For Marisa,

What can I say except,

"Thank you!"

Con affetto,

Randy

Methodology of Educational Measurement and Assessment

[1]This work was conducted while M. von Davier was employed with Educational Testing Service.

This book series collates key contributions to a fast-developing field of education research. It is an international forum for theoretical and empirical studies exploring new and existing methods of collecting, analyzing, and reporting data from educational measurements and assessments. Covering a high-profile topic from multiple viewpoints, it aims to foster a broader understanding of fresh developments as innovative software tools and new concepts such as competency models and skills diagnosis continue to gain traction in educational institutions around the world. Methodology of Educational Measurement and Assessment offers readers reliable critical evaluations, reviews and comparisons of existing methodologies alongside authoritative analysis and commentary on new and emerging approaches. It will showcase empirical research on applications, examine issues such as reliability, validity, and comparability, and help keep readers up to speed on developments in statistical modeling approaches. The fully peer-reviewed publications in the series cover measurement and assessment at all levels of education and feature work by academics and education professionals from around the world. Providing an authoritative central clearing-house for research in a core sector in education, the series forms a major contribution to the international literature.

More information about this series at http://www.springer.com/series/13206

Randy E. Bennett • Matthias von Davier
Editors

Advancing Human Assessment

The Methodological, Psychological and Policy Contributions of ETS

Editors
Randy E. Bennett
Educational Testing Service (ETS)
Princeton, NJ, USA

Matthias von Davier
National Board of Medical Examiners (NBME)
Philadelphia, PA, USA

ISSN 2367-170X ISSN 2367-1718 (electronic)
Methodology of Educational Measurement and Assessment
ISBN 978-3-319-58687-8 ISBN 978-3-319-58689-2 (eBook)
DOI 10.1007/978-3-319-58689-2

Library of Congress Control Number: 2017949698

Printed on acid-free paper

This Springer imprint is published by Springer Nature
The registered company is Springer International Publishing AG
The registered company address is: Gewerbestrasse 11, 6330 Cham, Switzerland

Foreword

Since its founding in 1947, Educational Testing Service (ETS), the world's largest private nonprofit testing organization, has conducted a significant and wide-ranging research program. The purpose of *Advancing Human Assessment: The Methodological, Psychological and Policy Contributions of ETS* is to review and synthesize in a single volume the extensive advances made in the fields of educational and psychological measurement, scientific psychology, and education policy and evaluation by researchers in the organization.

The individual chapters provide comprehensive reviews of work ETS researchers conducted to improve the science and practice of human assessment. Topics range from test fairness and validity to psychometric methodologies, statistics, and program evaluation. There are also reviews of ETS research in education policy, including the national and international assessment programs that contribute to policy formation. Finally, there are extensive treatments of research in cognitive, developmental, and personality psychology.

Many of the developments presented in these chapters have become de facto standards in human assessment, for example, item response theory (IRT), linking and equating, differential item functioning (DIF), and confirmatory factor analysis, as well as the design of large-scale group-score assessments and the associated analysis methodologies used in the National Assessment of Educational Progress (NAEP), the Programme for International Student Assessment (PISA), the Progress in International Reading Literacy Study (PIRLS), and the Trends in International Mathematics and Science Study (TIMSS).

The breadth and the depth of coverage the chapters provide are due to the fact that long-standing experts in the field, many of whom contributed to the developments described in the chapters, serve as lead chapter authors. These experts contribute insights that build upon decades of experience in research and in the use of best practices in educational measurement, evaluation, scientific psychology, and education policy.

The volume's editors, Randy E. Bennett and Matthias von Davier, are themselves distinguished ETS researchers.

Randy E. Bennett is the Norman O. Frederiksen chair in assessment innovation in the Research and Development Division at ETS. Since the 1980s, he has conducted research on integrating advances in the cognitive and learning sciences, measurement, and technology to create new forms of assessment intended to have positive impact on teaching and learning. For his work, he was given the ETS Senior Scientist Award in 1996, the ETS Career Achievement Award in 2005, and the Distinguished Alumni Award from Teachers College, Columbia University in 2016. He is the author of many publications including "Technology and Testing" (with Fritz Drasgow and Ric Luecht) in *Educational Measurement* (4th ed.). From 2007 to 2016, he led a long-term research and development activity at ETS called the Cognitively Based Assessment *of*, *for*, and *as* Learning (*CBAL*®) initiative, which created theory-based assessments designed to model good teaching and learning practice.

Matthias von Davier is a distinguished research scientist at the National Board of Medical Examiners in Philadelphia, PA. Until January 2017, he was a senior research director at ETS. He managed a group of researchers concerned with methodological questions arising in large-scale international comparative studies in education. He joined ETS in 2000 and received the ETS Research Scientist Award in 2006. He has served as the editor in chief of the *British Journal of Mathematical and Statistical Psychology* since 2013 and is one of the founding editors of the SpringerOpen journal *Large-Scale Assessments in Education*, which is sponsored by the International Association for the Evaluation of Educational Achievement (IEA) and ETS through the IEA-ETS Research Institute (IERI). His work at ETS involved the development of psychometric methodologies used in analyzing cognitive skills data and background data from large-scale educational surveys, such as the Organisation for Economic Co-operation and Development's PIAAC and PISA, as well as IEA's TIMSS and PIRLS. His work at ETS also included the development of extensions and of estimation methods for multidimensional models for item response data and the improvement of models and estimation methods for the analysis of data from large-scale educational survey assessments.

ETS is proud of the contributions its staff members have made to improving the science and practice of human assessment, and we are pleased to make syntheses of this work available in this volume.

Research and Development Division
Educational Testing Service,
Princeton, NJ, USA

Ida Lawrence

Preface

An edited volume on the history of the scientific contributions to educational research, psychology, and psychometrics made by the staff members of a nonprofit organization, Educational Testing Service (ETS), in Princeton, NJ, begs the questions: Why this self-inspection and who might benefit from that?

The answer can be found in current developments in these fields. Many of the advances that have occurred in psychometrics can be traced to the almost 70 years of work that transpired at ETS, and this legacy is true also for select areas in statistics, the analysis of education policy, and psychology.

When looking at other publications, be they conference proceedings, textbooks, or comprehensive collections like the *Handbook of Item Response Theory* (van der Linden 2016), the *Handbook of Test Development* (Lane et al. 2015), or *Educational Measurement* (4th ed.; Brennan 2006), one finds that many of the chapters were contributed by current and former ETS staff members, interns, or visiting scholars.

We believe that this volume can do more than summarize past achievements or collect and systematize contributions. A volume that compiles the scientific and policy work done at ETS in the years since 1947 also shows the importance, and the long-term effects, of a unique organizational form—the nonprofit measurement organization—and how that form can contribute substantially to advancing scientific knowledge, the way students and adults are assessed, and how learning outcomes and education policy more generally are evaluated.

Given the volume's purpose, we expect that it will be most attractive to those concerned with teaching, advancing, and practicing the diverse fields covered herein. It thus should make for an invaluable reference for those interested in the genesis of important lines of study that began or were significantly advanced at ETS.

Contributors to this volume are current and former ETS researchers who were asked to review and synthesize work around the many important themes that the organization explored over its history, including how these contributions have affected ETS and the field beyond. Each author brings his or her own perspective

---

This work was conducted while M. von Davier was employed with Educational Testing Service.

and writing style to the challenge of weaving together what in many cases constitutes a prodigious body of work. Some of the resulting accounts are thematically organized, while other authors chose more chronologically oriented approaches. In some chapters, individuals loom large simply because they had such significant impact on the topic at hand, while in other accounts, there were so many contributors that substantive themes offered a more sensible organizational structure. As a result, the chapters follow no single template, but each tells its own story in its own way.

The book begins with the reprint of a 2005 report by Randy E. Bennett that reviews the history of ETS, the special role that scientific research has played, and what that history might mean for the future.

With that opening chapter as context, Part I centers on ETS's contributions to the analytic tools employed in educational measurement. Chapters 2 and 3 by Tim Moses cover the basic statistical tools used by psychometricians and analysts to assess the quality of test items and test scores. These chapters focus on how ETS researchers invented new tools (e.g., confirmatory factor analysis), refined other tools (e.g., reliability indices), and developed versions of basic statistical quantities to make them more useful for assessing the quality of test and item scores for low- and high-stakes tests (e.g., differential item functioning procedures).

Chapter 4 by Neil J. Dorans and Gautam Puhan summarizes the vast literature developed by ETS researchers on one of the most fundamental procedures used to assure comparability of test scores across multiple test forms. This chapter on score linking is written with an eye toward the main purpose of this procedure, namely, to ensure fairness.

James E. Carlson and Matthias von Davier describe in Chap. 5 another staple of educational and psychological measurement that had significant roots at ETS. Their chapter on item response theory (IRT) traverses decades of work, focusing on the many developments and extensions of the theory, rather than summarizing the even more numerous applications of it.

In Chap. 6, Henry Braun describes important contributions to research on statistics at ETS. To name a few, ETS research contributed significantly to the methodology and discourse with regard to missing data imputation procedures, statistical prediction, and various aspects of causal inference.

The closing chapter in this part, Chap. 7 by Neil J. Dorans, covers additional topics relevant to ensuring fair assessments. The issues dealt with here go beyond that addressed in Chap. 4 and include procedures to assess differential item functioning (DIF) and give an overview of the different approaches used at ETS.

The chapters in Part II center on ETS's contributions to education policy and program evaluation. Two chapters cover contributions that built upon the developments described in Chaps. 2, 3, 4, 5, 6, and 7. Chapter 8 by Albert E. Beaton and John L. Barone describes work on large-scale, group-score assessments of school-based student populations. The chapter deals mainly with the National Assessment of Educational Progress (NAEP), describing the methodological approach and related developments. The following chapter by Irwin Kirsch, Mary Louise Lennon, Kentaro Yamamoto, and Matthias von Davier focuses on methods and procedures developed over 30 years at ETS to ensure relevance, comparability, and interpretability in large-scale, international assessments of adult literacy. It describes how methods developed for NAEP were extended to focus on new target populations and domains, as well as to link assessments over time, across countries and language versions, and between modes of delivery.

Chapter 10 discusses longitudinal studies and related methodological issues. This contribution by Donald A. Rock reviews the series of important longitudinal investigations undertaken by ETS, showing the need to carefully consider assumptions made in such studies and in the vertical linking of tests associated with them. The approaches taken offer solutions to methodological challenges associated with the measurement of growth and the interpretation of vertical scales.

Besides the book's opening chapter, the only other contribution not written for the current volume is Chap. 11, by Samuel Ball. The chapter was originally published by ETS in 1979. It describes the pinnacle of work at ETS on large program evaluation studies, including the classic investigation of the effects of Sesame Street. We include it because Ball directed much of that work and his account offers a unique, firsthand perspective.

Part II concludes with Chap. 12, a review by Richard J. Coley, Margaret E. Goertz, and Gita Z. Wilder of the extensive set of projects, primary and secondary analyses, and syntheses produced on education policy. Those endeavors have ranged widely, from school finance analyses to help build more equitable funding approaches to uncovering the complex of factors that contribute to achievement gaps.

Part III concerns ETS's contributions to research in psychology and consists of three chapters. ETS's work in personality, social, cognitive, and developmental psychology was extensive. Much of it, particularly in the early years, centered on theory development, as well as on the invention and improvement of assessment methodology. The first two chapters, Chap. 13 by Lawrence J. Stricker and Chap. 14 by Nathan Kogan, respectively, cover the very broad range of investigations conducted in cognitive, personality, and social psychology. Chap. 15 by Nathan Kogan, Lawrence J. Stricker, Michael Lewis, and Jeanne Brooks-Gunn documents the large and extremely productive research program around the social, cognitive, and psychological development of infants and children.

The final part concerns contributions to validity. Chapter 16 by Michael Kane and Brent Bridgeman reviews ETS staff members' seminal work on validity theory and practice, most notably that of Samuel Messick. Donald E. Powers' Chap. 17 delves into the historically contentious area of special test preparation, an activity that can threaten or enhance the validity of test scores. The part closes with Isaac I. Bejar's Chap. 18, a wide-ranging examination of constructed-response formats, with special attention to their validity implications.

We end the book with Chap. 19, a synthesis of the material covered.

This book would not have been possible without the vision of Henry Braun, who suggested the need for it; Ida Lawrence, who supported it; Lawrence J. Stricker, whose contributions of time, effort, advice, and thought were invaluable; and the authors, who gave of their time to document the accomplishments of their colleagues, past and present. Also invaluable was the help of Kim Fryer, whose editorial and managerial skills saw the project to a successful completion.

We hope that this collection will provide a review worthy of the developments that took place over the past seven decades at ETS.

Princeton, NJ, USA
September 2017

Randy E. Bennett
Matthias von Davier

References

Brennan, R. L. (Ed.). (2006). *Educational measurement* (4th ed.). Westport: Praeger.
Lane, S., Raymond, M. R., & Haladyna, T. M. (Eds.). (2015). *Handbook of test development* (2nd ed.). New York: Routledge.
van der Linden, W. J. (Ed.). (2016). *Handbook of item response theory: Models, statistical tools, and applications* (Vols. 1–3). Boca Raton: Chapman & Hall/CRC.

Contents

About the Editors

Randy E. Bennett is the Norman O. Frederiksen chair in assessment innovation in the Research and Development Division at Educational Testing Service in Princeton, NJ. Bennett's work has focused on integrating advances in cognitive science, technology, and educational measurement to create approaches to assessment that have positive impact on teaching and learning. From 1999 through 2005, he directed the NAEP Technology-Based Assessment Project, which included the first administration of computer-based performance assessments with nationally representative samples of school students and the first use of "clickstream," or logfile, data in such samples to measure the processes used in problem-solving. From 2007 to 2016, he directed an integrated research initiative titled *Cognitively Based Assessment of, for, and as Learning* (*CBAL*®), which focused on creating theory–based summative and formative assessment intended to model good teaching and learning practice. Randy Bennett is the president of the International Association for Educational Assessment (IAEA) (2016), an organization primarily constituted of governmental and nongovernmental nonprofit measurement organizations throughout the world, and president of the National Council on Measurement in Education (NCME) (2017–2018), whose members are individuals employed primarily in universities, testing organizations, state education departments, and school districts. He is a fellow of the American Educational Research Association.

Matthias von Davier is a distinguished research scientist at the National Board of Medical Examiners (NBME), in Philadelphia, PA. Until 2016, he was a senior research director in the Research and Development Division at Educational Testing Service (ETS) and codirector of the Center for Global Assessment at ETS, leading psychometric research and operations of the center. He earned his Ph.D. at the University of Kiel, Germany, in 1996, specializing in psychometrics. In the Center for Advanced Assessment at NBME, he works on psychometric methodologies for analyzing data from technology-based high-stakes assessments. He is one of the editors of the Springer journal *Large-Scale Assessments in Education*, which is jointly published by the International Association for the Evaluation of Educational Achievement (IEA) and ETS. He is also editor in chief of the *British Journal of*

Mathematical and Statistical Psychology (BJMSP) and coeditor of the Springer book series *Methodology of Educational Measurement and Assessment*. Dr. von Davier received the 2006 ETS Research Scientist Award and the 2012 NCME Bradley Hanson Award for contributions to educational measurement. His areas of expertise include topics such as item response theory, latent class analysis, diagnostic classification models, and, more broadly, classification and mixture distribution models, computational statistics, person-fit statistics, item-fit statistics, model checking, hierarchical extension of models for categorical data analysis, and analytical methodologies used in large-scale educational surveys.

Chapter 1
What Does It Mean to Be a Nonprofit Educational Measurement Organization in the Twenty-First Century?

Randy E. Bennett

The philosopher George Santayana (1905) said, "Those who cannot remember the past are condemned to repeat it" (p. 284). This quote is often called, "Santayana's Warning," because it is taken to mean that an understanding of history helps avoid having to relive previous mistakes. But the quote can also be read to suggest that, in order to make reasoned decisions about the future, we need to be always cognizant of where we have come from. This claim is especially true for a nonprofit organization because its continued existence is usually rooted in its founding purposes.

This chapter uses Educational Testing Service (ETS), the largest of the nonprofit educational measurement organizations, to illustrate that claim. The chapter is divided into four sections. First, the tax code governing the establishment and operation of educational nonprofits is reviewed. Second, the history around the founding of ETS is described. Third, the implications of ETS's past for its future are discussed. Finally, the main points of the paper are summarized.

This chapter was originally published in 2005 by Educational Testing Service.

R.E. Bennett (✉)
Educational Testing Service, Princeton, NJ, USA
e-mail: rbennett@ets.org

R.E. Bennett, M. von Davier (eds.), *Advancing Human Assessment*, Methodology of Educational Measurement and Assessment,
DOI 10.1007/978-3-319-58689-2_1

1.1 What Is an Educational Nonprofit?

The term *nonprofit* refers to how an organization is incorporated under state law. To be *federally* tax exempt, an educational nonprofit must become a 501(c)3 corporation.[1, 2] What is 501(c)3? It is a very important section in the Internal Revenue Code. The section is important because of what it does, and does not, allow educational nonprofits to do, as well as because of how the section came about.

Section 501(c)3 exempts certain types of organizations from federal income tax.[3] To qualify, an organization must meet certain discrete "tests." The tests are the organizational test, operational test, inurement test, lobbying restriction, electioneering prohibition, public benefit test, and public policy test (Harris 2004). Each of these tests is briefly reviewed in turn.

Under the Internal Revenue Code, to be exempt, an organization must be set up exclusively for one or more of the following purposes: charitable, religious, educational, scientific, literary, testing for public safety, fostering amateur national or international sports competition, or the prevention of cruelty to children or animals (Internal Revenue Service [IRS] 2003b). An entity meets this organizational test if its articles of incorporation limit its function to one or more exempt purposes (e.g., educational) and do not expressly allow the organization to engage, other than insubstantially, in activities that are not consistent with those purposes.

ETS's exempt purpose is educational and its organizing documents specify the activities it can pursue in keeping with that purpose. Paraphrasing the 2005 revision of the organization's charter and bylaws (ETS 2005), those activities are to:

- Conduct educational testing services,
- Counsel test users on measurement,
- Serve as a clearinghouse about research in testing,
- Determine the need for, encourage, and carry on research in major areas of assessment,
- Promote understanding of scientific educational measurement and the maintenance of the highest standards in testing,
- Provide teachers, parents, and students (including adults) with products and services to improve learning and decisions about opportunities,
- Enhance educational opportunities for minority and educationally disadvantaged students, and
- Engage in other advisory services and activities in testing and measurement from time to time.

[1] For purposes of this chapter, *nonprofit* and *501(c)3* corporation are used to mean the same thing, even though they are legally different.

[2] The terms *nonprofit* and *not-for-profit* are not legally distinct, at least not in the Internal Revenue Code.

[3] The IRS has 27 types of organizations that are tax exempt under 501(c), only one of which covers those institutions exempt under 501(c)3 (Internal Revenue Service [IRS] 2003a).

To meet the second—or operational—test, the organization must be run exclusively for one or more of the exempt purposes designated in its articles. The test is met if the organization's stated purpose and activities conform. Although Section 501(c)3 indicates that the organization must be operated "exclusively" for exempt purposes, the term *exclusively* has been interpreted by the IRS to mean "primarily" or "substantially." Thus, Section 501(c)3 does allow exempt organizations to engage in activities *un*related to their exempt purposes (IRS 2000). But those activities must not become "substantial" and tax must be paid on this unrelated business income.

Note that the operational test makes clear that engaging in unrelated activities to support the exempt purpose is, in itself, a *nonexempt purpose*, if it is done any more than *in*substantially (IRS n.d.-a).[4] To prevent such unrelated activities from becoming so substantial that they threaten tax-exempt status—as well as to allow outside investment and limit liability—an exempt organization may create for-profit subsidiaries.[5]

The inurement test is often cited as the fundamental difference between for-profit and nonprofit corporations. This test says that no part of the organization's net earnings may benefit any private individual. For example, there may be no stockholders and no distribution of net earnings, as in a dividend.

The lobbying restriction and electioneering prohibition mean, respectively, that no significant part of an organization's activities may consist of "carrying on propaganda or otherwise attempting to influence legislation…" and that an exempt organization may not participate or intervene in any political campaign for or against any candidate for public office (IRS n.d.-b).[6] Unlike the lobbying restriction, the electioneering prohibition is absolute.

To meet the public benefit test, the organization must operate for the advantage of public, rather than private, interests.[7] Private interests can be benefited, but only incidentally. Further, the principal beneficiaries of the organization's activities must be sufficiently numerous and well-defined, so that the community is, in some way, served.

Finally, there is the public policy test, which essentially says that an otherwise qualifying organization's "…purpose must not be so at odds with the common community conscience as to undermine any public benefit that might otherwise be conferred" (*Bob Jones University v. United States* 1983). The quintessential example is Bob Jones University, which lost its tax-exempt status as a result of racially

---

[4] There does not appear to be a statutory or regulatory definition of *substantial*. However, experts in nonprofit tax law often advise limiting gross unrelated business income to under 20% of gross revenue (e.g., *FAQ's—501(c)(3) Status* (n.d).

[5] The Chauncey Group International would be one example from ETS's history.

[6] The dollar limits associated with the lobbying restriction are defined by a relatively complex formula. See *Restrictions on Nonprofit Activities* (n.d.). Also see IRS (n.d.-c).

[7] This test differs from the inurement test in that the inurement test applies *only* to insiders—persons having a private interest in the organization's activities—whereas the "private interests" cited in the public benefit test apply more generally.

discriminatory practices that the IRS believed, and the Supreme Court affirmed, violated fundamental public policy.[8]

Organizations set up for educational purposes under 501(c)3 have several additional requirements (IRS 2003b). First, the "positions" they take must be educational. According to the IRS, "Advocacy of a particular position … may be educational if there is a sufficiently full and fair exposition of pertinent facts to permit an individual or the public to form an independent opinion or conclusion" (IRS 2003b, p. 25). Also, the method used by an organization to develop and present its views is a factor in determining if the organization is "educational."

What constitutes an "educational" method? The IRS says that the method is not educational when:

1. The presentation of viewpoints unsupported by facts is a significant part of the organization's communications.
2. The facts that purport to support the viewpoints are distorted.
3. The organization's presentations express conclusions more on the basis of emotion than objective evaluation. (IRS 2003b, p. 25)

That, then, is what 501(c)3 is about. But why did Congress decide to grant tax exemptions to certain organizations in the first place, thereby forgoing huge amounts of future revenue?

The statutory roots of 501(c)3 are commonly traced to the Tariff Act of 1894, which imposed a corporate income tax and exempted entities organized and conducted solely for charitable, religious, or educational purposes from having to pay it (Scrivner 2001). The congressional intent behind the exemption was to give preferential treatment because such organizations provided a benefit to society. Congress reaffirmed this view in the Revenue Act of 1938 when it said that tax exemption was based on the theory that the loss of revenue is compensated by relieving the government of a function it would otherwise have to perform (presumably because the for-profit sector would not, or should not be allowed to, perform it) and because of the benefits to the general welfare that the function would serve (*Bob Jones University v. United States* 1983).[9]

The Revenue Act of 1950 added unrelated business income tax rules, which were intended to eliminate unfair competition by taxing the unrelated activities of exempt

[8] The IRS revoked the tax-exempt status of Bob Jones University in 1975 even though the school had not violated any provision of 501(c)(3). The IRS revoked its tax-exempt status because the university had, on the basis of religious belief, at first, refused admission to Black students, and then only to Black students married within their own race. It then admitted Black students generally but enforced strict rules, including expulsion, against interracial dating. The university sued when its exempt status was revoked. The Supreme Court upheld the IRS decision by an 8–1 vote.

[9] Why shouldn't the for-profit sector supply some services? Because the need for profit may come into direct conflict with the intended public benefit behind the service. Some services require a disinterested party. See, for example, the inurement test, the lobbying restriction, and the electioneering prohibition, which are intended to distance the service provider from self-interest that could otherwise affect the provision of the service. Occupational and professional licensing and certification, which is often handled by private nonprofit associations, would be an example.

organizations in the same way as competing for-profit corporations were taxed (Scrivner 2001). The Internal Revenue Code of 1954 was a restructuring to the current numbering, which resulted in the section known today as "Section 501(c)3." Finally, the 1959 Regulations for the 1950 Act and the 1954 Code defined *charity* to more closely approach the English common-law definition (Scrivner 2001). That is, not only the relief of poverty, but also the advancement of education, religion, and other purposes beneficial to the community. So, legally, many 501(c)3 organizations like ETS are, in fact, "public charities."[10]

To summarize, in the words of the majority opinion rendered by the U.S. Supreme Court in *Bob Jones University v. United States* (1983), "In enacting ... 501(c)3, Congress sought to provide tax benefits to charitable organizations to encourage the development of private institutions that serve a useful public purpose or supplement or take the place of public institutions of the same kind."

Thus, Section 501(c)3 has its roots in the idea that the government might not be able to provide all the services the public needs, that the for-profit sector might not fill the gap, and that those organizations that do voluntarily address such social needs should be compensated through tax exemption.

How did ETS come to be a 501(c)3? The reasons for that lie fundamentally in how ETS came about. That story begins at the end of the nineteenth century, just prior to the establishment of the College Entrance Examination Board.

1.2 Where Did ETS Come From?

Prior to the founding of the College Entrance Examination Board (CEEB), admission to college and university in the United States was a disorganized, if not chaotic process (Fuess 1950). The Ivy League institutions each administered their own tests, which varied widely in subjects assessed, quality, and administration date. Wilson Farrand, principal of Newark Academy, summarized the disarray in entrance requirements as follows (cited in Fuess 1950, p. 17):

> Princeton requires Latin of candidates for one course, but not for the others. Yale demands it of all, Columbia of none. Princeton names five books of Caesar and four orations of Cicero; Yale names four books of Caesar and three books of Virgil ... Yale calls for Botany, Columbia for Physics and Chemistry, Princeton for no science. Princeton and Columbia demand both German and French, while Yale is satisfied with either. On the other hand, while Princeton and Columbia demand only American History, Yale calls also for that of England...

Other colleges annually reviewed and certified high schools so that they had a means of assuring the quality of the curriculum and of the recommendations coming from school principals (Fuess 1950; Hubin 1988). This certification system was

---

[10] All 501(c)3 organizations must be categorized under Section 509(a) as either private foundations or one of several types of public charity. See IRS (2003b, pp. 30–36) for the particulars of this classification.

a burden for both the colleges and the high schools. College staff had to physically visit each feeder school and each feeder school had to undergo multiple reviews annually, one for each receiving college.

To help rationalize admissions, the CEEB was established in 1900 through the Association of Colleges and Secondary Schools of the Middle States and Maryland, with 12 collegiate charter members, as well as representation from three secondary schools (Fuess 1950).[11] The Board's initial purpose was to create a single battery of centrally scored examinations and, in so doing, bring order and higher quality to the college preparatory curriculum. The first "College Boards," as they were called, were administered the following year, 1901.

The College Boards were a *weeklong* battery of essay tests in various content domains, which solved some problems but brought others to the fore (Hubin 1988). By the early 1920s, the CEEB membership, which now included secondary schools, was unhappy enough with the exams that it began publicly voicing concerns about their subjectivity in scoring, variation in difficulty from one administration (or form) to the next, and the narrowing effect the exams were having on the high school curriculum.[12]

As a consequence of its unhappiness, the Board commissioned two streams of investigation, with the idea being to supplement, not replace, the essay tests (Hubin 1988). The first stream of investigation focused on multiple-choice achievement tests and was led by Edward L. Thorndike and Ben D. Wood of Teachers College, Columbia University.

The second stream centered on multiple-choice intelligence tests derived from Yerkes' 1918 Army Alpha Test. The Board referred to this project as its "Psychological" examinations, and it was led by Carl Brigham of Princeton University.

As the Board was pursuing its twin investigations, the American Council on Education (ACE), began its own initiative to develop "psychological examinations." ACE was founded in 1918, as an association dominated by public universities. (The much smaller CEEB, in contrast, was composed primarily of private institutions.) In 1924, ACE commissioned Louis. L. Thurstone, of the Carnegie Institute of Technology, to create an admissions test based on Army Alpha.[13] ACE took this course of action because many public institutions were already using intelligence tests in admissions and ACE wanted to standardize this use (Hubin 1988).

Meanwhile, under the auspices of the CEEB, Brigham had by 1926 developed *his* "psychological" examination but under a new name because, Brigham, originally a eugenicist, no longer believed that intelligence tests measured native ability.

[11] The charter members were: Barnard College, Bryn Mawr College, Columbia University, Cornell University, Johns Hopkins University, New York University, Rutgers College, Swarthmore College, Union College, University of Pennsylvania, Vassar College, and the Woman's College of Baltimore (Fuess 1950).

[12] These issues, incidentally, remain with us in one way or another to this day.

[13] The Carnegie Institute would later merge with the Mellon Institute of Industrial Research to form Carnegie-Mellon University.

And in the first SAT® manual (Brigham 1926, cited in Hubin 1988, pp. 196–197), he was quite clear:

> The term 'scholastic aptitude' makes no stronger claim ... than that there is a tendency for ... scores in these tests to be associated positively with ... subsequent academic attainment.

Further:

> This additional test ... should be regarded merely as a supplementary record.

To evaluate Brigham's predictive validity claim, the CEEB began to administer the test experimentally that same year (Hubin 1988).

The story next moves to Harvard. The year was 1933 and the country was in the midst of its Great Depression. James Bryant Conant had just become president. As president, Conant found that applicants were being drawn primarily from a small number of northeastern preparatory schools and that the "College Boards" were still being used for admission (Hubin 1988). Conant disliked the "College Boards" because he saw them as nothing more than a measure of mastery of the prep school curriculum that couldn't be used to assess students coming from public schools. For Conant, Harvard admission was being based largely on ability to pay because if a student could not afford to attend prep school, that student was not going to do well on the College Boards and was not coming to Harvard.

Conant decided to address this problem by creating a scholarship program to increase economic and regional diversity (Hubin 1988). Note that Conant did not want to increase access to Harvard for everyone, but just for those with academic talent. How would one measure academic talent in the public schools? Obviously not with the College Boards, which were keyed toward the prep school curriculum. To find a solution, Conant turned to Henry Chauncey, his assistant dean of admissions. Chauncey contacted Carl Brigham, who was by this time associate director for research at the CEEB, on leave from Princeton University.

Chauncey arranged with the CEEB for Brigham's Scholastic Aptitude Test to be used for the award of Harvard scholarships in the very next year (1934), and by 1937, the "Scholarship Tests," as they were now being called, were used by 14 colleges. The Scholarship Tests were composed of two parts: the Scholastic Aptitude Test, as developed by Brigham, and a battery of achievement tests created by Ben Wood for ACE's Cooperative Test Service. This use of the Scholarship Tests was funded by the Carnegie Foundation for the Advancement of Teaching. In the following year, 1938, the Scholarship Tests were extended to use for university admissions.

Within the Scholarship Test project were collaborating the three organizations that would later form ETS: the College Entrance Examination Board, the American Council on Education, and the Carnegie Foundation for the Advancement of Teaching. This third organization, the Carnegie Foundation, was established by Andrew Carnegie in 1905 as an independent policy and research center to "... encourage, uphold, and dignify the profession of the teacher and the cause of higher education" (Carnegie Foundation for the Advancement of Teaching 2005). Around the time the Scholarship Tests were being put into place, William Learned of the Foundation was also encouraging development of the *GRE*® assessment as an experimental test for admissions to graduate liberal arts programs (Hubin 1988). The GRE consisted of an aptitude test, which was the SAT, and a battery of achievement tests created by Wood for ACE. The GRE was first administered experimentally at Harvard, Yale, Princeton, and Columbia in 1937, the very same year the Scholarship Tests were extended beyond Harvard.

Just as the GRE and the Scholarship Tests were being introduced, several key figures floated the idea of a unified testing organization (ETS 1992). In a 1937 speech at the annual meeting of the Educational Records Bureau, Conant advocated for formation of a "nationwide cooperative testing service" because he believed that standardized testing would be advanced by a substantial *research* program concentrated in one organization. Around the same time, Learned of Carnegie approached Wood of ACE about forming a "general examinations board" because Learned believed that testing could be made more efficient by eliminating competition and duplication, and because Learned thought that more resources would be available for *research* and development if a unified organization was formed (Hubin 1988).

George Mullins, CEEB Secretary, described the philosophical foundation that such an organization should have (ETS 1992, p. 9). He wrote:

> The organization should be built so that it can attack the problems of educational measurement scientifically … It should have no doctrine to sell, no propaganda concerning certain forms of tests to be spread … It must be an open-minded scientific organization if it is destined to give real service and therefore to endure.

So, clearly, *scientific research* was a principal motivation for the advocates of a unified testing organization. And, paradoxically, scientific research was also a motivation for those opposed. The strongest opponent was Carl Brigham (ETS 1992; Hubin 1988; Lemann 1999). What were Brigham's concerns?

First, Brigham did not believe that psychology or measurement was scientifically advanced enough to support the large-scale operational use of testing that a national agency would bring. He wrote (cited in Lemann 1999, p. 34):

> Practice has always outrun theory…this is a new field and…very little has been done which is right.

What was not right? A prime example comes from an earlier letter (cited in Lemann 1999, p. 33):

> The more I work in this field, the more I am convinced that psychologists have sinned greatly in sliding easily from the name of the test to the functions or trait measured.

This comment referred to the ease with which the psychologists of the day had concluded that IQ tests measured innate intelligence tied to ethnicity, a view he himself had taken but since, quite publicly, rejected (Hubin 1988).

A second reason for Brigham's opposition to a unified measurement organization was that he believed consolidation would slow the growth of measurement science and kill innovation in tests. In a letter to Conant (cited in ETS 1992, p. 6), he wrote:

> One of my complaints against the proposed organization is that although the word *research* will be mentioned many times in its charter, the very creation of powerful machinery to do more widely those things that are now being done badly will stifle research, discourage new developments, and establish existing methods, and even existing tests, as the correct ones.

What kind of innovation did Brigham wish to see? Among other things, Brigham was interested in connecting tests to teaching, learning, and cognition (Donlon 1979; Hubin 1988). Ideas in his 1932 book, A Study of Error, anticipated by a half-century what we, today, call *formative assessment*. Brigham believed that the mistakes students made in solving test items could provide a basis for instruction.

But according to Lemann (1999), what concerned Brigham most was that any organization that owned the rights to a particular examination would inescapably become more interested in marketing that test than in objectively researching its effectiveness and working toward its improvement. For Brigham, a strong research program, not just lip service, was essential because "The provision for extensive research will prevent degeneration into a sales and propaganda group…" (Brigham 1937, p. 756). Brigham's opposition was so strident, and his opinion so respected, that the idea for a consolidated organization was shelved—that is, until Brigham died in January of 1943 at the relatively young age 52.

Coincidentally, by the time Brigham died, the need for a unified testing agency had become more pronounced. In the 1940s, the CEEB was still primarily a regional membership organization, though it had grown from its original 12 institutions to more than 50 (Fuess 1950). While membership growth was certainly desirable, the Board found it difficult to integrate the increased operational testing activity it had to perform. The SAT was now equated and machine scored, making it both fairer and far more efficient to process than the College Boards. In 1941, the old College Boards were discontinued, which from William Learned's point of view, could have been no great loss, for he had earlier described them as "a few arbitrarily chosen questions with answers rated in terms of personal opinion by individuals of varying degrees of experience and competence" (cited in Hubin 1988, pg. 293).

The 1940s saw the CEEB's operational testing activities balloon. Initially, this growth came from large military contracts during the Second World War, most of which had nothing to do with educational testing and which, consequently, made the Board's membership of admissions officers quite uneasy (Hubin 1988). This activity was followed by an upsurge in college applicants after the war because of the GI Bill, which paid essentially the full cost of college for returning veterans.

Meanwhile, the Carnegie Foundation had its own concerns. In 1944, the GRE had gone operational and by 1946, the GRE unit had more staff than the Foundation proper (Hubin 1988). While the Foundation had hoped for such success, it was in the business of seeding development, not running established testing programs so, like the Board, it too was looking for a home for its "Big Test."

Finally, ACE's Cooperative Test Service (CTS) was operating at a loss (Lemann 1999). Moreover, ACE, which saw the CEEB as a competitor, did not want the CEEB to get the GRE. Thus, all three organizations had their own motivations for wanting a centralized testing agency. Carnegie took the next step by sponsoring a national commission, chaired by Conant, to recommend how to proceed.

When the Commission issued its recommendations, there was considerable negotiation, much of it facilitated by Henry Chauncey, over whether the agency should be a functionary of one of the three organizations or independent of them (ETS 1992; Lemann 1999). Finally, the three organizations agreed on an independent arrangement in which they would turn over their testing programs and a portion of their assets to the new ETS (ETS, 1992; Fuess 1950). Among other things, the CEEB contributed SAT operations, Law School Admission Test (LSAT) operations, 139 permanent employees, and its Princeton office. Carnegie gave the GRE, the Pre-Engineering Inventory, and 36 employees. ACE donated the Cooperative Test Service, which included NTE operations (and later full ownership), Thurstone's Psychological Examination, and 37 employees.[14]

In December 1947, ETS was granted a charter by the New York State Board of Regents. James Conant was designated chairman of the Board of Trustees. Henry Chauncey, now a director of the CEEB, was named ETS's first president, in recognition of the role he played in bringing the three organizations together.

Interestingly, neither the Board of Trustees, nor Henry Chauncey, forgot Carl Brigham's warnings about the need for research as a mechanism for driving innovation, helping maintain high standards, and, ultimately, for improving education. With respect to improving education, after its first meeting, the ETS Board of Trustees issued the following statement (cited in ETS 1992, p. 22):

> In view of the great need for research in all areas and the long-range importance of this work to the future development of sound educational programs, it is the hope of those who have brought the ETS into being that [through its research] it make fundamental contributions to the progress of education in the United States.

Henry Chauncey also made the connection between research and improvements to education. In his inaugural message to staff (cited in ETS 1992, pp. 21–22), he wrote:

> It is our ardent hope and confident expectation that the new organization will make important contributions to American education through developing and making available tests of the highest standards, by sponsoring distinguished research both on existing tests and on unexplored test areas, and by providing effective advisory services … to schools and colleges.

For Henry Chauncey, these were not just words. Over his tenure as ETS president, traditions for the conduct of research were established to forestall precisely the problems that Brigham feared. Chauncey put into place academic freedom to encourage independent thinking and constructive criticism of ETS's own tests, thereby fostering opportunities for the instruments' improvement. He put into place a policy requiring full publication of research results, unless the outcomes of a project were identified *in advance* as proprietary and confidential. He established this policy for two reasons. The first reason was to prevent the suppression of results unfavorable to ETS or its testing programs because, if suppression occurred, it would signal the organization's descent into producing "propaganda" to serve its

[14] Each organization also contributed capital in the form of cash or securities (Fuess 1950).

own ends.[15] A second reason for the publication policy was that Chauncey saw advancing the fields of education and measurement as a corporate obligation, and publication as a mechanism for achieving it.

Henry Chauncey also put into place a funding source that was purposely *not* controlled by the testing programs. He did this to encourage long-term, forward thinking projects unconstrained by short-term business needs. Last, he actively encouraged a wide-ranging agenda to push innovation along many measurement and education fronts simultaneously. Collectively, Chauncey's traditions had one other important effect. They created an environment that allowed him to attract—and retain—the brightest young research staff—Lord, Messick, Tucker, Gulliksen, Frederiksen—and, as a result, build ETS's reputation as a *scientific* educational measurement organization.

To summarize, Conant, Chauncey, Learned, and the others who spearheaded the creation of ETS saw the new organization as providing a *public service* that neither the government nor the private sector offered. In this way, the public service principles underlying the founding of ETS and those of the Internal Revenue Code met, with ETS being established as a nonprofit corporation.

Turning back to Santayana, "How might an organization like ETS use its past, in particular its public service roots, to help shape its future?" In other words, "How can such an organization succeed as a *nonprofit* educational measurement organization in the twenty-first century?"

1.3 What Does the Past Imply for the Future?

To answer these questions requires a close look at the idea of *mission*. Section 501(c)3 doesn't use that word. It uses the term *purpose* to describe the reason for which an exempt organization is operated. What's an organizational mission? It is, essentially, nothing more than a statement of purpose. Logically then, a nonprofit organization's mission must be aligned with its exempt purpose, otherwise, the organization could not remain exempt. ETS's mission is "... to help advance quality and equity in education by providing fair and valid assessments, research and related services," which is quite close in meaning to the inaugural words of the Trustees and of Henry Chauncey, as well as to ETS's chartered purpose.[16]

As noted earlier, the overwhelming majority of activities an exempt organization conducts must be consistent with its purposes under section 501(c)3. Such an

[15] It is interesting to note the repeated appearance of the word *propaganda* in the story of ETS. Brigham feared it, George Mullins (CEEB) hoped a unified organization would transcend it, the IRS enjoined educational nonprofits against it, and Henry Chauncey made sure to do something about it.

[16] ETS's chartered purpose reads as follows (ETS 2005): "The purposes for which the corporation is formed are to engage in, undertake, and carry on services, research and other activities in the field of educational testing and such other activities as may be appropriate to such purpose."

organization may not, as a primary operating strategy, use unrelated business to fund its mission (IRS n.d.-a). So, what additional strategies are available to fund the mission for such an organization?

One strategy often used by educational nonprofits with large endowments is to fund activities through investment proceeds. But this approach is not a long-term operating strategy for an organization like ETS whose relatively limited reserves are needed to drive innovation and cover unforeseen emergencies.

A second mechanism is through for-profit subsidiaries, which protect the exempt status of the parent from unrelated income and allow outside investment. ETS used this approach in the 1990s with the establishment of Chauncey Group International (CGI), K12 Works, and ETS Technologies. CGI was sold in 2004 to Thomson Prometric and the other two subsidiaries were incorporated into ETS proper a short time earlier. More recently ETS has created several new subsidiaries, including one for-profit, ETS Global, B.V.

The third strategy is to fund the mission *through* the mission—in essence, through new product and service development because, in ETS's case, the products that originally funded the mission—in particular, the SAT and GRE—are no longer providing the income growth needed to support continued innovation.

There is, of course, a catch to funding the mission through mission-related new product and service development. Obviously, these new products and services need to be financially successful. That is not the catch.

The catch is that, to be truly mission oriented, this new product and service development needs to be different in some fundamental way from the products and services that for-profits offer. Otherwise, what justifies *nonprofit* status? How can it be a *public service* if the for-profit sector is offering the very same thing?

How, then, does an organization like ETS differentiate itself?

ETS has typically tried to differentiate itself by providing higher levels of quality—or of integrity—than the for-profits offer. And when people's life-chances are at stake, that *is* a public service.

But there may be something more that it can do. That "something more" is to take on the big challenges the for-profits will not because the challenges are too hard—and they don't have the scientific capability to attack them—or because the return on investment isn't big enough or soon enough—and their need to satisfy shareholders won't allow it.

What are the big challenges? Arguably, the best examples can be found in the work of ETS's progenitors. Carl Brigham was interested in going beyond the SAT's selection function to using assessment to guide instruction. And what better time to pursue that goal than now? The relevant questions today are:

- Who needs help?
- With what specific content standards?
- How should they be taught those specific standards?

These questions are especially important to address in the case of minority and economically disadvantaged students, for whom closing the achievement gap has become a critical national issue.

For his part, Henry Chauncey had enormous interest in measuring new constructs to guide individuals in educational choices (Lemann 1999). He wanted to know:

- What are individuals good at?
- What are they interested in?
- How can those aptitudes and interests be guided for the good of the individual and society?

Finally, James Conant's original interest was using assessment to increase diversity. In today's terms:

- How can we better locate promising individuals from all backgrounds for purposes of encouraging them to pursue higher education?
- How can we help them further develop the skills they will need to succeed (including such new basics as English for second-language learners)?

The challenges that Brigham, Chauncey, and Conant posed well over 50 years ago remain largely unresolved today. If nonprofit measurement organizations like ETS can create financially successful new products and services that meet those challenges, they will be doing as they should: funding their mission by *doing* their mission.

But nonprofit organizations like ETS do not have a chance of realizing this goal without reorienting themselves dramatically. For one, they need to be more responsive to market needs. Being responsive means:

- Scientific research and development that solves problems customers care about,
- Business units that influence the choice of problems to be solved and that help shape those solutions into marketable products,
- Sales and marketing functions that effectively communicate what differentiates a nonprofit's products from competitor offerings, and that bring back intelligence about customer needs, product ideas, and potential product improvements,
- Technology delivery that can readily and cost-effectively incorporate advances, and
- Operational processes that execute accurately and efficiently.

However, in addition to being more responsive, such organizations need to be more *responsible*. Being responsible means:

- Products and marketing claims the organization can stand behind scientifically,
- Research that supports, but is not controlled by, the business units, and
- Research and development investments that are determined not only by financial return but also *mission return* (i.e., the positive impact on education the product or service is expected to have).

Ultimately, being "responsible" means continuing to champion integrity, quality, and scientific leadership in educational measurement and in education.

1.4 Summary

ETS is a nonprofit educational measurement organization with public service principles deeply rooted in its progenitors and in Section 501(c)3. Both its progenitors and the framers of the tax code were interested in the same thing: providing a social benefit.

A corollary of this history is that the purpose of a nonprofit measurement organization like ETS is *not* about making money. Its purpose is not even about making money to support its mission (because that, in itself, would be a nonexempt purpose). Its purpose is about *doing* its mission, making enough money in the process to continue doing it in better and bigger ways.

Third, it is reasonable to argue that, in the twenty-first century, renewed meaning can be brought to ETS's mission by attacking big challenges in education and measurement, such as using assessment to guide instruction, measuring new constructs to help individuals with educational choices, and using assessment to increase diversity. Interestingly, many of these big challenges are the same ones that motivated the organization's progenitors. Those individuals knew these were hard problems but they had faith that those problems could be solved. To solve them, ETS and organizations like it will need to act more responsively and more responsibly than ever before.

Make no mistake: If ETS succeeds in being responsive *without* being responsible, or in being responsible without being responsive, it will have failed. ETS, and kindred organizations, *must* do both. *Doing both* is what it means to be a nonprofit educational measurement organization in the twenty-first century.

Acknowledgments I am grateful to the following people for providing information, comments, or other assistance to me in developing the presentation on which this chapter was based: Isaac Bejar, Michal Beller, Julie Duminiak, Marisa Farnum, Eleanor Horne, Pat Kyllonen, Ernie Price, Larry Stricker, Rich Swartz, Stan Von Mayrhauser, Dylan Wiliam, and Ann Willard. However, all opinions contained herein, and any errors of fact, are my own.

References

Bob Jones University v. United States, 461 U.S. 574 (1983).

Brigham, C. C. (1937). The place of research in a testing organization. *School and Society, 46*, 756–759.

Carnegie Foundation for the Advancement of Teaching. (2005). *About the Carnegie Foundation.* Retrieved from http://web.archive.org/web/20041211123245/http://www.carnegiefoundation.org/AboutUs/index.htm

Donlon, T. F. (1979). Brigham's book. *The College Board Review, 113*, 24–30.

Educational Testing Service. (1992). *The origins of Educational Testing Service*. Princeton: Author.

Educational Testing Service. (2005). *Educational Testing Service: Charter and bylaws* (revised April 2005). Princeton: Author.

FAQ's—501(c)(3) status. (n.d.). Retrieved from http://web.archive.org/web/20050113093311/http://www.t-tlaw.com/lr-04.htm

Fuess, C. M. (1950). *The College Board: Its first fifty years*. New York: Columbia University Press.
Harris, J. A. (2004). *Requirements for federal income tax exemption under code section 501(c)3*. Retrieved from http://web.archive.org/web/20040107112241/http://www.zsrlaw.com/publications/articles/jah1999.htm
Hubin, D. R. (1988). *The Scholastic Aptitude Test: Its development and introduction, 1900–1947*. Retrieved from http://darkwing.uoregon.edu/~hubin/
Internal Revenue Service. (2000). *Tax on unrelated business income of exempt organizations* (IRS publication 598). Retrieved from http://web.archive.org/web/20030316230904/http://www.irs.gov/pub/irs-pdf/p598.pdf
Internal Revenue Service. (2003a). *IRS data book 1998–2001* Retrieved from http://web.archive.org/web/20050519035227/http://www.irs.gov/pub/irs-soi/01db22eo.xls
Internal Revenue Service. (2003b). *Tax exempt status for your organization* (IRS publication no. 557). Retrieved from http://web.archive.org/web/20040102203939/http://www.irs.gov/pub/irs-pdf/p557.pdf
Internal Revenue Service. (n.d.-a). *Unrelated business income tax—General rules*. Retrieved from http://web.archive.org/web/20050127092548/http://www.irs.gov/charities/article/0,,id=96104,00.html
Internal Revenue Service. (n.d.-b). *Section 501(c)3 organizations*. Retrieved from http://web.archive.org/web/20041118012537/http://www.irs.gov/publications/p557/ch03.html
Internal Revenue Service. (n.d.-c). *Political and lobbying activities*. Retrieved from http://web.archive.org/web/20050608095235/http://www.irs.gov/charities/charitable/article/0,,id=120703,00.html
Lemann, N. (1999). *The big test: The secret history of the American meritocracy*. New York: Farrar, Strauss and Giroux.
Restrictions on nonprofit activities for 501(c)(3) organizations. (n.d.). Retrieved from http://web.archive.org/web/20050207105910/http://www.alliance1.org/Public_Policy/IRS_rules.htm
Santayana, G. (1905). *The life of reason, reason in common sense*. New York: Scribner's.
Scrivner, G. N. (2001). A brief history of tax policy changes affecting charitable organizations. In J. S. Ott (Ed.), *The nature of the nonprofit sector* (pp. 126–142). Boulder: Westview.

Part I
ETS Contributions to Developing Analytic Tools for Educational Measurement

Chapter 2
A Review of Developments and Applications in Item Analysis

Tim Moses

This chapter summarizes contributions ETS researchers have made concerning the applications of, refinements to, and developments in item analysis procedures. The focus is on dichotomously scored items, which allows for a simplified presentation that is consistent with the focus of the developments and which has straightforward applications to polytomously scored items. Item analysis procedures refer to a set of statistical measures used by testing experts to review and revise items, to estimate the characteristics of potential test forms, and to make judgments about the quality of items and assembled test forms. These procedures and statistical measures have been alternatively characterized as conventional item analysis (Lord 1961, 1965a, b), traditional item analysis (Wainer 1989), analyses associated with classical test theory (Embretson and Reise 2000; Hambleton 1989; Tucker 1987; Yen and Fitzpatrick 2006), and simply item analysis (Gulliksen 1950; Livingston and Dorans 2004). This chapter summarizes key concepts of item analysis described in the sources cited. The first section describes item difficulty and discrimination indices. Subsequent sections review discussions about the relationships of item scores and test scores, visual displays of item analysis, and the additional roles item analysis methods have played in various psychometric contexts. The key concepts described in each section are summarized in Table 2.1.

T. Moses (✉)
College Board, New York, NY, USA
e-mail: tmoses@collegeboard.org

R.E. Bennett, M. von Davier (eds.), *Advancing Human Assessment*,
Methodology of Educational Measurement and Assessment,
DOI 10.1007/978-3-319-58689-2_2

Table 2.1 Summary key item analysis concepts

Item analysis concept	Motivation	Description of application to item analysis	Description of application(s) to other psychometric questions
Average item score ($\overline{x}_i$) and reference average item score ($\overline{x}_{i,2}$)	Index for summarizing item difficulty	Gulliksen (1950), Horst (1933), Lord and Novick (1968), Thurstone (1925), and Tucker (1987)	DIF (Dorans and Kulick 1986); item context/order (Dorans and Lawrence 1990; Moses et al. 2007)
Delta (Δ_i) and equated delta $\left[\hat{e}_2\left(\Delta_{i,1}\right)\right]$	Index for summarizing item difficulty with reduced susceptibility to score compression due to mostly high scores or mostly low scores	Brigham (1932), Gulliksen (1950), Holland and Thayer (1985), and Tucker (1987)	DIF (Holland and Thayer 1988); IRT comparisons (L. L. Cook et al. 1988)
Point biserial correlation $\left[\hat{r}_{\text{point biserial}}\left(x_i,y\right)\right]$	Index for summarizing item discrimination	Swineford (1936), Gulliksen (1950), and Lord and Novick (1968)	
Biserial correlation $\left[\hat{r}_{\text{biserial}}\left(x_i,y\right)\right]$	Index for summarizing item discrimination with reduced susceptibility to examinee group differences and to dichotomous scoring	Fan (1952), Pearson (1909), Tucker (1987), Turnbull (1946), and Lord and Novick (1968)	
r-Polyreg correlation $\left[\hat{r}_{\text{polyreg}}\left(x_i,y\right)\right]$	Index for summarizing item discrimination with reduced susceptibility to examinee group differences, dichotomous scoring, and the difficulties of estimating the biserial correlation	Lewis et al. (n.d.) and Livingston and Dorans (2004)	
Conditional average item score $\left(\overline{x}_{ik}\right)$ estimated from raw data	Obtain a detailed description of an item's functional relationship (difficulty and discrimination) with the criterion (usually a total test)	Thurstone (1925), Lord (1965a, b, 1970), and Wainer (1989)	DIF (Dorans and Holland 1993); IRT comparisons (Sinharay 2006)
Conditional average item scores $\left(\overline{x}_{ik}\right)$ estimated from raw data on percentile groupings of the total test scores	Obtain a detailed description of an item's functional relationship (difficulty and discrimination) for a total test with reduced susceptibility to sample fluctuations	Turnbull (1946), Tucker (1987), and Wainer (1989)	
Conditional average item scores $\left(\overline{x}_{ik}\right)$ estimated with kernel or other smoothing	Obtain a detailed description of an item's functional relationship (difficulty and discrimination) for a total test with reduced susceptibility to sample fluctuations	Ramsay (1991) and Livingston and Dorans (2004)	DIF (Moses et al. 2010); IRT comparisons (Moses 2016)

Note. DIF differential item functioning, *IRT* item response theory

2.1 Item Analysis Indices

In their discussions of item analysis, ETS researchers Lord and Novick (1968, p. 327) and, two decades later, Wainer (1989, p. 2) regarded items as the building blocks of a test form being assembled. The assembly of a high-quality test form depends on assuring that the individual building blocks are sound. Numerical indices can be used to summarize, evaluate, and compare a set of items, usually with respect to their difficulties and discriminations. Item difficulty and discrimination indices can also be used to check for potential flaws that may warrant item revision prior to item use in test form assembly. The most well-known and utilized difficulty and discrimination indices of item analysis were developed in the early twentieth century (W. W. Cook 1932; Guilford 1936; Horst 1933; Lentz et al. 1932; Long and Sandiford 1935; Pearson 1909; Symonds 1929; Thurstone 1925). Accounts of ETS scientists Tucker (1987, p. ii), Livingston and Dorans (2004) have described how historical item analysis indices have been applied and adapted at ETS from the mid-1940s to the present day.

2.1.1 *Item Difficulty Indices*

In their descriptions of item analyses, Gulliksen (1950) and Tucker (1987) listed two historical indices of item difficulty that have been the focus of several applications and adaptations at ETS. These item difficulty indices are defined using the following notation:

i is a subscript indexing the $i = 1$ to I items on Test Y,
j is a subscript indexing the $j = 1$ to N examinees taking Test Y,
x_{ij} indicates a score of 0 or 1 on the ith dichotomously scored Item i from examinee j (all N examinees have scores on all I items).

The most well-known item difficulty index is the average item score, or, for dichotomously scored items, the proportion of correct responses, the "p-value" or "P_+" (Gulliksen 1950; Hambleton 1989; Livingston and Dorans 2004; Lord and Novick 1968; Symonds 1929; Thurstone 1925; Tucker 1987; Wainer 1989):

$$\overline{x}_i = \frac{1}{N}\sum_{j}^{N} x_{ij}. \tag{2.1}$$

Estimates of the quantity defined in Eq. 2.1 can be obtained with several alternative formulas.[1] A more complex formula that is the basis of developments described in Sect. 2.2.1 can be obtained based on additional notation, where.

---

[1] Alternative expressions to the average item score computations shown in Eq. 2.1 are available in other sources. Expressions involving summations with respect to examinees are shown in Gulliksen (1950) and Lord and Novick (1968). More elaborate versions of Eq. 2.1 that address polytomously scored items and tests composed of both dichotomously and polytomously scored items have also been developed (J. Carlson, personal communication, November 6, 2013).

k is a subscript indexing the $k = 0$ to I possible scores of Test Y (y_k),
$\hat{p}_k$ is the observed proportion of examinees obtaining test score y_k,
$\overline{x}_{ik}$ is the average score on Item i for examinees obtaining test score y_k.

With the preceding notation, the average item score as defined in Eq. 2.1 can be obtained as

$$\overline{x}_i = \sum_k \hat{p}_k \, \overline{x}_{ik}.$$

Alternative item difficulty indices that use a transformation based on the inverse of the cumulative distribution function (CDF) of the normal distribution for the $\overline{x}_i$ in Eq. 2.1 have been proposed by ETS scientists (Gulliksen 1950; Horst 1933) and others (Symonds 1929; Thurstone 1925). The transformation based on the inverse of the CDF of the normal distribution is used extensively at ETS is the delta index developed by Brolyer (Brigham 1932; Gulliksen 1950):

$$\hat{\Delta}_i = 13 - 4\Phi^{-1}\left(\overline{x}_i\right), \tag{2.2}$$

where $\Phi^{-1}(p)$ represents the inverse of the standard normal cumulative distribution corresponding to the pth percentile. ETS scientists Gulliksen (1950, p. 368), Fan (1952, p. 1), Holland and Thayer (1985, p. 1), and Wainer (1989, p. 7) have described deltas as having features that differ from those of average item scores:

- The delta provides an increasing expression of an item's difficulty (i.e., is negatively associated with the average item score).
- The increments of the delta index are less compressed for very easy or very difficult items.
- The sets of deltas obtained for a test's items from two different examinee groups are more likely to be linearly related than the corresponding sets of average item scores.

Variations of the item difficulty indices in Eqs. 2.1 and 2.2 have been adapted and used in item analyses at ETS to address examinee group influences on item difficulty indices. These variations have been described both as actual item difficulty parameters (Gulliksen 1950, pp. 368–371) and as adjustments to existing item difficulty estimates (Tucker 1987, p. iii). One adjustment is the use of a linear function to transform the mean and standard deviation of a set of $\hat{\Delta}_i$ values from one examinee group to this set's mean and standard deviation from the examinee group of interest (Gulliksen 1950; Thurstone 1925, 1947; Tucker 1987):

$$\hat{e}_2\left(\hat{\Delta}_{i,1}\right) = \overline{\Delta}_{.,2} + \frac{\hat{\sigma}_{.,2}\left(\Delta\right)}{\hat{\sigma}_{.,1}\left(\Delta\right)}\left(\hat{\Delta}_{i,1} - \overline{\Delta}_{.,1}\right). \tag{2.3}$$

Equation 2.3 shows that the transformation of Group 1's item deltas to the scale of Group 2's deltas, $\hat{e}_2\left(\Delta_{i,1}\right)$, is obtained from the averages, $\overline{\Delta}_{.,1}$ and $\overline{\Delta}_{.,2}$, and standard deviations, $\hat{\sigma}_{.,1}\left(\Delta\right)$ and $\hat{\sigma}_{.,2}\left(\Delta\right)$, of the groups' deltas. The "mean sigma" adjustment in Eq. 2.3 has been exclusively applied to deltas (i.e., "delta equating"; Gulliksen 1950; Tucker 1987, p. ii) due to the higher likelihood of item deltas to reflect linear relationships between the deltas obtained from two examinee groups on the same set of items. Another adjustment uses Eq. 2.1 to estimate the average item scores for an examinee group that did not respond to those items but has available scores and $\hat{p}_k$ estimates on a total test (e.g., Group 2). Using Group 2's $\hat{p}_k$ estimates and the conditional average item scores from Group 1, which actually did respond to the items and also has scores on the same test as Group 2 (Livingston and Dorans 2004; Tucker 1987), the estimated average item score for Item i in Group 2 is

$$\overline{x}_{i,2} = \sum_k \hat{p}_{k,2}\, \overline{x}_{ik,1}. \tag{2.4}$$

The Group 2 adjusted or *reference* average item scores produced with Eq. 2.4 can be subsequently used with Eq. 2.2 to obtain delta estimates for Group 2.

Other measures have been considered as item difficulty indices in item analyses at ETS but have not been used as extensively as those in Eqs. 2.1, 2.2, 2.3, and 2.4. The motivation for considering the additional measures was to expand the focus of Eqs. 2.1, 2.2, and 2.3 beyond item difficulty to address the measurement heterogeneity that would presumably be reflected in relatively low correlations with other items, test scores, or assumed underlying traits (Gulliksen 1950, p. 369; Tucker 1948, 1987, p. iii). Different ways to incorporate items' biserial correlations (described in Sect. 2.1.2) have been considered, including the estimation of item–test regressions to identify the test score that predicts an average item score of 0.50 in an item (Gulliksen 1950). Other proposals to address items' measurement heterogeneity were attempts to incorporate heterogeneity indices into difficulty indices, such as by conducting the delta equating of Eq. 2.3 after dividing the items' deltas by the items' biserial correlations (Tucker 1948) and creating alternative item difficulty indices from the parameter estimates of three-parameter item characteristic curves (Tucker 1981). These additional measures did not replace delta equating in historical ETS practice, partly because of the computational and numerical difficulties in estimating biserial correlations (described later and in Tucker 1987, p. iii), accuracy loss due to computational difficulties in estimating item characteristic curves (Tucker 1981), and interpretability challenges (Tucker 1987, p. vi). Variations of the delta statistic in Eq. 2.2 have been proposed based on logistic cumulative functions rather than normal ogives (Holland and Thayer 1985). The potential benefits of logistic cumulative functions include a well-defined standard error estimate, odds ratio interpretations, and smoother and less biased estimation. These benefits have not been considered substantial enough to warrant a change to wide use of logistic cumulative functions, because the difference between the values of the logistic cumulative function and the normal ogive cumulative function is small

(Haley, cited in Birnbaum 1968, p. 399). In other ETS research by Olson, Scheuneman, and Grima (1989), proposals were made to study items' difficulties after exploratory and confirmatory approaches are used to categorize items into sets based on their content, context, and/or task demands.

2.1.2 *Item Discrimination Indices*

Indices of item discrimination summarize an item's relationship with a trait of interest. In item analysis, the total test score is almost always used as an approximation of the trait of interest. On the basis of the goals of item analysis to evaluate items, items that function well might be distinguished from those with flaws based on whether the item has a positive versus a low or negative association with the total score. One historical index of the item–test relationship applied in item analyses at ETS is the product moment correlation (Pearson 1895; see also Holland 2008; Traub 1997):

$$\hat{r}(x_i,y) = \frac{\hat{\sigma}(x_i,y)}{\hat{\sigma}(x_i)\hat{\sigma}(y)}, \tag{2.5}$$

where $\hat{\sigma}(x_i,y)$, $\hat{\sigma}(x_i)$, and $\hat{\sigma}(y)$ denote the estimated covariance and standard deviations of the item scores and test scores. For the dichotomously scored items of interest in this chapter, Eq. 2.5 is referred to as a point biserial correlation, which may be computed as

$$\hat{r}_{\text{point biserial}}(x_i,y) = \frac{\frac{1}{N}\sum_k N_k \overline{x}_{ik} y_k - \overline{x}_i \overline{y}}{\sqrt{\overline{x}_i(1-\overline{x}_i)}\hat{\sigma}(y)}, \tag{2.6}$$

where N and N_k denote the sample sizes for the total examinee group and for the subgroup of examinees obtaining total score y_k and $\overline{x}_i$ and $\overline{y}$ are the means of Item i and the test for the total examinee group. As described in Sect. 2.2.1, the point biserial correlation is a useful item discrimination index due to its direct relationship with respect to test score characteristics.

In item analysis applications, ETS researcher Swineford (1936) described how the point biserial correlation can be a "considerably lowered" (p. 472) measure of item discrimination when the item has an extremely high or low difficulty value. The biserial correlation (Pearson 1909) addresses the lowered point biserial correlation based on the assumptions that (a) the observed scores of Item i reflect an artificial dichotomization of a continuous and normally distributed trait (z), (b) y is normally distributed, and (c) the regression of y on z is linear. The biserial correlation can be estimated in terms of the point biserial correlation and is itself an estimate of the product moment correlation of z and y:

$$\hat{r}_{\text{biserial}}\left(x_i, y\right) = \hat{r}_{\text{point biserial}}\left(x_i, y\right)\frac{\sqrt{\overline{x}_i\left(1-\overline{x}_i\right)}}{\varphi\left(\hat{q}_i\right)} \approx \hat{r}_{zy}, \tag{2.7}$$

where $\varphi\left(\hat{q}_i\right)$ is the density of the standard normal distribution at $\hat{q}_i$ and where $\hat{q}_i$ is the assumed and estimated point that dichotomizes z into x_i (Lord and Novick 1968). Arguments have been made for favoring the biserial correlation estimate over the point biserial correlation as a discrimination index because the biserial correlation is not restricted in range due to Item i's dichotomization and because the biserial correlation is considered to be more invariant with respect to examinee group differences (Lord and Novick 1968, p. 343; Swineford 1936).

Despite its apparent advantages over the point biserial correlation (described earlier), ETS researchers and others have noted several drawbacks to the biserial correlation. Some of the potential drawbacks pertain to the computational complexities the $\varphi\left(\hat{q}_i\right)$ in Eq. 2.7 presented for item analyses conducted prior to modern computers (DuBois 1942; Tucker 1987). Theoretical and applied results revealed the additional problem that estimated biserial correlations could exceed 1 (and be lower than −1, for that matter) when the total test scores are not normally distributed (i.e., highly skewed or bimodal) and could also have high standard errors when the population value is very high (Lord and Novick 1968; Tate 1955a, b; Tucker 1987).

Various attempts have been made to address the difficulties of computing the biserial correlation. Prior to modern computers, these attempts usually involved different uses of punch card equipment (DuBois 1942; Tucker 1987). ETS researcher Turnbull (1946) proposed the use of percentile categorizations of the total test scores and least squares regression estimates of the item scores on the categorized total test scores to approximate Eq. 2.7 and also avoid its computational challenges. In other ETS work, lookup tables were constructed using the average item scores of the examinee groups falling below the 27th percentile or above the 73rd percentile on the total test and invoking bivariate normality assumptions (Fan 1952). Attempts to normalize the total test scores resulted in partially improved biserial correlation estimates but did not resolve additional estimation problems due to the discreteness of the test scores (Tucker 1987, pp. ii–iii, v). With the use of modern computers, Lord (1961) used simulations to evaluate estimation alternatives to Eq. 2.7, such as those proposed by Brogden (1949) and Clemens (1958). Other correlations based on maximum likelihood, ad hoc, and two-step (i.e., combined maximum likelihood and ad hoc) estimation methods have also been proposed and shown to have accuracies similar to each other in simulation studies (Olsson, Drasgow, and Dorans 1982).

The biserial correlation estimate eventually developed and utilized at ETS is from Lewis, Thayer, and Livingston (n.d.; see also Livingston and Dorans 2004). Unlike the biserial estimate in Eq. 2.7, the Lewis et al. method can be used with dichotomously or polytomously scored items, produces estimates that cannot exceed 1, and does not rely on bivariate normality assumptions. This correlation has been referred to as an *r*-polyreg correlation, an *r*-polyserial estimated by regression correlation (Livingston and Dorans 2004, p. 14), and an *r*-biserial correlation for dichotomously scored items. The correlation is based on the assumption that the item scores are determined by the examinee's position on an underlying latent continuous variable z. The distribution of z for candidates with a given criterion score y is assumed to be normal with mean $\beta_i y$ and variance 1, implying the following probit regression model:

$$P\left(x_i \leq 1 \middle| y\right) = P\left(z \leq_i \alpha \middle| y\right) = \varphi\left(a_i - \beta_i y\right), \tag{2.8}$$

where α_i is the value of z corresponding to $x_i = 1$, Φ is the standard normal cumulative distribution function, and a_i and β_i are intercept and slope parameters. Using the maximum likelihood estimate of β_i, the *r*-polyreg correlation can be computed as

$$\hat{r}_{\text{polyreg}}\left(x_i, y\right) = \frac{\sqrt{\hat{\beta}_i^2 \hat{\sigma}_y^2}}{\sqrt{\hat{\beta}_i^2 \hat{\sigma}_y^2 + 1}}, \tag{2.9}$$

where $\hat{\sigma}_y$ is the standard deviation of scores on criterion variable y and is estimated in the same group of examinees for which the polyserial correlation is to be estimated. In Olsson et al.'s (1982) terminology, the $\hat{r}_{\text{polyreg}}\left(x_i, y\right)$ correlation might be described as a two-step estimator that uses a maximum likelihood estimate of β_i and the traditional estimate of the standard deviation of y.

Other measures of item discrimination have been considered at ETS but have been less often used than those in Eqs. 2.5, 2.6, 2.7 and 2.9. In addition to describing relationships between total test scores and items' correct/incorrect responses, ETS researcher Myers (1959) proposed the use of biserial correlations to describe relationships between total test scores and distracter responses and between total test scores and not-reached responses. Product moment correlations are also sometimes used to describe and evaluate an item's relationships with other items (i.e., phi correlations; Lord and Novick 1968). Alternatives to phi correlations have been developed to address the effects of both items' dichotomizations (i.e., tetrachoric correlations; Lord and Novick 1968; Pearson 1909). Tetrachoric correlations have been used less extensively than phi correlations for item analysis at ETS, possibly due to their assumption of bivariate normality and their lack of invariance advantages (Lord and Novick 1968, pp. 347–349). Like phi correlations, tetrachoric correlations may also be infrequently used as item analysis measures because they describe the relationship of only two test items rather than an item and the total test.

2.2 Item and Test Score Relationships

Discussions of the relationships of item and test score characteristics typically arise in response to a perceived need to expand the focus of item indices. For example, in Sect. 2.1.2, item difficulty indices have been noted as failing to account for items' measurement heterogeneity (see also Gulliksen 1950, p. 369). Early summaries and lists of item indices (W. W. Cook 1932; Guilford 1936; Lentz et al. 1932; Long and Sandiford 1935; Pearson 1909; Richardson 1936; Symonds 1929), and many of the refinements and developments of these item indices from ETS, can be described with little coverage of their implications for test score characteristics. Even when test score implications have been covered in historical discussions, this coverage has usually been limited to experiments about how item difficulties relate to one or two characteristics of test scores (Lentz et al. 1932; Richardson 1936) or to "arbitrary indices" (Gulliksen 1950, p. 363) and "arbitrarily defined" laws and propositions (Symonds 1929, p. 482). In reviewing the sources cited earlier, Gulliksen (1950) commented that "the striking characteristic of nearly all the methods described is that no theory is presented showing the relationship between the validity or reliability of the total test and the method of item analysis suggested" (p. 363).

Some ETS contributions to item analysis are based on describing the relationships of item characteristics to test score characteristics. The focus on relationships of items and test score characteristics was a stated priority of Gulliksen's (1950) review of item analysis: "In developing and investigating procedures of item analysis, it would seem appropriate, first, to establish the relationship between certain item parameters and the parameters of the total test" (p. 364). Lord and Novick (1968) described similar priorities in their discussion of item analysis and indices: "In mental test theory, the basic requirement of an item parameter is that it have a definite (preferably a clear and simple) relationship to some interesting total-test-score parameter" (p. 328). The focus of this section's discussion is summarizing how the relationships of item indices and test form characteristics were described and studied by ETS researchers such as Green Jr. (1951), Gulliksen (1950), Livingston and Dorans (2004), Lord and Novick (1968), Sorum (1958), Swineford (1959), Tucker (1987), Turnbull (1946), and Wainer (1989).

2.2.1 Relating Item Indices to Test Score Characteristics

A test with scores computed as the sum of I dichotomously scored items has four characteristics that directly relate to average item scores and point biserial correlations of the items (Gulliksen 1950; Lord and Novick 1968). These characteristics include Test Y's mean (Gulliksen 1950, p. 367, Eq. 5; Lord and Novick 1968, p. 328, Eq. 15.2.3),

$$\bar{y} = \sum_i \bar{x}_i, \tag{2.10}$$

Test *Y*'s variance (Gulliksen 1950, p. 377, Equation 19; Lord and Novick 1968, p. 330, Equations 15.3.5 and 15.3.6),

$$\hat{\sigma}^2(y) = \sum_i \hat{r}_{\text{point biserial}}(x_i, y)\sqrt{\overline{x}_i(1-\overline{x}_i)}\hat{\sigma}(y) = \sum_i \hat{\sigma}(x_i, y) \tag{2.11}$$

Test *Y*'s alpha or KR-20 reliability (Cronbach 1951; Gulliksen 1950, pp. 378–379, Eq. 21; Kuder and Richardson 1937; Lord and Novick 1968, p. 331, Eq. 15.3.8),

$$\hat{r}\text{el}(y) = \left(\frac{I}{I-1}\right)\left\{1 - \frac{\sum_i \overline{x}_i(1-\overline{x}_i)}{\left[\sum_i \hat{r}_{\text{point biserial}}(x_i, y)\sqrt{\overline{x}_i(1-\overline{x}_i)}\right]^2}\right\}, \tag{2.12}$$

and Test *Y*'s validity as indicated by *Y*'s correlation with an external criterion, *W* (Gulliksen 1950, pp. 381–382, Eq. 24; Lord and Novick 1968, p. 332, Eq. 15.4.2),

$$\hat{r}_{wy} = \frac{\sum_i \hat{r}_{\text{point biserial}}(x_i, w)\sqrt{\overline{x}_i(1-\overline{x}_i)}}{\sum_i \hat{r}_{\text{point biserial}}(x_i, y)\sqrt{\overline{x}_i(1-\overline{x}_i)}}. \tag{2.13}$$

Equations 2.10–2.13 have several implications for the characteristics of an assembled test. The mean of an assembled test can be increased or reduced by including easier or more difficult items (Eq. 2.10). The variance and reliability of an assembled test can be increased or reduced by including items with higher or lower item–test correlations (Eqs. 2.11 and 2.12, assuming fixed item variances). The validity of an assembled test can be increased or reduced by including items with lower or higher item–test correlations (Eq. 2.13).

The test form assembly implications of Eqs. 2.10, 2.11, 2.12 and 2.13 have been the focus of additional research at ETS. Empirical evaluations of the predictions of test score variance and reliability from items' variances and correlations with test scores suggest that items' correlations with test scores have stronger influences than items' variances on test score variance and reliability (Swineford 1959). Variations of Eq. 2.12 have been proposed that use an approximated linear relationship to predict test reliability from items' biserial correlations with test scores (Fan, cited in Swineford 1959). The roles of item difficulty and discrimination have been described in further detail for differentiating examinees of average ability (Lord 1950) and for classifying examinees of different abilities (Sorum 1958). Finally, the correlation of a test and an external criterion shown in Eq. 2.13 has been used to develop methods of item selection and test form assembly based on maximizing test validity (Green 1951; Gulliksen 1950; Horst 1936).

2.2.2 Conditional Average Item Scores

In item analyses, the most detailed descriptions of relationships of items and test scores take the form of $\bar{x}_{ik}$, the average item score conditional on the kth score of total test Y (i.e., the discussion immediately following Eq. 2.1). ETS researchers have described these conditional average item scores as response curves (Livingston and Dorans 2004, p. 1), functions (Wainer 1989, pp. 19–20), item–test regressions (Lord 1965b, p. 373), and approximations to item characteristic curves (Tucker 1987, p. ii). Conditional average item scores tend to be regarded as one of the most fundamental and useful outputs of item analysis, because the $\bar{x}_{ik}$ are useful as the basis to calculate in item difficulty indices such as the overall average item score (the variation of Eq. 2.1), item difficulties estimated for alternative examinee groups (Eq. 2.4), and item discrimination indices such as the point biserial correlation (Eq. 2.6). Because the $1-\bar{x}_{ik}$ scores are also related to the difficulty and discrimination indices, the percentages of examinees choosing different incorrect (i.e., distracter) options or omitting the item making up the $1-\bar{x}_{ik}$ scores can provide even more information about the item. Item reviews based on conditional average item scores and conditional proportions of examinees choosing distracters and omitting the item involve relatively detailed presentations of individual items rather than tabled listings of all items' difficulty and discrimination indices for an entire test. The greater detail conveyed in conditional average item scores has prompted consideration of the best approaches to estimation and display of results.

The simplest and most direct approach to estimating and presenting $\bar{x}_{ik}$ and $1-\bar{x}_{ik}$ is based on the raw, unaltered conditional averages at each score of the total test. This approach has been considered in very early item analyses (Thurstone 1925) and also in more current psychometric investigations by ETS researchers Dorans and Holland (1993), Dorans and Kulick (1986), and Moses et al. (2010). Practical applications usually reveal that raw conditional average item scores are erratic and difficult to interpret without reference to measures of sampling instabilities (Livingston and Dorans 2004, p. 12).

Altered versions of $\bar{x}_{ik}$ and $1-\bar{x}_{ik}$ have been considered and implemented in operational and research contexts at ETS. Operational applications favored grouping total test scores into five or six percentile categories, with equal or nearly equal numbers of examinees, and reporting conditional average item scores and percentages of examinees choosing incorrect options across these categories (Tucker 1987; Turnbull 1946; Wainer 1989). Other, less practical alterations of the $\bar{x}_{ik}$ were considered in research contexts based on very large samples ($N > 100{,}000$), where, rather than categorizing the y_k scores, the $\bar{x}_{ik}$ values were only presented at total test scores with more than 50 examinees (Lore 1965b). Questions remained about how to present $\bar{x}_{ik}$ and $1-\bar{x}_{ik}$ at the uncategorized scores of the total test while also controlling for sampling variability (Wainer 1989, pp. 12–13).

Other research about item analysis has considered alterations of $\bar{x}_{ik}$ and $1-\bar{x}_{ik}$ (Livingston and Dorans 2004; Lord 1965a, b; Ramsay 1991). Most of these alterations involved the application of models and smoothing methods to reveal trends and eliminate irregularities due to sampling fluctuations in $\overline{x}_{ik}$ and $1-\overline{x}_{ik}$. Relatively strong mathematical models such as normal ogive and logistic functions have been found to be undesirable in theoretical discussions (i.e., the average slope of all test items' conditional average item scores does not reflect the normal ogive model; Lord 1965a) and in empirical investigations (Lord 1965b). Eventually,

> the developers of the ETS system chose a more flexible approach—one that allows the estimated response curve to take the shape implied by the data. Nonmonotonic curves, such as those observed with distracters, can be easily fit by this approach. (Livingston and Dorans 2004, p. 2)

This approach utilizes a special version of kernel smoothing (Ramsay 1991) to replace each $\overline{x}_{ik}$ or $1-\overline{x}_{ik}$ value with a weighted average of all $k = 0$ to I values:

$$KS\left(\overline{x}_{ik}\right)=\left(\sum_{l=0}^{I} w_{kl}\right)^{-1}\sum_{l=0}^{I} w_{kl}\overline{x}_{il}. \tag{2.14}$$

The w_{kl} values of Eq. 2.14 are Gaussian weights used in the averaging,

$$w_{kl}=\exp\left[\frac{-1}{2h}\frac{\left(y_l-y_k\right)^2}{\hat{\sigma}^2\left(y\right)}\right]n_l, \tag{2.15}$$

where exp denotes exponentiation, n_l is the sample size at test score y_l, and h is a kernel smoothing bandwidth parameter determining the extent of smoothing (usually set at $1.1N^{-0.2}$; Ramsay 1991). The rationale of the kernel smoothing procedure is to smooth out sampling irregularities by averaging adjacent $\overline{x}_{ik}$ values, but also to track the general trends in $\overline{x}_{ik}$ by giving the largest weights to the $\overline{x}_{ik}$ values at y scores closest to y_k and at y scores with relatively large conditional sample sizes, n_l. As indicated in the preceding Livingston and Dorans (2004) quote, the kernel smoothing in Eqs. 2.14 and 2.15 is also applied to the conditional percentages of examinees omitting and choosing each distracter that contribute to $1-\overline{x}_{ik}$. Standard errors and confidence bands of the raw and kernel-smoothed versions of $\overline{x}_{ik}$ values have been described and evaluated in Lewis and Livingston (2004) and Moses et al. (2010).

2.3 Visual Displays of Item Analysis Results

Presentations of item analysis results have reflected increasingly refined integrations of indices and conditional response information. In this section, the figures and discussions from the previously cited investigations are reviewed to trace the progression of item analysis displays from pre-ETS origins to current ETS practice.

The original item analysis example is Thurstone's (1925) scaling study for items of the Binet–Simon test, an early version of the Stanford–Binet test (Becker 2003; Binet and Simon 1905). The Binet–Simon and Stanford–Binet intelligence tests represent some of the earliest adaptive tests, where examiners use information they have about an examinee's maturity level (i.e., mental age) to determine where to begin testing and then administer only those items that are of appropriate difficulty for that examinee. The use of multiple possible starting points, and subsets of items, results in limited test administration time and maximized information obtained from each item but also presents challenges in determining how items taken by different examinees translate into a coherent scale of score points and of mental age (Becker 2003).

Thurstone (1925) addressed questions about the Binet–Simon test scales by developing and applying the item analysis methods described in this chapter to Burt's (1921) study sample of 2764 examinees' Binet–Simon test and item scores. Some steps of these analyses involved creating graphs of each of the test's 65 items' proportions correct, $\overline{x}_{ik}$, as a function of examinees' chronological ages, y. Then each item's "at par" (p. 444) age, y_k, is found such that 50% of examinees answered the item correctly, $\overline{x}_{ik} = 0.5$. Results of these steps for a subsample of the items were presented and analyzed in terms of plotted $\overline{x}_{ik}$ values (reprinted in Fig. 2.1).

Thurstone's (1925) analyses included additional steps for mapping all 65 items' at par ages to an item difficulty scale for 3.5-year-old examinees:

1. First the proportions correct of the items taken by 3-year-old, 4-year-old, …, 14-year-old examinees were converted into indices similar to the delta index shown in Eq. 2.2. That is, Thurstone's deltas were computed as $\Delta_{ik} = 0 - (1)\Phi^{-1}(\overline{x}_{ik})$, where the i subscript references the item and the k subscript references the age group responding to the item.

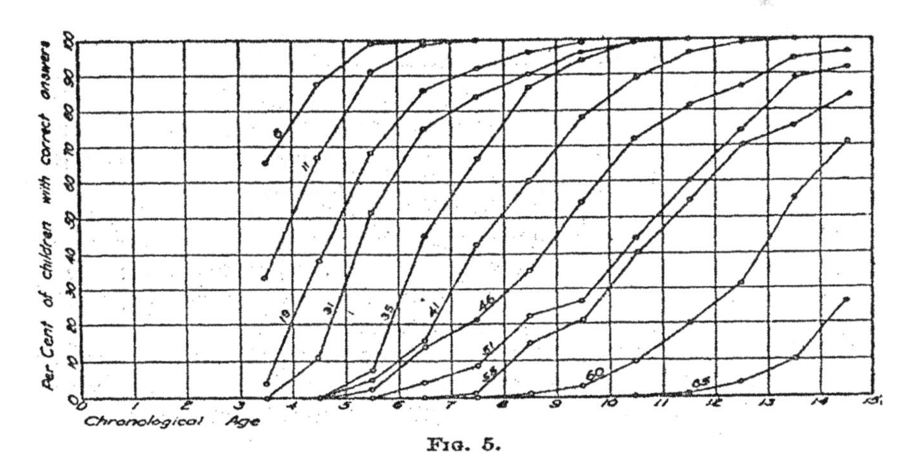

Fig. 2.1 Thurstone's (1925) Figure 5, which plots proportions of correct response (vertical axis) to selected items from the Binet–Simon test among children in successive age groups (horizontal axis)

2. For the sets of common items administered to two adjacent age groups (e.g., items administered to 8-year-old examinees and to 7-year-old examinees), the two sets of average item scores, $\bar{x}_{i7}$ and $\bar{x}_{i8}$, were converted into deltas, $\hat{\Delta}_{i7}$ and $\hat{\Delta}_{i8}$.
3. The means and standard deviations of the two sets of deltas from the common items administered to two adjacent age groups (e.g., 7- and 8-year-old examinees) were used with Eq. 2.3 to transform the difficulties of items administered to older examinees to the difficulty scale of items administered to the younger examinees,

$$\hat{e}_7\left(\hat{\Delta}_{i8}\right) = \bar{\Delta}_{.,7} + \frac{\hat{\sigma}_{.,7}(\Delta)}{\hat{\sigma}_{.,8}(\Delta)}\left(\hat{\Delta}_{i8} - \bar{\Delta}_{.,8}\right).$$

4. Steps 1–3 were repeated for the two sets of items administered to adjacent age groups from ages 3 to 14 years, with the purpose of developing scale transformations for the item difficulties observed for each age group to the difficulty scale of 3.5-year-old examinees.
5. The transformations obtained in Steps 1–4 for scaling the item difficulties at each age group to the difficulty scale of 3.5-year-old examinees were applied to items' $\hat{\Delta}_{ik}$ and $\bar{x}_{ik}$ estimates nearest to the items' at par ages. For example, with items at an at par age of 7.9, two scale transformations would be averaged, one for converting the item difficulties of 7-year-old examinees to the difficulty scale of 3.5-year-old examinees and another for converting the item difficulties of 8-year-old examinees to the difficulty scale of 3.5-year-old examinees. For items with different at par ages, the scale transformations corresponding to those age groups would be averaged and used to convert to the difficulty scale of 3.5-year-old examinees.

Thurstone (1925) used Steps 1–5 to map all 65 of the Binet–Simon test items to a scale and to interpret items' difficulties for 3.5-year-old examinees (Fig. 2.2). Items 1–7 are located to the left of the horizontal value of 0 in Fig. 2.2, indicating that these items are relatively easy (i.e., have $\bar{x}_{i3.5}$ values greater than 0.5 for the average 3.5-year-old examinee). Items to the right of the horizontal value of 0 in Fig. 2.2 are relatively difficult (i.e., have $\bar{x}_{i3.5}$ values less than 0.5 for the average 3.5-year-old examinee). The items in Fig. 2.2 at horizontal values far above 0 (i.e., greater than the mean item difficulty value of 0 for 3.5-year-old examinees by a given number of standard deviation units) are so difficult that they would not actually be administered to 3.5-year-old examinees. For example, Item 44 was actually administered to examinees 7 years old and older, but this item corresponds to a horizontal value of 5 in Fig. 2.2, implying that its proportion correct is estimated as 0.5 for 3.5-year-old examinees who are 5 standard deviation units more intelligent than the average 3.5-year-old examinee. The presentation in Fig. 2.2 provided empirical evidence that allowed Thurstone (1925) to describe the limitations of assembled forms of Burt–Simon items for measuring the intelligence of examinees at different ability levels and ages: "…the questions are unduly bunched at certain ranges and rather scarce at other ranges" (p. 448). The methods

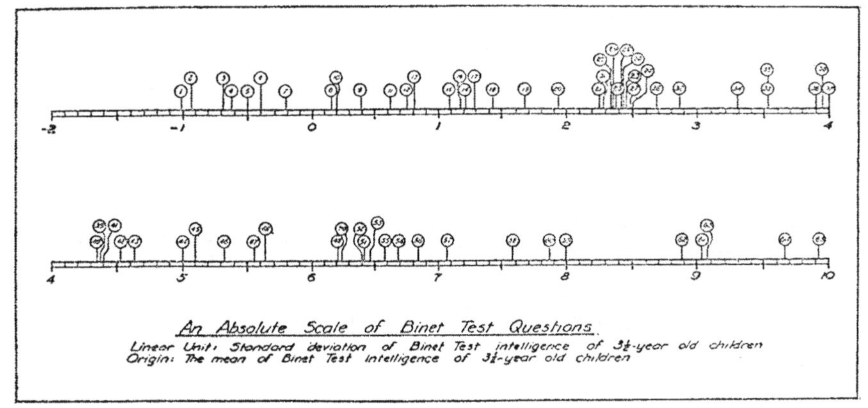

FIG. 6.

Fig. 2.2 Thurstone's (1925) Figure 6, which represents Binet–Simon test items' average difficulty on an absolute scale

Thurstone (1925) developed, and displayed in Figs. 2.1 and 2.2, were adapted and applied in item analysis procedures used at ETS (Gulliksen 1950, p. 368; Tucker 1987, p. ii).

Turnbull's (1946) presentation of item analysis results for an item from a 1946 College Entrance Examination Board test features an integration of tabular and graphical results, includes difficulty and discrimination indices, and also shows the actual multiple-choice item being analyzed (Fig. 2.3). The graph and table in Fig. 2.3 convey the same information, illustrating the categorization of the total test score into six categories with similar numbers of examinees ($n_k = 81$ or 82). Similar to Thurstone's conditional average item scores (Fig. 2.1), Turnbull's graphical presentation is based on a horizontal axis variable with few categories. The small number of categories limits sampling variability fluctuations in the conditional average item scores, but these categories are labeled in ways that conceal the actual total test scores corresponding to the conditional average item scores. In addition to presenting conditional average item scores, Turnbull's presentation reports conditional percentages of examinees choosing the item's four distracters. Wainer (1989, p. 10) pointed out that the item's correct option is not directly indicated but must be inferred to be the option with conditional scores that monotonically increase with the criterion categories. The item's overall average score (percentage choosing the right response) and biserial correlation, as well as initials of the staff who graphed and checked the results, are also included.

A successor of Turnbull's (1946) item analysis is the ETS version shown in Fig. 2.4 for a 1981 item from the *PSAT/NMSQT*® test (Wainer 1989).[2] The presentation in Fig. 2.4 is completely tabular, with the top table showing conditional sample

[2]In addition to the item analysis issues illustrated in Fig. 2.4 and in Wainer (1989), this particular item was the focus of additional research and discussion, which can be found in Wainer (1983).

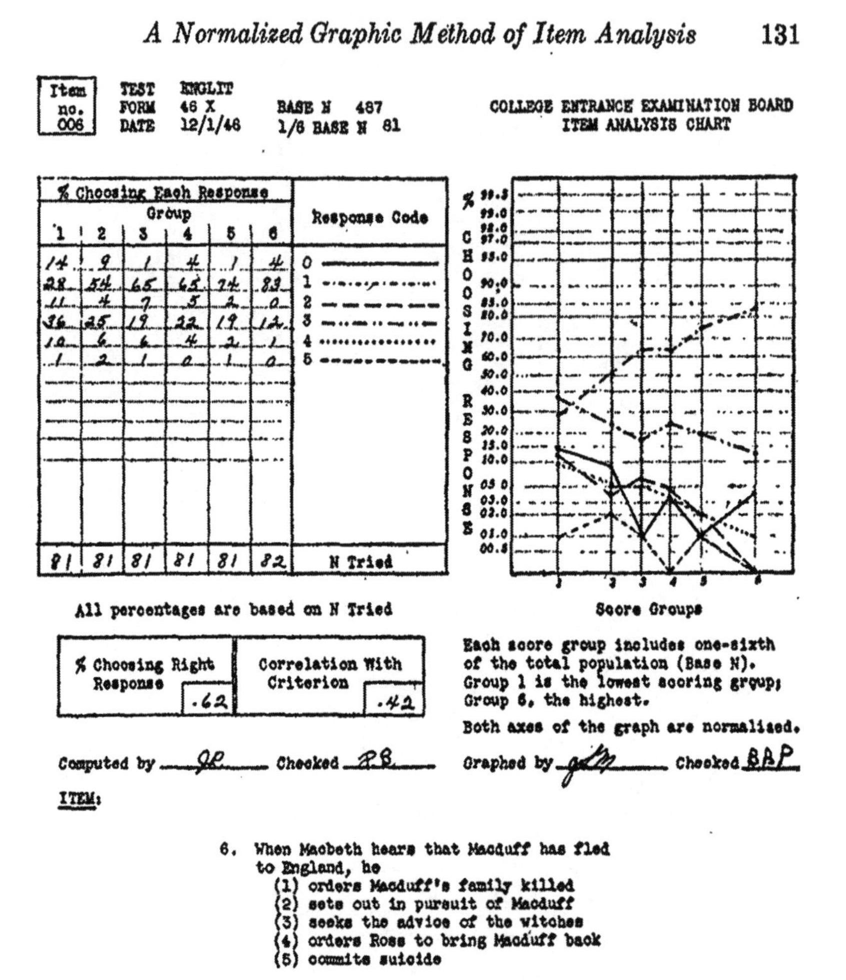

A Normalized Graphic Method of Item Analysis 131

Item no. 006 TEST ENGLIT FORM 46 X DATE 12/1/46 BASE N 487 1/6 BASE N 81

COLLEGE ENTRANCE EXAMINATION BOARD
ITEM ANALYSIS CHART

% Choosing Each Response						
Group						
1	2	3	4	5	6	Response Code
14	9	1	4	1	4	0
28	54	65	65	74	83	1
11	4	7	5	2	0	2
36	25	19	22	19	12	3
10	6	6	4	2	1	4
1	2	1	0	1	0	5
81	81	81	81	81	82	N Tried

All percentages are based on N Tried

% Choosing Right Response	Correlation With Criterion
.62	.42

Computed by JR Checked RB

ITEM:

Each score group includes one-sixth of the total population (Base N). Group 1 is the lowest scoring group; Group 6, the highest.

Both axes of the graph are normalized.

Graphed by gdm Checked BAP

6. When Macbeth hears that Macduff has fled to England, he
(1) orders Macduff's family killed
(2) sets out in pursuit of Macduff
(3) seeks the advice of the witches
(4) orders Ross to bring Macduff back
(5) commits suicide

FIGURE 1
Analysis of a sample item in English Literature

Fig. 2.3 Turnbull's (1946) Figure 1, which reports a multiple-choice item's normalized graph (*right*) and table (*left*) for all of its response options for six groupings of the total test score

sizes of examinees choosing the correct option, the distracters, and omitting the item, at five categories of the total test scores (Tucker 1987). The lower table in Fig. 2.4 shows additional overall statistics such as sample sizes and PSAT/NMSQT scores for the group of examinees choosing each option and the group omitting the item, overall average PSAT/NMSQT score for examinees reaching the item (M_{TOTAL}), observed deltas (Δ_O), deltas equated to a common scale using Eq. 2.3 (i.e., "equated deltas," Δ_E), percentage of examinees responding to the item (P_{TOTAL}), percentage of examinees responding correctly to the item (P_+), and the biserial correlation (r_{bis}). The lower table also includes an asterisk with the number of examinees choosing

ITEM NO. 44 TIS NO. 8012 TEST. MATH 2 FORM. 3CPT1 BASE N. 2930 DATE TABULATED. 2/12/81

EDUCATIONAL TESTING SERVICE

ITEM ANALYSIS

RESPONSE CODE	LOW N_1	N_2	N_3	N_4	HIGH N_5
OMIT	45	56	62	60	48
A	179	204	191	181	159
B	168	115	110	106	82
C	60	85	108	122	237
D	61	72	59	65	38
E	42	20	20	12	6
TOTAL	555	552	550	546	570

* DENOTES CORRECT RESPONSE

FORM	BASE N	OMIT	A	B	C	D	E	M_{TOTAL}	Δ_E SCALE	Δ_E	CRITERION
3CPT1	2930	271	914	581	612*	295	100	13.0	BOARD	15.2	IS050
TEST CODE	ITEM NO.	M_O	M_A	M_B	M_C	M_D	M_E	P_{TOTAL}	P_+	Δ_O	r_{bis}
MATH 2	44	13.0	12.8	12.0	15.0	12.4	10.8	0.95	0.22	16.1	0.36

Exhibit 1. Standard ETS Item Analysis Information Strip.

Fig. 2.4 Wainer's (1989) Exhibit 1, which illustrates a tabular display of classical item indices for a PSAT/NMSQT test's multiple-choice item's five responses and omitted responses from 1981

Option C to indicate that Option C is the correct option. Wainer used Turnbull's item presentation (Fig. 2.3) as a basis for critiquing the presentation of Fig. 2.4, suggesting that Fig. 2.4 could be improved by replacing the tabular presentation with a graphical one and also by including the actual item next to the item analysis results.

The most recent versions of item analyses produced at ETS are presented in Livingston and Dorans (2004) and reprinted in Figs. 2.5–2.7. These analysis presentations include graphical presentations of conditional percentages choosing the item's correct option, distracters, omits, and not-reached responses at individual uncategorized criterion scores. The dashed vertical lines represent percentiles of the score distribution where the user can choose which percentiles to show (in this case, the 20th, 40th, 60th, 80th, and 90th percentiles). The figures' presentations also incorporate numerical tables to present overall statistics for the item options and criterion scores as well as observed item difficulty indices, item difficulty indices equated using Eqs. 2.3 and 2.4 (labeled as Ref. in the figures), r-biserial correlations ($\hat{r}_{\text{polyreg}}\left(x_i, y\right)$; Eq. 2.9), and percentages of examinees reaching the item. Livingston and Dorans provided instructive discussion of how the item analysis presentations in Figs. 2.5–2.7 can reveal the typical characteristics of relatively easy items (Fig. 2.5), items too difficult for the intended examinee population (Fig. 2.6), and items exhibiting other problems (Fig. 2.7).

The results of the easy item shown in Fig. 2.5 are distinguished from those of the more difficult items in Figs. 2.6 and 2.7 in that the percentages of examinees choosing the correct option in Fig. 2.5 is 50% or greater for all examinees, and the percentages monotonically increase with the total test score. The items described in Figs. 2.6 and 2.7 exhibit percentages of examinees choosing the correct option that do not obviously rise for most criterion scores (Fig. 2.6) or do not rise more clearly than an intended incorrect option (Fig. 2.7). Livingston and Dorans (2004) interpreted Fig. 2.6 as indicative of an item that is too difficult for the examinees, where examinees do not clearly choose the correct option, Option E, at a higher rate than distracter C, except for the highest total test scores (i.e., the best performing exam-

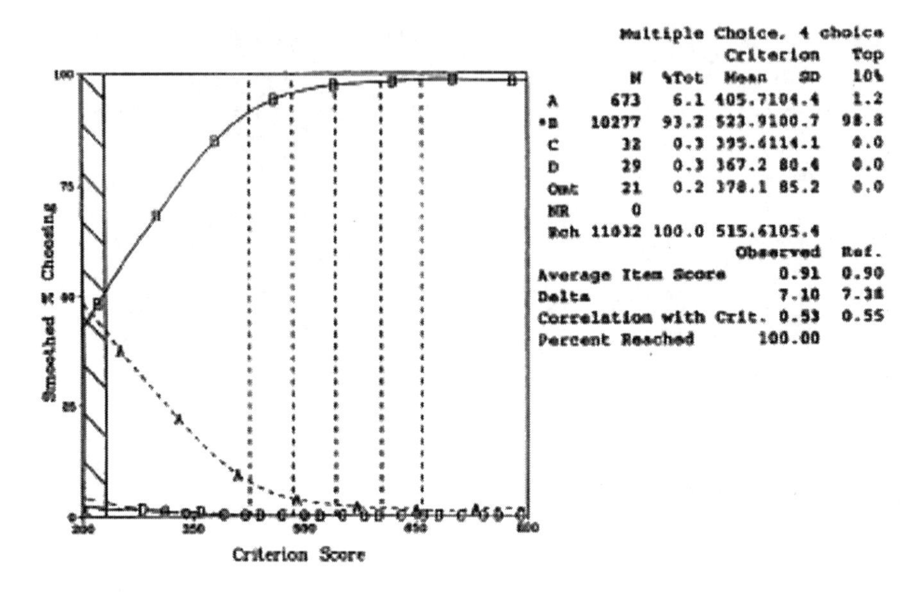

Fig. 2.5 Livingston and Dorans's (2004) Figure 1, which demonstrates classical item analysis results currently used at ETS, for a relatively easy item

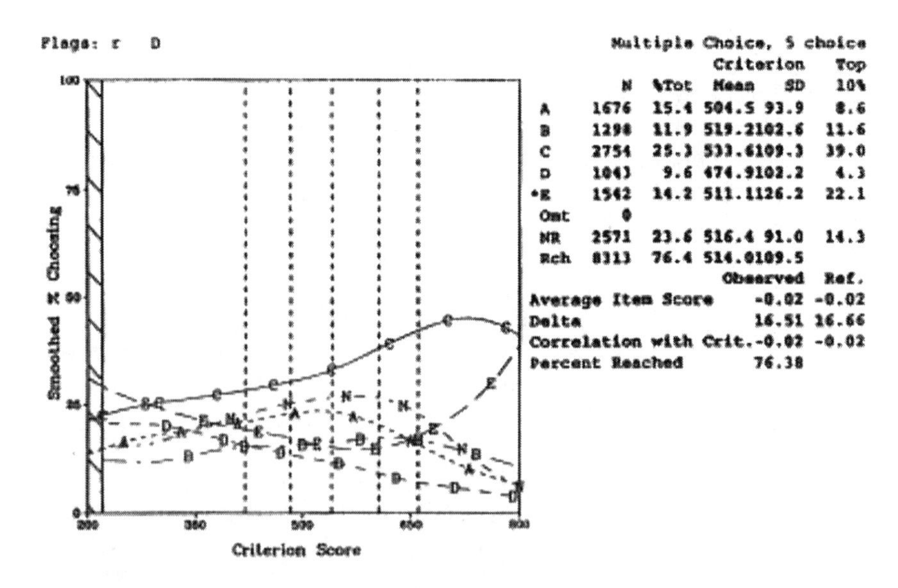

Fig. 2.6 Livingston and Dorans's (2004) Figure 5, which demonstrates classical item analysis results currently used at ETS, for a relatively difficult item

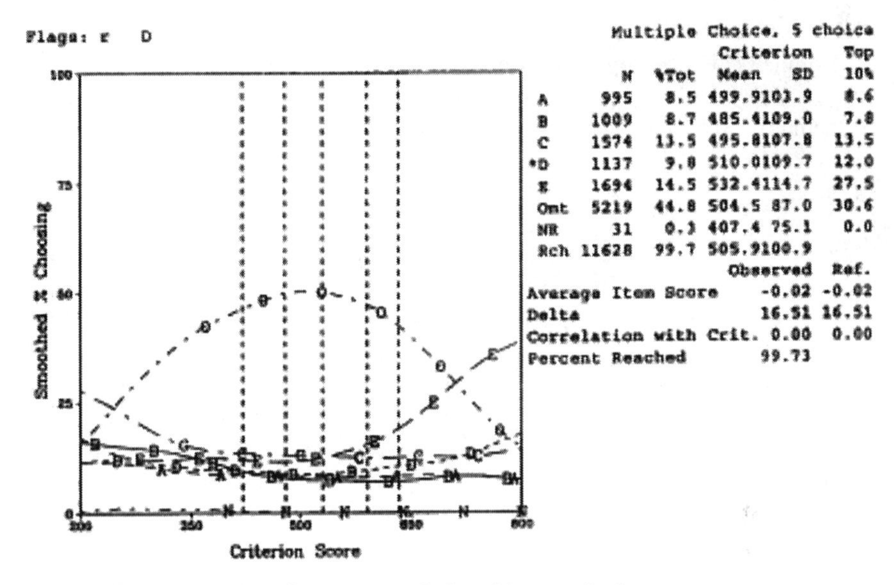

	N	%Tot	Criterion Mean	Criterion SD	Top 10%
A	995	8.5	499.9	103.9	8.6
B	1009	8.7	485.4	109.0	7.8
C	1574	13.5	495.8	107.8	13.5
*D	1137	9.8	510.0	109.7	12.0
E	1694	14.5	532.4	114.7	27.5
Omt	5219	44.8	504.5	87.0	30.6
NR	31	0.3	407.4	75.1	0.0
Rch	11628	99.7	505.9	100.9	

	Observed	Ref.
Average Item Score	-0.02	-0.02
Delta	16.51	16.51
Correlation with Crit.	0.00	0.00
Percent Reached	99.73	

***Figure 7*. An item that does not work for this population.**

Fig. 2.7 Livingston and Dorans's (2004) Figure 7, which demonstrates classical item analysis results currently used at ETS, for a problematic item

inees). Figure 2.7 is interpreted as indicative of an item that functions differently from the skill measured by the test (Livingston and Dorans 2004), where the probability of answering the item correctly is low for examinees at all score levels, where it is impossible to identify the correct answer (D) from the examinee response data, and where the most popular response for most examinees is to omit the item. Figures 2.6 and 2.7 are printed with statistical flags that indicate their problematic results, where the "*r*" flags indicate *r*-biserial correlations that are very low and even negative and the "*D*" flags indicate that high-performing examinees obtaining high percentiles of the criterion scores are more likely to choose one or more incorrect options rather than the correct option.

2.4 Roles of Item Analysis in Psychometric Contexts

2.4.1 Differential Item Functioning, Item Response Theory, and Conditions of Administration

The methods of item analysis described in the previous sections have been used for purposes other than informing item reviews and test form assembly with dichotomously scored multiple-choice items. In this section, ETS researchers' applications of item analysis to psychometric contexts such as differential item functioning

(DIF), item response theory (IRT), and evaluations of item order and context effects are summarized. The applications of item analysis in these areas have produced results that are useful supplements to those produced by the alternative psychometric methods.

2.4.2 Subgroup Comparisons in Differential Item Functioning

Item analysis methods have been applied to compare an item's difficulty for different examinee subgroups. These DIF investigations focus on "unexpected" performance differences for examinee subgroups that are matched in terms of their overall ability or their performance on the total test (Dorans and Holland 1993, p. 37). One DIF procedure developed at ETS is based on evaluating whether two subgroups' conditional average item scores differ from 0 (i.e., standardization; Dorans, and Kulick 1986):

$$\bar{x}_{ik,1} - \bar{x}_{ik,2} \neq 0, \quad k = 0, \ldots, I. \tag{2.16}$$

Another statistical procedure applied to DIF investigations is based on evaluating whether the odds ratios in subgroups for an item i differ from 1 (i.e., the Mantel–Haenszel statistic; Holland and Thayer 1988; Mantel and Haenszel 1959):

$$\frac{\bar{x}_{ik,1} / \left(1 - \bar{x}_{ik,1}\right)}{\bar{x}_{ik,2} / \left(1 - \bar{x}_{ik,2}\right)} \neq 1, \quad k = 0, \ldots, I. \tag{2.17}$$

Most DIF research and investigations focus on averages of Eq. 2.16 with respect to one "standardization" subgroup's total score distribution (Dorans and Holland 1993, pp. 48–49) or averages of Eq. 2.17 with respect to the combined subgroups' test score distributions (Holland and Thayer 1988, p. 134). Summary indices created from Eqs. 2.16 and 2.17 can be interpreted as an item's average difficulty difference for the two matched or standardized subgroups, expressed either in terms of the item's original scale (like Eq. 2.1) or in terms of the delta scale (like Eq. 2.2; Dorans and Holland 1993).

DIF investigations based on averages of Eqs. 2.16 and 2.17 have also been supplemented with more detailed evaluations, such as the subgroups' average item score differences at each of the total test scores indicated in Eq. 2.16. For example, Dorans and Holland (1993) described how the conditional average item score differences in Eq. 2.16 can reveal more detailed aspects of an item's differential functioning, especially when supplemented with conditional comparisons of matched subgroups' percentages choosing the item's distracters or of omitting the item. In ETS practice, conditional evaluations are implemented as comparisons of subgroups' conditional $\bar{x}_{ik}$ and $1 - \bar{x}_{ik}$ values after these values have been estimated with kernel smoothing (Eqs. 2.14 and 2.15). Recent research has shown that evaluations of differences in subgroups' conditional $\bar{x}_{ik}$ values can be biased when estimated with kernel smoothing and that more accurate subgroup comparisons of the conditional $\bar{x}_{ik}$ values can be obtained when estimated with logistic regression or loglinear models (Moses et al. 2010).

2.4.3 *Comparisons and Uses of Item Analysis and Item Response Theory*

Comparisons of item analysis and IRT with respect to methods, assumptions, and results have been an interest of early and contemporary psychometrics (Bock 1997; Embretson and Reise 2000; Hambleton 1989; Lord 1980; Lord and Novick 1968). These comparisons have also motivated considerations for updating and replacing item analysis procedures at ETS. In early years at ETS, potential IRT applications to item analysis were dismissed due to the computational complexities of IRT model estimation (Livingston and Dorans 2004) and also because of the estimation inaccuracies resulting from historical attempts to address the computational complexities (Tucker 1981). Some differences in the approaches' purposes initially slowed the adaptation of IRT to item analysis, as IRT methods were regarded as less oriented to the item analysis goals of item review and revision (Tucker 1987, p. iv). IRT models have also been interpreted to be less flexible in terms of reflecting the shapes of item response curves implied by actual data (Haberman 2009, p. 15; Livingston and Dorans 2004, p. 2).

This section presents a review of ETS contributions describing how IRT compares with item analysis. The contributions are reviewed with respect to the approaches' similarities, the approaches' invariance assumptions, and demonstrations of how item analysis can be used to evaluate IRT model fit. To make the discussions more concrete, the reviews are presented in terms of the following two-parameter normal ogive IRT model:

$$\text{prob}\left(x_i = 1 | \theta, a_i, b_i\right) = \int_{-\infty}^{a_i(\theta - b_i)} \frac{1}{\sqrt{2\pi}} \exp\left(\frac{-t^2}{2}\right) dt \tag{2.18}$$

where the probability of a correct response to dichotomously scored Item i is modeled as a function of an examinee's latent ability, θ, Item i's difficulty, b_i, and discrimination, a_i (Lord 1980). Alternative IRT models are reviewed by ETS researchers Lord (1980), Yen and Fitzpatrick (2006), and others (Embretson and Reise 2000; Hambleton 1989).

2.4.3.1 Similarities of Item Response Theory and Item Analysis

Item analysis and IRT appear to have several conceptual similarities. Both approaches can be described as predominantly focused on items and on the implications of items' statistics for assembling test forms with desirable measurement properties (Embretson and Reise 2000; Gulliksen 1950; Wainer 1989; Yen and Fitzpatrick 2006). The approaches have similar historical origins, as the Thurstone (1925) item scaling study that influenced item analysis (Gulliksen 1950; Tucker 1987) has also been described as an antecedent of IRT methods (Bock 1997, pp. 21–23; Thissen and Orlando 2001, pp. 79–83). The kernel smoothing methods used to depict conditional average item scores in item analysis (Eqs. 2.14 and 2.15) were originally developed as an IRT method that is nonparametric with respect to the shapes of its item response functions (Ramsay 1991, 2000).

In Lord and Novick (1968) and Lord (1980), the item difficulty and discrimination parameters of IRT models and item analysis are systematically related, and one can be approximated by a transformation of the other. The following assumptions are made to show the mathematical relationships (though these assumptions are not requirements of IRT models):

- The two-parameter normal ogive model in Eq. 2.18 is correct (i.e., no guessing).
- The regression of x_i on θ is linear with error variances that are normally distributed and homoscedastic.
- Variable θ follows a standard normal distribution.
- The reliability of total score y is high.
- Variable y is linearly related to θ.

With the preceding assumptions, the item discrimination parameter of the IRT model in Eq. 2.18 can be approximated from the item's biserial correlation as

$$a_i \approx \frac{r_{\text{biserial}}\left(x_i, y\right)}{\sqrt{1 - r_{\text{biserial}}\left(x_i, y\right)^2}}. \tag{2.19}$$

With the preceding assumptions, the item difficulty parameter of the IRT model in Eq. 2.18 can be approximated as

$$b_i \approx \frac{l\Delta_i}{r_{\text{biserial}}\left(x_i, y\right)}, \tag{2.20}$$

where $l\Delta_i$ is a linear transformation of the delta (Eq. 2.2). Although IRT does not require the assumptions listed earlier, the relationships in Eqs. 2.19 and 2.20 are used in some IRT estimation software to provide initial estimates in an iterative procedure to estimate $a_i$and b_i (Zimowski et al. 2003).

2.4.3.2 Comparisons and Contrasts in Assumptions of Invariance

One frequently described contrast of item analysis and IRT approaches is with respect to their apparent invariance properties (Embretson and Reise 2000; Hambleton 1989; Yen and Fitzpatrick 2006). A simplified statement of the question of interest is, When a set of items is administered to two not necessarily equal groups of examinees and then item difficulty parameters are estimated in the examinee groups using item analysis and IRT approaches, which approach's parameter estimates are more invariant to examinee group differences? ETS scientists Linda L. Cook, Daniel Eignor, and Hessy Taft (1988) compared the group sensitivities of item analysis deltas and IRT difficulty estimates after estimation and equating using achievement test data, sets of similar examinee groups, and other sets of dissimilar examinee groups. L. L. Cook et al.'s results indicate that equated deltas and IRT models' equated difficulty parameters are similar with respect to their stabilities and their potential for group dependence problems. Both approaches produced inaccurate estimates with very dissimilar examinee groups, results which are consistent with those of equating studies reviewed by ETS scientists L. L. Cook and Petersen (1987) and equating studies conducted by ETS scientists Lawrence and Dorans (1990), Livingston, Dorans, and Nancy Wright (1990), and Schmitt, Cook, Dorans, and Eignor (1990). The empirical results showing that difficulty estimates from item analysis and IRT can exhibit similar levels of group dependence tend to be underemphasized in psychometric discussions, which gives the impression that estimated IRT parameters are more invariant than item analysis indices (Embretson and Reise 2000, pp. 24–25; Hambleton 1989, p. 147; Yen and Fitzpatrick 2006, p. 111).

2.4.3.3 Uses of Item Analysis Fit Evaluations of Item Response Theory Models

Some ETS researchers have suggested the use of item analysis to evaluate IRT model fit (Livingston and Dorans 2004; Wainer 1989). The average item scores conditioned on the observed total test score, $\overline{x}_{ik}$, of interest in item analysis has been used as a benchmark for considering whether the normal ogive or logistic functions assumed in IRT models can be observed in empirical test data (Lord 1965a, b, 1970). One recent application by ETS scientist Sinharay (2006) utilized $\overline{x}_{ik}$ to describe and evaluate the fit of IRT models by considering how well the IRT models' posterior predictions of $\overline{x}_{ik}$ fit the $\overline{x}_{ik}$ values obtained from the raw data. Another recent investigation compared IRT models' $\overline{x}_{ik}$ values to those obtained from loglinear models of test score distributions (Moses 2016).

2.4.4 Item Context and Order Effects

A basic assumption of some item analyses is that items' statistical measures will be consistent if those items are administered in different contexts, locations, or positions (Lord and Novick 1968, p. 327). Although this assumption is necessary for supporting items' administration in adaptive contexts (Wainer 1989), examples in large-scale testing indicate that it is not always tenable (Leary and Dorans 1985; Zwick 1991). Empirical investigations of order and context effects on item statistics have a history of empirical evaluations focused on the changes in IRT estimates across administrations (e.g., Kingston and Dorans 1984). Other evaluations by ETS researchers Dorans and Lawrence (1990) and Moses et al. (2007) have focused on the implications of changes in item statistics on the total test score distributions from randomly equivalent examinee groups. These investigations have a basis in Gulliksen's (1950) attention to how item difficulty affects the distribution of the total test score (Eqs. 2.10 and 2.11). That is, the Dorans and Lawrence (1990) study focused on the changes in total test score means and variances that resulted from changes in the positions of items and intact sections of items. The Moses et al. (2007) study focused on changes in entire test score distributions that resulted from changes in the positions of items and from changes in the positions of intact sets of items that followed written passages.

2.4.5 Analyses of Alternate Item Types and Scores

At ETS, considerable discussion has been devoted to adapting and applying item analysis approaches to items that are not dichotomously scored. Indices of item difficulty and discrimination can be extended, modified, or generalized to account for examinees' assumed guessing tendencies and omissions (Gulliksen 1950; Lord and Novick 1968; Myers 1959). Average item scores (Eq. 2.1), point biserial correlations (Eq. 2.5), *r*-polyreg correlations (Eq. 2.9), and conditional average item scores have been adapted and applied in the analysis of polytomously scored items. Investigations of DIF based on comparing subgroups' average item scores conditioned on total test scores as in Eq. 2.16 have been considered for polytomously scored items by ETS researchers, including Dorans and Schmitt (1993), Moses et al. (2013), and Zwick et al. (1997). At the time of this writing, there is great interest in developing more innovative items that utilize computer delivery and are more interactive in how they engage examinees. With appropriate applications and possible additional refinements, the item analysis methods described in this chapter should have relevance for reviews of innovative item types and for attending to these items' potential adaptive administration contexts, IRT models, and the test forms that might be assembled from them.

Acknowledgments This manuscript was significantly improved from earlier versions thanks to reviews and suggestions from Jim Carlson, Neil Dorans, Skip Livingston and Matthias von Davier, and editorial work from Kim Fryer.

References

Becker, K. A. (2003). *History of the Stanford–Binet intelligence scales: Content and psychometrics* (Stanford–Binet intelligence scales, 5th Ed. Assessment Service Bulletin no. 1). Itasca: Riverside.

Binet, A., & Simon, T. (1905). Methodes nouvelles pour le diagnostic du nieveau intellectual anormoux [new methods for the diagnosis of levels of intellectual abnormality]. *L'Année Psychologique, 11*, 191–244. https://doi.org/10.3406/psy.1904.3675.

Birnbaum, A. (1968). Some latent trait models and their use in inferring an examinee's ability. In F. M. Lord & M. R. Novick (Eds.), *Statistical theories of mental test scores* (pp. 374–472). Reading: Addison-Wesley.

Bock, R. D. (1997). A brief history of item response theory. *Educational Measurement: Issues and Practice, 16*(4), 21–33. https://doi.org/10.1111/j.1745-3992.1997.tb00605.x.

Brigham, C. C. (1932). *A study of error*. New York: College Entrance Examination Board.

Brogden, H. E. (1949). A new coefficient: Application to biserial correlation and to estimation of selective efficiency. *Psychometrika, 14*, 169–182. https://doi.org/10.1007/BF02289151.

Burt, C. (1921). *Mental and scholastic tests*. London: King.

Clemens, W. V. (1958). An index of item-criterion relationship. *Educational and Psychological Measurement, 18*, 167–172. https://doi.org/10.1177/001316445801800118.

Cook, W. W. (1932). *The measurement of general spelling ability involving controlled comparisons between techniques*. Iowa City: University of Iowa Studies in Education.

Cook, L. L., & Petersen, N. S. (1987). Problems related to the use of conventional and item response theory equating methods in less than optimal circumstances. *Applied Psychological Measurement, 11*, 225–244. https://doi.org/10.1177/014662168701100302.

Cook, L. L., Eignor, D. R., & Taft, H. L. (1988). A comparative study of the effects of recency of instruction on the stability of IRT and conventional item parameter estimates. *Journal of Educational Measurement, 25*, 31–45. https://doi.org/10.1111/j.1745-3984.1988.tb00289.x.

Cronbach, L. J. (1951). Coefficient alpha and the internal structure of tests. *Psychometrika, 16*, 297–334. https://doi.org/10.1007/BF02310555.

Dorans, N. J., & Holland, P. W. (1993). DIF detection and description: Mantel–Haenszel and standardization. In P. W. Holland & H. Wainer (Eds.), *Differential item functioning* (pp. 35–66). Hillsdale: Erlbaum.

Dorans, N. J., & Kulick, E. (1986). Demonstrating the utility of the standardization approach to assessing unexpected differential item performance on the scholastic aptitude test. *Journal of Educational Measurement, 23*, 355–368. https://doi.org/10.1111/j.1745-3984.1986.tb00255.x.

Dorans, N. J., & Lawrence, I. M. (1990). Checking the statistical equivalence of nearly identical test editions. *Applied Measurement in Education, 3*, 245–254. https://doi.org/10.1207/s15324818ame0303_3.

Dorans, N. J., & Schmitt, A. P. (1993). Constructed response and differential item functioning: A pragmatic approach. In R. E. Bennett & W. C. Ward (Eds.), *Construction versus choice in cognitive measurement* (pp. 135–165). Hillsdale: Erlbaum.

DuBois, P. H. (1942). A note on the computation of biserial *r* in item validation. *Psychometrika, 7*, 143–146. https://doi.org/10.1007/BF02288074.

Embretson, S. E., & Reise, S. P. (2000). *Item response theory for psychologists*. Hillsdale: Erlbaum.

Fan, C.-T. (1952). *Note on construction of an item analysis table for the high-low-27-per-cent group method* (Research Bulletin no. RB-52-13). Princeton: Educational Testing Service. http://dx.doi.org/10.1002/j.2333-8504.1952.tb00227.x

Green, B. F., Jr. (1951). *A note on item selection for maximum validity* (Research Bulletin no. RB-51-17). Princeton: Educational Testing Service. http://dx.doi.org/10.1002/j.2333-8504.1951.tb00217.x

Guilford, J. P. (1936). *Psychometric methods*. New York: McGraw-Hill.

Gulliksen, H. (1950). *Theory of mental tests*. New York: Wiley. https://doi.org/10.1037/13240-000.

Haberman, S. J. (2009). *Use of generalized residuals to examine goodness of fit of item response models* (Research Report No. RR-09-15). Princeton: Educational Testing Service. http://dx.doi.org/10.1002/j.2333-8504.2009.tb02172.x

Hambleton, R. K. (1989). Principles and selected applications of item response theory. In R. L. Linn (Ed.), *Educational measurement* (3rd ed., pp. 147–200). Washington, DC: American Council on Education.

Holland, P. W. (2008, March). *The first four generations of test theory*. Paper presented at the ATP Innovations in Testing Conference, Dallas, TX.

Holland, P. W., & Thayer, D. T. (1985). *An alternative definition of the ETS delta scale of item difficulty* (Research Report No. RR-85-43). Princeton: Educational Testing Service. http://dx.doi.org/10.1002/j.2330-8516.1985.tb00128.x

Holland, P. W., & Thayer, D. T. (1988). Differential item performance and the Mantel–Haenszel procedure. In H. Wainer & H. I. Braun (Eds.), *Test validity* (pp. 129–145). Hillsdale: Erlbaum.

Horst, P. (1933). The difficulty of a multiple choice test item. *Journal of Educational Psychology, 24*, 229–232. https://doi.org/10.1037/h0073588.

Horst, P. (1936). Item selection by means of a maximizing function. *Psychometrika, 1*, 229–244. https://doi.org/10.1007/BF02287875.

Kingston, N. M., & Dorans, N. J. (1984). Item location effects and their implications for IRT equating and adaptive testing. *Applied Psychological Measurement, 8*, 147–154. https://doi.org/10.1177/014662168400800202.

Kuder, G. F., & Richardson, M. W. (1937). The theory of the estimation of test reliability. *Psychometrika, 2*, 151–160. https://doi.org/10.1007/BF02288391.

Lawrence, I. M., & Dorans, N. J. (1990). Effect on equating results of matching samples on an anchor test. *Applied Measurement in Education, 3*, 19–36. https://doi.org/10.1207/s15324818ame0301_3.

Leary, L. F., & Dorans, N. J. (1985). Implications for altering the context in which test items appear: A historical perspective on an immediate concern. *Review of Educational Research, 55*, 387–413. https://doi.org/10.3102/00346543055003387.

Lentz, T. F., Hirshstein, B., & Finch, J. H. (1932). Evaluation of methods of evaluating test items. *Journal of Educational Psychology, 23*, 344–350. https://doi.org/10.1037/h0073805.

Lewis, C., & Livingston, S. A. (2004). *Confidence bands for a response probability function estimated by weighted moving average smoothing*. Unpublished manuscript.

Lewis, C., Thayer, D., & Livingston, S. A. (n.d.). *A regression-based polyserial correlation coefficient*. Unpublished manuscript.

Livingston, S. A., & Dorans, N. J. (2004). *A graphical approach to item analysis* (Research Report No. RR-04-10). Princeton: Educational Testing Service. http://dx.doi.org/10.1002/j.2333-8504.2004.tb01937.x

Livingston, S. A., Dorans, N. J., & Wright, N. K. (1990). What combination of sampling and equating methods works best? *Applied Measurement in Education, 3*, 73–95. https://doi.org/10.1207/s15324818ame0301_6.

Long, J. A., & Sandiford, P. (1935). The validation of test items. *Bulletin of the Department of Educational Research, Ontario College of Education, 3*, 1–126.

Lord, F. M. (1950). *Properties of test scores expressed as functions of the item parameters* (Research Bulletin no. RB-50-56). Princeton: Educational Testing Service. http://dx.doi.org/10.1002/j.2333-8504.1950.tb00919.x

Lord, F. M. (1961). *Biserial estimates of correlation* (Research Bulletin no. RB-61-05). Princeton: Educational Testing Service. http://dx.doi.org/10.1002/j.2333-8504.1961.tb00105.x

Lord, F.M. (1965a). A note on the normal ogive or logistic curve in item analysis. *Psychometrika, 30*, 371–372. https://doi.org/10.1007/BF02289500

Lord, F.M. (1965b). An empirical study of item-test regression. *Psychometrika, 30*, 373–376. https://doi.org/10.1007/BF02289501

Lord, F.M. (1970). Item characteristic curves estimated without knowledge of their mathematical form—a confrontation of Birnbaum's logistic model. *Psychometrika, 35*, 43–50. https://doi.org/10.1007/BF02290592

Lord, F. M. (1980). *Applications of item response theory to practical testing problems*. Hillsdale: Erlbaum.

Lord, F. M., & Novick, M. R. (1968). *Statistical theories of mental test scores*. Reading: Addison-Wesley.

Mantel, N., & Haenszel, W. M. (1959). Statistical aspects of the analysis of data from retrospective studies of disease. *Journal of the National Cancer Institute, 22*, 719–748.

Moses, T. (2016). Estimating observed score distributions with loglinear models. In W. J. van der Linder & R. K. Hambleton (Eds.), *Handbook of item response theory* (2nd ed., pp. 71–85). Boca Raton: CRC Press.

Moses, T., Yang, W., & Wilson, C. (2007). Using kernel equating to check the statistical equivalence of nearly identical test editions. *Journal of Educational Measurement, 44*, 157–178. https://doi.org/10.1111/j.1745-3984.2007.00032.x.

Moses, T., Miao, J., & Dorans, N. J. (2010). A comparison of strategies for estimating conditional DIF. *Journal of Educational and Behavioral Statistics, 6*, 726–743. https://doi.org/10.3102/1076998610379135.

Moses, T., Liu, J., Tan, A., Deng, W., & Dorans, N. J. (2013). *Constructed response DIF evaluations for mixed format tests* (Research Report No. RR-13-33) Princeton: Educational Testing Service. http://dx.doi.org/10.1002/j.2333-8504.2013.tb02340.x

Myers, C. T. (1959). *An evaluation of the "not-reached" response as a pseudo-distracter* (Research Memorandum No. RM-59-06). Princeton: Educational Testing Service.

Olson, J. F., Scheuneman, J., & Grima, A. (1989). *Statistical approaches to the study of item difficulty* (Research Report No. RR-89-21). Princeton: Educational Testing Service. http://dx.doi.org/10.1002/j.2330-8516.1989.tb00136.x

Olsson, U., Drasgow, F., & Dorans, N. J. (1982). The polyserial correlation coefficient. *Psychometrika, 47*, 337–347. https://doi.org/10.1007/BF02294164.

Pearson, K. (1895). Contributions to the mathematical theory of evolution, II: Skew variation in homogeneous material. *Philosophical Transactions of the Royal Society, 186*, 343–414. https://doi.org/10.1098/rsta.1895.0010.

Pearson, K. (1909). On a new method for determining the correlation between a measured character a, and a character B. *Biometrika, 7*, 96–105. https://doi.org/10.1093/biomet/7.1-2.96.

Ramsay, J. O. (1991). Kernel smoothing approaches to nonparametric item characteristic curve estimation. *Psychometrika, 56*, 611–630. https://doi.org/10.1007/BF02294494.

Ramsay, J. O. (2000). *TESTGRAF: A program for the graphical analysis of multiple-choice test and questionnaire data* [Computer software and manual]. Retrieved from http://www.psych.mcgill.ca/faculty/ramsay/ramsay.html

Richardson, M. W. (1936). Notes on the rationale of item analysis. *Psychometrika, 1*, 69–76. https://doi.org/10.1007/BF02287926.

Schmitt, A. P., Cook, L. L., Dorans, N. J., & Eignor, D. R. (1990). Sensitivity of equating results to different sampling strategies. *Applied Measurement in Education, 3*, 53–71. https://doi.org/10.1207/s15324818ame0301_5.

Sinharay, S. (2006). Bayesian item fit analysis for unidimensional item response theory models. *British Journal of Mathematical and Statistical Psychology, 59*, 429–449. https://doi.org/10.1348/000711005X66888.

Sorum, M. (1958). *Optimum item difficulty for a multiple-choice test* (Research memorandum no. RM-58-06). Princeton: Educational Testing Service.

Swineford, F. (1936). Biserial *r* versus Pearson *r* as measures of test-item validity. *Journal of Educational Psychology, 27*, 471–472. https://doi.org/10.1037/h0052118.

Swineford, F. (1959, February). Some relations between test scores and item statistics. *Journal of Educational Psychology, 50*(1), 26–30. https://doi.org/10.1037/h0046332.

Symonds, P. M. (1929). Choice of items for a test on the basis of difficulty. *Journal of Educational Psychology, 20*, 481–493. https://doi.org/10.1037/h0075650.

Tate, R. F. (1955a). Applications of correlation models for biserial data. *Journal of the American Statistical Association, 50*, 1078–1095. https://doi.org/10.1080/01621459.1955.10501293.

Tate, R. F. (1955b). The theory of correlation between two continuous variables when one is dichotomized. *Biometrika, 42*, 205–216. https://doi.org/10.1093/biomet/42.1-2.205.

Thissen, D., & Orlando, M. (2001). Item response theory for items scored in two categories. In D. Thissen & H. Wainer (Eds.), *Test scoring* (pp. 73–140). Mahwah: Erlbaum.

Thurstone, L. L. (1925). A method of scaling psychological and educational tests. *Journal of Educational Psychology, 16*, 433–451. https://doi.org/10.1037/h0073357.

Thurstone, L. L. (1947). The calibration of test items. *American Psychologist, 3*, 103–104. https://doi.org/10.1037/h0057821.

Traub, R. E. (1997). Classical test theory in historical perspective. *Educational Measurement: Issues and Practice, 16*(4), 8–14. https://doi.org/10.1111/j.1745-3992.1997.tb00603.x.

Tucker, L. R. (1948). A method for scaling ability test items taking item unreliability into account. *American Psychologist, 3*, 309–310.

Tucker, L. R. (1981). *A simulation–Monte Carlo study of item difficulty measures delta and D.6* (Research Report No. RR-81-06). Princeton: Educational Testing Service. http://dx.doi.org/10.1002/j.2333-8504.1981.tb01239.x

Tucker, L. R. (1987). *Developments in classical item analysis methods* (Research Report No. RR-87-46). Princeton: Educational Testing Service. http://dx.doi.org/10.1002/j.2330-8516.1987.tb00250.x

Turnbull, W. W. (1946). A normalized graphic method of item analysis. *Journal of Educational Psychology, 37*, 129–141. https://doi.org/10.1037/h0053589.

Wainer, H. (1983). Pyramid power: Searching for an error in test scoring with 830,000 helpers. *American Statistician, 37*, 87–91. https://doi.org/10.1080/00031305.1983.10483095.

Wainer, H. (1989, Summer). The future of item analysis. *Journal of Educational Measurement, 26*, 191–208.

Yen, W. M., & Fitzpatrick, A. R. (2006). Item response theory. In R. L. Brennan (Ed.), *Educational measurement* (4th ed., pp. 111–153). Westport: American Council on Education and Praeger.

Zimowski, M. F., Muraki, E., Mislevy, R. J., & Bock, R. D. (2003). *BILOG-MG [computer software]*. Lincolnwood: Scientific Software International.

Zwick, R. (1991). Effects of item order and context on estimation of NAEP Reading proficiency. *Educational Measurement: Issues and Practice, 10*, 10–16. https://doi.org/10.1111/j.1745-3992.1991.tb00198.x.

Zwick, R., Thayer, D. T., & Mazzeo, J. (1997). *Describing and categorizing DIF in polytomous items* (Research Report No. RR-97-05). Princeton: Educational Testing Service. http://dx.doi.org/10.1002/j.2333-8504.1997.tb01726.x

Chapter 3
Psychometric Contributions: Focus on Test Scores

Tim Moses

This chapter is an overview of ETS psychometric contributions focused on test scores, in which issues about items and examinees are described to the extent that they inform research about test scores. Comprising this overview are Sect. 3.1 Test Scores as Measurements and Sect. 3.2 Test Scores as Predictors in Correlational and Regression Relationships. The discussions in these sections show that these two areas are not completely independent. As a consequence, additional contributions are the focus in Sect. 3.3 Integrating Developments About Test Scores as Measurements and Test Scores as Predictors. For each of these sections, some of the most important historical developments that predate and provide context for the contributions of ETS researchers are described.

3.1 Test Scores as Measurements

3.1.1 Foundational Developments for the Use of Test Scores as Measurements, Pre-ETS

By the time ETS officially began in 1947, the fundamental concepts of the classical theory of test scores had already been established. These original developments are usually traced to Charles Spearman's work in the early 1900s (Gulliksen 1950; Mislevy 1993), though Edgeworth's work in the late 1800s is one noteworthy predecessor (Holland 2008). Historical reviews describe how the major ideas of

A version of this chapter was originally published in 2013 by Educational Testing Service as a research report in the ETS R&D Scientific and Policy Contributions Series.

T. Moses (✉)
College Board, New York, NY, USA
e-mail: tmoses@collegeboard.org

R.E. Bennett, M. von Davier (eds.), *Advancing Human Assessment*,
Methodology of Educational Measurement and Assessment,
DOI 10.1007/978-3-319-58689-2_3

classical test theory, such as conceptions of test score averages and errors, were borrowed from nineteenth century astronomers and were probably even informed by Galileo's work in the seventeenth century (Traub 1997).

To summarize, the fundamental concepts of classical test theory are that an observed test score for examinee p on a particular form produced for test X, X'_p, can be viewed as the sum of two independent components: the examinee's true score that is assumed to be stable across all parallel forms of X, T_{Xp}, and a random error that is a function of the examinee and is specific to test form X', $E_{X'p}$,

$$X'_p = T_{Xp} + E_{X'p} \tag{3.1}$$

Classical test theory traditionally deals with the hypothetical scenario where examinee p takes an infinite number of parallel test forms (i.e., forms composed of different items but constructed to have identical measurement properties, X', X'', X''', …). As the examinee takes the infinite number of test administrations, the examinee is assumed to never tire from the repeated testing, does not remember any of the content in the test forms, and does not remember prior performances on the hypothetical test administrations. Under this scenario, classical test theory asserts that means of observed scores and errors for examinee p across all the X', X'', X'''… forms are

$$\mu\left(X'_p\right) = T_{Xp} \text{ } and \text{ } \mu\left(E_{X'p}\right) = 0, \tag{3.2}$$

and the conditional variance for examinee p across the forms is

$$\sigma^2_{X_p|T_{X_p}} = \sigma^2_{E_{X_p}} \tag{3.3}$$

The variance of the observed score turns out to be the sum of the true score variance and the error variance,

$$\sigma^2_X = \sigma^2_{T_X} + \sigma^2_{E_X}, \tag{3.4}$$

where the covariance of the true scores and errors, $\sigma_{T_X, E_{X_2}}$, is assumed to be zero. Research involving classical test theory often focuses on $\sigma^2_{T_X}$ and $\sigma^2_{E_X}$, meaning that considerable efforts have been devoted to developing approaches for estimating these quantities. The reliability of a test score can be summarized as a ratio of those variances,

$$rel\left(X\right) = \frac{\sigma^2_{T_X}}{\sigma^2_X} = 1 - \frac{\sigma^2_{E_X}}{\sigma^2_X} \tag{3.5}$$

Reliability indicates the measurement precision of a test form for the previously described hypothetical situation involving administrations of an infinite number of parallel forms given to an examinee group.

3.1.2 Overview of ETS Contributions

Viewed in terms of the historical developments summarized in the previous section, many psychometric contributions at ETS can be described as increasingly refined extensions of classical test theory. The subsections in Sect. 3.1 summarize some of the ETS contributions that add sophistication to classical test theory concepts. The summarized contributions have themselves been well captured in other ETS contributions that provide culminating and progressively more rigorous formalizations of classical test theory, including Gulliksen's (1950) *Theory of Mental Tests*, Lord and Novick's (1968) *Statistical Theories of Mental Test Scores*, and Novick's (1965) *The Axioms and Principal Results of Classical Test Theory*. In addition to reviewing and making specific contributions to classical test theory, the culminating formalizations address other more general issues such as different conceptualizations of observed score, true score, and error relationships (Gulliksen 1950), derivations of classical test theory resulting from statistical concepts of sampling, replications and experimental units (Novick 1965), and latent, platonic, and other interpretations of true scores (Lord and Novick 1968). The following subsections of this paper summarize ETS contributions about specific aspects of classical test theory. Applications of these contributions to improvements in the psychometric (measurement) quality of ETS tests are also described.

3.1.3 ETS Contributions About $\sigma_{E_X|T_{X_P}}$

The finding that σ_{E_X} (i.e., the standard error of measurement) may not indicate the actual measurement error for all examinees across all T_{Xp} values is an important, yet often forgotten contribution of early ETS researchers. The belief that classical test theory assumes that $\sigma^2_{E_X|T_{X_p}}$ is constant for all T_{Xp} values has been described as a common misconception (Haertel 2006), and appears to have informed misleading statements about the disadvantages of classical test theory relative to item response theory (e.g., Embretson and Reise 2000, p. 16).

In fact, the variability of the size of tests' conditional standard errors has been the focus of empirical study where actual tests were divided into two halves of equivalent difficulty and length (i.e., tau equivalent, described in Sect. 3.1.5.1), the standard deviation of the differences between the half test scores of examinees grouped by their total scores were computed, and a polynomial regression was fit to the estimated conditional standard errors on the total test scores and graphed (Mollenkopf 1949). By relating the coefficients of the polynomial regression to empirical test score distributions, Mollenkopf showed that conditional standard errors are usually larger near the center of the score distribution than at the tail and may only be expected to be constant for normally distributed and symmetric test-score distributions.

Another contribution to conditional standard error estimation involves assuming a binomial error model for number-correct scores (Lord 1955b, 1957a). If a test is regarded as a random sample of n dichotomously scored items, then the total score for an examinee with a particular true score, $T_{xp,}$ may be modeled as the sum of n draws from a binomial distribution with the probability of success on each draw equal to the average of their scores on the n items. The variance of the number-correct score under this model is binomial,

$$T_{Xp}\left(1-\frac{T_{Xp}}{n}\right). \tag{3.6}$$

The sample estimate of the conditional standard error can be computed by substituting observed scores for true scores and incorporating a correction for the use of the sample estimate of error variance,

$$\sqrt{\frac{X_p\left(n-X_p\right)}{n-1}}. \tag{3.7}$$

It is an estimator of the variance expected across hypothetical repeated measurements for each separate examinee where each measurement employs an independent sample of n items from an infinite population of such items. As such, it is appropriate for absolute or score-focused interpretations for each examinee.

An adjustment to Lord's (1955b, 1957a) conditional standard error for making relative interpretations of examinees' scores in relation to other examinees rather than with respect to absolute true score values was provided by Keats (1957). Noting that averaging Lord's $\frac{X_p\left(n-X_p\right)}{n-1}$ quantity produces the square of the overall standard error of measurement for the Kuder-Richardson Formula 21, $\sigma_{Xp}^2\left[1-rel_{21}\left(X\right)\right]$ (described in Sect. 3.1.5.2), Keats proposed a correction that utilizes the Kuder-Richardson Formula 21 reliability, $rel_{21}(X)$, and any other reliability estimate of interest, $\widehat{rel}\left(X\right)$. The conditional standard error estimate based on Keats' correction,

$$\sqrt{\frac{X_p\left(n-X_p\right)\left[1-\widehat{rel}\left(X\right)\right]}{\left(n-1\right)\left[1-rel_{21}\left(X\right)\right]}}, \tag{3.8}$$

produces a single standard error estimate for each observed score that is appropriate for tests consisting of equally weighted, dichotomously scored items.

3.1.4 Intervals for True Score Inference

One application of interest of standard errors of measurement in Sect. 3.1.3 is to true-score estimation, such as in creating confidence intervals for estimates of the true scores of examinees. Tolerance intervals around estimated true scores are attempts to locate the true score at a specified percentage of confidence (Gulliksen 1950). The confidence intervals around true scores formed from overall or conditional standard errors would be most accurate when errors are normally distributed (Gulliksen 1950, p. 17). These relatively early applications of error estimates to true score estimation are questionable, due in part to empirical investigations that suggest that measurement errors are more likely to be binomially distributed rather than normally distributed (Lord 1958a).

For number-correct or proportion-correct scores, two models that do not invoke normality assumptions are the beta-binomial strong true-score model (Lord 1965) and the four-parameter beta model (Keats and Lord 1962). The beta-binomial model builds on the binomial error model described in Sect. 3.1.3. If the observed test score of examinee p is obtained by a random sample of n items from some item domain, the mean item score is the probability of a correct response to each such randomly chosen item. This fact implies the binomial error model, that the observed score of examinee p follows a binomial distribution for the sum of n tries with the probability related to the mean for each trial (i.e., the average item score). The four-parameter beta-binomial model is a more general extension of the binomial error model, modeling the true-score distribution as a beta distribution linearly rescaled from the (0,1) interval to the (a,b) interval, $0 \leq a < b \leq 1$. Estimation for two-parameter and four-parameter beta-binomial models can be accomplished by the method of moments (Hanson 1991; Keats and Lord 1962, 1968, Chapter 23). The beta-binomial and four-parameter beta models have had widespread applicability, including not only the construction of tolerance intervals of specified percentages for the true scores of an examinee group (Haertel 2006; Lord and Stocking 1976), but also providing regression-based estimates of true scores (Lord and Novick 1968), and providing estimates of consistency and accuracy when examinees are classified at specific scores on a test (Livingston and Lewis 1995).

3.1.5 Studying Test Score Measurement Properties With Respect to Multiple Test Forms and Measures

3.1.5.1 Alternative Classical Test Theory Models

When the measurement properties of the scores of multiple tests are studied, approaches based on the classical test theory model and variations of this model typically begin by invoking assumptions that aspects of the test scores are identical. Strictly parallel test forms have four properties: They are built from identical test specifications, their observed score distributions are identical when administered to any (indefinitely large) population of examinees, they have equal covariances with one another (if there are more than two tests), and they have identical covariances with any other measure of the same or a different construct. Situations with multiple tests that have similar measurement properties but are not necessarily strictly parallel have been defined, and the definitions have been traced to ETS authors (Haertel 2006). In particular, Lord and Novick (1968, p. 48) developed a stronger definition of strictly parallel tests by adding to the requirement of equal covariances that the equality must hold for every subpopulation for which the test is to be used (also in Novick 1965). Test forms can be tau equivalent when each examinee's true score is constant across the forms while the error variances are unequal (Lord and Novick, p. 50). Test forms can be essentially tau equivalent when an examinee's true scores on the forms differ by an additive constant (Lord and Novick, p. 50). Finally, Haertel credits Jöreskog (1971b) for defining a weaker form of parallelism by dropping the requirement of equal true-score variances (i.e., congeneric test forms). That is, congeneric test forms have true scores that are perfectly and linearly related but with possibly unequal means and variances. Although Jöreskog is credited for the official definition of congeneric test form, Angoff (1953) and Kristof (1971) were clearly aware of this model when developing their reliability estimates summarized below.

3.1.5.2 Reliability Estimation

The interest in reliability estimation is often in assessing the measurement precision of a single test form. This estimation is traditionally accomplished by invoking classical test theory assumptions about two or more measures related to the form in question. The scenario in which reliability is interpreted as a measure of score precision when an infinite number of parallel test forms are administered to the same examinees under equivalent administration conditions (see Sect. 3.2.1) is mostly regarded as a hypothetical thought experiment rather than a way to estimate reliability empirically. In practice, reliability estimates are most often obtained as *internal consistency estimates*. This means the only form administered is the one for which reliability is evaluated and variances and covariances of multiple parts constructed from the individual items or half tests on the administered form are obtained while invoking classical test theory assumptions that these submeasures are parallel, tau equivalent, or congeneric.

Many of the popular reliability measures obtained as internal consistency estimates were derived by non-ETS researchers. One of these measures is the Spearman-Brown estimate for a test (X) divided into two strictly parallel halves (X_1 and X_2),

$$\frac{2\rho_{X1,X2}}{1+\rho_{X1,X2}}, \tag{3.9}$$

where $\rho_{X1,X2} = \frac{\sigma_{X1,X2}}{\sigma_{X1}\sigma_{X2}}$ is the correlation of X_1 and X_2 (Brown 1910; Spearman 1910). Coefficient alpha (Cronbach 1951) can be calculated by dividing a test into $i = 1, 2, \ldots, n$ parts assumed to be parallel,

$$\frac{n}{n-1}\left(\frac{\sigma_X^2 - \sum_i \sigma_{X,i}^2}{\sigma_X^2}\right) = \frac{n}{n-1}\left(1 - \frac{\sum_i \sigma_{X,i}^2}{\sigma_X^2}\right). \tag{3.10}$$

Coefficient alpha is known to be a general reliability estimate that produces previously proposed reliability estimates in special cases. For n parts that are all dichotomously scored items, coefficient alpha can be expressed as the Kuder-Richardson Formula 20 reliability (Kuder and Richardson 1937) in terms of the proportion of correct responses on the ith part, $\mu(X_i)$,

$$\frac{n}{n-1}\left(1 - \frac{\sum_i \mu(X_i)\left[1-\mu(X_i)\right]}{\sigma_X^2}\right). \tag{3.11}$$

The Kuder-Richardson Formula 21 ($rel_{21}(X)$) from Eq. 3.8 in Sect. 3.1.2) can be obtained as a simplification of Eq. 3.11, by replacing each $\mu(X_i)$ for the dichotomously scored items with the mean score on all the items, $\mu(X)$, resulting in

$$\frac{n}{n-1}\left(1 - \frac{\mu(X)\left[n-\mu(X)\right]}{n\sigma_X^2}\right). \tag{3.12}$$

Some ETS contributions to reliability estimation have been made in interpretive analyses of the above reliability approaches. The two Kuder-Richardson formulas have been compared and shown to give close results in practice (Lord 1959b), with the Kuder-Richardson Formula 21 estimate shown by Ledyard R Tucker (1949) always to be less than or equal to the Kuder-Richardson Formula 20 estimate. Cronbach (1951) described his coefficient alpha measure as equal to the mean of all possible split-half reliability estimates, and this feature has been pointed out as eliminating a source of error associated with the arbitrary choice of the split (Lord 1956). Lord (1955b) pointed out that the Kuder-Richardson Formula 21 reliability estimate requires an assumption that all item intercorrelations are equal and went on to show that an average of his binomial estimate of the squared standard errors of measurement can be used in the $1 - \frac{\sigma_{E_X}^2}{\sigma_X^2}$ reliability estimate in Eq. 3.5 to produce the Kuder-Richardson Formula 21 reliability estimate (i.e., the squared values in Eq. 3.7 can be averaged over examinees to estimate $\sigma_{E_X}^2$). Other ETS researchers have pointed out that if the part tests are not essentially tau equivalent, then coefficient alpha is a lower bound to the internal consistency reliability (Novick and Lewis 1967). The worry that internal consistency reliability estimates depend on how closely the parts are to parallel has prompted recommendations for constructing the parts, such as by grouping a test form's items based on their percent-correct score and biserial item-test correlations (Gulliksen 1950). Statistical sampling theory for coefficient alpha was developed by Kristof (1963b; and independently by Feldt 1965). If the coefficient alpha reliability is calculated for a test divided into n strictly parallel parts using a sample of N examinees, then a statistic based on coefficient alpha is distributed as a central F with $N - 1$ and $(n - 1)(N - 1)$ degrees of freedom. This result is exact only under the assumption that part-test scores follow a multivariate normal distribution with equal variances and with equal covariances (the compound symmetry assumption). Kristof (1970) presented a method for testing the significance of point estimates and for constructing confidence intervals for alpha calculated from the division of a test into $n = 2$ parts with unequal variances, under the assumption that the two part-test scores follow a bivariate normal distribution.

The ETS contributions to conditional error variance estimation from Sect. 3.1.2 have been cited as contributors to generalizability (G) theory. G theory uses analysis of variance concepts of experimental design and variance components to reproduce reliability estimates, such as coefficient alpha, and to extend these reliability estimates to address multiple sources of error variance and reliability estimates for specific administration situations (Brennan 1997; Cronbach et al. 1972). A description of the discussion of relative and absolute error variance and of applications of Lord's (1955b, 1957a) binomial error model results (see Sect. 3.1.2) suggested that these ETS contributions were progenitors to G theory:

> The issues Lord was grappling with had a clear influence on the development of G theory. According to Cronbach (personal communication, 1996), about 1957, Lord visited the Cronbach team in Urbana. Their discussions suggested that the error in Lord's formulation of the binomial error model (which treated one person at a time—that is, a completely nested design) could not be the same error as that in classical theory for a crossed design (Lord basically acknowledges this in his 1962 article.) This insight was eventually captured in the distinction between relative and absolute error in G theory, and it illustrated that errors of measurement are influenced by the choice of design. Lord's binomial error model is probably best known as a simple way to estimate conditional SEMs and as an important precursor to strong true score theory, but it is also associated with important insights that became an integral part of G theory. (Brennan 1997, p. 16)

Other ETS contributions have been made by deriving internal consistency reliability estimates based on scores from a test's parts that are not strictly parallel. This situation would seem advantageous because some of the more stringent assumptions required to achieve strictly parallel test forms can be relaxed. However, situations in which the part tests are not strictly parallel pose additional estimation challenges in that the two-part tests, which are likely to differ in difficulty, length, and so on, result in four unknown variances (the true score and error variances of the two parts) that must be estimated from three pieces of information (the variances and the covariance of the part scores). Angoff (1953; also Feldt 1975) addressed this challenge of reliability estimation by assuming that the part tests follow a congeneric model, so that even though the respective lengths of the part tests (i.e., true-score coefficients) cannot be directly estimated, the relative true-score variances and relative error variances of the parts can be estimated as functions of the difference in the effective test lengths of the parts. That is, if one part is longer or shorter than the other part by factor j, the proportional true scores of the first and second part differ by j, the proportional true-score variances differ by j^2, and the proportional error variances differ by j. These results suggest the following reliability coefficient referred to as the Angoff-Feldt coefficient (see Haertel 2006),

$$\frac{4\sigma\left(X_1, X_2\right)}{\sigma_X^2 - \frac{\left[\sigma_{X,1}^2 - \sigma_{X,2}^2\right]^2}{\sigma_X^2}} \tag{3.13}$$

Angoff also used his results to produce reliability estimates for a whole test, X, and an internal part, X_1,

$$\begin{aligned} rel\left(X\right) &= \frac{\rho_{X,X1}\sigma_X - \sigma_{X1}}{\rho_{X,X1}\left(\sigma_X - \rho_{X,X1}\sigma_{X1}\right)} \text{ and} \\ rel\left(X_1\right) &= \frac{\rho_{X,X1}\left(\rho_{X,X1}\sigma_X - \sigma_{X1}\right)}{\sigma_X - \rho_{X,X1}\sigma_{X1}}, \end{aligned} \tag{3.14}$$

and for a whole test X, and an external part not contained in X, Y,

$$\begin{aligned} rel\left(X\right) &= \frac{\rho_{X,Y}\left(\sigma_X + \rho_{X,Y}\sigma_Y\right)}{\sigma_Y + \rho_{X,Y}\sigma_X} \text{ and} \\ rel\left(Y\right) &= \frac{\rho_{X,Y}\left(\sigma_Y + \rho_{X,Y}\sigma_X\right)}{\sigma_X + \rho_{X,Y}\sigma_Y}. \end{aligned} \tag{3.15}$$

The same assumptions later used by Angoff and Feldt were employed in an earlier work by Horst (1951a) to generalize the Spearman-Brown split-half formula to produce a reliability estimate for part tests of unequal but known lengths. Reviews of alternative approaches to reliability estimation when the two-part test lengths are unknown have recommended the Angoff-Feldt estimate in most cases (Feldt 2002).

Kristof made additional contributions to reliability estimation by applying classical test theory models and assumptions (see Sect. 3.1.5.1) to tests divided into more than two parts. He demonstrated that improved statistical precision in reliability estimates could be obtained from dividing a test into more than two tau-equivalent parts (Kristof 1963b). By formulating test length as a parameter in a model for a population covariance matrix of two or more tests, Kristof (1971) described the estimation of test length and showed how to formulate confidence intervals for the relative test lengths. Finally, Kristof (1974) provided a solution to the problem of three congeneric parts of unknown length, where the reliability estimation problem is considered to be just identified, in that there are exactly as many variances and covariances as parameters to be estimated. Kristof's solution was shown to be at least as accurate as coefficient alpha and also gives stable results across alternative partitions. Kristof also addressed the problem of dividing a test into more than three parts of unknown effective test length where the solution is over-determined. Kristof's solution is obtained via maximum-likelihood and numerical methods.

3.1.5.3 Factor Analysis

Some well-known approaches to assessing the measurement properties of multiple tests are those based on factor-analysis models. Factor-analysis models are conceptually like multivariate versions of the classical test theory results in Sect. 3.1.1. Let $\boldsymbol{X}$ denote a q-by-1 column vector with the scores of q tests, $\boldsymbol{\mu}$ denote the q-by-1 vector of means for the q test forms in $\boldsymbol{X}$, $\boldsymbol{\Theta}$ denote a k-by-1 element vector of scores on k common factors, $k < q$, $\boldsymbol{\lambda}$ denote a q-by-k matrix of constants called factor loadings, and finally, let $\mathbf{v}$ denote a q-by-1 row vector of unique factors corresponding to the elements of $\boldsymbol{X}$. With these definitions, the factor-analytic model can be expressed as.

$$\boldsymbol{X} = \mu + \lambda\Theta + v, \tag{3.16}$$

and the covariance matrix of $\boldsymbol{X}$, $\boldsymbol{\Sigma}$, can be decomposed into a sum of q-by-q covariance matrices attributable to the common factors ($\boldsymbol{\lambda\Psi\lambda}'$, where Ψ is a k-by-k covariance matrix of the common factors, $\boldsymbol{\Theta}$) and $\boldsymbol{D}^2$ is a diagonal covariance matrix among the uncorrelated unique factors, $\mathbf{v}$,

$$\Sigma = \lambda\Psi\lambda' + \boldsymbol{D}^2. \tag{3.17}$$

The overall goal of factor analyses described in Eqs. 3.16 and 3.17 is to meaningfully explain the relationships among multiple test forms and other variables with a small number of common factors (i.e., $k << q$, meaning "k much less than q"). Since Spearman's (1904a) original factor analysis, motivations have been expressed for factor-analysis models that account for observed variables' intercorrelations using one, or very few, common factors. Spearman's conclusions from his factor analysis of scores from tests of abilities in a range of educational subjects (classics, French, English, Math, music, and musical pitch discrimination) and other scores from measures of sensory discrimination to light, sound, and weight were an important basis for describing a range of intellectual abilities in terms of a single, common, general factor:

> We reach the profoundly important conclusion that there really exists a something that we may provisionally term "General Sensory Discrimination" and similarly a "General Intelligence," and further that the functional correspondence between these two is not appreciably less than absolute. (Spearman 1904a, p. 272)

The predominant view regarding factor analysis is as a tool for describing the measurement properties of one or more tests in terms of factors hypothesized to underlie observed variables that comprise the test(s) (Cudeck and MacCallum 2007; Harman 1967; Lord and Novick 1968). Factor analysis models can be viewed as multivariate variations of the classical test theory model described in Sect. 3.1. In this sense, factor analysis informs a "psychometric school" of inquiry, which views a "...battery of tests as a selection from a large domain of tests that could be developed for the same psychological phenomenon and focused on the factors in this domain" (Jöreskog 2007, p. 47). Similar to the classical test theory assumptions, the means of **v** are assumed to be zero, and the variables' covariance matrix, $\boldsymbol{D}^2$, is diagonal, meaning that the unique factors are assumed to be uncorrelated. Somewhat different from the classical test theory model, the unique factors in **v** are not exactly error variables, but instead are the sum of the error factors and specific factors of the q variables. That is, the **v** factors are understood to reflect unreliability (error factors) as well as actual measurement differences (specific factors). The assumption that the **v** factors are uncorrelated implies that the observed covariances between the observed variables are attributable to common factors and loadings, $\boldsymbol{\lambda\Theta}$. The common factors are also somewhat different from the true scores of the variables because the factor-analysis model implies that the true scores reflect common factors as well as specific factors in **v**.

Many developments in factor analysis are attempts to formulate subjective aspects of model selection into mathematical, statistical, and computational solutions. ETS researchers have contributed several solutions pertaining to these interests, which are reviewed in Harman (1967) and in Lord and Novick (1968). In particular, iterative methods have been contrasted and developed for approximating the factor analysis model in observed data by Browne (1969) and Jöreskog (1965, 1967, 1969a; Jöreskog and Lawley 1968), including maximum likelihood, image factor analysis, and alpha factor analysis. An initially obtained factor solution is not uniquely defined, but can be transformed (i.e., rotated) in ways that result in different interpretations of how the factors relate to the observed variables and reproduce the variables' intercorrelations. Contributions by ETS scientists such as Pinzka, Saunders, and Jennrich include the development of different rotation methods that either allow the common factors to be correlated (oblique) or force the factors to remain orthogonal (Browne 1967, 1972a, b; Green 1952; Pinzka and Saunders 1954; Saunders 1953a). The most popular rules for selecting the appropriate number of common factors, k, are based on the values and graphical patterns of factors' eigenvalues, rules that have been evaluated and supported by simulation studies (Browne 1968; Linn 1968; Tucker et al. 1969). Methods for estimating statistical standard errors of estimated factor loadings have been derived (Jennrich 1973; Jennrich and Thayer 1973). Other noteworthy ETS contributions include mathematical or objective formalizations of interpretability in factor analysis (i.e., Thurstone's simple structure, Tucker 1955; Tucker and Finkbeiner 1981), correlation-like measures of the congruence or strength of association among common factors (Tucker 1951), and methods for postulating and simulating data that reflect a factor analysis model in terms of the variables common (major) factors and that also depart from the factor analysis model in terms of several intercorrelated unique (minor) factors (Tucker et al. 1969).

An especially important ETS contribution is the development and naming of confirmatory factor analysis, a method now used throughout the social sciences to address a range of research problems. This method involves fitting and comparing factor-analysis models with factorial structures, constraints, and values specified a priori and estimated using maximum-likelihood methods (Jöreskog 1969b; Jöreskog and Lawley 1968). Confirmatory factor analysis contrasts with the exploratory factor-analysis approaches described in the preceding paragraphs in that confirmatory factor-analysis models are understood to have been specified a priori with respect to the data. In addition, the investigator has much more control over the models and factorial structures that can be considered in confirmatory factor analysis than in exploratory factor analysis. Example applications of confirmatory factor analyses are investigations of the invariance of a factor-analysis solution across subgroups (Jöreskog 1971a) and evaluating test scores with respect to psychometric models (Jöreskog 1969a). These developments expanded factor analyses towards structural-equation modeling, where factors of the observed variables are not only estimated but are themselves used as predictors and outcomes in further analyses (Jöreskog 2007). The LISREL computer program, initially produced by Jöreskog at ETS, was one of the first programs made available to investigators for implementing maximum-likelihood estimation algorithms for confirmatory factor analysis and structural equation models (Jöreskog and van Thillo 1972).

3.1.6 Applications to Psychometric Test Assembly and Interpretation

The ETS contributions to the study of measurement properties of test scores reviewed in the previous sections can be described as relatively general contributions to classical test theory models and related factor-analysis models. Another set of developments has been more focused on applications of measurement theory concepts to the development, use, and evaluation of psychometric tests. These application developments are primarily concerned with building test forms with high measurement precision (i.e., high reliability and low standard errors of measurement).

The basic idea that longer tests are more reliable than shorter tests had been established before ETS (Brown 1910, Spearman 1910; described in Gulliksen 1950 and Mislevy 1993, 1997). ETS researchers developed more refined statements about test length, measurement precision, and scoring systems that maximize reliability. One example of these efforts was establishing that, like reliability, a test's overall standard error of measurement is also directly related to test length, both in theoretical predictions (Lord 1957a) and also in empirical verifications (Lord 1959b). Other research utilized factor-analysis methods to show how reliability for a test of dichotomous items can be maximized by weighting those items by their standardized component loadings on the first principal component (Lord 1958) and how the reliability of a composite can be maximized by weighting the scores for the composite's test battery according to the first principal axis of the correlations and reliabilities of the tests (Green 1950). Finally, conditions for maximizing the reliability of a composite were established, allowing for the battery of tests to have variable lengths and showing that summing the tests after they have been scaled to have equal standard errors of measurement would maximize composite reliability (Woodbury and Lord 1956).

An important limitation of many reliability estimation methods is that they pertain to overall or average score precision. Livingston and Lewis (1995) developed a method for score-specific reliability estimates rather than overall reliability, as score-specific reliability would be of interest for evaluating precision at one or more cut scores. The Livingston and Lewis method is based on taking a test with items not necessarily equally weighted or dichotomously scored and replacing this test with an idealized test consistent with some number of identical dichotomous items. An effective test length of the idealized test is calculated from the mean, variance, and reliability of the original test to produce equal reliability in the idealized test. Scores on the original test are linearly transformed to proportion-correct scores on the idealized test, and the four parameter beta-binomial model described previously is applied. The resulting analyses produce estimates of classification consistency when the same cut scores are used to classify examinees on a hypothetically administered alternate form and estimates of classification accuracy to describe the precision of the cut-score classifications in terms of the assumed true-score distribution.

Statistical procedures have been a longstanding interest for assessing whether two or more test forms are parallel or identical in some aspect of their measurement (i.e., the models in Sect. 3.1.5.1). The statistical procedures are based on evaluating the extent to which two or more test forms satisfy different measurement models when accounting for the estimation error due to inferring from the examinee sample at hand to a hypothetical population of examinees (e.g., Gulliksen 1950, Chapter 14; Jöreskog 2007). ETS researchers have proposed and developed several statistical procedures to assess multiple tests' measurement properties. Kristof (1969) presented iteratively computed maximum-likelihood estimation versions of the procedures described in Gulliksen for assessing whether tests are strictly parallel to also assess if tests are essentially tau equivalent. Procedures for assessing the equivalence of the true scores of tests based on whether their estimated true-score correlation equals 1 have been derived as a likelihood ratio significance test (Lord 1957b) and as F-ratio tests (Kristof 1973). Another F test was developed to assess if two tests differ only with respect to measurement errors, units, and origins of measurement (Lord 1973). A likelihood ratio test was derived for comparing two or more coefficient alpha estimates obtained from dividing two tests each into two part tests with equivalent error variances using a single sample of examinees (Kristof 1964). Different maximum likelihood and chi-square procedures have been developed for assessing whether tests have equivalent overall standard errors of measurement, assuming these tests are parallel (Green 1950), or that they are essentially tau equivalent (Kristof 1963a). Comprehensive likelihood ratio tests for evaluating the fit of different test theory models, including congeneric models, have been formulated within the framework of confirmatory factor-analysis models (Jöreskog 1969a).

3.2 Test Scores as Predictors in Correlational and Regression Relationships

This section describes the ETS contributions to the psychometric study of test scores that are focused on scores' correlations and regression-based predictions to criteria that are not necessarily parallel to the tests. The study of tests with respect to their relationships with criteria that are not necessarily alternate test forms means that test validity issues arise throughout this section and are treated primarily in methodological and psychometric terms. Although correlation and regression issues can be described as if they are parts of classical test theory (e.g., Traub 1997), they are treated as distinct from classical test theory's measurement concepts here because (a) the criteria with which the tests are to be related are often focused on observed scores rather than on explicit measurement models and (b) classical measurement concepts have specific implications for regression and correlation analyses, which are addressed in the next section. Section 3.1.1 reviews the basic correlational and regression developments established prior to ETS. Section 3.2.2 reviews ETS psychometric contributions involving correlation and regression analyses.

3.2.1 Foundational Developments for the Use of Test Scores as Predictors, Pre-ETS

The simple correlation describes the relationship of variables X and Y in terms of the standardized covariance of these variables, $\rho_{X,Y} = \frac{\sigma_{X,Y}}{\sigma_X \sigma_Y}$, and has been traced to the late 1800s work of Galton, Edgeworth, and Pearson (Holland 2008; Traub 1997). The X,Y correlation plays a central role in linear regression, the major concepts of which have been credited to the early nineteenth century work of Legendre, Gauss, and Laplace (Holland 2007). The correlation and regression methods establish a predictive relationship of Y's conditional mean to a linear function of X,

$$Y = \mu(Y|X) + \varepsilon = \mu_Y + \rho_{X,Y}\frac{\sigma_Y}{\sigma_X}(X - \mu_X) + \varepsilon \quad (3.18)$$

The prediction error, ε, in Eq. 3.18 describes the imprecision of the linear regression function as well as an X,Y correlation that is imperfect (i.e., less than 1). Prediction error is different from the measurement errors of X and Y that reflect unreliability, E_X and E_Y, (Sect. 3.1). The linear regression function in Eq. 3.18 is based on least-squares estimation because using this method results in the smallest possible value of $\sigma_\varepsilon^2 = \sigma_Y^2\left[1-\rho_{X,Y}^2\right]$. The multivariate version of Eq. 3.18 is based on predicting the conditional mean of Y from a combination of a set of q observable predictor variables,

$$\boldsymbol{Y} = \mathbf{X}\beta + \varepsilon = \widehat{Y} + \varepsilon, \tag{3.19}$$

where $\boldsymbol{Y}$ is an N-by-1 column vector of the N Y values in the data, $\widehat{Y} = \mathbf{X}\beta$ is an N-by-1 column vector of predicted values ($\widehat{Y}$), $\boldsymbol{X}$ is an N-by-q matrix of values on the predictor variables, $\boldsymbol{\beta}$ is a q-by-1 column vector of the regression slopes of the predictor variables (i.e., scaled semipartial correlations of Y and each X with the relationships to the other Xs partialed out of each X), and $\boldsymbol{\varepsilon}$ is an N-by-1 column vector of the prediction errors. The squared multiple correlation of Y and $\widehat{Y}$ predicted from the Xs in Eqs. 3.18 and 3.19 can be computed given the $\boldsymbol{\beta}$ parameters (or estimated using estimated parameters, $\widehat{\boldsymbol{\beta}}$) as,

$$\rho_{\widehat{Y},Y}^2 = \frac{\sum_{i=1}^{N}(\mathbf{X}_i\boldsymbol{\beta})^2 - \frac{1}{N}\left(\sum_{i=1}^{N}\mathbf{X}_i\boldsymbol{\beta}\right)^2}{\mathbf{Y}'\mathbf{Y} - \frac{1}{N}\left(\sum_{i=1}^{N}\mathbf{Y}_i\right)^2} = 1 - \frac{\sigma_\varepsilon^2}{\sigma_Y^2} \tag{3.20}$$

Early applications of correlation and regression concepts dealt with issues such as prediction in astronomy (Holland 2008; Traub 1997) and obtaining estimates of correlations that account for restrictions in the ranges and standard deviations of X and Y (Pearson 1903).

3.2.2 ETS Contributions to the Methodology of Correlations and Regressions and Their Application to the Study of Test Scores as Predictors

The following two subsections summarize ETS contributions about the sample-based aspects of estimated correlations and regressions. Important situations where relationships of tests to other tests and to criteria are of interest involve missing or incomplete data from subsamples of a single population and the feasibility of accounting for incomplete data of samples when those samples reflect distinct populations with preexisting differences. The third subsection deals with ETS contributions that focus directly on detecting group differences in the relationships of tests and what these group differences imply about test validity. The final section describes contributions pertaining to test construction such as determining testing time, weighting subsections, scoring items, and test length so as to maximize test validity.

3.2.2.1 Relationships of Tests in a Population's Subsamples With Partially Missing Data

Some contributions by ETS scientists, such as Gulliksen, Lord, Rubin, Thayer, Horst, and Moses, to test-score relationships have established the use of regressions for estimating test data and test correlations when subsamples in a dataset have partially missing data on the test(s) or the criterion. One situation of interest involves examinee subsamples, R and S, which are missing data on one of two tests, X and Y, but which have complete data on a third test, A. To address the missing data in this situation, regressions of each test onto test A can be used to estimate the means and standard deviations of X and Y for the subsamples with the missing data (Gulliksen 1950; Lord 1955a, c). For example, if group P takes tests X and A and subsample S takes only A, the mean and variance of the missing X scores of S can be estimated by applying the A-to-X regression of subsample R to the A scores of S using the sample statistics in

$$\mu_{X,S} = \mu_{X,R} - \rho_{X,A,R} \frac{\sigma_{X,R}}{\sigma_{A,R}} \left(\mu_{A,R} - \mu_{A,S} \right), \tag{3.21}$$

and

$$\sigma_{X,S}^2 = \sigma_{X,R}^2 - \left[\rho_{X,A,R} \frac{\sigma_{X,R}}{\sigma_{A,R}} \right]^2 \left[\sigma_{A,R}^2 - \sigma_{A,S}^2 \right]. \tag{3.22}$$

For the more general situation involving a group of standard tests given to an examinee group and one of several new tests administered to random subsamples in the overall group, correlations among all the new and standard tests can be estimated by establishing plausible values for the new tests' partial correlations of the new and standard tests and then using the intercorrelations of the standard tests to "uncondition" the partial correlations and obtain the complete set of simple correlations (Rubin and Thayer 1978, p. 5). Finally, for predicting an external criterion from a battery of tests, it is possible to identify the minimum correlation of an experimental test with the external criterion required to increase the multiple correlation of the battery with that criterion by a specified amount without knowing the correlation of the experimental test with the criterion (Horst 1951c). The fundamental assumption for all of the above methods and situations is that subsamples are randomly selected from a common population, so that other subsamples' correlations of their missing test with other tests and criteria can serve as reasonable estimates of the correlations for the subsamples with missing data.

Regressions and correlations have been regarded as optimal methods for addressing missing test score data in subsamples because under some assumed mathematical model (e.g., normally distributed bivariate or trivariate distributions), regression and correlation estimates maximize the fit of the complete and estimated missing data with the assumed model (Lord 1955a, c; Rubin and Thayer 1978). Thus regressions and correlations can sometimes be special cases of more general maximum-likelihood estimation algorithms for addressing missing data (e.g., the EM algorithm; Dempster et al. 1977). Similar to Lord's (1954b) establishment of linear regression estimates as maximum likelihood estimators for partially missing data, nonlinear regressions estimated with the usual regression methods have been shown to produce results nearly identical to those obtained by using the EM algorithm to estimate the same nonlinear regression models (Moses et al. 2011). It should be noted that the maximum-likelihood results apply to situations involving partially missing data and not necessarily to other situations where a regression equation estimated entirely in one subsample is applied to a completely different, second subsample that results in loss of prediction efficiency (i.e., a larger $\widehat{\sigma}^2(\varepsilon)$ for that second subsample; Lord 1950a).

3.2.2.2 Using Test Scores to Adjust Groups for Preexisting Differences

In practice, correlations and regressions are often used to serve interests such as assessing tests taken by subsamples that are likely due to pre-existing population differences that may not be completely explained by *X* or by the study being conducted. This situation can occur in quasi-experimental designs, observational studies, a testing program's routine test administrations, and analyses of selected groups. The possibilities by which preexisting group differences can occur imply that research situations involving preexisting group differences are more likely than subsamples that are randomly drawn from the same population and that have partially missing data (the situation of interest in Sect. 3.2.2.1). The use of correlation and regression for studying test scores and criteria based on examinees with preexisting group differences that have been matched with respect to other test scores has prompted both methodological proposals and discussions about the adequacy of correlation and regression methods for addressing such situations by ETS scientists such as Linn, Charles Werts, Nancy Wright, Dorans, Holland, Rosenbaum, and O'Connor.

Some problems of assessing the relationships among tests taken by groups with preexisting group differences involve a restricted or selected group that has been chosen based either on their criterion performance (explicit selection) or on some third variable (incidental selection, Gulliksen 1950). Selected groups would exhibit performance on tests and criteria that have restricted ranges and standard deviations, thereby affecting these groups' estimated correlations and regression equations. Gulliksen applied Pearson's (1903) ideas to obtain a estimated correlation, prediction error variance, or regression coefficients of the selected group after correcting these estimates for the range-restricted scores of the selected group on *X* and/or *Y*. These corrections for range restrictions are realized by using the *X* and/or *Y* standard deviations from an unselected group in place of those from the selected group.

Concerns have been raised about the adequacy of Gulliksen's (1950) corrections for the statistics of self-selected groups. In particular, the corrections may be inaccurate if the assumed regression model is incorrect (i.e., is actually nonlinear or if the error variance, $\sigma^2(\varepsilon)$, is not constant), or if the corrections are based on a purported selection variable that is not the actual variable used to select the groups (Linn 1967; Lord and Novick 1968). Cautions have been expressed for using the corrections involving selected and unselected groups when those two groups have very different standard deviations (Lord and Novick 1968). The issue of accurately modeling the selection process used to establish the selected group is obviously relevant when trying to obtain accurate prediction estimates (Linn 1983; Linn and Werts 1971; Wright and Dorans 1993).

The use of regressions to predict criterion *Y*'s scores from groups matched on *X* is another area where questions have been raised about applications for groups with preexisting differences. In these covariance analyses (i.e., ANCOVAs), the covariance-adjusted means of the two groups on *Y* are compared, where the adjustment is obtained by applying an *X*-to-*Y* regression using both groups' data to estimate the regression slope ($\rho_{X,Y,R+S}\frac{\sigma_{Y,R+S}}{\sigma_{X,R+S}}$) and each group's means ($\mu_{Y,R}$, $\mu_{Y,S}$, $\mu_{X,R}$ and $\mu_{X,S}$) in the estimation and comparison of the groups' intercepts,

$$\mu_{Y,R} - \mu_{Y,S} - \rho_{X,Y,R+S}\frac{\sigma_{Y,R+S}}{\sigma_{X,R+S}}\left(\mu_{X,R} - \mu_{X,S}\right). \tag{3.23}$$

The application of the covariance analyses of Eq. 3.23 to adjust the *Y* means for preexisting group differences by matching the groups on *X* has been criticized for producing results that can, under some circumstances, contradict analyses of average difference scores, $\mu_{Y,R} - \mu_{Y,S} - (\mu_{X,R} - \mu_{X,S})$, (Lord 1967). In addition, covariance analyses have been described as inadequate for providing an appropriate adjustment for the preexisting group differences that are confounded with the study groups and not completely due to *X* (Lord 1969). Attempts have been made to resolve the problems of covariance analysis for groups with preexisting differences. For instance, Novick (1983) elaborated on the importance of making appropriate assumptions about the subpopulation to which individuals are exchangeable members, Holland and Rubin (1983) advised investigators to make their untestable assumptions about causal inferences explicit, and Linn and Werts (1973) emphasized research designs that provide sufficient information about the measurement errors of the variables. Analysis strategies have also been recommended to account for and explain the preexisting group differences with more than one variable using multiple regression (O'Connor 1973), Mahalanobis distances (Rubin 1980), a combination of Mahalanobis distances and regression (Rubin 1979), and propensity-score matching methods (Rosenbaum and Rubin 1984, 1985).

3.2.2.3 Detecting Group Differences in Test and Criterion Regressions

Some ETS scientists such as Schultz, Wilks, Cleary, Frederiksen, and Melville have developed and applied statistical methods for comparing the regression functions of groups. Developments for statistically comparing regression lines of groups tend to be presented in terms of investigations in which the assessment of differences in regressions of groups is the primary focus. Although these developments can additionally be described as informing the developments in the previous section (e.g., establishing the most accurate regressions to match groups from the same population or different populations), these developments tend to describe the applications of matching groups and adjusting test scores as secondary interests. To the extent that groups are found to differ with respect to X,Y correlations, the slopes and/or intercepts of their $Y|X$ regressions and so on, other ETS developments interpret these differences as reflecting important psychometric characteristics of the test(s). Thus these developments are statistical, terminological, and applicative.

Several statistical strategies have been developed for an investigation with the primary focus of determining whether regressions differ by groups. Some statistical significance procedures are based on directly comparing aspects of groups' regression functions to address sequential questions. For example, some strategies center on assessing differences in the regression slopes of two groups and, if the slope differences are likely to be zero, assessing the intercept differences of the groups based on the groups' parallel regression lines using a common slope (Schultz and Wilks 1950). More expansive and general sequential tests involve likelihood ratio and F-ratio tests to sequentially test three hypotheses: first, whether the prediction error variances of the groups are equal; then, whether the regression slopes of the groups are equal (assuming equal error variances), and finally, whether the regression intercepts of the groups are equal (assuming equal error variances and regression slopes; Gulliksen and Wilks 1950). Significance procedures have also been described to consider how the correlation from the estimated regression model in Eq. 3.18, based only on X, might be improved by incorporating a group membership variable, G, as a moderator (i.e., moderated multiple regression; Saunders 1953b),

$$\begin{bmatrix} Y_1 \\ Y_1 \\ . \\ . \\ Y_N \end{bmatrix} = \begin{bmatrix} 1_1 & X_1 & G_1 & X_1G_1 \\ 1_2 & X_2 & G_2 & X_2G_2 \\ . & & & \\ . & & & \\ 1_N & X_N & G_N & X_NG_N \end{bmatrix} \begin{bmatrix} \beta_0 \\ \beta_X \\ \beta_G \\ \beta_{XG} \end{bmatrix} + \begin{bmatrix} e_1 \\ e_1 \\ . \\ . \\ e_N \end{bmatrix}. \tag{3.24}$$

Other statistical procedures for assessing group differences include extensions of the Johnson-Neyman procedure for establishing regions of predictor-variable values in which groups significantly differ in their expected criterion scores (Potthoff 1964) and iterative, exploratory procedures for allowing the regression weights of individuals to emerge in ways that maximize prediction accuracy (Cleary 1966a).

The previously described statistical procedures for assessing group differences in regressions have psychometric implications for the tests used as predictors in those regressions. These implications have sometimes been described in terms of test use in which differential predictability investigations have been encouraged that determine the subgroups for which a test is most highly correlated with a criterion and, therefore, most accurate as a predictor of it (Frederiksen and Melville 1954). Other investigators have made particularly enduring arguments that if subgroups are found for which the predictions of a test for a criterion in a total group's regression are inaccurate, the use of that test as a predictor in the total group regression is biased for that subgroup (Cleary 1966b). The statistical techniques in this section, such as moderated multiple regression (Saunders 1953b) for assessing differential predictability and Cleary's test bias,[1] help to define appropriate and valid uses for tests.

3.2.2.4 Using Test Correlations and Regressions as Bases for Test Construction

Interest in test validity has prompted early ETS developments concerned with constructing, scoring, and administering tests in ways that maximized tests' correlations with an external criterion). In terms of test construction, ETS authors such as Gulliksen, Lord, Novick, Horst, Green, and Plumlee have proposed simple, mathematically tractable versions of the correlation between a test and criterion that might be maximized based on item selection (Gulliksen 1950; Horst 1936). Although the correlations to be maximized are different, the Gulliksen and Horst methods led to similar recommendations that maximum test validity can be approximated by selecting items based on the ratio of correlations of items with the criterion and with the total test (Green 1954). Another aspect of test construction addressed in terms of validity implications is the extent to which multiple-choice tests lead to validity reductions relative to open-ended tests (i.e., tests with items that do not present examinees with a set of correct and incorrect options) because of the probability of chance success in multiple-choice items (Plumlee 1954). Validity implications have also been described in terms of the decrement in validity that results when items are administered and scored as the sum of the correct responses of examinees rather than through formulas designed to discourage guessing and to correct examinee scores for random guessing (Lord 1963).

For situations in which a battery of tests are administered under fixed total testing time, several ETS contributions have considered how to determine the length of

[1]Although the summary of Cleary's (1966b) work in this chapter uses the *test bias* phrase actually used by Cleary, it should be acknowledged that more current descriptions of Cleary's regression applications favor different phrases such as prediction bias, overprediction, and underprediction (e.g., Bridgeman et al. 2008). The emphasis of current descriptions on prediction accuracy allows for distinctions to be made between tests that are not necessarily biased but that may be used in ways that result in biased predictions.

each test in ways that maximize the multiple correlation of the battery with an external criterion. These developments have origins in Horst (1951b), but have been extended to a more general and sophisticated solution by Woodbury and Novick (1968). Further extensions deal with computing the composite scores of the battery as the sum of the scores of the unweighted tests in the battery rather than based on the regression weights (Jackson and Novick 1970). These methods have been extensively applied and compared to suggest situations in which validity gains might be worthwhile for composites formed from optimal lengths and regression weights (Novick and Thayer 1969).

3.3 Integrating Developments About Test Scores as Measurements and Test Scores as Predictors

The focus of this section is on ETS contributions that integrate and simultaneously apply measurement developments in Sect. 3.1 and the correlational and regression developments in Sect. 3.2. As previously stated, describing measurement and correlational concepts as if they are completely independent is an oversimplification. Some of the reliability estimates in Sect. 3.1 explicitly incorporate test correlations. In Sect. 3.2, a review of algorithms by Novick and colleagues for determining the lengths of tests in a battery that maximize validity utilize classical test theory assumptions and test reliabilities, but ultimately produce regression and multiple correlation results based on the observed test and criterion scores (Jackson and Novick 1970; Novick and Thayer 1969; Woodbury and Novick 1968). The results by Novick and his colleagues are consistent with other results that have shown that observed-score regressions such as Eq. 3.18 can serve as optimal predictors of the true scores of a criterion (Holland and Hoskens 2003). What distinguishes this section's developments is that measurement, correlational, and regression concepts are integrated in ways that lead to fundamentally unique results.

Integrations of measurement concepts into correlations and regressions build upon historical developments that predate ETS. Spearman's (1904b, 1910) use of classical test theory assumptions to derive an X, Y correlation disattenuated for X and Y's measurement errors (assumed to be independent) is one major influence,

$$\frac{\rho_{X,Y}}{\sqrt{rel(X)rel(Y)}}. \tag{3.25}$$

Kelley's (1923, 1947) regression estimate of the true scores of a variable from its observed scores is another influence,

$$\hat{T}_{Xp} = rel(X)X_p + \left[1 - rel(X)\right]\mu(X) \tag{3.26}$$

Equations 3.25 and 3.26 suggest that some types of analyses that utilize observed scores to compute correlations and regressions can be inaccurate due to measurement errors of *Y*, *X*, or the combination of *Y*, *X*, and additional predictor variables (Moses 2012). Examples of analyses that can be rendered inaccurate when *X* is unreliable are covariance analyses that match groups based on *X* (Linn and Werts 1973) and differential prediction studies that evaluate *X's* bias (Linn and Werts 1971). Lord (1960a) developed an approach for addressing unreliable *X* scores in covariance analyses. In Lord's formulations, the standard covariance analysis model described in Eq. 3.23 is altered to produce an estimate of the covariance results that might be obtained based on a perfectly reliable *X*,

$$\mu_{Y,R} - \mu_{Y,S} - \widehat{\beta}_{T_X}\left(\mu_{X,R} - \mu_{X,S}\right), \tag{3.27}$$

where $\widehat{\beta}_{T_X}$ is estimated as slope disattenuated for the unreliability of *X* based on the classical test theory assumption of *X* having measurement errors independent of measurement errors for *Y*,

$$\widehat{\beta}_{T_X} = \frac{N_R\sigma_{X,Y,R} + N_S\sigma_{X,Y,S}}{N_R rel_R\left(X\right)\sigma^2_{X,R} + N_S rel_S\left(X\right)\sigma^2_{X,S}}\left[1 - \frac{k\left(k-w\right)}{\left(N_R + N_S\right)w^2}\right], \tag{3.28}$$

where

$$k = \frac{N_R\sigma^2_{X,R} + N_S\sigma^2_{X,S}}{N_R + N_S},\ w = \frac{N_R rel_R\left(X\right)\sigma^2_{X,R} + N_S rel_S\left(X\right)\sigma^2_{X,S}}{N_R + N_S},$$

and the bracketed term in Eq. 3.28 is a correction for sampling bias. Large sample procedures are used to obtain a sample estimate of the slope in Eq. 3.28 and produce a statistical significance procedure for evaluating Eq. 3.27.

Another ETS contribution integrating measurement, correlation, and regression is in the study of change (Lord 1962a). Regression procedures are described as valuable for estimating the changes of individuals on a measure obtained in a second time period, *Y*, while controlling for the initial statuses of the individuals in a first time period, *X*, $Y - X$. Noting that measurement errors can both deflate and inflate regression coefficients with respect to true differences, Lord proposed a multiple regression application to estimate true change from the observed measures, making assumptions that the measurement errors of *X* and *Y* are independent and have the same distributions,

$$\widehat{T}_Y - \widehat{T}_X = \mu\left(Y\right) + \widehat{\beta}_{Y|X}\left[Y - \mu\left(Y\right)\right] - \mu\left(X\right) - \widehat{\beta}_{X|Y}\left[X - \mu\left(X\right)\right], \tag{3.29}$$

where the regression coefficients incorporate disattenuation for the unreliabilities of *X* and *Y*,

$$\hat{\beta}_{Y|X} = \frac{rel(Y) - \rho_{X,Y}^2 - [1 - rel(X)]\rho_{X,Y}\sigma_X / \sigma_Y}{1 - \rho_{X,Y}^2}, \quad (3.30)$$

$$\hat{\beta}_{X|Y} = \frac{rel(X) - \rho_{X,Y}^2 - [1 - rel(Y)]\rho_{X,Y}\sigma_Y / \sigma_X}{1 - \rho_{X,Y}^2}. \quad (3.31)$$

Lord also showed that the reliability of the observed change can be estimated as follows (related to the Lord-McNemar estimate of true change, Haertel 2006),

$$rel(Y - X) = \frac{rel(Y)\sigma_Y^2 + rel(X)\sigma_X^2 - 2\rho_{X,Y}\sigma_X\sigma_Y}{\sigma_Y^2 + \sigma_X^2 - 2\rho_{X,Y}\sigma_X\sigma_Y}. \quad (3.32)$$

Another ETS contribution, by Shelby Haberman, considers the question of whether subscores should be reported. This question integrates correlational and measurement concepts to determine if the true scores of subscore X are better estimated in regressions on the observed scores of the subscore (such as Eq. 3.26), the observed scores of total test Y, or a combination of the X and Y observed scores (Haberman 2008). Extending the results of Lord and Novick (1968) and Holland and Hoskens (2003), versions of the prediction error variance for an X-to-Y regression, $\sigma_\varepsilon^2 = \sigma_Y^2\left[1 - \rho_{X,Y}^2\right]$, are produced for the prediction in Eq. 3.26 of the subscore's true score from its observed score,

$$rel(X)\sigma_X^2\left[1 - rel(X)\right], \quad (3.33)$$

and for the prediction from the observed total score, Y,

$$rel(X)\sigma_X^2\left[1 - \rho_{T_X,Y}^2\right] \quad (3.34)$$

The prediction error variance for the regression of the true scores of X on both X and Y is obtained in extensions of Eqs. 3.33 and 3.34,

$$rel(X)\sigma_X^2\left[1 - rel(X)\right]\left[1 - \rho_{Y,T_X \cdot X}^2\right] \quad (3.35)$$

where $\rho_{Y,T_X \cdot X}$ is the partial correlation of the true score of X and the observed score of Y given the observed score of X. Estimates of the correlations in Eqs. 3.34 and 3.35 are obtained somewhat like the disattenuated correlation in Eq. 3.25, but with modifications to account for subscore X being contained within total score Y (i.e., violations of the classical test theory assumptions of X and Y having independent measurement errors).

Comparisons of the prediction error variances from Eqs. 3.33, 3.34, and 3.35 produce an indication for when the observed subscore has value for reporting (i.e., when Eq. 3.33 is less than Eqs. 3.34 and 3.35, such as when the subscore has high reliability and a moderate correlation with the total test score). Comparisons of Eqs. 3.33, 3.34 and 3.35 can also suggest when the total test score is a more accurate reflection of the true subscore (i.e., when Eq. 3.34 is less than Eq. 3.33, such as when the subscore has low reliability and/or a high correlation with the total test score). Haberman's (2008) applications to real data from testing programs suggested that the use of the observed scores of the total test is generally more precise than the use of the observed scores of the subscore and also is usually not appreciably worse than the combination of the observed scores of the subscore and the total test.

The final ETS contributions summarized in this section involve true-score estimation methods that are more complex than Kelley's (1923, 1947) linear regression (Eq. 3.26). Some of these more complex true-score regression estimates are based on the tau equivalent classical test theory model, in which frequency distributions are obtained from two or more tests assumed to be tau equivalent and these tests' distributions are used to infer several moments of the tests' true-score and error distributions (i.e., means, variances, skewness, kurtosis, and conditional versions of these; Lord 1959a). Other true-score regression estimates are based on invoking binomial assumptions about a single test's errors and beta distribution assumptions about that test's true scores (Keats and Lord 1962; Lord 1965). These developments imply regressions of true scores on observed scores that are not necessarily linear, though linearity does result when the true scores follow a beta distribution and the observed scores follow a negative hypergeometric distribution. The regressions reflect relationships among true scores and errors that are more complex than assumed in classical test theory, in which the errors are not independent of the true scores and for which attention cannot be restricted only to means, variances, and covariances. Suggested applications for these developments include estimating classification consistency and accuracy (Livingston and Lewis 1995), smoothing observed test score distributions (Hanson and Brennan 1990; Kolen and Brennan 2004), producing interval estimates for true scores (Lord and Novick 1968), predicting test norms (Lord 1962b), and predicting the bivariate distribution of two tests assumed to be parallel (Lord and Novick 1968).

3.4 Discussion

The purpose of this chapter was to summarize more than 60 years of ETS psychometric contributions pertaining to test scores. These contributions were organized into a section about the measurement properties of tests and developments of classical test theory, another section about the use of tests as predictors in correlational and regression relationships, and a third section based on integrating and applying measurement theories and correlational and regression analyses to address test-score issues. Work described in the third section on the integrations of measurement and correlational concepts and their consequent applications, is especially relevant to the operational work of psychometricians on ETS testing programs. Various integrations and applications are used when psychometricians assess a testing program's alternate test forms with respect to their measurement and prediction properties, equate alternate test forms (Angoff 1971; Kolen and Brennan 2004), and employ adaptations of Cleary's (1966b) test bias[2] approach to evaluate the invariance of test equating functions (Dorans and Holland 2000; Myers 1975). Other applications are used to help testing programs face increasing demand for changes that might be supported with psychometric methods based on the fundamental measurement and regression issues about test scores covered in this chapter.

One unfortunate aspect of this undertaking is the large number of ETS psychometric contributions that were not covered. These contributions are difficult to describe in terms of having a clear and singular focus on scores or other issues, but they might be accurately described as studies of the interaction of items and test scores. The view of test scores as a sum of items suggests several ways in which an item's characteristics influence test-score characteristics. Some ETS contributions treat item and score issues almost equally and interactively in describing their relationships, having origins in Gulliksen's (1950) descriptions of how item statistics influence test score means, standard deviations, reliability, and validity. ETS researchers such as Swineford, Lord, and Novick have clarified Gulliksen's descriptions through empirically estimated regression functions that predict test score standard deviations and reliabilities from correlations of items and test scores, through item difficulty statistics (Swineford 1959), and through mathematical functions derived to describe the influence of items with given difficulty levels on the moments of test-score distributions (Lord 1960b; Lord and Novick 1968). Other mathematical functions describe the relationships of the common factor of the items to the discrimination, standard error of measurement, and expected scores of the test (Lord 1950b). Using item response theory (IRT) methods that focus primarily on items rather than scores, ETS researchers (see the chapter on ETS contributions to IRT in this volume) have explained the implications of IRT item models for test-score characteristics, showing how observed test score distributions can be estimated from IRT models (Lord and Wingersky 1984) and showing how classical test theory results can be directly obtained from some IRT models (Holland and Hoskens 2003).

The above contributions are not the only ones dealing with interactions between scores, items, and/or fairness. Similarly, advances such as differential item functioning (DIF) can be potentially described with respect to items, examinees, and item-examinee interactions. Developments such as IRT and its application to adaptive testing can be described in terms of items and using item parameters to estimate examinees' abilities as the examinees interact with and respond to the items. ETS

[2] Although the summary of Cleary's (1966b) work in this chapter uses the *test bias* phrase actually used by Cleary, it should be acknowledged that more current descriptions of Cleary's regression applications favor different phrases such as prediction bias, overprediction, and underprediction (e.g., Bridgeman et al. 2008). The emphasis of current descriptions on prediction accuracy allows for distinctions to be made between tests that are not necessarily biased but that may be used in ways that result in biased predictions.

contributions to DIF and to IRT are just two of several additional areas of psychometrics summarized in other chapters (Carlson and von Davier, Chap. 5, this volume; Dorans, Chap. 7, this volume).

References

Angoff, W. H. (1953). Test reliability and effective test length. *Psychometrika, 18*, 1–14. https://doi.org/10.1007/BF02289023

Angoff, W. H. (1971). Scales, norms, and equivalent scores. In R. L. Thorndike (Ed.), *Educational measurement* (2nd ed., pp. 508–600). Washington, DC: American Council on Education.

Brennan, R. L. (1997). A perspective on the history of generalizability theory. *Educational Measurement: Issues and Practice, 16*(4), 14–20. https://doi.org/10.1111/j.1745-3992.1997.tb00604.x

Bridgeman, B., Pollack, J. M., & Burton, N. W. (2008). Predicting grades in college courses: A comparison of multiple regression and percent succeeding approaches. *Journal of College Admission, 199*, 19–25.

Brown, W. (1910). Some experimental results in the correlation of mental abilities. *British Journal of Psychology, 3,* 296–322. https://doi.org/10.1111/j.2044-8295.1910.tb00207.x

Browne, M. W. (1967). On oblique procrustes rotation. *Psychometrika, 32*, 125–132. https://doi.org/10.1007/BF02289420

Browne, M. W. (1968). A comparison of factor analytic techniques. *Psychometrika, 33,* 267–334. https://doi.org/10.1007/BF02289327

Browne, M. W. (1969) Fitting the factor analysis model. *Psychometrika, 34*, 375. https://doi.org/10.1007/BF02289365

Browne, M. W. (1972a). Oblique rotation to a partially specified target. *British Journal of Mathematical and Statistical Psychology, 25,* 207–212. https://doi.org/10.1111/j.2044-8317.1972.tb00492.x

Browne, M. W. (1972b). Orthogonal rotation to a partially specified target. *British Journal of Mathematical and Statistical Psychology, 25,* 115–120. https://doi.org/10.1111/j.2044-8317.1972.tb00482.x

Cleary, T. A. (1966a). An individual differences model for multiple regression. *Psychometrika, 31*, 215–224. https://doi.org/10.1007/BF02289508

Cleary, T. A. (1966b). *Test bias: Validity of the Scholastic Aptitude Test for Negro and White students in integrated colleges* (Research Bulletin No. RB-66-31). Princeton: Educational Testing Service. http://dx.doi.org/10.1002/j.2333-8504.1966.tb00529.x

Cronbach, L. J. (1951). Coefficient alpha and the internal structure of tests. *Psychometrika, 16,* 297–334. https://doi.org/10.1007/BF02310555

Cronbach, L. J., Gleser, G. C., Nanda, H., & Rajaratnam, N. (1972). *The dependability of behavioral measurements: Theory of generalizability for scores and profiles*. New York: Wiley.

Cudeck, R., & MacCallum, R. C. (2007). *Factor analysis at 100: Historical developments and future directions*. Mahwah: Erlbaum.

Dempster, A. P., Laird, N. M., & Rubin, D. B. (1977). Maximum likelihood from incomplete data via the EM algorithm. *Journal of the Royal Statistical Society, Series B, 39*, 1–22.

Dorans, N. J., & Holland, P. W. (2000). Population invariance and the equitability of tests: Basic theory and the linear case. *Journal of Educational Measurement, 37*, 281–306.

Embretson, S. E., & Reise, S. P. (2000). *Item response theory for psychologists*. Hillsdale: Erlbaum.

Feldt, L. S. (1965). The approximate sampling distribution of Kuder-Richardson reliability coefficient twenty. *Psychometrika, 30,* 357–370. https://doi.org/10.1007/BF02289499

Feldt, L. S. (1975). Estimation of the reliability of a test divided into two parts of unequal length. *Psychometrika, 40,* 557–561. https://doi.org/10.1007/BF02291556

Feldt, L. S. (2002). Reliability estimation when a test is split into two parts of unknown effective length. *Applied Measurement in Education, 15,* 295–308. https://doi.org/10.1207/S15324818AME1503_4

Frederiksen, N., & Melville, S.D. (1954). Differential predictability in the use of test scores. *Educational and Psychological Measurement, 14,* 647–656. https://doi.org/10.1177/001316445401400040

Green, B. F., Jr. (1950). A test of the equality of standard errors of measurement. *Psychometrika, 15,* 251–257. https://doi.org/10.1007/BF02289041

Green, B. F., Jr. (1952). The orthogonal approximation of an oblique structure in factor analysis. *Psychometrika, 17,* 429–440. https://doi.org/10.1007/BF02288918

Green, B. F., Jr. (1954). A note on item selection for maximum validity. *Educational and Psychological Measurement, 14,* 161–164. https://doi.org.10.1177/001316445401400116

Gulliksen, H. (1950). *Theory of mental tests*. New York: Wiley. https://doi.org/10.1037/13240-000

Gulliksen, H., & Wilks, S. S. (1950). Regression tests for several samples. *Psychometrika, 15,* 91–114. https://doi.org/10.1007/BF02289195

Haberman, S. J. (2008). When can subscores have value? *Journal of Educational and Behavioral Statistics, 33,* 204–229. https://doi.org/10.3102/1076998607302636

Haertel, E. H. (2006). Reliability. In R. L. Brennan (Ed.), *Educational measurement* (4th ed., pp. 65–110). Westport: American Council on Education and Praeger.

Hanson, B. A. (1991). *Method of moments estimates for the four-parameter beta compound binomial model and the calculation of classification consistency indexes* (Research Report No. 91–5). Iowa City: American College Testing Program.

Hanson, B. A., & Brennan, R. L. (1990). An investigation of classification consistency indexes estimated under alternative strong true score models. *Journal of Educational Measurement, 27,* 345–359. https://doi.org/10.1111/j.1745-3984.1990.tb00753.x

Harman, H. H. (1967). *Modern factor analysis* (3rd ed.). Chicago: University of Chicago Press.

Holland, P. W. (2007). A framework and history for score linking. In N. J. Dorans, M. Pommerich, & P. W. Holland (Eds.), *Linking and aligning scores and scales* (pp. 5–30). New York: Springer. https://doi.org/10.1007/978-0-387-49771-6_2

Holland, P. W. (2008, March). *The first four generations of test theory*. Presentation at the ATP Innovations in Testing Conference, Dallas, TX.

Holland, P. W., & Hoskens, M. (2003). Classical test theory as a first-order item response theory: Applications to true-score prediction from a possibly nonparallel test. *Psychometrika, 68,* 123–149. https://doi.org/10.1007/BF02296657

Holland, P. W., & Rubin, D. B. (1983). On Lord's paradox. In H. Wainer & S. Messick (Eds.), *Principals of modern psychological measurement: A festschrift for Frederic M. Lord* (pp. 3–25). Hillsdale: Erlbaum.

Horst, P. (1936). Item selection by means of a maximizing function. *Psychometrika, 1,* 229–244. https://doi.org/10.1007/BF02287875

Horst, P. (1951a). Estimating total test reliability from parts of unequal length. *Educational and Psychological Measurement, 11,* 368–371. https://doi.org/10.1177/001316445101100306

Horst, P. (1951b). Optimal test length for maximum battery validity. Psychometrika, 16, 189–202. https://doi.org/10.1007/BF02289114

Horst, P. (1951c). The relationship between the validity of a single test and its contribution to the predictive efficiency of a test battery. *Psychometrika, 16,* 57–66. https://doi.org/10.1007/BF02313427

Jackson, P. H., & Novick, M. R. (1970). Maximizing the validity of a unit-weight composite as a function of relative component lengths with a fixed total testing time. *Psychometrika, 35,* 333–347. https://doi.org/10.1007/BF02310793

Jennrich, R. I. (1973). Standard errors for obliquely rotated factor loadings. *Psychometrika, 38,* 593–604. https://doi.org/10.1007/BF02291497

Jennrich, R. I., & Thayer, D. T. (1973). A note on Lawley's formulas for standard errors in maximum likelihood factor analysis. *Psychometrika, 38,* 571–592. https://doi.org/10.1007/BF02291495

Jöreskog, K. G. (1965). *Image factor analysis* (Research Bulletin No RB-65-05). Princeton: Educational Testing Service. http://dx.doi.org/10.1002/j.2333-8504.1965.tb00134.x

Jöreskog, K. G. (1967). Some contributions to maximum likelihood factor analysis. *Psychometrika, 32,* 443–482. https://doi.org/10.1007/BF02289658

Jöreskog, K. G. (1969a). Efficient estimation in image factor analysis. *Psychometrika, 34,* 51–75. https://doi.org/10.1007/BF02290173

Jöreskog, K. G. (1969b). A general approach to confirmatory maximum likelihood factor analysis. *Psychometrika, 34,* 183–202. https://doi.org/10.1007/BF02289343

Jöreskog, K.G. (1971a). Simultaneous factor analysis in several populations. *Psychometrika, 36,* 409–426. https://doi.org/10.1007/BF02291366

Jöreskog, K.G. (1971b). Statistical analysis of sets of congeneric tests. *Psychometrika, 36,* 109–133. https://doi.org/10.1007/BF02291393

Jöreskog, K. G. (2007). Factor analysis and its extensions. In R. Cudeck & R. C. MacCallum (Eds.), *Factor analysis at 100: Historical developments and future directions* (pp. 47–77). Mahwah: Erlbaum.

Jöreskog, K. G., & Lawley, D. N. (1968). New methods in maximum likelihood factor analysis. *British Journal of Mathematical and Statistical Psychology, 21,* 85–96. https://doi.org/10.1111/j.2044-8317.1968.tb00399.x

Jöreskog, K. G., & van Thillo, M. (1972). *LISREL: A general computer program for estimating a linear structural equation system involving multiple indicators of unmeasured variables* (Research Bulletin No. RB-72-56). Princeton: Educational Testing Service. http://dx.doi.org/10.1002/j.2333-8504.1972.tb00827.x

Keats, J. A. (1957). Estimation of error variances of test scores. *Psychometrika, 22,* 29–41. https://doi.org/10.1007/BF02289207

Keats, J. A., & Lord, F. M. (1962). A theoretical distribution for mental test scores. *Psychometrika, 27,* 59–72. https://doi.org/10.1007/BF02289665

Kelley, T. L. (1923). *Statistical methods.* New York: Macmillan.

Kelley, T. L. (1947). *Fundamentals of statistics.* Cambridge, MA: Harvard University Press.

Kolen, M. J., & Brennan, R. L. (2004). *Test equating, scaling, and linking: Methods and practices* (2nd ed.). New York: Springer. https://doi.org/10.1007/978-1-4757-4310-4

Kristof, W. (1963a). Statistical inferences about the error variance. Psychometrika, 28, 129–143. https://doi.org/10.1007/BF02289611

Kristof, W. (1963b). The statistical theory of stepped-up reliability coefficients when a test has been divided into several equivalent parts. *Psychometrika, 28,* 221–238. https://doi.org/10.1007/BF02289571

Kristof, W. (1964). Testing differences between reliability coefficients. *British Journal of Statistical Psychology, 17,* 105–111. https://doi.org/10.1111/j.2044-8317.1964.tb00253.x

Kristof, W. (1969). Estimation of true score and error variance for tests under various equivalence assumptions. *Psychometrika, 34,* 489–507. https://doi.org/10.1007/BF02290603

Kristof, W. (1970). On the sampling theory of reliability estimation. *Journal of Mathematical Psychology, 7,* 371–377. https://doi.org/10.1016/0022-2496(70)90054-4

Kristof, W. (1971). On the theory of a set of tests which differ only in length. *Psychometrika, 36,* 207–225. https://doi.org/10.1007/BF02297843

Kristof, W. (1973). Testing a linear relation between true scores of two measures. *Psychometrika, 38,* 101–111. https://doi.org/10.1007/BF02291178

Kristof, W. (1974). Estimation of reliability and true score variance from a split of a test into three arbitrary parts. *Psychometrika, 39,* 491–499. https://doi.org/10.1007/BF02291670

Kuder, G. F., & Richardson, M. W. (1937). The theory of the estimation of test reliability. *Psychometrika, 2,* 151–160. https://doi.org/10.1007/BF02288391

Linn, R. L. (1967). *Range restriction problems in the validation of a guidance test battery* (Research Bulletin No. RB-67-08). Princeton: Educational Testing Service. http://dx.doi.org/10.1002/j.2333-8504.1967.tb00149.x

Linn, R. L. (1968). A Monte Carlo approach to the number of factors problem. *Psychometrika, 33,* 33–71. https://doi.org/10.1007/BF02289675

Linn, R. L. (1983). Predictive bias as an artifact of selection procedures. In H. Wainer & S. Messick (Eds.), *Principals of modern psychological measurement: A festschrift for Frederic M. Lord* (pp. 27–40). Hillsdale: Erlbaum.

Linn, R. L., & Werts, C. E. (1971). Considerations for studies of test bias. *Journal of Educational Measurement, 8,* 1–4. https://doi.org/10.1007/BF02289675

Linn, R. L., & Werts, C. E. (1973). Errors of inference due to errors of measurement. *Educational and Psychological Measurement, 33*, 531–543. https://doi.org/10.1177/001316447303300301

Livingston, S. A., & Lewis, C. (1995). Estimating the consistency and accuracy of classifications based on test scores. *Journal of Educational Measurement, 32,* 179–197. https://doi.org/10.1111/j.1745-3984.1995.tb00462.x

Lord, F. M. (1950a). *Efficiency of prediction when a regression equation from one sample is used in a new sample* (Research Bulletin No. RB-50-40). Princeton: Educational Testing Service. http://dx.doi.org/10.1002/j.2333-8504.1950.tb00478.x

Lord, F. M. (1950b). *Properties of test scores expressed as functions of the item parameters* (Research Bulletin No. RB-50-56). Princeton: Educational Testing Service. http://dx.doi.org/10.1002/j.2333-8504.1950.tb00919.x

Lord, F. M. (1955a). Equating test scores—A maximum likelihood solution. *Psychometrika, 20,* 193–200. https://doi.org/10.1007/BF02289016

Lord, F. M. (1955b). *Estimating test reliability* (Research Bulletin No. RB-55-07). Princeton: Educational Testing Service. http://dx.doi.org/10.1002/j.2333-8504.1955.tb00054.x

Lord, F. (1955c). Estimation of parameters from incomplete data. *Journal of the American Statistical Association, 50*, 870–876. https://doi.org/10.2307/2281171

Lord, F. M. (1956). Sampling error due to choice of split in split-half reliability coefficients. *Journal of Experimental Education, 24,* 245–249. https://doi.org/10.1080/00220973.1956.11010545

Lord, F. M. (1957a). Do tests of the same length have the same standard errors of measurement? *Educational and Psychological Measurement, 17*, 510–521. https://doi.org/10.1177/001316445701700407

Lord, F. M. (1957b). A significance test for the hypothesis that two variables measure the same trait except for errors of measurement. *Psychometrika, 22*, 207–220. https://doi.org/10.1007/BF02289122

Lord, F. M. (1958). Some relations between Guttman's principal components of scale analysis and other psychometric theory. *Psychometrika, 23*, 291–296. https://doi.org/10.1007/BF02289779

Lord, F. M. (1959a). Statistical inferences about true scores. *Psychometrika, 24*, 1–17. https://doi.org/10.1007/BF02289759

Lord, F. M. (1959b). Tests of the same length do have the same standard error of measurement. *Educational and Psychological Measurement, 19,* 233–239. https://doi.org/10.1177/001316445901900208

Lord, F. M. (1960). An empirical study of the normality and independence of errors of measurement in test scores. *Psychometrika, 25*, 91–104. https://doi.org/10.1007/BF02288936

Lord, F. M. (1960a). Large-sample covariance analysis when the control variable is fallible. *Journal of the American Statistical Association, 55,* 307–321. https://doi.org/10.1080/01621459.1960.10482065

Lord, F. M. (1960b). Use of true-score theory to predict moments of univariate and bivariate observed score distributions. *Psychometrika, 25*, 325–342. https://doi.org/10.1007/BF02289751

Lord, F. M. (1962a). *Elementary models for measuring change*. (Research Memorandum No. RM-62-05). Princeton: Educational Testing Service.

Lord. F. M. (1962b). Estimating norms by item-sampling. *Educational and Psychological Measurement, 22*, 259–267. https://doi.org/10.1177/001316446202200202

Lord, F. M. (1963). Formula scoring and validity. *Educational and Psychological Measurement, 23*, 663–672. https://doi.org/10.1177/001316446302300403

Lord, F. M. (1965). A strong true score theory with applications. *Psychometrika, 30,* 239–270. https://doi.org/10.1007/BF02289490

Lord, F. M. (1967). A paradox in the interpretation of group comparisons. *Psychological Bulletin, 68*, 304–305.

Lord, F. M. (1969). Statistical adjustments when comparing preexisting groups. *Psychological Bulletin, 72*, 336–337. https://doi.org/10.1037/h0028108

Lord, F. M. (1973). Testing if two measuring procedures measure the same dimension. *Psychological Bulletin, 79*, 71–72. https://doi.org/10.1037/h0033760

Lord, F. M., & Novick, M. R. (1968). *Statistical theories of mental test scores*. Reading: Addison-Wesley.

Lord, F. M., & Stocking, M. (1976). An interval estimate for making statistical inferences about true score. *Psychometrika, 41,* 79–87. https://doi.org/10.1007/BF02291699

Lord, F. M., & Wingersky, M. S. (1984). Comparison of IRT true-score and equipercentile observed-score "equatings." *Applied Psychological Measurement, 8,* 453–461. https://doi.org/10.1177/014662168400800409

Mislevy, R. J. (1993). Foundations of a new test theory. In N. Frederiksen, R. J. Mislevy, & I. I. Bejar (Eds.), *Test theory for a new generation of tests* (pp. 19–39). Hillsdale: Erlbaum.

Mollenkopf, W. G. (1949). Variation of the standard error of measurement. *Psychometrika, 14,* 189–229. https://doi.org/10.1007/BF02289153

Moses, T. (2012). Relationships of measurement error and prediction error in observed-score regression. *Journal of Educational Measurement, 49,* 380–398. https://doi.org/10.1111/j.1745-3984.2012.00182.x

Moses, T., Deng, W., & Zhang, Y.-L. (2011). Two approaches for using multiple anchors in NEAT equating. *Applied Psychological Measurement, 35*, 362–379. https://doi.org/10.1177/0146621611405510

Myers, C. T. (1975). *Test fairness: A comment on fairness in statistical analysis* (Research Bulletin No. RB-75-12). Princeton: Educational Testing Service. http://dx.doi.org/10.1002/j.2333-8504.1975.tb01051.x

Novick, M. R. (1965). *The axioms and principal results of classical test theory* (Research Bulletin No. RB-65-02). Princeton: Educational Testing Service. http://dx.doi.org/10.1002/j.2333-8504.1965.tb00132.x

Novick, M. R. (1983). The centrality of Lord's paradox and exchangeability for all statistical inference. In H. Wainer & S. Messick (Eds.), *Principals of modern psychological measurement: A festschrift for Frederic M. Lord* (pp. 41–53). Hillsdale: Erlbaum.

Novick, M. R., & Lewis, C. (1967). Coefficient alpha and the reliability of composite measurements. *Psychometrika, 32,* 1–13. https://doi.org/10.1007/BF02289400

Novick, M. R., & Thayer, D. T. (1969). *Some applications of procedures for allocating testing time* (Research Bulletin No. RB-69-01). Princeton: Educational Testing Service. http://dx.doi.org/10.1002/j.2333-8504.1969.tb00161.x

O'Connor, E. F. (1973). *Unraveling Lord's paradox: The appropriate use of multiple regression analysis in quasi-experimental research* (Research Bulletin No. RB-73-53). Princeton: Educational Testing Service. http://dx.doi.org/10.1002/j.2333-8504.1973.tb00839.x

Pearson, K. (1903). Mathematical contributions to the theory of evolution. XI. On the influence of natural selection on the variability and correlation of organs. *Philosophicali. Transactions 200-A,* 1–66. London: Royal Society

Pinzka, C., & Saunders, D. R. (1954). *Analytic rotation to simple structure: II. Extension to an oblique solution* (Research Bulletin No. RB-54-31). Princeton: Educational Testing Service. http://dx.doi.org/10.1002/j.2333-8504.1954.tb00487.x

Plumlee, L. B. (1954). Predicted and observed effect of chance on multiple-choice test validity. *Psychometrika, 19,* 65–70. https://doi.org/10.1007/BF02288994

Potthoff, R. F. (1964). On the Johnson-Neyman technique and some extensions thereof. *Psychometrika, 29,* 241–256. https://doi.org/10.1007/BF02289721

Rosenbaum, P. R., & Rubin, D. B. (1984). Reducing bias in observational studies using subclassification on the propensity score. *Journal of the American Statistical Association, 79,* 516–524. https://doi.org/10.1080/01621459.1984.10478078

Rosenbaum, P. R., & Rubin, D. B. (1985). Constructing a control group using multivariate matched sampling methods that incorporate the propensity score. *American Statistician, 39,* 33–8. https://doi.org/10.1080/00031305.1985.10479383

Rubin, D. B. (1979). Using multivariate matched sampling and regression adjustment to control bias in observational studies. *Journal of the American Statistical Association, 74*, 318–328. https://doi.org/10.2307/2286330

Rubin, D. B. (1980). Bias reduction using Mahalanobis-metric matching. *Biometrics, 36*, 293–298. https://doi.org/10.2307/2529981

Rubin, D. B., & Thayer, D. (1978). Relating tests given to different samples. *Psychometrika, 43,* 1–10. https://doi.org/10.1007/BF02294084

Saunders, D. R. (1953a). *An analytic method for rotation to orthogonal simple structure* (Research Bulletin No. RB-53-10). Princeton: Educational Testing Service. http://dx.doi.org/10.1002/j.2333-8504.1953.tb00890.x

Saunders, D. R. (1953b). *Moderator variables in prediction, with special reference to freshman engineering grades and the strong vocational interest blank* (Research Bulletin No. RB-53-23). Princeton: Educational Testing Service. http://dx.doi.org/10.1002/j.2333-8504.1953.tb00238.x

Schultz, D. G., & Wilks, S. S. (1950). *A method for adjusting for lack of equivalence in groups* (Research Bulletin No. RB-50-59). Princeton: Educational Testing Service. http://dx.doi.org/10.1002/j.2333-8504.1950.tb00682.x

Spearman, C. (1904a). General intelligence objectively determined and measured. *American Journal of Psychology, 15,* 201–293. https://doi.org/10.2307/1412107

Spearman, C. (1904b). The proof and measurement of association between two things. *American Journal of Psychology, 15,* 72–101. https://doi.org/10.2307/1412159

Spearman, C. (1910). Correlation calculated from faulty data. *British Journal of Psychology, 3,* 271–295. https://doi.org/10.1111/j.2044-8295.1910.tb00206.x

Swineford, F. (1959). Some relations between test scores and item statistics. *Journal of Educational Psychology, 50*, 26–30. https://doi.org/10.1037/h0046332

Traub, R. E. (1997). Classical test theory in historical perspective. *Educational Measurement: Issues and Practice, 16*(4), 8–14. https://doi.org/10.1111/j.1745-3992.1997.tb00603.x

Tucker, L. R. (1949). A note on the estimation of test reliability by the Kuder-Richardson formula (20). *Psychometrika, 14,* 117–119. https://doi.org/10.1007/BF02289147

Tucker, L. R. (1951). *A method for synthesis of factor analysis studies* (Personnel Research Section Report No. 984). Washington, DC: Department of the Army.

Tucker, L. R. (1955). The objective definition of simple structure in linear factor analysis. *Psychometrika, 20,* 209–225. https://doi.org/10.1007/BF02289018

Tucker, L. R., & Finkbeiner, C.T. (1981). *Transformation of factors by artificial personal probability functions* (Research Report No. RR-81-58). Princeton: Educational Testing Service. http://dx.doi.org/10.1002/j.2333-8504.1981.tb01285.x

Tucker, L. R., Koopman, R. F., & Linn, R. L. (1969). Evaluation of factor analytic research procedures by means of simulated correlation matrices. *Psychometrika, 34,* 421–459. https://doi.org/10.1007/BF02290601

Woodbury, M. A., & Lord, F. M. (1956). The most reliable composite with a specified true score. *British Journal of Statistical Psychology, 9*, 21–28. https://doi.org/10.1111/j.2044-8317.1956.tb00165.x

Woodbury, M. A., & Novick, M. R. (1968). Maximizing the validity of a test battery as a function of relative test lengths for a fixed total testing time. *Journal of Mathematical Psychology, 5*, 242–259. https://doi.org/10.1016/0022-2496(68)90074-6

Wright, N. K., & Dorans, N. J. (1993). *Using the selection variable for matching or equating* (Research Report No. RR-93-04). Princeton: Educational Testing Service. http://dx.doi.org/10.1002/j.2333-8504.1993.tb01515.x

Chapter 4
Contributions to Score Linking Theory and Practice

Neil J. Dorans and Gautam Puhan

Test score equating is essential for testing programs that use multiple editions of the same test and for which scores on different editions are expected to have the same meaning. Different editions may be built to a common blueprint and be designed to measure the same constructs, but they almost invariably differ somewhat in their psychometric properties. If one edition were more difficult than another, test takers would tend to receive lower scores on the harder form. Score equating seeks to eliminate the effects on scores of these unintended differences in test form difficulty. Score equating is necessary to be fair to test takers.

ETS statisticians and psychometricians have contributed indirectly or directly to the wealth of material in the chapters on score equating or on score linking that have appeared in the four editions of *Educational Measurement*. ETS's extensive involvement with the score equating chapters of these editions of *Educational Measurement* highlights the impact that ETS has had in this important area of psychometrics.

At the time of publication, each of the four editions of *Educational Measurement* represented the state of the art in domains that are essential to the purview of the National Council on Measurement in Education. Experts in each domain wrote a chapter in each edition. Harold Gulliksen was one of the key contributors to the Flanagan (1951) chapter on units, scores, and norms that appeared in the first edition. Several of the issues and problems raised in that first edition are still current, which shows their persistence. Angoff (1971), in the second edition, provided a comprehensive introduction to scales, norms, and test equating. Petersen et al. (1989) introduced new material developed since the Angoff chapter. Holland and Dorans (2006) included a brief review of the history of test score linking. In addition to test equating, Holland and Dorans (2006) discussed other ways that scores on different tests are connected or linked together.

N.J. Dorans (✉) • G. Puhan
Educational Testing Service, Princeton, NJ, USA
e-mail: ndorans@ets.org

R.E. Bennett, M. von Davier (eds.), *Advancing Human Assessment*,
Methodology of Educational Measurement and Assessment,
DOI 10.1007/978-3-319-58689-2_4

The purpose of this chapter is to document ETS's involvement with score linking theory and practice. This chapter is not meant to be a book on score equating and score linking.[1] Several books on equating exist; some of these have been authored by ETS staff, as is noted in the last section of this chapter. We do not attempt to summarize all extant research and development pertaining to score equating or score linking. We focus on efforts conducted by ETS staff. We do not attempt to pass judgment on research or synthesize it. Instead, we attempt to describe it in enough detail to pique the interest of the reader and help point him or her in the right direction for further exploration on his or her own. We presume that the reader is familiar enough with the field so as not to be intimidated by the vocabulary that has evolved over the years in this area of specialization so central to ETS's mission to foster fairness and quality.

The particular approach to tackling this documentation task is to cluster studies around different aspects of score linking. Section 4.1 lists several examples of score linking to provide a motivation for the extent of research on score linking. Section 4.2 summarizes published efforts that provide conceptual frameworks of score linking or examples of scale aligning. Section 4.3 deals with data collection designs and data preparation issues. In Sect. 4.4, the focus is on the various procedures that have been developed to link or equate scores. Research describing processes for evaluating the quality of equating results is the focus of Sect. 4.5. Studies that focus on comparing different methods are described in Sect. 4.6. Section 4.7 is a brief chronological summary of the material covered in Sects. 4.2, 4.3, 4.4, 4.5 and 4.6. Section 4.8 contains a summary of the various books and chapters that ETS authors have contributed on the topic of score linking. Section 4.9 contains a concluding comment.

4.1 Why Score Linking Is Important

Two critical ingredients are needed to produce test scores: the test and those who take the test, the test takers. Test scores depend on the blueprint or specifications used to produce the test. The specifications describe the construct that the test is supposed to measure, how the items or components of the test contribute to the measurement of this construct (or constructs), the relative difficulty of these items for the target population of test takers, and how the items and test are scored. The definition of the target population of test takers includes who qualifies as a member of that population and is preferably accompanied by an explanation of why the test

[1] The term *linking* is often used in an IRT context to refer to procedures for aligning item parameter and proficiency metrics from one calibration to another, such as those described by M. von Davier and A. A. von Davier (2007). We do not consider this type of IRT linking in this chapter; it is treated in the chapter by Carlson and von Davier (Chap. 5, this volume). We do, however, address IRT true-score linking in Sect. 4.6.4 and IRT preequating in Sect. 4.4.4.

is appropriate for these test takers and examples of appropriate and inappropriate use.

Whenever scores from two different tests are going to be compared, there is a need to link the scales of the two test scores. The goal of scale aligning is to transform the scores from two different tests onto a common scale. The types of linkages that result depend on whether the test scores being linked measure different constructs or similar constructs, whether the tests are similar or dissimilar in difficulty, and whether the tests are built to similar or different test specifications. We give several practical examples in the following.

When two or more tests that measure different constructs are administered to a common population, the scores for each test may be transformed to have a common distribution for the target population of test takers (i.e., the reference population). The data are responses from (a) administering all the tests to the same sample of test takers or (b) administering the tests to separate, randomly equivalent samples of test takers from the same population. In this way, all of the tests are taken by equivalent groups of test takers from the reference population. One way to define comparable scores is in terms of comparable percentiles in the reference population.

Even though the scales on the different tests are made comparable in this narrow sense, the tests do measure different constructs. The recentering of the *SAT® I* test scale is an example of this type of scale aligning (Dorans 2002a, b). The scales for the SAT Verbal (SAT-V) and SAT Mathematical (SAT-M) scores were redefined so as to give the scaled scores on the SAT-V and SAT-M the same distribution in a reference population of students tested in 1990. The recentered score scales enable a student whose SAT-M score is higher than his or her SAT-V score to conclude that he or she did in fact perform better on the mathematical portion than on the verbal portion, at least in relation to the students tested in 1990.

Tests of skill subjects (e.g., reading) that are targeted for different school grades may be viewed as tests of similar constructs that are intended to differ in difficulty—those for the lower grades being easier than those for the higher grades. It is often desired to put scores from such tests onto a common overall scale so that progress in a given subject, such as mathematics or reading, can be tracked over time. A topic such as mathematics or reading, when considered over a range of school grades, has several subtopics or dimensions. At different grades, potentially different dimensions of these subjects are relevant and tested. For this reason, the constructs being measured by the tests for different grade levels may differ somewhat, but the tests are often similar in reliability.

Sometimes tests that measure the same construct have similar levels of difficulty but differ in reliability (e.g., length). The classic case is scaling the scores of a short form of a test onto the scale of its full or long form.

Sometimes tests to be linked all measure similar constructs, but they are constructed according to different specifications. In most cases, they are similar in test length and reliability. In addition, they often have similar uses and may be taken by the same test takers for the same purpose. Score linking adds value to the scores on both tests by expressing them as if they were scores on the other test. Many colleges and universities accept scores on either the ACT or SAT for the purpose of admissions decisions, and they often have more experience interpreting the results from one of these tests than the other.

Test equating is a necessary part of any testing program that produces new test forms and for which the uses of these tests require the meaning of the score scale be maintained over time. Although they measure the same constructs and are usually built to the same test specifications or test blueprint, different editions or forms of a test almost always differ somewhat in their statistical properties. For example, one form may be harder than another, so without adjustments, test takers would be expected to receive lower scores on this harder form. A primary goal of test equating for testing programs is to eliminate the effects on scores of these unintended differences in test form difficulty. The purpose of equating test scores is to allow the scores from each test to be used interchangeably, as if they had come from the same test. This purpose puts strong requirements on the tests and on the method of score linking. Most of the research described in the following pages focused on this particular form of scale aligning, known as *score equating*.

In the remaining sections of this chapter, we focus on score linking issues for tests that measure characteristics at the level of the individual test taker. Large-scale assessments, which are surveys of groups of test takers, are described in Beaton and Barone (Chap. 8, this volume) and Kirsh et al. (Chap. 9, this volume).

4.2 Conceptual Frameworks for Score Linking

Holland and Dorans (2006) provided a framework for classes of score linking that built on and clarified earlier work found in Mislevy (1992) and Linn (1993). Holland and Dorans (2006) made distinctions between different types of linkages and emphasized that these distinctions are related to how linked scores are used and interpreted. A link between scores on two tests is a transformation from a score on one test to a score on another test. There are different types of links, and the major difference between these types is not procedural but interpretative. Each type of score linking uses either equivalent groups of test takers or common items for linkage purposes. It is essential to understand why these types differ because they can be confused in practice, which can lead to violations of the standards that guide professional practice. Section 4.2.1 describes frameworks used for score linking. Section 4.2.2 contains a discussion of score equating frameworks.

4.2.1 Score Linking Frameworks

Lord (1964a, b) published one of the early articles to focus on the distinction between test forms that are actually or rigorously parallel and test forms that are nominally parallel—those that are built to be parallel but fall short for some reason.

This distinction occurs in most frameworks on score equating. Lord (1980) later went on to say that equating was either unnecessary (rigorously parallel forms) or impossible (everything else).

Mislevy (1992) provided one of the first extensive treatments of different aspects of what he called linking of educational assessments: *equating, calibration, projection, statistical moderation,* and *social moderation.*

Dorans (1999) made distinctions between three types of linkages or score correspondences when evaluating linkages among SAT scores and ACT scores. These were equating, scaling, and prediction. Later, in a special issue of *Applied Psychological Measurement*, edited by Pommerich and Dorans (2004), he used the terms *equating*, *concordance*, and *expectation* to refer to these three types of linkings and provided means for determining which one was most appropriate for a given set of test scores (Dorans 2004b). This framework was elaborated on by Holland and Dorans (2006), who made distinctions between *score equating, scale aligning*, and *predicting*, noting that scale aligning was a broad category that could be further subdivided into subcategories on the basis of differences in the construct assessed, test difficulty, test reliability, and population ability.

Many of the types of score linking cited by Mislevy (1992) and Dorans (1999, 2004b) could be found in the broad area of scale aligning, including concordance, vertical linking, and calibration. This framework was adapted for the public health domain by Dorans (2007) and served as the backbone for the volume on linking and aligning scores and scales by Dorans et al. (2007).

4.2.2 Equating Frameworks

Dorans et al. (2010a) provided an overview of the particular type of score linking called score equating from a perspective of best practices. After defining equating as a special form of score linking, the authors described the most common data collection designs used in the equating of test scores, some common observed-score equating functions, common data-processing practices that occur prior to computations of equating functions, and how to evaluate an equating function.

A.A. von Davier (2003, 2008) and A.A. von Davier and Kong (2005), building on the unified statistical treatment of score equating, known as *kernel equating*, that was introduced by Holland and Thayer (1989) and developed further by A.A. von Davier et al. (2004b), described a new unified framework for linear equating in a nonequivalent groups anchor test design. They employed a common parameterization to show that three linear methods, Tucker, Levine observed score, and chained,[2] can be viewed as special cases of a general linear function. The concept of a method function was introduced to distinguish among the possible forms that a linear equating function might take, in general, and among the three equating methods, in particular. This approach included a general formula for the standard error of equating

[2] These equating methods are described in Sect. 4.4.

for all linear equating functions in the nonequivalent groups anchor test design and advocated the use of the standard error of equating difference (SEED) to investigate if the observed differences in the equating functions are statistically significant.

A.A. von Davier (2013) provided a conceptual framework that encompassed traditional observed-score equating methods, kernel equating methods, and item response theory (IRT) observed-score equating, all of which produce one equating function between two test scores, along with local equating or local linking, which can produce a different linking function between two test scores given a score on a third variable (Wiberg et al. 2014). The notion of multiple conversions between two test scores is a source of controversy (Dorans 2013; Gonzalez and von Davier 2013; Holland 2013; M. von Davier et al. 2013).

4.3 Data Collection Designs and Data Preparation

Data collection and preparation are prerequisites to score linking.

4.3.1 Data Collection

Numerous data collection designs have been used for score linking. To obtain unbiased estimates of test form difficulty differences, all score equating methods must control for differential ability of the test-taker groups employed in the linking process. Data collection procedures should be guided by a concern for obtaining equivalent groups, either directly or indirectly. Often, two different, nonstrictly parallel tests are given to two different groups of test takers of unequal ability. Assuming that the samples are large enough to ignore sampling error, differences in the distributions of the resulting scores can be due to one or both of two factors. One factor is the relative difficulty of the two tests, and the other is the relative ability of the two groups of test takers on these tests. Differences in difficulty are what test score equating is supposed to take care of; difference in ability of the groups is a confounding factor that needs to be eliminated before the equating process can take place.

In practice, two distinct approaches address the separation of test difficulty and group ability differences. The first approach is to use a common population of test takers so that there are no ability differences. The other approach is to use an anchor measure of the construct being assessed by the tests to be equated. Ideally, the data should come from a large representative sample of motivated test takers that is divided in half either randomly or randomly within strata to achieve equivalent groups. Each half of this sample is administered either the new form or the old form of a test. It is typical to assume that all samples are random samples from populations of interest, even though, in practice, this may be only an approximation. When the same test takers take both tests, we achieve direct control over differential test-taker ability. In practice, it is more common to use two equivalent samples of test takers from a common population instead of identical test takers.

The second approach assumes that performance on a set of common items or an anchor measure can quantify the ability differences between two distinct, but not necessarily equivalent, samples of test takers. The use of an anchor measure can lead to more flexible data collection designs than those that require common test takers. However, the use of anchor measures requires users to make various assumptions that are not needed when the test takers taking the tests are either the same or from equivalent samples. When there are ability differences between new and old form samples, the various statistical adjustments for ability differences often produce different results because the methods make different assumptions about the relationships of the anchor test score to the scores to be equated. In addition, assumptions are made about the invariance of item characteristics across different locations within the test.

Some studies have attempted to link scores on tests in the absence of either common test material or equivalent groups of test takers. Dorans and Middleton (2012) used the term *presumed linking* to describe these situations. These studies are not discussed here.

It is generally considered good practice to have the anchor test be a mini-version of the total tests being equated. That means it should have the same difficulty and similar content. Often an external anchor is not available, and internal anchors are used. In this case, context effects become a possible issue. To minimize these effects, anchor (or common) items are often placed in the same location within each test. When an anchor test is used, the items should be evaluated via procedures for assessing whether items are functioning in the same way in both the old and new form samples. All items on both total tests are evaluated to see if they are performing as expected. If they are not, it is often a sign of a quality-control problem. More information can be found in Holland and Dorans (2006).

When there are large score differences on the anchor test between samples of test takers given the two different test forms to be equated, equating based on the nonequivalent-groups anchor test design can often become problematic. Accumulation of potentially biased equating results can occur over a chain of prior equatings and lead to a shift in the meaning of numbers on the scores scale.

In practice, the true equating function is never known, so it is wise to look at several procedures that make different assumptions or that use different data. Given the potential impact of the final score conversion on all participants in an assessment process, it is important to check as many factors that can cause problems as possible. Considering multiple conversions is one way to do this.

Whereas many sources, such as Holland and Dorans (2006), have focused on the structure of data collection designs, the amount of data collected has a substantial effect on the usefulness of the resulting equatings. Because it is desirable for the statistical uncertainty associated with test equating to be much smaller than the other sources of variation in test results, it is important that the results of test equating be based on samples that are large enough to ensure this. This fact should always be kept in mind when selecting a data collection design. Section 4.4 describes procedures that have been developed to deal with the threats associated with small samples.

4.3.2 Data Preparation Activities

Prior to equating and other forms of linking, several steps can be taken to improve the quality of the data. These best practices of data preparation often deal with sample selection, smoothing score distributions, excluding outliers, repeaters, and so on. These issues are the focus of the next four parts of this section.

4.3.2.1 Sample Selection

Before conducting the equating analyses, testing programs often filter the data based on certain heuristics. For example, a testing program may choose to exclude test takers who do not attempt a certain number of items on the test. Other programs might exclude test takers based, for example, on repeater status. ETS researchers have conducted studies to examine the effect of such sample selection practices on equating results. Liang et al. (2009) examined whether nonnative speakers of the language in which the test is administered should be excluded and found that this may not be an issue as long as the proportion of nonnative speakers does not change markedly across administrations. Puhan (2009b, 2011c) studied the impact of repeaters in the equating samples and found in the data he examined that inclusion or exclusion of repeaters had very little impact on the final equating results. Similarly, Yang et al. (2011) examined the effect of repeaters on score equating and found no significant effects of repeater performance on score equating for the exam being studied. However, Kim and Walker (2009a, b) found in their study that when the repeater subgroup was subdivided based on the particular form test takers took previously, subgroup equating functions substantially differed from the total-group equating function.

4.3.2.2 Weighted Samples

Dorans (1990c) edited a special issue of *Applied Measurement in Education* that focused on the topic of equating with samples matched on the anchor test score (Dorans 1990a). The studies in that special issue used simulations that varied in the way in which real data were manipulated to produce simulated samples of test takers. These and related studies are described in Sect. 4.6.3.

Other authors used demographic data to achieve a form of matching. Livingston (2014a) proposed the demographically adjusted groups procedure, which uses demographic information about the test takers to transform the groups taking the two different test forms into groups of equal ability by weighting the test takers unequally. Results indicated that although this procedure adjusts for group differences, it does not reduce the ability difference between the new and old form samples enough to warrant use.

Qian et al. (2013) used techniques for weighting observations to yield a weighted sample distribution that is consistent with the target population distribution to achieve true-score equatings that are more invariant across administrations than those obtained with unweighted samples.

Haberman (2015) used adjustment by minimum discriminant information to link test forms in the case of a nonequivalent-groups design in which there are no satisfactory common items. This approach employs background information other than scores on individual test takers in each administration so that weighted samples of test takers form pseudo-equivalent groups in the sense that they resemble samples from equivalent groups.

4.3.2.3 Smoothing

Irregularities in score distributions can produce irregularities in the equipercentile equating adjustment that might not generalize to different groups of test takers because the methods developed for continuous data are applied to discrete data. Therefore it is generally advisable to presmooth the raw-score frequencies in some way prior to equipercentile equating.

The idea of smoothing score distributions prior to equating goes far back to the 1950s. Karon and Cliff (1957) proposed the Cureton–Tukey procedure as a means for reducing sampling error by mathematically smoothing the sample score data before equating. However, the differences among the linear equating method, the equipercentile equating method with no smoothing of the data, and the equipercentile equating method after smoothing by the Cureton–Tukey method were not statistically significant. Nevertheless, this was an important idea, and although Karon and Cliff's results did not show the benefits of smoothing, currently most testing programs using equipercentile equating use some form of pre- or postsmoothing to obtain more stable equating results.

Ever since the smoothing method using loglinear models was adapted by ETS researchers in the 1980s (for details, see Holland and Thayer 1987; Rosenbaum and Thayer 1987) smoothing has been an important component of the equating process. The new millennium saw a renewed interest in smoothing research. Macros using the statistical analysis software SAS loglinear modeling routines were developed at ETS to facilitate research on smoothing (Moses and von Davier 2006, 2013; Moses et al. 2004). A series of studies were conducted to assess selection strategies (e.g., strategies based on likelihood ratio tests, equated score difference tests, Akaike information criterion (AIC) for univariate and bivariate loglinear smoothing models and their effects on equating function accuracy (Moses 2008a, 2009; Moses and Holland 2008, 2009a, b, c, 2010a, b).

Studies also included comparisons of traditional equipercentile equating with various degrees of presmoothing and kernel equating (Moses and Holland 2007) and smoothing approaches for composite scores (Moses 2014) as well as studies that compared smoothing with pseudo-Bayes probability estimates (Moses and Oh 2009).

There has also been an interest in smoothing in the context of systematic irregularities in the score distributions that are due to scoring practice and scaling issues (e.g., formula scoring, impossible scores) rather than random irregularities (J. Liu et al. 2009b; Puhan et al. 2008b, 2010).

4.3.2.4 Small Samples and Smoothing

Presmoothing the data before conducting an equipercentile equating has been shown to reduce error in small-sample equating. For example, Livingston and Feryok (1987) and Livingston (1993b) worked with small samples and found that presmoothing substantially improved the equating results obtained from small samples. Puhan (2011a, b), based on the results of an empirical study, however, concluded that although presmoothing can reduce random equating error, it is not likely to reduce equating bias caused by using an unrepresentative small sample and presented other alternatives to the small-sample equating problem that focused more on improving data collection (see Sect. 4.4.5).

4.4 Score Equating and Score Linking Procedures

Many procedures for equating tests have been developed by ETS researchers. In this section, we consider equating procedures such as linear, equipercentile equating, kernel equating, and IRT true-score linking.[3] Equating procedures developed to equate new forms under special circumstances (e.g., preequating and small-sample equating procedures) are also considered in this section.

[3] We have chosen to use the term *linking* instead of *equating* when it comes to describing the IRT true-score approach that is in wide use. This linking procedure defines the true-score equating that exists between true scores on Test X and true scores on Test Y, which are perfectly related to each other, as both are monotonic transformations of the same IRT proficiency estimate. Typically, this true-score equating is applied to observed scores as if they were true scores. This application produces an observed-score linking that is not likely to yield equated scores, however, as defined by Lord (1980) or Holland and Dorans (2006); hence our deliberate use of linking instead of equating.

4.4.1 Early Equating Procedures

Starting in the 1950s, ETS researchers have made substantial contributions to the equating literature by proposing new methods for equating, procedures for improving existing equating methods, and procedures for evaluating equating results.

Lord (1950) provided a definition of comparability wherein the score scales of two equally reliable tests are considered comparable with respect to a certain group of test takers if the score distributions of the two tests are identical for this group. He provided the basic formulas for equating means and standard deviations (in six different scenarios) to achieve comparability of score scales. Tucker (1951) emphasized the need to establish a formal system within which to consider scaling error due to sampling. Using simple examples, he illustrated possible ways of defining the scaling error confidence range and setting a range for the probability of occurrence of scaling errors due to sampling that would be considered within normal operations. Techniques were developed to investigate whether regressions differ by groups. Schultz and Wilks (1950) presented a technique to adjust for the lack of equivalence in two samples. This technique focused on the intercept differences from the two group regressions of total score onto anchor score obtained under the constraint that the two regressions had the same slope. Koutsopoulos (1961) presented a linear practice effect solution for a counterbalanced case of equating, in which two equally random groups (alpha and beta) take two forms, X and Y, of a test, alpha in the order X, Y and beta in the order Y, X. Gulliksen (1968) presented a variety of solutions for determining the equivalence of two measures, ranging from a criterion for strict interchangeability of scores to factor methods for comparing multifactor batteries of measures and multidimensional scaling. Boldt (1972) laid out an alternative approach to linking scores that involved a principle for choosing objective functions whose optimization would lead to a selection of conversion constants for equating.

Angoff (1953) presented a method of equating test forms of the American Council on Education (ACE) examination by using a miniature version of the full test as an external anchor to equate the test forms. Fan and Swineford (1954) and Swineford and Fan (1957) introduced a method based on item difficulty estimates to equate scores administered under the nonequivalent anchor test design, which the authors claimed produced highly satisfactory results, especially when the two groups taking the two forms were quite different in ability.

Assuming that the new and old forms are equally reliable, Lord (1954, 1955) derived maximum likelihood estimates of the population mean and standard deviation, which were then substituted into the basic formula for linear equating.

Levine (1955) developed two linear equating procedures for the common-item nonequivalent population design. Levine observed-score equating relates observed scores on a new form to the scale of observed scores on an old form. Levine true-score equating equates true scores. Approximately a half-century later, A.A. von Davier et al. (2007) introduced an equipercentile version of the Levine linear observed-score equating function, which is based on assumptions about true scores.

Based on theoretical and empirical results, Chen (2012) showed that linear IRT observed-score linking and Levine observed-score equating for the anchor test design are closely related despite being based on different methodologies. Chen and Livingston (2013) presented a new equating method for the nonequivalent groups with anchor test design: poststratification equating based on true anchor scores. The linear version of this method is shown to be equivalent, under certain conditions, to Levine observed-score equating.

4.4.2 True-Score Linking

As noted in the previous section, Levine (1955) also developed the so-called Levine true-score equating procedure that equates true scores.

Lord (1975) compared equating methods based on item characteristic curve (ICC) theory, which he later called item response theory (IRT) in Lord (1980), with nonlinear conventional methods and pointed out the effectiveness of ICC-based methods for increasing stability of the equating near the extremes of the data, reducing scale drift, and preequating. Lord also included a chapter on IRT preequating. (A review of research related to IRT true-score linking appears in Sect. 4.6.4.)

4.4.3 Kernel Equating and Linking With Continuous Exponential Families

As noted earlier, Holland and Thayer (1989) introduced the kernel method of equating score distributions. This new method included both linear and standard equipercentile methods as special cases and could be applied under most equating data collection designs.

Within the Kernel equating framework, Chen and Holland (2010) developed a new curvilinear equating for the nonequivalent groups with anchor test (NEAT) design which they called curvilinear Levine observed score equating.

In the context of equivalent-groups design, Haberman (2008a) introduced a new way to continuize discrete distribution functions using exponential families of functions. Application of this linking method was also considered for the single-group design (Haberman 2008b) and the nonequivalent anchor test design (Haberman and Yan 2011). For the nonequivalent groups with anchor test design, this linking method produced very similar results to kernel equating and equipercentile equating with loglinear presmoothing.

4.4.4 Preequating

Preequating has been tried for several ETS programs over the years. Most notably, the computer-adaptive testing algorithm employed for the *GRE*® test, the *TOEFL*® test, and GMAT examination in the 1990s could be viewed as an application of IRT preequating. Since the end of the twentieth century, IRT preequating has been used for the *CLEP*® examination and with the GRE revised General Test introduced in 2011. This section describes observed-score preequating procedures. (The results of several studies that used IRT preequating can be found in Sect. 4.6.5.)

In the 1980s, section preequating was used with the GMAT examination. A preequating procedure was developed for use with small-volume tests, most notably the *PRAXIS*® assessments. This approach is described in Sect. 4.4.5. Holland and Wightman (1982) described a preliminary investigation of a linear section preequating procedure. In this statistical procedure, data collected from equivalent groups via the nonscored variable or experimental section(s) of a test were combined across tests to produce statistics needed for linear preequating of a form composed of these sections. Thayer (1983) described the maximum likelihood estimation procedure used for estimating the joint covariance matrix for sections of tests given to distinct samples of test takers, which was at the heart of the section preequating approach.

Holland and Thayer (1981) applied this procedure to the GRE test and obtained encouraging results. Holland and Thayer (1984, 1985) extended the theory behind section preequating to allow for practice effects on both the old and new forms and, in the process, provided a unified account of the procedure. Wightman and Wightman (1988) examined the effectiveness of this approach when there is only one variable or experimental section of the test, which entailed using different missing data techniques to estimate correlations between sections.

After a long interlude, section preequating with a single variable section was studied again. Guo and Puhan (2014) introduced a method for both linear and nonlinear preequating. Simulations and a real-data application showed the proposed method to be fairly simple and accurate. Zu and Puhan (2014) examined an observed-score preequating procedure based on empirical item response curves, building on work done by Livingston in the early 1980s. The procedure worked reasonably well in the score range that contained the middle 90th percentile of the data, performing as well as the IRT true-score equating procedure.

4.4.5 Small-Sample Procedures

In addition to proposing new methods for test equating in general, ETS researchers have focused on equating under special circumstances, such as equating with very small samples. Because equating with very small samples tends to be less stable, researchers have proposed new approaches that aim to produce more stable equating results under small-sample conditions. For example, Kim et al. (2006, 2007, 2008c, 2011) proposed the synthetic linking function (which is a weighted average of the small-sample equating and the identity function) for small samples and conducted several empirical studies to examine its effectiveness in small-sample conditions. Similarly, the circle-arc equating method, which constrains the equating curve to pass through two prespecified endpoints and an empirically determined middle point, was also proposed for equating with small samples (Livingston and Kim 2008, 2009, 2010a, b) and evaluated in empirical studies by Kim and Livingston (2009, 2010). Finally, Livingston and Lewis (2009) proposed the empirical Bayes approach for equating with small samples whereby prior information comes from equatings of other test forms, with an appropriate adjustment for possible differences in test length. Kim et al. (2008d, 2009) conducted resampling studies to evaluate the effectiveness of the empirical Bayes approach with small samples and found that this approach tends to improve equating accuracy when the sample size is 25 or fewer, provided the prior equatings are accurate.

The studies summarized in the previous paragraph tried to incorporate modifications to existing equating methods to improve equating under small-sample conditions. Their efficacy depends on the correctness of the strong assumptions that they employ to affect their proposed solutions (e.g., the appropriateness of the circle arc or the identity equatings).

Puhan (2011a, b) presented other alternatives to the small-sample equating problem that focused more on improving data collection. One approach would be to implement an equating design whereby data conducive to improved equatings can be collected to help with the small-sample equating problem. An example of such a design developed at ETS is the single-group nearly equivalent test design, or the SiGNET design (Grant 2011), which introduces a new form in stages rather than all at once. The SiGNET design has two primary merits. First, it facilitates the use of a single-group equating design that has the least random equating error of all designs, and second, it allows for the accumulation of data to equate the new form with a larger sample. Puhan et al. (2008a, 2009) conducted a resampling study to compare equatings under the SiGNET and common-item equating designs and found lower equating error for the SiGNET design than for the common-item equating design in very small sample size conditions (e.g., $N = 10$).

4.5 Evaluating Equatings

In this part, we address several topics in the evaluation of links formed by scale alignment or by equatings. Section 4.5.1 describes research on assessing the sampling error of linking functions. In Sect. 4.5.2, we summarize research dealing with measures of the effect size for assessing the invariance of equating and scale-aligning functions over subpopulations of a larger population. Section 4.5.3 is concerned with research that deals with scale continuity.

4.5.1 Sampling Stability of Linking Functions

All data based linking functions are statistical estimates, and they are therefore subject to sampling variability. If a different sample had been taken from the target population, the estimated linking function would have been different. A measure of statistical stability gives an indication of the uncertainty in an estimate that is due to the sample selected. In Sect. 4.5.1.1, we discuss the standard error of equating (SEE). Because the same methods are also used for concordances, battery scaling, vertical scaling, calibration, and some forms of anchor scaling, the SEE is a relevant measure of statistical accuracy for these cases of test score linking as well as for equating.

In Sects. 4.5.1.1 and 4.5.1.2, we concentrate on the basic ideas and large-sample methods for estimating standard error. These estimates of the SEE and related measures are based on the delta method. This means that they are justified as standard error estimates only for large samples and may not be valid in small samples.

4.5.1.1 The Standard Error of Equating

Concern about the sampling error associated with different data collection designs for equating has occupied ETS researchers since the 1950s (e.g., Karon 1956; Lord 1950). The SEE is the oldest measure of the statistical stability of estimated linking functions. The SEE is defined as the conditional standard deviation of the sampling distribution of the equated score for a given raw score over replications of the equating process under similar conditions. We may use the SEE for several purposes. It gives a direct measure of how consistently the equating or linking function is estimated. Using the approximate normality of the estimate, the SEE can be used to form confidence intervals. In addition, comparing the SEE for various data collection designs can indicate the relative advantage some designs have over others for particular sample sizes and other design factors. This can aid in the choice of a data collection design for a specific purpose.

The SEE can provide us with statistical caveats about the instability of linkings based on small samples. As the size of the sample(s) increases, the SEE will decrease. With small samples, there is always the possibility that the estimated linking function is a poor representation of the population linking function.

The earliest work on the SEE is found in Lord (1950) and reproduced in Angoff (1971). These papers were concerned with linear-linking methods and assumed normal distributions of scores. Zu and Yuan (2012) examined estimates for linear equating methods under conditions of nonnormality for the nonequivalent-groups design. Lord (1982b) derived the SEE for the equivalent- and single-group designs for the equipercentile function using linear interpolation for continuization of the linking functions. However, these SEE calculations for the equipercentile function did not take into account the effect of presmoothing, which can produce reductions in the SEE in many cases, as demonstrated by Livingston (1993a). Liou and Cheng

(1995) gave an extensive discussion (including estimation procedures) of the SEE for various versions of the equipercentile function that included the effect of presmoothing. Holland et al. (1989) and Liou et al. (1996, 1997) discussed the SEE for kernel equating for the nonequivalent-groups anchor test design.

A.A. von Davier et al. (2004b) provided a system of statistical accuracy measures for kernel equating for several data collection designs. Their results account for four factors that affect the SEE: (a) the sample sizes; (b) the effect of presmoothing; (c) the data collection design; and (d) the form of the final equating function, including the method of continuization. In addition to the SEE and the SEED (described in Sect. 4.5.1.2), they recommend the use of percent relative error to summarize how closely the moments of the equated score distribution match the target score distribution that it is striving to match. A.A. von Davier and Kong (2005) gave a similar analysis for linear equating in the non-equivalent-groups design.

Lord (1981) derived the asymptotic standard error of a true-score equating by IRT for the anchor test design and illustrated the effect of anchor test length on this SEE. Y. Liu et al. (2008) compared a Markov chain Monte Carlo (MCMC) method and a bootstrap method in the estimation of standard errors of IRT true-score linking. Grouped jackknifing was used by Haberman et al. (2009) to evaluate the stability of equating procedures with respect to sampling error and with respect to changes in anchor selection with illustrations involving the two-parameter logistic (2PL) IRT model.

4.5.1.2 The Standard Error of Equating Difference Between Two Linking Functions

Those who conduct equatings are often interested in the stability of differences between linking functions. A.A. von Davier et al. (2004b) were the first to explicitly consider the standard error of the distribution of the difference between two estimated linking functions, which they called the SEED. For kernel equating methods, using loglinear models to presmooth the data, the same tools used for computing the SEE can be used for the SEED for many interesting comparisons of kernel equating functions. Moses and Zhang (2010, 2011) extended the notion of the SEED to comparisons between kernel linear and traditional linear and equipercentile equating functions, as well.

An important use of the SEED is to compare the linear and nonlinear versions of kernel equating. von Davier et al. (2004b) combined the SEED with a graphical display of the plot of the difference between the two equating functions. In addition to the difference, they added a band of ±2SEED to put a rough bound on how far the two equating functions could differ due to sampling variability. When the difference curve is outside of this band for a substantial number of values of the X-scores, this is evidence that the differences between the two equating functions exceed what might be expected simply due to sampling error. The ±2SEED band is narrower for larger sample sizes and wider for smaller sample sizes.

Duong and von Davier (2012) illustrated the flexibility of the observed-score equating framework and the availability of the SEED in allowing practitioners to compare statistically the equating results from different weighting schemes for distinctive subgroups of the target population.

In the special situation where we wish to compare an estimated equating function to another nonrandom function, for example, the identity function, the SEE plays the role of the SEED. Dorans and Lawrence (1988, 1990) used the SEE to create error bands around the difference plot to determine whether the equating between two section orders of a test was close enough to the identity. Moses (2008a, 2009) examined a variety of approaches for selecting equating functions for the equivalent-groups design and recommended that the likelihood ratio tests of loglinear models and the equated score difference tests be used together to assess equating function differences overall and also at score levels. He also encouraged a consideration of the magnitude of equated score differences with respect to score reporting practices.

In addition to the statistical significance of the difference between the two linking functions (the SEED), it is also useful to examine whether this difference has any important consequences for reported scores. This issue was addressed by Dorans and Feigenbaum (1994) in their notion of a difference that matters (DTM). They called a difference in reported score points a DTM if the testing program considered it to be a difference worth worrying about. This, of course, depends on the test and its uses. If the DTM that is selected is smaller than 2 times an appropriate SEE or SEED, then the sample size may not be sufficient for the purposes that the equating is intended to support.

4.5.2 *Measures of the Subpopulation Sensitivity of Score Linking Functions*

Neither the SEE nor the SEED gives any information about how different the estimated linking function would be if the data were sampled from other populations of test takers. Methods for checking the sensitivity of linking functions to the population on which they are computed (i.e., subpopulation invariance checks) serve as diagnostics for evaluating links between tests (especially those that are intended to be test equatings). The most common way that population invariance checks are made is on subpopulations of test takers within the larger population from which the samples are drawn. Subgroups such as male and female are often easily identifiable in the data. Other subgroups are those based on ethnicity, region of the country, and so on. In general, it is a good idea to select subgroups that are known to differ in their performance on the tests in question.

Angoff and Cowell (1986) examined the population sensitivity of linear conversions for the GRE Quantitative test (GRE-Q) and the specially constituted GRE Verbal-plus-Quantitative test (GREV+Q) using equivalent groups of approximately

13,000 taking each form. The data clearly supported the assumption of population invariance for GRE-Q but not quite so clearly for GREV+Q.

Dorans and Holland (2000a, b) developed general indices of population invariance/sensitivity of linking functions for the equivalent groups and single-group designs. To study population invariance, they assumed that the target population is partitioned into mutually exclusive and exhaustive subpopulations. A.A. von Davier et al. (2004a) extended that work to the nonequivalent-groups anchor test design that involves two populations, both of which are partitioned into similar subpopulations.

Moses (2006, 2008b) extended the framework of kernel equating to include the standard errors of indices described in Dorans and Holland (2000a, b). The accuracies of the derived standard errors were evaluated with respect to empirical standard errors.

Dorans (2004a) edited a special issue of the *Journal of Educational Measurement*, titled "Assessing the Population Sensitivity of Equating Functions," that examined whether equating or linking functions relating test scores achieved population invariance. A. A. von Davier et al. (2004a) extended the work on subpopulation invariance done by Dorans and Holland (2000a, b) for the single-population case to the two-population case, in which the data are collected on an anchor test as well as the tests to be equated. Yang (2004) examined whether the multiple-choice (MC) to composite linking functions of the *Advanced Placement*® examinations remain invariant over subgroups by region. Dorans (2004c) examined population invariance across gender groups and placed his investigation within a larger fairness context by introducing score equity analysis as another facet of fair assessment, a complement to differential item functioning and differential prediction.

A.A. von Davier and Liu (2007) edited a special issue of *Applied Psychological Measurement*, titled "Population Invariance," that built on and extended prior research on population invariance and examined the use of population invariance measures in a wide variety of practical contexts. A.A. von Davier and Wilson (2008) examined IRT models applied to Advanced Placement exams with both MC and constructed-response (CR) components. M. Liu and Holland (2008) used Law School Admission Test (LSAT) data to extend the application of population invariance methods to subpopulations defined by geographic region, whether test takers applied to law school, and their law school admission status. Yang and Gao (2008) investigated the population invariance of the one-parameter IRT model used with the testlet-based computerized exams that are part of CLEP. Dorans et al. (2008) examined the role that the choice of anchor test plays in achieving population invariance of linear equatings across male and female subpopulations and test administrations.

Rijmen et al. (2009) compared two methods for obtaining the standard errors of two population invariance measures of equating functions. The results indicated little difference between the standard errors found by the delta method and the grouped jackknife method.

Dorans and Liu (2009) provided an extensive illustration of the application of score equity assessment (SEA), a quality-control process built around the use of population invariance indices, to the SAT-M exam. Moses et al. (2009, 2010b) developed a SAS macro that produces Dorans and Liu's (2009) prototypical SEA analyses, including various tabular and graphical analyses of the differences between scaled score conversions from one or more subgroups and the scaled score conversion based on a total group. J. Liu and Dorans (2013) described how SEA can be used as a tool to assess a critical aspect of construct continuity, the equivalence of scores, whenever planned changes are introduced to testing programs. They also described how SEA can be used as a quality-control check to evaluate whether tests developed to a static set of specifications remain within acceptable tolerance levels with respect to equitability.

Kim et al. (2012) illustrated the use of subpopulation invariance with operational data indices to assess whether changes to the test specifications affected the equatability of a redesigned test to the current test enough to change the meaning of points on the score scale. Liang et al. (2009), also reported in Sinharay et al. (2011b), used SEA to examine the sensitivity of equating procedures to increasing numbers of nonnative speakers in equating samples.

4.5.3 Consistency of Scale Score Meaning

In an ideal world, measurement is flawless, and score scales are properly defined and well maintained. Shifts in performance on a test reflect shifts in the ability of test-taker populations, and any variability in the raw-to-scale conversions across editions of a test is minor and due to random sampling error. In an ideal world, many things need to mesh. Reality differs from the ideal in several ways that may contribute to scale inconsistency, which, in turn, may contribute to the appearance or actual existence of scale drift. Among these sources of scale inconsistency are inconsistent or poorly defined test-construction practices, population changes, estimation error associated with small samples of test takers, accumulation of errors over a long sequence of test administrations, inadequate anchor tests, and equating model misfit. Research into scale continuity has become more prevalent in the twenty-first century. Haberman and Dorans (2011) made distinctions among different sources of variation that may contribute to score-scale inconsistency. In the process of delineating these potential sources of scale inconsistency, they indicated practices that are likely either to contribute to inconsistency or to attenuate it.

Haberman (2010) examined the limits placed on scale accuracy by sample size, number of administrations, and number of forms to be equated. He demonstrated analytically that a testing program with a fixed yearly volume is likely to experience more substantial scale drift with many small-volume administrations than with fewer large volume administrations. As a consequence, the comparability of scores across different examinations is likely to be compromised from many small-volume administrations. This loss of comparability has implications for some modes of continuous testing. Guo (2010) investigated the asymptotic accumulative SEE for linear equating methods under the nonequivalent groups with anchor test design. This tool measures the magnitude of equating errors that have accumulated over a series of equatings.

Lee and Haberman (2013) demonstrated how to use harmonic regression to assess scale stability. Lee and von Davier (2013) presented an approach for score-scale monitoring and assessment of scale drift that used quality-control charts and time series techniques for continuous monitoring, adjustment of customary variations, identification of abrupt shifts, and assessment of autocorrelation.

With respect to the SAT scales established in the early 1940s, Modu and Stern (1975) indicated that the reported score scale had drifted by almost 14 points for the verbal section and 17 points for the mathematics section between 1963 and 1973. Petersen et al. (1983) examined scale drift for the verbal and mathematics portions of the SAT and concluded that for reasonably parallel tests, linear equating was adequate, but for tests that differed somewhat in content and length, 3PL IRT-based methods lead to greater stability of equating results. McHale and Ninneman (1994) assessed the stability of the SAT scale from 1973 to 1984 and found that the SAT-V score scale showed little drift. Furthermore, the results from the Mathematics scale were inconsistent, and therefore the stability of this scale could not be determined.

With respect to the revised SAT scales introduced in 1995, Guo et al. (2012) examined the stability of the SAT Reasoning Test score scales from 2005 to 2010. A 2005 old form was administered along with a 2010 new form. Critical Reading and Mathematics score scales experienced, at most, a moderate upward scale drift that might be explained by an accumulation of random equating errors. The Writing score scale experienced a significant upward scale drift, which might reflect more than random error.

Scale stability depends on the number of items or sets of items used to link tests across administrations. J. Liu et al. (2014) examined the effects of using one, two, or three anchor tests on scale stability of the SAT from 1995 to 2003. Equating based on one old form produced persistent scale drift and also showed increased variability in score means and standard deviations over time. In contrast, equating back to two or three old forms produced much more stable conversions and had less variation.

Guo et al. (2013) advocated the use of the conditional standard error of measurement when assessing scale deficiencies as measured by gaps and clumps, which were defined in Dorans et al. (2010b).

Using data from a teacher certification program, Puhan (2007, 2009a) examined scale drift for parallel equating chains and a single long chain. Results of the study indicated that although some drift was observed, the effect on pass or fail status of test takers was not large.

Cook (1988) explored several alternatives to the scaling procedures traditionally used for the College Board Achievement Tests. The author explored additional scaling covariates that might improve scaling results for tests that did not correlate highly with the SAT Reasoning Test, possible respecification of the sample of students used to scale the tests, and possible respecification of the hypothetical scaling population.

4.6 Comparative Studies

As new methods or modifications to existing methods for data preparation and analysis continued to be developed at ETS, studies were conducted to evaluate the new approaches. These studies were diverse and included comparisons between newly developed methods and existing methods, chained versus poststratification methods, comparisons of equatings using different types of anchor tests, and so on. In this section we attempt to summarize this research in a manner that parallels the structure employed in Sects. 4.3 and 4.4. In Sect. 4.6.1, we address research that focused on data collection issues, including comparisons of equivalent-groups equating and anchor test equating and comparisons of the various anchor test equating procedures. Section 4.6.2 contains research pertaining to anchor test properties. In Sect. 4.6.3, we consider research that focused on different types of samples of test takers. Next, in Sect. 4.6.4, we consider research that focused on IRT equating. IRT preequating is considered in Sect. 4.6.5. Then some additional topics are addressed. Section 4.6.6 considers equating tests with CR components. Equating of subscores is considered in Sect. 4.6.7, whereas Sect. 4.6.8 considers equating in the presence of multidimensional data. Because several of the studies addressed in Sect. 4.6 used simulated data, we close with a caveat about the strengths and limitations of relying on simulated data in Sect. 4.6.9.

4.6.1 Different Data Collection Designs and Different Methods

Comparisons between different equating methods (e.g., chained vs. poststratification methods) and different equating designs (e.g., equivalent groups vs. nonequivalent groups with anchor test design) have been of interest for many ETS researchers. (Comparisons that focused on IRT linking are discussed in Sect. 4.6.4.)

Kingston and Holland (1986) compared alternative equating methods for the GRE General Test. They compared the equivalent-groups design with two other designs (i.e., nonequivalent groups with an external anchor test and equivalent groups with a preoperational section) and found that the equivalent groups with preoperational section design produced fairly poor results compared to the other designs.

After Holland and Thayer introduced kernel equating in 1989, Livingston (1993b) conducted a study to compare kernel equating with traditional equating methods and concluded that kernel equating and equipercentile equating based on smoothed score distributions produce very similar results, except at the low end of the score scale, where the kernel results were slightly more accurate. However, much of the research work at ETS comparing kernel equating with traditional equating methods happened after A.A. von Davier et al. (2004b) was published. For example, A.A. von Davier et al. (2006) examined how closely the kernel equating (KE) method approximated the results of other observed-score equating methods under the common-item equating design and found that the results from kernal equating (KE) and the other methods were quite similar. Similarly, results from a study by Mao et al. (2006) indicated that the differences between KE and the traditional equating methods are very small (for most parts of the score scale) for both the equivalent-groups and common-item equating design. J. Liu and Low (2007, 2008) compared kernel equating with analogous traditional equating methods and concluded that KE results are comparable to the results of other methods. Similarly, Grant et al. (2009) compared KE with traditional equating methods, such as Tucker, Levine, chained linear, and chained equipercentile methods, and concluded that the differences between KE and traditional equivalents were quite small. Finally, Lee and von Davier (2008) compared equating results based on different kernel functions and indicated that the equated scores based on different kernel functions do not vary much, except for extreme scores.

There has been renewed interest in chained equating (CE) versus poststratification equating (PSE) research in the new millennium. For example, Guo and Oh (2009) evaluated the frequency estimation (FE) equating method, a PSE method, under different conditions. Based on their results, they recommended FE equating when neither the two forms nor the observed conditional distributions are very different. Puhan (2010a, b) compared Tucker, chained linear, and Levine observed equating under conditions where the new and old form samples were either similar in ability or not and where the tests were built to the same set of content specifications and concluded that, for most conditions, chained linear equating produced fairly accurate equating results. Predictions from both PSE and CE assumptions were compared using data from a special study that used a fairly novel approach (Holland et al. 2006, 2008). This research used real data to simulate tests built to the same set of content specifications and found that that both CE and PSE make very similar predictions but that those of CE are slightly more accurate than those of PSE, especially where the linking function is nonlinear. In a somewhat similar vein as the preceding studies, Puhan (2012) compared Tucker and chained linear equating in two scenarios. In the first scenario, known as rater comparability scoring and equating, chained linear equating produced more accurate results. Note that although rater comparability scoring typically results in a single-group equating design, the study evaluated a special case in which the rater comparability scoring data were used under a common-item equating design. In the second situation, which used a common-item equating design where the new and old form samples were randomly equivalent, Tucker equating produced more accurate results. Oh and Moses (2012) investigated differences between uni- and bidirectional approaches to chained equipercentile equating and concluded that although the bidirectional results were slightly less erratic and smoother, both methods, in general, produce very similar results.

4.6.2 The Role of the Anchor

Studies have examined the effect of different types of anchor tests on test equating, including anchor tests that are different in content and statistical characteristics. For example, Echternacht (1971) compared two approaches (i.e., using common items or scores from the GRE Verbal and Quantitative measures as the anchor) for equating the GRE Advanced tests. Results showed that both approaches produce equating results that are somewhat different from each other. DeMauro (1992) examined the possibility of equating the *TWE*® test by using TOEFL as an anchor and concluded that using TOEFL as an anchor to equate the TWE is not appropriate.

Ricker and von Davier (2007) examined the effects of external anchor test length on equating results for the common-item equating design. Their results indicated that bias tends to increase in the conversions as the anchor test length decreases, although FE and kernel poststratification equating are less sensitive to this change than other equating methods, such as chained equipercentile equating. Zu and Liu (2009, 2010) compared the effect of discrete and passage-based anchor items on common-item equating results and concluded that anchor tests that tend to have more passage-based items than discrete items result in larger equating errors, especially when the new and old samples differ in ability. Liao (2013) evaluated the effect of speededness on common-item equating and concluded that including an item set toward the end of the test in the anchor affects the equating in the anticipated direction, favoring the group for which the test is less speeded.

Moses and Kim (2007) evaluated the impact of unequal reliability on test equating methods in the common-item equating design and noted that unequal and/or low reliability inflates equating function variability and alters equating functions when there is an ability difference between the new and old form samples.

Sinharay and Holland (2006a, b) questioned conventional wisdom that an anchor test used in equating should be a statistical miniature version of the tests to be equated. They found that anchor tests with a spread of item difficulties less than that of a total test (i.e., a midi test) seem to perform as well as a mini test (i.e., a miniature version of the full test), thereby suggesting that the requirement of the anchor test to mimic the statistical characteristics of the total test may not be optimal. Sinharay et al. (2012) also demonstrated theoretically that the mini test may not be the optimal anchor test with respect to the anchor test–total test correlation. Finally, several empirical studies by J. Liu et al. (2009a, 2011a, b) also found that the midi anchor performed as well or better than the mini anchor across most of the score scale, except the top and bottom, which is where inclusion or exclusion of easy or hard items might be expected to have an effect.

For decades, new editions of the SAT were equated back to two past forms using the nonequivalent-groups anchor test design (Holland and Dorans 2006). Successive new test forms were linked back to different pairs of old forms. In 1994, the SAT equatings began to link new forms back to four old forms. The rationale for this new scheme was that with more links to past forms, it is easier to detect a poor past conversion function, and it makes the final new conversion function less reliant on any particular older equating function. Guo et al. (2011) used SAT data collected from 44 administrations to investigate the effect of accumulated equating error in equating conversions and the effect of the use of multiple links in equating. It was observed that the single-link equating conversions drifted further away from the operational four-link conversions as equating results accumulated over time. In addition, the single-link conversions exhibited an instability that was not obvious for the operational data. A statistical random walk model was offered to explain the mechanism of scale drift in equating caused by random equating error. J. Liu et al. (2014) tried to find a balance point where the needs for equating, control of item/form exposure, and pretesting could be satisfied. Three equating scenarios were examined using real data: equating to one old form, equating to two old forms, or equating to three old forms. Equating based on one old form produced persistent score drift and showed increased variability in score means and standard deviations over time. In contrast, equating back to two or three old forms produced much more stable conversions and less variation in means and standard deviations. Overall, equating based on multiple linking designs produced more consistent results and seemed to limit scale drift.

Moses et al. (2010a, 2011) studied three different ways of using two anchors that link the same old and new form tests in the common-item equating design. The overall results of this study suggested that when using two anchors, the poststratification approach works better than the imputation and propensity score matching approaches. Poststratification also produced more accurate SEEDs, quantities that are useful for evaluating competing equating and scaling functions.

4.6.3 Matched-Sample Equating

Equating based on samples with identical anchor score distributions was viewed as a potential solution to the variability seen across equating methods when equating samples of test takers were not equivalent (Dorans 1990c). Cook et al. (1988) discussed the need to equate achievement tests using samples of students who take the new and old forms at comparable points in the school year. Stocking et al. (1988) compared equating results obtained using representative and matched samples and concluded that matching equating samples on the basis of a fallible measure of ability is not advisable for any equating method, except possibly the Tucker equating method. Lawrence and Dorans (1988) compared equating results obtained using a representative old-form sample and an old-form sample matched to the new-form sample (matched sample) and found that results for the five studied equating methods tended to converge under the matched sample condition.

Lawrence and Dorans (1990), using the verbal anchor to create differences from the reference or base population and the pseudo-populations, demonstrated that the poststratification methods did best and the true-score methods did slightly worse than the chained method when the same verbal anchor was used for equating. Eignor et al. (1990a, b) used an IRT model to simulate data and found that the weakest results were obtained for poststratification on the basis of the verbal anchor and that the true-score methods were slightly better than the chained method. Livingston et al. (1990) used SAT-M scores to create differences in populations and examined the equating of SAT-V scores via multiple methods. The poststratification method produced the poorest results. They also compared equating results obtained using representative and matched samples and found that the results for all equating methods in the matched samples were similar to those for the Tucker and FE methods in the representative samples. In a follow-up study, Dorans and Wright (1993) compared equating results obtained using representative samples, samples matched on the basis of the equating set, and samples matched on the basis of a selection variable (i.e., a variable along which subpopulations differ) and indicated that matching on the selection variable improves accuracy over matching on the equating test for all methods. Finally, a study by Schmitt et al. (1990) indicated that matching on an anchor test score provides greater agreement among the results of the various equating procedures studied than were obtained under representative sampling.

4.6.4 Item Response Theory True-Score Linking

IRT true-score linking[4] was first used with TOEFL in 1979. Research on IRT-based linking methods received considerable attention in the 1980s to examine their applicability to other testing programs. ETS researchers have focused on a wide variety of research topics, including studies comparing non-IRT observed-score and IRT-based linking methods (including IRT true-score linking and IRT observed-score equating methods), studies comparing different IRT linking methods, studies examining the consequences of violation of assumptions on IRT equating, and so on. These studies are summarized here.

Marco et al. (1983a) examined the adequacy of various linear and curvilinear (observed-score methods) and ICC (one- and three-parameter logistic) equating models when certain sample and test characteristics were systematically varied. They found the 3PL model to be most consistently accurate. Using TOEFL data, Hicks (1983, 1984) evaluated three IRT variants and three conventional equating methods (Tucker, Levine and equipercentile) in terms of scale stability and found that the true-score IRT linking based on scaling by fixing the *b* parameters produces the least discrepant results. Lord and Wingersky (1983, 1984) compared IRT true-score linking with equipercentile equating using observed scores and concluded that the two methods yield almost identical results.

Douglass et al. (1985) studied the extent to which three approximations to the 3PL model could be used in item parameter estimation and equating. Although

[4] Several of the earlier studies cited in this section used the phrase IRT equating to describe the application of an IRT true-score equating function to linking two sets of observed scores. We are using the word linking because this procedure does not ensure that the linked scores are interchangeable in the sense described by Lord (1980) and Holland and Dorans (2006).

these approximations yielded accurate results (based on their circular equating criteria), the authors recommended further research before these methods are used operationally. Boldt (1993) compared linking based on the 3PL IRT model and a modified Rasch model (common nonzero lower asymptote) and concluded that the 3PL model should not be used if sample sizes are small. Tang et al. (1993) compared the performance of the computer programs LOGIST and BILOG (see Carlson and von Davier, Chap. 5, this volume, for more on these programs) on TOEFL 3PL IRT-based linking. The results indicated that the BILOG estimates were closer to the true parameter values in small-sample conditions. In a simulation study, Y. Li (2012) examined the effect of drifted (i.e., items performing differently than the remaining anchor items) polytomous anchor items on the test characteristic curve (TCC) linking and IRT true-score linking. Results indicated that drifted polytomous items have a relatively large impact on the linking results and that, in general, excluding drifted polytomous items from the anchor results in an improvement in equating results.

Kingston et al. (1985) compared IRT linking to conventional equating of the GMAT and concluded that violation of local independence had a negligible effect on the linking results. Cook and Eignor (1985) indicated that it was feasible to use IRT to link the four College Board Achievement tests used in their study. Similarly, McKinley and Kingston (1987) investigated the use of IRT linking for the GRE Subject Test in Mathematics and indicated that IRT linking was feasible for this test. McKinley and Schaefer (1989) conducted a simulation study to evaluate the feasibility of using IRT linking to reduce test form overlap of the GRE Subject Test in Mathematics. They compared double-part IRT true-score linking (i.e., linking to two old forms) with 20-item common-item blocks to triple-part linking (i.e., linking to three old forms) with 10-item common-item blocks. On the basis of the results of their study, they suggested using more than two links.

Cook and Petersen (1987) summarized a series of ETS articles and papers produced in the 1980s that examined how equating is affected by sampling errors, sample characteristics, and the nature of anchor items, among other factors. This summary added greatly to our understanding of the uses of IRT and conventional equating methods in suboptimal situations encountered in practice. Cook and Eignor (1989, 1991) wrote articles and instructional modules that provided a basis for understanding the process of score equating through the use of IRT. They discussed the merits of different IRT equating approaches.

A.A. von Davier and Wilson (2005, 2007) used data from the *Advanced Placement Program®* examinations to investigate the assumptions made by IRT true-score linking method and discussed the approaches for checking whether these assumptions are met for a particular data set. They provided a step-by-step check of how well the assumptions of IRT true-score linking are met. They also compared equating results obtained using IRT as well as traditional methods and showed that IRT and chained equipercentile equating results were close for most of the score range.

D. Li et al. (2012) compared the IRT true-score equating to chained equipercentile equating and observed that the sample variances for the chained equipercentile equating were much smaller than the variances for the IRT true-score equating, except at low scores.

4.6.5 Item Response Theory Preequating Research

In the early 1980s, IRT was evaluated for its potential in preequating tests developed from item pools. Bejar and Wingersky (1981) conducted a feasibility study for preequating the TWE and concluded that the procedure did not exhibit problems beyond those already associated with using IRT on this exam. Eignor (1985) examined the extent to which item parameters estimated on SAT-V and SAT-M pretest data could be used for equating purposes. The preequating results were mixed; three of the four equatings examined were marginally acceptable at best. Hypotheses for these results were posited by the author. Eignor and Stocking (1986) studied these hypotheses in a follow-up investigation and concluded that there was a problem either with the SAT-M data or the way in which LOGIST calibrated items under the 3PL model. Further hypotheses were generated. Stocking and Eignor (1986) investigated these results further and concluded that difference in ability across samples and multidimensionality may have accounted for the lack of item parameter invariance that undermined the preequating effort. While the SAT rejected the use of preequating on the basis of this research, during the 1990s, other testing programs moved to test administration and scoring designs, such as computer-adaptive testing, that relied on even more restrictive invariance assumptions than those that did not hold in the SAT studies.

Gao et al. (2012) investigated whether IRT true-score preequating results based on a Rasch model agreed with equating results based on observed operational data (postequating) for CLEP. The findings varied from subject to subject. Differences among the equating results were attributed to the manner of pretesting, contextual/order effects, or the violations of IRT assumptions. Davey and Lee (2011) examined the potential effect of item position on item parameter and ability estimates for the GRE revised General Test, which would use preequating to link scores obtained via its two-stage testing model. In an effort to mitigate the impact of position effects, they recommended that questions be pretested in random locations throughout the test. They also recommended considering the impact of speededness in the design of the revised test because multistage tests are more subject to speededness compared to linear forms of the same length and testing time.

4.6.6 Equating Tests With Constructed-Response Items

Large-scale testing programs often include CR as well as MC items on their tests. Livingston (2014b) listed some characteristics of CR tests (i.e., small number of tasks and possible raw scores, tasks that are easy to remember and require judgment for scoring) that cause problems when equating scores obtained from CR tests. Through the years, ETS researchers have tried to come up with innovative solutions to equating CR tests effectively.

When a CR test form is reused, raw scores from the two administrations of the form may not be comparable due to two different sets of raters among other reasons. The solution to this problem requires a rescoring, at the new administration, of test-taker responses from a previous administration. The scores from this "rescoring" are used as an anchor for equating, and this process is referred to as rater comparability scoring and equating (Puhan 2013b). Puhan (2013a, b) challenged conventional wisdom and showed theoretically and empirically that the choice of target population weights (for poststratification equating) has a predictable impact on final equating results obtained under the rater comparability scoring and equating scenario. The same author also indicated that chained linear equating produces more accurate equating results than Tucker equating under this equating scenario (Puhan 2012).

Kim et al. (2008a, b, 2010a, b) have compared various designs for equating CR-only tests, such as using an anchor test containing either common CR items or rescored common CR items or an external MC test and an equivalent-groups design incorporating rescored CR items (no anchor test). Results of their studies showed that the use of CR items without rescoring results in much larger bias than the other designs. Similarly, they have compared various designs for equating tests containing both MC and CR items such as using an anchor test containing only MC items, both MC and CR items, both MC and rescored CR items, and an equivalent-groups design incorporating rescored CR items (no anchor test). Results of their studies indicated that using either MC items alone or a mixed anchor without CR item rescoring results in much larger bias than the other two designs and that the equivalent-groups design with rescoring results in the smallest bias. Walker and Kim (2010) examined the use of an all-MC anchor for linking mixed-format tests containing both MC and CR items in a nonequivalent-groups design. They concluded that a MC-only anchor could effectively link two such test forms if either the MC or CR portion of the test measured the same knowledge and skills and if the relationship between the MC portion and the total test remained constant across the new and reference linking groups.

Because subpopulation invariance is considered a desirable property for equating relationships, Kim and Walker (2009b, 2012a) examined the appropriateness of the anchor composition in a mixed-format test, which includes both MC and CR items, using subpopulation invariance indices. They found that the mixed anchor was a better choice than the MC-only anchor to achieve subpopulation invariance between males and females. Muraki et al. (2000) provided an excellent summary describing issues and developments in linking performance assessments and included comparisons of common linking designs (single group, equivalent groups, nonequivalent groups) and linking methodologies (traditional and IRT).

Myford et al. (1995) pilot-tested a quality-control procedure for monitoring and adjusting for differences in reader performance and discussed steps that might enable different administrations of the TWE to be equated. Tan et al. (2010)

compared equating results using different sample sizes and equating designs (i.e., single group vs. common-item equating designs) to examine the possibility of reducing the rescoring sample. Similarly, Kim and Moses (2013) conducted a study to evaluate the conditions under which single scoring for CR items is as effective as double scoring in a licensure testing context. Results of their study indicated that under the conditions they examined, the use of single scoring would reduce scoring time and cost without increasing classification inconsistency. Y. Li and Brown (2013) conducted a rater comparability scoring and equating study and concluded that raters maintained the same scoring standards across administrations for the CRs in the *TOEFL iBT®* test Speaking and Writing sections. They recommended that the TOEFL iBT program use this procedure as a tool to periodically monitor Speaking and Writing scoring.

Some testing programs require all test takers to complete the same common portion of a test but offer a choice of essays in another portion of the test. Obviously there can be a fairness issue if the different essays vary in difficulty. ETS researchers have come up with innovative procedures whereby the scores on the alternate questions can be adjusted based on the estimated total group mean and standard deviation or score distribution on each alternate question (Cowell 1972; Rosenbaum 1985). According to Livingston (1988), these procedures tend to make larger adjustments when the scores to be adjusted are less correlated with scores on the common portion. He therefore suggested an adjustment procedure that makes smaller adjustments when the correlation between the scores to be adjusted and the scores on the common portion is low. Allen et al. (1993) examined Livingston's proposal, which they demonstrate to be consistent with certain missing data assumptions, and compared its adjustments to those from procedures that make different kinds of assumptions about the missing data that occur with essay choice.

In an experimental study, Wang et al. (1995) asked students to identify which items within three pairs of MC items they would prefer to answer, and the students were required to answer both items in each of the three pairs. The authors concluded that allowing choice will only produce fair tests when it is not necessary to allow choice. Although this study used tests with MC items only and involved small numbers of items and test takers, it attempted to answer via an experiment a question similar to what the other, earlier discussed studies attempted to answer, namely, making adjustments for test-taker choice among questions.

The same authors attempted to equate tests that allowed choice of questions by using existing IRT models and the assumption that the ICCs for the items obtained from test takers who chose to answer them are the same as the ICCs that would be obtained from the test takers who did not answer them (Wainer et al. 1991, 1994). Wainer and Thissen (1994) discussed several issues pertaining to tests that allow a choice to test takers. They provided examples where equating such tests is impossible and where allowing choice does not necessarily elicit the test takers' best performance.

4.6.7 Subscores

The demand for subscores has been increasing for a number of reasons, including the desire of candidates who fail the test to know their strengths and weaknesses in different content areas and because of mandates by legislatures to report subscores. Furthermore, states and academic institutions such as colleges and universities want a profile of performance for their graduates to better evaluate their training and focus on areas that need remediation. However, for subscores to be reported operationally, they should be comparable across the different forms of a test. One way to achieve comparability is to equate the subscores.

Sinharay and Haberman (2011a, b) proposed several approaches for equating augmented subscores (i.e., a linear combination of a subscore and the total score) under the nonequivalent groups with anchor test design. These approaches only differ in the way the anchor score is defined (e.g., using subscore, total score or augmented subscore as the anchor). They concluded that these approaches performed quite accurately under most practical situations, although using the total score or augmented subscore as the anchor performed slightly better than using only the subscore as the anchor. Puhan and Liang (2011a, b) considered equating subscores using internal common items or total scaled scores as the anchor and concluded that using total scaled scores as the anchor is preferable, especially when the internal common items are small.

4.6.8 Multidimensionality and Equating

The call for CR items and subscores on MC tests reflects a shared belief that a total score based on MC items underrepresents the construct of interest. This suggests that more than one dimension may exist in the data.

ETS researchers such as Cook et al. (1985) examined the relationship between violations of the assumption of unidimensionality and the quality of IRT true-score equating. Dorans and Kingston (Dorans and Kingston 1985; Kingston and Dorans 1982) examined the consequences of violations of unidimensionality assumptions on IRT equating and noted that although violations of unidimensionality may have an impact on equating, the effect may not be substantial. Using data from the LSAT, Camilli et al. (1995) examined the effect of multidimensionality on equating and concluded that violations of unidimensionality may not have a substantial impact on estimated item parameters and true-score equating tables. Dorans et al. (2014) did a comparative study where they varied content structure and correlation between underlying dimensions to examine their effect on latent-score and observed-score linking results. They demonstrated analytically and with simulated data that score equating is possible with multidimensional tests, provided the tests are parallel in content structure.

4.6.9 A Caveat on Comparative Studies

Sinharay and Holland (2008, 2010a, b) demonstrated that the equating method with explicit or implicit assumptions most consistent with the model used to generate the data performs best with those simulated data. When they compared three equating methods—the FE equipercentile equating method, the chained equipercentile equating method, and the IRT observed-score equating method—each one worked best in data consistent with its assumptions. The chained equipercentile equating method was never the worst performer. These studies by Sinharay and Holland provide a valuable lens from which to view the simulation studies summarized in Sect. 4.6 whether they used data simulated from a model or real test data to construct simulated scenarios: The results of the simulation follow from the design of the simulation. As Dorans (2014) noted, simulation studies may be helpful in studying the strengths and weakness of methods but cannot be used as a substitute for analysis of real data.

4.7 The Ebb and Flow of Equating Research at ETS

In this section, we provide a high-level summary of the ebb and flow of equating research reported in Sects. 4.2, 4.3, 4.5, and 4.6. We divide the period from 1947, the birth of ETS, through 2015 into four periods: (a) before 1970, (b) 1970s to mid-1980s, (c) mid-1980s to 2000, and (d) 2001–2015.

4.7.1 Prior to 1970

As might be expected, much of the early research on equating was procedural as many methods were introduced, including those named after Tucker and Levine (Sect. 4.4.1). Lord attended to the SEE (Sect. 4.5.1.1). There were early efforts to smooth data from small samples (Sect. 4.3.2.3). With the exception of work done by Lord in 1964, distinctions between equating and other forms of what is now called score linking did not seem to be made (Sect. 4.2.1).

4.7.2 The Year 1970 to the Mid-1980s

Equating research took on new importance in the late 1970s and early 1980s as test disclosure legislation led to the creation of many more test forms in a testing program than had been needed in the predisclosure period. This required novel data collection designs and led to the investigation of preequating approaches. Lord introduced his equating requirements (Sect. 4.2.1) and concurrently introduced IRT score linking methods, which became the subject of much research (Sects. 4.4.2 and 4.6.4). Lord estimated the SEE for IRT (Sect. 4.5.1.1). IRT preequating research was prevalent and generally discouraging (Sect. 4.6.5). Holland and his colleagues introduced section preequating (section 4.4.4) as another preequating solution to the problems posed by the test disclosure legislation.

4.7.3 The Mid-1980s to 2000

Equating research was more dormant in this period, as first differential item functioning and then computer-adaptive testing garnered much of the research funding at ETS. While some work was motivated by practice, such as matched-sample equating research (Sect. 4.6.3) and continued investigations of IRT score linking (Sect. 4.6.4), there were developments of theoretical import. Most notable among these were the development of kernel equating by Holland and his colleagues (Sects. 4.4.3 and 4.6.1), which led to much research about its use in estimating standard errors (Sect. 4.5.1.1). Claims made by some that scores from a variety of sources could be used interchangeably led to the development of cogent frameworks for distinguishing between different kinds of score linkings (Sect. 4.2.1). The role of dimensionality in equating was studied (Sect. 4.6.8).

4.7.4 The Years 2002–2015

The twenty-first century witnessed a surge of equating research. The kernel equating method and its use in estimating standard errors was studied extensively (Sects. 4.4.3, 4.5.1, 4.5.2, and 4.6.1). A new equating method was proposed by Haberman (Sect. 4.4.3).

Data collection and preparation received renewed interest in the areas of sample selection (Sect. 4.3.2.1) and weighting of samples (Sect. 4.3.2.2). A considerable amount of work was done on smoothing (Sect. 4.3.2.3), mostly by Moses and Holland and their colleagues. Livingston and Puhan and their colleagues devoted much attention to developing small-sample equating methods (Sect. 4.4.5).

CE was the focus of many comparative investigations (Sect. 4.6.1). The anchor continued to receive attention (Sect. 4.6.2). Equating subscores became an important issue as there were more and more calls to extract information from less and less (Sect. 4.6.7). The comparability problems faced by reliance on subjectively scored CR items began to be addressed (Sect. 4.6.6). The role of dimensionality in equating was examined again (Sect. 4.6.8).

Holland and Dorans provided a detailed framework for classes of linking (Sect. 4.2.1) as a further response to calls for linkages among scores from a variety of sources. Central to that framework was the litmus test of population invariance, which led to an area of research that uses equating to assess the fairness of test scores across subgroups (Sect. 4.5.2).

4.8 Books and Chapters

Books and chapters can be viewed as evidence that the authors are perceived as possessing expertise that is worth sharing with the profession. We conclude this chapter by citing the various books and chapters that have been authored by ETS staff in the area of score linking, and then we allude to work in related fields and forecast our expectation that ETS will continue to work the issues in this area.

An early treatment of score equating appeared in Gulliksen (1950), who described, among other things, Ledyard R Tucker's proposed use of an anchor test to adjust for differences in the abilities of samples. Tucker proposed this approach to deal with score equating problems with the SAT that occurred when the SAT started to be administered more than once a year to test takers applying to college. Books that dealt exclusively with score equating did not appear for more than 30 years, until the volume edited by ETS researchers Holland and Rubin (1982) was published. The 1980s was the first decade in which much progress was made in score equating research, spearheaded in large part by Paul Holland and his colleagues.

During the 1990s, ETS turned its attention first toward differential item functioning (Dorans, Chap. 7, this volume) and then toward CR and computer-adaptive testing. The latter two directions posed particular challenges to ensuring comparability of measurements, leaning more on strong assumptions than on an empirical basis. After a relatively dormant period in the 1990s, score equating research blossomed in the twenty-first century. Holland and his colleagues played major roles in this rebirth. The Dorans and Holland (2000a, b) article on the population sensitivity of score linking functions marked the beginning of a renaissance of effort on score equating research at ETS.

With the exception of early chapters by Angoff (1967, 1971), most chapters on equating prior to 2000 appeared between 1981 and 1990. Several appeared in the aforementioned Holland and Rubin (1982). Angoff (1981) provided a summary of procedures in use at ETS up until that time. Braun and Holland (1982) provided a formal mathematical framework to examine several observed-score equating procedures used at ETS at that time. Cowell (1982) presented an early application of IRT true-score linking, which was also described in a chapter by Lord (1982a). Holland and Wightman (1982) described a preliminary investigation of a linear section pre-equating procedure. Petersen et al. (1982) summarized the linear equating portion of a massive simulation study that examined linear and curvilinear methods of anchor test equating, ranging from widely used methods to rather obscure methods. Some anchors were external (did not count toward the score), whereas others were internal. They examined different types of content for the internal anchor. Anchors varied in difficulty. In addition, equating samples were randomly equivalent, similar, or dissimilar in ability. Rock (1982) explored how equating could be represented from the perspective of confirmatory factor analysis. Rubin (1982) commented on the chapter by Braun and Holland, whereas Rubin and Szatrowski (1982) critiqued the preequating chapter.

ETS researchers contributed chapters related to equating and linking in edited volumes other than Holland and Rubin's (1982). Angoff (1981) discussed equating and equity in a volume on new directions in testing and measurement circa 1980. Marco (1981) discussed the efforts of test disclosure on score equating in a volume on coaching, disclosure, and ethnic bias. Marco et al. (1983b) published the curvilinear equating analogue to their linear equating chapter that appeared in Holland and Rubin (1982) in a volume on latent trait theory and computer-adaptive testing. Cook and Eignor (1983) addressed the practical considerations associated with using IRT to equate or link test scores in a volume on IRT. Dorans (1990b) produced a chapter on scaling and equating in a volume on computer-adaptive testing edited by Wainer et al. (1990). Angoff and Cook (1988) linked scores across languages by relating the SAT to the College Board *PAA*™ test in a chapter on access and assessment for Hispanic students.

Since 2000, ETS authors have produced several books on the topics of score equating and score linking, including two quite different books, the theory-oriented unified statistical treatment of score equating by A.A. von Davier et al. (2004b) and an introduction to the basic concepts of equating by Livingston (2004). A.A. von Davier et al. (2004b) focused on a single method of test equating (i.e., kernel equating) in a unifying way that introduces several new ideas of general use in test equating. Livingston (2004) is a lively and straightforward account of many of the major issues and techniques. Livingston (2014b) is an updated version of his 2004 publication.

In addition to these two equating books were two edited volumes, one by Dorans et al. (2007) and one by A.A. von Davier (2011c). ETS authors contributed several chapters to both of these volumes.

There were six integrated parts to the volume *Linking and Aligning Scores and Scales* by Dorans et al. (2007). The first part set the stage for the remainder of the volume. Holland (2007) noted that linking scores or scales from different tests has a history about as long as the field of psychometrics itself. His chapter included a typology of linking methods that distinguishes among predicting, scaling, and equating. In the second part of the book, Cook (2007) considered some of the daunting challenges facing practitioners and discussed three major stumbling blocks encountered when attempting to equate scores on tests under difficult conditions: characteristics of the tests to be equated, characteristics of the groups used for equating, and characteristics of the anchor tests. A. A. von Davier (2007) addressed potential future directions for improving equating practices and included a brief introduction to kernel equating and issues surrounding assessment of the population sensitivity of equating functions. Educational testing programs in a state of transition were considered in the third part of the volume. J. Liu and Walker (2007) addressed score linking issues associated with content changes to a test. Eignor (2007) discussed linkings between test scores obtained under different modes of administration, noting why scores from computer-adaptive tests and paper-and-pencil tests cannot be considered equated. Concordances between tests built for a common purpose but in different ways were discussed by Dorans and Walker (2007) in a whimsical chapter that was part of the fourth part of the volume, which dealt with concordances. Yen (2007) examined the role of vertical scaling in the pre–No Child Left Behind (NCLB) era and the NCLB era in the fifth part, which was dedicated to vertical scaling. The sixth part dealt with relating the results obtained by surveys of educational achievement that provide aggregate results to tests designed to assess individual test takers. Braun and Qian (2007) modified and evaluated a procedure developed to link state standards to the National Assessment of Educational Progress scale and illustrated its use. In the book's postscript, Dorans et al. (2007) peered into the future and speculated about the likelihood that more and more linkages of dubious merit would be sought.

The A.A. von Davier (2011c) volume titled *Statistical Models for Test Equating, Scaling and Linking*, which received the American Educational Research Association 2013 best publication award, covered a wide domain of topics. Several chapters in the book addressed score linking and equating issues. In the introductory chapter of the book, A.A. von Davier (2011a) described the equating process as a feature of complex statistical models used for measuring abilities in standardized assessments and proposed a framework for observed-score equating methods. Dorans et al. (2011) emphasized the practical aspects of the equating process, the need for a solid data collection design for equating, and the challenges involved in applying specific equating procedures. Carlson (2011) addressed how to link vertically the results of tests that are constructed to intentionally differ in difficulty and content and that are taken by groups of test takers who differ in ability. Holland and Strawderman (2011) described a procedure that might be considered for averaging equating conversions that come from linkings to multiple old forms. Livingston and Kim (2011) addressed different approaches to dealing with the problems associated with equating test scores in small samples. Haberman (2011b) described the use of exponential families for continuizing test score distributions. Lee and von Davier (2011) discussed how various continuous variables with distributions (normal, logistic, and uniform) can be used as kernels to continuize test score distributions. Chen et al. (2011) described new hybrid models within the kernel equating framework, including a nonlinear version of Levine linear equating. Sinharay et al. (2011a) presented a detailed investigation of the untestable assumptions behind two popular nonlinear equating methods used with a nonequivalent-groups design. Rijmen et al. (2011) applied the SEE difference developed by A.A. von Davier et al. (2004b) to the full vector of equated raw scores and constructed a test for testing linear hypotheses about the equating results. D. Li et al. (2011) proposed the use of time series methods for monitoring the stability of reported scores over a long sequence of administrations.

ETS researchers contributed chapters related to equating and linking in edited volumes other than Dorans et al. (2007) and A. A. von Davier (2011c). Dorans (2000) produced a chapter on scaling and equating in a volume on computer-adaptive testing edited by Wainer et al. (2000). In a chapter in a volume dedicated to examining the adaptation of tests from one language to another, Cook and Schmitt-Cascallar (2005) reviewed different approaches to establishing score linkages on tests that are administered in different languages to different populations and critiqued three attempts to link the English-language SAT to the Spanish-language PAA over a 25-year period, including Angoff and Cook (1988) and Cascallar and Dorans (2005). In volume 26 of the *Handbook of Statistics*, dedicated to psychometrics and edited by Rao and Sinharay (2007), Holland et al. (2007) provided an introduction to test score equating, its data collection procedures, and methods used for equating. They also presented sound practices in the choice and evaluation of equating designs and functions and discussed challenges often encountered in practice.

Dorans and Sinharay (2011) edited a volume dedicated to feting the career of Paul Holland, titled *Looking Back*, in which the introductory chapter by Haberman (2011a) listed score equating as but one of Holland's many contributions. Three chapters on score equating were included in that volume. These three authors joined Holland and other ETS researchers in promoting the rebirth of equating research at ETS. Moses (2011) focused on one of Holland's far-reaching applications: his application of loglinear models as a smoothing method for equipercentile equating. Sinharay (2011) discussed the results of several studies that compared the performances of the poststratification equipercentile and chained equipercentile equating methods. Holland was involved in several of these studies. In a book chapter, A. A. von Davier (2011b) focused on the statistical methods available for equating test forms from standardized educational assessments that report scores at the individual level.

4.9 Concluding Comment

Lord (1980) stated that score equating is either not needed or impossible. Scores will be compared, however. As noted by Dorans and Holland (2000a),

> The comparability of measurements made in differing circumstances by different methods and investigators is a fundamental pre-condition for all of science. Psychological and educational measurement is no exception to this rule. Test equating techniques are those statistical and psychometric methods used to adjust scores obtained on different tests measuring the same construct so that they are comparable. (p. 281)

Procedures will attempt to facilitate these comparisons.

As in any scientific endeavor, instrument preparation and data collection are critical. With large equivalent groups of motivated test takers taking essentially parallel forms, the ideal of "no need to equate" is within reach. Score equating methods converge. As samples get small or contain unmotivated test takers or test takers with preknowledge of the test material, or as test takers take un-pretested tests that differ in content and difficulty, equating will be elusive. Researchers in the past have suggested solutions for suboptimal conditions. They will continue to do so in the future. We hope this compilation of studies will be valuable for future researchers who grapple with the inevitable less-than-ideal circumstances they will face when linking score scales or attempting to produce interchangeable scores via score equating.

References

Allen, N. L., Holland, P. W., & Thayer, D. T. (1993). *The optional essay problem and the hypothesis of equal difficulty* (Research Report No. RR-93-40). Princeton: Educational Testing Service. http://dx.doi.org/10.1002/j.2333-8504.1993.tb01551.x

Angoff, W. H. (1953). *Equating of the ACE psychological examinations for high school students* (Research Bulletin No. RB-53-03). Princeton: Educational Testing Service. http://dx.doi.org/10.1002/j.2333-8504.1953.tb00887.x

Angoff, W. H. (1967). Technical problems of obtaining equivalent scores on tests. In W. A. Mehrens & R. L. Ebel (Eds.), *Principles of educational and psychological measurement: A book of selected readings* (pp. 84–86). Chicago: Rand McNally.

Angoff, W. H. (1971). Scales, norms and equivalent scores. In R. L. Thorndike (Ed.), *Educational measurement* (2nd ed., pp. 508–600). Washington, DC: American Council on Education.

Angoff, W. H. (1981). Equating and equity. *New Directions for Testing and Measurement, 9*, 15–20.

Angoff, W. H., & Cook, L. L. (1988). *Equating the scores on the "Prueba de Apitud Academica" and the "Scholastic Aptitude Test"* (College Board Report No. 88-2). New York: College Board. http://dx.doi.org/10.1002/j.2330-8516.1988.tb00259.x

Angoff, W. H., & Cowell, W. R. (1986). An examination of the assumption that the equating of parallel forms is population-independent. *Journal of Educational Measurement, 23*, 327–345. https://doi.org/10.1111/j.1745-3984.1986.tb00253.x

Bejar, I. I., & Wingersky, M. S. (1981). *An application of item response theory to equating the Test of Standard Written English* (Research Report No. RR-81-35). Princeton: Educational Testing Service.

Boldt, R. F. (1972). *Anchored scaling and equating: Old conceptual problems and new methods* (Research Bulletin No. RB-72-28). Princeton: Educational Testing Service. http://dx.doi.org/10.1002/j.2333-8504.1972.tb01025.x

Boldt, R. F. (1993). *Simulated equating using several item response curves* (Research Report No. RR-93-57). Princeton: Educational Testing Service. http://dx.doi.org/10.1002/j.2333-8504.1993.tb01568.x

Braun, H. I., & Holland, P. W. (1982). Observed-score test equating: A mathematical analysis of some ETS equating procedures. In P. W. Holland & D. B. Rubin (Eds.), *Test equating* (pp. 9–49). New York: Academic Press.

Braun, H. I., & Qian, J. (2007). An enhanced method for mapping state standards onto the NAEP scale. In N. J. Dorans, M. Pommerich, & P. W. Holland (Eds.), *Linking and aligning scores and scales* (pp. 313–338). New York: Springer. https://doi.org/10.1007/978-0-387-49771-6_17

Camilli, G., Wang, M.-M., & Fesq, J. (1995). The effects of dimensionality on equating the Law School Admission Test. *Journal of Educational Measurement, 32*, 79–96. https://doi.org/10.1111/j.1745-3984.1995.tb00457.x

Carlson, J. (2011). Statistical models for vertical linking. In A. A. von Davier (Ed.), *Statistical models for test equating, scaling, and linking* (pp. 59–70). New York: Springer. https://doi.org/10.1007/978-0-387-98138-3_4

Cascallar, A. S., & Dorans, N. J. (2005). Linking scores from tests of similar content given in different languages: An illustration involving methodological alternatives. *International Journal of Testing, 5*, 337–356. https://doi.org/10.1207/s15327574ijt0504_1

Chen, H. (2012). A comparison between linear IRT observed score equating and Levine observed score equating under the generalized kernel equating framework. *Journal of Educational Measurement, 49*, 269–284. https://doi.org/10.1111/j.1745-3984.2012.00175.x

Chen, H., & Holland, P. (2010). New equating methods and their relationships with Levine observed score linear equating under the kernel equating framework. *Psychometrika, 75*, 542–557. https://doi.org/10.1007/s11336-010-9171-7

Chen, H., & Livingston, S. A. (2013). *Poststratification equating based on true anchor scores and its relationship to Levine observed score equating* (Research Report No. RR-13-11). Princeton: Educational Testing Service. http://dx.doi.org/10.1002/j.2333-8504.2013.tb02318.x

Chen, H., Livingston, S. A., & Holland, P. W. (2011). Generalized equating functions for NEAT designs. In A. A. von Davier (Ed.), *Statistical models for test equating, scaling, and linking* (pp. 89–107). New York: Springer. https://doi.org/10.1007/978-0-387-98138-3_12

Cook, L. L. (1988). *Achievement test scaling* (Research Report No. RR-88-34). Princeton: Educational Testing Service. http://dx.doi.org/10.1002/j.2330-8516.1988.tb00290.x

Cook, L. L. (2007). Practical problems in equating test scores: A practitioner's perspective. In N. J. Dorans, M. Pommerich, & P. W. Holland (Eds.), *Linking and aligning scores and scales* (pp. 73–88). New York: Springer. https://doi.org/10.1007/978-0-387-49771-6_5

Cook, L. L., & Eignor, D. R. (1983). Practical considerations regarding the use of item response theory to equate tests. In R. K. Hambleton (Ed.), *Applications of item response theory* (pp. 175–195). Vancouver: Educational Research Institute of British Columbia.

Cook, L. L., & Eignor, D. R. (1985). *An investigation of the feasibility of applying item response theory to equate achievement tests* (Research Report No. RR-85-31). Princeton: Educational Testing Service. http://dx.doi.org/10.1002/j.2330-8516.1985.tb00116.x

Cook, L. L., & Eignor, D. R. (1989). Using item response theory in test score equating. *International Journal of Educational Research, 13*, 161–173. https://doi.org/10.1016/0883-0355(89)90004-9

Cook, L. L., & Eignor, D. R. (1991). IRT equating methods. *Educational Measurement: Issues and Practice, 10*, 37–45. https://doi.org/10.1111/j.1745-3992.1991.tb00207.x

Cook, L. L., & Petersen, N. S. (1987). Problems related to the use of conventional and item response theory equating methods in less than optimal circumstances. *Applied Psychological Measurement, 11*, 225–244. https://doi.org/10.1177/014662168701100302

Cook, L. L., & Schmitt-Cascallar, A. P. (2005). Establishing score comparability for tests given in different languages. In R. Hambleton, P. F. Meranda, & C. D. Spielberger (Eds.), *Adapting educational and psychological tests for cross-cultural assessment* (pp. 171–192). Mahwah: Erlbaum.

Cook, L. L., Dorans, N. J., Eignor, D. R., & Petersen, N. S. (1985). *An assessment of the relationship between the assumption of unidimensionality and the quality of IRT true-score equating* (Research Report No. RR-85-30). Princeton: Educational Testing Service. http://dx.doi.org/10.1002/j.2330-8516.1985.tb00115.x

Cook, L. L., Eignor, D. R., & Schmitt, A. P. (1988). *The effects on IRT and conventional achievement test equating results of using equating samples matched on ability* (Research Report No. RR-88-52). Princeton: Educational Testing Service. http://dx.doi.org/10.1002/j.2330-8516.1988.tb00308.x

Cowell, W. R. (1972). *A technique for equating essay question scores* (Statistical Report No. SR-72-70). Princeton: Educational Testing Service.

Cowell, W. R. (1982). Item-response-theory pre-equating in the TOEFL testing program. In P. W. Holland & D. B. Rubin (Eds.), *Test equating* (pp. 149–161). New York: Academic Press.

Davey, T., & Lee, Y.-H. (2011). *Potential impact of context effects on the scoring and equating of the multistage GRE revised General Test* (Research Report No. RR-11-26). Princeton: Educational Testing Service. http://dx.doi.org/10.1002/j.2333-8504.2011.tb02262.x

DeMauro, G. E. (1992). *An investigation of the appropriateness of the TOEFL test as a matching variable to equate TWE topics* (Research Report No. RR-92-26). Princeton: Educational Testing Service. http://dx.doi.org/10.1002/j.2333-8504.1992.tb01457.x

Dorans, N. J. (1990a). The equating methods and sampling designs. *Applied Measurement in Education, 3*, 3–17. https://doi.org/10.1207/s15324818ame0301_2

Dorans, N. J. (1990b). Scaling and equating. In H. Wainer, N. J. Dorans, R. Flaugher, B. F. Green, R. Mislevy, L. Steinberg, & D. Thissen (Eds.), *Computerized adaptive testing: A primer* (pp. 137–160). Hillsdale: Erlbaum.

Dorans, N. J. (Ed.). (1990c). Selecting samples for equating: To match or not to match [Special issue]. *Applied Measurement in Education, 3*(1).

Dorans, N. J. (1999). *Correspondences between ACT and SAT I scores* (College Board Report No. 99-1). New York: College Board. https://doi.org/10.1002/j.2333-8504.1999.tb01800.x

Dorans, N. J. (2000). Scaling and equating. In H. Wainer, N. J. Dorans, D. Eignor, R. Flaugher, B. F. Green, R. Mislevy, L. Steinberg, & D. Thissen (Eds.), *Computerized adaptive testing: A primer* (2nd ed., pp. 135–158). Hillsdale: Erlbaum.

Dorans, N. J. (2002a). *The recentering of SAT scales and its effects on score distributions and score interpretations* (College Board Research Report No. 2002-11). New York: College Board. https://doi.org/10.1002/j.2333-8504.2002.tb01871.x

Dorans, N. J. (2002b). Recentering the SAT score distributions: How and why. *Journal of Educational Measurement, 39*(1), 59–84. https://doi.org/10.1111/j.1745-3984.2002.tb01135.x

Dorans, N. J. (Ed.). (2004a). Assessing the population sensitivity of equating functions. [Special issue]. *Journal of Educational Measurement, 41*(1).

Dorans, N. J. (2004b). Equating, concordance and expectation. *Applied Psychological Measurement, 28*, 227–246. https://doi.org/10.1177/0146621604265031

Dorans, N. J. (2004c). Using population invariance to assess test score equity. *Journal of Educational Measurement, 41*, 43–68. https://doi.org/10.1111/j.1745-3984.2004.tb01158.x

Dorans, N. J. (2007). Linking scores from multiple health outcome instruments. *Quality of Life Research, 16*(S1), S85–S94. https://doi.org/10.1007/s11136-006-9155-3

Dorans, N. J. (2013). On attempting to do what Lord said was impossible: Commentary on van der Linden's conceptual issues in observed-score equating. *Journal of Educational Measurement, 50*, 304–314. https://doi.org/10.1111/jedm.12017

Dorans, N. J. (2014). *Simulate to understand models, not nature* (Research Report No. RR-14-16). Princeton: Educational Testing Service. http://dx.doi.org/10.1002/ets2.12013

Dorans, N. J., & Feigenbaum, M. D. (1994). Equating issues engendered by changes to the SAT and *PSAT/NMSQT®*. In I. M. Lawrence, N. J. Dorans, M. D. Feigenbaum, N. J. Feryok, A. P. Schmitt, & N. K. Wright (Eds.), *Technical issues related to the introduction of the new SAT and PSAT/NMSQT* (Research Memorandum No. RM-94-10). Princeton: Educational Testing Service.

Dorans, N. J., & Holland, P. W. (2000a). Population invariance and the equatability of tests: Basic theory and the linear case. *Journal of Educational Measurement, 37*, 281–306. https://doi.org/10.1111/j.1745-3984.2000.tb01088.x

Dorans, N. J., & Holland, P. W. (2000b). *Population invariance and the equatability of tests: Basic theory and the linear case* (Research Report No. RR-00-19). Princeton: Educational Testing Service. http://dx.doi.org/10.1002/j.2333-8504.2000.tb01842.x

Dorans, N. J., & Kingston, N. M. (1985). The effects of violations of unidimensionality on the estimation of item and ability parameters and on item response theory equating of the GRE verbal scale. *Journal of Educational Measurement, 22*, 249–262. https://doi.org/10.1111/j.1745-3984.1985.tb01062.x

Dorans, N. J., & Lawrence, I. M. (1988). *Checking the equivalence of nearly identical test forms* (Research Report No. RR-88-06). Princeton: Educational Testing Service. http://dx.doi.org/10.1002/j.2330-8516.1988.tb00262.x

Dorans, N. J., & Lawrence, I. M. (1990). Checking the statistical equivalence of nearly identical test editions. *Applied Measurement in Education, 3*, 245–254. https://doi.org/10.1207/s15324818ame0303_3

Dorans, N. J., & Liu, J. (2009). *Score equity assessment: Development of a prototype analysis using SAT Mathematics test data across several administrations* (Research Report No. RR-09-08). Princeton: Educational Testing Service. http://dx.doi.org/10.1002/j.2333-8504.2009.tb02165.x

Dorans, N. J., & Middleton, K. (2012). Addressing the extreme assumptions of presumed linkings. *Journal of Educational Measurement, 49*, 1–18. https://doi.org/10.1111/j.1745-3984.2011.00157.x

Dorans, N. J., & Sinharay, S. (Eds.). (2011). *Looking back: Proceedings of a conference in honor of Paul W. Holland*. New York: Springer. https://doi.org/10.1007/978-1-4419-9389-2

Dorans, N. J., & Walker, M. E. (2007). Sizing up linkages. In N. J. Dorans, M. Pommerich, & P. W. Holland (Eds.), *Linking and aligning scores and scales* (pp. 179–198). New York: Springer. https://doi.org/10.1007/978-0-387-49771-6_10

Dorans, N. J., & Wright, N. K. (1993). *Using the selection variable for matching or equating* (Research Report No. RR-93-04). Princeton: Educational Testing Service. https://doi.org/10.1002/j.2333-8504.1993.tb01515.x

Dorans, N. J., Pommerich, M., & Holland, P. W. (Eds.). (2007). *Linking and aligning scores and scales*. New York: Springer.

Dorans, N. J., Liu, J., & Hammond, S. (2008). Anchor test type and population invariance: An exploration across subpopulations and test administrations. *Applied Psychological Measurement, 32*, 81–97. https://doi.org/10.1177/0146621607311580

Dorans, N. J., Moses, T. P., & Eignor, D. R. (2010a). *Principles and practices of test score equating* (Research Report No. RR-10-29). Princeton: Educational Testing Service. http://dx.doi.org/10.1002/j.2333-8504.2010.tb02236.x

Dorans, N. J., Liang, L., & Puhan, G. (2010b). *Aligning scales of certification tests* (Research Report No. RR-10-07). Princeton: Educational Testing Service. http://dx.doi.org/10.1002/j.2333-8504.2010.tb02214.x

Dorans, N. J., Moses, T. P., & Eignor, D. R. (2011). Equating test scores: Towards best practices. In A. A. von Davier (Ed.), *Statistical models for scaling, equating and linking* (pp. 21–42). New York: Springer. https://doi.org/10.1007/978-0-387-98138-3_2

Dorans, N. J., Lin, P., Wang, W., & Yao, L. (2014). *The invariance of latent and observed linking functions in the presence of multiple latent test-taker dimensions* (Research Report No. RR-14-41). Princeton: Educational Testing Service. http://dx.doi.org/10.1002/ets2.12041

Douglass, J. B., Marco, G. L., & Wingersky, M. S. (1985). *An evaluation of three approximate item response theory models for equating test scores* (Research Report No. RR-85-46). Princeton: Educational Testing Service. https://doi.org/10.1002/j.2330-8516.1985.tb00131.x

Duong, M., & von Davier, A. A. (2012). Observed-score equating with a heterogeneous target population. *International Journal of Testing, 12*, 224–251. https://doi.org/10.1080/15305058.2011.620725

Echternacht, G. (1971). *Alternate methods of equating GRE advanced tests* (GRE Board Professional Report No. GREB No. 69-2P). Princeton: Educational Testing Service.

Eignor, D. R. (1985). *An investigation of the feasibility and practical outcomes of the pre-equating of the SAT Verbal and Mathematical sections* (Research Report No. RR-85-10). Princeton: Educational Testing Service. http://dx.doi.org/10.1002/j.2330-8516.1985.tb00095.x

Eignor, D. R. (2007). Linking scores derived under different modes of test administration. In N. J. Dorans, M. Pommerich, & P. W. Holland (Eds.), *Linking and aligning scores and scales* (pp. 135–159). New York: Springer. https://doi.org/10.1007/978-0-387-49771-6_8

Eignor, D. R., & Stocking, M. L. (1986). *An investigation of the possible causes of the inadequacy of IRT pre-equating* (Research Report No. RR-86-14). Princeton: Educational Testing Service. http://dx.doi.org/10.1002/j.2330-8516.1986.tb00169.x

Eignor, D. R., Stocking, M. L., & Cook, L. L. (1990a). *The effect on observed- and true-score equating procedures of matching on a fallible criterion: A simulation with test variation* (Research Report No. RR-90-25). Princeton: Educational Testing Service. http://dx.doi.org/10.1002/j.2333-8504.1990.tb01361.x

Eignor, D. R., Stocking, M. L., & Cook, L. L. (1990b). Simulation results of the effects on linear and curvilinear observed- and true-score equating procedures of matching on a fallible criterion. *Applied Measurement in Education, 3*, 37–55. https://doi.org/10.1207/s15324818ame0301_4

Fan, C. T., & Swineford, F. (1954). *A method of score conversion through item statistics* (Research Bulletin No. RB-54-16). Princeton: Educational Testing Service. http://dx.doi.org/10.1002/j.2333-8504.1954.tb00243.x

Flanagan, J. C. (1951). Units, scores, and norms. In E. F. Lindquist (Ed.), *Educational measurement* (pp. 695–763). Washington DC: American Council on Education.

Gao, R., He, W., & Ruan, C. (2012). *Does preequating work? An investigation into a preequated testlet-based college placement exam using postadministration data* (Research Report No. RR-12-12). Princeton: Educational Testing Service. http://dx.doi.org/10.1002/j.2333-8504.2012.tb02294.x

Gonzalez, J., & von Davier, M. (2013). Statistical models and inference for the true equating transformation in the context of local equating. *Journal of Educational Measurement, 50*, 315–320. https://doi.org/10.1111/jedm.12018

Grant, M. C. (2011). *The single group with nearly equivalent tests (SiGNET) design for equating very small volume multiple-choice tests* (Research Report No. RR-11-31). Princeton: Educational Testing Service. http://dx.doi.org/10.1002/j.2333-8504.2011.tb02267.x

Grant, M. C., Zhang, Y., & Damiano, M. (2009). *An evaluation of kernel equating: Parallel equating with classical methods in the SAT Subject tests program* (Research Report No. RR-09-06). Princeton: Educational Testing Service. http://dx.doi.org/10.1002/j.2333-8504.2009.tb02163.x

Gulliksen, H. (1950). *Theory of mental test scores*. New York: Wiley. https://doi.org/10.1037/13240-000

Gulliksen, H. (1968). Methods for determining equivalence of measures. *Psychological Bulletin, 70*, 534–544. https://doi.org/10.1037/h0026721

Guo, H. (2010). Accumulative equating error after a chain of linear equatings. *Psychometrika, 75*, 438–453. https://doi.org/10.1007/s11336-010-9160-x

Guo, H., & Oh, H.-J. (2009). *A study of frequency estimation equipercentile equating when there are large ability differences* (Research Report No. RR-09-45). Princeton: Educational Testing Service. http://dx.doi.org/10.1002/j.2333-8504.2009.tb02202.x

Guo, H., & Puhan, G. (2014). Section pre-equating under the equivalent groups design without IRT. *Journal of Educational Measurement, 51*, 301–317. https://doi.org/10.1111/jedm.12049

Guo, H., Liu, J., Dorans, N., & Feigenbaum, M. (2011). *Multiple linking in equating and random scale drift* (Research Report No. RR-11-46). Princeton: Educational Testing Service. http://dx.doi.org/10.1002/j.2333-8504.2011.tb02282.x

Guo, H., Liu, J., Curley, E., Dorans, N., & Feigenbaum, M. (2012). *The stability of the score scale for the SAT Reasoning Test from 2005–2012* (Research Report No. RR-12-15). Princeton: Educational Testing Service. http://dx.doi.org/10.1002/j.2333-8504.2012.tb02297.x

Guo, H., Puhan, G., & Walker, M. E. (2013). *A criterion to evaluate the individual raw-to-scale equating conversions* (Research Report No. RR-13-05). Princeton: Educational Testing Service. http://dx.doi.org/10.1002/j.2333-8504.2013.tb02312.x

Haberman, S. J. (2008a). *Continuous exponential families: An equating tool* (Research Report No. RR-08-05). Princeton: Educational Testing Service. http://dx.doi.org/10.1002/j.2333-8504.2008.tb02091.x

Haberman, S. J. (2008b). *Linking with continuous exponential families: Single-group designs* (Research Report No. RR-08-61). Princeton: Educational Testing Service. http://dx.doi.org/10.1002/j.2333-8504.2008.tb02147.x

Haberman, S. (2010). *Limits on the accuracy of linking* (Research Report No. RR-10-22). Princeton: Educational Testing Service. http://dx.doi.org/10.1002/j.2333-8504.2010.tb02229.x

Haberman, S. J. (2011a). The contributions of Paul Holland. In N. J. Dorans & S. Sinharay (Eds.), *Looking back: Proceedings of a conference in honor of Paul W. Holland* (pp. 3–17). New York: Springer. https://doi.org/10.1007/978-1-4419-9389-2_1

Haberman, S. J. (2011b). Using exponential families for equating. In A. A. von Davier (Ed.), *Statistical models for scaling, equating and linking* (pp. 125–140). New York: Springer.

Haberman, S. J. (2015). Pseudo-equivalent groups and linking. *Journal of Educational and Behavioral Statistics, 40*, 254–273. https://doi.org/10.3102/1076998615574772

Haberman, S. J., & Dorans, N. J. (2011). *Sources of scale inconsistency* (Research Report No. RR-11-10). Princeton: Educational Testing Service. http://dx.doi.org/10.1002/j.2333-8504.2011.tb02246.x

Haberman, S. J., & Yan, D. (2011). *Use of continuous exponential families to link forms via anchor tests* (Research Report No. RR-11-11), Princeton: Educational Testing Service. http://dx.doi.org/10.1002/j.2333-8504.2011.tb02247.x

Haberman, S. J., Lee, Y.-H., & Qian, J. (2009). *Jackknifing techniques for evaluation of equating accuracy* (Research Report No. RR-09-39). Princeton: Educational Testing Service. http://dx.doi.org/10.1002/j.2333-8504.2009.tb02196.x

Hicks, M. M. (1983). True score equating by fixed b's scaling: A flexible and stable equating alternative. *Applied Psychological Measurement, 7*, 255–266. https://doi.org/10.1177/014662168300700302

Hicks, M. M. (1984). *A comparative study of methods of equating TOEFL test scores* (Research Report No. RR-84-20). Princeton: Educational Testing Service. http://dx.doi.org/10.1002/j.2330-8516.1984.tb00060.x

Holland, P. W. (2007). A framework and history for score linking. In N. J. Dorans, M. Pommerich, & P. W. Holland (Eds.), *Linking and aligning scores and scales* (pp. 5–30). New York: Springer. https://doi.org/10.1007/978-0-387-49771-6_2

Holland, P. W. (2013). Comments on van der Linden's critique and proposal for equating. *Journal of Educational Measurement, 50*, 286–294. https://doi.org/10.1111/jedm.12015

Holland, P. W., & Dorans, N. J. (2006). Linking and equating. In R. L. Brennan (Ed.), *Educational measurement* (4th ed., pp. 187–220). Westport, CT: American Council on Education and Praeger.

Holland, P. W., & Rubin, D. B. (Eds.). (1982). *Test equating*. New York: Academic Press.

Holland, P. W., & Strawderman, W. (2011). How to average equating functions, if you must. In A. A. von Davier (Ed.), *Statistical models for test equating, scaling, and linking* (pp. 89–107). New York: Springer.

Holland, P. W., & Thayer, D. T. (1981). *Section pre-equating the graduate record examination* (Program Statistics Research Technical Report No. 81-51). Princeton: Educational Testing Service. https://doi.org/10.1002/j.2333-8504.1981.tb01278.x

Holland, P. W., & Thayer, D. T. (1984). *Section pre-equating in the presence of practice effects* (Research Report No. RR-84-07). Princeton: Educational Testing Service. http://dx.doi.org/10.1002/j.2330-8516.1984.tb00047.x

Holland, P. W., & Thayer, D. T. (1985). Section pre-equating in the presence of practice effects. *Journal of Educational Statistics, 10*, 109–120. https://doi.org/10.2307/1164838

Holland, P. W., & Thayer, D. T. (1987). *Notes on the use of log-linear model for fitting discrete probability distribution* (Research Report No. RR-87-31). Princeton: Educational Testing Service. http://dx.doi.org/10.1002/j.2330-8516.1987.tb00235.x

Holland, P. W., & Thayer, D. T. (1989). *The kernel method of equating score distributions* (Program Statistics Research Technical Report No. 89-84).Princeton: Educational Testing Service. https://doi.org/10.1002/j.2333-8504.1981.tb01278.x

Holland, P. W., & Wightman, L. E. (1982). Section pre-equating. A preliminary investigation. In P. W. Holland & D. B. Rubin (Eds.), *Testing equating* (pp. 217–297). New York: Academic Press.

Holland, P. W., King, B. F., & Thayer, D. T. (1989). *The standard error of equating for the kernel method of equating score distributions* (Research Report No. RR-89-06). Princeton: Educational Testing Service. http://dx.doi.org/10.1002/j.2330-8516.1989.tb00332.x

Holland, P. W., von Davier, A. A., Sinharay, S., & Han, N. (2006). *Testing the untestable assumptions of the chain and poststratification equating methods for the NEAT design* (Research Report No. RR-06-17). Princeton: Educational Testing Service. http://dx.doi.org/10.1002/j.2333-8504.2006.tb02023.x

Holland, P. W., Dorans, N. J., & Petersen, N. S. (2007). Equating test scores. In C. R. Rao & S. Sinharay (Eds.), *Handbook of statistics: Vol. 26, Psychometrics* (pp. 169–203). Amsterdam: Elsevier.

Holland, P. W., Sinharay, S., von Davier, A. A., & Han, N. (2008). An approach to evaluating the missing data assumptions of the chain and post-stratification equating methods for the NEAT design. *Journal of Educational Measurement, 45*(1), 17–43. https://doi.org/10.1111/j.1745-3984.2007.00050.x

Karon, B. P. (1956). The stability of equated test scores. *Journal of Experimental Education, 24*, 181–195. https://doi.org/10.1080/00220973.1956.11010539

Karon, B. P., & Cliff, R. H. (1957). *The Cureton–Tukey method of equating test scores* (Research Bulletin No. RB-57-06). Princeton: Educational Testing Service. http://dx.doi.org/10.1002/j.2333-8504.1957.tb00072.x

Kim, S., & Livingston, S. A. (2009). *Methods of linking with small samples in a common-item design: An empirical comparison* (Research Report No. RR-09-38). Princeton: Educational Testing Service. http://dx.doi.org/10.1002/j.2333-8504.2009.tb02195.x

Kim, S., & Livingston, S. A. (2010). Comparisons among small sample equating methods in a common-item design. *Journal of Educational Measurement, 47*, 286–298. https://doi.org/10.1111/j.1745-3984.2010.00114.x

Kim, S., & Moses, T. P. (2013). Determining when single scoring for constructed-response items is as effective as double scoring in mixed-format licensure tests. *International Journal of Testing, 13*, 314–328. https://doi.org/10.1080/15305058.2013.776050

Kim, S., & Walker, M. E. (2009a). *Effect of repeaters on score equating in a large scale licensure test* (Research Report No. RR-09-27). Princeton: Educational Testing Service. http://dx.doi.org/10.1002/j.2333-8504.2009.tb02184.x

Kim, S., & Walker, M. E. (2009b). *Evaluating subpopulation invariance of linking functions to determine the anchor composition for a mixed-format test* (Research Report No. RR-09-36). Princeton: Educational Testing Service. http://dx.doi.org/10.1002/j.2333-8504.2009.tb02193.x

Kim, S., & Walker, M. E. (2012a). Determining the anchor composition for a mixed-format test: Evaluation of subpopulation invariance of linking functions. *Applied Measurement in Education, 25*, 178–195. https://doi.org/10.1080/08957347.2010.524720

Kim, S., & Walker, M. E. (2012b). Examining repeater effects on chained equipercentile equating with common anchor items. *Applied Measurement in Education, 25*, 41–57. https://doi.org/10.1080/08957347.2012.635481

Kim, S., von Davier, A. A., & Haberman, S. J. (2006). *An alternative to equating with small samples in the non-equivalent groups anchor test design* (Research Report No. RR-06-27). Princeton: Educational Testing Service. http://dx.doi.org/10.1002/j.2333-8504.2006.tb02033.x

Kim, S., von Davier, A. A., & Haberman, S. (2007). *Investigating the effectiveness of a synthetic linking function on small sample equating* (Research Report No. RR-07-37). Princeton: Educational Testing Service. http://dx.doi.org/10.1002/j.2333-8504.2007.tb02079.x

Kim, S., Walker, M. E., & McHale, F. (2008a). *Comparisons among designs for equating constructed response tests* (Research Report No. RR-08-53). Princeton: Educational Testing Service. http://dx.doi.org/10.1002/j.2333-8504.2008.tb02139.x

Kim, S., Walker, M. E., & McHale, F. (2008b). *Equating of mixed-format tests in large-scale assessments* (Research Report No. RR-08-26). Princeton: Educational Testing Service. http://dx.doi.org/10.1002/j.2333-8504.2008.tb02112.x

Kim, S., von Davier, A. A., & Haberman, S. (2008c). Small-sample equating using a synthetic linking function. *Journal of Educational Measurement, 45*, 325–342. https://doi.org/10.1111/j.1745-3984.2008.00068.x

Kim, S., Livingston, S. A., & Lewis, C. (2008d). *Investigating the effectiveness of collateral information on small-sample equating.* (Research Report No. RR-08-52). Princeton: Educational Testing Service. http://dx.doi.org/10.1002/j.2333-8504.2008.tb02138.x

Kim, S., Livingston, S. A., & Lewis, C. (2009). *Evaluating sources of collateral information on small-sample equating* (Research Report No. RR-09-14). Princeton: Educational Testing Service. http://dx.doi.org/10.1002/j.2333-8504.2009.tb02171.x

Kim, S., Walker, M. E., & McHale, F. (2010a). Comparisons among designs for equating mixed-format tests in large-scale assessments. *Journal of Educational Measurement, 47*, 36–53. https://doi.org/10.1111/j.1745-3984.2009.00098.x

Kim, S., Walker, M. E., & McHale, F. (2010b). Investigation the effectiveness of equating designs for constructed response tests in large scale assessment. *Journal of Educational Measurement, 47*, 186–201. https://doi.org/10.1111/j.1745-3984.2010.00108.x

Kim, S., von Davier, A. A., & Haberman, S. (2011). Practical application of a synthetic linking function on small sample equating. *Applied Measurement in Education, 24*, 95–114. http://dx.doi.org/10.1080/08957347.2011.554601

Kim, S., Walker, M. E., & Larkin, K. (2012). Examining possible construct changes to a licensure test by evaluating equating requirements. *International Journal of Testing, 12*, 365–381. https://doi.org/10.1080/15305058.2011.645974

Kingston, N. M., & Dorans, N. J. (1982). *The feasibility of using item response theory as a psychometric model for the GRE aptitude test* (Research Report No. RR-82-12). Princeton: Educational Testing Service. http://dx.doi.org/10.1002/j.2333-8504.1982.tb01298.x

Kingston, N. M., & Holland, P. W. (1986). *Alternative methods for equating the GRE general test* (Research Report No. RR-86-16). Princeton: Educational Testing Service. http://dx.doi.org/10.1002/j.2330-8516.1986.tb00171.x

Kingston, N. M., Leary, L. F., & Wightman, L. E. (1985). *An exploratory study of the applicability of item response theory methods to the Graduate Management Admission Test* (Research Report No. RR-85-34). Princeton: Educational Testing Service. http://dx.doi.org/10.1002/j.2330-8516.1985.tb00119.x

Koutsopoulos, C. J. (1961). *A linear practice effect solution for the counterbalanced case of equating* (Research Bulletin No. RB-61-19). Princeton: Educational Testing Service. http://dx.doi.org/10.1002/j.2333-8504.1961.tb00287.x

Lawrence, I. M., & Dorans, N. J. (1988). *A comparison of observed score and true score equating methods for representative samples and samples matched on an anchor test* (Research Report No. RR-88-23). Princeton: Educational Testing Service. http://dx.doi.org/10.1002/j.2330-8516.1988.tb00279.x

Lawrence, I. M., & Dorans, N. J. (1990). Effect on equating results of matching samples on an anchor test. *Applied Measurement in Education, 3*, 19–36. https://doi.org/10.1207/s15324818ame0301_3

Lee, Y.-H., & Haberman, S. H. (2013). Harmonic regression and scale stability. *Psychometrika, 78*, 815–829. https://doi.org/10.1007/s11336-013-9337-1

Lee, Y.-H., & von Davier, A. A. (2008). *Comparing alternative kernels for the kernel method of test equating: Gaussian, logistic, and uniform kernels* (Research Report No. RR-08-12). Princeton: Educational Testing Service. http://dx.doi.org/10.1002/j.2333-8504.2008.tb02098.x

Lee, Y.-H., & von Davier, A. A. (2011). Equating through alternative kernels. In A. A. von Davier (Ed.), *Statistical models for test equating, scaling, and linking* (pp. 159–173). New York: Springer. https://doi.org/10.1007/978-0-387-98138-3_10

Lee, Y.-H., & von Davier, A. A. (2013). Monitoring scale scores over time via quality control charts, model-based approaches, and time series techniques. *Psychometrika, 78*, 557–575. https://doi.org/10.1007/s11336-013-9317-5

Levine, R. (1955). *Equating the score scales of alternate forms administered to samples of different ability* (Research Bulletin No. RB-55-23). Princeton: Educational Testing Service. http://dx.doi.org/10.1002/j.2333-8504.1955.tb00266.x

Li, D., Li, S., & von Davier, A. A. (2011). Applying time-series analysis to detect scale drift. In A. A. von Davier (Ed.), *Statistical models for test equating, scaling, and linking* (pp. 327–346). New York: Springer. https://doi.org/10.1007/978-0-387-98138-3_20

Li, D., Jiang, Y., & von Davier, A. A. (2012). The accuracy and consistency of a series of IRT true score equatings. *Journal of Educational Measurement, 49*, 167–189. https://doi.org/10.1111/j.1745-3984.2012.00167.x

Li, Y. (2012). *Examining the impact of drifted polytomous anchor items on test characteristic curve (TCC) linking and IRT true score equating* (Research Report No. RR-12-09). Princeton: Educational Testing Service. http://dx.doi.org/10.1002/j.2333-8504.2012.tb02291.x

Li, Y., & Brown, T. (2013). *A trend-scoring study for the TOEFL iBT Speaking and Writing Sections* (Research Memorandum No. RM-13-05). Princeton: Educational Testing Service.

Liang, L., Dorans, N. J., & Sinharay, S. (2009). *First language of examinees and its relationship to equating* (Research Report No. RR-09-05). Princeton: Educational Testing Service. http://dx.doi.org/10.1002/j.2333-8504.2009.tb02162.x

Liao, C. (2013). *An evaluation of differential speededness and its impact on the common item equating of a reading test* (Research Memorandum No. RM-13-02). Princeton: Educational Testing Service.

Linn, R. L. (1993). Linking results of distinct assessments. *Applied Measurement in Education, 6*, 83–102. https://doi.org/10.1207/s15324818ame0601_5

Liou, M., & Cheng, P. E. (1995). Asymptotic standard error of equipercentile equating. *Journal of Educational and Behavioral Statistics, 20*, 259–286. https://doi.org/10.3102/10769986020003259

Liou, M., Cheng, P. E., & Johnson, E. G. (1996). *Standard errors of the kernel equating methods under the common-item design* (Research Report No. RR-96-11). Princeton: Educational Testing Service. http://dx.doi.org/10.1002/j.2333-8504.1996.tb01689.x

Liou, M., Cheng, P. E., & Johnson, E. G. (1997). Standard errors of the kernel equating methods under the common-item design. *Applied Psychological Measurement, 21*, 349–369. https://doi.org/10.1177/01466216970214005

Liu, J., & Dorans, N. J. (2013). Assessing a critical aspect of construct continuity when test specifications change or test forms deviate from specifications. *Educational Measurement: Issues and Practice, 32*, 15–22. https://doi.org/10.1111/emip.12001

Liu, M., & Holland, P. W. (2008). Exploring population sensitivity of linking functions across three law school admission test administrations. *Applied Psychological Measurement, 32*, 27–44. https://doi.org/10.1177/0146621607311576.

Liu, J., & Low, A. C. (2007). *An exploration of kernel equating using SAT data: Equating to a similar population and to a distant population* (Research Report No. RR-07-17). Princeton: Educational Testing Service. http://dx.doi.org/10.1002/j.2333-8504.2007.tb02059.x

Liu, J., & Low, A. (2008). A comparison of the kernel equating method with the traditional equating methods using SAT data. *Journal of Educational Measurement, 45*, 309–323. https://doi.org/10.1111/j.1745-3984.2008.00067.x

Liu, J., & Walker, M. E. (2007). Score linking issues related to test content changes. In N. J. Dorans, M. Pommerich, & P. W. Holland (Eds.), *Linking and aligning scores and scales* (pp. 109–134). New York: Springer. https://doi.org/10.1007/978-0-387-49771-6_7

Liu, J., Sinharay, S., Holland, P., Curley, E., & Feigenbaum, M. (2009a). *The effect of different types of anchor test on observed score equating* (Research Report No. RR-09-41). Princeton: Educational Testing Service. http://dx.doi.org/10.1002/j.2333-8504.2009.tb02198.x

Liu, J., Moses, T. P., & Low, A. C. (2009b). *Evaluation of the effects of loglinear smoothing models on equating functions in the presence of structured data irregularities* (Research Report No. RR-09-22). Princeton: Educational Testing Service. http://dx.doi.org/10.1002/j.2333-8504.2009.tb02179.x

Liu, J., Sinharay, S., Holland, P. W., Curley, E., & Feigenbaum, M. (2011a). Observed score equating using a mini-version anchor and an anchor with less spread of difficulty: A comparison study. *Educational and Psychological Measurement, 71*, 346–361. https://doi.org/10.1177/0013164410375571

Liu, J., Sinharay, S., Holland, P., Curley, E., & Feigenbaum, M. (2011b). Test score equating using a mini-version anchor and a midi anchor: A case study using SAT data. *Journal of Educational Measurement, 48*, 361–379. https://doi.org/10.1111/j.1745-3984.2011.00150.x

Liu, J., Guo, H., & Dorans, N. J. (2014). *A comparison of raw-to-scale conversion consistency between single- and multiple-linking using a nonequivalent groups anchor test design*

(Research Report No. RR-14-13). Princeton: Educational Testing Service. http://dx.doi.org/10.1002/ets2.12014

Liu, M., & Holland, P. W. (2008). Exploring population sensitivity of linking functions across three law school admission test administrations. *Applied Psychological Measurement, 32*, 27–44. https://doi.org/10.1177/0146621607311576.

Liu, Y., Shultz, E. M., & Yu, L. (2008). Standard error estimation of 3PL IRT true score equating with an MCMC method. *Journal of Educational and Behavioral Statistics, 33*, 257–278. https://doi.org/10.3102/1076998607306076.

Livingston, S. A. (1988). *Adjusting scores on examinations offering a choice of essay questions* (Research Report No. RR-88-64). Princeton: Educational Testing Service. http://dx.doi.org/10.1002/j.2330-8516.1988.tb00320.x

Livingston, S. A. (1993a). *An empirical tryout of kernel equating* (Research Report No. RR-93-33). Princeton: Educational Testing Service. http://dx.doi.org/10.1002/j.2333-8504.1993.tb01544.x

Livingston, S. A. (1993b). Small-sample equating with log-linear smoothing. *Journal of Educational Measurement, 30*, 23–39. https://doi.org/10.1111/j.1745-3984.1993.tb00420.x.

Livingston, S. A. (2004). *Equating test scores (without IRT)*. Princeton: Educational Testing Service.

Livingston, S. A. (2014a). *Demographically adjusted groups for equating test scores* (Research Report No. RR-14-30). Princeton: Educational Testing Service. http://dx.doi.org/10.1002/ets2.12030

Livingston, S. A. (2014b). *Equating test scores (without IRT)* (2nd ed.). Princeton: Educational Testing Service.

Livingston, S. A., & Feryok, N. J. (1987). *Univariate versus bivariate smoothing in frequency estimation equating* (Research Report No. RR-87-36). Princeton: Educational Testing Service. http://dx.doi.org/10.1002/j.2330-8516.1987.tb00240.x

Livingston, S. A., & Kim, S. (2008). *Small sample equating by the circle-arc method* (Research Report No. RR-08-39). Princeton: Educational Testing Service. http://dx.doi.org/10.1002/j.2333-8504.2008.tb02125.x

Livingston, S. A., & Kim, S. (2009). The circle-arc method for equating in small samples. *Journal of Educational Measurement, 46*, 330–343. https://doi.org/10.1111/j.1745-3984.2009.00084.x

Livingston, S. A., & Kim, S. (2010a). *An empirical comparison of methods for equating with randomly equivalent groups of 50 to 400 test takers* (Research Report No. RR-10-05). Princeton: Educational Testing Service. http://dx.doi.org/10.1002/j.2333-8504.2010.tb02212.x

Livingston, S. A., & Kim, S. (2010b). Random-groups equating with samples of 50 to 400 test takers. *Journal of Educational Measurement, 47*, 175–185. https://doi.org/10.1111/j.1745-3984.2010.00107.x

Livingston, S. A., & Kim, S. (2011). New approaches to equating with small samples. In A. A. von Davier (Ed.), *Statistical models for test equating, scaling, and linking* (pp. 109–122). New York: Springer. https://doi.org/10.1007/978-0-387-98138-3_7

Livingston, S. A., & Lewis, C. (2009). *Small-sample equating with prior information* (Research Report No. RR-09-25). Princeton: Educational Testing Service. http://dx.doi.org/10.1002/j.2333-8504.2009.tb02182.x

Livingston, S. A., Dorans, N. J., & Wright, N. K. (1990). What combination of sampling and equating methods works best? *Applied Measurement in Education, 3*, 73–95. https://doi.org/10.1207/s15324818ame0301_6

Lord, F. M. (1950). *Notes on comparable scales for test scores* (Research Bulletin No. RB-50-48). Princeton: Educational Testing Service. http://dx.doi.org/10.1002/j.2333-8504.1950.tb00673.x

Lord, F. M. (1954). *Equating test scores: The maximum likelihood solution for a common item equating problem* (Research Bulletin No. RB-54-01). Princeton: Educational Testing Service. http://dx.doi.org/10.1002/j.2333-8504.1954.tb00040.x

Lord, F. M. (1955). Equating test scores: A maximum likelihood solution. *Psychometrika, 20*, 193–200. https://doi.org/10.1007/BF02289016

Lord, F. M. (1964a). Nominally and rigorously parallel test forms. *Psychometrika, 29*, 335–345. https://doi.org/10.1007/BF02289600

Lord, F. M. (1964b). *Rigorously and nonrigorously parallel test forms* (Research Bulletin No. RB-64-14). Princeton: Educational Testing Service. http://dx.doi.org/10.1002/j.2333-8504.1964.tb00323.x

Lord, F. M. (1975). *A survey of equating methods based on item characteristic curve theory* (Research Bulletin No. RB-75-13). Princeton: Educational Testing Service. http://dx.doi.org/10.1002/j.2333-8504.1975.tb01052.x

Lord, F. M. (1980). *Applications of item response theory to practical testing problems.* Hillsdale: Lawrence Erlbaum Associates.

Lord, F. M. (1981). *Standard error of an equating by item response theory* (Research Report No. RR-81-49). Princeton: Educational Testing Service. http://dx.doi.org/10.1002/j.2333-8504.1981.tb01276.x

Lord, F. M. (1982a). Item response theory and equating: A technical summary. In P. W. Holland & D. B. Rubin (Eds.), *Test equating* (pp. 141–148). New York: Academic Press. https://doi.org/10.2307/1164642

Lord, F. M. (1982b). The standard error of equipercentile equating. *Journal of Educational Statistics, 7*, 165–174. https://doi.org/10.2307/1164642

Lord, F. M., & Wingersky, M. S. (1983). *Comparison of IRT observed-score and true-score equatings* (Research Report No. RR-83-26). Princeton: Educational Testing Service. http://dx.doi.org/10.1002/j.2330-8516.1983.tb00026.x

Lord, F. M., & Wingersky, M. S. (1984). Comparison of IRT true-score and equipercentile observed score equatings. *Applied Psychological Measurement, 8*, 453–461. https://doi.org/10.1177/014662168400800409

Mao, X., von Davier, A. A., & Rupp, S. (2006). *Comparisons of the kernel equating method with the traditional equating methods on PRAXIS data* (Research Report No. RR-06-30). Princeton: Educational Testing Service. http://dx.doi.org/10.1002/j.2333-8504.2006.tb02036.x

Marco, G. L. (1981). Equating tests in an era of test disclosure. In B. F. Green (Ed.), *New directions for testing and measurement: Issues in testing—coaching, disclosure, and ethnic bias* (pp. 105–122). San Francisco: Jossey-Bass.

Marco, G. L., Petersen, N. S., & Stewart, E. E. (1983a). *A large-scale evaluation of linear and curvilinear score equating models* (Research Memorandum No. RM-83-02). Princeton: Educational Testing Service.

Marco, G. L., Stewart, E. E., & Petersen, N. S. (1983b). A test of the adequacy of curvilinear score equating models. In D. J. Weiss (Ed.), *New horizons in testing: Latent trait test theory and computerized adaptive testing* (pp. 147–177). New York: Academic Press. https://doi.org/10.1016/B978-0-12-742780-5.50018-4

McHale, F. J., & Ninneman, A. M. (1994). *The stability of the score scale for the Scholastic Aptitude Test from 1973 to 1984* (Statistical Report No. SR-94-27). Princeton: Educational Testing Service.

McKinley, R. L., & Kingston, N. M. (1987). *Exploring the use of IRT equating for the GRE Subject Test in Mathematics* (Research Report No. RR-87-21). Princeton: Educational Testing Service. http://dx.doi.org/10.1002/j.2330-8516.1987.tb00225.x

McKinley, R. L., & Schaefer, G. (1989). *Reducing test form overlap of the GRE Subject Test in Mathematics using IRT triple-part equating* (Research Report No. RR-89-08). Princeton: Educational Testing Service. http://dx.doi.org/10.1002/j.2330-8516.1989.tb00334.x

Mislevy, R. J. (1992). *Linking educational assessments: Concepts, issues, methods, and prospects* (Policy Information Report). Princeton: Educational Testing Service.

Modu, C. C., & Stern, J. (1975). *The stability of the SAT-Verbal score scale* (Research Bulletin No. RB-75-09). Princeton: Educational Testing Service. http://dx.doi.org/10.1002/j.2333-8504.1975.tb01048.x

Moses, T. P. (2006). *Using the kernel method of test equating for estimating the standard errors of population invariance measures* (Research Report No. RR-06-20). Princeton: Educational Testing Service. http://dx.doi.org/10.1002/j.2333-8504.2006.tb02026.x

Moses, T. P. (2008a). *An evaluation of statistical strategies for making equating function selections* (Research Report No. RR-08-60). Princeton: Educational Testing Service. http://dx.doi.org/10.1002/j.2333-8504.2008.tb02146.x

Moses, T. P. (2008b). Using the kernel method of test equating for estimating the standard errors of population invariance measures. *Journal of Educational and Behavioral Statistics, 33*, 137–157. https://doi.org/10.3102/1076998607302634

Moses, T. P. (2009). A comparison of statistical significance tests for selecting equating functions. *Applied Psychological Measurement, 33*, 285–306. https://doi.org/10.1177/0146621608321757

Moses, T. P. (2011). Log-linear models as smooth operators: Holland's statistical applications and their practical uses. In N. J. Dorans & S. Sinharay (Eds.), *Looking back: Proceedings of a conference in honor of Paul W. Holland* (pp. 185–202). New York: Springer. https://doi.org/10.1007/978-1-4419-9389-2_10

Moses, T. P. (2014). Alternative smoothing and scaling strategies for weighted composite scores. *Educational and Psychological Measurement, 74*, 516–536. https://doi.org/10.1177/0013164413507725

Moses, T. P., & Holland, P. (2007). *Kernel and traditional equipercentile equating with degrees of presmoothing* (Research Report No. RR-07-15). Princeton: Educational Testing Service. http://dx.doi.org/10.1002/j.2333-8504.2007.tb02057.x

Moses, T. P., & Holland, P. W. (2008). *The influence of strategies for selecting loglinear smoothing models on equating functions* (Research Report No. RR-08-25). Princeton: Educational Testing Service. http://dx.doi.org/10.1002/j.2333-8504.2008.tb02111.x

Moses, T. P., & Holland, P. W. (2009a). *Alternative loglinear smoothing models and their effect on equating function accuracy* (Research Report No. RR-09-48). Princeton: Educational Testing Service. http://dx.doi.org/10.1002/j.2333-8504.2009.tb02205.x

Moses, T. P., & Holland, P. W. (2009b). *Selection strategies for bivariate loglinear smoothing models and their effects on NEAT equating functions* (Research Report No. RR-09-04). Princeton: Educational Testing Service. http://dx.doi.org/10.1002/j.2333-8504.2009.tb02161.x

Moses, T. P., & Holland, P. W. (2009c). Selection strategies for univariate loglinear smoothing models and their effects on equating function accuracy. *Journal of Educational Measurement, 46*, 159–176. https://doi.org/10.1111/j.1745-3984.2009.00075.x

Moses, T. P., & Holland, P. W. (2010a). A comparison of statistical selection strategies for univariate and bivariate loglinear smoothing models. *British Journal of Mathematical and Statistical Psychology, 63*, 557–574. https://doi.org/10.1348/000711009X478580

Moses, T. P., & Holland, P. W. (2010b). The effects of selection strategies for bivariate loglinear smoothing models on NEAT equating functions. *Journal of Educational Measurement, 47*(1), 76–91. https://doi.org/10.1111/j.1745-3984.2009.00100.x

Moses, T. P., & Kim, S. (2007). *Reliability and the nonequivalent groups with anchor test design* (Research Report No. RR-07-16). Princeton: Educational Testing Service. http://dx.doi.org/10.1002/j.2333-8504.2007.tb02058.x

Moses, T. P., & Oh, H. (2009). *Pseudo Bayes estimates for test score distributions and chained equipercentile equating* (Research Report No. RR-09-47). Princeton: Educational Testing Service. http://dx.doi.org/10.1002/j.2333-8504.2009.tb02204.x

Moses, T. P., & von Davier, A. A. (2006). *A SAS macro for loglinear smoothing: Applications and implications* (Research Report No. RR-06-05). Princeton: Educational Testing Service. http://dx.doi.org/10.1002/j.2333-8504.2006.tb02011.x

Moses, T. P., & von Davier, A. A. (2013). A SAS IML macro for loglinear smoothing. *Applied Psychological Measurement, 35*, 250–251. https://doi.org/10.1177/0146621610369909

Moses, T. P., & Zhang, W. (2010). *Research on standard errors of equating differences* (Research Report No. RR-10-25). Princeton: Educational Testing Service. http://dx.doi.org/10.1002/j.2333-8504.2010.tb02232.x

Moses, T. P., & Zhang, W. (2011). Standard errors of equating differences: Prior developments, extensions, and simulations. *Journal of Educational and Behavioral Statistics, 36*, 779–803. https://doi.org/10.3102/1076998610396892

Moses, T. P., von Davier, A. A., & Casabianca, J. (2004). *Loglinear smoothing: An alternative numerical approach using SAS* (Research Report No. RR-04-27). Princeton: Educational Testing Service. http://dx.doi.org/10.1002/j.2333-8504.2004.tb01954.x

Moses, T. P., Liu, J., & Dorans, N. J. (2009). *Systematized score equity assessment in SAS* (Research Memorandum No. RM-09-08). Princeton: Educational Testing Service.

Moses, T. P., Deng, W., & Zhang, Y.-L. (2010a). *The use of two anchors in nonequivalent groups with anchor test (NEAT) equating* (Research Report No. RR-10-23). Princeton: Educational Testing Service. http://dx.doi.org/10.1002/j.2333-8504.2010.tb02230.x

Moses, T. P., Liu, J., & Dorans, N. J. (2010b). Systemized SEA in SAS. *Applied Psychological Measurement, 34*, 552–553. https://doi.org/10.1177/0146621610369909

Moses, T. P., Deng, W., & Zhang, Y.-L. (2011). Two approaches for using multiple anchors in NEAT equating: A description and demonstration. *Applied Psychological Measurement, 35*, 362–379. https://doi.org/10.1177/0146621611405510

Muraki, E., Hombo, C. M., & Lee, Y.-W. (2000). Equating and linking of performance assessments. *Applied Psychological Measurement, 24*, 325–337. https://doi.org/10.1177/01466210022031787

Myford, C., Marr, D. B., & Linacre, J. M. (1995). *Reader calibration and its potential role in equating for the Test of Written English* (Research Report No. RR-95-40). Princeton: Educational Testing Service. http://dx.doi.org/10.1002/j.2333-8504.1995.tb01674.x

Oh, H., & Moses, T. P. (2012). Comparison of the one- and bi-direction chained equipercentile equating. *Journal of Educational Measurement, 49*, 399–418. https://doi.org/10.1111/j.1745-3984.2012.00183.x

Petersen, N. S., Marco, G. L., & Stewart, E. E. (1982). A test of the adequacy of linear score equating models. In P. W. Holland & D. B. Rubin (Eds.), *Test equating* (pp. 71–135). New York: Academic Press.

Petersen, N. S., Cook, L. L., & Stocking, M. L. (1983). IRT versus conventional equating methods: A comparative study of scale stability. *Journal of Educational Statistics, 8*, 137–156. https://doi.org/10.2307/1164922

Petersen, N. S., Kolen, M. J., & Hoover, H. D. (1989). Scaling, norming and equating. In R. L. Linn (Ed.), *Educational measurement* (3rd ed., pp. 221–262). New York: Macmillan.

Pommerich, M., & Dorans, N. J. (Eds.). (2004). Concordance. [Special issue]. *Applied Psychological Measurement, 28*(4).

Puhan, G. (2007). *Scale drift in equating on a test that employs cut scores* (Research Report No. RR-07-34). Princeton: Educational Testing Service. http://dx.doi.org/10.1002/j.2333-8504.2007.tb02076.x

Puhan, G. (2009a). Detecting and correcting scale drift in test equating: An illustration from a large scale testing program. *Applied Measurement in Education, 22*, 79–103. https://doi.org/10.1080/08957340802558391

Puhan, G. (2009b). *What effect does the inclusion or exclusion of repeaters have on test equating?* (Research Report No. RR-09-19). Princeton: Educational Testing Service. http://dx.doi.org/10.1002/j.2333-8504.2009.tb02176.x

Puhan, G. (2010a). *Chained versus post stratification equating: An evaluation using empirical data* (Research Report No. RR-10-06). Princeton: Educational Testing Service. http://dx.doi.org/10.1002/j.2333-8504.2010.tb02213.x

Puhan, G. (2010b). A comparison of chained linear and poststratification linear equating under different testing conditions. *Journal of Educational Measurement, 47*, 54–75. https://doi.org/10.1111/j.1745-3984.2009.00099.x

Puhan, G. (2011a). *Can smoothing help when equating with unrepresentative small samples* (Research Report No. RR-11-09). Princeton: Educational Testing Service. http://dx.doi.org/10.1002/j.2333-8504.2011.tb02245.x

Puhan, G. (2011b). Futility of log linear smoothing when equating with unrepresentative small samples. *Journal of Educational Measurement, 48*, 274–292. https://doi.org/10.1111/j.1745-3984.2011.00147.x

Puhan, G. (2011c). Impact of inclusion or exclusion of repeaters on test equating. *International Journal of Testing, 11*, 215–230. https://doi.org/10.1080/15305058.2011.555575

Puhan, G. (2012). Tucker versus chained linear equating in two equating situations—Rater comparability scoring and randomly equivalent groups with an anchor. *Journal of Educational Measurement, 49*, 313–330. https://doi.org/10.1111/j.1745-3984.2012.00177.x.

Puhan, G. (2013a). *Choice of target population weights in rater comparability scoring and equating* (Research Report No. RR-13-03). Princeton: Educational Testing Service. http://dx.doi.org/10.1002/j.2333-8504.2013.tb02310.x

Puhan, G. (2013b). Rater comparability scoring and equating: Does choice of target population weights matter in this context? *Journal of Educational Measurement, 50*, 374–380. https://doi.org/10.1111/jedm.12023

Puhan, G., & Liang, L. (2011a). Equating subscores under the non-equivalent anchor test (NEAT) design. *Educational Measurement: Issues and Practice, 30*(1), 23–35. https://doi.org/10.1111/j.1745-3992.2010.00197.x

Puhan, G., & Liang, L. (2011b). *Equating subscores using total scaled scores as the anchor test* (Research Report No. RR-11-07). Princeton: Educational Testing Service. http://dx.doi.org/10.1002/j.2333-8504.2011.tb02243.x

Puhan, G., Moses, T. P., Grant, M., & McHale, F. (2008a). *An alternative data collection design for equating with very small samples* (Research Report No. RR-08-11). Princeton: Educational Testing Service. http://dx.doi.org/10.1002/j.2333-8504.2008.tb02097.x

Puhan, G., von Davier, A. A., & Gupta, S. (2008b). *Impossible scores resulting in zero frequencies in the anchor test: Impact on smoothing and equating* (Research Report No. RR-08-10). Princeton: Educational Testing Service. http://dx.doi.org/10.1002/j.2333-8504.2008.tb02096.x

Puhan, G., Moses, T. P., Grant, M., & McHale, F. (2009). Small sample equating using a single group nearly equivalent test (SiGNET) design. *Journal of Educational Measurement, 46*, 344–362. https://doi.org/10.1111/j.1745-3984.2009.00085.x

Puhan, G., von Davier, A. A., & Gupta, S. (2010). A brief report on how impossible scores impact smoothing and equating. *Educational and Psychological Measurement, 70*, 953–960. https://doi.org/10.1177/0013164410382731

Qian, J., von Davier, A. A., & Jiang, Y. (2013). *Weighting test samples in IRT linking and equating: Toward an improved sampling design for complex equating* (Research Report No. RR-13-39). Princeton: Educational Testing Service. http://dx.doi.org/10.1002/j.2333-8504.2013.tb02346.x

Rao, C. R., & Sinharay, S. (Eds.). (2007). *Psychometrics* (Handbook of statistics, Vol. 26). Amsterdam: Elsevier.

Ricker, K. L., & von Davier, A. A. (2007). *The impact of anchor test lengths on equating results in a nonequivalent groups design* (Research Report No. RR-07-44). Princeton: Educational Testing Service. http://dx.doi.org/10.1002/j.2333-8504.2007.tb02086.x

Rijmen, F., Manalo, J., & von Davier, A. A. (2009). Asymptotic and sampling-based standard errors for two population invariance measures in the linear equating case. *Applied Psychological Measurement, 33*, 222–237. https://doi.org/10.1177/0146621608323927

Rijmen, F., Qu, Y., & von Davier, A. A. (2011). Hypothesis testing of equating differences in the kernel equating framework. In A. A. von Davier (Ed.), *Statistical models for test equating, scaling, and linking* (pp. 317–326). New York: Springer. https://doi.org/10.1007/978-0-387-98138-3_19

Rock, D. R. (1982). Equating using the confirmatory factor analysis model. In P. W. Holland & D. B. Rubin (Eds.), *Test equating* (pp. 247–257). New York: Academic Press.

Rosenbaum, P. R. (1985). *A generalization of adjustment, with an application to the scaling of essay scores* (Research Report No. RR-85-02). Princeton: Educational Testing Service. http://dx.doi.org/10.1002/j.2330-8516.1985.tb00087.x

Rosenbaum, P. R., & Thayer, D. T. (1987). Smoothing the joint and marginal distributions of scored two-way contingency tables in test equating. *British Journal of Mathematical and Statistical Psychology, 40*, 43–49. https://doi.org/10.1111/j.2044-8317.1987.tb00866.x

Rubin, D. B. (1982). Discussion of "Partial orders and partial exchangeability in test theory." In P. W. Holland & D. B. Rubin (Eds.), *Test equating* (pp. 339–341). New York: Academic Press.

Rubin, D. B., & Szatrowski, T. (1982). Discussion of "Section pre-equating: A preliminary investigation." In P. W. Holland & D. B. Rubin (Eds.), *Test equating* (pp. 301–306). New York: Academic Press.

Schmitt, A. P., Cook, L. L., Dorans, N. J., & Eignor, D. R. (1990). Sensitivity of equating results to different sampling strategies. *Applied Measurement in Education, 3*, 53–71. https://doi.org/10.1207/s15324818ame0301_5

Schultz, D. G., & Wilks, S. S. (1950). *A method for adjusting for lack of equivalence in groups* (Research Bulletin No. RB-50-59). Princeton: Educational Testing Service. http://dx.doi.org/10.1002/j.2333-8504.1950.tb00682.x

Sinharay, S. (2011). Chain equipercentile equating and frequency estimation equipercentile equating: Comparisons based on real and stimulated data. In N. J. Dorans & S. Sinharay (Eds.), *Looking back: Proceedings of a conference in honor of Paul W. Holland* (pp. 203–219). New York: Springer. https://doi.org/10.1007/978-1-4419-9389-2_11

Sinharay, S., & Haberman, S. J. (2011a). Equating of augmented subscores. *Journal of Educational Measurement, 48*, 122–145. https://doi.org/10.1111/j.1745-3984.2011.00137.x

Sinharay, S., & Haberman, S. J. (2011b). *Equating of subscores and weighted averages under the NEAT design* (Research Report No. RR-11-01). Princeton: Educational Testing Service. http://dx.doi.org/10.1002/j.2333-8504.2011.tb02237.x

Sinharay, S., & Holland, P. W. (2006a). *Choice of anchor test in equating* (Research Report No. RR-06-35). Princeton: Educational Testing Service. http://dx.doi.org/10.1002/j.2333-8504.2006.tb02040.x

Sinharay, S., & Holland, P. W. (2006b). *The correlation between the scores of a test and an anchor test* (Research Report No. RR-06-04). Princeton: Educational Testing Service. http://dx.doi.org/10.1002/j.2333-8504.2006.tb02010.x

Sinharay, S., & Holland, P. W. (2008). *The missing data assumption of the NEAT design and their implications for test equating* (Research Report No. RR-09-16). Princeton: Educational Testing Service. http://dx.doi.org/10.1002/j.2333-8504.2009.tb02173.x

Sinharay, S., & Holland, P. W. (2010a). A new approach to comparing several equating methods in the context of the NEAT design. *Journal of Educational Measurement, 47*, 261–285. https://doi.org/10.1111/j.1745-3984.2010.00113.x

Sinharay, S., & Holland, P. W. (2010b). The missing data assumption of the NEAT design and their implications for test equating. *Psychometrika, 75*, 309–327. https://doi.org/10.1007/s11336-010-9156-6

Sinharay, S., Holland, P. W., & von Davier, A. A. (2011a). Evaluating the missing data assumptions of the chain and poststratification equating methods. In A. A. von Davier (Ed.), *Statistical models for test equating, scaling, and linking* (pp. 281–296). New York: Springer. https://doi.org/10.1007/978-0-387-98138-3_17

Sinharay, S., Dorans, N. J., & Liang, L. (2011b). First language of examinees and fairness assessment procedures. *Educational Measurement: Issues and Practice, 30*, 25–35. https://doi.org/10.1111/j.1745-3992.2011.00202.x

Sinharay, S., Haberman, S., Holland, P. W., & Lewis, C. (2012). *A note on the choice of an anchor test in equating* (Research Report No. RR-12-14). Princeton: Educational Testing Service. http://dx.doi.org/10.1002/j.2333-8504.2012.tb02296.x

Stocking, M. L., & Eignor, D. R. (1986). *The impact of different ability distributions on IRT pre-equating* (Research Report No. RR-86-49). Princeton: Educational Testing Service. http://dx.doi.org/10.1002/j.2330-8516.1986.tb00204.x

Stocking, M. L., Eignor, D. R., & Cook, L. L. (1988). *Factors affecting the sample invariant properties of linear and curvilinear observed- and true-score equating procedures* (Research Report No. RR-88-41). Princeton: Educational Testing Service. http://dx.doi.org/10.1002/j.2330-8516.1988.tb00297.x

Swineford, F., & Fan, C. T. (1957). A method of score conversion through item statistics. *Psychometrika, 22*, 185–188. https://doi.org/10.1007/BF02289053

Tan, X., Ricker, K., & Puhan, G. (2010). *Single versus double scoring of trend responses in trend score equating with constructed response tests* (Research Report No. RR-10-12). Princeton: Educational Testing Service. http://dx.doi.org/10.1002/j.2333-8504.2010.tb02219.x

Tang, L. K., Way, W. D., & Carey, P. A. (1993). *The effect of small calibration sample sizes on TOEFL IRT-based equating* (Research Report No. RR-93-59). Princeton: Educational Testing Service. http://dx.doi.org/10.1002/j.2333-8504.1993.tb01570.x

Thayer, D. T. (1983). Maximum likelihood estimation of the joint covariance matrix for sections of tests given to distinct samples with application to test equating. *Psychometrika, 48*, 293–297. https://doi.org/10.1007/BF02294023

Tucker, L. (1951). *Notes on the nature of gamble in test score scaling* (Research Bulletin No. RB-51-27). Princeton: Educational Testing Service. http://dx.doi.org/10.1002/j.2333-8504.1951.tb00226.x

von Davier, A. A. (2003). *Notes on linear equating methods for the non-equivalent-groups design* (Research Report No. RR-03-24). Princeton: Educational Testing Service. http://dx.doi.org/10.1002/j.2333-8504.2003.tb01916.x

von Davier, A. A. (2007). Potential solutions to practical equating issues. In N. J. Dorans, M. Pommerich, & P. W. Holland (Eds.), *Linking and aligning scores and scales* (pp. 89–106). New York: Springer. https://doi.org/10.1007/978-0-387-49771-6_6

von Davier, A. A. (2008). New results on the linear equating methods for the non-equivalent-groups design. *Journal of Educational and Behavioral Statistics, 33*, 186–203. https://doi.org/10.3102/1076998607302633

von Davier, A. A. (2011a). A statistical perspective on equating test scores. In A. A. von Davier (Ed.), *Statistical models for test equating, scaling and linking* (pp. 1–17). New York: Springer. https://doi.org/10.1007/978-0-387-98138-3

von Davier, A. A. (2011b). An observed-score equating framework. In N. J. Dorans & S. Sinharay (Eds.), *A festschrift for Paul W. Holland* (pp. 221–237). New York: Springer. https://doi.org/10.1007/978-1-4419-9389-2_12

von Davier, A. A. (Ed.). (2011c). *Statistical models for test equating, scaling and linking*. New York: Springer. https://doi.org/10.1007/978-0-387-98138-3

von Davier, A. A. (2013). Observed score equating: An overview. *Psychometrika, 78*, 605–623. https://doi.org/10.1007/s11336-013-9319-3

von Davier, A. A., & Kong, N. (2005). A unified approach to linear equating for the non-equivalent groups design. *Journal of Educational and Behavioral Statistics, 30*, 313–334. https://doi.org/10.3102/10769986030003313

von Davier, A. A., & Liu, M. (Eds.). (2007). Population invariance. [Special issue]. *Applied Psychological Measurement, 32*(1).

von Davier, A. A., & Wilson, C. (2007). IRT true-score test equating: A guide through assumptions and applications. *Educational and Psychological Measurement, 67*, 940–957. https://doi.org/10.1177/0013164407301543

von Davier, A. A., & Wilson, C. (2008). Investigation the population sensitivity assumption of item response theory true-score equating across two subgroups of examinees and two test formats. *Applied Psychological Measurement, 32*, 11–26. https://doi.org/10.1177/0146621607311560

von Davier, A. A., Holland, P. W., & Thayer, D. T. (2004a). The chain and post-stratification methods of observed-score equating: Their relationship to population invariance. *Journal of Educational Measurement, 41*, 15–32. https://doi.org/10.1111/j.1745-3984.2004.tb01156.x

von Davier, A. A., Holland, P. W., & Thayer, D. T. (2004b). *The kernel method of test equating*. New York: Springer. https://doi.org/10.1007/b97446

von Davier, A. A., Holland, P. W., Livingston, S. A., Casabianca, J., Grant, M. C., & Martin, K. (2006). *An evaluation of the kernel equating method: A special study with pseudo-tests constructed from real test data* (Research Report No. RR-06-02). Princeton: Educational Testing Service. http://dx.doi.org/10.1002/j.2333-8504.2006.tb02008.x

von Davier, A.A., Fournier-Zajac, S., & Holland, P. W. (2007). *An equipercentile version of the Levine linear observed score equating function using the methods of kernel equating* (Research Report No. RR-07-14). Princeton: Educational Testing Service. http://dx.doi.org/10.1002/j.2333-8504.2007.tb02056.x

von Davier, M., & von Davier, A. A. (2007). A unified approach to IRT scale linkage and scale transformations. *Methodology, 3*, 115–124. https://doi.org/10.1027/1614-2241.3.3.115.

von Davier, M., Gonzalez, J., & von Davier, A. A. (2013). Local equating using the Rasch model, the OPLM, and the 2PL IRT model—or—What is it anyway if the model captures everything there is to know about the test takers? *Journal of Educational Measurement, 50*, 295–303. https://doi.org/10.1111/jedm.12016

Wainer, H., & Thissen, D. (1994). On examinee choice in educational testing. *Review of Educational Research, 64*, 159–195. https://doi.org/10.3102/00346543064001159

Wainer, H., Dorans, N. J., Flaugher, R., Green, B. F., Mislevy, R., Sternberg, L., & Thissen, D. (1990). *Computerized adaptive testing: A primer*. Hillsdale: Lawrence Erlbaum Associates.

Wainer, H., Wang, X.-B., & Thissen, D. (1991). *How well can we equate test forms that are constructed by the examinees* (Research Report No. RR-91-57). Princeton: Educational Testing Service. http://dx.doi.org/10.1002/j.2333-8504.1991.tb01424.x

Wainer, H., Wang, X.-B., & Thissen, D. (1994). How well can we compare scores on test forms that are constructed by examinee choice? *Journal of Educational Measurement, 31*, 183–199. https://doi.org/10.1111/j.1745-3984.1994.tb00442.x

Wainer, H., Dorans, N. J., Eignor, D., Flaugher, R., Green, B. F., Mislevy, R., Sternberg, L., & Thissen, D. (2000). *Computerized adaptive testing: A primer* (2nd ed.). Hillsdale: Erlbaum.

Walker, M. E., & Kim, S. (2010). *Examining two strategies to link mixed-format tests using multiple-choice anchors* (Research Report No. RR-10-18). Princeton: Educational Testing Service.. https://doi.org/10.1002/j.2333-8504.2010.tb02225.x

Wang, X.-B., Wainer, H., & Thissen, D. (1995). On the viability of some untestable assumptions in equating exams that allow examinee choice. *Applied Measurement in Education, 8*, 211–225. https://doi.org/10.1207/s15324818ame0803_2

Wiberg, M., van der Linden, W., & von Davier, A. A. (2014). Local observed-score kernel equating. *Journal of Educational Measurement, 51*, 57–74. https://doi.org/10.1111/jedm.12034

Wightman, L. E., & Wightman, L. F. (1988). *An empirical investigation of one variable section pre-equating* (Research Report No. RR-88-37). Princeton: Educational Testing Service. http://dx.doi.org/10.1002/j.2330-8516.1988.tb00293.x

Yang, W.-L. (2004). Sensitivity of linkings between *AP*® multiple-choice scores and composite scores to geographical region: An illustration of checking for population invariance. *Journal of Educational Measurement, 41*, 33–41. https://doi.org/10.1111/j.1745-3984.2004.tb01157.x

Yang, W.-L., & Gao, R. (2008). Invariance of score linkings across gender groups for forms of a testlet-based college-level examination program examination. *Applied Psychological Measurement, 32*, 45–61. https://doi.org/10.1177/0146621607311577

Yang, W.-L., Bontya, A. M., & Moses, T. P. (2011). *Repeater effects on score equating for a graduate admissions exam* (Research Report No. RR-11-17). Princeton: Educational Testing Service. http://dx.doi.org/10.1002/j.2333-8504.2011.tb02253.x

Yen, W. (2007). Vertical scaling and No Child Left Behind. In N. J. Dorans, M. Pommerich, & P. W. Holland (Eds.), *Linking and aligning scores and scales* (pp. 273–283). New York: Springer. https://doi.org/10.1007/978-0-387-49771-6_15

Zu, J., & Liu, J. (2009). *Comparison of the effects of discrete anchor items and passage-based anchor items on observed-score equating results* (Research Report No. RR-09-44). Princeton: Educational Testing Service. http://dx.doi.org/10.1002/j.2333-8504.2009.tb02201.x

Zu, J., & Liu, J. (2010). Observed score equating using discrete and passage-based anchor items. *Journal of Educational Measurement, 47*, 395–412. https://doi.org/10.1111/j.1745-3984.2010.00120.x

Zu, J., & Puhan, G. (2014). Pre-equating with empirical item characteristic curves: An observed-score pre-equating method. *Journal of Educational Measurement, 51*, 281–300. https://doi.org/10.1111/jedm.12047

Zu, J., & Yuan, K. (2012). Standard error of linear observed-score equating for the neat design with nonnormally distributed data. *Journal of Educational Measurement, 49*, 190–213. https://doi.org/10.1111/j.1745-3984.2012.00168.x

Chapter 5
Item Response Theory

James E. Carlson and Matthias von Davier

Item response theory (IRT) models, in their many forms, are undoubtedly the most widely used models in large-scale operational assessment programs. They have grown from negligible usage prior to the 1980s to almost universal usage in large-scale assessment programs, not only in the United States, but in many other countries with active and up-to-date programs of research in the area of psychometrics and educational measurement.

Perhaps the most important feature leading to the dominance of IRT in operational programs is the characteristic of estimating individual item locations (difficulties) and test-taker locations (abilities) separately, but on the same scale, a feature not possible with classical measurement models. This estimation allows for tailoring tests through judicious item selection to achieve precise measurement for individual test takers (e.g., in computerized adaptive testing, CAT) or for defining important cut points on an assessment scale. It also provides mechanisms for placing different test forms on the same scale (linking and equating). Another important characteristic of IRT models is local independence: for a given location of test takers on the scale, the probability of success on any item is independent of that of every other item on that scale. This characteristic is the basis of the likelihood function used to estimate test takers' locations on the scale.

Few would doubt that ETS researchers have contributed more to the general topic of IRT than individuals from any other institution. In this chapter we briefly review most of those contributions, dividing them into sections by decades of publication. Of course, many individuals in the field have changed positions between

---

This chapter was originally published in 2013 as a research report in the ETS R&D Scientific and Policy Contributions Series.

This work was conducted while M. von Davier was employed with Educational Testing Service.

J.E. Carlson (✉) • M. von Davier
Educational Testing Service, Princeton, NJ, USA
e-mail: jcarlson@ets.org

R.E. Bennett, M. von Davier (eds.), *Advancing Human Assessment*,
Methodology of Educational Measurement and Assessment,
DOI 10.1007/978-3-319-58689-2_5

different testing agencies and universities over the years, some having been at ETS during more than one period of time. This chapter includes some contributions made by ETS researchers before taking a position at ETS, and some contributions made by researchers while at ETS, although they have since left. It is also important to note that IRT developments at ETS were not made in isolation. Many contributions were collaborations between ETS researchers and individuals from other institutions, as well as developments that arose from communications with others in the field.

5.1 Some Early Work Leading up to IRT (1940s and 1950s)

Tucker (1946) published a precursor to IRT in which he introduced the term *item characteristic curve*, using the normal ogive model (Green 1980).[1] Green stated:

> Workers in IRT today are inclined to reference Birnbaum in Novick and Lord [sic] when needing historical perspective, but, of course Lord's 1955 monograph, done under Tuck's direction, precedes Birnbaum, and Tuck's 1946 paper precedes practically everybody. He used normal ogives for item characteristic curves, as Lord did later. (p. 4)

Some of the earliest work leading up to a complete specification of IRT was carried out at ETS during the 1950s by Lord and Green. Green was one of the first two psychometric fellows in the joint doctoral program of ETS and Princeton University. Note that the work of Lord and Green was completed prior to Rasch's (1960) publication describing and demonstrating the one-parameter IRT model, although in his preface Rasch mentions modeling data in the mid-1950s, leading to what is now referred to as the Rasch model. Further background on the statistical and psychometric underpinnings of IRT can be found in the work of a variety of authors, both at and outside of ETS (Bock 1997; Green 1980; Lord 1952a, b, 1953).[2]

Lord (1951, 1952a, 1953) discussed test theory in a formal way that can be considered some of the earliest work in IRT. He introduced and defined many of the now common IRT terms such as item characteristic curves (ICCs), test characteristic curves (TCCs), and standard errors conditional on latent ability.[3] He also

[1] Green stated that Tucker was at Princeton and ETS from 1944 to 1960; as head of statistical analysis at ETS, Tucker was responsible for setting up the statistical procedures for test and item analysis, as well as equating.

[2] These journal articles by Green and Lord are based on their Ph.D. dissertations at Princeton University, both presented in 1951.

[3] Lord (1980a, p. 19) attributes the term *local independence* to Lazarsfeld (1950) and mentions that Lazarsfeld used the term *trace line* for a curve like the ICC. Rasch (1960) makes no mention of the earlier works referred to by Lord so we have to assume he was unaware of them or felt they were not relevant to his research direction.

discussed what we now refer to as local independence and the invariance of item parameters (not dependent on the ability distribution of the test takers). His 1953 article is an excellent presentation of the basics of IRT, and he also mentions the relevance of works specifying mathematical forms of ICCs in the 1940s (by Lawley, by Mosier, and by Tucker), and in the 1950s, (by Carroll, by Cronbach & Warrington, and by Lazarsfeld).

The emphasis of Green (1950a, b, 1951a, b, 1952) was on analyzing item response data using latent structure (LS) and latent class (LC) models. Green (1951b) stated:

> Latent Structure Analysis is here defined as a mathematical model for describing the interrelationships of items in a psychological test or questionnaire on the basis of which it is possible to make some inferences about hypothetical fundamental variables assumed to underlie the responses. It is also possible to consider the distribution of respondents on these underlying variables. This study was undertaken to attempt to develop a general procedure for applying a specific variant of the latent structure model, the latent class model, to data. (abstract)

He also showed the relationship of the latent structure model to factor analysis (FA)

> The general model of latent structure analysis is presented, as well as several more specific models. The generalization of these models to continuous manifest data is indicated. It is noted that in one case, the generalization resulted in the fundamental equation of linear multiple factor analysis. (abstract)

The work of Green and Lord is significant for many reasons. An important one is that IRT (previously referred to as latent trait, or LT, theory) was shown by Green to be directly related to the models he developed and discussed. Lord (1952a) showed that if a single latent trait is normally distributed, fitting a linear FA model to the tetrachoric correlations of the items yields a unidimensional normal-ogive model for the item response function.

5.2 More Complete Development of IRT (1960s and 1970s)

During the 1960s and 1970s, Lord (1964, 1965a, b, 1968a, b, 1970) expanded on his earlier work to develop IRT more completely, and also demonstrated its use on operational test scores (including early software to estimate the parameters). Also at this time, Birnbaum (1967) presented the theory of logistic models and Ross (1966) studied how actual item response data fit Birnbaum's model. Samejima (1969)[4] published her development of the graded response (GR) model suitable for polytomous data. The theoretical developments of the 1960s culminated in some of

[4] Samejima produced this work while at ETS. She later developed her GR models more fully while holding university positions.

the most important work on IRT during this period, much of it assembled into Lord and Novick's (1968) *Statistical Theories of Mental Test Scores* (which also includes contributions of Birnbaum: Chapters 17, 18, 19, and 20). Also Samejima's continuing work on graded response models, was further developed (1972) while she held academic positions.

An important aspect of the work at ETS in the 1960s was the development of software, particularly by Wingersky, Lord, and Andersen (Andersen 1972; Lord, 1968a; Lord and Wingersky 1973) enabling practical applications of IRT. The LOGIST computer program (Lord et al. 1976; see also Wingersky 1983) was the standard IRT estimation software used for many years in many other institutions besides ETS. Lord (1975b) also published a report in which he evaluated LOGIST estimates using artificial data. Developments during the 1950s were limited by a lack of such software and computers sufficiently powerful to carry out the estimation of parameters. In his 1968 publication, Lord presented a description and demonstration of the use of maximum likelihood (ML) estimation of the ability and item parameters in the three-parameter logistic (3PL) model, using *SAT*® items. He stated, with respect to ICCs:

> The problems of estimating such a curve for each of a large number of items simultaneously is one of the problems that has delayed practical application of Birnbaum's models since they were first developed in 1957. The first step in the present project (see Appendix B) was to devise methods for estimating three descriptive parameters simultaneously for each item in the Verbal test. (1968a, p. 992)

Lord also discussed and demonstrated many other psychometric concepts, many of which were not put into practice until fairly recently due to the lack of computing power and algorithms. In two publications (1965a, b) he emphasized that ICCs are the functions relating probability of response to the underlying latent trait, not to the total test score, and that the former and not the latter can follow a cumulative normal or logistic function (a point he originally made much earlier, Lord 1953). He also discussed (1968a) optimum weighting in scoring and information functions of items from a Verbal SAT test form, as well as test information, and relative efficiency of tests composed of item sets having different psychometric properties. A very interesting fact is that Lord (1968a, p. 1004) introduced and illustrated multistage tests (MTs), and discussed their increased efficiency relative to "the present Verbal SAT" (p. 1005). What we now refer to as *router* tests in using MTs, Lord called *foretests*. He also introduced *tailor-made tests* in this publication (and in Lord 1968c) and discussed how they would be administered using computers. Tailor-made tests are now, of course, commonly known as computerized adaptive tests (CATs); as suggested above, MTs and CATs were not employed in operational testing programs until fairly recently, but it is fascinating to note how long ago Lord introduced these notions and discussed and demonstrated the potential increase in efficiency of assessments achievable with their use. With respect to CATs Lord stated:

> The detailed strategy for selecting a sequence of items that will yield the most information about the ability of a given examinee has not yet been worked out. It should be possible to work out such a strategy on the basis of a mathematical model such as that used here, however. (1968a, p. 1005)

In this work, Lord also presented a very interesting discussion (1968a, p. 1007) on improving validity by using the methods described and illustrated. Finally, in the appendix, Lord derived the ML estimators (MLEs) of the item parameters and, interestingly points out the fact, well known today, that MLEs of the 3PL lower asymptote or *c* parameter, are often "poorly determined by the data" (p. 1014). As a result, he fixed these parameters for the easier items in carrying out his analyses.

During the 1970s Lord produced a phenomenal number of publications, many of them related to IRT, but many on other psychometric topics. On the topics related to IRT alone, he produced six publications besides those mentioned above; these publications dealt with such diverse topics as individualized testing (1974b), estimating power scores from tests that used improperly timed administration (1973), estimating ability and item parameters with missing responses (1974a), the ability scale (1975c), practical applications of item characteristic curves (1977), and equating methods (1975a). In perusing Lord's work, including Lord and Novick (1968), the reader should keep in mind that he discussed many item response methods and functions using classical test theory (CTT) as well as what we now call IRT. Other work by Lord includes discussions of item characteristic curves and information functions without, for example, using normal ogive or logistic IRT terminology, but the methodology he presented dealt with the theory of item response data. During this period, Erling Andersen visited ETS and during his stay developed one of the seminal papers on testing goodness of fit for the Rasch model (Andersen 1973). Besides the work of Lord, during this period ETS staff produced many publications dealing with IRT, both methodological and application oriented. Marco (1977), for example, described three studies indicating how IRT can be used to solve three relatively intractable testing problems: designing a multipurpose test, evaluating a multistage test, and equating test forms using pretest statistics. He used data from various College Board testing programs and demonstated the use of the information function and relative efficiency using IRT for preequating. Cook (Hambleton and Cook 1977) coauthored an article on using LT models to analyze educational test data. Hambleton and Cook described a number of different IRT models and functions useful in practical applications, demonstrated their use, and cited computer programs that could be used in estimating the parameters. Kreitzberg et al. (1977) discussed potential advantages of CAT, constraints and operational requirements, psychometric and technical developments that make it practical, and its advantages over conventional paper-and-pencil testing. Waller (1976) described a method of estimating Rasch model parameters eliminating the effects of random guessing, without using a computer, and reported a Monte Carlo study on the performance of the method.

5.3 Broadening the Research and Application of IRT (the 1980s)

During this decade, psychometricians, with leadership from Fred Lord, continued to develop the IRT methodology. Also, of course, computer programs for IRT were further developed. During this time many ETS measurement professionals were engaged in assessing the use of IRT models for scaling dichotomous item response data in operational testing programs. In many programs, IRT linking and equating procedures were compared with conventional methods, to inform programs about whether changing these methods should be considered.

5.3.1 Further Developments and Evaluation of IRT Models

In this section we describe further psychometric developments at ETS, as well as research studies evaluating the models, using both actual test and simulated data.

Lord continued to contribute to IRT methodology with works by himself as well as coauthoring works dealing with unbiased estimators of ability parameters and their parallel forms reliability (1983d), a four-parameter logistic model (Barton and Lord 1981), standard errors of IRT equating (1982), IRT parameter estimation with missing data (1983a), sampling variances and covariances of IRT parameter estimates (Lord and Wingersky 1982), IRT equating (Stocking and Lord 1983), statistical bias in ML estimation of IRT item parameters (1983c), estimating the Rasch model when sample sizes are small (1983b), comparison of equating methods (Lord and Wingersky 1984), reducing sampling error (Wingersky and Lord 1984), conjunctive and disjunctive item response functions (1984), ML and Bayesian parameter estimation in IRT (1986), and confidence bands for item response curves with Pashley (Lord and Pashley 1988).

Although Lord was undoubtedly the most prolific ETS contributor to IRT during this period, other ETS staff members made many contributions to IRT. Holland (1981), for example, wrote on the question, "When are IRT models consistent with observed data?" and Cressie and Holland (1983) examined how to characterize the manifest probabilities in LT models. Holland and Rosenbaum (1986) studied monotone unidimensional latent variable models. They discussed applications and generalizations and provided a numerical example. Holland (1990b) also discussed the *Dutch identity* as a useful tool for studying IRT models and conjectured that a quadratic form based on the identity is a limiting form for log manifest probabilities for all smooth IRT models as test length tends to infinity (but see Zhang and Stout 1997, later in this chapter). Jones discussed the adequacy of LT models (1980) and robustness tools for IRT (1982).

Wainer and several colleagues published articles dealing with standard errors in IRT (Wainer and Thissen 1982), review of estimation in the Rasch model for "longish tests" (Gustafsson et al. 1980), fitting ICCs with spline functions (Winsberg et al. 1984), estimating ability with wrong models and inaccurate parameters (Jones et al. 1984), evaluating simulation results of IRT ability estimation (Thissen and Wainer 1984; Thissen et al. 1984), and confidence envelopes for IRT (Thissen and Wainer 1990). Wainer (1983) also published an article discussing IRT and CAT, which he described as a coming technological revolution. Thissen and Wainer (1985) followed up on Lord's earlier work, discussing the estimation of the c parameter in IRT. Wainer and Thissen (1987) used the 1PL, 2PL, and 3PL models to fit simulated data and study accuracy and efficiency of robust estimators of ability. For short tests, simple models and robust estimators best fit the data, and for longer tests more complex models fit well, but using robust estimation with Bayesian priors resulted in substantial shrinkage. Testlet theory was the subject of Wainer and Lewis (1990).

Mislevy has also made numerous contributions to IRT, introducing Bayes modal estimation (1986b) in 1PL, 2PL, and 3PL IRT models, providing details of an expectation-maximization (EM) algorithm using two-stage modal priors, and in a simulation study, demonstrated improvement in estimation. Additionally he wrote on Bayesian treatment of latent variables in sample surveys (Mislevy 1986a). Most significantly, Mislevy (1984) developed the first version of a model that would later become the standard analytic approach for the National Assessment of Educational Progress (NAEP) and virtually all other large scale international survey assessments (see also Beaton and Barone's Chap. 8 and Chap. 9 by Kirsch et al. in this volume on the history of adult literacy assessments at ETS). Mislevy (1987a) also introduced application of empirical Bayes procedures, using auxililary information about test takers, to increase the precision of item parameter estimates. He illustrated the procedures with data from the Profile of American Youth survey. He also wrote (1988) on using auxilliary information about items to estimate Rasch model item difficulty parameters and authored and coauthored other papers, several with Sheehan, dealing with use of auxiliary/collateral information with Bayesian procedures for estimation in IRT models (Mislevy 1988; Mislevy and Sheehan 1989b; Sheehan and Mislevy 1988). Another contribution Mislevy made (1986c) is a comprehensive discussion of FA models for test item data with reference to relationships to IRT models and work on extending currently available models. Mislevy and Sheehan (1989a) discussed consequences of uncertainty in IRT linking and the information matrix in latent variable models. Mislevy and Wu (1988) studied the effects of missing responses and discussed the implications for ability and item parameter estimation relating to alternate test forms, targeted testing, adaptive testing, time limits, and omitted responses. Mislevy also coauthored a book chapter describing a hierarchical IRT model (Mislevy and Bock 1989).

Many other ETS staff members made important contributions. Jones (1984a, b) used asymptotic theory to compute approximations to standard errors of Bayesian and robust estimators studied by Wainer and Thissen. Rosenbaum wrote on testing the local independence assumption (1984) and showed (1985) that the observable distributions of item responses must satisfy certain constraints when two groups of test takers have generally different ability to respond correctly under a unidimensional

IRT model. Dorans (1985) contributed a book chapter on item parameter invariance. Douglass et al. (1985) studied the use of approximations to the 3PL model in item parameter estimation and equating. Methodology for comparing distributions of item responses for two groups was contributed by Rosenbaum (1985). McKinley and Mills (1985) compared goodness of fit statistics in IRT models, and Kingston and Dorans (1985) explored item-ability regressions as a tool for model fit.

Tatsuoka (1986) used IRT in developing a probabilistic model for diagnosing and classifying cognitive errors. While she held a postdoctoral fellowship at ETS, Lynne Steinberg coathored (Thissen and Steinberg 1986) a widely used and cited taxonomy of IRT models, which mentions, among other contributions, that the expressions they use suggest additional, as yet undeveloped, models. One explicitly suggested is basically the two-parameter partial credit (2PPC) model developed by Yen (see Yen and Fitzpatrick 2006) and the equivalent generalized partial credit (GPC) model developed by Muraki (1992a), both some years after the Thissen-Steinberg article. Rosenbaum (1987) developed and applied three nonparametric methods for comparisons of the shapes of two item characteristic surfaces. Stocking (1989) developed two methods of online calibration for CAT tests and compared them in a simulation using item parameters from an operational assessment. She also (1990) conducted a study on calibration using different ability distributions, concluding that the best estimation for applications that are highly dependent on item parameters, such as CAT and test construction, resulted when the calibration sample contained widely dispersed abilities. McKinley (1988) studied six methods of combining item parameter estimates from different samples using real and simulated item response data. He stated, "results support the use of covariance matrix-weighted averaging and a procedure that involves sample-size-weighted averaging of estimated item characteristic curves at the center of the ability distribution." (abstract). McKinley also (1989a) developed and evaluated with simulated data a confirmatory multidimensional IRT (MIRT) model. Yamamoto (1989) developed HYBRID, a model combining IRT and LC analysis, and used it to "present a structure of cognition by a particular response vector or set of them" (abstract). The software developed by Yamamoto was also used in a paper by Mislevy and Verhelst (1990) that presented an approach to identifying latent groups of test takers. Folk (Folk and Green 1989) coauthored a work on adaptive estimation when the unidimensionality assumption of IRT is violated.

5.3.2 *IRT Software Development and Evaluation*

With respect to IRT software, Mislevy and Stocking (1987) provided a guide to use of the LOGIST and BILOG computer programs that was very helpful to new users of IRT in applied settings. Mislevy, of course, was one of the developers of BILOG (Mislevy and Bock 1983). Wingersky (1987), the primary developer of LOGIST, developed and evaluated, with real and artificial data, a one-stage version of LOGIST for use when estimates of item parameters but not test-taker abilities are required.

Item parameter estimates were not as good as those from LOGIST, and the one-stage software did not reduce computer costs when there were missing data in the real dataset. Stocking (1989) conducted a study of estimation errors and relationship to properties of the test or item set being calibrated; she recommended improvements to the methods used in the LOGIST and BILOG programs. Yamamoto (1989) produced the HYBIL software for the HYBRID model and mixture IRT we referred to above. Both HYBIL and BILOG utilize marginal ML estimation, whereas LOGIST uses joint ML estimation methods.

5.3.3 Explanation, Evaluation, and Application of IRT Models

During this decade ETS scientists began exploring the use of IRT models with operational test data and producing works explaining IRT models for potential users. Applications of IRT were seen in many ETS testing programs.

Lord's book, *Applications of Item Response Theory to Practical Testing Problems* (1980a), presented much of the current IRT theory in language easily understood by many practitioners. It covered basic concepts, comparison to CTT methods, relative efficiency, optimal number of choices per item, flexilevel tests, multistage tests, tailored testing, mastery testing, estimating ability and item parameters, equating, item bias, omitted responses, and estimating true score distributions. Lord (1980b) also contributed a book chapter on practical issues in tailored testing.

Bejar illustrated use of item characteristic curves in studying dimensionality (1980), and he and Wingersky (1981, 1982) applied IRT to the Test of Standard Written English, concluding that using the 3PL model and IRT preequating "did not appear to present problems" (abstract). Kingston and Dorans (1982) applied IRT to the *GRE*® Aptitude Test, stating that "the most notable finding in the analytical equatings was the sensitivity of the precalibration design to practice effects on analytical items … this might present a problem for any equating design" (abstract). Kingston and Dorans (1982a) used IRT in the analysis of the effect of item position on test taker responding behavior. They also (1982b) compared IRT and conventional methods for equating the GRE Aptitude Test, assessing the reasonableness of the assumptions of item response theory for GRE item types and test taker populations, and finding that the IRT precalibration design was sensitive to practice effects on analytical items. In addition, Kingston and Dorans (1984) studied the effect of item location on IRT equating and adaptive testing, and Dorans and Kingston (1985) studied effects of violation of the unidimensionality assumption on estimation of ability and item parameters and on IRT equating with the GRE Verbal Test, concluding that there were two highly correlated verbal dimensions that had an effect on equating, but that the effect was slight. Kingston et al. (1985) compared IRT to conventional equating of the Graduate Management Admission Test (GMAT) and concluded that violation of local independence of this test had little effect on the equating results (they cautioned that further study was necessary before using other IRT-based procedures with the test). McKinley and Kingston (1987) investigated using IRT equating for the GRE Subject Test in Mathematics and also studied the unidimensionality and model fit assumptions, concluding that the test was reasonably unidimensional and the 3PL model provided reasonable fit to the data.

Cook, Eignor, Petersen and colleagues wrote several explanatory papers and conducted a number of studies of application of IRT on operational program data, studying assumptions of the models, and various aspects of estimation and equating (Cook et al. 1985a, c, 1988a, b; Cook and Eignor 1985, 1989; Eignor 1985; Stocking 1988). Cook et al. (1985b, 1988c) examined effects of curriculum (comparing results for students tested before completing the curriculum with students tested after completing it) on stability of CTT and IRT difficulty parameter estimates, effects on equating, and the dimensionality of the tests. Cook and colleagues (Wingersky et al. 1987), using simulated data based on actual SAT item parameter estimates, studied the effect of anchor item characteristics on IRT true-score equating.

Kreitzberg and Jones (1980) presented results of a study of CAT using the Broad-Range Tailored Test and concluded,"computerized adaptive testing is ready to take the first steps out of the laboratory environment and find its place in the educational community" (abstract). Scheuneman (1980) produced a book chapter on LT theory and item bias. Hicks (1983) compared IRT equating with fixed versus estimated parameters and three "conventional" equating methods using *TOEFL*® test data, concluding that fixing the *b* parameters to pretest values (essentially this is what we now call preequating) is a "very acceptable option." She followed up (1984) with another study in which she examined controlling for native language and found this adjustment resulted in increased stability for one test section but a decrease in another section. Peterson, Cook, and Stocking (1983) studied several equating methods using SAT data and found that for reasonably parallel tests, linear equating methods perform adequately, but when tests differ somewhat in content and length, methods based on the three-parameter logistic IRT model lead to greater stability of equating results. In a review of research on IRT and conventional equating procedures, Cook and Petersen (1987) discussed how equating methods are affected by sampling error, sample characteristics, and anchor item characteristics, providing much useful information for IRT users.

Cook coauthored a book chapter (Hambleton and Cook 1983) on robustness of IRT models, including effects of test length and sample size on precision of ability estimates. Several ETS staff members contributed chapters to that same edited book on applications of item response theory (Hambleton 1983). Bejar (1983) contributed an introduction to IRT and its assumptions; Wingersky (1983) a chapter on the LOGIST computer program; Cook and Eignor (1983) on practical considerations for using IRT in equating. Tatsuoka coauthored on appropriateness indices (Harnisch and Tatsuoka 1983); and Yen wrote on developing a standardized test with the 3PL model (1983); both Tatsuoka and Yen later joined ETS.

Lord and Wild (1985) compared the contribution of the four verbal item types to measurement accuracy of the GRE General Test, finding that the reading comprehension item type measures something slightly different from what is measured by sentence completion, analogy, or antonym item types. Dorans (1986) used IRT to study the effects of item deletion on equating functions and the score distribution on the SAT, concluding that reequating should be done when an item is dropped. Kingston and Holland (1986) compared equating errors using IRT and several other equating methods, and several equating designs, for equating the GRE General Test, with varying results depending on the specific design and method. Eignor and Stocking (Eignor and Stocking 1986; Stocking and Eignor 1986) conducted two studies to investigate whether calibration or linking methods might be reasons for poor equating results on the SAT. In the first study they used actual data, and in the second they used simulations, concluding that a combination of differences in true mean ability and multidimensionality were consistent with the real data. Eignor et al. (1986) studied the potential of a new plotting procedures for assessing fit to the 3PL model using SAT and TOEFL data. Wingersky and Sheehan (1986) also wrote on fit to IRT models, using regressions of item scores onto observed (number correct) scores rather than the previously used method of regressing onto estimated ability.

Bejar (1990), using IRT, studied an approach to psychometric modeling that explicitly incorporates information on the mental models test takers use in solving an item, and concluded that it is not only workable, but also necessary for future developments in psychometrics. Kingston (1986) used full information FA to estimate difficulty and discrimination parameters of a MIRT model for the GMAT, finding there to be dominant first dimensions for both the quantitative and verbal measures. Mislevy (1987b) discussed implications of IRT developments for teacher certification. Mislevy (1989) presented a case for a new test theory combining modern cognitive psychology with modern IRT. Sheehan and Mislevy (1990) wrote on the integration of cognitive theory and IRT and illustrated their ideas using the Survey of Young Adult Literacy data. These ideas seem to be the first appearance of a line of research that continues today. The complexity of these models, built to integrate cognitive theory and IRT, evolved dramatically in the twenty-first century due to rapid increase in computational capabilities of modern computers and developments in understanding problem solving. Lawrence coauthored a paper (Lawrence and Dorans 1988) addressing the sample invariance properties of four equating methods with two types of test-taker samples (matched on anchor test score distributions or taken from different administrations and differing in ability). Results for IRT, Levine, and equipercentile methods differed for the two types of samples, whereas the Tucker observed score method did not. Henning (1989) discussed the appropriateness of the Rasch model for multiple-choice data, in response to an article that questioned such appropriateness. McKinley (1989b) wrote an explanatory article for potential users of IRT. McKinley and Schaeffer (1989) studied an IRT equating method for the GRE designed to reduce the overlap on test forms. Bejar et al. (1989), in a paper on methods used for patient management items in medical licensure testing, outlined recent developments and introduced a procedure that integrates those developments with IRT. Boldt (1989) used LC analysis to study the dimensionality of the TOEFL and assess whether different dimensions were necessary to fit models to diverse groups of test takers. His findings were that a single dimension LT model fits TOEFL data well but "suggests the use of a restrictive assumption of proportionality of item response curves" (p. 123).

In 1983, ETS assumed the primary contract for NAEP, and ETS psychometricians were involved in designing analysis procedures, including the use of an IRT-based latent regression model using ML estimation of population parameters from observed item responses without estimating ability parameters for test takers (e.g., Mislevy 1984, 1991). Asymptotic standard errors and tests of fit, as well as approximate solutions of the integrals involved, were developed in Mislevy's 1984 article. With leadership from Messick (Messick 1985; Messick et al. 1983), a large team of ETS staff developed a complex assessment design involving new analysis procedures for direct estimation of average achievement of groups of students. Zwick (1987) studied whether the NAEP reading data met the unidimensionality assumption underlying the IRT scaling procedures. Mislevy (1991) wrote on making inferences about latent variables from complex samples, using IRT proficiency estimates as an example and illustrating with NAEP reading data. The innovations introduced include the linking of multiple test forms using IRT, a task that would be virtually impossible without IRT-based methods, as well as the intregration of IRT with a regression-based population model that allows the prediction of an ability prior, given background data collected in student questionnaires along with the cogntive NAEP tests.

5.4 Advanced Item Response Modeling: The 1990s

During the 1990s, the use of IRT in operational testing programs expanded considerably. IRT methodology for dichotomous item response data was well developed and widely used by the end of the 1980s. In the early years of the 1990s, models for polytomous item response data were developed and began to be used in operational programs. Muraki (1990) developed and illustrated an IRT model for fitting a polytomous item response theory model to Likert-type data. Muraki (1992a) also developed the GPC model, which has since become one of the most widely used models for polytomous IRT data. Concomitantly, before joining ETS, Yen[5] developed the 2PPC model that is identical to the GPC, differing only in the parameterization incorporated into the model. Muraki (1993) also produced an article detailing the IRT information functions for the GPC model. Chang and Mazzeo (1994) discussed item category response functions (ICRFs) and the item response functions (IRFs), which are weighted sums of the ICRFs, of the partial credit and graded response models. They showed that if two polytomously scored items have the same IRF, they must have the same number of categories that have the same ICRFs. They also discussed theoretical and practical implications. Akkermans and Muraki (1997) studied and described characteristics of the item information and discrimination functions for partial credit items.

[5]Developed in 1991 (as cited in Yen and Fitzpatrick 2006), about the same time as Muraki was developing the GPC model.

In work reminiscent of the earlier work of Green and Lord, Gitomer and Yamamoto (1991) described HYBRID (Yamamoto 1989), a model that incorporates both LT and LC components; these authors, however, defined the latent classes by a cognitive analysis of the understanding that individuals have for a domain. Yamamoto and Everson (1997) also published a book chapter on this topic. Bennett et al. (1991) studied new cognitively sensitive measurement models, analyzing them with the HYBRID model and comparing results to other IRT methodology, using partial-credit data from the GRE General Test. Works by Tatsuoka (1990, 1991) also contributed to the literature relating IRT to cognitive models. The integration of IRT and a person-fit measure as a basis for rule space, as proposed by Tatsuoka, allowed in-depth examinations of items that require multiple skills. Sheehan (1997) developed a tree-based method of proficiency scaling and diagnostic assessment and applied it to developing diagnostic feedback for the SAT I Verbal Reasoning Test. Mislevy and Wilson (1996) presented a version of Wilson's Saltus model, an IRT model that incorporates developmental stages that may involve discontinuities. They also demonstrated its use with simulated data and an example of mixed number subtraction.

The volume *Test Theory for a New Generation of Tests* (Frederiksen et al. 1993) presented several IRT-based models that anticipated a more fully integrated approach providing information about measurement qualities of items as well as about complex latent variables that align with cognitive theory. Examples of these advances are the chapters by Yamamoto and Gitomer (1993) and Mislevy (1993a).

Bradlow (1996) discussed the fact that, for certain values of item parameters and ability, the information about ability for the 3PL model will be negative and has consequences for estimation—a phenomenon that does not occur with the 2PL. Pashley (1991) proposed an alternative to Birnbaum's 3PL model in which the asymptote parameter is a linear component within the logit of the function. Zhang and Stout (1997) showed that Holland's (1990b) conjecture that a quadratic form for log manifest probabilities is a limiting form for all smooth unidimensional IRT models does not always hold; these authors provided counterexamples and suggested that only under strong assumptions can this conjecture be true.

Holland (1990a) published an article on the sampling theory foundations of IRT models. Stocking (1990) discussed determining optimum sampling of test takers for IRT parameter estimation. Chang and Stout (1993) showed that, for dichotomous IRT models, under very general and nonrestrictive nonparametric assumptions, the posterior distribution of test taker ability given dichotomous responses is approximately normal for a long test. Chang (1996) followed up with an article extending this work to polytomous responses, defining a global information function, and he showed the relationship of the latter to other information functions.

Mislevy (1991) published on randomization-based inference about latent variables from complex samples. Mislevy (1993b) also presented formulas for use with Bayesian ability estimates. While at ETS as a postdoctoral fellow, Roberts coauthored works on the use of unfolding[6] (Roberts and Laughlin 1996). A parametric IRT model for unfolding dichotomously or polytomously scored responses, called the graded unfolding model (GUM), was developed; a subsequent recovery simulation showed that reasonably accurate estimates could be obtained. The applicability of the GUM to common attitude testing situations was illustrated with real data on student attitudes toward capital punishment. Roberts et al. (2000) described the generalized GUM (GGUM), which introduced a parameter to the model, allowing for variation in discrimination across items; they demonstrated the use of the model with real data.

Wainer and colleagues wrote further on testlet response theory, contributing to issues of reliability of testlet-based tests (Sireci et al. 1991). These authors also developed, and illustrated using operational data, statistical methodology for detecting differential item functioning (DIF) in testlets (Wainer et al. 1991). Thissen and Wainer (1990) also detailed and illustrated how *confidence envelopes* could be formed for IRT models. Bradlow et al. (1999) developed a Bayesian IRT model for testlets and compared results with those from standard IRT models using a released SAT dataset. They showed that degree of precision bias was a function of testlet effects and the testlet design. Sheehan and Lewis (1992) introduced, and demonstrated with actual program data, a procedure for determining the effect of testlet nonequivalence on the operating characteristics of a computerized mastery test based on testlets.

Lewis and Sheehan (1990) wrote on using Bayesian decision theory to design computerized mastery tests. Contributions to CAT were made in a book, *Computer Adaptive Testing: A Primer*, edited by Wainer et al. (1990a) with chapters by ETS psychometricians: "Introduction and History" (Wainer 1990), "Item Response Theory, Item Calibration and Proficiency Estimation" (Wainer and Mislevy 1990); "Scaling and Equating" (Dorans 1990); "Testing Algorithms" (Thissen and Mislevy 1990); "Validity" (Steinberg et al. 1990); "Item Pools" (Flaugher 1990); and "Future Challenges" (Wainer et al. 1990b). Automated item selection (AIS) using IRT was the topic of two publications (Stocking et al. 1991a, b). Mislevy and Chang (2000) introduced a term to the expression for probability of response vectors to deal with item selection in CAT, and to correct apparent incorrect response pattern probabilities in the context of adaptive testing. Almond and Mislevy (1999) studied graphical modeling methods for making inferences about multifaceted skills and models in an IRT CAT environment, and illustrated in the context of language testing.

In an issue of an early volume of *Applied Measurement in Education*, Eignor et al. (1990) expanded on their previous studies (Cook et al. 1988b) comparing IRT

---

[6]Unfolding models are proximity IRT models developed for assessments with binary disagree-agree or graded disagree-agree responses. Responses on these assessments are not necessarily cumulative and one cannot assume that higher levels of the latent trait will lead to higher item scores and thus to higher total test scores. Unfolding models predict item scores and total scores on the basis of the distances between the test taker and each item on the latent continuum (Roberts n.d.).

equating with several non-IRT methods and with different sampling designs. In another article in that same issue, Schmitt et al. (1990) reported on the sensitivity of equating results to sampling designs; Lawrence and Dorans (1990) contributed with a study of the effect of matching samples in equating with an anchor test; and Livingston et al. (1990) also contributed on sampling and equating methodolgy to this issue.

Zwick (1990) published an article showing when IRT and Mantel-Haenszel definitions of DIF coincide. Also in the DIF area, Dorans and Holland (1992) produced a widely disseminated and used work on the Mantel-Haenszel (MH) and standardization methodologies, in which they also detailed the relationship of the MH to IRT models. Their methodology, of course, is the mainstay of DIF analyses today, at ETS and at other institutions. Muraki (1999) described a stepwise DIF procedure based on the multiple group PC model. He illustrated the use of the model using NAEP writing trend data and also discussed item parameter drift. Pashley (1992) presented a graphical procedure, based on IRT, to display the location and magnitude of DIF along the ability continuum.

MIRT models, although developed earlier, were further developed and illustrated with operational data during this decade; McKinley coauthored an article (Reckase and McKinley 1991) describing the discrimination parameter for these models. Muraki and Carlson (1995) developed a multidimensional graded response (MGR) IRT model for polytomously scored items, based on Samejima's normal ogive GR model. Relationships to the Reckase-McKinley and FA models were discussed, and an example using NAEP reading data was presented and discussed. Zhang and Stout (1999a, b) described models for detecting dimensionality and related them to FA and MIRT.

Lewis coauthored publications (McLeod and Lewis 1999; McLeod et al. 2003) with a discussion of person-fit measures as potential ways of detecting memorization of items in a CAT environment using IRT, and introduced a new method. None of the three methods showed much power to detect memorization. Possible methods of altering a test when the model becomes inappropriate for a test taker were discussed.

5.4.1 IRT Software Development and Evaluation

During this period, Muraki developed the PARSCALE computer program (Muraki and Bock 1993) that has become one of the most widely used IRT programs for polytomous item response data. At ETS it has been incorporated into the GENASYS software used in many operational programs to this day. Muraki (1992b) also developed the RESGEN software, also widely used, for generating simulated polytomous and dichotomous item response data.

Many of the research projects in the literature reviewed here involved development of software for estimation of newly developed or extended models. Some examples involve Yamamoto's (1989) HYBRID model, the MGR model (Muraki and Carlson 1995) for which Muraki created the POLYFACT software, and the Saltus model (Mislevy and Wilson 1996) for which an EM algorithm-based program was created.

5.4.2 *Explanation, Evaluation, and Application of IRT Models*

In this decade ETS researchers continued to provide explanations of IRT models for users, to conduct research evaluating the models, and to use them in testing programs in which they had not been previously used. The latter activity is not emphasized in this section as it was for sections on previous decades because of the sheer volume of such work and the fact that it generally involves simply applying IRT to testing programs, whereas in previous decades the research made more of a contribution, with recommendations for practice in general. Although such work in the 1990s contributed to improving the methodology used in specific programs, it provided little information that can be generalized to other programs. This section, therefore covers research that is more generalizable, although illustrations may have used specific program data.

Some of this research provided new information about IRT scaling. Donoghue (1992), for example, described the common misconception that the partial credit and GPC IRT model item category functions are symmetric, helping explain characteristics of items in these models for users of them. He also (1993) studied the information provided by polytomously scored NAEP reading items and made comparisons to information provided by dichotomously scored items, demonstrating how other users can use such information for their own programs. Donoghue and Isham (1998) used simulated data to compare IRT and other methods of detecting item parameter drift. Zwick (1991), illustrating with NAEP reading data, presented a discussion of issues relating to two questions: "What can be learned about the effects of item order and context on invariance of item parameter estimates?" and "Are common-item equating methods appropriate when measuring trends in educational growth?" Camili et al. (1993) studied scale shrinkage in vertical equating, comparing IRT with equipercentile methods using real data from NAEP and another testing program. Using IRT methods, variance decreased from fall to spring testings, and also from lower- to upper-grade levels, whereas variances have been observed to increase across grade levels for equipercentile equating. They discussed possible reasons for scale shrinkage and proposed a more comprehensive, model-based approach to establishing vertical scales. Yamamoto and Everson (1997) estimated IRT parameters using TOEFL data and Yamamoto's extended HYBRID model (1989), which uses a combination of IRT and LC models to characterize when test takers switch from ability-based to random responses. Yamamoto studied effects of time limits on speededness, finding that this model estimated the parameters more accurately than the usual IRT model. Yamamoto and Everson (1995) using three different sets of actual test data, found that the HYBRID model successfully determined the switch point in the three datasets. Liu coauthored (Lane et al.

1995) an article in which mathematics performance-item data were used to study the assumptions of and stability over time of item parameter estimates using the GR model. Sheehan and Mislevy (1994) used a tree-based analysis to examine the relationship of three types of item attributes (constructed-response [CR] vs. multiple choice [MC], surface features, aspects of the solution process) to operating characteristics (using 3PL parameter estimates) of computer-based *PRAXIS®* mathematics items. Mislevy and Wu (1996) built on their previous research (1988) on estimation of ability when there are missing data due to assessment design (alternate forms, adaptive testing, targeted testing), focusing on using Bayesian and direct likelihood methods to estimate ability parameters.

Wainer et al. (1994) examined, in an IRT framework, the comparability of scores on tests in which test takers choose which CR prompts to respond to, and illustrated using the College Board *Advanced Placement®* Test in Chemistry.

Zwick et al. (1995) studied the effect on DIF statistics of fitting a Rasch model to data generated with a 3PL model. The results, attributed to degredation of matching resulting from Rasch model ability estimation, indicated less sensitive DIF detection.

In 1992, special issues of the *Journal of Educational Measurement* and the *Journal of Educational Statistics* were devoted to methodology used by ETS in NAEP, including the NAEP IRT methodology. Beaton and Johnson (1992), and Mislevy et al. (1992b) detailed how IRT is used and combined with the plausible values methodology to estimate proficiencies for NAEP reports. Mislevy et al. (1992a) wrote on how population characteristics are estimated from sparse matrix samples of item responses. Yamamoto and Mazzeo (1992) described IRT scale linking in NAEP.

5.5 IRT Contributions in the Twenty-First Century

5.5.1 Advances in the Development of Explanatory and Multidimensional IRT Models

Multidimensional models and dimensionality considerations continued to be a subject of research at ETS, with many more contributions than in the previous decades. Zhang (2004) proved that, when simple structure obtains, estimation of unidimensional or MIRT models by joint ML yields identical results, but not when marginal ML is used. He also conducted simulations and found that, with small numbers of items, MIRT yielded more accurate item parameter estimates but the unidimensional approach prevailed with larger numbers of items, and that when simple structure does not hold, the correlations among dimensions are overestimated.

A genetic algorithm was used by Zhang (2005b) in the maximization step of an EM algorithm to estimate parameters of a MIRT model with complex, rather than simple, structure. Simulated data suggested that this algorithm is a promising approach to estimation for this model. Zhang (2007) also extended the theory of conditional covariances to the case of polytomous items, providing a theoretical foundation for study of dimensionality. Several estimators of conditional covariance were constructed, including the case of complex incomplete designs such as those used in NAEP. He demonstrated use of the methodology with NAEP reading assessment data, showing that the dimensional structure is consistent with the purposes of reading that define NAEP scales, but that the degree of multidimensionality is weak in those data.

Haberman et al. (2008) showed that MIRT models can be based on ability distributions that are multivariate normal or multivariate polytomous, and showed, using empirical data, that under simple structure the two cases yield comparable results in terms of model fit, parameter estimates, and computing time. They also discussed numerical methods for use with the two cases.

Rijmen wrote two papers dealing with methodology relating to MIRT models, further showing the relationship between IRT and FA models. As discussed in the first section of this chapter, such relationships were shown for more simple models by Bert Green and Fred Lord in the 1950s. In the first (2009) paper, Rijmen showed how an approach to full information ML estimation can be placed into a graphical model framework, allowing for derivation of efficient estimation schemes in a fully automatic fashion. This avoids tedious derivations, and he demonstrated the approach with the bifactor and a MIRT model with a second-order dimension. In the second paper, (2010) Rijmen studied three MIRT models for testlet-based tests, showing that the second-order MIRT model is formally equivalent to the testlet model, which is a bifactor model with factor loadings on the specific dimensions restricted to being proportional to the loadings on the general factor.

M. von Davier and Carstensen (2007) edited a book dealing with multivariate and mixture distribution Rasch models, including extensions and applications of the models. Contributors to this book included: Haberman (2007b) on the interaction model; M. von Davier and Yamamoto (2007) on mixture distributions and hybrid Rasch models; Mislevy and Huang (2007) on measurement models as narrative structures; and Boughton and Yamamoto (2007) on a hybrid model for test speededness.

Antal (2007) presented a coordinate-free approach to MIRT models, emphasizing understanding these models as extensions of the univariate models. Based on earlier work by Rijmen et al. (2003), Rijmen et al. (2013) described how MIRT models can be embedded and understood as special cases of generalized linear and nonlinear mixed models.

Haberman and Sinharay (2010) studied the use of MIRT models in computing subscores, proposing a new statistical approach to examining when MIRT model subscores have added value over total number correct scores and subscores based on CTT. The MIRT-based methods were applied to several operational datasets, and results showed that these methods produce slightly more accurate scores than CTT-based methods.

Rose et al. (2010) studied IRT modeling of nonignorable missing item responses in the context of large-scale international assessments, comparing using CTT and simple IRT models, the usual two treatments (missing item responses as wrong, or as not administered), with two MIRT models. One model used indicator variables as a dimension to designate where missing responses occurred, and the other was a multigroup MIRT model with grouping based on a within-country stratification by the amount of missing data. Using both simulated and operational data, they demonstrated that a simple IRT model ignoring missing data performed relatively well when the amount of missing data was moderate, and the MIRT-based models only outperformed the simple models with larger amounts of missingness, but they yielded estimates of the correlation of missingness with ability estimates and improved the reliability of the latter.

van Rijn and Rijmen (2015) provided an explanation of a "paradox" that in some MIRT models answering an additional item correctly can result in a decrease in the test taker's score on one of the latent variables, previously discussed in the psychometric literature. These authors showed clearly how it occurs and also pointed out that it does not occur in testlet (restricted bifactor) models.

ETS researchers also continued to develop CAT methodology. Yan et al. (2004b) introduced a nonparametric tree-based algorithm for adaptive testing and showed that it may be superior to conventional IRT methods when the IRT assumptions are not met, particularly in the presence of multidimensionality. While at ETS, Weissman coauthored an article (Belov et al. 2008) in which a new CAT algorithm was developed and tested in a simulation using operational test data. Belov et al. showed that their algorithm, compared to another algorithm incorporating content constraints had lower maximum item exposure rates, higher utilization of the item pool, and more robust ability estimates when high (low) ability test takers performed poorly (well) at the beginning of testing.

The second edition of *Computerized Adaptive Testing: A Primer* (Wainer et al. 2000b) was published and, as in the first edition (Wainer et al. 1990a), many chapters were authored or coauthored by ETS researchers (Dorans 2000; Flaugher 2000; Steinberg et al. 2000; Thissen and Mislevy 2000; Wainer 2000; Wainer et al. 2000c; Wainer and Eignor 2000; Wainer and Mislevy 2000). Xu and Douglas (2006) explored the use of nonparametric IRT models in CAT; derivatives of ICCs required by the Fisher information criterion might not exist for these models, so alternatives based on Shannon entropy and Kullback-Leibler information (which do not require derivatives) were proposed. For long tests these methods are equivalent to the maximum Fisher information criterion, and simulations showed them to perform similarly, and much better than random selection of items.

Diagnostic models for assessment including cognitive diagnostic (CD) assessment, as well as providing diagnostic information from common IRT models, continued to be an area of research by ETS staff. Yan et al. (2004a), using a mixed number subtraction dataset, and cognitive research originally developed by Tatsuoka and her colleagues, compared several models for providing diagnostic information on score reports, including IRT and other types of models, and characterized the kinds of problems for which each is suited. They provided a general Bayesian psychometric framework to provide a common language, making it easier to appreciate the differences. M. von Davier (2008a) presented a class of general diagnostic (GD) models that can be estimated by marginal ML algorithms; that allow for both dichotomous and polytomous items, compensatory and noncompensatory models; and subsume many common models including unidimensional and multidimensional Rasch models, 2PL, PC and GPC, facets, and a variety of skill profile models. He demonstrated the model using simulated as well as TOEFL iBT data.

Xu (2007) studied monotonicity properties of the GD model and found that, like the GPC model, monotonicity obtains when slope parameters are restricted to be equal, but does not when this restriction is relaxed, although model fit is improved. She pointed out that trade offs between these two variants of the model should be considerred in practice. M. von Davier (2007) extended the GD model to a hierarchical model and further extended it to the mixture general diagnostic (MGD) model (2008b), which allows for estimation of diagnostic models in multiple known populations as well as discrete unknown, or not directly observed mixtures of populations.

Xu and von Davier (2006) used a MIRT model specified in the GD model framework with NAEP data and verified that the model could satisfactorily recover parameters from a sparse data matrix and could estimate group characteristics for large survey data. Results under both single and multiple group assumptions and comparison with the NAEP model results were also presented. The authors suggested that it is possible to conduct cognitive diagnosis for NAEP proficiency data. Xu and von Davier (2008b) extended the GD model, employing a log-linear model to reduce the number of parameters to be estimated in the latent skill distribution. They extended that model (2008a) to allow comparison of constrained versus nonconstrained parameters across multiple populations, illustrating with NAEP data.

M. von Davier et al. (2008) discussed models for diagnosis that combine features of MIRT, FA, and LC models. Hartz and Roussos (2008)[7] wrote on the fusion model for skills diagnosis, indicating that the development of the model produced advancements in modeling, parameter estimation, model fitting methods, and model fit evaluation procedures. Simulation studies demonstrated the accuracy of the estimation procedure, and effectiveness of model fitting and model fit evaluation procedures. They concluded that the model is a promising tool for skills diagnosis that merits further research and development.

Linking and equating also continue to be important topics of ETS research. In this section the focus is research on IRT-based linking/equating methods. M. von Davier and von Davier (2007, 2011) presented a unified approach to IRT scale linking and transformation. Any linking procedure is viewed as a restriction on the item parameter space, and then rewriting the log-likelihood function together with implementation of a maximization procedure under linear or nonlinear restrictions accomplishes the linking. Xu and von Davier (2008c) developed an IRT linking approach for use with the GD model and applied the proposed approach to NAEP data. Holland and Hoskens (2002) developed an approach viewing CTT as a first-order version of IRT and the latter as detailed elaborations of CTT, deriving general results for the prediction of true scores from observed scores, leading to a new view

[7]While these authors were not ETS staff members, this report was completed under the auspices of the External Diagnostic Research Team, supported by ETS.

of linking tests not designed to be linked. They illustrated the theory using simulated and actual test data. M. von Davier et al. (2011) presented a model that generalizes approaches by Andersen (1985), and Embretson (1991), respectively, to utilize MIRT in a multiple-population longitudinal context to study individual and group-level learning trajectories.

Research on testlets continued to be a focus at ETS, as well as research involving item families. Wang et al. (2002) extended the development of testlet models to tests comprising polytomously scored and/or dichotomously scored items, using a fully Bayesian method. They analyzed data from the Test of Spoken English (TSE) and the North Carolina Test of Computer Skills, concluding that the latter exhibited significant testlet effects, whereas the former did not. Sinharay et al. (2003) used a Bayesian hierarchical model to study item families, showing that the model can take into account the dependence structure built into the families, allowing for calibration of the family rather than the individual items. They introduced the family expected response function (FERF) to summarize the probability of a correct response to an item randomly generated from the family, and suggested a way to estimate the FERF.

Wainer and Wang (2000) conducted a study in which TOEFL data were fitted to an IRT testlet model, and for comparative purposes to a 3PL model. They found that difficulty parameters were estimated well with either model, but discrimination and lower asymptote parameters were biased when conditional independence was incorrectly assumed. Wainer also coauthored book chapters explaining methodology for testlet models (Glas et al. 2000; Wainer et al. 2000a).

Y. Li et al. (2010) used both simulated data and operational program data to compare the parameter estimation, model fit, and estimated information of testlets comprising both dichotomous and polytomous items. The models compared were a standard 2PL/GPC model (ignoring local item dependence within testlets) and a general dichotomous/polytomous testlet model. Results of both the simulation and real data analyses showed little difference in parameter estimation but more difference in fit and information. For the operational data, they also made comparisons to a MIRT model under a simple structure constraint, and this model fit the data better than the other two models.

Roberts et al. (2002) in a continuation of their research on the GGUM, studied the characteristics of marginal ML and expected a posteriori (EAP) estimates of item and test-taker parameter estimates, respectively. They concluded from simulations that accurate estimates could be obtained for items using 750–1000 test takers and for test takers using 15–20 items.

Checking assumptions, including the fit of IRT models to both the items and test takers of a test, is another area of research at ETS during this period. Sinharay and Johnson (2003) studied the fit of IRT models to dichotomous item response data in the framework of Bayesian posterior model checking. Using simulations, they studied a number of discrepancy measures and suggest graphical summaries as having a potential to become a useful psychometric tool. In further work on this model checking (Sinharay 2003, 2005, 2006; Sinharay et al. 2006) they discussed the model-checking technique, and IRT model fit in general, extended some aspects of it, demonstrated it with simulations, and discussed practical applications. Deng coauthored (de la Torre and Deng 2008) an article proposing a modification of the standardized log likelihood of the response vector measure of person fit in IRT models, taking into account test reliability and using resampling methods. Evaluating the method, they found type I error rates were close to the nominal and power was good, resulting in a conclusion that the method is a viable and promising approach.

Based on earlier work during a postdoctoral fellowship at ETS, M. von Davier and Molenaar (2003) presented a person-fit index for dichotomous and polytomous IRT and latent structure models. Sinharay and Lu (2008) studied the correlation between fit statistics and IRT parameter estimates; previous researchers had found such a correlation, which was a concern for practitioners. These authors studied some newer fit statistics not examined in the previous research, and found these new statistics not to be correlated with the item parameters. Haberman (2009b) discussed use of generalized residuals in the study of fit of 1PL and 2PL IRT models, illustrating with operational test data.

Mislevy and Sinharay coauthored an article (Levy et al. 2009) on posterior predictive model checking, a flexible family of model-checking procedures, used as a tool for studying dimensionality in the context of IRT. Factors hypothesized to influence dimensionality and dimensionality assessment are couched in conditional covariance theory and conveyed via geometric representations of multidimensionality. Key findings of a simulation study included support for the hypothesized effects of the manipulated factors with regard to their influence on dimensionality assessment and the superiority of certain discrepancy measures for conducting posterior predictive model checking for dimensionality assessment.

Xu and Jia (2011) studied the effects on item parameter estimation in Rasch and 2PL models of generating data from different ability distributions (normal distribution, several degrees of generalized skew normal distributions), and estimating parameters assuming these different distributions. Using simulations, they found for the Rasch model that the estimates were little affected by the fitting distribution, except for fitting a normal to an extremely skewed generating distribution; whereas for the 2PL this was true for distributions that were not extremely skewed, but there were computational problems (unspecified) that prevented study of extremely skewed distributions.

M. von Davier and Yamamoto (2004) extended the GPC model to enable its use with discrete mixture IRT models with partially missing mixture information. The model includes LC analysis and multigroup IRT models as special cases. An application to large-scale assessment mathematics data, with three school types as groups and 20% of the grouping data missing, was used to demonstrate the model.

M. von Davier and Sinharay (2010) presented an application of a stochastic approximation EM algorithm using a Metropolis-Hastings sampler to estimate the parameters of an item response latent regression (LR) model. These models extend IRT to a two-level latent variable model in which covariates serve as predictors of the conditional distribution of ability. Applications to data from NAEP were presented, and results of the proposed method were compared to results obtained using the current operational procedures.

Haberman (2004) discussed joint and conditional ML estimation for the dichotomous Rasch model, explored conditions for consistency and asymptotic normality, investigated effects of model error, estimated errors of prediction, and developed generalized residuals. The same author (Haberman 2005a) showed that if a parametric model for the ability distribution is not assumed, the 2PL and 3PL (but not 1PL) models have identifiability problems that impose restrictions on possible models for the ability distribution. Haberman (2005b) also showed that LC item response models with small numbers of classes are competitive with IRT models for the 1PL and 2PL cases, showing that computations are relatively simple under these conditions. In another report, Haberman (2006) applied adaptive quadrature to ML estimation for IRT models with normal ability distributions, indicating that this method may achieve significant gains in speed and accuracy over other methods.

Information about the ability variable when an IRT model has a latent class structure was the topic of Haberman (2007a) in another publication. He also discussed reliability estimates and sampling and provided examples. Expressions for bounds on log odds ratios involving pairs of items for unidimensional IRT models in general, and explicit bounds for 1PL and 2Pl models were derived by Haberman, Holland, and Sinharay (2007). The results were illustrated through an example of their use in a study of model-checking procedures. These bounds can provide an elementary basis for assessing goodness of fit of these models. In another publication, Haberman (2008) showed how reliability of an IRT scaled score can be estimated and that it may be obtained even though the IRT model may not be valid.

Zhang (2005a) used simulated data to investigate whether Lord's bias function and weighted likelihood estimation method for IRT ability with known item parameters would be effective in the case of unknown parameters, concluding that they may not be as effective in that case. He also presented algorithms and methods for obtaining the global maximum of a likelihood, or weighted likelihood (WL), function.

Lewis (2001) produced a chapter on expected response functions (ERFs) in which he discussed Bayesian methods for IRT estimation. Zhang and Lu (2007) developed a new corrected weighted likelihood (CWL) function estimator of ability in IRT models based on the asymptotic formula of the WL estimator; they showed via simulation that the new estimator reduces bias in the ML and WL estimators, caused by failure to take into account uncertainty in item parameter estimates. Y.-H. Lee and Zhang (2008) further studied this estimator and Lewis' ERF estimator under various conditions of test length and amount of error in item parameter estimates. They found that the ERF reduced bias in ability estimation under all conditions and the CWL under certain conditions.

Sinharay coedited a volume on psychometrics in the *Handbook of Statistics* (Rao and Sinharay 2007), and contributions included chapters by: M. von Davier et al. (2007) describing recent developments and future directions in NAEP statistical procedures; Haberman and von Davier (2007) on models for cognitively based skills; von Davier and Rost (2007) on mixture distribution IRT models; Johnson et al. (2007) on hierarchical IRT models; Mislevy and Levy (2007) on Bayesian approaches; Holland et al. (2007) on equating, including IRT.

D. Li and Oranje (2007) compared a new method for approximating standard error of regression effects estimates within an IRT-based regression model, with the imputation-based estimator used in NAEP. The method is based on accounting for complex samples and finite populations by Taylor series linearization, and these authors formally defined a general method, and extended it to multiple dimensions. The new method was compared to the NAEP imputation-based method.

Antal and Oranje (2007) described an alternative numerical integration applicable to IRT and emphasized its potential use in estimation of the LR model of NAEP. D. Li, Oranje, and Jiang (2007) discussed parameter recovery and subpopulation proficiency estimation using the hierarchical latent regression (HLR) model and made comparisons with the LR model using simulations. They found the regression effect estimates were similar for the two models, but there were substantial differences in the residual variance estimates and standard errors, especially when there was large variation across clusters because a substantial portion of variance is unexplained in LR.

M. von Davier and Sinharay (2004) discussed stochastic estimation for the LR model, and Sinharay and von Davier (2005) extended a bivariate approach that represented the gold standard for estimation to allow estimation in more than two dimensions. M. von Davier and Sinharay (2007) presented a Robbins-Monro type stochastic approximation algorithm for LR IRT models and applied this approach to NAEP reading and mathematics data.

5.6 IRT Software Development and Evaluation

Wang et al. (2001, 2005) produced SCORIGHT, a program for scoring tests composed of testlets. M. von Davier (2008a) presented stand-alone software for multidimensional discrete latent trait (MDLT) models that is capable of marginal ML estimation for a variety of multidimensional IRT, mixture IRT, and hierarchical IRT models, as well as the GD approach. Haberman (2005b) presented a stand-alone general software for MIRT models. Rijmen (2006) presented a MATLAB toolbox utilizing tools from graphical modeling and Bayesian networks that allows estimation of a range of MIRT models.

5.6.1 Explanation, Evaluation, and Application of IRT Models

For the fourth edition of *Educational Measurement* edited by Brennan, authors Yen and Fitzpatrick (2006) contributed the chapter on IRT, providing a great deal of information useful to both practictioners and researchers. Although other ETS staff were authors or coauthors of chapters in this book, they did not focus on IRT methodology, per se.

Muraki et al. (2000) presented IRT methodology for psychometric procedures in the context of performance assessments, including description and comparison of many IRT and CTT procedures for scaling, linking, and equating. Tang and Eignor (2001), in a simulation, studied whether CTT item statistics could be used as collateral information along with IRT calibration to reduce sample sizes for pretesting TOEFL items, and found that CTT statistics, as the only collateral information, would not do the job.

Rock and Pollack (2002) investigated model-based methods (including IRT-based methods), and more traditional methods of measuring growth in prereading and reading at the kindergarten level, including comparisons between demographic groups. They concluded that the more traditional methods may yield uninformative if not incorrect results.

Scrams et al. (2002) studied use of item variants for continuous linear computer-based testing. Results showed that calibrated difficulty parameters of analogy and antonym items from the GRE General Test were very similar to those based on variant family information, and, using simulations, they showed that precision loss in ability estimation was less than 10% in using parameters estimated from expected response functions based only on variant family information.

A study comparing linear, fixed common item, and concurrent parameter estimation equating methods in capturing growth was conducted and reported by Jodoin et al. (2003). A. A. von Davier and Wilson studied the assumptions made at each step of calibration through IRT true-score equating and methods of checking whether the assumptions are met by a dataset. Operational data from the *AP*® Calculus AB exam were used as an illustration. Rotou et al. (2007) compared the measurement precision, in terms of reliability and conditional standard error of measurement (CSEM), of multistage (MS), CAT, and linear tests, using 1PL, 2PL, and 3PL IRT models. They found the MS tests to be superior to CAT and linear tests for the 1PL and 2PL models, and performance of the MS and CAT to be about the same, but better than the linear for the 3PL case.

Liu et al. (2008) compared the bootstrap and Markov chain Monte Carlo (MCMC) methods of estimation in IRT true-score equating with simulations based on operational testing data. Patterns of standard error estimates for the two methods were similar, but the MCMC produced smaller bias and mean square errors of equating. G. Lee and Fitzpatrick (2008), using operational test data, compared IRT equating by the Stocking-Lord method with and without fixing the *c* parameters. Fixing the *c* parameters had little effect on parameter estimates of the nonanchor items, but a considerable effect at the lower end of the scale for the anchor items. They suggeted that practitioners consider using the fixed-*c* method.

A regression procedure was developed by Haberman (2009a) to simultaneously link a very large number of IRT parameter estimates obtained from a large number of test forms, where each form has been separately calibrated and where forms can be linked on a pairwise basis by means of common items. An application to 2PL and GPC model data was also presented. Xu et al. (2011) presented two methods of using nonparametric IRT models in linking, illustrating with both simulated and operational datasets. In the simulation study, they showed that the proposed methods recover the true linking function when parametric models do not fit the data or when there is a large discrepancy in the populations.

Y. Li (2012), using simulated data, studied the effects, for a test with a small number of polytomous anchor items, of item parameter drift on TCC linking and IRT true-score equating. Results suggest that anchor length, number of items with drifting parameters, and magnitude of the drift affected the linking and equating results. The ability distributions of the groups had little effect on the linking and equating results. In general, excluding drifted polytomous anchor items resulted in an improvement in equating results.

D. Li et al. (2012) conducted a simulation study of IRT equating of six forms of a test, comparing several equating transformation methods and separate versus concurrent item calibration. The characteristic curve methods yielded smaller biases and smaller sampling errors (or accumulation of errors over time) so the former were concluded to be superior to the latter and were recommended in practice.

Livingston (2006) described IRT methodology for item analysis in a book chapter in *Handbook of Test Development* (Downing and Haladyna 2006). In the same publication, Wendler and Walker (2006) discussed IRT methods of scoring, and Davey and Pitoniak (2006) discussed designing CATs, including use of IRT in scoring, calibration, and scaling.

Almond et al. (2007) described Bayesian network models and their application to IRT-based CD modeling. The paper, designed to encourage practitioners to learn to use these models, is aimed at a general educational measurement audience, does not use extensive technical detail, and presents examples.

5.6.2 The Signs of (IRT) Things to Come

The body of work that ETS staff has contributed to in the development and applications of IRT, MIRT, and comprehensive integrated models based on IRT has been documented in multiple published monographs and edited volumes. At the point of writing this chapter, the history is still in the making; there are three more edited volumes that would have not been possible without the contributions of ETS researchers reporting on the use of IRT in various applications. More specifically:

- *Handbook of Item Response Theory* (second edition) contains chapters by Shelby Haberman, John Mazzeo, Robert Mislevy, Tim Moses, Frank Rijmen, Sandip Sinharay, and Matthias von Davier.
- *Computerized Multistage Testing: Theory and Applications* (edited by Duanli Yan, Alina von Davier, & Charlie Lewis, 2014) contains chapters by Isaac Bejar, Brent Bridgeman, Henry Chen, Shelby Haberman, Sooyeon Kim, Ed Kulick,

Yi-Hsuan Lee, Charlie Lewis, Longjuan Liang, Skip Livingston, John Mazzeo, Kevin Meara, Chris Mills, Andreas Oranje, Fred Robin, Manfred Steffen, Peter van Rijn, Alina von Davier, Matthias von Davier, Carolyn Wentzel, Xueli Xu, Kentaro Yamamoto, Duanli Yan, and Rebecca Zwick.

- *Handbook of International Large Scale International Assessment* (edited by Leslie Rutkowski, Matthias von Davier, & David Rutkowski, 2013) contains chapters by Henry Chen, Eugenio Gonzalez, John Mazzeo, Andreas Oranje, Frank Rijmen, Matthias von Davier, Jonathan Weeks, Kentaro Yamamoto, and Lei Ye.

5.7 Conclusion

Over the past six decades, ETS has pushed the envelope of modeling item response data using a variety of latent trait models that are commonly subsumed under the label IRT. Early developments, software tools, and applications allowed insight into the particular advantages of approaches that use item response functions to make inferences about individual differences on latent variables. ETS has not only provided theoretical developments, but has also shown, in large scale applications of IRT, how these methodologies can be used to perform scale linkages in complex assessment designs, and how to enhance reporting of results by providing a common scale and unbiased estimates of individual or group differences.

In the past two decades, IRT, with many contributions from ETS researchers, has become an even more useful tool. One main line of development has connected IRT to cognitive models and integrated measurement and structural modeling. This integration allows for studying questions that cannot be answered by secondary analyses using simple scores derived from IRT- or CTT-based approaches. More specifically, differential functioning of groups of items, the presence or absence of evidence that suggests that multiple diagnostic skill variables can be identified, and comparative assessment of different modeling approaches are part of what the most recent generation of multidimensional explanatory item response models can provide.

ETS will continue to provide cutting edge research and development on future IRT-based methodologies, and continues to play a leading role in the field, as documented by the fact that nine chapters of the *Handbook of Item Response Theory (second edition)* are authored by ETS staff. Also, of course, at any point in time, including the time of publication of this work, there are numerous research projects being conducted by ETS staff, and for which reports are being drafted, reviewed, or submitted for publication. By the timeaa this work is published, there will undoubtedly be additional publications not included herein.

References

Akkermans, W., & Muraki, E. (1997). Item information and discrimination functions for trinary PCM items. *Psychometrika, 62,* 569–578. https://doi.org/10.1007/BF02294643

Almond, R. G., & Mislevy, R. J. (1999). Graphical models and computerized adaptive testing. *Applied Psychological Measurement, 23,* 223–237. https://doi.org/10.1177/01466219922031347

Almond, R. G., DiBello, L. V., Moulder, B., & Zapata-Rivera, J.-D. (2007). Modeling diagnostic assessments with Bayesian networks. *Journal of Educational Measurement, 44*, 341–359.

Andersen, E. B. (1972). *A computer program for solving a set of conditional maximum likelihood equations arising in the Rasch model for questionnaires* (Research Memorandum No. RM-72-06). Princeton: Educational Testing Service.

Andersen, E. B. (1973). A goodness of fit test for the Rasch model. *Psychometrika, 38,* 123–140. https://doi.org/10.1007/BF02291180

Andersen, E. B. (1985). Estimating latent correlations between repeated testings. *Psychometrika, 50*, 3–16. https://doi.org/10.1007/BF02294143

Antal, T. (2007). *On multidimensional item response theory: A coordinate-free approach* (Research Report No. RR-07-30). Princeton: Educational Testing Service. https://doi.org/10.1002/j.2333-8504.2007.tb02072.x

Antal, T., & Oranje, A. (2007). *Adaptive numerical integration for item response theory* (Research Report No. RR-07-06). Princeton: Educational Testing Service. https://doi.org/10.1002/j.2333-8504.2007.tb02048.x

Barton, M. A., & Lord, F. M. (1981). *An upper asymptote for the three-parameter logistic item-response model* (Research Report No. RR-81-20). Princeton: Educational Testing Service. https://doi.org/10.1002/j.2333-8504.1981.tb01255.x

Beaton, A. E., & Johnson, E. G. (1992). Overview of the scaling methodology used in the National Assessment. *Journal of Educational Measurement, 29,* 163–175. https://doi.org/10.1111/j.1745-3984.1992.tb00372.x

Bejar, I. I. (1980). A procedure for investigating the unidimensionality of achievement tests based on item parameter estimates. *Journal of Educational Measurement, 17,* 283–296. https://doi.org/10.1111/j.1745-3984.1980.tb00832.x

Bejar, I. I. (1983). Introduction to item response models and their assumptions. In R. K. Hambleton (Ed.), *Applications of item response theory* (pp. 1–23). Vancouver: Educational Research Institute of British Columbia.

Bejar, I. I. (1990). A generative analysis of a three-dimensional spatial task. *Applied Psychological Measurement, 14,* 237–245. https://doi.org/10.1177/014662169001400302

Bejar, I. I., & Wingersky, M. S. (1981). *An application of item response theory to equating the Test of Standard Written English* (Research Report No. RR-81-35). Princeton: Educational Testing Service.

Bejar, I. I., & Wingersky, M. S. (1982). A study of pre-equating based on item response theory. *Applied Psychological Measurement, 6*, 309–325. https://doi.org/10.1177/014662168200600308

Bejar, I. I., Braun, H. I., & Carlson, S. B. (1989). *Psychometric foundations of testing based on patient management problems* (Research Memorandum No. RM-89-02). Princeton: Educational Testing Service.

Belov, D., Armstrong, R. D., & Weissman, A. (2008). A Monte Carlo approach for adaptive testing with content constraints. *Applied Psychological Measurement, 32,* 431–446. https://doi.org/10.1177/0146621607309081

Bennett, R. E., Sebrechts, M. M., & Yamamoto, K. (1991). *Fitting new measurement models to GRE General Test constructed-response item data* (Research Report No. RR-91-60). Princeton: Educational Testing Service. https://doi.org/10.1002/j.2333-8504.1991.tb01427.x

Birnbaum, A. (1967). *Statistical theory for logistic mental test models with a prior distribution of ability* (Research Bulletin No. RB-67-12). Princeton: Educational Testing Service. https://doi.org/10.1002/j.2333-8504.1967.tb00363.x

Bock, R. D. (1997). A brief history of item response theory. *Educational Measurement: Issues and Practice, 16*(4), 21–33. https://doi.org/10.1111/j.1745-3992.1997.tb00605.x
Boldt, R. F. (1989). Latent structure analysis of the Test of English as a Foreign Language. *Language Testing, 6,* 123–142. https://doi.org/10.1177/026553228900600201
Boughton, K., & Yamamoto, K. (2007). A HYBRID model for test speededness. In M. von Davier & C. H. Carstensen (Eds.), *Multivariate and mixture distribution Rasch models: Extensions and applications* (pp. 147–156). New York: Springer. https://doi.org/10.1007/978-0-387-49839-3_9
Bradlow, E. T. (1996). Negative information and the three-parameter logistic model. *Journal of Educational and Behavioral Statistics, 21*, 179–185.
Bradlow, E. T., Wainer, H., & Wang, X. (1999). A Bayesian random effects model for testlets. *Psychometrika, 64,* 153–168. https://doi.org/10.1007/BF02294533
Camilli, G., Yamamoto, K., & Wang, M.-M. (1993). Scale shrinkage in vertical equating. *Applied Psychological Measurement, 17,* 379–388. https://doi.org/10.1177/014662169301700407
Chang, H.-H. (1996). The asymptotic posterior normality of the latent trait for polytomous IRT models. *Psychometrika, 61,* 445–463. https://doi.org/10.1007/BF02294549
Chang, H.-H., & Mazzeo, J. (1994). The unique correspondence of the item response function and item category response functions in polytomously scored item response models. *Psychometrika, 59,* 391–404. https://doi.org/10.1007/BF02296132
Chang, H.-H., & Stout, W. (1993). The asymptotic posterior normality of the latent trait in an IRT model. *Psychometrika, 58,* 37–52. https://doi.org/10.1007/BF02294469
Cook, L. L., & Eignor, D. R. (1983). Practical considerations regarding the use of item response theory to equate tests. In R. Hambleton (Ed.), *Applications of item response theory* (pp. 175–195). Vancouver: Educational Research Institute of British Columbia.
Cook, L. L., & Eignor, D. R. (1985). *An investigation of the feasibility of applying item response theory to equate achievement tests* (Research Report No. RR-85-31). Princeton: Educational Testing Service. https://doi.org/10.1002/j.2330-8516.1985.tb00116.x
Cook, L. L., & Eignor, D. R. (1989). Using item response theory in test score equating. *International Journal of Educational Research, 13*, 161–173.
Cook, L. L., & Petersen, N. S. (1987). Problems related to the use of conventional and item response theory equating methods in less than optimal circumstances. *Applied Psychological Measurement, 11,* 225–244. https://doi.org/10.1177/014662168701100302
Cook, L. L., Dorans, N. J., Eignor, D. R., & Petersen, N. S. (1985a). *An assessment of the relationship between the assumption of unidimensionality and the quality of IRT true-score equating* (Research Report No. RR-85-30) . Princeton: Educational Testing Service. https://doi.org/10.1002/j.2330-8516.1985.tb00115.x
Cook, L. L., Eignor, D. R., & Taft, H. L. (1985b). *A comparative study of curriculum effects on the stability of IRT and conventional item parameter estimates* (Research Report No. RR-85-38). Princeton: Educational Testing Service. https://doi.org/10.1002/j.2330-8516.1985.tb00123.x
Cook, L. L., Eignor, D. R., & Petersen, N. S. (1985c). *A study of the temporal stability of IRT item parameter estimates* (Research Report No. RR-85-45). Princeton: Educational Testing Service. https://doi.org/10.1002/j.2330-8516.1985.tb00130.x
Cook, L. L., Eignor, D. R., & Schmitt, A. P. (1988a). *The effects on IRT and conventional achievement test equating results of using equating samples matched on ability* (Research Report No. RR-88-52). Princeton: Educational Testing Service. https://doi.org/10.1002/j.2330-8516.1988.tb00308.x
Cook, L. L., Dorans, N. J., & Eignor, D. R. (1988b). An assessment of the dimensionality of three SAT-Verbal test editions. *Journal of Educational Statistics, 13,* 19–43. https://doi.org/10.2307/1164949
Cook, L. L., Eignor, D. R., & Taft, H. L. (1988c). A comparative study of the effects of recency of instruction on the stability of IRT and conventional item parameter estimates. *Journal of Educational Measurement, 25,* 31–45. https://doi.org/10.1111/j.1745-3984.1988.tb00289.x
Cressie, N., & Holland, P. W. (1983). Characterizing the manifest probabilities of latent trait models. *Psychometrika, 48*, 129–141. https://doi.org/10.1007/BF02314681

Davey, T., & Pitoniak, M. (2006). Designing computerized adaptive tests. In S. M. Downing & T. M. Haladyna (Eds.), *Handbook of test development* (pp. 543–573). Mahwah: Erlbaum.

de la Torre, J., & Deng, W. (2008). Improving person-fit assessment by correcting the ability estimate and its reference distribution. *Journal of Educational Measurement, 45,* 159–177. https://doi.org/10.1111/j.1745-3984.2008.00058.x

Donoghue, J. R. (1992). *On a common misconception concerning the partial credit and generalized partial credit polytomous IRT models* (Research Memorandum No. RM-92-12). Princeton: Educational Testing Service.

Donoghue, J. R. (1993). *An empirical examination of the IRT information in polytomously scored reading items* (Research Report No. RR-93-12). Princeton: Educational Testing Service. https://doi.org/10.1002/j.2333-8504.1993.tb01523.x

Donoghue, J. R., & Isham, S. P. (1998). A comparison of procedures to detect item parameter drift. *Applied Psychological Measurement, 22,* 33–51. https://doi.org/10.1177/01466216980221002

Dorans, N. J. (1985). Item parameter invariance: The cornerstone of item response theory. In K. M. Rowland & G. R. Ferris (Eds.), *Research in personnel and human resources management* (Vol. 3, pp. 55–78). Greenwich: JAI Press.

Dorans, N. J. (1986). The impact of item deletion on equating conversions and reported score distributions. *Journal of Educational Measurement, 23,* 245–264. https://doi.org/10.1111/j.1745-3984.1986.tb00250.x

Dorans, N. J. (1990). Scaling and equating. In H. Wainer, N. J. Dorans, R. Flaugher, B. F. Green, R. J. Mislevy, L. Steinberg, & D. Thissen (Eds.), *Computerized adaptive testing: A primer* (pp. 137–160). Hillsdale: Erlbaum.

Dorans, N. J. (2000). Scaling and equating. In H. Wainer, N. J. Dorans, D. Eignor, R. Flaugher, B. F. Green, R. J. Mislevy, L. Steinberg, & D. Thissen (Eds.), *Computerized adaptive testing: A primer* (pp. 135–158). Mahwah: Erlbaum.

Dorans, N. J., & Holland, P. W. (1992). *DIF detection and description: Mantel-Haenszel and standardization* (Research Report No. RR-92-10). Princeton: Educational Testing Service. https://doi.org/10.1002/j.2333-8504.1992.tb01440.x

Dorans, N. J., & Kingston, N. M. (1985). The effects of violations of unidimensionality on the estimation of item and ability parameters and on item response theory equating of the GRE Verbal Scale. *Journal of Educational Measurement, 22,* 249–262. https://doi.org/10.1111/j.1745-3984.1985.tb01062.x

Douglass, J. B., Marco, G. L., & Wingersky, M. S. (1985). *An evaluation of three approximate item response theory models for equating test scores* (Research Report No. RR-85-46). Princeton: Educational Testing Service. https://doi.org/10.1002/j.2330-8516.1985.tb00131.x

Downing, S. M., & Haladyna, T. M. (Eds.). (2006). *Handbook of test development.* Mahwah: Erlbaum.

Eignor, D. R. (1985). *An investigation of the feasibility and practical outcomes of pre-equating the SAT Verbal and Mathematical sections* (Research Report No. RR-85-10). Princeton: Educational Testing Service. https://doi.org/10.1002/j.2330-8516.1985.tb00095.x

Eignor, D. R., & Stocking, M. L. (1986). *An investigation of possible causes for the inadequacy of IRT pre-equating* (Research Report No. RR-86-14). Princeton: Educational Testing Service. https://doi.org/10.1002/j.2330-8516.1986.tb00169.x

Eignor, D. R., & Stocking, M. L. (1986). *The impact of different ability distributions on IRT pre-equating, Research Report No. RR-86-49.* Princeton: Educational Testing Service. https://doi.org/10.1002/j.2330-8516.1986.tb00204.x

Eignor, D. R., Golub-Smith, M. L., & Wingersky, M. S. (1986). *Application of a new goodness-of-fit plot procedure to SAT and TOEFL item type data* (Research Report No. RR-86-47). Princeton: Educational Testing Service. https://doi.org/10.1002/j.2330-8516.1986.tb00202.x

Eignor, D. R., Stocking, M. L., & Cook, L. L. (1990). Simulation results of effects on linear and curvilinear observed- and true-score equating procedures of matching on a fallible criterion. *Applied Measurement in Education, 3,* 37–52. https://doi.org/10.1207/s15324818ame0301_4

Embretson, S. E. (1991). A multidimensional latent trait model for measuring learning and change. *Psychometrika, 56*, 495–515. https://doi.org/10.1007/BF02294487

Flaugher, R. (1990). Item pools. In H. Wainer, N. J. Dorans, R. Flaugher, B. F. Green, R. J. Mislevy, L. Steinberg, & D. Thissen (Eds.), *Computerized adaptive testing: A primer* (pp. 41–63). Hillsdale: Erlbaum.

Flaugher, R. (2000). Item pools. In H. Wainer, N. J. Dorans, D. Eignor, R. Flaugher, B. F. Green, R. J. Mislevy, L. Steinberg, & D. Thissen (Eds.), *Computerized adaptive testing: A primer* (pp. 37–59). Mahwah: Erlbaum.

Folk, V. G., & Green, B. F. (1989). Adaptive estimation when the unidimensionality assumption of IRT is violated. *Applied Psychological Measurement, 13,* 373–389. https://doi.org/10.1177/014662168901300404

Frederiksen, N., Mislevy, R. J., & Bejar, I. I. (Eds.). (1993). *Test theory for a new generation of tests*. Hillsdale: Erlbaum.

Gitomer, D. H., & Yamamoto, K. (1991). Performance modeling that integrates latent trait and class theory. *Journal of Educational Measurement, 28,* 173–189. https://doi.org/10.1111/j.1745-3984.1991.tb00352.x

Glas, C. A. W., Wainer, H., & Bradlow, E. T. (2000). MML and EAP estimation in testlet-based adaptive testing. In W. J. van der Linden & C. A. W. Glas (Eds.), *Computerized adaptive testing: Theory and practice* (pp. 271–287). Dordrecht: Kluwer. https://doi.org/10.1007/0-306-47531-6_14

Green, B. F., Jr. (1950a). *A proposal for a comparative study of the measurement of attitude* (Research Memorandum no. RM-50-20). Princeton: Educational Testing Service.

Green, B. F., Jr. (1950b). *A proposal for an empirical evaluation of the latent class model of latent structure analysis* (Research Memorandum No. RM-50-26). Princeton: Educational Testing Service.

Green, B. F., Jr. (1951a). A general solution for the latent class model of latent structure analysis. *Psychometrika, 16,* 151–166. https://doi.org/10.1007/BF02289112

Green, B. F., Jr. (1951b). *Latent class analysis: A general solution and an empirical evaluation* (Research Bulletin No. RB-51-15). Princeton: Educational Testing Service. https://doi.org/10.1002/j.2333-8504.1951.tb00215.x

Green, B. F., Jr. (1952). Latent structure analysis and its relation to factor analysis. *Journal of the American Statistical Association, 47,* 71–76. https://doi.org/10.1080/01621459.1952.10501155

Green, B. F., Jr. (1980, April). *Ledyard R Tucker's affair with psychometrics: The first 45 years.* Paper presented at a special symposium in honor of Ledyard R Tucker. Champaign: The University of Illinois.

Gustafsson, J.-E., Morgan, A. M. B., & Wainer, H. (1980). A review of estimation procedures for the Rasch model with an eye toward longish tests. *Journal of Educational Statistics, 5*, 35–64.

Haberman, S. J. (2004). *Joint and conditional maximum likelihood estimation for the Rasch model of binary responses* (Research Report No. RR-04-20). Princeton: Educational Testing Service. https://doi.org/10.1002/j.2333-8504.2004.tb01947.x

Haberman, S. J. (2005a). *Identifiability of parameters in item response models with unconstrained ability distributions* (Research Report No. RR-05-24). Princeton: Educational Testing Service. https://doi.org/10.1002/j.2333-8504.2005.tb02001.x

Haberman, S. J. (2005b). *Latent-class item response models* (Research Report No. RR-05-28). Princeton: Educational Testing Service. https://doi.org/10.1002/j.2333-8504.2005.tb02005.x

Haberman, S. J. (2006). *Adaptive quadrature for item response models* (Research Report No. RR-06-29). Princeton: Educational Testing Service. https://doi.org/10.1002/j.2333-8504.2006.tb02035.x

Haberman, S. J. (2007a). *The information a test provides on an ability parameter* (Research Report No. RR-07-18). Princeton: Educational Testing Service. https://doi.org/10.1002/j.2333-8504.2007.tb02060.x

Haberman, S. J. (2007b). The interaction model. In M. von Davier & C. H. Carstensen (Eds.), *Multivariate and mixture distribution Rasch models: Extensions and applications* (pp. 201–216). New York: Springer. https://doi.org/10.1007/978-0-387-49839-3_13

Haberman, S. J. (2008). *Reliability of scaled scores* (Research Report No. RR-08-70). Princeton: Educational Testing Service. https://doi.org/10.1002/j.2333-8504.2008.tb02156.x

Haberman, S. J. (2009a). *Linking parameter estimates derived from an item response model through separate calibrations* (Research Report No. RR-09-40). Princeton: Educational Testing Service. https://doi.org/10.1002/j.2333-8504.2009.tb02197.x

Haberman, S. J. (2009b). *Use of generalized residuals to examine goodness of fit of item response models* (Research Report No. RR-09-15). Princeton: Educational Testing Service. https://doi.org/10.1002/j.2333-8504.2009.tb02172.x

Haberman, S. J., & Sinharay, S. (2010). Reporting of subscores using multidimensional item response theory. *Psychometrika, 75,* 209–227. https://doi.org/10.1007/s11336-010-9158-4

Haberman, S. J., & von Davier, M. (2007). Some notes on models for cognitively based skills. In C. R. Rao & S. Sinharay (Eds.), *Handbook of statistics: Vol 26. Psychometrics* (pp. 1031–1038). Amsterdam: Elsevier.

Haberman, S. J., Holland, P. W., & Sinharay, S. (2007). Limits on log odds ratios for unidimensional item response theory models. *Psychometrika, 72,* 551–561. https://doi.org/10.1007/s11336-007-9009-0

Haberman, S. J., von Davier, M., & Lee, Y.-H. (2008). *Comparison of multidimensional item response models: Multivariate normal ability distributions versus multivariate polytomous ability distributions* (Research Report No. RR-08-45). Princeton: Educational Testing Service. https://doi.org/10.1002/j.2333-8504.2008.tb02131.x

Hambleton, R. K. (1983). *Applications of item response theory*. Vancouver: Educational Research Institute of British Columbia.

Hambleton, R. K., & Cook, L. L. (1977). Latent trait models and their use in the analysis of educational test data. *Journal of Educational Measurement, 14,* 75–96. https://doi.org/10.1111/j.1745-3984.1977.tb00030.x

Hambleton, R. K., & Cook, L. L. (1983). Robustness of item response models and effects of test length and sample size on the precision of ability estimates. In D. J. Weiss (Ed.), *New horizons in testing: Latent trait test theory and computerized adaptive testing* (pp. 31–49). New York: Academic Press. https://doi.org/10.1016/B978-0-12-742780-5.50010-X

Harnisch, D. L., & Tatsuoka, K. K. (1983). A comparison of appropriateness indices based on item response theory. In R. K. Hambleton (Ed.), *Applications of item response theory* (pp. 104–122). Vancouver: Educational Research Institute of British Columbia.

Hartz, S., & Roussos, L. (2008). *The fusion model for skills diagnosis: Blending theory with practicality* (Research Report No. RR-08-71). Princeton: Educational Testing Service. https://doi.org/10.1002/j.2333-8504.2008.tb02157.x

Henning, G. (1989). Does the Rasch model really work for multiple-choice items? Take another look: A response to Divgi. *Journal of Educational Measurement, 26,* 91–97. https://doi.org/10.1111/j.1745-3984.1989.tb00321.x

Hicks, M. M. (1983). True score equating by fixed b's scaling: A flexible and stable equating alternative. *Applied Psychological Measurement, 7,* 255–266. https://doi.org/10.1177/014662168300700302

Hicks, M. M. (1984). *A comparative study of methods of equating TOEFL test scores* (Research Report No. RR-84-20). Princeton: Educational Testing Service. https://doi.org/10.1002/j.2330-8516.1984.tb00060.x

Holland, P. W. (1981). When are item response models consistent with observed data? *Psychometrika, 46,* 79–92. https://doi.org/10.1007/BF02293920

Holland, P. (1990a). On the sampling theory foundations of item response theory models. *Psychometrika, 55,* 577–601. https://doi.org/10.1007/BF02294609

Holland, P. (1990b). The Dutch identity: A new tool for the study of item response theory models. *Psychometrika, 55,* 5–18. https://doi.org/10.1007/BF02294739

Holland, P. W., & Hoskens, M. (2002). *Classical test theory as a first-order item response theory: Application to true-score prediction from a possibly nonparallel test* (Research Report No. RR-02-20). Princeton: Educational Testing Service. https://doi.org/10.1002/j.2333-8504.2002.tb01887.x

Holland, P. W., & Rosenbaum, P. R. (1986). Conditional association and unidimensionality in monotone latent variable models. *Annals of Statistics, 14,* 1523–1543. https://doi.org/10.1214/aos/1176350174

Holland, P., Dorans, N., & Petersen, N. (2007). Equating. In C. R. Rao & S. Sinharay (Eds.), *Handbook of statistics: Vol. 26. Psychometrics* (pp. 169–204). Amsterdam: Elsevier.

Jodoin, M. G., Keller, L. A., & Swaminathan, H. (2003). A comparison of linear, fixed common item, and concurrent parameter estimation equating procedures in capturing academic growth. *The Journal of Experimental Education, 71,* 229–250. https://doi.org/10.1080/00220970309602064

Johnson, M., Sinharay, S., & Bradlow, E. T. (2007). Hierarchical item response theory models. In C. R. Rao & S. Sinharay (Eds.), *Handbook of statistics: Vol. 26. Psychometrics* (pp. 587–605). Amsterdam: Elsevier.

Jones, D. H. (1980). *On the adequacy of latent trait models* (Program Statistics Research Technical Report No. 80–08). Princeton: Educational Testing Service.

Jones, D. H. (1982). *Tools of robustness for item response theory* (Research Report No. RR-82-41). Princeton: Educational Testing Service.

Jones, D. H. (1984a). *Asymptotic properties of the robustified jackknifed estimator* (Research Report No. RR-84-41). Princeton: Educational Testing Service. https://doi.org/10.1002/j.2330-8516.1984.tb00081.x

Jones, D. H. (1984b). *Bayesian estimators, robust estimators: A comparison and some asymptotic results* (Research Report No. RR-84-42). Princeton: Educational Testing Service. https://doi.org/10.1002/j.2330-8516.1984.tb00082.x

Jones, D. H., Kaplan, B. A., & Wainer, H. (1984). *Estimating ability with three item response models when the models are wrong and their parameters are inaccurate* (Research Report No. RR-84-26). Princeton: Educational Testing Service. https://doi.org/10.1002/j.2330-8516.1984.tb00066.x

Kingston, N. M. (1986). *Assessing the dimensionality of the GMAT Verbal and Quantitative measures using full information factor analysis* (Research Report No. RR-86-13). Princeton: Educational Testing Service. https://doi.org/10.1002/j.2330-8516.1986.tb00168.x

Kingston, N. M., & Dorans, N. J. (1982a). *The effect of the position of an item within a test on item responding behavior: An analysis based on item response theory* (Research Report No. RR-82-22). Princeton: Educational Testing Service. https://doi.org/10.1002/j.2333-8504.1982.tb01308.x

Kingston, N. M., & Dorans, N. J. (1982b). *The feasibility of using item response theory as a psychometric model for the GRE Aptitude Test* (Research Report No. RR-82-12). Princeton: Educational Testing Service. https://doi.org/10.1002/j.2333-8504.1982.tb01298.x

Kingston, N. M., & Dorans, N. J. (1984). Item location effects and their implications for IRT equating and adaptive testing. *Applied Psychological Measurement, 8,* 147–154. https://doi.org/10.1177/014662168400800202

Kingston, N. M., & Dorans, N. J. (1985). The analysis of item-ability regressions: An exploratory IRT model fit tool. *Applied Psychological Measurement, 9,* 281–288. https://doi.org/10.1177/014662168500900306

Kingston, N. M., & Holland, P. W. (1986). *Alternative methods of equating the GRE General Test* (Research Report No. RR-86-16). Princeton: Educational Testing Service. https://doi.org/10.1002/j.2330-8516.1986.tb00171.x

Kingston, N. M., Leary, L. F., & Wightman, L. E. (1985). *An exploratory study of the applicability of item response theory methods to the Graduate Management Admission Test* (Research Report No. RR-85-34). Princeton: Educational Testing Service.

Kreitzberg, C. B., & Jones, D. H. (1980). *An empirical study of the broad range tailored test of verbal ability* (Research Report No. RR-80-05). Princeton: Educational Testing Service. https://doi.org/10.1002/j.2333-8504.1980.tb01195.x

Kreitzberg, C. B., Stocking, M. L., & Swanson, L. (1977). *Computerized adaptive testing: The concepts and its potentials* (Research Memorandum No. RM-77-03). Princeton: Educational Testing Service.

Lane, S., Stone, C. A., Ankenmann, R. D., & Liu, M. (1995). Examination of the assumptions and properties of the graded item response model: An example using a mathematics performance assessment. *Applied Measurement in Education, 8,* 313–340. https://doi.org/10.1207/s15324818ame0804_3

Lawrence, I. M., & Dorans, N. J. (1988). *A comparison of observed score and true score equating methods for representative samples and samples matched on an anchor test* (Research Report No. RR-88-23). Princeton: Educational Testing Service. https://doi.org/10.1002/j.2330-8516.1988.tb00279.x

Lawrence, I. M., & Dorans, N. J. (1990). Effect on equating results of matching samples on an anchor test. *Applied Measurement in Education, 3,* 19–36. https://doi.org/10.1207/s15324818ame0301_3

Lazarsfeld, P. F. (1950). The logical and mathematical foundation of latent structure analysis. In S. A. Stouffer, L. Guttman, E. A. Suchman, P. F. Lazarsfeld, S. A. Star, & J. A. Clausen (Eds.), *Studies in social psychology in World War II: Vol. 4. Measurement and prediction* (pp. 362–472). Princeton: Princeton University Press.

Lee, G., & Fitzpatrick, A. R. (2008). A new approach to test score equating using item response theory with fixed c-parameters. *Asia Pacific Education Review, 9,* 248–261. https://doi.org/10.1007/BF03026714

Lee, Y.-H., & Zhang, J. (2008). *Comparing different approaches of bias correction for ability estimation in IRT models* (Research Report No. RR-08-13). Princeton: Educational Testing Service. https://doi.org/10.1002/j.2333-8504.2008.tb02099.x

Levy, R., Mislevy, R. J., & Sinharay, S. (2009). Posterior predictive model checking for multidimensionality in item response theory. *Applied Psychological Measurement, 33,* 519–537. https://doi.org/10.1177/0146621608329504

Lewis, C. (2001). Expected response functions. In A. Boomsma, M. A. J. van Duijn, & T. A. B. Snijders (Eds.), *Essays on item response theory* (pp. 163–171). New York: Springer. https://doi.org/10.1007/978-1-4613-0169-1_9

Lewis, C., & Sheehan, K. M. (1990). Using Bayesian decision theory to design a computerized mastery test. *Applied Psychological Measurement, 14,* 367–386. https://doi.org/10.1177/014662169001400404

Li, D., & Oranje, A. (2007). *Estimation of standard error of regression effects in latent regression models using Binder's linearization* (Research Report No. RR-07-09). Princeton: Educational Testing Service. https://doi.org/10.1002/j.2333-8504.2007.tb02051.x

Li, D., Oranje, A., & Jiang, Y. (2007). *Parameter recovery and subpopulation proficiency estimation in hierarchical latent regression models* (Research Report No. RR-07-27). Princeton: Educational Testing Service. https://doi.org/10.1002/j.2333-8504.2007.tb02069.x

Li, D., Jiang, Y., & von Davier, A. A. (2012). The accuracy and consistency of a series of IRT true score equatings. *Journal of Educational Measurement, 49,* 167–189. https://doi.org/10.1111/j.1745-3984.2012.00167.x

Li, Y. (2012). *Examining the impact of drifted polytomous anchor items on test characteristic curve (TCC) linking and IRT true score equating* (Research Report No. RR-12-09). Princeton: Educational Testing Service. https://doi.org/10.1002/j.2333-8504.2012.tb02291.x

Li, Y., Li, S., & Wang, L. (2010). *Application of a general polytomous testlet model to the reading section of a large-scale English language assessment* (Research Report No. RR-10-21). Princeton: Educational Testing Service. https://doi.org/10.1002/j.2333-8504.2010.tb02228.x

Liu, Y., Schulz, E. M., & Yu, L. (2008). Standard error estimation of 3PL IRT true score equating with an MCMC method. *Journal of Educational and Behavioral Statistics, 33,* 257–278. https://doi.org/10.3102/1076998607306076

Livingston, S. A. (2006). Item analysis. In S. M. Downing & T. M. Haladyna (Eds.), *Handbook of test development* (pp. 421–441). Mahwah: Erlbaum.

Livingston, S. A., Dorans, N. J., & Wright, N. K. (1990). What combination of sampling and equating methods works best? *Applied Measurement in Education, 3,* 73–95. https://doi.org/10.1207/s15324818ame0301_6

Lord, F. M. (1951). *A theory of test scores and their relation to the trait measured* (Research Bulletin No. RB-51-13). Princeton: Educational Testing Service. https://doi.org/10.1002/j.2333-8504.1951.tb00922.x

Lord, F. M. (1952a). *A theory of test scores* (Psychometric Monograph No. 7). Richmond: Psychometric Corporation.

Lord, F. M. (1952b). *The scale proposed for the academic ability test* (Research Memorandum No. RM-52-03). Princeton: Educational Testing Service.

Lord, F. M. (1953). The relation of test score to the trait underlying the test. *Educational and Psychological Measurement, 13,* 517–549. https://doi.org/10.1177/001316445301300401

Lord, F. M. (1964). *A strong true score theory, with applications* (Research Bulletin No. RB-64-19). Princeton: Educational Testing Service. https://doi.org/10.1002/j.2333-8504.1951.tb00922.x

Lord, F. M. (1965a). An empirical study of item-test regression. *Psychometrika, 30,* 373–376. https://doi.org/10.1007/BF02289501

Lord, F. M. (1965b). A note on the normal ogive or logistic curve in item analysis. *Psychometrika, 30,* 371–372. https://doi.org/10.1007/BF02289500

Lord, F. M. (1968a). An analysis of the Verbal Scholastic Aptitude Test using Birnbaum's three-parameter logistic model. *Educational and Psychological Measurement, 28,* 989–1020. https://doi.org/10.1177/001316446802800401

Lord, F. M. (1968b). *Some test theory for tailored testing* (Research Bulletin No. RB-68-38). Princeton: Educational Testing Service. https://doi.org/10.1002/j.2333-8504.1968.tb00562.x

Lord, F. M. (1970). Item characteristic curves estimated without knowledge of their mathematical form—A confrontation of Birnbaum's logistic model. *Psychometrika, 35,* 43–50. https://doi.org/10.1007/BF02290592

Lord, F. M. (1973). Power scores estimated by item characteristic curves. *Educational and Psychological Measurement, 33,* 219–224. https://doi.org/10.1177/001316447303300201

Lord, F. M. (1974a). Estimation of latent ability and item parameters when there are omitted responses. *Psychometrika, 39,* 247–264. https://doi.org/10.1007/BF02291471

Lord, F. M. (1974b). Individualized testing and item characteristic curve theory. In D. H. Krantz, R. C. Atkinson, R. D. Luce, & P. Suppes (Eds.), *Contemporary developments in mathematical psychology* (Vol. II, pp. 106–126). San Francisco: Freeman.

Lord, F. M. (1975a). *A survey of equating methods based on item characteristic curve theory* (Research Bulletin No. RB-75-13). Princeton: Educational Testing Service. https://doi.org/10.1002/j.2333-8504.1975.tb01052.x

Lord, F. M. (1975b). *Evaluation with artificial data of a procedure for estimating ability and item characteristic curve parameters* (Research Bulletin No. RB-75-33). Princeton: Educational Testing Service. https://doi.org/10.1002/j.2333-8504.1975.tb01073.x

Lord, F. M. (1975c). The 'ability' scale in item characteristic curve theory. *Psychometrika, 40,* 205–217. https://doi.org/10.1007/BF02291567

Lord, F. M. (1977). Practical applications of item characteristic curve theory. *Journal of Educational Measurement, 14,* 117–138. https://doi.org/10.1111/j.1745-3984.1977.tb00032.x

Lord, F. M. (1980a). *Applications of item response theory to practical testing problems*. Hillsdale: Erlbaum.

Lord, F. M. (1980b). Some how and which for practical tailored testing. In L. J. T. van der Kamp, W. F. Langerak, & D. N. M. de Gruijter (Eds.), *Psychometrics for educational debates* (pp. 189–205). New York: Wiley.

Lord, F. M. (1982). Standard error of an equating by item response theory. *Applied Psychological Measurement, 6,* 463–472. https://doi.org/10.1177/014662168200600407

Lord, F. M. (1983a). Maximum likelihood estimation of item response parameters when some responses are omitted. *Psychometrika, 48,* 477–482. https://doi.org/10.1007/BF02293689

Lord, F. M. (1983b). Small *N* justifies Rasch model. In D. J. Weiss (Ed.), *New horizons in testing: Latent trait test theory and computerized adaptive testing* (pp. 51–61). New York: Academic Press. https://doi.org/10.1016/B978-0-12-742780-5.50011-1

Lord, F. M. (1983c). Statistical bias in maximum likelihood estimation of item parameters. *Psychometrika, 48,* 425–435. https://doi.org/10.1007/BF02293684

Lord, F. M. (1983d). Unbiased estimators of ability parameters, of their variance, and of their parallel-forms reliability. Psychometrika, 48, 233–245. https://doi.org/10.1007/BF02294018

Lord, F. M. (1984). *Conjunctive and disjunctive item response functions* (Research Report No. RR-84-45). Princeton: Educational Testing Service. https://doi.org/10.1002/j.2330-8516.1984.tb00085.x

Lord, F. M. (1986). Maximum likelihood and Bayesian parameter estimation in item response theory. *Journal of Educational Measurement, 23,* 157–162. https://doi.org/10.1111/j.1745-3984.1986.tb00241.x

Lord, F. M., & Novick, M. R. (1968). *Statistical theories of mental test scores.* Reading: Addison-Wesley.

Lord, F. M., & Pashley, P. (1988). *Confidence bands for the three-parameter logistic item response curve* (Research Report No. RR-88-67). Princeton: Educational Testing Service. https://doi.org/10.1002/j.2330-8516.1988.tb00323.x

Lord, F. M., & Wild, C. L. (1985). *Contribution of verbal item types in the GRE General Test to accuracy of measurement of the verbal scores* (Research Report No. RR-85-29). Princeton: Educational Testing Service. https://doi.org/10.1002/j.2330-8516.1985.tb00114.x

Lord, F. M., & Wingersky, M. S. (1973). *A computer program for estimating examinee ability and item characteristic curve parameters* (Research Memorandum No. RM-73-02). Princeton: Educational Testing Service.

Lord, F. M., & Wingersky, M. S. (1982). *Sampling variances and covariances of parameter estimates in item response theory* (Research Report No. RR-82-33). Princeton: Educational Testing Service. https://doi.org/10.1002/j.2333-8504.1982.tb01318.x

Lord, F. M., & Wingersky, M. S. (1983). *Comparison of IRT observed-score and true-score 'equatings'* (Research Report No. RR-83-26). Princeton: Educational Testing Service. https://doi.org/10.1002/j.2330-8516.1983.tb00026.x

Lord, F. M., & Wingersky, M. S. (1984). Comparison of IRT true-score and equipercentile observed-score "equatings". *Applied Psychological Measurement, 8*, 453–461. https://doi.org/10.1177/014662168400800409

Lord, F. M., Wingersky, M. S., & Wood, R. L. (1976). *LOGIST: A computer program for estimating examinee ability and item characteristic curve parameters* (Research Memorandum No. RM-76-06). Princeton: Educational Testing Service.

Marco, G. (1977). Item characteristic curve solutions to three intractable testing problems. *Journal of Educational Measurement, 14*, 139–160. https://doi.org/10.1111/j.1745-3984.1977.tb00033.x

McKinley, R. L. (1988). A comparison of six methods for combining multiple IRT item parameter estimates. *Journal of Educational Measurement, 25,* 233–246. https://doi.org/10.1111/j.1745-3984.1988.tb00305.x

McKinley, R. L. (1989a). *Confirmatory analysis of test structure using multidimensional item response theory* (Research Report No. RR-89-31). Princeton: Educational Testing Service. https://doi.org/10.1002/j.2330-8516.1989.tb00145.x

McKinley, R. L. (1989b). Methods plainly speaking: An introduction to item response theory. *Measurement and Evaluation in Counseling and Development, 22*, 37–57.

McKinley, R. L., & Kingston, N. M. (1987). *Exploring the use of IRT equating for the GRE subject test in mathematics* (Research Report No. RR-87-21). Princeton: Educational Testing Service. https://doi.org/10.1002/j.2330-8516.1987.tb00225.x

McKinley, R. L., & Mills, C. N. (1985). A comparison of several goodness-of-fit statistics. *Applied Psychological Measurement, 9,* 49–57. https://doi.org/10.1177/014662168500900105

McKinley, R. L., & Schaeffer, G. A. (1989). *Reducing test form overlap of the GRE Subject Test in Mathematics using IRT triple-part equating* (Research Report No. RR-89-08). Princeton: Educational Testing Service. https://doi.org/10.1002/j.2330-8516.1989.tb00334.x

McLeod, L. D., & Lewis, C. (1999). Detecting item memorization in the CAT environment. *Applied Psychological Measurement, 23,* 147–160. https://doi.org/10.1177/01466219922031275

McLeod, L. D., Lewis, C., & Thissen, D. (2003). A Bayesian method for the detection of item preknowledge in computerized adaptive testing. *Applied Psychological Measurement, 27,* 121–137. https://doi.org/10.1177/0146621602250534

Messick, S. J. (1985). *The 1986 NAEP design: Changes and challenges* (Research Memorandum No. RM-85-02). Princeton: Educational Testing Service.

Messick, S. J., Beaton, A. E., Lord, F. M., Baratz, J. C., Bennett, R. E., Duran, R. P., … Wainer, H. (1983). *National Assessment of Educational Progress reconsidered: A new design for a new era* (NAEP Report No. 83–01). Princeton: Educational Testing Service.

Mislevy, R. J. (1984). Estimating latent distributions. *Psychometrika, 49,* 359–381. https://doi.org/10.1007/BF02306026

Mislevy, R. (1985). *Inferences about latent populations from complex samples* (Research Report No. RR-85-41). Princeton: Educational Testing Service. https://doi.org/10.1002/j.2330-8516.1985.tb00126.x

Mislevy, R. (1986a). *A Bayesian treatment of latent variables in sample surveys* (Research Report No. RR-86-01). Princeton: Educational Testing Service. https://doi.org/10.1002/j.2330-8516.1986.tb00155.x

Mislevy, R. (1986b). Bayes modal estimation in item response models. *Psychometrika, 51,* 177–195. https://doi.org/10.1007/BF02293979

Mislevy, R. (1986c). Recent developments in the factor analysis of categorical variables. *Journal of Educational Statistics, 11,* 3–31. https://doi.org/10.2307/1164846

Mislevy, R. (1987a). Exploiting auxiliary information about examinees in the estimation of item parameters. *Applied Psychological Measurement, 11,* 81–91. https://doi.org/10.1177/014662168701100106

Mislevy, R. (1987b). Recent developments in item response theory with implications for teacher certification. *Review of Research in Education, 14,* 239–275. https://doi.org/10.2307/1167313

Mislevy, R. (1988). Exploiting auxiliary information about items in the estimation of Rasch item parameters. *Applied Psychological Measurement, 12,* 281–296. https://doi.org/10.1177/014662168801200306

Mislevy, R. (1989). *Foundations of a new test theory* (Research Report No. RR-89-52). Princeton: Educational Testing Service. https://doi.org/10.1002/j.2333-8504.1982.tb01336.x

Mislevy, R. (1991). Randomization-based inference about latent variables from complex samples. *Psychometrika, 56,* 177–196 https://doi.org/10.1007/BF02294457

Mislevy, R. (1993a). Foundations of a new test theory. In N. Frederiksen, R. J. Mislevy, & I. I. Bejar (Eds.), *Test theory for a new generation of tests* (pp. 19–39). Hillsdale: Erlbaum.

Mislevy, R. (1993b). Some formulas for use with Bayesian ability estimates. *Educational and Psychological Measurement, 53,* 315–328. https://doi.org/10.1177/0013164493053002002

Mislevy, R. J., & Bock, R. D. (1983). *BILOG: Item analysis and test scoring with binary logistic models [Computer software].* Mooresville: Scientific Software.

Mislevy, R., & Bock, R. D. (1989). A hierarchical item-response model for educational testing. In R. D. Bock (Ed.), *Multilevel analysis of educational data* (pp. 57–75). San Diego: Academic Press.

Mislevy, R., & Chang, H.-H. (2000). Does adaptive testing violate local independence? *Psychometrika, 65,* 149–156. https://doi.org/10.1007/BF02294370

Mislevy, R. J., & Huang, C.-W. (2007). Measurement models as narrative structures. In M. von Davier & C. H. Carstensen (Eds.), *Multivariate and mixture distribution Rasch model: Extensions and applications* (pp.15–35). New York: Springer. https://doi.org/10.1007/978-0-387-49839-3_2

Mislevy, R., & Levy, R. (2007). Bayesian psychometric modeling from an evidence centered design perspective. In C. R. Rao & S. Sinharay (Eds.), *Handbook of statistics: Vol. 26. Psychometrics* (pp. 839–866). Amsterdam: Elsevier.

Mislevy, R. J., & Sheehan, K. M. (1989a). Information matrices in latent-variable models. *Journal of Educational Statistics, 14,* 335–350. https://doi.org/10.2307/1164943

Mislevy, R. J., & Sheehan, K. M. (1989b). The role of collateral information about examinees in item parameter estimation. *Psychometrika, 54,* 661–679. https://doi.org/10.1007/BF02296402

Mislevy, R., & Stocking, M. L. (1987). *A consumer's guide to LOGIST and BILOG* (Research Report No. RR-87-43). Princeton: Educational Testing Service. https://doi.org/10.1002/j.2330-8516.1987.tb00247.x

Mislevy, R., & Verhelst, N. (1990). Modeling item responses when different subjects employ different solution strategies. *Psychometrika, 55,* 195–215. https://doi.org/10.1007/BF02295283

Mislevy, R. J., & Wilson, M. (1996). Marginal maximum likelihood estimation for a psychometric model of discontinuous development. *Psychometrika, 61,* 41–71. https://doi.org/10.1007/BF02296958

Mislevy, R. J., & Wu, P.-K. (1988). *Inferring examinee ability when some item responses are missing* (Research Report No. RR-88-48). Princeton: Educational Testing Service. https://doi.org/10.1002/j.2330-8516.1988.tb00304.x

Mislevy, R. J., & Wu, P.-K. (1996). *Missing responses and IRT ability estimation: Omits, choice, time limits, and adaptive testing* (Research Report No. RR-96-30). Princeton: Educational Testing Service. https://doi.org/10.1002/j.2333-8504.1996.tb01708.x

Mislevy, R. J., Beaton, A. E., Kaplan, B. A., & Sheehan, K. M. (1992a). Estimating population characteristics from sparse matrix samples of item responses. *Journal of Educational Measurement, 29,* 133–161. https://doi.org/10.1111/j.1745-3984.1992.tb00371.x

Mislevy, R. J., Johnson, E. G., & Muraki, E. (1992b). Scaling procedures in NAEP. *Journal of Educational Statistics, 17,* 131–154. https://doi.org/10.2307/1165166

Muraki, E. (1990). Fitting a polytomous item response model to Likert-type data. *Applied Psychological Measurement, 14,* 59–71. https://doi.org/10.1177/014662169001400106

Muraki, E. (1992a). A generalized partial credit model: Application of an EM algorithm. *Applied Psychological Measurement, 16,* 159–176. https://doi.org/10.1177/014662169201600206

Muraki, E. (1992b). *RESGEN item response generator* (Research Report No. RR-92-07). Princeton: Educational Testing Service.

Muraki, E. (1993). Information functions of the generalized partial credit model. *Applied Psychological Measurement, 17,* 351–363. https://doi.org/10.1177/014662169301700403

Muraki, E. (1999). Stepwise analysis of differential item functioning based on multiple-group partial credit model. *Journal of Educational Measurement, 36,* 217–232. https://doi.org/10.1111/j.1745-3984.1999.tb00555.x

Muraki, E., & Bock, R. D. (1993). *PARSCALE: IRT-based test scoring and item analysis for graded items and rating scales* [Computer program]. Chicago: Scientific Software.

Muraki, E., & Carlson, J. E. (1995). Full information factor analysis for polytomous item responses. *Applied Psychological Measurement, 19,* 73–90. https://doi.org/10.1177/014662169501900109

Muraki, E., Hombo, C. M., & Lee, Y.-W. (2000). Equating and linking of performance assessments. *Applied Psychological Measurement, 24,* 325–337. https://doi.org/10.1177/01466210022031787

Pashley, P. J. (1991). *An alternative three-parameter logistic item response model* (Research Report No. RR-91-10). Princeton: Educational Testing Service. https://doi.org/10.1002/j.2333-8504.1991.tb01376.x

Pashley, P. J. (1992). *Graphical IRT-based DIF analyses* (Research Report No. RR-92-66). Princeton: Educational Testing Service. https://doi.org/10.1002/j.2333-8504.1992.tb01497.x

Peterson, N. S., Cook, L. L., & Stocking, M. L. (1983). IRT versus conventional equating methods: A comparative study of scale stability. *Journal of Educational Statistics, 8*, 137–156.

Rao, C. R., & Sinharay, S. (Eds.). (2007). *Handbook of statistics: Vol. 26. Psychometrics*. Amsterdam: Elsevier.

Rasch, G. (1960). *Probabilistic models for some intelligence and attainment tests*. Copenhagen: Nielsen & Lydiche.

Reckase, M. D., & McKinley, R. L. (1991). The discriminating power of items that measure more than one dimension. *Applied Psychological Measurement, 15,* 361–373. https://doi.org/10.1177/014662169101500407

Rijmen, F. (2006). *BNL: A Matlab toolbox for Bayesian networks with logistic regression nodes* (Technical Report). Amsterdam: VU University Medical Center.

Rijmen, F. (2009). *Efficient full information maximum likelihood estimation for multidimensional IRT models* (Research Report No. RR-09-03). Princeton: Educational Testing Service. https://doi.org/10.1002/j.2333-8504.2009.tb02160.x

Rijmen, F. (2010). Formal relations and an empirical comparison among the bi-factor, the testlet, and a second-order multidimensional IRT model. *Journal of Educational Measurement, 47,* 361–372. https://doi.org/10.1111/j.1745-3984.2010.00118.x

Rijmen, F., Tuerlinckx, F., De Boeck, P., & Kuppens, P. (2003). A nonlinear mixed model framework for item response theory. *Psychological Methods, 8,* 185–205. https://doi.org/10.1037/1082-989X.8.2.185

Rijmen, F., Jeon, M., von Davier, M., & Rabe-Hesketh, S. (2013). A general psychometric approach for educational survey assessments: Flexible statistical models and efficient estimation methods. In L. Rutkowski, M. von Davier, & D. Rutkowski (Eds.), *A handbook of international large-scale assessment data analysis*. London: Chapman & Hall.

Roberts, J. S. (n.d.). *Item response theory models for unfolding*. Retrieved from http://www.psychology.gatech.edu/unfolding/Intro.html

Roberts, J. S., & Laughlin, J. E (1996). A unidimensional item response model for unfolding from a graded disagree-agree response scale. *Applied Psychological Measurement, 20,* 231–255. https://doi.org/10.1177/014662169602000305

Roberts, J. S., Donoghue, J. R., & Laughlin, L. E. (2000). A general item response theory model for unfolding unidimensional polytomous responses. *Applied Psychological Measurement, 24,* 3–32. https://doi.org/10.1177/01466216000241001

Roberts, J. S., Donoghue, J. R., & Laughlin, L. E. (2002). Characteristics of MML/EAP parameter estimates in the generalized graded unfolding model. *Applied Psychological Measurement, 26,* 192–207. https://doi.org/10.1177/01421602026002006

Rock, D. A., & Pollack, J. M. (2002). *A model-based approach to measuring cognitive growth in pre-reading and reading skills during the kindergarten year* (Research Report No. RR-02-18). Princeton: Educational Testing Service. https://doi.org/10.1002/j.2333-8504.2002.tb01885.x

Rose, N., von Davier, M., & Xu, X. (2010). *Modeling nonignorable missing data with item response theory (IRT)* (Research Report No. RR-10-11). Princeton: Educational Testing Service. https://doi.org/10.1002/j.2333-8504.2010.tb02218.x

Rosenbaum, P. R. (1984). *Testing the local independence assumption in item response theory* (Research Report No. RR-84-09). Princeton: Educational Testing Service. https://doi.org/10.1002/j.2330-8516.1984.tb00049.x

Rosenbaum, P. R. (1985). Comparing distributions of item responses for two groups. *British Journal of Mathematical and Statistical Psychology, 38,* 206–215. https://doi.org/10.1111/j.2044-8317.1985.tb00836.x

Rosenbaum, P. R. (1987). Comparing item characteristic curves. *Psychometrika, 52,* 217–233. https://doi.org/10.1007/BF02294236

Ross, J. (1966). An empirical study of a logistic mental test model. *Psychometrika, 31,* 325–340. https://doi.org/10.1007/BF02289466

Rotou, O., Patsula, L. N., Steffen, M., & Rizavi, S. M. (2007). *Comparison of multistage tests with computerized adaptive and paper-and-pencil tests* (Research Report No. RR-07-04). Princeton: Educational Testing Service. https://doi.org/10.1002/j.2333-8504.2007.tb02046.x

Samejima, F. (1969). Estimation of latent ability using a response pattern of graded scores *Psychometrika*, *34*(4, Whole Pt. 2). https://doi.org/10.1007/BF03372160

Samejima, F. (1972). A general model for free-response data. *Psychometrika*, *37*(1, Whole Pt. 2).

Scheuneman, J. D. (1980). Latent trait theory and item bias. In L. J. T. van der Kamp, W. F. Langerak, & D. N. M. de Gruijter (Eds.), *Psychometrics for educational debates* (pp. 140–151). New York: Wiley.

Schmitt, A. P., Cook, L. L., Dorans, N. J., & Eignor, D. R. (1990). Sensitivity of equating results to different sampling strategies. *Applied Measurement in Education, 3,* 53–71. https://doi.org/10.1207/s15324818ame0301_5

Scrams, D. J., Mislevy, R. J., & Sheehan, K. M. (2002). *An analysis of similarities in item functioning within antonym and analogy variant families* (Research Report No. RR-02-13). Princeton: Educational Testing Service. https://doi.org/10.1002/j.2333-8504.2002.tb01880.x

Sheehan, K. M. (1997). A tree-based approach to proficiency scaling and diagnostic assessment. *Journal of Educational Measurement, 34,* 333–352. https://doi.org/10.1111/j.1745-3984.1997.tb00522.x

Sheehan, K. M., & Lewis, C. (1992). Computerized mastery testing with nonequivalent testlets. *Applied Psychological Measurement, 16,* 65–76. https://doi.org/10.1177/014662169201600108

Sheehan, K. M., & Mislevy, R. J. (1988). *Some consequences of the uncertainty in IRT linking procedures* (Research Report No. RR-88-38). Princeton: Educational Testing Service. https://doi.org/10.1002/j.2330-8516.1988.tb00294.x

Sheehan, K. M., & Mislevy, R. (1990). Integrating cognitive and psychometric models to measure document literacy. *Journal of Educational Measurement, 27,* 255–272 https://doi.org/10.1111/j.1745-3984.1990.tb00747.x

Sheehan, K. M., & Mislevy, R. J. (1994). *A tree-based analysis of items from an assessment of basic mathematics skills* (Research Report No. RR-94-14). Princeton: Educational Testing Service. https://doi.org/10.1002/j.2333-8504.1994.tb01587.x

Sinharay, S. (2003). *Practical applications of posterior predictive model checking for assessing fit of common item response theory models* (Research Report No. RR-03-33). Princeton: Educational Testing Service. https://doi.org/10.1002/j.2333-8504.2003.tb01925.x

Sinharay, S. (2005). Assessing fit of unidimensional item response theory models using a Bayesian approach. *Journal of Educational Measurement, 42,* 375–394. https://doi.org/10.1111/j.1745-3984.2005.00021.x

Sinharay, S. (2006). Bayesian item fit analysis for unidimensional item response theory models. *British Journal of Mathematical and Statistical Psychology, 59,* 429–449. https://doi.org/10.1348/000711005X66888

Sinharay, S., & Johnson, M. S. (2003). *Simulation studies applying posterior predictive model checking for assessing fit of the common item response theory models* (Research Report No. RR-03-28). Princeton: Educational Testing Service. https://doi.org/10.1002/j.2333-8504.2003.tb01920.x

Sinharay, S., & Lu, Y. (2008). A further look at the correlation between item parameters and item fit statistics. *Journal of Educational Measurement, 45,* 1–15. https://doi.org/10.1111/j.1745-3984.2007.00049.x

Sinharay, S., & von Davier, M. (2005). *Extension of the NAEP BRGOUP program to higher dimensions* (Research Report No. RR-05-27). Princeton: Educational Testing Service. https://doi.org/10.1002/j.2333-8504.2005.tb02004.x

Sinharay, S., Johnson, M. S., & Williamson, D. (2003). *An application of a Bayesian hierarchical model for item family calibration* (Research Report No. RR-03-04). Princeton: Educational Testing Service. https://doi.org/10.1002/j.2333-8504.2003.tb01896.x

Sinharay, S., Johnson, M. S., & Stern, H. S. (2006). Posterior predictive assessment of item response theory models. *Applied Psychological Measurement, 30,* 298–321. https://doi.org/10.1177/0146621605285517

Sireci, S. G., Thissen, D., & Wainer, H. (1991). On the reliability of testlet-based tests. *Journal of Educational Measurement, 28,* 237–247. https://doi.org/10.1111/j.1745-3984.1991.tb00356.x

Steinberg, L., Thissen, D., & Wainer, H. (1990). Validity. In H. Wainer, N. J. Dorans, R. Flaugher, B. F. Green, R. J. Mislevy, L. Steinberg, & D. Thissen (Eds.), *Computerized adaptive testing: A primer* (pp. 187–231). Hillsdale: Erlbaum.

Steinberg, L., Thissen, D., & Wainer, H. (2000). Validity. In H. Wainer, N. J. Dorans, D. Eignor, R. Flaugher, B. F. Green, R. J. Mislevy, L. Steinberg, & D. Thissen (Eds.), *Computerized adaptive testing: A primer* (2nd ed., pp. 185–229). Mahwah: Erlbaum.

Stocking, M. L. (1988). *Scale drift in on-line calibration* (Research Report No. RR-88-28). Princeton: Educational Testing Service. https://doi.org/10.1002/j.2330-8516.1988.tb00284.x

Stocking, M. L. (1989). *Empirical estimation errors in item response theory as a function of test properties* (Research Report No. RR-89-05). Princeton: Educational Testing Service. https://doi.org/10.1002/j.2330-8516.1989.tb00331.x

Stocking, M. L. (1990). Specifying optimum examinees for item parameter estimation in item response theory. *Psychometrika, 55,* 461–475. https://doi.org/10.1007/BF02294761

Stocking, M. L., & Lord, F. M. (1983). Developing a common metric in item response theory. *Applied Psychological Measurement, 7,* 201–210. https://doi.org/10.1177/014662168300700208

Stocking, M. L., Swanson, L., & Pearlman, M. (1991a). *Automatic item selection (AIS) methods in the ETS testing environment* (Research Memorandum No. RM-91-05). Princeton: Educational Testing Service.

Stocking, M. L., Swanson, L., & Pearlman, M. (1991b). *Automated item selection using item response theory* (Research Report No. RR-91-09). Princeton: Educational Testing Service. https://doi.org/10.1002/j.2333-8504.1991.tb01375.x

Tang, K. L., & Eignor, D. R. (2001). *A study of the use of collateral statistical information in attempting to reduce TOEFL IRT item parameter estimation sample sizes* (Research Report No. RR-01-11). Princeton: Educational Testing Service. https://doi.org/10.1002/j.2333-8504.2001.tb01853.x

Tatsuoka, K. K. (1986). Diagnosing cognitive errors: Statistical pattern classification based on item response theory. *Behaviormetrika, 13,* 73–86. https://doi.org/10.2333/bhmk.13.19_73

Tatsuoka, K. K. (1990). Toward an integration of item-response theory and cognitive error diagnosis. In N. Frederiksen, R. Glaser, A. Lesgold, & M. G. Shafto (Eds.), *Diagnostic monitoring of skill and knowledge acquisition* (pp. 453–488). Hillsdale: Erlbaum.

Tatsuoka, K. K. (1991). *A theory of IRT-based diagnostic testing* (Office of Naval Research Report). Princeton: Educational Testing Service.

Thissen, D., & Mislevy, R. J. (1990). Testing algorithms. In H. Wainer, N. J. Dorans, R. Flaugher, B. F. Green, R. J. Mislevy, L. Steinberg, & D. Thissen (Eds.), *Computerized adaptive testing: A primer* (pp. 103–135). Hillsdale: Erlbaum.

Thissen, D., & Mislevy, R. J. (2000). Testing algorithms. In H. Wainer, N. J. Dorans, D. Eignor, R. Flaugher, B. F. Green, R. J. Mislevy, L. Steinberg, & D. Thissen (Eds.), *Computerized adaptive testing: A primer* (2nd ed., pp. 101–131). Mahwah: Erlbaum.

Thissen, D., & Steinberg, L. (1986). A taxonomy of item response models. *Psychometrika, 51,* 567–577. https://doi.org/10.1007/BF02295596

Thissen, D., & Wainer, H. (1984). *The graphical display of simulation results with applications to the comparison of robust IRT estimators of ability* (Research Report No. RR-84-36). Princeton: Educational Testing Service. https://doi.org/10.1002/j.2330-8516.1984.tb00076.x

Thissen, D., & Wainer, H. (1985). *Some supporting evidence for Lord's guideline for estimating "c" theory* (Research Report No. RR-85-15). Princeton: Educational Testing Service. https://doi.org/10.1002/j.2330-8516.1985.tb00100.x

Thissen, D., & Wainer, H. (1990). Confidence envelopes for item response theory. *Journal of Educational Statistics, 15,* 113–128. https://doi.org/10.2307/1164765

Thissen, D., Wainer, H., & Rubin, D. (1984). *A computer program for simulation evaluation of IRT ability estimators* (Research Report No. RR-84-37). Princeton: Educational Testing Service. https://doi.org/10.1002/j.2330-8516.1984.tb00077.x

Tucker, L. R. (1946). Maximum validity of a test with equivalent items. *Psychometrika, 11,* 1–13. https://doi.org/10.1007/BF02288894

van Rijn, P., & Rijmen, F. (2015). On the explaining-away phenomenon in multivariate latent variable models. British Journal of Mathematical and Statistical Psychology, 68, 1–22. https://doi.org/10.1111/bmsp.12046

von Davier, A. A., & Wilson, C. (2005). *A didactic approach to the use of IRT true-score equating model* (Research Report No. RR-05-26). Princeton: Educational Testing Service.

von Davier, M. (2007). *Hierarchical general diagnostic models* (Research Report No. RR-07-19). Princeton: Educational Testing Service. https://doi.org/10.1002/j.2333-8504.2007.tb02061.x

von Davier, M. (2008a). A general diagnostic model applied to language testing data. *British Journal of Mathematical and Statistical Psychology, 61,* 287–307. https://doi.org/10.1348/000711007X193957

von Davier, M. (2008b). The mixture general diagnostic model. In G. R. Hancock & K. M. Samuelsen (Eds.), *Advances in latent variable mixture models* (pp. 255–274). Charlotte: Information Age Publishing.

von Davier, M., & Carstensen, C. H. (Eds.). (2007). *Multivariate and mixture distribution Rasch models: Extensions and applications*. New York: Springer. https://doi.org/10.1007/978-0-387-49839-3

von Davier, M. & Molenaar, I. W. (2003). A person-fit index for polytomous Rasch models, latent class models, and their mixture generalizations. *Psychometrika, 68,* 213–228. https://doi.org/10.1007/BF02294798

von Davier, M., & Rost, J. (2007). Mixture distribution item response models. In C. R. Rao & S. Sinharay (Eds.), *Handbook of statistics: Vol. 26. Psychometrics* (pp. 643–661). Amsterdam: Elsevier. https://doi.org/10.1007/978-0-387-49839-3

von Davier, M., & Sinharay, S. (2004). *Application of the stochastic EM method to latent regression models* (Research Report No. RR-04-34). Princeton: Educational Testing Service. https://doi.org/10.1002/j.2333-8504.2004.tb01961.x

von Davier, M., & Sinharay, S. (2007). An importance sampling EM algorithm for latent regression models. *Journal of Educational and Behavioral Statistics, 32,* 233–251. https://doi.org/10.3102/1076998607300422

von Davier, M., & Sinharay, S. (2010). Stochastic approximation methods for latent regression item response models. *Journal of Educational and Behavioral Statistics, 35,* 174–193. https://doi.org/10.3102/1076998609346970

von Davier, M., & von Davier, A. A. (2007). A unified approach to IRT scale linking and scale transformation. *Methodology: European Journal of Research Methods for the Behavioral and Social Sciences, 3,* 115–124. https://doi.org/10.1027/1614-2241.3.3.115

von Davier, M., & von Davier, A. A. (2011). A general model for IRT scale linking and scale transformation. In A. A. von Davier (Ed.), *Statistical models for test equating, scaling, and linking* (pp. 225–242). New York: Springer. https://doi.org/10.1007/978-0-387-98138-3

von Davier, M., & Yamamoto, K. (2004). Partially observed mixtures of IRT models: An extension of the generalized partial-credit model. *Applied Psychological Measurement, 28,* 389–406. https://doi.org/10.1177/0146621604268734

von Davier, M., & Yamamoto, K. (2007). Mixture distribution and HYBRID Rasch models. In M. von Davier & C. H. Carstensen (Eds.), *Multivariate and mixture distribution Rasch models: Extensions and applications* (pp. 99–115). New York: Springer. https://doi.org/10.1007/978-0-387-49839-3

von Davier, M., Sinharay, S., Oranje, A., & Beaton, A. (2007). The statistical procedures used in National Assessment of Educational Progress: Recent developments and future directions. In C. R. Rao & S. Sinharay S. (Eds.), *Handbook of statistics: Vol. 26. Psychometrics* (pp. 1039–1055). Amsterdam: Elsevier.

von Davier, M., DiBello, L., & Yamamoto, K. (2008). Reporting test outcomes using models for cognitive diagnosis. In J. Hartig, E. Klieme, & D. Leutner (Eds.), *Assessment of competencies in educational contexts* (pp. 151–174). Cambridge, MA: Hogrefe & Huber.

von Davier, M., Xu, X., & Carstensen, C. H. (2011). Measuring growth in a longitudinal large-scale assessment with a general latent variable model. *Psychometrika, 76,* 318–336. https://doi.org/10.1007/s11336-011-9202-z

Wainer, H. (1983). On item response theory and computerized adaptive tests: The coming technological revolution in testing. *Journal of College Admission, 28*, 9–16.

Wainer, H. (1990). Introduction and history. In H. Wainer, N. J. Dorans, R. Flaugher, B. F. Green, R. J. Mislevy, L. Steinberg, & D. Thissen (Eds.), *Computerized adaptive testing: A primer* (pp. 1–21). Hillsdale: Erlbaum.

Wainer, H. (2000). Introduction and history. In H. Wainer, N. J. Dorans, D. Eignor, R. Flaugher, B. F. Green, R. J. Mislevy, L. Steinberg, & D. Thissen (Eds.), *Computerized adaptive testing: A primer* (2nd ed., pp. 1–21). Mahwah: Erlbaum.

Wainer, H., & Eignor, D. (2000). Caveats, pitfalls and unexpected consequences of implementing large-scale computerized testing. In H. Wainer, N. J. Dorans, D. Eignor, R. Flaugher, B. F. Green, R. J. Mislevy, L. Steinberg, & D. Thissen (Eds.), *Computer adaptive testing: A primer* (2nd ed., pp. 271–299). Mahwah: Erlbaum.

Wainer, H., & Lewis, C. (1990). Toward a psychometrics for testlets. *Journal of Educational Measurement, 27*, 1–14. https://doi.org/10.1111/j.1745-3984.1990.tb00730.x

Wainer, H., & Mislevy, R. (1990). Item response theory, item calibration, and proficiency estimation. In H. Wainer, N. J. Dorans, D. R. Flaugher, B. F. Green, R. J. Mislevy, L. Steinberg, & D. Thissen (Eds.), *Computer adaptive testing: A primer* (pp. 65–102). Hillsdale: Erlbaum.

Wainer, H., & Mislevy, R. (2000). Item response theory, item calibration, and proficiency estimation. In H. Wainer, N. J. Dorans, D. Eignor, R. Flaugher, B. F. Green, R. J. Mislevy, L. Steinberg, & D. Thissen (Eds.), *Computer adaptive testing: A primer* (2nd ed., pp. 61–100). Mahwah: Erlbaum.

Wainer, H., & Thissen, D. (1982). Some standard errors in item response theory. *Psychometrika, 47,* 397–412. https://doi.org/10.1007/BF02293705

Wainer, H., & Thissen, D. (1987). Estimating ability with the wrong model. *Journal of Educational Statistics, 12,* 339–368. https://doi.org/10.2307/1165054

Wainer, H., & Wang, X. (2000). Using a new statistical model for testlets to score TOEFL. *Journal of Educational Measurement, 37,* 203–220. https://doi.org/10.1111/j.1745-3984.2000.tb01083.x

Wainer, H., Dorans, N. J., Flaugher, R., Green, B. F., Mislevy, R. J., Steinberg, L., & Thissen, D. (Eds.). (1990a). *Computer adaptive testing: A primer*. Hillsdale: Erlbaum.

Wainer, H., Dorans, N. J., Green, B. F., Mislevy, R. J., Steinberg, L., & Thissen, D. (1990b). Future challenges. In H. Wainer, N. J. Dorans, R. Flaugher, B. F. Green, R. J. Mislevy, L. Steinberg, & D. Thissen (Eds.), *Computer adaptive testing: A primer* (pp. 233–270). Hillsdale: Erlbaum.

Wainer, H., Sireci, S. G., & Thissen, D. (1991) Differential testlet functioning: Definitions and detection. *Journal of Educational Measurement, 28,* 197–219. https://doi.org/10.1111/j.1745-3984.1991.tb00354.x

Wainer, H., Wang, X.-B., & Thissen, D. (1994). How well can we compare scores on test forms that are constructed by examinees choice? *Journal of Educational Measurement, 31*, 183–199. https://doi.org/10.1111/j.1745-3984.1994.tb00442.x

Wainer, H., Bradlow, E. T., & Du, Z. (2000a). Testlet response theory: An analog for the 3PL model useful in testlet-based adaptive testing. In W. J. van der Linden & C. A. W. Glas (Eds.), *Computerized adaptive testing: Theory and practice* (pp. 245–269). Dordrecht: Kluwer. https://doi.org/10.1007/0-306-47531-6_13

Wainer, H., Dorans, N. J., Eignor, D., Flaugher, R., Green, B. F., Mislevy, R. J., Steinberg, L., & Thissen, D. (Eds.). (2000b). *Computer adaptive testing: A primer* (2nd ed.). Mahwah: Erlbaum.

Wainer, H., Dorans, N. J., Green, B. F., Mislevy, R. J., Steinberg, L., & Thissen, D. (2000c). Future challenges. In H. Wainer, N. J. Dorans, D. Eignor, R. Flaugher, B. F. Green, R. J. Mislevy,

L. Steinberg, & D. Thissen (Eds.), *Computer adaptive testing: A primer* (2nd ed., pp. 231–269). Mahwah: Erlbaum.

Waller, M. I. (1976). *Estimating parameters in the Rasch model: Removing the effects of random guessing theory* (Research Bulletin No. RB-76-08). Princeton: Educational Testing Service. https://doi.org/10.1002/j.2333-8504.1976.tb01094.x

Wang, X., Bradlow, E. T., & Wainer, H. (2001). *User's guide for SCORIGHT (version 1.2): A computer program for scoring tests built of testlets* (Research Report No. RR-01-06). Princeton: Educational Testing Service.

Wang, X., Bradlow, E. T., & Wainer, H. (2002). A general Bayesian model for testlets: Theory and applications. *Applied Psychological Measurement, 26*, 109–128. https://doi.org/10.1177/0146621602026001007

Wang, X., Bradlow, E. T., & Wainer, H. (2005). *User's guide for SCORIGHT (version 3.0): A computer program for scoring tests built of testlets including a module for covariate analysis* (Research Report No. RR-04-49). Princeton: Educational Testing Service. https://doi.org/10.1002/j.2333-8504.2004.tb01976.x

Wendler, C. L. W., & Walker, M. E. (2006). Practical issues in designing and maintaining multiple test forms for large-scale programs. In S. M. Downing & T. M. Haladyna (Eds.), *Handbook of test development* (pp. 445–467). Mahwah: Erlbaum.

Wingersky, M. S. (1983). LOGIST: A program for computing maximum likelihood procedures for logistic test models. In R. K. Hambleton (Ed.), *Applications of item response theory* (pp. 45–56). Vancouver: Educational Research Institute of British Columbia.

Wingersky, M. S. (1987). *One-stage LOGIST* (Research Report No. RR-87-45). Princeton: Educational Testing Service. https://doi.org/10.1002/j.2330-8516.1987.tb00249.x

Wingersky, M. S., & Lord, F. M. (1984). An investigation of methods for reducing sampling error in certain IRT procedures. *Applied Psychological Measurement, 8,* 347–364. https://doi.org/10.1177/014662168400800312

Wingersky, M. S., & Sheehan, K. M. (1986). *Using estimated item observed-score regressions to test goodness-of-fit of IRT models* (Research Report No. RR-86-23). Princeton: Educational Testing Service. https://doi.org/10.1002/j.2330-8516.1986.tb00178.x

Wingersky, M. S., Cook, L. L., & Eignor, D. R. (1987). *Specifying the characteristics of linking items used for item response theory item calibration* (Research Report No. RR-87-24). Princeton: Educational Testing Service. https://doi.org/10.1002/j.2330-8516.1987.tb00228.x

Winsberg, S., Thissen, D., & Wainer, H. (1984). *Fitting item characteristic curves with spline functions* (Research Report No. RR-84-40). Princeton: Educational Testing Service. https://doi.org/10.1002/j.2330-8516.1984.tb00080.x

Xu, X. (2007). *Monotone properties of a general diagnostic model* (Research Report No. RR-07-25). Princeton: Educational Testing Service. https://doi.org/10.1002/j.2333-8504.2007.tb02067.x

Xu, X., & Douglas, J. (2006). Computerized adaptive testing under nonparametric IRT models. *Psychometrika, 71,* 121–137. https://doi.org/10.1007/s11336-003-1154-5

Xu, X., & Jia, Y. (2011). *The sensitivity of parameter estimates to the latent ability distribution* (Research Report No. RR-11-41). Princeton: Educational Testing Service. https://doi.org/10.1002/j.2333-8504.2011.tb02277.x

Xu, X., & von Davier, M. (2006). *Cognitive diagnostics for NAEP proficiency data* (Research Report No. RR-06-08). Princeton: Educational Testing Service. https://doi.org/10.1002/j.2333-8504.2006.tb02014.x

Xu, X., & von Davier, M. (2008a). *Comparing multiple-group multinomial log-linear models for multidimensional skill distributions in the general diagnostic model* (Research Report No. RR-08-35). Princeton: Educational Testing Service. https://doi.org/10.1002/j.2333-8504.2008.tb02121.x

Xu, X., & von Davier, M. (2008b). *Fitting the structured general diagnostic model to NAEP* data (Research Report No. RR-08-27). Princeton: Educational Testing Service. https://doi.org/10.1002/j.2333-8504.2008.tb02113.x

Xu, X., & von Davier, M. (2008c). Linking for the general diagnostic model. *IERI Monograph Series: Issues and Methodologies in Large-Scale Assessments, 1*, 97–111.
Xu, X., Douglas, J., & Lee, Y.-S. (2011). Linking with nonparametric IRT models. In A. A. von Davier (Ed.), *Statistical models for test equating, scaling, and linking* (pp. 243–258). New York: Springer.
Yamamoto, K. (1989). *HYBRID model of IRT and latent class models* (Research Report No. RR-89-41). Princeton: Educational Testing Service. https://doi.org/10.1002/j.2333-8504.1982.tb01326.x
Yamamoto, K., & Everson, H. T. (1995). *Modeling the mixture of IRT and pattern responses by a modified hybrid model* (Research Report No. RR-95-16). Princeton: Educational Testing Service. https://doi.org/10.1002/j.2333-8504.1995.tb01651.x
Yamamoto, K., & Everson, H. T. (1997). Modeling the effects of test length and test time on parameter estimation using the HYBRID model. In J. Rost & R. Langeheine (Eds.), *Applications of latent trait class models in the social sciences* (pp. 89–99). New York: Waxmann.
Yamamoto, K., & Gitomer, D. H. (1993). Application of a HYBRID model to a test of cognitive skill representation. In N. Frederiksen, R. J. Mislevy, & I. I. Bejar (Eds.), *Test theory for a new generation of tests* (pp. 275–295). Hillsdale: Erlbaum.
Yamamoto, K., & Mazzeo, J. (1992). Item response theory scale linking in NAEP. *Journal of Educational Statistics, 17,* 155–173. https://doi.org/10.2307/1165167
Yan, D., Almond, R. G., & Mislevy, R. J. (2004a). *A comparison of two models for cognitive diagnosis* (Research Report No. RR-04-02). Princeton: Educational Testing Service. https://doi.org/10.1002/j.2333-8504.2004.tb01929.x
Yan, D., Lewis, C., & Stocking, M. L. (2004b). Adaptive testing with regression trees in the presence of multidimensionality. *Journal of Educational and Behavioral Statistics, 29,* 293–316. https://doi.org/10.3102/10769986029003293
Yen, W. M. (1983). Use of the three-parameter logistic model in the development of a standardized achievement test. In R. K. Hambleton (Ed.), *Applications of item response theory* (pp. 123–141). Vancouver: Educational Research Institute of British Columbia.
Yen, W. M., & Fitzpatrick, A. R. (2006). Item response theory. In R. L. Brennan (Ed.), *Educational measurement* (4th ed., pp. 111–153). Westport: American Council on Education and Praeger Publishers.
Zhang, J. (2004). *Comparison of unidimensional and multidimensional approaches to IRT parameter estimation* (Research Report No. RR-04-44). Princeton: Educational Testing Service. https://doi.org/10.1002/j.2333-8504.2004.tb01971.x
Zhang, J. (2005a). *Bias correction for the maximum likelihood estimate of ability* (Research Report No. RR-05-15). Princeton: Educational Testing Service. https://doi.org/10.1002/j.2333-8504.2005.tb01992.x
Zhang, J. (2005b). *Estimating multidimensional item response models with mixed structure* (Research Report No. RR-05-04). Princeton: Educational Testing Service. http://dx.doi.org/10.1002/j.2333-8504.2005.tb01981.x
Zhang, J. (2007). Conditional covariance theory and DETECT for polytomous items. *Psychometrika, 72,* 69–91. https://doi.org/10.1007/s11336-004-1257-7
Zhang, J., & Lu, T. (2007). *Refinement of a bias-correction procedure for the weighted likelihood estimator of ability* (Research Report No. RR-07-23). Princeton: Educational Testing Service. https://doi.org/10.1002/j.2333-8504.2007.tb02065.x
Zhang, J., & Stout, W. (1997). On Holland's Dutch identity conjecture. *Psychometrika, 62,* 375–392. https://doi.org/10.1007/BF02294557
Zhang, J. & Stout, W. (1999a). Conditional covariance structure of generalized compensatory multidimensional items. *Psychometrika, 64,* 129–152. https://doi.org/10.1007/BF02294532
Zhang, J. & Stout, W. (1999b). The theoretical detect index of dimensionality and its application to approximate simple structure. *Psychometrika, 64,* 213–249. https://doi.org/10.1007/BF02294536

Zwick, R. J. (1987). Assessing the dimensionality of NAEP reading data. *Journal of Educational Measurement, 24,* 293–308. https://doi.org/10.1111/j.1745-3984.1987.tb00281.x
Zwick, R. J. (1990). When do item response function and Mantel-Haenszel definitions of differential item functioning coincide? *Journal of Educational Statistics, 15,* 185–197. https://doi.org/10.2307/1165031
Zwick, R. J. (1991). Effects of item order and context on estimation of NAEP reading proficiency. *Educational Measurement: Issues and Practice, 10*(3), 10–16. https://doi.org/10.1111/j.1745-3992.1991.tb00198.x
Zwick, R. J., Thayer, D. T., & Wingersky, M. S. (1995). Effect of Rasch calibration on ability and DIF estimation in computer-adaptive tests. *Journal of Educational Measurement, 32,* 341–363. https://doi.org/10.1111/j.1745-3984.1995.tb00471.x

Chapter 6
Research on Statistics

Henry Braun

Since its founding in 1947, ETS has supported research in a variety of areas—a fact attested to by the many different chapters comprising this volume. As a private, nonprofit organization known primarily for its products and services related to standardized testing, it comes as no surprise that ETS conducted extensive research in educational measurement and psychometrics, which together provide the scientific foundations for the testing industry. This work is documented in the chapters in this book. At the same time, a good part of educational measurement and perhaps most of psychometrics can be thought of as drawing upon—and providing an impetus for extending—work in theoretical and applied statistics. Indeed, many important developments in statistics are to be found in the reports alluded to above.

One may ask, therefore, if there is a need for a separate chapter on statistics. The short answer is yes. The long answer can be found in the rest of the chapter. A review of the ETS Research Report (RR) series and other archival materials reveals that a great deal of research in both theoretical and applied statistics was carried out at ETS, both by regular staff members and by visitors. Some of the research was motivated by longstanding problems in statistics, such as the Behrens-Fisher problem or the problem of simultaneous inference, and some by issues arising at ETS during the course of business. Much of this work is distinguished by both its depth and generality. Although a good deal of statistics-related research is treated in other chapters, much is not.

The purpose of this chapter, then, is to tell *a* story of statistics research at ETS. It is not *the* story, as it is not complete; rather, it is structured in terms of a number of major domains and, within each domain, a roughly chronological narrative of key highlights. As will be evident, the boundaries between domains are semipermeable so that the various narratives sometimes intermix. Consequently, reference will also

H. Braun (✉)
Boston College, Chestnut Hill, MA, USA
e-mail: braunh@bc.edu

R.E. Bennett, M. von Davier (eds.), *Advancing Human Assessment*,
Methodology of Educational Measurement and Assessment,
DOI 10.1007/978-3-319-58689-2_6

be made to topic coverage in other chapters. The writing of this chapter was made more challenging by the fact that some important contributions made by ETS researchers or by ETS visitors (supported by ETS) did not appear in the RR series but in other technical report series and/or in the peer-reviewed literature. A good faith effort was made to identify some of these contributions and include them as appropriate.

The chapter begins with a treatment of classic linear models, followed by sections on latent regression, Bayesian methods, and causal inference. It then offers shorter treatments of a number of topics, including missing data, complex samples, and data displays. A final section offers some closing thoughts on the statistical contributions of ETS researchers over the years.

6.1 Linear Models

Linear models, comprising such techniques as regression, analysis of variance, and analysis of covariance, are the workhorses of applied statistics. Whether offering convenient summaries of data patterns, modeling data to make predictions, or even serving as the basis for inferring causal relationships, they are both familiar tools and the source of endless questions and puzzles that have fascinated statisticians for more than a century. Research on problems related to linear models goes back to ETS's earliest days and continues even today.

From the outset, researchers were interested in the strength of the relationship between scores on admissions tests and school performance as measured by grades. The best known example, of course, is the relationship between *SAT*® test scores and performance in the first year of college. The strength of the relationship was evidence of the predictive validity of the test, with predictive validity being one component of the *validity trinity*.[1] From this simple question, many others arose: How did the strength of the relationship change when other predictors (e.g., high school grades) were included in the model? What was the impact of restriction of range on the observed correlations, and to what extent was differential restriction of range the cause of the variation in validity coefficients across schools? What could explain the year-to-year volatility in validity coefficients for a given school, and how could it be controlled? These and other questions that arose over the years provided the impetus for a host of methodological developments that have had an impact on general statistical practice. The work at ETS can be divided roughly into three categories: computation, inference, and prediction.

[1] The validity trinity comprises content validity, criterion-related validity, and predictive validity.

6.1.1 Computation

In his doctoral dissertation, Beaton (1964) developed the sweep operator, which was one of the first computational algorithms to take full advantage of computer architecture to improve statistical calculations with respect to both speed and the size of the problem that could be handled. After coming to ETS, Beaton and his colleagues developed F4STAT, an expandable subroutine library to carry out statistical calculations that put ETS in the forefront of statistical computations. More on F4STAT can be found in Beaton and Barone (Chap. 8, this volume). (It is worth noting that, over the years, the F4STAT system has been expanded and updated to more current versions of FORTRAN and is still in use today.) Beaton et al. (1972) considered the problem of computational accuracy in regression. Much later, Longford, in a series of reports (Longford 1987a, b, 1993), addressed the problem of obtaining maximum likelihood estimates in multilevel models with random effects. Again, accuracy and speed were key concerns. (Other aspects of multilevel models are covered in Sect. 6.2.3). A contribution to robust estimation of regression models was authored by Beaton and Tukey (1974).

6.1.2 Inference

The construction of confidence intervals with specific confidence coefficients is another problem that appears throughout the RR series, with particular attention to the setting of simultaneous confidence intervals when making inferences about multiple parameters, regression planes, and the like. One of the earliest contributions was by Abelson (1953) extending the Neyman-Johnson technique for regression. Aitkin (1973) made further developments. Another famous inference problem, the Behrens-Fisher problem, attracted the attention of Potthoff (1963, 1965), who devised Scheffé-type tests. Beaton (1981) used a type of permutation test approach to offer a way to interpret the coefficients of a least squares fit in the absence of random sampling. This was an important development, as many of the data sets subjected to regression analysis do not have the required pedigree and, yet, standard inferential procedures are applied nonetheless. A. A. von Davier (2003a) treated the problem of comparing regression coefficients in large samples. Related work can be found in Moses and Klockars (2009).

A special case of simultaneous inference arises in analysis of variance (ANOVA) when comparisons among different levels of a factor are of interest and control of the overall error rate is desired. This is known as the problem of multiple comparisons, and many procedures have been devised. Braun and Tukey (1983) proposed a new procedure and evaluated its operating characteristics. Zwick (1993) provided a comprehensive review of multiple comparison procedures. Braun (1994) edited Volume VIII of *The Collected Works of John W. Tukey*, a volume dedicated to Tukey's work in the area of simultaneous inference. Especially noteworthy in this collection is that Braun, in collaboration with ETS colleagues Kaplan, Sheehan, and Wang, prepared a corrected, complete version of the never-published manuscript (1953) by Tukey titled *The Problem of Multiple Comparisons* (1994), which set the stage for the modern treatment of simultaneous inference. A review of Tukey's contributions to simultaneous inference was presented in Benjamini and Braun (2003).

6.1.3 Prediction

Most of the standardized tests that ETS was and is known for are intended for use in admissions to higher education. A necessary, if not sufficient, justification for their utility is their predictive validity; that is, for example, that scores on the SAT are strongly correlated with first year averages (FYA) in college and, more to the point, that they possess explanatory power above and beyond that available with the use of other quantitative measures, such as high school grades. Another important consideration is that the use of the test does not inappropriately disadvantage specific subpopulations. (A more general discussion of validity can be found in Chap. 16 by Kane and Bridgeman, this volume. See also Kane 2013). Another aspect of test fairness, differential prediction, is discussed in the chapter by Dorans and Puhan (Chap. 4, this volume).

Consequently, the study of prediction equations and, more generally, prediction systems has been a staple of ETS research. Most of the validity studies conducted at ETS were done under the auspices of particular programs and the findings archived in the report series of those programs. At the same time, ETS researchers were continually trying to improve the quality and utility of validity studies through developing new methodologies.

Saunders (1952) investigated the use of the analysis of covariance (ANCOVA) in the study of differential prediction. Rock (1969) attacked a similar problem using the notion of moderator variables. Browne (1969) published a monograph that proposed measures of predictive accuracy, developed estimates of those measures, and evaluated their operating characteristics.

Tucker established ETS's test validity procedures and supervised their implementation until his departure to the University of Illinois. He published some of the earliest ETS work in this area (1957, 1963). His first paper proposed a procedure to simplify the prediction problem with many predictors by constructing a smaller number of composite predictors. The latter paper, titled *Formal Models for a Central Prediction System*, tackled a problem that bedeviled researchers in this area. The problem can be simply stated: Colleges receive applications from students attending many different high schools, each with its own grading standards. Thus, high school grade point averages (HSGPA) are not comparable even when they are reported on a common scale. Consequently, including HSGPA in a single prediction equation without any adjustment necessarily introduces noise in the system and induces bias in the estimated regression coefficients. Standardized test scores, such as the SAT, are on a common scale—a fact that surely contributes to their strong correlation with FYA. Tucker's monograph discusses three approaches to constructing composite predictors based on placing multiple high school grades on a common scale for purposes of predicting college grades. This work, formally published in Tucker (1963), led to further developments, which were reviewed by Linn (1966) and, later, by Young and Barrett (1992). More recently, Zwick (2013) and Zwick and Himelfarb (2011) conducted further analyses of HSGPA as a predictor of FYA, with a focus on explaining why HSGPA tends to overpredict college performance for students from some demographic subgroups.

Braun and Szatrowski (1984a, b) investigated a complementary prediction problem. When conducting a typical predictive validity study at an institution, the data are drawn from those students who matriculate and obtain a FYA. For schools that use the predictor in the admissions process, especially those that are at least moderately selective, the consequence is a restriction of range for the predictor and an attenuated correlation. Although there are standard corrections for restriction of range, they rest on untestable assumptions. At the same time, unsuccessful applicants to selective institutions likely attend other institutions and obtain FYAs at those institutions. The difficulty is that FYAs from different institutions are not on a common scale and cannot be used to carry out an *ideal validity study* for a single institution in which the prediction equation is estimated on, for example, all applicants.

Using data from the Law School Admissions Council, Braun and Szatrowski (1984a, b) were able to link the FYA grade scales for different law schools to a single, common scale and, hence, carry out institutional validity studies incorporating data from nearly all applicants. The resulting fitted regression planes differed from the standard estimates in expected ways and were in accord with the fitted planes obtained through an Empirical Bayes approach. During the 1980s, there was considerable work on using Empirical Bayes methods to improve the accuracy and stability of prediction equations. (These are discussed in the section on Bayes and Empirical Bayes.)

A longstanding concern with predictive validity studies, especially in the context of college admissions, is the nature of the criterion. In many colleges, freshmen enroll in a wide variety of courses with very different grading standards. Consequently, first year GPAs are rather heterogeneous, which has a complex impact on the observed correlations with predictors. This difficulty was tackled by Ramist et al. (1990). They investigated predictive validity when course-level grades (rather than FYAs) were employed as the criterion. Using this more homogeneous criterion yielded rather different results for the correlations with SAT alone, HSGPA alone, and SAT with HSGPA. Patterns were examined by subject and course rigor, as was variation across the 38 colleges in the study. This approach was further pursued by Lewis et al. (1994) and by Bridgeman et al. (2008).

Over the years, Willingham maintained an interest in investigating the differences between grades and test scores, especially with respect to differential predictive validity (Willingham et al. 2002). Related contributions include Lewis and Willingham (1995) and Haberman (2006). The former showed how restriction of range can affect estimates of *gender bias* in prediction and proposed some strategies for generating improved estimates. The latter was concerned with the bias in predicting multinomial responses and the use of different penalty functions in reducing that bias.

Over the years, ETS researchers also published volumes that explored aspects of test validity and test use, with some attention to methodological considerations. Willingham (1988) considered issues in testing *handicapped people* (a term now replaced by the term *students with disabilities*) for the SAT and *GRE*® programs. The chapter in that book by Braun et al. (1988) studied the predictive validity for those testing programs for students with different disabilities. Willingham and Cole (1997) examined testing issues in gender-related fairness, with some attention to the implications for predictive validity.

6.1.4 Latent Regression

Latent regression methods were introduced at ETS by Mislevy (1984) for use in the National Assessment of Educational Progress (NAEP) and are further described in Sheehan and Mislevy (1989), Mislevy (1991), and Mislevy et al. (1992). An overview of more recent developments is given in M. von Davier et al. (2006) and M. von Davier and Sinharay (2013). Mislevy's key insight was that NAEP was not intended to, and indeed was prohibited from, reporting scores at the individual level. Instead, scores were to be reported at various levels of aggregation, either by political jurisdiction or by subpopulation of students. By virtue of the matrix sampling design of NAEP, the amount of data available for an individual student is relatively sparse. Consequently, the estimation bias in statistics of interest may be considerable, but can be reduced through application of latent regression techniques. With latent regression models, background information on students is combined with their responses to cognitive items to yield unbiased estimates of score distributions at the subpopulation level—provided that the characteristics used to define the subpopulations are included in the latent regression model. This topic is also dealt with in the chapter by Beaton and Barone (Chap. 8, this volume), especially in Appendix A; the chapter by Kirsch et al. (Chap. 9, this volume) describes assessments of literacy skills in adult populations that use essentially the same methodologies.

In NAEP, the fitting of a latent regression model results in a family of posterior distributions. To generate plausible values, five members of the family are selected at random, and from each a single random draw is made.[2] The plausible values are used to produce estimates of the target population parameters and to estimate the measurement error components of the total variance of the estimates. Note that latent regression models are closely related to empirical Bayes models.

Latent regression models are very complex and, despite more than 25 years of use, many questions remain. In particular, there are attempts to simplify the

[2] In the series of international surveys of adult skills, 10 PV are generated for each respondent.

estimation procedure without increasing the bias. Comparisons of the ETS approach with so-called direct estimation methods were carried out by M. von Davier (2003b). ETS researchers continue to refine the models and the estimation techniques (Li and Oranje 2007; Li et al. 2007; M. von Davier and Sinharay 2010). Goodness-of-fit issues are addressed in Sinharay et al. (2009). In that paper, the authors apply a strategy analogous to Bayesian posterior model checking to evaluate the quality of the fit of a latent regression model and apply the technique to NAEP data.

6.2 Bayesian Methods

Bayesian inference comes in many different flavors, depending on the type of probability formalism that is employed. The main distinction between Bayesian inference and classical, frequentist inference (an amalgam of the approaches of Fisher and Neyman) is that, in the former, distribution parameters of interest are treated as random quantities, rather than as fixed quantities. The Bayesian procedure requires specification of a so-called prior distribution, based on information available before data collection. Once relevant data are collected, they can be combined with the prior distribution to yield a so-called posterior distribution which represents current belief about the likely values of the parameter. This approach can be directly applied to evaluating competing hypotheses, so that one can speak of the posterior probabilities associated with different hypotheses—these are the conditional probabilities of the hypotheses, given prior beliefs and the data collected. As many teachers of elementary (and not so elementary) statistics are aware, these are the kinds of interpretations that many ascribe (incorrectly) to the results of a frequentist analysis.

Over the last 50 years, the Bayesian approach to statistical inference has gained more adherents, particularly as advances in computer hardware/software have made Bayesian calculations more feasible. Both theoretical developments and successful applications have moved Bayesian and quasi-Bayesian methods closer to normative statistical practice. In this respect, a number of ETS researchers have made significant contributions in advancing the Bayesian approach, as well as providing a Bayesian perspective on important statistical issues. This section is organized into three sections: Bayes for classical models, later Bayes, and empirical Bayes.

6.2.1 *Bayes for Classical Models*

Novick was an early proponent of Bayes methods and a prolific contributor to the Bayesian analysis of classical statistical and psychometric models. Building on earlier work by Bohrer (1964) and Lindley (1969b, c, 1970), Novick and colleagues tackled estimation problems in multiple linear regression with particular attention to applications to predictive validity (Novick et al. 1971, 1972; Novick and Thayer 1969). These studies demonstrated the superior properties of Bayesian regression estimates when many models were to be estimated. The advantage of *borrowing strength* across multiple contexts anticipated later work by Rubin and others who employed Empirical Bayes methods. Rubin and Stroud (1977) continued this work by treating the problem of Bayesian estimation in unbalanced multivariate analysis of variance (MANOVA) designs.

Birnbaum (1969) presented a Bayesian formulation of the logistic model for test scores, which was followed by Lindley (1969a) and Novick and Thayer (1969), who studied the Bayesian estimation of true scores. Novick et al. (1971) went on to develop a comprehensive Bayesian analysis of the classical test theory model addressing such topics as reliability, validity, and prediction.

During this same period, there were contributions of a more theoretical nature as well. For example, Novick (1964) discussed the differences between the subjective probability approach favored by Savage and the logical probability approach favored by Jefferies, arguing for the relative advantages of the latter. Somewhat later, Rubin (1975) offered an example of where Bayesian and standard frequentist inferences can differ markedly. Rubin (1979a) provided a Bayesian analysis of the bootstrap procedure proposed by Efron, which had already achieved some prominence. Rubin showed that the bootstrap could be represented as a Bayesian procedure—but with a somewhat unusual prior distribution.

6.2.2 Later Bayes

The development of graphical models and associated inference networks found applications in intelligent tutoring systems. The Bayesian formulation is very natural, since prior probabilities on an individual's proficiency profile could be obtained from previous empirical work or simply based on plausible (but not necessarily correct) assumptions about the individual. As the individual attempts problems, data accumulates, the network is updated, and posterior probabilities are calculated. These posterior probabilities can be used to select the next problem in order to optimize some criterion or to maximize the information with respect to a subset of proficiencies.

At ETS, early work on intelligent tutoring systems was carried out by Gitomer and Mislevy under a US Air Force contract to develop a tutoring system for troubleshooting hydraulic systems on F-15s. The system, called HYDRIVE, was one of the first to employ rigorous probability models in the analysis of sequential data. The model is described in Mislevy and Gitomer (1995), building on previous work by Mislevy (1994a, b). Further developments can be found in Almond et al. (2009).

Considerable work in the Bayesian domain concerns issues of either computational efficiency or model validation. Sinharay (2003a, b, 2006) has made contributions to both. In particular, the application of posterior predictive model checking to Bayesian measurement models promises to be an important advance in refining these models. At the same time, ETS researchers have developed Bayesian formulations of hierarchical models (Johnson and Jenkins 2005) and extensions to testlet theory (Wang et al. 2002).

6.2.3 Empirical Bayes

The term *empirical Bayes* (EB) actually refers to a number of different strategies to eat the Bayesian omelet without breaking the Bayesian eggs; that is, EB is intended to reap the benefits of a Bayesian analysis without initially fully specifying a Bayesian prior. Braun (1988) described some of the different methods that fall under this rubric. We have already noted fully Bayesian approaches to the estimation of prediction equations. Subsequently, Rubin (1980d) proposed an EB strategy to deal with a problem that arose from the use of standardized test scores and school grades in predicting future performance; namely, the prediction equation for a particular institution (e.g., a law school) would often vary considerably from year to year—a phenomenon that caused some concern among admissions officers. Although the causes of this volatility, such as sampling variability and differential restriction of range, were largely understood, they did not lead immediately to a solution.

Rubin's version of EB for estimating many multiple linear regression models (as would be the case in a validity study of 100+ law schools) postulated a multivariate normal prior distribution, but did not specify the parameters of the prior. These were estimated through maximum likelihood along with estimates of the regression coefficients for each institution. In this setting, the resulting EB estimate of the regression model for a particular institution can be represented as a weighted combination of the ordinary least squares (OLS) estimate (based on the data from that institution only) and an overall estimate of the regression (aggregating data across institutions), with the weights proportional to the relative precisions of the two estimates. Rubin showed that, in comparison to the OLS estimate, the EB estimates yielded better prediction for the following year and much lower year-to-year volatility. This work led to changes in the validity study services provided by ETS to client programs.

Braun et al. (1983) extended the EB method to the case where the OLS estimate did not necessarily exist because of insufficient data. This problem can arise in prediction bias studies when the focal group is small and widely scattered among institutions. Later, Braun and Zwick (1993) developed an EB approach to estimating survival curves in a validity study in which the criterion was graduate degree attainment. EB or shrinkage-type estimators are now quite commonly applied in various contexts and are mathematically equivalent to the multilevel models that are used to analyze nested data structures.

6.3 Causal Inference

Causal inference in statistics is concerned with using data to elucidate the causal relationships among different factors. Of course, causal inference holds an important place in the history and philosophy of science. Early statistical contributions centered on the role of randomization and the development of various experimental designs to obtain the needed data most efficiently. In the social sciences, experiments are often not feasible, and various alternative designs and analytic strategies have been devised. The credibility of the causal inferences drawn from those designs has been an area of active research. ETS researchers have made important contributions to both the theoretical and applied aspects of this domain.

With respect to theory, Rubin (1972, 1974b, c), building on earlier work by Neyman, proposed a model for inference from randomized studies that utilized the concept of *potential outcomes*. That is, in comparing two treatments, ordinarily an individual can be exposed to only one of the treatments, so that only one of the two potential outcomes can be observed. Thus, the treatment effect on an individual is inestimable. However, if individuals are randomly allocated to treatments, an unbiased estimate of the average treatment effect can be obtained. He also made explicit the conditions under which causal inferences could be justified.

Later, Rubin (1978a) tackled the role of randomization in Bayesian inference for causality. This was an important development because, until then, many Bayesians argued that randomization was irrelevant to the Bayesian approach. Rubin's argument (in part) was that with a randomized design, Bayesian procedures were not only simpler, but also less sensitive to specification of the prior distribution. He also further explicated the crucial role of the stable unit treatment value assumption (SUTVA) in causal inference. This assumption asserts that the outcome of exposing a unit (e.g., an individual) to a particular treatment does not depend on which other units are exposed to that treatment. Although the SUTVA may be unobjectionable in some settings (e.g., agricultural or industrial experiments), in educational settings it is less plausible and argues for caution in interpreting the results.

Holland and Rubin (1980, 1987) clarified the statistical approach to causal inference. In particular, they emphasized the importance of manipulability; that is, the putative *causal agent* should have at least two possible states. Thus, the investigation of the differential effectiveness of various instructional techniques is a reasonable undertaking since, in principle, students could be exposed to any one of the techniques. On the other hand, an individual characteristic like gender or race cannot be treated as a causal agent, since ordinarily it is not subject to manipulation. (On this point, see also Holland, 2003). They go on to consider these issues in the context of retrospective studies, with consideration of estimating causal effects in various subpopulations defined in different ways.

Lord (1967) posed a problem involving two statisticians who drew radically different conclusions from the same set of data. The essential problem lies in attempting to draw causal conclusions from an analysis of covariance applied to nonexperimental data. The resulting longstanding conundrum, usually known as

Lord's Paradox, engendered much confusion. Holland and Rubin (1983) again teamed up to resolve the paradox, illustrating the power of the application of the Neyman-Rubin model, with careful consideration of the assumptions underlying different causal inferences.

In a much-cited paper, Holland (1986) reviewed the philosophical and epistemological foundations of causal inference and related them to the various statistical approaches that had been proposed to analyze experimental or quasi-experimental data, as well as the related literature on causal modeling. An invitational conference that touched on many of these issues was held at ETS, with the proceedings published in Wainer (1986). Holland (1987) represents a continuation of his work on the foundations of causal inference with a call for the measurement of effects rather than the deduction of causes. Holland (1988) explored the use of path analysis and recursive structural equations in causal inference, while Holland (1993) considered Suppes' theory of causality and related it to the statistical approach based on randomization.

As noted above, observational studies are much more common in the social sciences than are randomized experimental designs. In a typical observational study, units are exposed to treatments through some nonrandom mechanism that is often denoted by the term *self-selection* (whether or not the units actually exercised any discretion in the process). The lack of randomization means that the ordinary estimates of average treatment effects may be biased due to the initial nonequivalence of the groups. If the treatment groups are predetermined, one bias-reducing strategy involves matching units in different treatment groups on a number of observed covariates, with the hope that the resulting matched groups are approximately equivalent on all relevant factors except for the treatments under study. Were that the case, the observed average differences between the matched treatment groups would be approximately unbiased estimates of the treatment effects. Sometimes, an analysis of covariance is conducted instead of matching and, occasionally, both are carried out. These strategies raise some obvious questions. Among the most important are: What are the best ways to implement the matching and how well do they work? ETS researchers have made key contributions to answering both questions.

Rubin (1974b, c, 1980a) investigated various approaches to matching simultaneously on multiple covariates and, later, he considered combined strategies of matching and regression adjustment (1979b). Subsequently, Rosenbaum and Rubin (1985a) investigated the bias due to incomplete matching and suggested strategies for minimizing the number of unmatched treatment cases. Rosenbaum and Rubin (1983b) published a seminal paper on matching using propensity scores. Propensity scores facilitate multifactor matching through construction of a scalar index such that matching on this index typically yields samples that are well-matched on all the factors contributing to the index. Further developments and explications can be found in Rosenbaum and Rubin (1984, 1985b), as well as the now substantial literature that has followed. In 1986, the previously mentioned ETS-sponsored conference (Wainer 1986) examined the topic of inference from self-selected samples. The focus was a presentation by James Heckman on his model-based approach to the problem, with comments and critiques by a number of statisticians. A particular concern was the sensitivity of the findings to an untestable assumption about the value of a correlation parameter.

More generally, with respect to the question of how well a particular strategy works, one approach is to vary the assumptions and determine (either analytically or through simulation) how much the estimated treatment effects change as a result. In many situations, such sensitivity analyses can yield very useful information. Rosenbaum and Rubin (1983a) pioneered an empirical approach that involved assuming the existence of an unobserved binary covariate that accounts for the residual selection bias and incorporating this variable into the statistical model used for adjustment. By varying the parameters associated with this variable, it is possible to generate a response surface that depicts the sensitivity of the estimated treatment effect as a function of these parameters. The shape of the surface near the *naïve* estimate offers a qualitative sense of the confidence to be placed in its magnitude and direction.

This approach was extended by Montgomery et al. (1986) in the context of longitudinal designs. They showed that if there are multiple observations on the outcome, then under certain stability assumptions it is possible to obtain estimates of the parameters governing the unobserved binary variable and, hence, obtain a point estimate of the treatment effect in the expanded model.

More recently, education policy makers have seized on using indicators derived from student test scores as a basis for holding schools and teachers accountable. Under No Child Left Behind, the principal indicator is the percent of students meeting a state-determined proficiency standard. Because of the many technical problems with such status-based indicators, interest has shifted to indicators related to student progress. Among the most popular are the so-called value-added models (VAM) that attempt to isolate the specific contributions that schools and teachers make to their students' learning. Because neither students nor teachers are randomly allocated to schools (or to each other), this is a problem of causal inference (i.e., attribution of responsibility) from an observational study with a high degree of self-selection. The technical and policy issues were explicated in Braun (2005a, b) and in Braun and Wainer (2007). A comparison of the results of applying different VAMs to the same data was considered in Braun, Qu, and Trapani (2008).

6.4 Missing Data

The problem of missing data is ubiquitous in applied statistics. In a longitudinal study of student achievement, for example, data can be missing because the individual was not present at the administration of a particular assessment. In other cases, relevant data may not have been recorded, recorded but lost, and so on. Obviously, the existence of missing data complicates both the computational and inferential aspects of analysis. Adjusting calculation routines to properly take account of missing values can be challenging. Simple methods, such as deleting cases with missing data or filling in the missing values with some sort of average, can be wasteful, bias-inducing, or both. Standard inferences can also be suspect when there are missing values if they do not take account of how the data came to be missing. Thus, characterizing the process by which the *missingness* occurs is key to making credible inferences, as well as appropriate uses of the results. Despite the fact that ETS's testing programs and other activities generate oceans of data, problems of missing data are common, and ETS researchers have made fundamental contributions to addressing these problems.

Both Lord (1955) and Gulliksen (1956) tackled specific estimation problems in the presence of missing data. This tradition was continued by Rubin (1974a, 1976b, c). In this last report, concerned with fitting regression models, he considered how patterns of missingness of different potential predictors, along with multiple correlations, can be used to guide the selection of a prediction model. This line of research culminated in the celebrated paper by Dempster et al. (1977) that introduced, and elaborated on, the expectation-maximization (EM) algorithm for obtaining maximum likelihood estimates in the presence of missing data. The EM algorithm is an iterative estimation procedure that converges to the maximum likelihood estimate(s) of model parameters under broad conditions. Since that publication, the EM algorithm has become the tool of choice for a wide range of problems, with many researchers developing further refinements and modifications over the years. An ETS contribution is due to M. von Davier and Sinharay (2007), in which they develop a stochastic EM algorithm that is applied to latent regression problems.

Of course, examples of applications of EM abound. One particular genre involves embedding a complete data problem (for which obtaining maximum likelihood estimates is difficult or computationally intractable) in a larger missing data problem to which EM can be readily applied. Rubin and Szatrowski (1982) employed this strategy to obtain estimates in the case of multivariate normal distributions with patterned covariance matrices. Rubin and Thayer (1982) applied the EM algorithm to estimation problems in factor analysis. A more expository account of the EM algorithm and its applications can be found in Little and Rubin (1983).

With respect to inference, Rubin (1973, 1976b) investigated the conditions under which estimation in the presence of missing data would yield unbiased parameter estimates. The concept of *missing at random* was defined and its implications investigated in both the frequentist and Bayesian traditions. Further work on *ignorable nonresponse* was conducted in the context of sample surveys (see the next section).

6.5 Complex Samples

The problem of missing data, usually termed *nonresponse*, is particularly acute in sample surveys and is the cause of much concern with respect to estimation bias—both of the parameters of interest and their variances. Nonresponse can take many forms, from the complete absence of data to having missing values for certain variables (which may vary from individual to individual). Rubin (1978b) represents an early contribution using a Bayesian approach to address a prediction problem in which all units had substantial background data recorded but more than a quarter had no data on the dependent variables of interest. The method yields a pseudo-confidence interval for the population average.

Subsequently, Rubin (1980b, c) developed the multiple imputations methodology for dealing with nonresponse. This approach relies on generating posterior distributions for the missing values, based on prior knowledge (if available) and relevant auxiliary data (if available). Random draws from the posterior distribution are then used to obtain estimates of population quantities, as well as estimates of the component of error due to the added uncertainty contributed by the missing data. This work ultimately led to two publications that have had a great impact on the field (Rubin 1987; Rubin et al. 1983). Note that the multiple imputations methodology, combined with latent regression, is central to the estimation strategy in NAEP (Beaton and Barone, Chap. 8, this volume).

A related missing data problem arises in NAEP as the result of differences among states in the proportions of sampled students, either with disabilities or who are English-language learners, who are exempted from sitting for the assessment. Since these differences can be quite substantial, McLaughlin (2000) pointed out that these gaps likely result in biased comparisons between states on NAEP achievement. The suggested solution was to obtain so-called *full-population estimates* based on model assumptions regarding the performance of the excluded students. Braun et al. (2010) attacked the problem by investigating whether the observed differences in exemption rates could be explained by relevant differences in the focal subpopulations. Concluding that was not the case, they devised a new approach to obtaining full-population estimates and developed an agenda to guide further research and policy. Since then, the National Assessment Governing Board has imposed stricter limits on exemption rates.

Of course, missing data is a perennial problem in all surveys. ETS has been involved in a number of international large-scale assessment surveys, including those sponsored by the Organization for Economic Cooperation and Development (e.g., Program for International Student Assessment—PISA, International Adult Literacy Survey – IALS, Program for the International Assessment of Adult Competencies—PIAAC) and by the International Association for the Evaluation of Educational Achievement (e.g., Trends in International Mathematics and Science Study—TIMSS, Progress in International Reading Literacy Study—PIRLS). Different strategies for dealing with missing (or omitted) data have been advanced, especially for the cognitive items. An interesting and informative comparison of different approaches was presented by Rose et al. (2010). In particular, they compared deterministic rules with model-based rules using different item response theory (IRT) models.

6.6 Data Displays

An important tool in the applied statistician's kit is the use of graphical displays, a precept strongly promoted by Tukey in his work on exploratory data analysis. Plotting data in different ways can reveal patterns that are not evident in the usual summaries generated by standard statistical software. Moreover, good displays not only can suggest directions for model improvement, but also may uncover possible data errors.

No one at ETS took this advice more seriously than Wainer. An early effort in this direction can be found in Wainer and Thissen (1981). In subsequent years, he wrote a series of short articles in *The American Statistician* and *Chance* addressing both what to do—and what not to do—in displaying data. See, for example, Wainer (1984, 1993, 1996). During and subsequent to his tenure at ETS, Wainer also was successful in reaching a broader audience through his authorship of a number of well-received books on data display (1997, 2005, 2009).

6.7 Conclusion

This chapter is the result of an attempt to span the range of statistical research conducted at ETS over nearly 70 years, with the proviso that much of that research is covered in other chapters sponsored by this initiative. In the absence of those chapters, this one would have been much, much longer. To cite but one example, Holland and Thayer (1987, 2000) introduced a new approach to smoothing empirical score distributions based on employing a particular class of log-linear models. This innovation was motivated by problems arising in equipercentile equating and led to methods that were much superior to the ones used previously—superior with respect to accuracy, quantification of uncertainty, and asymptotic consistency. This work is described in more detail in Dorans and Puhan (Chap. 4, this volume). In short, only a perusal of many other reports can fully reflect the body of statistical research at ETS.

From ETS's founding, research has been a cornerstone of the organization. In particular, it has always offered a rich environment for statisticians and other quantitatively minded individuals. Its programs and activities generate enormous amounts of data that must be organized, described, and analyzed. Equally important, the various uses proposed for the data often raise challenging issues in computational efficiency, methodology, causality, and even philosophy. To address these issues, ETS has been fortunate to attract and retain (at least for a time) many exceptional individuals, well-trained in statistics and allied disciplines, eager to apply their skills to a wide range of problems, and effective collaborators. That tradition continues with attendant benefits to both ETS and the research community at large.

Acknowledgments The author acknowledges with thanks the superb support offered by Katherine Shields of Boston College. Katherine trolled the ETS ReSEARCHER database at http://search.ets.org/researcher/, made the first attempt to cull relevant reports, and assisted in the organization of the chapter. Thanks also for comments and advice to: Al Beaton, Randy Bennett, Brent Bridgeman, Neil Dorans, Shelby Haberman, Paul Holland, and the editors.

References

Abelson, R. P. (1953). A note on the Neyman-Johnson technique. *Psychometrika, 18*, 213–218. https://doi.org/10.1007/BF02289058

Aitkin, M. A. (1973). Fixed-width confidence intervals in linear regression with applications to the Johnson-Neyman technique. *British Journal of Mathematical and Statistical Psychology, 26*, 261–269. https://doi.org/10.1111/j.2044-8317.1973.tb00521.x

Almond, R. G., Mulder, J., Hemat, L. A., & Yan, D. (2009). Bayesian network models for local dependence among observable outcome variables. *Journal of Educational and Behavioral Statistics, 34*, 491–521.

Beaton, A. E. (1964). *The use of special matrix operators in statistical calculus.* Unpublished doctoral dissertation, Harvard University, Cambridge, MA.

Beaton, A. E. (1981). *Interpreting least squares without sampling assumptions* (Research Report No. RR-81-38). Princeton: Educational Testing Service. https://doi.org/10.1002/j.2333-8504.1981.tb01265.x

Beaton, A. E., Rubin, D. B., & Barone, J. L. (1972). The acceptability of regression solutions: Another look at computational accuracy. *Journal of the American Statistical Association, 71*, 158–168. https://doi.org/10.1080/01621459.1976.10481507

Beaton, A. E., & Tukey, J. W. (1974). The fitting of power series, meaning polynomials, illustrated on band–spectroscopic data. *Technometrics, 16*, 147–185. https://doi.org/10.1080/00401706.1974.10489171

Benjamini, Y., & Braun, H. I. (2003). John W. Tukey's contributions to multiple comparisons. *Annals of Statistics, 30*, 1576–1594.

Birnbaum, A. (1969). Statistical theory for logistic mental test models with a prior distribution of ability. *Journal of Mathematical Psychology, 6*, 258–276. https://doi.org/10.1016/0022-2496(69)90005-4

Bohrer, R. E. (1964). *Bayesian analysis of linear models: Fixed effects* (Research Bulletin No. RB-64-46). Princeton: Educational Testing Service. https://doi.org/10.1002/j.2333-8504.1964.tb00516.x

Braun, H. I. (1988). *Empirical Bayes methods: A tool for exploratory analysis* (Research Report No. RR-88-25). Princeton: Educational Testing Service. https://doi.org/10.1002/j.2330-8516.1988.tb00281.x

Braun, H. I. (Ed.). (1994). *The collected works of John W. Tukey: Vol. VIII. Multiple comparisons*. New York: Chapman & Hall, Inc..

Braun, H. I. (2005a). *Using student progress to evaluate teachers: A primer on value-added models* (Policy Information Report). Princeton: Educational Testing Service.

Braun, H. I. (2005b). Value-added modeling: What does due diligence require? In R. Lissitz (Ed.), *Value added models in education: Theory and applications* (pp. 19–39). Maple Grove: JAM Press.

Braun, H. I., Jones, D. H., Rubin, D. B., & Thayer, D. T. (1983). Empirical Bayes estimation of coefficients in the general linear model from data of deficient rank. *Psychometrika, 48*, 171–181. https://doi.org/10.1007/BF02294013

Braun, H. I., Qu, Y., & Trapani, C. (2008). *Robustness of a value-added assessment of school effectiveness* (Research Report No. RR-08-22). Princeton: Educational Testing Service. https://doi.org/10.1002/j.2333-8504.2008.tb02108.x

Braun, H. I., Ragosta, M., & Kaplan, B. (1988). Predictive validity. In W. W. Willingham (Ed.), *Testing handicapped people* (pp. 109–132). Boston: Allyn and Bacon.
Braun, H. I., & Szatrowski, T. H. (1984a). The scale-linkage algorithm: Construction of a universal criterion scale for families of institutions. *Journal of Educational Statistics, 9*, 311–330. https://doi.org/10.2307/1164744
Braun, H. I., & Szatrowski, T. H. (1984b). Validity studies based on a universal criterion scale. *Journal of Educational Statistics, 9*, 331–344. https://doi.org/10.2307/1164745
Braun, H. I., & Tukey, J. W. (1983). Multiple comparisons through orderly partitions: The maximum subrange procedure. In H. Wainer & S. Messick (Eds.), *Principals of modern psychological measurement: A festschrift for Frederic M. Lord* (pp. 55–65). Hillsdale: Erlbaum.
Braun, H. I., & Wainer, H. (2007). Value-added modeling. In C. R. Rao & S. Sinharay (Eds.), *Handbook of statistics: Vol. 27. Psychometrics* (pp. 867–892). Amsterdam: Elsevier.
Braun, H. I., Zhang, J., & Vezzu, S. (2010). An investigation of bias in reports of the National Aassessment of Educational Progress. *Educational Evaluation and Policy Analysis, 32*, 24–43. https://doi.org/10.3102/0162373709351137
Braun, H. I., & Zwick, R. J. (1993). Empirical Bayes analysis of families of survival curves: Applications to the analysis of degree attainment. *Journal of Educational Statistics, 18*, 285–303.
Bridgeman, B., Pollack, J. M., & Burton, N. W. (2008). Predicting grades in college courses: A comparison of multiple regression and percent succeeding approaches. *Journal of College Admission, 199*, 19–25.
Browne, M. W. (1969). *Precision of prediction* (Research Bulletin No. RB-69-69). Princeton: Educational Testing Service. https://doi.org/10.1002/j.2333-8504.1969.tb00748.x
Dempster, A. P., Laird, N., & Rubin, D. B. (1977). Maximum likelihood from incomplete data via the EM algorithm. *The Journal of the Royal Statistical Society, Series B, 39*, 1–38.
Gulliksen, H. O. (1956). A least squares solution for paired comparisons with incomplete data. *Psychometrika, 21*, 125–134. https://doi.org/10.1007/BF02289093
Haberman, S. J. (2006). Bias in estimation of misclassification rates. *Psychometrika, 71*, 387–394. https://doi.org/10.1007/s11336-004-1145-6
Holland, P. W. (1986). Statistics and causal inference. *Journal of the American Statistical Association, 81*, 945–960. https://doi.org/10.1080/01621459.1986.10478354
Holland, P. W. (1987). *Which comes first, cause or effect?* (Research Report No. RR-87-08). Princeton: Educational Testing Service. https://doi.org/10.1002/j.2330-8516.1987.tb00212.x
Holland, P. W. (1988). *Causal inference, path analysis and recursive structural equations models* (Research Report No. RR-88-14). Princeton: Educational Testing Service. https://doi.org/10.1002/j.2330-8516.1988.tb00270.x
Holland, P. W. (1993). *Probabilistic causation without probability* (Research Report No. RR-93-19). Princeton: Educational Testing Service. https://doi.org/10.1002/j.2333-8504.1993.tb01530.x
Holland, P. W. (2003). *Causation and race* (Research Report No. RR-03-03). Princeton: Educational Testing Service. https://doi.org/10.1002/j.2333-8504.2003.tb01895.x
Holland, P. W., & Rubin, D. B. (1980). *Causal inference in prospective and retrospective studies*. Washington, DC: Education Resources Information Center.
Holland, P. W., & Rubin, D. B. (1983). On Lord's paradox. In H. Wainer & S. Messick (Eds.), *Principals of modern psychological measurement: A festschrift for Frederick M. Lord* (pp. 3–25). Hillsdale: Erlbaum.
Holland, P. W., & Rubin, D. B. (1987). *Causal inference in retrospective studies* (Research Report No. RR-87-07). Princeton: Educational Testing Service. https://doi.org/10.1002/j.2330-8516.1987.tb00211.x
Holland, P. W., & Thayer, D. T. (1987). *Notes on the use of log-linear models for fitting discrete probability distributions* (Research Report No. RR-87-31). Princeton: Educational Testing Service. https://doi.org/10.1002/j.2330-8516.1987.tb00235.x
Holland, P. W., & Thayer, D. T. (2000). Univariate and bivariate loglinear models for discrete test score distributions. *Journal of Educational and Behavioral Statistics, 25*, 133–183. https://doi.org/10.3102/10769986025002133

Johnson, M. S., & Jenkins, F. (2005). *A Bayesian hierarchical model for large-scale educational surveys: An application to the National Assessment of Educational Progress* (Research Report No. RR-04-38). Princeton: Educational Testing Service. https://doi.org/10.1002/j.2333-8504.2004.tb01965.x

Kane, M. T. (2013). Validating the interpretations and uses of test scores. *Journal of Educational Measurement, 50*, 1–73. https://doi.org/10.1111/jedm.12000

Lewis, C., McCamley-Jenkins, L., & Ramist, L. (1994). *Student group differences in predicting college grades: Sex, language, and ethnic groups* (Research Report No. RR-94-27). Princeton: Educational Testing Service. https://doi.org/10.1002/j.2333-8504.1994.tb01600.x

Lewis, C., & Willingham, W. W. (1995). *The effects of sample restriction on gender differences* (Research Report No. RR-95-13). Princeton: Educational Testing Service. https://doi.org/10.1002/j.2333-8504.1995.tb01648.x

Li, D., & Oranje, A. (2007). *Estimation of standard error of regression effects in latent regression models using Binder's linearization* (Research Report No. RR-07-09). Princeton: Educational Testing Service. https://doi.org/10.1002/j.2333-8504.2007.tb02051.x

Li, D., Oranje, A., & Jiang, Y. (2007). *Parameter recovery and subpopulation proficiency estimation in hierarchical latent regression models* (Research Report No. RR-07-27). Princeton: Educational Testing Service. https://doi.org/10.1002/j.2333-8504.2007.tb02069.x

Lindley, D. V. (1969a). *A Bayesian estimate of true score that incorporates prior information* (Research Bulletin No. RB-69-75). Princeton: Educational Testing Service. https://doi.org/10.1002/j.2333-8504.1969.tb00754.x

Lindley, D. V. (1969b). *A Bayesian solution for some educational prediction problems* (Research Bulletin No. RB-69-57). Princeton: Educational Testing Service. https://doi.org/10.1002/j.2333-8504.1969.tb00735.x

Lindley, D. V. (1969c). *A Bayesian solution for some educational prediction problems, II* (Research Bulletin No. RB-69-91). Princeton: Educational Testing Service. https://doi.org/10.1002/j.2333-8504.1969.tb00770.x

Lindley, D. V. (1970). *A Bayesian solution for some educational prediction problems, III* (Research Bulletin No. RB-70-33). Princeton: Educational Testing Service. https://doi.org/10.1002/j.2333-8504.1970.tb00591.x

Linn, R. L. (1966). Grade adjustments for prediction of academic performance: A review. *Journal of Educational Measurement, 3*, 313–329. https://doi.org/10.1111/j.1745-3984.1966.tb00897.x

Little, R. J. A., & Rubin, D. B. (1983). On jointly estimating parameters and missing data by maximizing the complete-data likelihood. *American Statistician, 37*, 218–220. https://doi.org/10.1080/00031305.1983.10483106

Longford, N. T. (1987a). A fast scoring algorithm for maximum likelihood estimation in unbalanced mixed models with nested random effects. *Biometrika, 74*, 817–827. https://doi.org/10.1093/biomet/74.4.817

Longford, N. T. (1987b). *Fisher scoring algorithm for variance component analysis with hierarchically nested random effects* (Research Report No. RR-87-32). Princeton: Educational Testing Service. https://doi.org/10.1002/j.2330-8516.1987.tb00236.x

Longford, N. T. (1993). *Logistic regression with random coefficients* (Research Report No. RR-93-20). Princeton: Educational Testing Service. https://doi.org/10.1002/j.2333-8504.1993.tb01531.x

Lord, F. M. (1955). Estimation of parameters from incomplete data. *Journal of the American Statistical Association, 50*, 870–876. https://doi.org/10.1080/01621459.1955.10501972

Lord, F. M. (1967). A paradox in the interpretation of group comparisons. *Psychological Bulletin, 68*, 304–305. https://doi.org/10.1037/h0025105

McLaughlin, D. (2000). *Protecting state NAEP trends from changes in SD/LEP inclusion rates* (Technical report). Palo Alto: American Institutes for Research.

Mislevy, R. J. (1984). Estimating latent distributions. *Psychometrika, 49*, 359–381. https://doi.org/10.1007/BF02306026

Mislevy, R. J. (1991). Randomization-based inference about latent variables from complex samples. *Psychometrika, 56*, 177–196. https://doi.org/10.1007/BF02294457

Mislevy, R. J. (1994a). *Information-decay pursuit of dynamic parameters in student models* (Research Memorandum No. RM-94-14-ONR). Princeton: Educational Testing Service.
Mislevy, R. J. (1994b). *Virtual representation of IID observations in Bayesian belief networks* (Research Memorandum No. RM-94-13-ONR). Princeton: Educational Testing Service.
Mislevy, R. J., & Gitomer, D. H. (1995). The role of probability-based inference in an intelligent tutoring system. *User Modeling and User-Adapted Interaction, 5*, 253–282. https://doi.org/10.1007/BF01126112
Mislevy, R. J., Johnson, E. G., & Muraki, E. (1992). Scaling procedures in NAEP. *Journal of Educational Statistics, 17*, 131–154. https://doi.org/10.2307/1165166
Montgomery, M. R., Richards, T., & Braun, H. I. (1986). Child health, breast-feeding, and survival in Malaysia: A random-effects logit approach. *Journal of the American Statistical Association, 81*, 297–309. https://doi.org/10.1080/01621459.1986.10478273
Moses, T., & Klockars, A. (2009). *Strategies for testing slope differences* (Research Report No. RR-09-32). Princeton: Educational Testing Service. https://doi.org/10.1002/j.2333-8504.2009.tb02189.x
Novick, M. R. (1964). *On Bayesian logical probability* (Research Bulletin No. RB-64-22). Princeton: Educational Testing Service. https://doi.org/10.1002/j.2333-8504.1964.tb00330.x
Novick, M. R., Jackson, P. H., & Thayer, D. T. (1971). Bayesian inference and the classical test theory model: Reliability and true scores. *Psychometrika, 36*, 261–288. https://doi.org/10.1007/BF02297848
Novick, M. R., Jackson, P. H., Thayer, D. T., & Cole, N. S. (1972). Estimating multiple regressions in m-groups: A cross-validation study. *British Journal of Mathematical and Statistical Psychology, 25*, 33–50. https://doi.org/10.1111/j.2044-8317.1972.tb00476.x
Novick, M. R., & Thayer, D. T. (1969). *A comparison of Bayesian estimates of true score* (Research Bulletin No. RB-69-74). Princeton: Educational Testing Service. https://doi.org/10.1002/j.2333-8504.1969.tb00753.x
Potthoff, R. F. (1963). *Illustrations of some Scheffe-type tests for some Behrens-Fisher-type regression problems* (Research Bulletin No. RB-63-36). Princeton: Educational Testing Service. https://doi.org/10.1002/j.2333-8504.1963.tb00502.x
Potthoff, R. F. (1965). Some Scheffe-type tests for some Behrens-Fisher-type regression problem. *Journal of the American Statistical Association, 60*, 1163–1190. https://doi.org/10.1080/01621459.1965.10480859
Ramist, L., Lewis, C., & McCamley, L. (1990). Implications of using freshman GPA as the criterion for the predictive validity of the SAT. In W. W. Willingham, C. Lewis, R. Morgan, & L. Ramist (Eds.), *Predicting college grades: An analysis of institutional trends over two decades* (pp. 253–288). Princeton: Educational Testing Service.
Rock, D. A. (1969). *The identification and utilization of moderator effects in prediction systems* (Research Bulletin No. RB-69-32). Princeton: Educational Testing Service. https://doi.org/10.1002/j.2333-8504.1969.tb00573.x
Rose, N., von Davier, M., & Xu, X. (2010). *Modeling nonignorable missing data with item response theory (IRT)* (Research Report No. RR-10-11). Princeton: Educational Testing Service. https://doi.org/10.1002/j.2333-8504.2010.tb02218.x
Rosenbaum, P. R., & Rubin, D. B. (1983a). Assessing sensitivity to an unobserved binary covariate in an observational study with binary outcome. *The Journal of the Royal Statistical Society, Series B, 45*, 212–218.
Rosenbaum, P. R., & Rubin, D. B. (1983b). *The bias due to incomplete matching* (Research Report No. RR-83-37). Princeton: Educational Testing Service.
Rosenbaum, P. R., & Rubin, D. B. (1984). Reducing bias in observational studies using subclassification on the propensity score. *Journal of the American Statistical Association, 79*, 516–524. https://doi.org/10.1080/01621459.1984.10478078
Rosenbaum, P. R., & Rubin, D. B. (1985a). The bias due to incomplete matching. *Biometrics, 41*, 103–116. https://doi.org/10.2307/2530647

Rosenbaum, P. R., & Rubin, D. B. (1985b). Constructing a control group using multivariate matched sampling methods that incorporate the propensity score. *The American Statistician, 39*, 33–38. https://doi.org/10.1080/00031305.1985.10479383

Rubin, D. B. (1972). *Estimating causal effects of treatments in experimental and observational studies* (Research Bulletin No. RB-72-39). Princeton: Educational Testing Service. https://doi.org/10.1002/j.2333-8504.1972.tb00631.x

Rubin, D. B. (1973). *Missing at random: What does it mean?* (Research Bulletin No. RB-73-02). Princeton: Educational Testing Service. https://doi.org/10.1002/j.2333-8504.1973.tb00198.x

Rubin, D. B. (1974a). Characterizing the estimation of parameters in incomplete data problems. *Journal of the American Statistical Association, 69*, 467–474. https://doi.org/10.1080/01621459.1974.10482976

Rubin, D. B. (1974b). Multivariate matching methods that are equal percent bias reducing, I: Some examples. *Biometrics, 32*, 109–120. https://doi.org/10.2307/2529342

Rubin, D. B. (1974c). Multivariate matching methods that are equal percent bias reducing, II: Maximums on bias reduction for fixed sample sizes. *Biometrics, 32*, 121–132. https://doi.org/10.2307/2529343

Rubin, D. B. (1975). *A note on a simple problem in inference* (Research Bulletin No. RB-75-20). Princeton: Educational Testing Service. https://doi.org/10.1002/j.2333-8504.1975.tb01061.x

Rubin, D. B. (1976a). Comparing regressions when some predictor values are missing. *Technometrics, 18*, 201–205. https://doi.org/10.1080/00401706.1976.10489425

Rubin, D. B. (1976b). Inference and missing data. *Biometrika, 63*, 581–592.

Rubin, D. B. (1976c). Noniterative least squares estimates, standard errors, and F-tests for any analysis of variance with missing data. *The Journal of the Royal Statistical Society, 38*, 270–274.

Rubin, D. B. (1978a). Bayesian inference for causal effects: The role of randomization. *The Annals of Statistics, 6*(1), 34–58. https://doi.org/10.1214/aos/1176344064

Rubin, D. B. (1978b). Multiple imputations in sample surveys—A phenomenological Bayesian approach to nonresponse. *The Proceedings of the Survey Research Methods Section of the American Statistical Association*, 20–34.

Rubin, D. B. (1979a). *The Bayesian bootstrap* (Program Statistical Report No. PSRTR-80-03). Princeton: Educational Testing Service.

Rubin, D. B. (1979b). Using multivariate matched sampling and regression to control bias in observational studies. *Journal of the American Statistical Association, 74*, 318–328. https://doi.org/10.2307/2286330

Rubin, D. B. (1980a). Bias reduction using Mahalanobis metric matching. *Biometrics, 36*, 293–298. https://doi.org/10.2307/2529981

Rubin, D. B. (1980b). *Handling nonresponse in sample surveys by multiple imputations*. Washington, DC: U.S. Department of Commerce, Bureau of the Census.

Rubin, D. B. (1980c). Illustrating the use of multiple imputations to handle nonresponse in sample surveys. In *Proceedings of the 42nd session of the International Statistical Institute, 1979* (Book 2, pp. 517–532). The Hague: The International Statistical Institute.

Rubin, D. B. (1980d). Using empirical Bayes techniques in the law school validity studies. *Journal of the American Statistical Association, 75*, 801–816. https://doi.org/10.1080/01621459.1980.10477553

Rubin, D. B. (1987). *Multiple imputation for nonresponse in surveys*. New York: Wiley. https://doi.org/10.1002/9780470316696

Rubin, D., & Stroud, T. (1977). The calculation of the posterior distribution of the cell means in a two-way unbalanced MANOVA. *Journal of the Royal Statistical Society. Series C (Applied Statistics), 26*, 60–66. https://doi.org/10.2307/2346868

Rubin, D. B., & Szatrowski, T. H. (1982). Finding maximum likelihood estimates of patterned covariance matrices by the EM algorithm. *Biometrika, 69*, 657–660. https://doi.org/10.1093/biomet/69.3.657

Rubin, D. B., & Thayer, D. T. (1982). EM algorithms for ML factor analysis. *Psychometrika, 47*, 69–76. https://doi.org/10.1007/BF02293851

Rubin, D. B., Madow, W. G., & Olkin, I. (Eds.). (1983). *Incomplete data in sample surveys: Vol. 2. Theory and bibliographies*. New York: Academic Press.

Saunders, D. R. (1952). *The "ruled surface regression" as a starting point in the investigation of "differential predictability"* (Research Memorandum No. RM-52-18). Princeton: Educational Testing Service.

Sheehan, K. M., & Mislevy, R. J. (1989). Information matrices in latent-variable models. *Journal of Educational Statistics, 14*, 335–350.

Sinharay, S. (2003a). *Assessing convergence of the Markov chain Monte Carlo algorithms: A review* (Research Report No. RR-03-07). Princeton: Educational Testing Service. https://doi.org/10.1002/j.2333-8504.2003.tb01899.x

Sinharay, S. (2003b). *Practical applications of posterior predictive model checking for assessing fit of the common item response theory models* (Research Report No. RR-03-33). Princeton: Educational Testing Service. https://doi.org/10.1002/j.2333-8504.2003.tb01925.x

Sinharay, S. (2006). Model diagnostics for Bayesian networks. *Journal of Educational and Behavioral Statistics, 31*, 1–33. https://doi.org/10.3102/10769986031001001

Sinharay, S., Guo, Z., von Davier, M., & Veldkamp, B. P. (2009). *Assessing fit of latent regression models* (Research Report No. RR-09-50). Princeton: Educational Testing Service. https://doi.org/10.1002/j.2333-8504.2009.tb02207.x

Tucker, L. R. (1957). *Computation procedure for transformation of predictor variables to a simplified regression structure* (Research Memorandum No. RM-57-01). Princeton: Educational Testing Service.

Tucker, L. R. (1963). *Formal models for a central prediction system* (Psychometric Monograph No. 10). Richmond: Psychometric Corporation.

Tukey, J. W. (1953). *The problem of multiple comparisons*. Unpublished manuscript.

Tukey, J. W. (1994). The problem of multiple comparisons. In H. I. Braun (Ed.), *The collected works of John Tukey: Vol. VIII. Multiple comparisons* (pp. 1–300). New York: Chapman & Hall.

von Davier, A. A. (2003a). *Large sample tests for comparing regression coefficients in models with normally distributed variables* (Research Report No. RR-03-19). Princeton: Educational Testing Service. https://doi.org/10.1002/j.2333-8504.2003.tb01911.x

von Davier, M. (2003b). *Comparing conditional and marginal direct estimation of subgroup distributions* (Research Report No. RR-03-02). Princeton: Educational Testing Service. https://doi.org/10.1002/j.2333-8504.2003.tb01894.x

von Davier, M., & Sinharay, S. (2007). An importance sampling EM algorithm for latent regression models. *Journal of Educational and Behavioral Statistics, 32*, 233–251. https://doi.org/10.3102/1076998607300422

von Davier, M., Sinharay, S., Oranje., A. & Beaton, A. (2006). Statistical procedures used in the National Assessment of Educational Progress (NAEP): Recent developments and future directions. In C. R. Rao & S. Sinharay (Eds.), *Handbook of statistics: Vol. 26. Psychometrics* (pp. 1039–1055). Amsterdam: Elsevier.

von Davier, M., & Sinharay, S. (2010). Stochastic approximation methods for latent regression item response models. *Journal of Educational and Behavioral Statistics, 35*, 174–193. https://doi.org/10.3102/1076998609346970

von Davier, M., & Sinharay, S. (2013). Analytics in international large-scale assessments: Item response theory and population models. In L. Rutkowski, M. von Davier, & D. Rutkowski (Eds.), *Handbook of international large-scale assessment: Background, technical issues, and methods of data analysis* (pp. 155–174). New York: CRC Press.

Wainer, H. (1984). How to display data badly. *The American Statistician, 38*, 137–147. https://doi.org/10.1080/00031305.1984.10483186

Wainer, H. (Ed.). (1986). *Drawing inferences from self-selected samples*. New York: Springer. https://doi.org/10.1007/978-1-4612-4976-4

Wainer, H. (1993). Graphical answers to scientific questions. *Chance, 6*(4), 48–50. https://doi.org/10.1080/09332480.1993.10542398

Wainer, H. (1996). Depicting error. *The American Statistician, 50*, 101–111. https://doi.org/10.1080/00031305.1996.10474355
Wainer, H. (1997). *Visual revelations: Graphical tales of fate and deception from Napoleon Bonaparte to Ross Perot*. New York: Copernicus Books.
Wainer, H. (2005). *Graphic discovery: A trout in the milk and other visual adventures*. Princeton: Princeton University Press.
Wainer, H. (2009). *Picturing the uncertain world: How to understand, communicate and control uncertainty through graphical display*. Princeton: Princeton University Press.
Wainer, H., & Thissen, D. (1981). Graphical data analysis. In M. R. Rosenzweig & L. W. Porter (Eds.), *Annual review of psychology* (pp. 191–241). Palo Alto: Annual Reviews. https://doi.org/10.1177/014662160202600l007
Wang, X., Bradlow, E. T., & Wainer, H. (2002). A general Bayesian model for testlets: Theory and applications. *Applied Psychological Measurement, 26*, 109–128. https://doi.org/10.1177/0146621602026001007
Willingham, W. W. (Ed.). (1988). *Testing handicapped people*. Boston: Allyn and Bacon.
Willingham, W. W., & Cole, N. S. (1997). *Gender and fair assessment*. Mahwah, NJ: Erlbaum.
Willingham, W. W., Pollack, J., & Lewis, C. (2002). Grades and test scores: Accounting for observed differences. *Journal of Educational Measurement, 39*, 1–37. https://doi.org/10.1111/j.1745-3984.2002.tb01133.x
Young, J. W., & Barrett, C. A. (1992). Analyzing high school transcripts to improve prediction of college performance. *Journal of College Admission, 137*, 25–29.
Zwick, R. J. (1993). Pairwise comparison procedures for one-way analysis of variance designs. In G. Keren & C. Lewis (Eds.), *A handbook for data analysis in the behavioral sciences: Methodological issues* (pp. 43–71). Hillsdale: Erlbaum.
Zwick, R. J. (2013). *Disentangling the role of high school grades, SAT scores, and SES in predicting college achievement* (Research Report No. RR-13-09). Princeton: Educational Testing Service. https://doi.org/10.1002/j.2333-8504.2013.tb02316.x
Zwick, R. J., & Himelfarb, I. (2011). The effect of high school socioeconomic status on the predictive validity of SAT scores and high school grade-point average. *Journal of Educational Measurement, 48*, 101–121. https://doi.org/10.1111/j.1745-3984.2011.00136.x

Chapter 7
Contributions to the Quantitative Assessment of Item, Test, and Score Fairness

Neil J. Dorans

ETS was founded in 1947 as a not-for-profit organization (Bennett, Chap. 1, this volume). Fairness concerns have been an issue at ETS almost since its inception. William Turnbull (1949, 1951a, b), who in 1970 became the second president of ETS, addressed the Canadian Psychological Association on socioeconomic status and predictive test scores. He made a cogent argument for rejecting the notion that differences in subgroup performance on a test means that a test score is biased. He also advocated the comparison of prediction equations as a means of assessing test fairness. His article was followed by a number of articles by ETS staff on the issue of differential prediction. By the 1980s, under the direction of its third president, Gregory Anrig, ETS established the industry standard for fairness assessment at the item level (Holland and Wainer 1993; Zieky 2011). This century, fairness analyses have begun to focus on relationships between tests that purport to measure the same thing in the same way across different subgroups (Dorans and Liu 2009; Liu and Dorans 2013).

In this chapter, I review quantitative fairness procedures that have been developed and modified by ETS staff over the past decades. While some reference is made to events external to ETS, the focus is on ETS, which has been a leader in fairness assessment. In the first section, Fair Prediction of a Criterion, I consider differential prediction and differential validity, procedures that examine whether test scores predict a criterion, such as performance in college, across different subgroups in a similar manner. The bulk of this review is in the second section, Differential Item Functioning (DIF), which focuses on item-level fairness, or

This chapter was originally published in 2013 by Educational Testing Service as a research report in the ETS R&D Scientific and Policy Contributions Series. ETS staff continue to contribute to the literature on fairness. See Dorans and Cook (2016) for some examples.

N.J. Dorans (✉)
Educational Testing Service, Princeton, NJ, USA
e-mail: ndorans@ets.org

R.E. Bennett, M. von Davier (eds.), *Advancing Human Assessment*,
Methodology of Educational Measurement and Assessment,
DOI 10.1007/978-3-319-58689-2_7

DIF. Then in the third section, Fair Linking of Test Scores, I consider research pertaining to whether tests built to the same set of specifications produce scores that are related in the same way across different gender and ethnic groups. In the final section, Limitations of Quantitative Fairness Assessment Procedures, limitations of these procedures are mentioned.

7.1 Fair Prediction of a Criterion

Turnbull (1951a) concluded his early ETS treatment of fairness with the following statement: "Fairness, like its amoral brother, validity, resides not in tests or test scores but in the relation to its uses" (p. 4–5).

While several ETS authors had addressed the relative lower performance of minority groups on tests of cognitive ability and its relationship to grades (e.g., Campbell 1964), Cleary (1968) conducted one of the first differential prediction studies. That study has been widely cited and critiqued. A few years after the Cleary article, the field was replete with differential validity studies, which focus on comparing correlation coefficients, and differential prediction studies, which focus on comparing regression functions, in large part because of interest engendered by the Supreme Court decision *Griggs v. Duke Power Co.* in 1971. This decision included the terms *business necessity* and *adverse impact*, both of which affected employment testing. Adverse impact is a substantially different rate of selection in hiring, promotion, transfer, training, or other employment-related decisions for any race, sex, or ethnic group. Business necessity can be used by an employer as a justification for using a selection mechanism that appears to be neutral with respect to sex, race, national origin, or religious group even though it excludes members of one sex, race, national origin, or religious group at a substantially higher rate than members of other groups. The employer must prove that the selection requirement having the adverse impact is job related and consistent with business necessity. In other words, in addition to avoiding the use of race/ethnic/gender explicitly as part of the selection process, the selection instrument had to have demonstrated predictive validity for its use. Ideally, this validity would be the same for all subpopulations.

Linn (1972) considered the implications of the Griggs decision for test makers and users. A main implication was that there would be a need for empirical demonstrations that test scores predict criterion performance, such as how well one does on the job. (In an educational context, test scores may be used with other information to predict the criterion of average course grade). Reliability alone would not be an adequate justification for use of test scores. Linn also noted that for fair prediction to hold, the prediction model must include all the appropriate variables in the model. Otherwise misspecification of the model can give the appearance of statistical bias. The prediction model should include all the predictors needed to predict Y, and the functional form used to combine the predictors should be the correct one. The reliabilities of the predictors also were noted to play a role. These limitations with differential validity and differential predictions studies were cogently summarized in four pages by Linn and Werts (1971). One of the quandaries faced by researchers that was not noted in this 1971 study is that some of the variables that contribute to prediction are variables over which a test taker has little control, such as gender, race, parent's level of education and income, and even zip code. Use of variables such as zip code to predict grades in an attempt to eliminate differential prediction would be unfair.

Linn (1975) later noted that differential prediction analyses should be preferred to differential validity studies because differences in predictor or criterion variability can produce differential validity even when the prediction model is fair. Differential prediction analyses examine whether the same prediction models hold across different groups. Fair prediction or selection requires invariance of prediction equations across groups,

$$R(Y|\mathbf{X},G=1)=R(Y|\mathbf{X},G=2)=\ldots=R(Y|\mathbf{X},G=g),$$

where R is the symbol for the function used to predict Y, the criterion score, from $\mathbf{X}$, the predictor. G is a variable indicating subgroup membership.

Petersen and Novick (1976) compared several models for assessing fair selection, including the regression model (Cleary 1968), the constant ratio model (Thorndike 1971), the conditional probability model (Cole 1973), and the constant probability model (Linn 1973) in the lead article in a special issue of the *Journal of Educational Measurement* dedicated to the topic of fair selection. They demonstrated that the regression, or Cleary, model, which is a differential prediction model, was a preferred model from a logical perspective in that it was consistent with its converse (i.e., fair selection of applicants was consistent with fair rejection of applicants). In essence, the Cleary model examines whether the regression of the criterion onto the predictor space is invariant across subpopulations.

Linn (1976) in his discussion of the Petersen and Novick (1976) analyses noted that the quest to achieve fair prediction is hampered by the fact that the criterion in many studies may itself be unfairly measured. Even when the correct equation is correctly specified and the criterion is measured well in the full population, invariance may not hold in subpopulations because of selection effects. Linn (1983) described how predictive bias may be an artifact of selection procedures. Linn used a simple case to illustrate his point. He posited that a single predictor X and linear model were needed to predict Y in the full population P. To paraphrase his argument, assume that a very large sample is drawn from P based on a selection variable U that might depend on X in a linear way. Errors in the prediction of Y from X and U from X are thus also linearly related because of their mutual dependence on X. Linn showed that the sample regression for the selected sample, R ($Y|X$, G) equals the regression in the full unselected population if the correlation between X and U is zero, or if errors in prediction of Y from X and U from X are uncorrelated.

Myers (1975) criticized the regression model because regression effects can produce differences in intercepts when two groups differ on X and Y and the predictor is unreliable, a point noted by Linn and Werts (1971). Myers argued for a linking or scaling model for assessing fairness. He noted that his approach made sense when

X and Y were measures of the same construct, but admitted that scaling test scores to grades or vice versa had issues. This brief report by Myers can be viewed as a remote harbinger of work on the population invariance of score linking functions done by Dorans and Holland (2000), Dorans (2004), Dorans and Liu (2009), and Liu and Dorans (2013).

As can be inferred from the studies above, in particular Linn and Werts (1971) and Linn (1975, 1983), there are many ways in which a differential prediction study can go awry, and even more ways that differential validity studies can be problematic.

7.2 Differential Item Functioning (DIF)

During the 1980s, the focus in the profession shifted to DIF studies. Although interest in item bias studies began in the 1960s (Angoff 1993), it was not until the 1980s that interest in fair assessment at the item level became widespread. During the 1980s, the measurement profession engaged in the development of item level models for a wide array of purposes. DIF procedures developed as part of that shift in attention from the score to the item.

Moving the focus of attention to prediction of item scores, which is what DIF is about, represented a major change from focusing primarily on fairness in a domain, where so many factors could spoil the validity effort, to a domain where analyses could be conducted in a relatively simple, less confounded way. While factors such as multidimensionality can complicate a DIF analysis, as described by Shealy and Stout (1993), they are negligible compared to the many influences that can undermine a differential prediction study, as described in Linn and Werts (1971). In a DIF analysis, the item is evaluated against something designed to measure a particular construct and something that the test producer controls, namely a test score.

Around 100 ETS research bulletins, memoranda, or reports have been produced on the topics of item fairness, DIF, or item bias. The vast majority of these studies were published in the late 1980s and early 1990s. The major emphases of these reports can be sorted into categories and are treated in subsections of this section: Differential Item Functioning Methods, Matching Variable Issues, Study Group Definitions, and Sample Size and Power Issues. The DIF methods section begins with some definitions followed by a review of procedures that were suggested before the term DIF was introduced. Most of the section then describes the following procedures: Mantel-Haenszel (MH), standardization (STAND), item response theory (IRT), and SIBTEST.

7.2.1 Differential Item Functioning (DIF) Methods

Two reviews of DIF methods were conducted by ETS staff: Dorans and Potenza (1994), which was shortened and published as Potenza and Dorans (1995), and Mapuranga et al. (2008), which then superseded Potenza and Dorans. In the last of these reviews, the criteria for classifying DIF methods were (a) definition of null DIF, (b) definition of the studied item score, (c) definition of the matching variable, and (d) the variable used to define groups.

Null DIF is the absence of DIF. One definition of null DIF, observed score null DIF, is that all individuals with the same score on a test of the shared construct measured by that item should have the same proportions answering the item correctly regardless of whether they are from the reference or focal group. The latent variable definition of null DIF can be used to compare the performance of focal and reference subgroups that are matched with respect to a latent variable. An observed difference in average item scores between two groups that may differ in their distributions of scores on the matching variable is referred to as *impact*. With impact, we compare groups that may or may not be comparable with respect to the construct being measured by the item; using DIF, we compare item scores on groups that are comparable with respect to an estimate of their standing on that construct.

The *studied item score* refers to the scoring rule used for the items being studied for DIF. Studied items are typically[1] scored as correct/incorrect (i.e., binary) or scored using more than two response categories (i.e., polytomous). The *matching variable* is a variable used in the process of comparing the reference and focal groups (e.g., total test score or subscore) so that comparable groups are formed. In other words, matching is a way of establishing score equivalence between groups that are of interest in DIF analyses. The matching variable can either be an observed score or an estimate of the unobserved latent variable consistent with a specific model for item performance, and can be either a univariate or multivariate variable.

In most DIF analyses, a single focal group is compared to a single reference group where the subgroup-classification variable (gender, race, geographic location, etc.) is referred to as the *grouping variable*. This approach ignores potential interactions between types of subgroups, (e.g., male/female and ethnic/racial). Although it might be better to analyze all grouping variables for DIF simultaneously (for statistical and computational efficiency), most DIF methods compare only two groups at a time. While convention is often the reason for examining two groups at a time, small sample size sometimes makes it a necessity.

The remainder of this section describes briefly the methods that have been developed to assess what has become known as DIF. After reviewing some early work, I turn to the two methods that are still employed operationally here at ETS: the MH method and the STAND method. After briefly discussing IRT methods, I mention

[1] All options can be treated as nominally scored, which could be useful in cases where the focus is on differential functioning on options other than the key (distractors).

the SIBTEST method. Methods that do not fit into any of these categories are addressed in what seems to be the most relevant subsection.

7.2.1.1 Early Developments: The Years Before Differential Item Functioning (DIF) Was Defined at ETS

While most of the focus in the 1960s and 1970s was on the differential prediction issue, several researchers turned their attention to item-level fairness issues. Angoff (1993) discussed several, but not all of these efforts. Cardall and Coffman (1964) and Cleary and Hilton (1966, 1968) defined *item bias*, the phrase that was commonly used before DIF was introduced, as an item-by-subgroup interaction. Analysis of variance was used by both studies of DIF. Identifying individual problem items was not the goal of either study.

Angoff and Sharon (1974) also employed an analysis of variance (ANOVA) method, but by then the transformed item difficulty (TID) or delta-plot method had been adopted for item bias research. Angoff (1972) introduced this approach, which was rooted in Thurstone's absolute scaling model. This method had been employed by Tucker (1951) in a study of academic ability on vocabulary items and by Gulliksen (1964) in a cross-national study of occupation prestige. This method uses an inverse normal transformation to convert item proportion-correct values for two groups to normal deviates that are expect to form an ellipse. Items that deviate from the ellipse exhibit the item difficulty by group interaction that is indicative of what was called item bias. Angoff and Ford (1971, 1973) are the standard references for this approach.

The delta-plot method ignores differences in item discrimination. If items differ in their discriminatory power and the groups under study differ in terms of proficiency, then items will exhibit item-by-group interactions even when there are no differences in item functioning. This point was noted by several scholars including Lord (1977) and affirmed by Angoff (1993). As a consequence, the delta-plot method is rarely used for DIF assessment, except in cases where small samples are involved.

Two procedures may be viewed as precursors of the eventual move to condition directly on total score that was adopted by the STAND (Dorans and Kulick 1983) and MH (Holland and Thayer 1988) DIF approaches. Stricker (1982) recommended a procedure that looks for DIF by examining the partial correlation between group membership and item score with the effect of total test score removed. Scheuneman (1979) proposed a test statistic that looked like a chi-square. This method was shown by Baker (1981) and others to be affected inappropriately by sample size and to possess no known sampling distribution.

The late 1980s and the early 1990s were the halcyon days of DIF research and development at ETS and in the profession. Fairness was of paramount concern, and practical DIF procedures were developed and implemented (Dorans and Holland 1993; Zieky 1993). In October 1989, ETS and the Air Force Human Resources Laboratory sponsored a DIF conference that was held at ETS in October 1989. The papers presented at that conference, along with a few additions, were collected in the volume edited by Holland and Wainer (1993), known informally as the DIF book. It contains some of the major work conducted in this early DIF era, including several chapters about MH and STAND. The chapter by Dorans and Holland (1993) is the source of much of the material in the next two sections, which describe the MH and STAND procedures in some detail because they have been used operationally at ETS since that time. Dorans (1989) is another source that compares and contrasts these two DIF methods.

7.2.1.2 Mantel-Haenszel (MH): Original Implementation at ETS

In their seminal paper, Mantel and Haenszel (1959) introduced a new procedure for the study of matched groups. Holland and Thayer (1986, 1988) adapted the procedure for use in assessing DIF. This adaptation, the MH method, is used at ETS as the primary DIF detection device. The basic data used by the MH method are in the form of *M* 2-by-2 contingency tables or one large three dimensional 2-by-2-by-M table, where *M* is the number of levels of the matching variable.

Under rights-scoring for the items in which responses are coded as either correct or incorrect (including omissions), proportions of rights and wrongs on each item in the target population can be arranged into a contingency table for each item being studied. There are two levels for group: the focal group (f) that is the focus of analysis, and the reference group (r) that serves as a basis for comparison for the focal group. There are also two levels for item response: right (R) or wrong (W), and there are *M* score levels on the matching variable, (e.g., total score). Finally, the item being analyzed is referred to as the studied item. The 2 (groups)-by-2 (item scores)-by-M (score levels) contingency table (see Table 7.1) for each item can be viewed in *2-by-2 slices*.

The null DIF hypothesis for the MH method can be expressed as

$$H_0 : \left[R_{rm} / W_{rm}\right] = \left[R_{fm} / W_{fm}\right] \ \forall m = 1,\ldots,M.$$

In other words, the odds of getting the item correct at a given level of the matching variable is the same in both the focal group and the reference group portions of the population, and this equality holds across all M levels of the matching variable.

Table 7.1 2-by-2-by-M contingency table for an item, viewed in a 2-by-2 Slice

	Item score		
Group	Right	Wrong	Total
Focal group (f)	R_{fm}	W_{fm}	N_{fm}
Reference group (r)	R_{rm}	W_{rm}	N_{rm}
Total group (t)	R_{tm}	W_{tm}	N_{tm}

In their original work, Mantel and Haenszel (1959) developed a chi-square test of the null hypothesis against a particular alternative hypothesis known as the constant odds ratio hypothesis,

$$H_a : [R_{rm} / W] = \alpha \left[R_{fm} / W_{fm} \right] \forall m = 1, \ldots, M, and\, \alpha \neq 1.$$

Note that when $\alpha = 1$, the alternative hypothesis reduces to the null DIF hypothesis. The parameter α is called the *common odds ratio* in the M 2-by-2 tables because under H_a, the value of α is the odds ratio common for all m.

Holland and Thayer (1988) reported that the MH approach is the test possessing the most statistical power for detecting departures from the null DIF hypothesis that are consistent with the constant odds-ratio hypothesis.

Mantel and Haenszel (1959) also provided an estimate of the constant odds – ratio that ranges from 0 to ∞, for which a value of 1 can be taken to indicate null DIF. This odds-ratio metric is not particularly meaningful to test developers who are used to working with numbers on an item difficulty scale. In general, odds are converted to ln[odds-ratio] because the latter is symmetric around zero and easier to interpret.

At ETS, test developers use item difficulty estimates in the *delta metric*, which has a mean of 13 and a standard deviation of 4. Large values of delta correspond to difficult items, while easy items have small values of delta. Holland and Thayer (1985) converted the estimate of the common odds ratio, α_{MH}, into a difference in deltas via:

$$MH\,\mathrm{D} - DIF = -2.35 \ln \left[\alpha MH \right].$$

Note that positive values of MH D-DIF favor the focal group, while negative values favor the reference group. An expression for the standard error of for MH D-DIF was provided in Dorans and Holland (1993).

7.2.1.3 Subsequent Developments With the Mantel-Haenszel (MH) Approach

Subsequent to the operational implementation of the MH approach to DIF detection by ETS in the late 1980s (Zieky 1993, 2011), there was a substantial amount of DIF research conducted by ETS staff through the early 1990s. Some of this research was presented in Holland and Wainer (1993); other presentations appeared in journal articles and ETS Research Reports. This section contains a partial sampling of research conducted primarily on the MH approach.

Donoghue et al. (1993) varied six factors in an IRT-based simulation of DIF in an effort to better understand the properties of the MH and STAND (to be described in the next section) effect sizes and their standard errors. The six factors varied were level of the IRT discrimination parameter, the number of DIF items in the matching variable, the amount of DIF on the studied item, the difficulty of the studied item, whether the studied item was included in the matching variable, and the number of items in the matching variable. Donoghue et al. found that both the MH and STAND methods had problems detecting IRT DIF in items with nonzero lower asymptotes. Their two major findings were the need to have enough items in the matching variable to ensure reliable matching for either method, and the need to include the studied item in the matching variable in MH analysis. This study thus provided support for the analytical argument for inclusion of the studied item that had been made by Holland and Thayer (1986). As will be seen later, Zwick et al. (1993a), Zwick (1990), Lewis (1993), and Tan et al. (2010) also addressed the question of inclusion of the studied item.

Longford et al. (1993) demonstrated how to use a random-effect or variance-component model to aggregate DIF results for groups of items. In particular they showed how to combine DIF estimates from several administrations to obtain variance components for administration differences for DIF within an item. In their examples, they demonstrated how to use their models to improve estimations within an administration, and how to combine evidence across items in randomized DIF studies. Subsequently, ETS researchers have employed Bayesian methods with the goal of pooling data across administrations to yield more stable DIF estimates within an administration. These approaches are discussed in the section on sample size and power issues.

Allen and Holland (1993) used a missing data framework to address the missing data problem in DIF analyses where "no response" to the self-reported group identification question is large, a common problem in applied settings. They showed how MH and STAND statistics can be affected by different assumptions about nonresponses.

Zwick and her colleagues examined DIF in the context of computer adaptive testing (CAT) in which tests are tailored to the individual test taker on the basis of his or her response to previous items. Zwick et al. (1993b) described in great detail a simulation study in which they examined the performance of MH and STAND procedures that had been modified for use with data collected adaptively. The modification to the DIF procedures involved replacing the standard number-right matching variable with a matching variable based on IRT, which was obtained by converting a maximum likelihood estimate of ability to an expected number-right true score on all items in the reference pool. Examinees whose expected true scores fell in the same one-unit intervals were considered to be matched. They found that DIF statistics computed in this way for CAT were similar to those obtained with the traditional matching variable of performance on the total test. In addition they found that pretest DIF statistics were generally well behaved, but the MH DIF statistics tended to have larger standard errors for the pretest items than for the CAT items.

Zwick et al. (1994) addressed the effect of using alternative matching methods for pretest items. Using a more elegant matching procedure did not lead to a reduction of the MH standard errors and produced DIF measures that were nearly identical to those from the earlier study. Further investigation showed that the MH standard errors tended to be larger when items were administered to examinees with a wide ability range, whereas the opposite was true of the standard errors of the

STAND DIF statistic. As reported in Zwick (1994), there may be a theoretical explanation for this phenomenon.

CAT can be thought of as a very complex form of item sampling. The sampling procedure used by the National Assessment of Educational Progress (NAEP) is another form of complex sampling. Allen and Donoghue (1996) used a simulation study to examine the effect of complex sampling of items on the measurement of DIF using the MH DIF procedure. Data were generated using a three-parameter logistic (3PL) IRT model according to the balanced incomplete block design. The length of each block of items and the number of DIF items in the matching variable were varied, as was the difficulty, discrimination, and presence of DIF in the studied item. Block, booklet, pooled booklet, and other approaches to matching on more than the block, were compared to a complete data analysis using the transformed log-odds on the delta scale. The pooled booklet approach was recommended for use when items are selected for examinees according to a balanced incomplete block (BIB) data collection design.

Zwick et al. (1993a) noted that some forms of performance assessment may in fact be more likely to tap construct-irrelevant factors than multiple-choice items are. The assessment of DIF can be used to investigate the effect on subpopulations of the introduction of performance tasks. Two extensions of the MH procedure were explored: the test of conditional association proposed by Mantel (1963) and the generalized statistic proposed by Mantel and Haenszel (1959). Simulation results showed that, for both inferential procedures, the studied item should be included in the matching variable, as in the dichotomous case. Descriptive statistics that index the magnitude of DIF, including that proposed by Dorans and Schmitt (1991; described below) were also investigated.

7.2.1.4 Standardization (STAND)

Dorans (1982) reviewed item bias studies that had been conducted on data from the *SAT®* exam in the late 1970s, and concluded that these studies were flawed because either DIF was confounded with lack of model fit or it was contaminated by impact as a result of *fat matching*, the practice of grouping scores into broad categories of roughly comparable ability. A new method was needed. Dorans and Kulick (1983, 1986) developed the STAND approach after consultation with Holland. The formulas in the following section can be found in these articles and in Dorans and Holland (1993) and Dorans and Kulick (2006).

Standardization's (STAND's) Definition of Differential Item Functioning (DIF)

An item exhibits DIF when the expected performance on an item differs for matched examinees from different groups. Expected performance can be estimated by non-parametric item-test regressions. Differences in empirical item-test regressions are indicative of DIF.

The first step in the STAND analysis is to use all available item response data in the target population of interest to estimate nonparametric item-test regressions in the reference group and in the focal group. Let $E_f(Y|X)$ define the empirical item-test regression for the focal group f, and let $E_r(Y|X)$ define the empirical item-test regression for the reference group r, where Y is the item-score variable and X is the matching variable. For STAND, the definition of null-DIF conditions on an observed score is $E_r(Y|X)=E_r(Y|X)$ Plots of difference in empirical item-test regressions, focal minus reference, provide visual descriptions of DIF in fine detail for binary as well as polytomously scored items. For illustrations of nonparametric item-test regressions and differences for an actual SAT item that exhibits considerable DIF, see Dorans and Kulick (1986).

Standardization's (STAND's) Primary Differential Item Functioning (DIF) Index

While plots described DIF directly, there was a need for some numerical index that targets suspect items for close scrutiny while allowing acceptable items to pass swiftly through the screening process. For each score level, the focal group supplies specific weights that are used for each individual D_m before accumulating the weighted differences across score levels to arrive at a summary item-discrepancy index, $STD-EISDIF$, which is defined as:

$$STD-EISDIF = \frac{\sum_{m=1}^{M} N_{fm} * E_f\left(Y|X=m\right)}{\sum_{m=1}^{M} N_{fm}} - \frac{\sum_{m=1}^{M} N_{fm} * E_r\left(Y|X=m\right)}{\sum_{m=1}^{M} N_{fm}}$$

or simplified

$$STD-EISDIF = \frac{\sum_{m=1}^{M} N_{fm}\left(*E_f\left(Y|X=m\right) - E_r\left(Y|X=m\right)\right)}{\sum_{m=1}^{M} N_{fm}}$$

where $N_{fm} / \sum_{m=1}^{M} N_{fm}$ is the weighting factor at score level X_m supplied by the focal group to weight differences in expected item performance observed in the focal group $E_f(Y|X)$ and expected item performance observed in the reference group $E_r(Y|X)$.

In contrast to impact, in which each group has its relative frequency serve as a weight at each score level, STAND uses a standard or common weight on both $E_f(Y|X)$ and $E_r(Y|X)$, namely $N_{fm} / \sum_{m=1}^{M} N_{fm}$. The use of the same weight on both $E_f(Y|X)$ and $E_r(Y|X)$ is the essence of the STAND approach. Use of N_{fm} means that $STD-EISDIF$ equals the difference between the observed performance of the focal group on the item and the predicted performance of selected reference group members who are matched in ability to the focal group members. This difference can be derived very simply; see Dorans and Holland (1993).

Extensions to Standardization (STAND)

The generalization of the STAND methodology to all response options including omission and not reached is straightforward and is known as standardized distractor analysis (Dorans and Schmitt 1993; Dorans et al. 1988, 1992). It is as simple as replacing the keyed response with the option of interest in all calculations. For example, a standardized response-rate analysis on Option A would entail computing the proportions choosing A in both the focal and reference groups. The next step is to compute differences between these proportions at each score level. Then these individual score-level differences are summarized across score levels by applying some standardized weighting function to these differences to obtain $STD-DIF(A)$, the standardized difference in response rates to Option A. In a similar fashion one can compute standardized differences in response rates for Options B, C, D, and E, and for nonresponses as well. This procedure is used routinely at ETS.

Application of the STAND methodology to counts of examinees at each level of the matching variable who did not reach the item results in a standardized not-reached difference. For items at the end of a separately timed section of a test, these standardized differences provide measurement of the differential speededness of a test. *Differential speededness* refers to the existence of differential response rates between focal group members and matched reference group members to items appearing at the end of a section. Schmitt et al. (1993) reported that excluding examinees who do not reach an item from the calculation of the DIF statistic for that item partially compensates for the effects of item location on the DIF estimate.

Dorans and Schmitt (1991) proposed an extended version of STAND for ordered polytomous data. This extension has been used operationally with NAEP data since the early 1990s. This approach, called standardized mean difference (*SMD*) by Zwick et al. (1993a), provides an average DIF value for describing DIF on items with ordered categories. At each matching score level, there exist distributions of ordered item scores, I, for both the focal group (e.g., females) and the reference group (e.g., males). The expected item scores for each group at each matching score level can be computed by using the frequencies to obtain a weighted average of the score levels. The difference between these expected items scores for the focal and reference groups, $STD-EISDIF$, is the DIF statistic. Zwick and Thayer (1996) provide standard errors for *SMD* (or $STD-EISDIF$).

7.2.1.5 Item Response Theory (IRT)

DIF procedures differ with respect to whether the matching variable is explicitly an observed score (Dorans and Holland 1993) or implicitly a latent variable (Thissen et al. 1993). Observed score DIF and DIF procedures based on latent variables do not measure the same thing, and both are not likely to measure what they strive to measure, which is DIF with respect to the construct that the item purports to measure. The observed score procedures condition on an observed score, typically the score reported to a test taker, which contains measurement error and clearly differs from a pure measure of the construct of interest, especially for test scores of inadequate reliability. The latent variable approaches in essence condition on an unobservable that the test is purportedly measuring. As such they employ what Meredith and Millsap (1992) would call a measurement invariance definition of null DIF, while methods like MH and STAND employ a prediction invariance definition, which may be viewed as inferior to measurement invariance from a theoretical perspective. On the other hand, procedures that purport to assess measurement invariance employ a set of assumptions; in essence they are assessing measurement invariance under the constraint that the model they assume to be true is in fact true.

The observed score methods deal with the fact that an unobservable is unknowable by replacing the null hypothesis of measurement invariance (i.e., the items measure the construct of interest in the same way in the focal and reference groups with a prediction invariance assumption and use the data directly to assess whether expected item score is a function of observed total score in the same way across groups). The latent variable approaches retain the measurement invariance hypothesis and use the data to estimate and compare functional forms of the measurement model relating item score to a latent variable in the focal and reference groups. The assumptions embodied in these functional forms may or may not be correct, however, and model misfit might be misconstrued as a violation of measurement invariance, as noted by Dorans and Kulick (1983). For example applying the Rasch (1960) model to data fit by the two-parameter logistic (2PL) model would flag items with lower IRT slopes as having DIF favoring the lower scoring group, while items with higher slopes would favor the higher scoring group.

Lord (1977, 1980) described early efforts to assess DIF from a latent trait variable perspective. Lord recommended a statistical significance test on the joint difference between the IRT difficulty and discrimination parameters between the two groups under consideration. Thissen et al. (1993) discussed Lord's procedure and described the properties of four other procedures that used IRT. All these methods used statistical significance testing. They also demonstrated how the IRT methods can be used to assess differential distractor functioning. Thissen et al. remains a very informative introduction and review of IRT methods circa 1990.

Pashley (1992) suggested a method for producing simultaneous confidence bands for the difference between item response curves. After these bands have been plotted, the size and regions of DIF can be easily identified. Wainer (1993) provided an IRT-based effect size of amount of DIF that is based on the STAND weighting system that allows one to weight difference in the item response functions (IRF) in a manner that is proportional to the density of the ability distribution.

Zwick et al. (1994) and Zwick et al. (1995) applied the Rasch model to data simulated according to the 3PL model. They found that the DIF statistics based on the Rasch model were highly correlated with the DIF values associated with the generated data, but that they tended to be smaller in magnitude. Hence the Rasch model did not detect DIF as well, which was attributed to degradation in the accuracy of matching. Expected true scores from the Rasch-based computer-adaptive test tended to be biased downward, particularly for lower-ability examinees. If the Rasch model had been used to generate the data, different results would probably have been obtained.

Wainer et al. (1991) developed a procedure for examining DIF in collections of related items, such as those associated with a reading passage. They called this DIF for a set of items a *testlet DIF*. This methodology paralleled the IRT-based likelihood procedures mentioned by Thissen et al. (1993).

Zwick (1989, 1990) demonstrated that the null definition of DIF for the MH procedure (and hence STAND and other procedures employing observed scores as matching variables) and the null hypothesis based on IRT are different because the latter compares item response curves, which in essence condition on unobserved ability. She also demonstrated that the item being studied for DIF should be included in the matching variable if MH is being used to identify IRT DIF.

7.2.1.6 SIBTEST

Shealy and Stout (1993) introduced a general model-based approach to assessing DIF and other forms of differential functioning. They cited the STAND approach as a progenitor. From a theoretical perspective, SIBTEST is elegant. It sets DIF within a general multidimensional model of item and test performance. Unlike most IRT approaches, which posit a specific form for the item response model (e.g., a 2PL model), SIBTEST does not specify a particular functional form. In this sense it is a nonparametric IRT model, in principle, in which the null definition of STAND involving regressions onto observed scores is replaced by one involving regression onto true scores,

$$\varepsilon_f\left(Y|T_x\right) = \varepsilon_r\left(Y|T_x\right),$$

where T_x represents a true score for$X$. As such, SIBTEST employs a measurement invariance definition of null DIF, while STAND employs a prediction invariance definition (Meredith and Millsap 1992).

Chang et al. (1995, 1996) extended SIBTEST to handle polytomous items. Two simulation studies compared the modified SIBTEST procedure with the generalized Mantel (1963) and SMD or STAND procedures. The first study compared the procedures under conditions in which the generalized Mantel and SMD procedures had been shown to perform well (Zwick et al. 1993a. Results of Study 1 suggested that SIBTEST performed reasonably well, but that the generalized Mantel and SMD procedures performed slightly better. The second study used data simulated under conditions in which observed-score DIF methods for dichotomous items had not performed well (i.e., a short nonrepresentative matching test). The results of Study 2 indicated that, under these conditions, the modified SIBTEST procedure provided better control of impact-induced Type I error inflation with respect to detecting DIF (as defined by SIBTEST) than the other procedures.

Zwick et al. (1997b) evaluated statistical procedures for assessing DIF in polytomous items. Three descriptive statistics – the SMD (Dorans and Schmitt 1991) and two procedures based on SIBTEST (Shealy and Stout 1993) were considered, along with five inferential procedures: two based on SMD, two based on SIBTEST, and one based on the Mantel (1963) method. The DIF procedures were evaluated through applications to simulated data, as well as to empirical data from ETS tests. The simulation included conditions in which the two groups of examinees had the same ability distribution and conditions in which the group means differed by one standard deviation. When the two groups had the same distribution, the descriptive index that performed best was the SMD. When the two groups had different distributions, a modified form of the SIBTEST DIF effect-size measure tended to perform best. The five inferential procedures performed almost indistinguishably when the two groups had identical distributions. When the two groups had different distributions and the studied item was highly discriminating, the SIBTEST procedures showed much better Type I error control than did the SMD and Mantel methods, particularly with short tests. The power ranking of the five procedures was inconsistent; it depended on the direction of DIF and other factors. The definition of DIF employed was the IRT definition, measurement invariance, not the observed score definition, prediction invariance.

Dorans (2011) summarized differences between SIBTEST and its progenitor, STAND. STAND uses observed scores to assess whether the item-test regressions are the same across focal and reference groups. On its surface, the SIBTEST DIF method appears to be more aligned with measurement models. This method assumes that examinee group differences influence DIF or test form difficulty differences more than can be observed in unreliable test scores. SIBTEST adjusts the observed data toward what is suggested to be appropriate by the measurement model. The degree to which this adjustment occurs depends on the extent that these data are unreliable. To compensate for unreliable data on the individual, SIBTEST regresses observed performance on the test to what would be expected for the focal or reference group on the basis of the ample data that show that race and gender are related to item performance. SIBTEST treats true score estimation as a prediction problem, introducing bias to reduce mean squared error. In essence, the SIBTEST method uses subgroup-specific true score estimates as a surrogate for the true score that is defined in the classical test theory model. If SIBTEST regressed all test takers to the same mean it would not differ from STAND.

7.2.2 Matching Variable Issues

Dorans and Holland (1993) laid out an informal research agenda with respect to observed score DIF. The matching variable was one area that merited investigation. Inclusion of the studied item in the matching variable and refinement or purification of the criterion were mentioned. Dimensionality and DIF was, and remains, an important factor; DIF procedures presume that all items measure the same construct in the same way across all groups.

Donoghue and Allen (1993) examined two strategies for forming the matching variable for the MH DIF procedure; "thin" matching on total test score was compared to forms of "thick" matching, pooling levels of the matching variable. Data were generated using a 3PL IRT model with a common guessing parameter. Number of subjects and test length were manipulated, as were the difficulty, discrimination, and presence/absence of DIF in the studied item. For short tests (five or ten items), thin matching yielded very poor results, with a tendency to falsely identify items as possessing DIF against the reference group. The best methods of thick matching yielded outcome measure values closer to the expected value for non-DIF items and a larger value than thin matching when the studied item possessed DIF. Intermediate-length tests yielded similar results for thin matching and the best methods of thick matching.

The issue of whether or not to include the studied item in the matching variable was investigated by many researchers from the late 1980s to early 1990s. Holland and Thayer (1988) demonstrated mathematically that when the data were consistent with the Rasch model, it was necessary to include the studied item in a purified rights-scored matching criterion in order to avoid biased estimates of DIF (of the measurement invariance type) for that studied item. Inclusion of the studied item removes the dependence of the item response on group differences in ability distributions. Zwick (1990) and Lewis (1993) developed this idea further to illustrate the applicability of this finding to more general item response models. Both authors proved mathematically that the benefit in bias correction associated with including the studied item in the matching criterion held true for the binomial model, and they claimed that the advantage of including the studied item in the matching criterion would not be evident for any IRT model more complex than the Rasch model.

Donoghue et al. (1993) evaluated the effect of including/excluding the studied item under the 3PL IRT model. In their simulation, they fixed the discrimination parameters for all items in a simulated test in each studied condition and fixed the guessing parameter for all conditions, but varied the difficulty (b) parameters for different items for each studied condition. Although the 3PL model was used to simulate data, only the b-parameter was allowed to vary. On the basis of their study, they recommended including the studied item in the matching variable when the MH procedure is used for DIF detection. They also recommended that short tests not be used for matching variables.

Zwick et al. (1993a) extended the scope of their DIF research to performance tasks. In their study, multiple-choice (MC) items and performance tasks were simulated using the 3PL model and the partial-credit model, respectively. The MC items were simulated to be free of DIF and were used as the matching criterion. The performance tasks were simulated to be the studied items with or without DIF. They found that the item should be included in the matching criterion.

Zwick (1990) analytically examined item inclusion for models more complex than the Rasch. Her findings apply to monotone IRFs with local independence for the case where the IRFs on the matching items were assumed identical for the two groups. If the studied item is excluded from the matching variable, the MH null hypothesis will not hold in general even if the two groups had the same IRF for the studied item. It is assured to hold only if the groups have the same ability distribution. If the ability distributions are ordered, the MH will show DIF favoring the higher group (generalization of Holland and Thayer's [1988] Rasch model findings). Even if the studied item is included, the MH null hypothesis will not hold in general. It is assured to hold only if the groups have the same ability distribution or if the Rasch model holds. Except in these special situations, the MH can produce a conclusion of DIF favoring either the focal or reference group.

Tan et al. (2010) studied the impact of including/excluding the studied item in the matching variable on bias in DIF estimates under conditions where the assumptions of the Rasch model were violated. Their simulation study varied different magnitudes of DIF and different group ability distributions, generating data from a 2PL IRT model and a multidimensional IRT model. Results from the study showed that including the studied item leads to less biased DIF estimates and more appropriate Type I error rate, especially when group ability distributions are different. Systematic biased estimates in favor of the high ability group were consistently found across all simulated conditions when the studied item was excluded from the matching criterion.

Zwick and Ercikan (1989) used bivariate matching to examine DIF on the NAEP history assessment, conditioning on number-right score and historical period studied. Contrary to expectation, the additional conditioning did not lead to a reduction in the number of DIF items.

Pomplun et al. (1992) evaluated the use of bivariate matching to study DIF with formula-scored tests, where item inclusion cannot be implemented in a straightforward fashion. Using SAT Verbal data with large and small samples, both male-female and black-white group comparisons were investigated. MH D-DIF values and DIF category classifications based on bivariate matching on rights score and nonresponse were compared with MH D-DIF values and categories based on rights-scored and formula-scored matching criteria. When samples were large, MH D-DIF values based on the bivariate matching criterion were ordered very similarly to MH D-DIF values based on the other criteria. However, with small samples the MH D-DIF values based on the bivariate matching criterion displayed only moderate correlations with MH D-DIF values from the other criteria.

7.2.3 Study Group Definition

Another area mentioned by Dorans and Holland (1993) was the definition of the focal and reference groups. Research has continued in this area as well.

Allen and Wainer (1989) noted that the accuracy of procedures that are used to compare the performance of different groups of examinees on test items obviously depends upon the correct classification of members in each examinee group. They argued that because the number of nonrespondents to questions of ethnicity is often of the same order of magnitude as the number of identified members of most minority groups, it is important to understand the effect of nonresponse on DIF results. They examined the effect of nonresponse to questions of ethnic identity on the measurement of DIF for SAT Verbal items using the MH procedure. They demonstrated that efforts to obtain more complete ethnic identifications from the examinees would lead to more accurate DIF analyses.

DIF analyses are performed on target populations. One of the requirements for inclusion in the analysis sample is that the test taker has sufficient skill in the language of the test. Sinharay (2009b) examined how an increase in the proportion of examinees who report that English is not their first language would affect DIF results if they were included in the DIF analysis sample of a large-scale assessment. The results varied by group. In some combinations of focal/reference groups, the magnitude of DIF was not appreciably affected by whether DIF was performed on examinees whose first language was not English. In other groups, first language status mattered. The results varied by type of test as well. In addition, the magnitude of DIF for some items was substantially affected by whether the DIF was performed on examinees whose first language was not English.

Dorans and Holland (1993) pointed out that in traditional one-way DIF analysis, deleting items due to DIF can have unintended consequences on the focal group. DIF analysis performed on gender and on ethnicity/race alone ignores the potential interactions between the two main effects. Additionally, Dorans and Holland suggested applying a "melting-pot" DIF method wherein the total group would function as the reference group and each gender-by-ethnic subgroup would serve sequentially as a focal group. Zhang et al. (2005) proposed a variation on the melting-pot approach called DIF dissection. They adapted the STAND methodology so that the reference group was defined to be the total group, while each of the subgroups independently acted as a focal group. They argued that using a combination of all groups as the reference group and each combination of gender and ethnicity as a focal group produces more accurate, though potentially less stable, findings than using a simple majority group approach. As they hypothesized, the deletion of a sizable DIF item had its greatest effect on the mean score of the focal group that had the most negative DIF according to the DIF dissection method. In addition, the study also found that the DIF values obtained by the DIF procedure reliably predicted changes in scaled scores after item deletion.

7.2.4 Sample Size and Power Issues

From its inaugural use as an operational procedure, DIF has had to grapple with sample size considerations (Zieky 1993). The conflict between performing as many DIF analyses as possible and limiting the analysis to those cases where there is sufficient power to detect DIF remains as salient as ever.

Lyu et al. (1995) developed a smoothed version of STAND, which merged kernel smoothing with the traditional STAND DIF approach, to examine DIF for student produced response (SPR) items on the SAT I Math at both the item and testlet levels. Results from the smoothed item-level DIF analysis showed that regular multiple-choice items have more variability in DIF values than SPRs.

Bayesian methods are often resorted to when small sample sizes limit the potential power of a statistical procedure. Bayesian statistical methods can incorporate, in the form of a prior distribution, existing information on the inference problem at hand, leading to improved estimation, especially for small samples for which the posterior distribution is sensitive to the choice of prior distribution. Zwick et al. (1997a, 1999) developed an empirical Bayes (EB) enhancement to MH DIF analysis in which they assumed that the MH statistics were normally distributed and that the prior distribution of underlying DIF parameters was also normal. They used the posterior distribution of DIF parameters to make inferences about the item's true DIF status and the posterior predictive distribution to predict the item's future observed status. DIF status was expressed in terms of the probabilities associated with each of the five DIF levels defined by the ETS classification system (Zieky 1993). The EB method yielded more stable DIF estimates than did conventional methods, especially in small samples. The EB approach also conveyed information about DIF stability in a more useful way by representing the state of knowledge about an item's DIF status as probabilistic.

Zwick et al. (2000) investigated a DIF flagging method based on loss functions. The approach built on their earlier research that involved the development of an EB enhancement to MH DIF analysis. The posterior distribution of DIF parameters was estimated and used to obtain the posterior expected loss for the proposed approach and for competing classification rules. Under reasonable assumptions about the relative seriousness of Type I and Type II errors, the loss-function-based DIF detection rule was found to perform better than the commonly used ETS DIF classification system, especially in small samples.

Zwick and Thayer (2002) used a simulation to investigate the applicability to computerized adaptive test data of an EB DIF analysis method developed by (Zwick et al. 1997a, 1999) and showed that the performance of the EB DIF approach to be quite promising, even in extremely small samples. When combined with a loss-function-based decision rule, the EB method is better at detecting DIF than conventional approaches, but it has a higher Type I error rate.

The EB method estimates the prior mean and variance from the current data and uses the same prior information for all the items. For most operational tests, however, a large volume of past data is available, and for any item appearing in a current test, a number of similar items are often found to have appeared in past operational administrations of the test. Conceptually, it should be possible to incorporate that past information into a prior distribution in a Bayesian DIF analysis. Sinharay (2009a) developed a full Bayesian (FB) DIF estimation method that used this type of past information. The FB Bayesian DIF analysis method was shown to be an improvement over existing methods in a simulation study.

Zwick et al. (2000) proposed a Bayesian updating (BU) method that may avert the shrinkage associated with the EB and FB approaches. Zwick et al. (2012) implemented the BU approach and compared it to the EB and FB approaches in both simulated and empirical data. They maintained that the BU approach was a natural way to accumulate all known DIF information about an item while mitigating the tendency to shrink DIF toward zero that characterized the EB and FB approaches.

Smoothing is another alternative used for dealing with small sample sizes. Yu et al. (2008) applied smoothing techniques to frequency distributions and investigated the impact of smoothed data on MH DIF detection in small samples. Eight sample-size combinations were randomly drawn from a real data set were replicated 80 times to produce stable results. Loglinear smoothing was found to provide slight-to-moderate improvements in MH DIF estimation with small samples.

Puhan, Moses, Yu, and Dorans (Puhan et al. 2007, 2009) examined the extent to which loglinear smoothing could improve the accuracy of SIBTEST DIF estimates in small samples of examinees. Examinee responses from a certification test were used. Separate DIF estimates for seven small-sample-size conditions were obtained using unsmoothed and smoothed score distributions. Results indicated that for most studied items smoothing the raw score distributions reduced random error and bias of the DIF estimates, especially in the small-sample-size conditions.

7.3 Fair Linking of Test Scores

Scores on different forms or editions of a test that are supposed to be used interchangeably should be related to each other in the same way across different subpopulations. Score equity assessment (SEA) uses subpopulation invariance of linking functions across important subpopulations to assess the degree of interchangeability of scores.

Test score equating is a statistical process that produces scores considered comparable enough across test forms to be used interchangeably. Five requirements are often regarded as basic to all test equating (Dorans and Holland 2000). One of the most basic requirements of score equating is that equating functions should be subpopulation invariant (Dorans and Holland 2000; Holland and Dorans 2006). That is, they should not be influenced by the subpopulation of examinees on which they are computed. The same construct and equal reliability requirements are prerequisites for subpopulation invariance. One way to demonstrate that two test forms are not equatable is to show that the equating functions used to link their scores are not invariant across different subpopulations of examinees. Lack of invariance in a linking function indicates that the differential difficulty of the two test forms is not consistent across different groups. The invariance can hold if the relative difficulty changes as a function of score level in the same way across subpopulations. If, however, the relative difficulty of the two test forms interacts with group membership or an interaction among score level, difficulty, and group is present, then invariance does not hold. SEA uses the subpopulation invariance of linking functions across important subgroups (e.g., gender groups and other groups, sample sizes permitting) to assess the degree of score exchangeability.

In an early study, Angoff and Cowell (1985, 1986) examined the invariance of equating scores on alternate forms of the *GRE®* quantitative test for various populations, including gender, race, major, and ability. Angoff and Cowell conducted equatings for each of the populations and compared the resulting conversions to each other and to differences that would be expected given the standard errors of equating. Differences in the equatings were found to be within that expected given sampling error. Angoff and Cowell concluded that population invariance was supported.

Dorans and Holland (2000) included several examples of linkings that are invariant (e.g., SAT Mathematics to SAT Mathematics and SAT Verbal to SAT Verbal, and SAT Mathematics to ACT Mathematics) as well as ones that are not (e.g., verbal to mathematics, and linkings between non-math ACT subscores and SAT Verbal). Equatability indexes are used to quantify the degree to which linkings are subpopulation invariant.

Since 2000, several evaluations of population invariance have been performed. Yang (2004) examined whether the linking functions that relate multiple-choice scores to composite scores based on weighted sums of multiple choice and constructed response scores for selected *Advanced Placement®* (*AP®*) exams remain invariant over subgroups by geographical region. The study focused on two questions: (a) how invariant were cut-scores across regions and (b) whether the small sample size for some regional groups presented particular problems for assessing linking invariance. In addition to using the subpopulation invariance indexes to evaluate linking functions, Yang also evaluated the invariance of the composite score thresholds for determining final AP grades. Dorans (2004) used the population sensitivity of linking functions to assess score equity for two AP exams.

Dorans et al. (2008) used population sensitivity indexes with SAT data to evaluate how consistent linear equating results were across males and females. Von Davier and Wilson (2008) examined the population invariance of IRT equating for an AP exam. Yang and Gao (2008) looked at invariance of linking computer-administered *CLEP®* data across gender groups.

SEA has also been used as a tool to evaluate score interchangeability when a test is revised (Liu and Dorans 2013). Liu et al. (2006) and Liu and Walker (2007) used SEA tools to examine the invariance of linkages across the old and new versions of the SAT using data from a major field trail conducted in 2003. This check was followed by SEA analyses conducted on operational data (see studies cited in Dorans and Liu 2009).

All these examples, as well as others such as Dorans et al. (2003), are illustrations of using SEA to assess the fairness of a test score by examining the degree to which the linkage between scores is invariant across subpopulations. In some of these illustrations, such as one form of SAT Mathematics with another form of SAT Mathematics, the expectation of score interchangeability was very high since alternate forms of this test are designed to be parallel in both content and difficulty. There are cases, however, where invariance was expected but did not hold. Cook et al. (1988), for example, found that the linking function between two biology exams depended on whether the equating was with students in a December administration, where most of the examinees were seniors who had not taken a biology course for some time, versus a June administration, where most of the examinees had just completed a biology course. This case, which has become an exemplar of lack of invariance where invariance would be expected, is discussed in detail by Cook (2007) and Peterson (2007). Invariance cannot be presumed to occur simply because tests are built to the same blueprint. The nature of the population can be critical, especially when diverse subpopulations are involved. For most testing programs, analysis that focuses on the invariance of equating functions should be conducted to confirm the fairness of the assembly process.

7.4 Limitations of Quantitative Fairness Assessment Procedures

First, not all fairness considerations can be reduced to quantitative evaluation. Because this review was limited to quantitative fairness procedures, it was limited in scope. With this important caveat in mind, this section will discuss limitations with the classes of procedures that have been examined.

Fair prediction is difficult to achieve. Differential prediction studies are difficult to complete effectively because there are so many threats to the subpopulation invariance of regression equations. Achieving subpopulation invariance of regressions is difficult because of selection effects, misspecification errors, predictor unreliability, and criterion issues. Any attempt to assess whether a prediction equation is invariant across subpopulations such as males and females must keep these confounding influences in mind.

To complicate validity assessment even more, there are as many external criteria as there are uses of a score. Each use implies a criterion against which the test's effectiveness can be assessed. The process of validation via prediction studies is an unending yet necessary task.

DIF screening is and has been possible to do. But it could be done better. Zwick (2012) reviewed the status of ETS DIF analysis procedures, focusing on three aspects: (a) the nature and stringency of the statistical rules used to flag items, (b) the minimum sample size requirements that are currently in place for DIF analysis, and (c) the efficacy of criterion refinement. Recommendations were made with respect to improved flagging rules, minimum sample size requirements, and procedures for combining data across administrations. Zwick noted that refinement of the matching criterion improves detection rates when DIF is primarily in one direction but can depress detection rates when DIF is balanced.

Most substantive DIF research studies that have tried to explain DIF have used observational data and the generation of post-hoc explanations for why items were flagged for DIF. The chapter by O'Neill and McPeek (1993) in the Holland and Wainer (1993) DIF book is a good example of this approach. As both those authors and Bond (1993) noted, this type of research with observed data is fraught with peril because of the highly selected nature of the data examined, namely items that have been flagged for DIF. In the same section of the DIF book, Schmitt et al. (1993) provided a rare exemplar on how to evaluate DIF hypotheses gleaned from observational data with experimental evaluations of the hypotheses via a carefully designed and executed experimental manipulation of item properties followed by a proper data analysis.

DIF can be criticized for several reasons. An item is an unreliable measure of the construct of interest. Performance on an item is susceptible to many influences that have little to do with the purpose of the item. An item, by itself, can be used to support a variety of speculations about DIF. It is difficult to figure out why DIF occurs. The absence of DIF is not a prerequisite for fair prediction. In addition, DIF analysis tells little about the effects of DIF on reported scores.

SEA focuses on invariance at the reported score level where inferences are made about the examinee. SEA studies based on counterbalanced single-group designs are likely to give the cleanest results about the invariance of score linking functions because it is a data collection design that allows for the computation of correlations between tests across subpopulations.

This chapter focused primarily on studies that focused on methodology and that were conducted by ETS staff members. As a result, many DIF and differential prediction studies that used these methods have been left out and need to be summarized elsewhere. As noted, qualitative and philosophical aspects of fairness have not been considered.

In addition, ETS has been the leader in conducting routine DIF analyses for over a quarter of century. This screening for DIF practice has made it difficult to find items that exhibit the high degree of DIF depicted on the cover of the Winter 2012 issue of *Educational Measurement: Issues and Practices,* an item that Dorans (2012) cited as a vintage example of DIF. Although item scores exhibit less DIF than they did before due diligence made DIF screening an operational practice, a clear need remains for continued research in fairness assessment. This includes improved methods for detecting evidence of unfairness and the use of strong data collection designs that allow researchers to arrive at a clearer understanding of sources of unfairness.

References

Allen, N. L., & Donoghue, J. R. (1996). Applying the Mantel-Haenszel procedure to complex samples of items. *Journal of Educational Measurement, 33*, 231–251. https://doi.org/10.1111/j.1745-3984.1996.tb00491.x

Allen, N., & Holland, P. H. (1993). A model for missing information about the group membership of examinees in DIF studies. In P. W. Holland & H. Wainer (Eds.), *Differential item functioning* (pp. 241–252). Hillsdale: Erlbaum.

Allen, N. L., & Wainer, H. (1989). *Nonresponse in declared ethnicity and the identification of differentially functioning items* (Research Report No. RR-89-47). Princeton: Educational Testing Service. http://dx.doi.org/10.1002/j.2333-8504.1982.tb01331.x

Angoff, W. H. (1972, September). *A technique for the investigation of cultural differences*. Paper presented at the meeting of the American Psychological Association, Honolulu.

Angoff, W. H. (1993). Perspectives on differential item functioning methodology. In P. W. Holland & H. Wainer (Eds.), *Differential item functioning* (pp. 3–23). Hillsdale: Erlbaum.

Angoff, W. H., & Cowell, W. R. (1985). *An examination of the assumption that the equating of parallel forms is population-independent* (Research Report No. RR-85-22). Princeton: Educational Testing Service. http://dx.doi.org/10.1002/j.2330-8516.1985.tb00107.x

Angoff, W. H., & Cowell, W. R. (1986). An examination of the assumption that the equating of parallel forms is population-independent. *Journal of Educational Measurement, 23*, 327–345. https://doi.org/10.1111/j.1745-3984.1986.tb00253.x

Angoff, W. H., & Ford, S. F. (1971). *Item-race interaction on a test of scholastic aptitude* (Research Bulletin No. RB-71-59). Princeton: Educational Testing Service. http://dx.doi.org/10.1002/j.2333-8504.1971.tb00812.x

Angoff, W. H., & Ford, S. (1973). Item-race interaction on a test of scholastic aptitude. *Journal of Educational Measurement, 10*, 95–105. https://doi.org/10.1111/j.1745-3984.1973.tb00787.x

Angoff, W. H., & Sharon, A. (1974). The evaluation of differences in test performance of two or more groups. *Educational and Psychological Measurement, 34*, 807–816. https://doi.org/10.1177/001316447403400408

Baker, F. B. (1981). A criticism of Scheuneman's item bias technique. *Journal of Educational Measurement, 18*, 59–62.

Bond, L. (1993). Comments on the O'Neill & McPeek chapter. In P. W. Holland & H. Wainer (Eds.), *Differential item functioning* (pp. 277–279). Hillsdale: Erlbaum.

Campbell, J. T. (1964). *Testing of culturally different groups* (Research Bulletin No. RB-64–34). Princeton: Educational Testing Service. http://dx.doi.org/10.1002/j.2333-8504.1964.tb00506.x

Cardall, C., & Coffman, W. E. (1964). *A method for comparing the performance of different groups on the items in a test* (Research Bulletin No. RB-64-61). Princeton: Educational Testing Service.

Chang, H.-H., Mazzeo, J., & Roussos, L. (1995). *Detecting DIF for polytomously scored items: An adaptation of the SIBTEST procedure* (Research Report No. RR-95-05). Princeton: Educational Testing Service. http://dx.doi.org/10.1002/j.2333-8504.1995.tb01640.x

Chang, H.-H., Mazzeo, J., & Roussos, L. (1996). Detecting DIF for polytomously scored items: An adaptation of the SIBTEST procedure. *Journal of Educational Measurement, 33*, 333–354. https://doi.org/10.1111/j.1745-3984.1996.tb00496.x

Cleary, T. A. (1968). Test bias: Prediction of grades of Negro and White students in integrated colleges. *Journal of Educational Measurement, 5*, 115–124. https://doi.org/10.1111/j.1745-3984.1968.tb00613.x

Cleary, T. A., & Hilton, T. L. (1966). *An investigation of item bias* (Research Bulletin No. RB-66-17). Princeton: Educational Testing Service. http://dx.doi.org/10.1002/j.2333-8504.1966.tb00355.x

Cleary, T. A., & Hilton, T. J. (1968). An investigation of item bias. *Educational and Psychological Measurement, 5*, 115–124.

Cole, N. S. (1973). Bias in selection. *Journal of Educational Measurement, 5*, 237–255. https://doi.org/10.1177/001316446802800106

Cook, L. L. (2007). Practical problems in equating test scores: A practitioner's perspective. In N. J. Dorans, M. Pommerich, & P. W. Holland (Eds.), *Linking and aligning scores and scales* (pp. 73–88). New York: Springer-Verlag. https://doi.org/10.1007/978-0-387-49771-6_5

Cook, L. L., Eignor, D. R., & Taft, H. L. (1988). A comparative study of the effects of recency of instruction on the stability of IRT and conventional item parameter estimates. *Journal of Educational Measurement, 25*, 31–45. https://doi.org/10.1111/j.1745-3984.1988.tb00289.x

Donoghue, J. R., & Allen, N. L. (1993). "Thin" versus "thick" in the Mantel-Haenszel procedure for detecting DIF. *Journal of Educational Statistics, 18*, 131–154. https://doi.org/10.2307/1165084

Donoghue, J. R., Holland, P. W., & Thayer, D. T. (1993). A Monte Carlo study of factors that affect the Mantel-Haenszel and standardization measures of differential item functioning. In P. W. Holland & H. Wainer (Eds.), *Differential item functioning* (pp. 137–166). Hillsdale: Erlbaum.

Dorans, N. J. (1982). *Technical review of item fairness studies: 1975–1979* (Statistical Report No. SR-82-90). Princeton: Educational Testing Service.

Dorans, N. J. (1989). Two new approaches to assessing differential item functioning: Standardization and the Mantel-Haenszel method. *Applied Measurement in Education, 2*, 217–233. https://doi.org/10.1207/s15324818ame0203_3

Dorans, N. J. (2004). Using subpopulation invariance to assess test score equity. *Journal of Educational Measurement, 41*, 43–68. https://doi.org/10.1111/j.1745-3984.2004.tb01158.x

Dorans, N. J. (2011). Holland's advice during the fourth generation of test theory: Blood tests can be contests. In N. J. Dorans & S. Sinharay (Eds.), *Looking back: Proceedings of a conference in honor of Paul W. Holland* (pp. 259–272). New York: Springer-Verlag.

Dorans, N. J. (2012). The contestant perspective on taking tests: Emanations from the statue within. *Educational Measurement: Issues and Practice, 31*(4), 20–37. https://doi.org/10.1111/j.1745-3992.2012.00250.x

Dorans, N. L., & Cook, L. L. (Eds.). (2016). *NCME application of educational assessment and measurement: Volume 3. Fairness in educational assessment and measurement*. New York: Routledge.

Dorans, N. J., & Holland, P. W. (1993). DIF detection and description: Mantel-Haenszel and standardization. In P. W. Holland & H. Wainer (Eds.), *Differential item functioning* (pp. 35–66). Hillsdale: Erlbaum.

Dorans, N. J., & Holland, P. W. (2000). Population invariance and the equatability of tests: Basic theory and the linear case. *Journal of Educational Measurement, 37*, 281–306. https://doi.org/10.1111/j.1745-3984.2000.tb01088.x

Dorans, N. J., & Kulick, E. (1983). *Assessing unexpected differential item performance of female candidates on SAT and TSWE forms administered in December 1977: An application of the standardization approach* (Research Report No. RR-83-09). Princeton: Educational Testing Service. http://dx.doi.org/10.1002/j.2330-8516.1983.tb00009.x

Dorans, N. J., & Kulick, E. (1986). Demonstrating the utility of the standardization approach to assessing unexpected differential item performance on the scholastic aptitude test. *Journal of Educational Measurement, 23*, 355–368. https://doi.org/10.1111/j.1745-3984.1986.tb00255.x

Dorans, N. J., & Kulick, E. (2006). Differential item functioning on the mini-mental state examination: An application of the Mantel-Haenzel and standardization procedures. *Medical Care, 44*(11), S107–S114. https://doi.org/10.1097/01.mlr.0000245182.36914.4a

Dorans, N. J., & Liu, J. (2009). *Score equity assessment: Development of a prototype analysis using SAT Mathematics test data across several administrations* (Research Report No. RR-09-08). Princeton: Educational Testing Service. https://doi.org/10.1002/j.2333-8504.2009.tb02165.x

Dorans, N. J., & Potenza, M. T. (1994). *Equity assessment for polytomously scored items: A taxonomy of procedures for assessing differential item functioning* (Research Report No. RR-94-49). Princeton: Educational Testing Service. https://doi.org/10.1002/j.2333-8504.1994.tb01622.x

Dorans, N. J., & Schmitt, A. P. (1991). *Constructed response and differential item functioning: A pragmatic approach* (Research Report No. RR-91-47). Princeton: Educational Testing Service. https://doi.org/10.1002/j.2333-8504.1991.tb01414.x

Dorans, N. J., & Schmitt, A. P. (1993). Constructed response and differential item functioning: A pragmatic approach. In R. E. Bennett & W. C. Ward (Eds.), *Construction versus choice in cognitive measurement* (pp. 135–165). Hillsdale: Erlbaum.

Dorans, N. J., Schmitt, A. P., & Bleistein, C. A. (1988). *The standardization approach to assessing differential speededness* (Research Report No. RR-88-31). Princeton: Educational Testing Service. https://doi.org/10.1002/j.2330-8516.1988.tb00287.x

Dorans, N. J., Schmitt, A. P., & Bleistein, C. A. (1992). The standardization approach to assessing comprehensive differential item functioning. *Journal of Educational Measurement, 29*, 309–319. https://doi.org/10.1111/j.1745-3984.1992.tb00379.x

Dorans, N. J., Holland, P. W., Thayer, D. T., & Tateneni, K. (2003). Invariance of score linking across gender groups for three Advanced Placement Program examinations. In N. J. Dorans (Ed.), *Population invariance of score linking: Theory and applications to Advanced Placement Program examinations* (Research Report No. RR-03-27, pp. 79–118). Princeton: Educational Testing Service. http://dx.doi.org/10.1002/j.2333-8504.2003.tb01919.x

Dorans, N. J., Liu, J., & Hammond, S. (2008). Anchor test type and population invariance: An exploration across subpopulations and test administrations. *Applied Psychological Measurement, 32*, 81–97. https://doi.org/10.1177/0146621607311580

Griggs v. Duke Power Company, 401 U.S. 424 (1971).

Gulliksen, H. O. (1964). Intercultural studies of attitudes. In N. Frederiksen & H. O. Gulliksen (Eds.), *Contributions to mathematical psychology* (pp. 61–108). New York: Holt, Rinehart & Winston.

Holland, P. W., & Dorans, N. J. (2006). Linking and equating. In R. L. Brennan (Ed.), *Educational measurement* (4th ed., pp. 187–220). Westport: American Council on Education and Praeger.

Holland, P. W., & Thayer, D. T. (1985). *An alternative definition of the ETS delta scale of item difficulty* (Research Report No. RR-85-43). Princeton: Educational Testing Service. https://doi.org/10.1002/j.2330-8516.1985.tb00128.x

Holland, P. W., & Thayer, D. T. (1986). *Differential item functioning and the Mantel-Haenszel procedure* (Research Report No. RR-86-31). Princeton: Educational Testing Service. https://doi.org/10.1002/j.2330-8516.1986.tb00186.x

Holland, P. W., & Thayer, D. T. (1988). Differential item performance and the Mantel-Haenszel procedure. In H. Wainer & H. Braun (Eds.), *Test validity* (pp. 129–145). Hillsdale: Erlbaum.

Holland, P. W., & Wainer, H. (1993). *Differential item functioning*. Hillsdale: Erlbaum.

Lewis, C. (1993). Bayesian methods for the analysis of variance. In G. Keren & C. Lewis (Eds.), *A handbook for data analysis in the behavioral sciences: Vol. II. Statistical issues* (pp. 233–256). Hillsdale: Erlbaum.

Linn, R. L. (1972). *Some implications of the Griggs decision for test makers and users* (Research Memorandum No. RM-72-13). Princeton: Educational Testing Service.

Linn, R. L. (1973). Fair test use in selection. *Review of Educational Research, 43*, 139–161. https://doi.org/10.3102/00346543043002139

Linn, R. L. (1975). *Test bias and the prediction of grades in law school* (Report No. LSAC-75-01), Newtown: Law School Admissions Council.

Linn, R. L. (1976). In search of fair selection procedures. *Journal of Educational Measurement, 13*, 53–58. https://doi.org/10.1111/j.1745-3984.1976.tb00181.x

Linn, R. L. (1983). Predictive bias as an artifact of selection procedures. In H. Wainer & S. Messick (Eds.), *Principals of modern psychological measurement: A festschrift for Frederic M. Lord* (pp. 27–40). Hillsdale: Erlbaum.

Linn, R. L., & Werts, C. E. (1971). Considerations for studies of test bias. *Journal of Educational Measurement, 8*, 1–4. https://doi.org/10.1111/j.1745-3984.1971.tb00898.x

Liu, J., & Dorans, N. J. (2013). Assessing a critical aspect of construct continuity when test specifications change or test forms deviate from specifications. *Educational Measurement: Issues and Practice, 32*(1), 15–22. https://doi.org/10.1111/emip.12001

Liu, J., & Walker, M. E. (2007). Score linking issues related to test content changes. In N. J. Dorans, M. Pommerich, & P. W. Holland (Eds.), *Linking and aligning scores and scales* (pp. 109–134). New York: Springer-Verlag. https://doi.org/10.1007/978-0-387-49771-6_7

Liu, J., Cahn, M., & Dorans, N. J. (2006). An application of score equity assessment: Invariance of linking of new SAT to old SAT across gender groups. *Journal of Educational Measurement, 43*, 113–129. https://doi.org/10.1111/j.1745-3984.2006.00008.x

Longford, N. T., Holland, P. W., & Thayer, D. T. (1993). Stability of the MH D-DIF statistics across populations. In P. W. Holland & H. Wainer (Eds.), *Differential item functioning* (pp. 171–196). Hillsdale: Erlbaum.

Lord, F. M. (1977). A study of item bias, using item characteristic curve theory. In Y. H. Poortinga (Ed.), *Basic problems in cross-cultural psychology* (pp. 19–29). Amsterdam: Swets and Zeitlinger.

Lord, F. M. (1980). *Applications of item response theory to practical testing problems*. Hillsdale: Erlbaum.

Lyu, C. F., Dorans, N. J., & Ramsay, J. O. (1995). *Smoothed standardization assessment of testlet level DIF on a math free-response item type* (Research Report No. RR-95-38). Princeton: Educational Testing Service. https://doi.org/10.1002/j.2333-8504.1995.tb01672.x

Mantel, N. (1963). Chi-square tests with one degree of freedom; extensions of the Mantel-Haenszel procedure. *Journal of the American Statistical Association, 58*, 690–700. https://doi.org/10.1080/01621459.1963.10500879

Mantel, N., & Haenszel, W. (1959). Statistical aspects of the analysis of data from retrospective studies of disease. *Journal of the National Cancer Institute, 22*, 719–748.

Mapuranga, R., Dorans, N. J., & Middleton, K. (2008). *A review of recent developments in differential item functioning* (Research Report No. RR-08-43). Princeton: Educational Testing Service. https://doi.org/10.1002/j.2333-8504.2008.tb02129.x

Meredith, W., & Millsap, R. E. (1992). On the misuse of manifest variables in the detection of measurement bias. *Psychometrika, 57*, 289–311. https://doi.org/10.1007/BF02294510

Myers, C. T. (1975). *Test fairness: A comment on fairness in statistical analysis* (Research Bulletin No. RB-75-12). Princeton: Educational Testing Service. https://doi.org/10.1002/j.2333-8504.1975.tb01051.x

O'Neill, K. O., & McPeek, W. M. (1993). Item and test characteristics that are associated with differential item functioning. In P. W. Holland & H. Wainer (Eds.), *Differential item functioning* (pp. 255–276). Hillsdale: Erlbaum.

Pashley, P. J. (1992). *Graphical IRT-based DIF analysis* (Research Report No. RR-92-66). Princeton: Educational Testing Service. https://doi.org/10.1002/j.2333-8504.1992.tb01497.x

Petersen, N. S., & Novick, M. R. (1976). An evaluating of some models for culture-fair selection. *Journal of Educational Measurement, 13*, 3–29. https://doi.org/10.1111/j.1745-3984.1976.tb00178.x

Peterson, N. S. (2007). Equating: Best practices and challenges to best practices. In N. J. Dorans, M. Pommerich, & P. W. Holland (Eds.), *Linking and aligning scores and scales* (pp. 59–72). New York: Springer-Verlag. https://doi.org/10.1007/978-0-387-49771-6_4

Pomplun, M., Baron, P. A., & McHale, F. J. (1992). *An initial evaluation of the use of bivariate matching in DIF analyses for formula scored tests* (Research Report No. RR-92-63). Princeton: Educational Testing Service. https://doi.org/10.1002/j.2333-8504.1992.tb01494.x

Potenza, M. T., & Dorans, N. J. (1995). DIF assessment for polytomously scored items: A framework for classification and evaluation. *Applied Psychological Measurement, 19*, 23–37. https://doi.org/10.1177/014662169501900104

Puhan, G., Moses, T. P., Yu, L., & Dorans, N. J. (2007). *Small-sample DIF estimation using log-linear smoothing: A SIBTEST application* (Research Report No. RR-07-10). Princeton: Educational Testing Service. https://doi.org/10.1002/j.2333-8504.2007.tb02052.x

Puhan, G., Moses, T. P., Yu, L., & Dorans, N. J. (2009). Using log-linear smoothing to improve small-sample DIF estimation. *Journal of Educational Measurement, 46*, 59–83. https://doi.org/10.1111/j.1745-3984.2009.01069.x

Rasch, G. (1960). *Probabilistic models for some intelligence and attainment tests*. Copenhagen: Danmarks Paedagogiske Institut.

Scheuneman, J. D. (1979). A method of assessing bias in test items. *Journal of Educational Measurement, 16*, 143–152. https://doi.org/10.1111/j.1745-3984.1979.tb00095.x

Schmitt, A. P., Holland, P. W., & Dorans, N. J. (1993). Evaluating hypotheses about differential item functioning. In P. W. Holland & H. Wainer (Eds.), *Differential item functioning* (pp. 281–315). Hillsdale: Erlbaum.

Shealy, R., & Stout, W. (1993). A model-based standardization approach that separates true bias/DIF from group ability differences and detects test bias/DIF as well as item bias/DIF. *Psychometrika, 58*, 159–194. https://doi.org/10.1007/BF02294572

Sinharay, S., Dorans, N. J., Grant, M. C., & Blew, E. O. (2009a). Using past data to enhance small-sample DIF estimation: A Bayesian approach. *Journal of Educational and Behavioral Statistics, 34*, 74–96. https://doi.org/10.3102/1076998607309021

Sinharay, S., Dorans, N. J., & Liang, L. (2009b). *First language of examinees and its relationship to differential item functioning* (Research Report No. RR-09-11). Princeton: Educational Testing Service. https://doi.org/10.1002/j.2333-8504.2009.tb02162.x

Stricker, L. J. (1982). Identifying test items that perform differently in population subgroups: A partial correlation index. *Applied Psychological Measurement, 6*, 261–273. https://doi.org/10.1177/014662168200600302

Tan, X., Xiang, B., Dorans, N. J., & Qu, Y. (2010). *The value of the studied item in the matching criterion in differential item functioning (DIF) analysis* (Research Report No. RR-10-13). Princeton: Educational Testing Service. https://doi.org/10.1002/j.2333-8504.2010.tb02220.x

Thissen, D., Steinberg, L., & Wainer, H. (1993). Detection of differential item functioning using the parameters of item response models. In P. W. Holland & H. Wainer (Eds.), *Differential item functioning* (pp. 67–113). Hillsdale: Erlbaum.

Thorndike, R. L. (1971). Concepts of culture-fairness. *Journal of Educational Measurement, 8*, 63–70. https://doi.org/10.1111/j.1745-3984.1971.tb00907.x

Tucker, L. R. (1951). *Academic ability test* (Research Memorandum No. RM-51-17). Princeton: Educational Testing Service.

Turnbull, W. W. (1949). Influence of cultural background on predictive test scores. In *Proceedings of the ETS invitational conference on testing problems* (pp. 29–34). Princeton: Educational Testing Service.

Turnbull, W. W. (1951a). *Socio-economic status and predictive test scores* (Research Memorandum No. RM-51-09). Princeton: Educational Testing Service.

Turnbull, W. W. (1951b). Socio-economic status and predictive test scores. *Canadian Journal of Psychology, 5*, 145–149. https://doi.org/10.1037/h0083546

von Davier, A. A., & Wilson, C. (2008). Investigating the population sensitivity assumption of item response theory true-score equating across two subgroups of examinees and two test formats. *Applied Psychological Measurement, 32*, 11–26. https://doi.org/10.1177/0146621607311560

Wainer, H. (1993). Model-based standardized measurement of an item's differential impact. In P. W. Holland & H. Wainer (Eds.), *Differential item functioning* (pp. 123–135). Hillsdale: Erlbaum.

Wainer, H., Sireci, S. G., & Thissen, D. (1991). Differential testlet functioning: Definitions and detection. *Journal of Educational Measurement, 28*, 197–219. https://doi.org/10.1111/j.1745-3984.1991.tb00354.x

Yang, W.-L. (2004). Sensitivity of linkings between AP multiple-choice scores and composite scores to geographical region: An illustration of checking for population invariance. *Journal of Educational Measurement, 41*, 33–41. https://doi.org/10.1111/j.1745-3984.2004.tb01157.x

Yang, W.-L., & Gao, R. (2008). Invariance of score linkings across gender groups for forms of a testlet-based college-level examination Program examination. *Applied Psychological Measurement, 32*, 45–61. https://doi.org/10.1177/0146621607311577

Yu, L., Moses, T., Puhan, G., & Dorans, N. J. (2008). *DIF detection with small samples: Applying smoothing techniques to frequency distributions in the Mantel-Haenszel procedure* (Research Report No. RR-08-44. Princeton: Educational Testing Service. https://doi.org/10.1002/j.2333-8504.2008.tb02130.x

Zhang, Y., Dorans, N. J., & Mathews-Lopez, J. (2005). *Using DIF dissection method to assess effects of item deletion* (Research Report No. 2005-10). New York: The College Board. https://doi.org/10.1002/j.2333-8504.2005.tb02000.x

Zieky, M. (1993). Practical questions in the use of DIF statistics in test development. In P. W. Holland & H. Wainer (Eds.), *Differential item functioning* (pp. 337–347). Hillsdale: Erlbaum.

Zieky, M. (2011). The origins of procedures for using differential item functioning statistics at Educational Testing Service. In N. J. Dorans & S. Sinharay (Eds.), *Looking back: Proceedings of a conference in honor of Paul W. Holland* (pp. 115–127). New York: Springer-Verlag. https://doi.org/10.1007/978-1-4419-9389-2_7

Zwick, R. (1989). *When do item response function and Mantel-Haenszel definitions of differential item functioning coincide* (Research Report No. RR-89-32). Princeton: Educational Testing Service. https://doi.org/10.1002/j.2330-8516.1989.tb00146.x

Zwick, R. (1990). When do item response function and Mantel-Haenszel definitions of differential item functioning coincide? *Journal of Educational Statistics, 15*, 185–197. https://doi.org/10.2307/1165031

Zwick, R. (1994). *The effect of the probability of correct response on the variability of measures of differential item functioning* (Research Report No. RR-94-44). Princeton: Educational Testing Service. https://doi.org/10.1002/j.2333-8504.1994.tb01617.x

Zwick, R. (2012). *A review of ETS differential item functioning assessment procedures: Flagging rules, minimum sample size requirements, and criterion refinement* (Research Report No. RR-12-08). Princeton: Educational Testing Service. https://doi.org/10.1002/j.2333-8504.2012.tb02290.x

Zwick, R., & Ercikan, K. (1989). Analysis of differential item functioning in the NAEP history assessment. *Journal of Educational Measurement, 26*, 55–66. https://doi.org/10.1111/j.1745-3984.1989.tb00318.x

Zwick, R., & Thayer, D. T. (1996). Evaluating the magnitude of differential item functioning in polytomous items. *Journal of Educational and Behavioral Statistics, 21*, 187–201. https://doi.org/10.3102/10769986021003187

Zwick, R., & Thayer, D. T. (2002). Application of an empirical Bayes enhancement of Mantel-Haenszel DIF analysis to a computerized adaptive test. *Applied Psychological Measurement, 26*, 57–76. https://doi.org/10.1177/0146621602026001004

Zwick, R., Donoghue, J. R., & Grima, A. (1993a). Assessment of differential item functioning for performance tasks. *Journal of Educational Measurement, 30*, 233–251. https://doi.org/10.1111/j.1745-3984.1993.tb00425.x

Zwick, R., Thayer, D. T., & Wingersky, M. (1993b). *A simulation study of methods for assessing differential item functioning in computer-adaptive tests* (Research Report No. RR-93-11). Princeton: Educational Testing Service. https://doi.org/10.1002/j.2333-8504.1993.tb01522.x

Zwick, R., Thayer, D. T., & Wingersky, M. (1994). A simulation study of methods for assessing differential item functioning in computerized adaptive tests. *Applied Psychological Measurement, 18*, 121–140. https://doi.org/10.1177/014662169401800203

Zwick, R., Thayer, D. T., & Wingersky, M. (1995). Effect of Rasch calibration on ability and DIF estimation in computer-adaptive tests. *Journal of Educational Measurement, 32*, 341–363. https://doi.org/10.1177/014662169401800203

Zwick, R., Thayer D. T., & Lewis, C. (1997a). *An investigation of the validity of an empirical Bayes approach to Mantel-Haenszel DIF analysis* (Research Report No. RR-97-21). Princeton: Educational Testing Service. https://doi.org/10.1002/j.2333-8504.1997.tb01742.x

Zwick, R., Thayer, D. T., & Mazzeo, J. (1997b). Descriptive and inferential procedures for assessing DIF in polytomous items. *Applied Measurement in Education, 10*, 321–344. https://doi.org/10.1207/s15324818ame1004_2

Zwick, R., Thayer, D. T., & Lewis, C. (1999). An empirical Bayes approach to Mantel-Haenszel DIF analysis. *Journal of Educational Measurement, 36*, 1–28. https://doi.org/10.1111/j.1745-3984.1999.tb00543.x

Zwick, R., Thayer, D. T., & Lewis, C. (2000). Using loss functions for DIF detection: An empirical Bayes approach. *Journal of Educational and Behavioral Statistics, 25*, 225–247. https://doi.org/10.3102/10769986025002225

Zwick, R., Ye, L., & Isham, S. (2012). Improving Mantel–Haenszel DIF estimation through Bayesian updating. *Journal of Educational and Behavioral Statistics, 37*, 601–629. https://doi.org/10.3102/1076998611431085

Part II
ETS Contributions to Education Policy and Evaluation

Chapter 8
Large-Scale Group-Score Assessment

Albert E. Beaton and John L. Barone

Large-scale group assessments are widely used to inform educational policymakers about the needs and accomplishments of various populations and subpopulations. The purpose of this section is to chronicle the ETS technical contributions in this area.

Various types of data have been used to describe demographic groups, and so we must limit the coverage here. We will consider only assessments that have important measurements, such as educational achievement tests, and also have population-defining variables such as racial/ethnic, gender, and other policy-relevant variables, such as the number of hours watching TV or mathematics courses taken. The assessed population must be large, such as the United States as a whole, or an individual state.

The design of group assessments is conceptually simple: define the population and measurement instruments and then test all students in the population. For example, if a high school exit examination is administered to all high school graduates, then finding differences among racial/ethnic groupings or academic tracks is straightforward. However, if the subgroup differences are the only matter of interest, then this approach would be expensive and consume a substantial amount of student time.

To take advantage of the fact that only group and subgroup comparisons are needed, large-scale group assessments make use of sampling theory. There are two sampling areas:

- Population to be measured: Scientific samples are selected so that the population and its subpopulations can be measured to the degree required.

A.E. Beaton
Boston College, Walnut Hill, MA, USA

J.L. Barone (✉)
Educational Testing Service, Princeton, NJ, USA
e-mail: jbarone@ets.org

R.E. Bennett, M. von Davier (eds.), *Advancing Human Assessment*,
Methodology of Educational Measurement and Assessment,
DOI 10.1007/978-3-319-58689-2_8

- Subject domain to be measured: The subject area domains may be many (e.g., reading, writing, and mathematics) and may have subareas (e.g., algebra, geometry, computational skills).

Population sampling involves selecting a sample of students that is large enough to produce estimates with sufficiently small standard errors. The domain sampling determines the breadth of measurement within a subject area. These decisions determine the costs and feasibility of the assessment.

It is informative to note the similarities and differences of group and individual assessments. Individual assessments have been in use for a long time. Some examples:

- The Army Alpha examination, which was administered to recruits in World War I.
- The *SAT*® and ACT examinations that are administered to applicants for selected colleges.

Such tests are used for important decisions about the test takers and thus must be sufficiently reliable and valid for their purposes.

As defined here, group tests are intended for population and subpopulation descriptions and not for individual decision making. As such, the tests need not measure an individual accurately as long as the target population or subpopulations parameters are well estimated.

Both group and individual assessments rely on available technology from statistics, psychometrics, and computer science. The goals of the assessment determine what technical features are used or adapted. In turn, new assessment often requires the development of enhanced technology.

For group assessments, the goal is to select the smallest sample size that will meet the assessment's measurement standards. Small subpopulations (e.g., minority students) may be oversampled to ensure a sufficient number for accurate measurement, and then sampling weights are computed so that population estimates can be computed appropriately.

Domain sampling is used to ensure that the assessment instruments cover a wide range of a subject area. Item sampling is used to create different test forms. In this way, the content of a subject-matter domain can be covered while individual students respond to a small sample of test items from the total set.

In short, group assessment typically sacrifices tight individual assessment to reduce the number of students measured and the amount of time each measured student participates in the assessment.

8.1 Organization of This Chapter

There are many different ways to present the many and varied contributions of ETS to large-scale group assessments. We have chosen to do so by topic. Topics may be considered as milestones or major events in the development of group technology. We have listed the topics chronologically to stress the symbiotic relationship of information needs and technical advancements. The information demands spur technical developments, and they in turn spur policy maker demands for information. This chapter begins by looking at the early 1960s, when the use of punch cards and IBM scoring machines limited the available technology. It leads up to the spread of large-scale group technology in use around the world.

In Sect. 8.2, Overview of Technological Contributions, 12 topics are presented. These topics cover the last half-century of development in this field, beginning with early assessments in the 1960s. ETS has had substantial influence in many but not all of these topics. All topics are included to show the contributions of other organizations to this field. Each topic is described in a few paragraphs. Some important technical contributions are mentioned but not fully described. The point here is to give an overview of large-scale group assessments and the various forces that have produced the present technology.

In Sect. 8.3, ETS and Large-Scale Assessment, gives the details of technical contributions. Each topic in Sect. 8.2 is given an individual subsection in Sect. 8.3. These subsections describe the topic in some detail. Section 8.3 is intended to be technical—but not too technical. The names of individual contributors are given along with references and URLs. Interested readers will find many opportunities to gain further knowledge of the technical contributions.

Topics will vary substantially in amount of space devoted to them depending on the degree of ETS contribution. In some cases, a topic is jointly attributable to an ETS and a non-ETS researcher.

Finally, there is an appendix, which describes in some detail the basic psychometric model used in the National Assessment of Educational Progress (NAEP). This also contains a record of the many years of comparing alternative methods for ways to improve the present methodology.

8.2 Overview of Technological Contributions

The following section is intended to give an overview of the evolving technology of large-scale group assessments. It is divided into 12 topics that describe the major factors in the development of group assessment technology. The topics are introduced chronologically, although their content may overlap considerably; for example, the topic on longitudinal studies covers 40 years. Each topic is followed by a detailed description in the next section that contains individual contributions, the names of researchers, references, and URLs. We intend for the reader to view the Overview and then move to other sections where more detail is available.

8.2.1 Early Group Assessments

The early days of group assessments brings back memories of punch cards and IBM scoring machines. Two pioneering assessments deserve mention:

- Project TALENT: The launching of Sputnik by the Soviet Union in 1957 raised concern about the quantity and quality of science education in the United States. Were there enough students studying science to meet future needs? Were students learning the basic ideas and applications of science? To answer these and other questions, Congress passed the National Defense Education Act (NDEA) in 1958.[1] To gather more information, Project TALENT was funded, and a national sample of high school students was tested in 1960. This group assessment was conducted by the American Institutes for Research.
- IEA Mathematics Assessment: At about the same time, International Association for the Evaluation of Educational Achievement (IEA) was formed and began gathering information for comparing various participating countries.

  ETS was not involved in either of these studies.

8.2.2 NAEP's Conception

In 1963, Francis Keppel was appointed the United States Commissioner of Education. He found that the commissioner was required to report annually on the progress of education in the United States. To this end, he wrote Ralph Tyler, who was then the director of the Institute for Advanced Studies in the Behavioral Sciences, for ideas on how this might be done. Tyler responded with a memorandum that became the beginning of the NAEP.

[1] U. S. Congress. National Defense Education Act of 1958, P.L. 85-864. 85th Congress, September 2, 1958. Washington, DC: GPO.U. S. Congress. The NDEA was signed into law on September 2, 1958 and provided funding to United States education institutions at all levels.

8.2.3 Educational Opportunities Survey (EOS)

Among the many facets of the Civil Rights Act of 1964[2] was the commissioning of a survey of the equality of educational opportunity in the United States. Although the EOS study did report on various inputs to the educational system, it focused on the output of education as represented by the test scores of various racial/ethnic groups in various regions of the country. The final report of this EOS, which is commonly known as the Coleman report (Coleman et al. 1966) has been heralded as one of the most influential studies ever done in education (Gamoran and Long 2006).

ETS was the prime contractor for this study. The project demonstrated that a large-scale study could be designed, administered, analyzed, interpreted, and published in a little over a year.

8.2.4 NAEP'S Early Assessments

The first phase of NAEP began with a science assessment in 1969. This assessment had many innovative features, such as matrix sampling, administration by tape recorder, and jackknife standard error estimates. In its early days, NAEP was directed by the Education Commission of the States.

8.2.5 Longitudinal Studies

The EOS report brought about a surge of commentaries in Congress and the nation's courts, as well as in the professional journals, newspapers, and magazines (e.g., Bowles and Levin 1968; Cain and Watts 1968). Different commentators often reached different interpretations of the same data (Mosteller et al. 2010; Viadero 2006). Harvard University sponsored a semester-long faculty seminar on the equality of educational opportunity that produced a number of new analyses and commentaries (Mosteller and Moynihan 1972). It soon became apparent that more data and, in particular, student growth data were necessary to address some of the related policy questions. The result was the start of a series of longitudinal studies.

[2] Civil Rights Act of 1964, P.L. No. 88-352, 78 Stat. 241 (July 2, 1964).

8.2.6 Scholastic Aptitude Test (SAT) Score Decline

In the early 1970s, educational policymakers and the news media noticed that the average SAT scores had been declining monotonically from a high point in 1964. To address this phenomenon, the College Board formed a blue ribbon panel, which was chaired by Willard Wirtz, a former Secretary of Labor. The SAT decline data analysis for this panel required linking Project Talent and the National Longitudinal Study[3] (NLS-72) data. ETS researchers developed partitioning analysis for this study. The panel submitted a report titled *On Further Examination: Report of the Advisory Panel on the Scholastic Aptitude Test Score Decline* (Wirtz 1977).

8.2.7 Calls for Change

The improvement of the accuracy and timeliness of large-scale group assessments brought about requests for more detailed policy information. The 1980s produced several reports that suggested further extensions of and improvement in the available data on educational issues. Some reports were particularly influential:

- The Wirtz and Lapointe (1982) report made suggestions for improvement of NAEP item development and reporting methods.
- The *Nation at Risk* report (National Commission on Excellence in Education 1983) decried the state of education in the United States and suggested changes in the governance of NAEP.

8.2.7.1 The Wall Charts

Secretary of Education, Terrence Bell, wanted information to allow comparison of educational policies in different states. In 1984, he released his *wall charts,* presenting a number of educational statistics for each state, and challenged the educational community to come up with a better state indicator of student achievement. These reports presented challenges to NAEP and other information collection systems.

[3] The National Longitudinal Study of the high school class of 1972 was the first longitudinal study funded by the United States Department of Education's National Center for Education Statistics (NCES).

8.2.8 NAEP's New Design

In 1983, the National Institute of Education released a request for proposals for the NAEP grant. ETS won this competition. The general design has been published by Messick et al. (1983) with the title, *A New Design for a New Era*. Archie Lapointe was the executive director of this effort.

Implementing the new design was challenging. The NAEP item pool had been prepared by the previous contractor, Education Commission of the States, but needed to be organized for balanced incomplete block (BIB) spiraling. Foremost was the application of item response theory (IRT), which was largely developed at ETS by Lord (see, for example, Carlson and von Davier, Chap. 5, this volume). IRT was used to summarize a host of item data into a single scale. The sample design needed to change to allow both age and grade sampling. The sample design also needed to be modified for bridge studies (studies designed to link newer forms to older forms of an assessment), which were needed to ensure maintenance of existing trends.

The implementation phase brought about opportunities for improving the assessment results. The first assessment under the new design occurred in the 1983–1984 academic year and assessed reading and writing. A vertical reading scale was developed so that students at various age and grade levels could be compared. Scale anchoring was developed to describe what students knew and could do at different points on the scale. Since the IRT methods at that time could handle only right/wrong items, the average response method (ARM) was developed for the writing items, which had graded responses. The approach to standard errors using the jackknife method used replicate weights to simplify computations using standard statistical systems.

The implementation was not without problems. It was intended to use the LOGIST program (Wood et al. 1976) to create maximum likelihood scores for individual students. However, this method was unacceptable, since it could not produce scores for students who answered all items correctly or scored below the chance level. Instead, a marginal maximum likelihood program (BILOG; Mislevy and Bock 1982) was used. This method produced a likelihood distribution for each student, and five plausible values were randomly chosen from those distributions. Mislevy (1985) has shown that plausible values can produce consistent estimates of group parameters and their standard errors.

Another problem occurred in the 1985–1986 NAEP assessment, in which reading, mathematics, and science were assessed. The results in reading were anomalous. Intensive investigations into the reading results produced a report by Beaton and Zwick (1990).

ETS's technical staff has continued to examine and improve the assessment technology. When graded responses were developed for IRT, the PARSCALE program (Muraki and Bock 1997) replaced the ARM program for scaling writing data. Of special interest is the examination of alternative methods for estimating population distributions. A detailed description of alternative methods and their evaluation is provided in the appendix.

The introduction of IRT into NAEP was extremely important in the acceptance and use of NAEP reports. The 1983–1984 unidimensional reading scale led the way and was followed by multidimensional scales in mathematics, science, and reading itself. These easy to understand and use scales facilitated NAEP interpretation.

8.2.9 NAEP's Technical Dissemination

Under its new design, NAEP produced a series of reports to present the findings of completed assessments. These reports were intended for policymakers and the general public. The reports featured graphs and tables to show important findings for different racial/ethnic and gender groupings. The publication of these reports was announced at press conferences, along with press releases. This method ensured that NAEP results would be covered in newspapers, magazines, and television broadcasts.

NAEP has also been concerned with describing its technology to interested professionals. This effort has included many formal publications:

- *A New Design for a New Era* (Messick et al. 1983), which describes the aims and technologies that were included in the ETS proposal.
- Textual reports that described in detail the assessment process.
- Descriptions of NAEP technology in professional journals.
- Research reports and memoranda that are available to the general public.
- A NAEP Primer that is designed to help secondary analysts get started in using NAEP data.

The new design included public-use data files for secondary analysis, and such files have been prepared for each NAEP assessment since 1983. However, these files were not widely used because of the considerable intellectual commitment that was necessary to understand the NAEP design and computational procedures. To address the need of secondary analysts, ETS researchers developed a web-based analysis system, the NAEP Data Explorer, which allows the user to recreate the published tables or revise them if needed. The tables and the associated standard errors are computed using the full NAEP database and appropriate algorithms. In short, powerful analyses can be computed using simple commands.[4]

This software is necessarily limited in appropriate ways; that is, in order to protect individual privacy, the user cannot identify individual schools or students. If a table has cells representing very small samples, the program will refuse to compute the table. However, the database sample is large, and such small cells rarely occur.

For more sophisticated users, there is a series of data tools that help the user to select a sample that is appropriate for the policy question at issue. This program can

[4] This software is freely available at http://nces.ed.gov/nationsreportcard/naepdata/

produce instructions for use with available statistical systems such as SAS or SPSS. For these users, a number of programs for latent regression analyses are also provided. These programs may be used under licenses from ETS.

8.2.10 National Assessment Governing Board

The National Assessment Governing Board was authorized by an amendment to the Elementary and Secondary Education Act in 1988. The amendment authorized the Governing Board to set NAEP policies, schedules, and subject area assessment frameworks. The governing board made important changes in the NAEP design that challenged the ETS technical staff.

The major change was allowing assessment results to be reported by individual states so that the performance of students in various states could be compared. Such reporting was not permitted in previous assessments. At first, state participation was voluntary, so that a sample of students from nonparticipating states was needed to provide a full national sample. ETS ran several studies to assess the effects of changing from a single national sample to national data made up from summarizing various state results.

Comparing state results led to concern about differing states exclusion procedures. NAEP had developed tight guidelines for the exclusion of students with disabilities or limited English ability. However, differing state laws and practices resulted in differences in exclusion rates. To address this problem, two different technologies for adjusting state results were proposed and evaluated at a workshop of the National Institute of Statistical Sciences.

The No Child Left behind Act (2002) required that each state provide standards for student performance in reading and mathematics at several grade levels. Using NAEP data as a common measure, ETS studied the differences in the percentages of students at different performance levels (e.g., proficient) in different states.

On another level, the Governing Board decided to define aspirational achievement levels for student performance, thus replacing the scale anchoring already in practice in NAEP. ETS did not contribute to this project; however, the method used to define aspirational levels was originally proposed by William Angoff, an ETS researcher.

At around the same time, ETS researchers looked into the reliability of item ratings (ratings obtained through human scoring of open-ended or constructed student responses to individual assessment items).This resulted in a review of the literature and recommendations for future assessments.

ETS has also explored the use of computer-based assessment models. This work used models for item generation as well as item response evaluation. An entire writing assessment was developed and administered. The possibilities for future assessments are exciting.

The appropriateness of the IRT model became an important issue in international assessments, where different students respond in different languages. It is possible that the IRT models will fit well in one culture but not in another. The issue was faced directly when Puerto Rican students were assessed using NAEP items that were translated into Spanish. The ETS technical staff came up with a method for testing whether or not the data in an assessment fit the IRT model. This approach has been extended for latent regression analyses.

8.2.11 NAEP's International Effects

Beginning with the 1988 International Assessment of Educational Progress (IAEP) school-based assessment, under the auspices of ETS and the United Kingdom's National Foundation for Educational Research, the ETS NAEP technologies for group assessment were readily adapted and extended into international settings. In 1994, ETS in collaboration with Statistics Canada conducted the International Adult literacy Survey (IALS), the world's first internationally comparative survey of adult skills. For the past 20 years, ETS group software has been licensed for use for the Trends in International Mathematics and Science Study (TIMSS), and for the past 15 years for the Progress in International Reading Literacy Study (PIRLS). As the consortium and technology lead for the 2013 Programme for the International Assessment of Adult Competencies (PIAAC), and the 2015 Program for International Student Assessment (PISA), ETS continues its research efforts to advance group assessment technologies—advances that include designing and developing instruments, delivery platforms, and methodology for computer-based delivery and multistage adaptive testing.

8.2.12 Other ETS Technical Contributions

ETS has a long tradition of research in the fields of statistics, psychometrics, and computer science. Much of this work is not directly associated with projects such as those mentioned above. However, much of this work involves understanding and improving the tools used in actual projects. Some examples of these technical works are described briefly here and the details and references are given in the next section of this paper.

F4STAT is a flexible and efficient statistical system that made the implementation of assessment data analysis possible. Development of the system began in 1964 and has continued over many following years.

One of the basic tools of assessment data analysis is multiple regressions. ETS has contributed to this field in a number of ways:

- Exploring methods of fitting robust regression statistics using power series.
- Exploring the accuracy of regression algorithms.
- Interpreting least squares without sampling assumptions.

ETS has also contributed to the area of latent regression analysis.

8.3 ETS and Large-Scale Assessment

8.3.1 Early Group Assessments

8.3.1.1 Project Talent

Project Talent was a very large-scale group assessment that reached for a scientific sample of 5% of the students in American high schools in 1960. In the end, Project Talent collected data on more than 440,000 students in Grades 9 through 12, attending more than 1,300 schools. The students were tested in various subject areas such as mathematics, science, and reading comprehension. The students were also administered three questionnaires that included items on family background, personal and educational experiences, aspirations for future education and vocation, and interests in various occupations and activities. The students were followed up by mail questionnaires after high school graduation. ETS was not involved in this project.[5]

8.3.1.2 First International Mathematics Study (FIMS)

At about the same time, the IEA was formed and began an assessment of mathematical competency in several nations including the United States. The IEA followed up this assessment with assessments in different subject areas at different times. Although ETS was not involved in the formative stage of international assessments it did contribute heavily to the design and implementation of the third mathematics and science study (TIMSS) in 1995.[6]

8.3.2 NAEP's Conception

The original design was created by Ralph Tyler and Princeton professor John Tukey. For more detailed information see *The Nation's Report Card: Evolutions and Perspectives* (Jones and Olkin 2004).

[5] More information is available at http://www.projecttalent.org/

[6] See http://nces.ed.gov/timss/

8.3.3 Educational Opportunities Survey

The Civil Rights Act of 1964 was a major piece of legislation that affected the American educational system. Among many other things, the act required that the U.S. Office of Education undertake a survey of the equality of educational opportunity for different racial and ethnic groups. The act seemed to require measuring the effectiveness of inputs to education such as the qualifications of teachers and the number of books in school libraries. Ultimately, it evolved into what we would consider today to be a value-added study that estimated the effect of school input variables on student performance as measured by various tests. The final report of the EOS, *The Equality of Educational Opportunity* (Coleman et al. 1966), has been hailed as one of the most influential reports in American education (Gamoran and Long 2006).

The survey was conducted under the direction of James Coleman, then a professor at Johns Hopkins University, and an advisory committee of prominent educators. NCES performed the sampling, and ETS received the contract to conduct the survey. Albert Beaton organized and directed the data analysis for ETS. John Barone had key responsibilities for data analysis systems development and application. This massive project, one of the largest of its kind, had a firm end date: July 1, 1966. Mosteller and Moynihan (1972) noted that the report used data from "some 570,000 school pupils" and "some 60,000 teachers" and gathered elaborate "information on the facilities available in some 4,000 schools."

The analysis of the EOS data involved many technical innovations and adaptations: foremost, the analysis would have been inconceivable without F4STAT.[7] The basic data for the surveyed grades (Grades 1, 3, 6, 9, and 12) and their teachers' data were placed on a total of 43 magnetic tapes and computer processing took 3 to 4 hours per analysis per grade—a formidable set of data and analyses given the computer power available at the time. With the computing capacity needed for such a project exceeding what ETS had on hand, mainframe computers in the New York area were used. Beaton (1968) provided details of the analysis.

The modularity of F4STAT was extremely important in the data analysis. Since the commercially available computers used a different operating system, a module had to be written to bridge this gap. A separate module was written to enter, score, and check the data for each grade so that the main analysis programs remained the same while the modules varied. Modules were added to the main programs to create publishable tables in readable format.

The data analysis involved fitting a regression model using the variables for students, their backgrounds, and schools that was collected in the survey. The dependent variables were test scores, such as those from a reading or mathematics test. The sampling weights were computed as the inverse of the probability of selection. Although F4STAT allowed for sampling weights, the sampling weights summed to the population size, not the sample size, which inappropriately reduced the error

[7] F4STAT is described in the next section.

estimates, and so sampling errors were not published.[8] John Tukey, a professor at Princeton University, was a consultant on this project. He discussed with Coleman and Beaton the possibility of using the jackknife method of error estimation. The jackknife method requires several passes over slightly modified data sets, which was impossible within the time and resource constraints. It was decided to produce self-weighting samples of 1,000 for each racial/ethnic grouping at each grade. Linear regression was used in further analyses.

After the EOS report was published, George Mayeske of the U.S. Office of Planning, Budgeting, and Evaluation organized further research into the equality of educational opportunity. Alexander Mood, then Assistant Commissioner of NCES, suggested using commonality analysis. Commonality analysis was first suggested in papers by Newton and Spurell (1967a, b). Beaton (1973a) generalized the algorithm and detailed its advantages and limitations. John Barone analyzed the EOS data using the commonality technique. This resulted in books by Mayeske et al. (1972, 1973a, b), and Mayeske and Beaton (1975).

The Mayeske analyses separated the total variance of student performance into "within-school" and "among-school" components. Regressions were run separately for within- and among-school components. This approach was a precursor to hierarchical linear modeling, which came later (Bryk and Raudenbush 1992).

Criterion scaling was also an innovation that resulted from experiences with the EOS. Large-scale analysis of variance becomes tedious when the number of levels or categories is large and the numbers of observations in the cells are irregular. Coding category membership by indicator or dummy variables may become impractically large. For example, coding all of the categorical variables for the ninth-grade students used in the Coleman report would entail 600 indicator variables; including all possible interactions would involve around 10^{75} such variables, a number larger than the number of grains of sand in the Sahara Desert.

To address this problem, Beaton (1969) developed *criterion scaling*. Let us say that there is a criterion or dependent variable that is measured on a large number of students who are grouped into a number of categories. We wish to test the hypothesis that the expected value of a criterion variable is the same for all categories. For example, let us say we have mathematics scores for students in a large number of schools and we wish to test the hypothesis that the school means are equal. We can create a criterion variable by giving each student in a school the average score of all students in that school. The regression of the individual mathematics scores on the criterion variable produced the results of a simple analysis of variance. The criterion variable can be used for many other purposes. This method and its advantages and limitations were described by Pedhazur (1997), who also included a numerical example.

[8] Later, F4STAT introduced a model that made the sum of the weights equal to the sample size.

8.3.4 NAEP's Early Assessments

The early NAEP assessments were conducted under the direction of Ralph Tyler and Princeton professor John Tukey. The Education Commission of the States was the prime administrator, with the sampling and field work done by a subcontract with the Research Triangle Institute.

The early design of NAEP had many interesting features:

- Sampling by student age, not grade. The specified ages were 9-, 13-, and 17-year-olds, as well as young adults. Out of school 17-year-olds were also sampled.
- Use of matrix sampling to permit a broad coverage of the subject area. A student was assigned a booklet that required about an hour to complete. Although all students in an assessment session were assigned the same booklet, the booklets varied from school to school.
- Administration by tape recorder. In all subject areas except reading, the questions were read to the students through a tape recording, so that the effect of reading ability on the subject areas would be minimized.
- Results were reported by individual items or by the average percentage correct over various subject matter areas.
- The jackknife method was used to estimate sampling variance in NAEP's complex sampling design.

For more extensive discussion of the design see Jones and Olkin (2004).

ETS was not involved in the design and analysis of these data sets, but did have a contract to write some assessment items. Beaton was a member of the NAEP computer advisory committee. ETS analyzed these data later as part of its trend analyses.

8.3.5 Longitudinal Studies

The EOS reported on the status of students at a particular point in time but did not address issues about future accomplishments or in-school learning. Many educational policy questions required information about growth or changes in student accomplishments. This concern led to the funding and implementation of a series of longitudinal studies.

ETS has made many important contributions to the methodology and analysis technology of longitudinal assessments. Continual adaptation occurred as the design of longitudinal studies responded to different policy interests and evolving technology. This is partially exemplified by ETS contributions addressing multistage adaptive testing (Cleary et al. 1968; Lord 1971), IRT intersample cross-walking to produce comparable scales, and criterion-referenced proficiency levels as indicators of student proficiency. Its expertise has been developed by the longitudinal study group, which was founded by Thomas Hilton, and later directed by Donald Rock, and then by Judy Pollack. We will focus here on the national longitudinal studies sponsored by the U.S. Department of Education[9].

The first of the national studies was the National Longitudinal Study of the Class of 1972[10] (Rock et al. 1985) which was followed by a series of somewhat different studies. The first study examined high school seniors who were followed up after graduation. The subsequent studies measured high school accomplishments as well as postsecondary activities. The policy interests then shifted to the kindergarten and elementary years. The change in student populations being studied shows the changes in the policymakers' interests.

Rock (Chap. 10, this volume) presented a comprehensive 4-decade history of ETS's research contributions and role in modeling and developing psychometric procedures for measuring change in large-scale longitudinal assessments. He observed that many of these innovations in the measurement of change profited from research solutions developed by ETS for NAEP.

In addition to the national studies, ETS has been involved in other longitudinal studies of interest:

- Study of the accomplishments of U.S. Air Force members 25 years after enlistment. The study (Thorndike and Hagen 1959) was done in collaboration with the National Bureau for Economic Research. Beaton (1975) developed and applied econometric modeling methods to analyze this database.
- The Parent Child Development Center (PCDC) study[11] of children from birth through the elementary school years. This study was unique in that the children were randomly assigned *in utero* to treatment or control groups. In their final evaluation report, Bridgeman, Blumenthal, and Andrews (Bridgeman et al. 1981) indicated that replicable program effects were obtained.

8.3.6 SAT Score Decline

In the middle of the 1970s, educational policymakers and news media were greatly concerned with the decline in average national SAT scores. From 1964 to the mid-1970s, the average score had dropped a little every year. To study the phenomenon, the College Board appointed a blue ribbon commission led by Willard Wirtz, a former U.S. Secretary of Labor.

The question arose as to whether the SAT decline was related to lower student ability or to changes in the college-entrant population. ETS researchers proposed a

[9] National Longitudinal studies were originally sponsored by the U.S. Office of Education. That office evolved into the present Department of Education.

[10] Thomas Hilton was the principal investigator; Hack Rhett and Albert Beaton contributed to the proposal and provided team leadership in the first year.

[11] Samuel Messick and Albert Beaton served on the project's steering committee. Thomas Hilton of the ETS Developmental Research Division was the Project Director. Samuel Ball and Brent Bridgeman directed the PCDC evaluation.

design to partition the decline in average SAT scores into components relating to shifts in student performance, shifts in student populations, and their interaction. To do so required that comparable national tests be available to separate the college-bound SAT takers from the other high school students. The only available national tests at that time were the tests from Project Talent and from NLS-72 . A carefully designed study linking the tests was administered to make the test scores equivalent.

8.3.6.1 Improvisation of Linking Methods

The trouble was that the reliabilities of the tests were different. The Project Talent test had 49 items and a higher reliability than the NLS-72 20-item test. The SAT mean was substantially higher for the top 10% of the Project Talent scores than of the NLS-72 scores, as would be expected from the different reliabilities. Improving the reliability of the NLS-72 test was impossible; as Fred Lord wisely noted that, if it were possible to convert a less reliable test to a reliable one, there would be no point to making reliable tests. No equating could do so.

The study design required that the two tests have equal—but not perfect—reliability. If we could not raise the reliability of the NLS-72 test, we could lower the reliability of the Project Talent test. We did so by adding a small random normal deviate to each Project Talent score where the standard deviation of the normal deviate was calculated to give the adjusted Project Talent scores the same reliability as the NLS-72 scores. When this was done, the SAT means for the top two 10% samples were within sampling error.

8.3.6.2 Partitioning Analysis

Partitioning analysis (Beaton et al. 1977) was designed for this study. Many scientific studies explore the differences among population means. If the populations are similar, then the comparisons are straightforward. However, if they differ, the mean comparisons are problematic. Partitioning analysis separates the difference between two means into three parts: proficiency effect, population effect, and joint effect. The proficiency effect is the change in means attributable to changes in student ability, the population effect is the part attributable to population changes, and the joint effect is the part attributable to the way that the population and proficiency work together. Partitioning analysis makes it simple to compute a well-known statistic, the standardized mean, which estimates what the mean would have been if the percentages of the various subgroups had remained the same.

In the SAT study, partitioning analysis showed that most of the decline in SAT means was attributable to population shifts, not changes in performance of those at particular levels of the two tests. What had happened is that the SAT-taking population had more than doubled in size, with more students going to college; that is, democratizing college attendance resulted in persons of lower ability entering the college-attending population.

Partitioning analysis would be applied again in future large-scale-assessment projects. For example, to explore the NAEP 1985–1986 reading anomaly (discussed later in this chapter), and also in a special study and resulting paper, *Partitioning NAEP Trend Data* (Beaton and Chromy 2007), that was commissioned by the NAEP validity studies panel. The SAT project also led to a book by Hilton on merging large databases (Hilton 1992).

8.3.7 Call for Change

The early 1980s produced three reports that influenced the NAEP design and implementation:

- The Wirtz and Lapointe (1982) report *Measuring the Quality of Education: A Report on Assessing Educational Progress* commended the high quality of the NAEP design but suggested changes in the development of test items and in the reporting of results.
- The report of the National Commission on Excellence in Education (NCEE), titled *A Nation at Risk: The Imperative for Educational Reform* (NCEE 1983), decried the state of education in the United States.
- Terrence Bell, then Secretary of Education, published wall charts, which contained a number of statistics for individual states. Included among the statistics were the average SAT and ACT scores for these states. Realizing that the SAT and ACT statistics were representative of college-bound students only, he challenged the education community to come up with better statistics of student attainment.

8.3.8 NAEP's New Design

The NAEP is the only congressionally mandated, regularly administered assessment of the performance of students in American schools. NAEP has assessed proficiency in many school subject areas (e.g., reading, mathematics, science) at different ages and grades, and at times young adults. NAEP is not a longitudinal study, since individual students are not measured as they progress in schooling; instead, NAEP assesses the proficiency of a probability sample of students at targeted school levels. Progress is measured by comparing the proficiencies of eighth-grade students to students who were eighth graders in past assessments.

In 1983, ETS competed for the NAEP grant and won. Westat was the subcontractor for sampling and field operations. The design that ETS proposed is published in

A New Design for a New Era (Messick et al. 1983).[12] The new design had many innovative features:

- *IRT scaling.* IRT scaling was introduced to NAEP as a way to summarize the data in a subject area (e.g., reading). This will be discussed below.
- *BIB spiraling.* BIB spiraling was introduced to address concerns about the dimensionality of NAEP testing data. To assess a large pool of items while keeping the testing time for an individual student to less than an hour, BIB spiraling involved dividing the item pool into individually timed (e.g., 15-minute) blocks and assigning the blocks to assessment booklets so that each item is paired with each other item in some booklet. In this way, the correlation between each pair of items is estimable. This method was suggested by Beaton and implemented by James Ferris. The idea was influenced by the work of Geoffrey Beall[13] on lattice designs (Beall and Ferris 1971) while he was at ETS.
- *Grade and age ("grage") sampling.* Previous NAEP samples were defined by age. ETS added overlapping grade samples so that results could be reported either by age or by grade.
- *"Bridge" studies.* These studies were introduced to address concerns about maintaining the already existing trend data. Bridge studies were created to link the older and newer designs. Building the bridge involved collecting randomly equivalent samples under both designs.

Implementing a new, complex design in a few months is challenging and fraught with danger but presents opportunities for creative developments. The most serious problem was the inability to produce maximum likelihood estimates of proficiency for the students who answered all their items correctly or answered below the chance level. Because reading and writing blocks were combined in some assessment booklets, many students were given only a dozen or so reading items. The result was that an unacceptable proportion of students had extreme, nonestimable, reading scores. The problem was exacerbated by the fact that the proportion of high and low scorers differed by racial/ethnic groups, which would compromise any statistical conclusions. No classical statistical methods addressed this problem adequately. The maximum likelihood program LOGIST (Wingersky et al. 1982; Wingersky 1983), could not be used.

Mislevy (1985) noted that NAEP did not need individual student scores; it needed only estimates of the distribution of student performance for different subpopulations such as gender or racial/ethnic groupings. In fact, it was not permissible or desirable to report individual scores. Combining the recent developments in

[12] Archie Lapointe was executive director. Original staff members included Samuel Messick as coordinator with the NAEP Design and Analysis Committee, Albert Beaton as director of data analysis, John Barone as director of data analysis systems, John Fremer as director of test development, and Jules Goodison as director of operations. Ina Mullis later moved from Education Commission of the States (the previous NAEP grantee) to ETS to become director of test development.

[13] Geoffrey Beall was an eminent retired statistician who was given working space and technical support by ETS. James Ferris did the programming for Beall's work.

marginal maximum likelihood available in the BILOG program (Mislevy and Bock 1982) and the missing data theory of Rubin (1977, 1987), he was able to propose consistent estimates of various group performances.

A result of the estimation process was the production of plausible values, which are used in the computations. Although maximum likelihood estimates could not be made for some students, estimation of the likelihood of a student receiving any particular score was possible for all. To remove bias in estimates, the distribution was "conditioned" using the many reporting and other variables that NAEP collected. A sample of five plausible values was selected at random from these distributions in making group estimates. von Davier et al. (2009) discussed plausible values and why they are useful.

The development of IRT estimation techniques led to addressing another problem. At that time, IRT allowed only right/wrong items, whereas the NAEP writing data were scored using graded responses. It was intended to present writing results one item at a time. Beaton and Johnson (1990) developed the ARM to scale the writing data. Essentially, the plausible value technology was applied to linear models.

In 1988, the National Council for Measurement in Education (NCME) gave its Award for Technical Contribution to Educational Measurement to ETS researchers Robert Mislevy, Albert Beaton, Eugene Johnson, and Kathleen Sheehan for the development of the plausible values methodology in the NAEP. The development of NAEP estimation procedures over time is detailed in the appendix.

The NAEP analysis plan included using the jackknife method for estimating standard errors, as in past NAEP assessments. However, the concept of replicate weights was introduced to simplify the computations. Essentially, the jackknife method involves pairing the primary sampling units and then systematically removing one of each pair and doubling the weight of the other. This process is done separately for each pair, resulting in half as many replicate weights as primary sampling units in the full sample. The replicate weights make it possible to compute the various population estimates using a regression program that uses sampling weights.

Another problem was reporting what students in American schools know and can do, which is the purpose of the assessment. The scaling procedures summarize the data across a subject area such as reading in general or its subscales. To describe the meaning of scales, scale anchoring was developed (Beaton and Allen 1992). In so doing, several anchor points on the scale were selected at about a standard deviation apart. At each point, items were selected that a large percentage of students at that point could correctly answer and most students at the next lower point could not. At the lowest level, items were selected only on the probability of answering the item correctly. These discriminating items were then interpreted and generalized as anchor descriptors. The scale-anchoring process and descriptors were a precursor to what would become the National Assessment Governing Board's achievement levels for NAEP.

Of special interest to NAEP was the question of dimensionality, that is, whether a single IRT scale could encapsulate the important information about student proficiency in an area such as reading. In fact the BIB spiraling method was developed and applied to the 1983–1984 NAEP assessment precisely to address this question.

Rebecca Zwick (1987a, b) addressed this issue. Three methods were applied to the 1984 reading data: principal components analysis, full-information factor analysis (Bock et al. 1988), and a test of unidimensionality, conditional independence, and monotonicity based on contingency tables (Rosenbaum 1984). Results were consistent with the assumption of unidimensionality. A complicating factor in these analyses was the structure of the data that resulted from NAEP's BIB design. A simulation was conducted to investigate the impact of using the BIB-spiraled data in dimensionality analyses. Results from the simulated BIB data were similar to those from the complete data. The *Psychometrika* paper (Zwick 1987b), which describes some unique features of the correlation matrix of dichotomous Guttman items, was a spin-off of the NAEP research. Additional studies of dimensionality were performed by Carlson and Jirele (1992) and Carlson (1993).

Dimensionality has taken on increased importance as new uses are proposed for large-scale assessment data. Future survey design and analysis methods are evolving over time to address dimensionality as well as new factors that may affect the interpretation of assessment results. Some important factors are the need to ensure that the psychometric models incorporate developments in theories of how students learn, how changes in assessment frameworks affect performance, and how changes in the use of technology and integrated tasks affect results. Addressing these factors will require new psychometric models. These models will need to take into account specified relationships between tasks and underlying content domains, the cognitive processes required to solve these tasks, and the multilevel structure of the assessment sample. These models may also require development and evaluation of alternative estimation methods. Continuing efforts to further develop these methodologies include a recent methodological research project that is being conducted by ETS researchers Frank Rijmen and Matthias von Davier and is funded by the U.S. Department of Education's Institute of Education Sciences. This effort, through the application of a combination of general latent variable model frameworks (Rijmen et al. 2003; von Davier 2010) with new estimation methods based on stochastic (von Davier and Sinharay 2007, 2010) as well as a graphical model framework approach (Rijmen 2011), will offer a contribution to the research community that applies to NAEP as well as to other survey assessments.

The 1986 assessment produced unacceptable results, which have been referred to as the *reading anomaly*. The average score for 12th grade students fell by an estimated 2 years of growth, which could not have happened in the 2 years since the last assessment. The eighth grade students showed no decline, and the fourth graders showed a slight decline. This reading anomaly brought about a detailed exploration of possible explanations. Although a single factor was not isolated, it was concluded that many small changes produced the results. The results were published in a book by Beaton and Zwick (1990), who introduced the maxim "If you want to measure change, don't change the measure."

Further research was published by Zwick (1991). This paper summarized the key analyses described in the Beaton and Zwick reading anomaly report, focusing on the effects of changes in item position.

While confidence intervals for scaled scores are relatively straightforward, a substantial amount of research investigates confidence intervals for percentages (Brown et al. 2001; Oranje 2006a). NAEP utilizes an adjustment proposed by Satterthwaite (1941) to calculate effective degrees of freedom. However, Johnson and Rust (1993) detected through simulation that Satterthwaite's formula tends to underestimate effective degrees of freedom, which could cause the statistical tests to be too conservative. Qian (1998) conducted further simulation studies to support Johnson and Rust's conclusion. He also pointed out the instability associated with Satterthwaite's estimator.

8.3.9 NAEP's Technical Dissemination

An important contribution of ETS to large-scale group assessments is the way in which NAEP's substantive results and technology have been documented and distributed to the nation. This first part of this section will describe the many ways NAEP has been documented in publications. This will be followed by a discussion of the public-use data files and simple ways to perform secondary analyses using the NAEP data. The final section will present a description of some of the software available for advanced secondary analysts.

8.3.9.1 Documentation of NAEP Procedures and Results

ETS considered that communicating the details of the NAEP design and implementation was very important, and thus communication was promised in its winning proposal. This commitment led to a long series of publications, such as the following:

- *A New Design for a New Era* (Messick et al. 1983), which was a summary of the winning ETS NAEP proposal, including the many innovations that it planned to implement.
- The NAEP *Report Cards,* which give the results of NAEP assessments in different subject areas and different years. The first of these reports was *The Reading Report Card: Progress Toward Excellence in Our Schools, Trends in Reading over Four National Assessments, 1971–1984* (NAEP 1985).[14]
- NAEP Technical Reports,[15] which contain detailed information about sampling, assessment construction, administration, weighting, and psychometric methods. Beginning with the 2000 assessment, technical information has been published directly on the web.

---

[14]A full listing of such reports can be found at http://nces.ed.gov/pubsearch/getpubcats.asp?sid=031. These reports are complemented by press conferences.

[15]See http://nces.ed.gov/nationsreportcard/tdw/

- In 1992, two academic journal issues were dedicated to NAEP technology: *Journal of Educational Statistics*, Vol. 17, No. 2 (Summer, 1992) and *Journal of Educational Measurement*, Vol. 29, No. 2 (June, 1992).
- ETS has produced a series of reports to record technical contributions in NAEP. These scholarly works are included in the ETS Research publication series, peer reviewed by ETS staff and made available to the general public. A searchable database of such reports is available at http://search.ets.org/researcher/. Many of these reports are later published in professional journals.
- *The NAEP Primer*, written by Beaton and Gonzalez (1995) and updated extensively by Beaton et al. (2011).

8.3.9.2 NAEP's Secondary-Use Data and Web Tools

The NAEP staff has made extensive efforts to make its data available to secondary analysts. To encourage such uses, the NAEP design of 1983–1984 included public-use data files to make the data available. At that time, persons interested in secondary data analysis needed to receive a license from NCES before they were allowed to use the data files to investigate new educational policy issues. They could also check published statistics and explore alternative technologies. The public-use data files were designed to be used in commonly available statistical systems such as SPSS and SAS; in fact the choice of the plausible values technique was chosen in part over direct estimation methods to allow the data files tapes to use the rectangular format that was in general use at that time. Such files were produced for the 1984, 1986, and 1988 assessments.

The public-use data files did not bring about as much secondary analysis as hoped for. The complex technology introduced in NAEP, such as plausible values and replicate sampling weights, was intimidating. The data files contain very large numbers of students and school variables. To use the database properly required a considerable investment in comprehending the NAEP designs and analysis plans. The intellectual cost of using the public-use data files had discouraged many potential users.

In 1988, Congressional legislation authorized NAEP state assessments, beginning in 1990. Because of increased confidentiality concerns, the legislation precluded the issuing of public-use data files going forward. This action brought about a number of different approaches to data availability. The strict rules required by the Privacy Act (1974) made maintaining privacy more challenging. We will describe a few approaches to this problem in which ETS has played an important role.

Simple, Easily Available Products There are many potential users for the published NAEP graphs and tables and also for simple or complex variations on published outputs. Potential users include NAEP report writers and NAEP state coordinators, but also include educational policy makers, newspaper reporters, educational researchers, and interested members of the general public. To make the NAEP data available to such potential users, there was a need for computer programs that were easy to use but employed the best available algorithms to help the users perform statistical analyses.

To respond to this need, ETS has developed and maintains web-based data tools for the purpose of analyzing large-scale assessment data. The foremost of these tools is the NAEP Data Explorer (NDE), whose principal developers at ETS were Alfred Rogers and Stephen Szyszkiewicz. NDE allows anyone with access to the Internet to navigate through the extensive, rich NAEP data archive and to produce results and reports that adhere to strict statistical, reporting, and technical standards. The user simply locates NDE on the web and, after electronically signing a user's agreement, is asked to select the data of interest: NAEP subject area; year(s) of assessment; states or other jurisdictions to be analyzed; and the correlates to be used in the analysis.[16]

NDE serves two sets of audiences: internal users (e.g., NAEP report writers and state coordinators) and the general public. NDE can be used by novice users and also contains many features appropriate for advanced users. Opening this data source to a much wider audience greatly increases the usefulness and transparency of NAEP. With a few clicks of a mouse, interested persons can effortlessly search a massive database, perform an analysis, and develop a report within a few minutes.

However, the NDE has its limitations. The NDE uses the full NAEP database and results from the NDE will be the same as those published by NAEP but, to ensure privacy, the NDE user is not allowed to view individual or school responses. The availability of statistical techniques is thus limited. NDE will refuse to compute statistics that might compromise individual responses, as might occur, for example, in a table in which the statistics in one or more cells are based on very small samples.

ETS has addressed making its data and techniques available through the *NAEP Primer* (Beaton et al. 2011). This publication for researchers provides much greater detail on how to access and analyze NAEP data, as well as an introduction to the available analysis tools and instruction on their use. A mini-sample of real data that have been approved for public use enables secondary analysts to familiarize themselves with the procedures before obtaining a license to a full data set. A NAEP-like data set is included for exploring the examples in the primer text.[17]

Full-Power, Licensed Products As mentioned above, using the NAEP database requires a substantial intellectual commitment. Keeping the NAEP subject areas, years, grades, and so forth straight is difficult and tedious. To assist users in the management of NAEP secondary-use data files, ETS developed the NAEP Data Toolkit. Alfred Rogers at ETS was the principal developer of the toolkit, which provides a data management application, NAEPEX, and procedures for performing two-way cross-tabulation and regression analysis. NAEPEX guides the user through the process of selecting samples and data variables of interest for analysis and

[16] The NDE is available free of charge at http://nces.ed.gov/nationsreportcard/naepdata/

[17] The primer is available at http://nces.ed.gov/nationsreportcard/researchcenter/datatools2.aspx

creates an extract data file or a set of SAS or SPSS control statements, which define the data of interest to the appropriate analysis system.[18]

Computational Analysis Tools Used for NAEP In addition to NAEPEX, ETS has developed a number of computer programs for more advanced users. These programs are intended to improve user access, operational ease, and computational efficiency in analyzing and reporting information drawn from the relatively large and complex large-scale assessment data sets. Continual development, enhancement, and documentation of applicable statistical methods and associated software tools are important and necessary. This is especially true given the ever increasing demand for—and scrutiny of—the surveys. Although initial large-scale assessment reports are rich and encyclopedic, there is great value in focused secondary analyses for interpretation, enhancing the value of the information, and formulation of policy. Diverse user audiences seeking to conduct additional analyses need to be confident in the methodologies, the computations, and in their ability to replicate, verify, and extend findings. The following presents a brief overview of several research-oriented computational analysis tools that have been developed and are available for both initial large-scale assessment operation and secondary research and analysis.

The methods and software required to perform direct estimation of group population parameters without introducing plausible values has developed substantially over the years. To analyze and report on the 1984 NAEP reading survey, ETS researchers and analysts developed the first operational version of the GROUP series of computer programs that estimate latent group effects. The GROUP series of programs is in continual development and advancement as evolving methods are incorporated. In addition to producing direct estimates of group differences, these programs may also produce plausible values based on Rubin's (1987) multiple imputations procedures for missing data. The output provides consistent estimates of population characteristics in filled-in data sets that enhance the ability to correctly perform secondary analyses with specialized software.

The separate programs in the GROUP series were later encapsulated into the DESI (Direct Estimation Software Interactive: ETS 2007; Gladkova et al. 2005) suite. DESI provides an intuitive graphical user interface (GUI) for ease of access and operation of the GROUP programs. The computational and statistical kernel of DESI can be applied to a broad range of problems, and the suite is now widely used in national and international large-scale assessments. WESVAR, developed at Westat, and the AM software program, developed at the American Institutes for Research (AIR) by Cohen (1998), also address direct estimation in general and are used primarily for analyzing data from complex samples, especially large-scale assessments such as NAEP. Descriptions and comparison of DESI and AM are found in papers by von Davier (2003) and Donoghue et al. (2006a). Sinharay and von Davier (2005) and von Davier and Sinharay (2007) discussed research around issues dealing with high performance statistical computing for large data sets found

[18]The NAEP Data Toolkit is available upon request from NAEP via http://nces.ed.gov/nationsreportcard/researchcenter/datatools2.aspx

in international assessments. Von Davier et al. (2006) presented an overview of large-scale assessment methodology and outlined steps for future extensions.

8.3.10 National Assessment Governing Board

The Elementary and Secondary Education act of 1988 authorized the national assessment governing board to set NAEP policies, schedules, and subject area assessment frameworks. This amendment made some important changes to the NAEP design. The main change was to allow assessment results to be reported by individual states so that the performance of students in various states could be compared. Such reporting was not permitted in previous assessments. This decision increased the usefulness and importance of NAEP. Reporting individual state results was introduced on a trial basis in 1990 and was approved as a permanent part of NAEP in 1996. Due to the success of individual state reporting, NAEP introduced separate reports for various urban school districts in 2002. These changes in NAEP reporting required vigilance to ensure that the new expanded assessments did not reduce the integrity of NAEP.

Several investigations were conducted to ensure the comparability and appropriateness of statistics over years and assessment type. Some of these are discussed in the sections below.

8.3.10.1 Comparability of State and National Estimate

At first, individual state reporting was done on a voluntary basis. The participating states needed large samples so that state subpopulations could be measured adequately. To maintain national population estimates, a sample of students from nonparticipating states was also collected. The participating and nonparticipating states' results were then merged with properly adjusted sampling weights. This separate sample for nonparticipating states became moot when all states participated as a result of the No Child Left Behind Act of 2002.

Two studies (Qian and Kaplan 2001; Qian et al. 2003) investigated the changes. The first described an analysis to ensure quality control of the combined national and state data. The second described the analyses directed at three main issues relevant to combining NAEP samples:

- Possible discrepancies in results between the combined sample and the current national sample.
- The effects of combined samples on the results of significance tests in comparisons, such as comparisons for reporting groups within the year and trend comparisons across years.
- The necessity of poststratification to adjust sample strata population estimates to the population values used in sample selection.

The findings of these studies showed that the combined samples will provide point estimates of population parameters similar to those from the national samples. Few substantial differences existed between combined and national estimates. In addition, the standard errors were smaller in the combined samples. With combined samples, there was a greater number of statistically significant differences in subpopulation comparisons within and across assessment years. The analysis also showed little difference between the results of nonpoststratified combined samples and those of poststratified combined samples.

8.3.10.2 Full Population Estimation

The publication of NAEP results for individual states allowed for comparisons of student performance. When more than one year was assessed in a subject area, estimation of trends in that area is possible. Trend comparisons are made difficult, since the published statistics are affected not only by the proficiency of students but also by the differences in the sizes of the subpopulations that are assessed. Early state trend results tended to show that states that excluded a larger percentage of students tended to have larger increases in reported average performance. This finding led to the search for full population estimates.

Although NAEP might like to estimate the proficiency of all students within an assessed grade, doing so is impractical. NAEP measurement tools cannot accurately measure the proficiency of some students with disabilities or students who are English language learners. While accommodations are made to include students with disabilities, such as allowing extra assessment time or use of braille booklets, some students are excluded. Despite strict rules for inclusion in NAEP, state regulations and practices vary somewhat and thus affect the comparability of state results.

To address this issue, Beaton (2000) suggested using a full population median, which Paul Holland renamed *bedian*. The bedian assumes only that the excluded students would do less well than the median of the full student population, and adjusts the included student median accordingly. McLaughlin (2000, 2005) proposed a regression approach by imputing excluded students' proficiencies from other available data. McLaughlin's work was further developed by Braun et al. (2008).

The National Institute of Statistical Sciences held a workshop on July 10–12, 2000, titled *NAEP Inclusion Strategies*. This workshop focused on comparing the full population statistics proposed by Beaton and McLaughlin. Included in its report is a detailed comparison by Holland (2000) titled "Notes on Beaton's and McLaughlin's Proposals."

8.3.11 Mapping State Standards Onto NAEP

The No Child Left Behind Act of 2002 required all states to set performance standards in reading and mathematics for Grades 3–8 and also for at least one grade in high school. The act, however, left to states the responsibility of determining the curriculum, selecting the assessments, and setting challenging academic standards. The result was that, in a particular grade, a standard such as *proficient* was reached by substantially different proportions of students in different states.

To understand the differences in state standards, ETS continued methodological development of an approach originally proposed by McLaughlin (1998) for making useful comparisons among state standards. It is assumed that the state assessments and NAEP assessment reflect similar content and have comparable structures, although they differ in test and item formats as well as standard-setting procedures. The Braun and Qian (2007) modifications involved (a) a shift from a school-based to a student-based strategy for estimating NAEP equivalent to a state standard, and (b) the derivation of a more refined estimate of the variance of NAEP parameter estimates by taking into account the NAEP design in the calculation of sampling error and by obtaining an estimate of the contribution of measurement error.

Braun and Qian applied the new methodology to four sets of data: (a) Year 2000 state mathematics tests and the NAEP 2000 mathematics assessments for Grades 4 and 8, and (b) Year 2002 state reading tests and the NAEP 2002 reading assessments for Grades 4 and 8. The study found that for both mathematics and reading, there is a strong negative linear relationship across states between the proportions meeting the standard and the apparent stringency of the standard as indicated by its NAEP equivalent. The study also found that the location of the NAEP score equivalent of a state's proficiency standard is not simply a function of the placement of the state's standard on the state's own test score scale. Rather, it also depends on the curriculum delivered to students across the state and the test's coverage of that curriculum with respect to both breadth and depth, as well as the relationship of both to the NAEP framework and the NAEP assessment administered to students. Thus, the variation among states' NAEP equivalent scores reflects the interaction of multiple factors, which can complicate interpretation of the results.

8.3.11.1 Testing Model Fit

IRT technology assumes that a student's response to an assessment item is dependent upon the students' ability, the item parameters of a known mathematical model, and an error term. The question arises as to how well the actual assessment data fit the assumed model. This question is particularly important in international assessments and also in any assessment where test items are translated into different languages. It is possible that the IRT model may fit well in one language but not well in another. For this reason, ETS applied an innovative model-fitting analysis for comparing Puerto Rican students with mainland students. The Puerto Rican students responded to NAEP questions that were translated into Spanish.

The method for analyzing model fit was suggested by Albert Beaton (2003). The model was explored by Kelvin Gregory when he was at ETS. John Donoghue suggested using standardized errors in the comparison process. The method requires that the data set from an assessment has been analyzed using IRT and its results are available. Using the estimated student abilities and item parameters, a large number (e.g., 1000) of randomly equivalent data sets are created under the assumption of local independence. Statistics from the actual sample are then compared to the distribution of statistics from the randomly equivalent data sets. Large differences between the actual and randomly equivalent statistics indicate misfit. This approach indicates the existence of items or persons that do not respond as expected by the IRT model.

Additional research and procedures for assessing the fit of latent regression models was discussed by Sinharay et al. (2010). Using an operational NAEP data set, they suggested and applied a simulation-based model-fit procedure that investigated whether the latent regression model adequately predicted basic statistical summaries.

8.3.11.2 Aspirational Performance Standards

The National Assessment Governing Board decided to create achievement levels that were intended as goals for student performance. The levels were for *basic*, *proficient*, and *advanced*. Although ETS staff did not have a hand in implementing these levels, the standard-setting procedure of ETS researcher William Angoff (1971) was used in the early stages of the standard setting.

8.3.12 Other ETS Contributions

The ETS research staff continued to pursue technical improvements in NAEP under the auspices of the governing board, including those discussed in the following sections.

8.3.12.1 Rater Reliability in NAEP

Donoghue et al. (2006b) addressed important issues in rater reliability and the potential applicability of rater effects models for NAEP. In addition to a detailed literature review of statistics used to monitor and evaluate within- and across-year rater reliability, they proposed several alternative statistics. They also extensively discussed IRT-based rater-effect approaches to modeling rater leniency, and provided several novel developments by applying signal detection theory in these models.

8.3.12.2 Computer-Based Assessment in NAEP

A key step towards computer-based testing in NAEP was a series of innovative studies in writing, mathematics, and critical reasoning in science and in technology-rich environments. The 2011 writing assessment was the first to be fully computer-based. Taking advantage of digital technologies enabled tasks to be delivered in audio and video multimedia formats. Development and administration of computer-delivered interactive computer tasks (ICTs) for the 2009 science assessment enabled measurement of science knowledge, processes, and skills that are not measurable in other modes. A mathematics online study in 2001 (Bennett et al. 2008) used both automated scoring and automatic item generation principles to assess mathematics for fourth and eighth graders on computers. This study also investigated the use of adaptive testing principles in the NAEP context. As of this writing, a technology and engineering literacy assessment is being piloted that assesses literacy as the capacity to use, understand, and evaluate technology, as well as to understand technological principles and strategies needed to develop solutions and achieve goals. The assessment is completely computer-based and engages students through the use of multimedia presentations and interactive simulations.

8.3.12.3 International Effects

The ETS methodology for group assessments has quickly spread around the world. At least seven major international studies have used or adapted the technology:

- School-based assessments
- The International Assessment of Educational Progress (IAEP)
- Trends in Mathematics and Science Study (TIMSS)
- Progress in International Reading Literacy Study (PIRLS)
- The Program for International Student Assessment (PISA 2015)
- Household-Based Adult Literacy Assessments
- The International Adult Literacy Study (IALS)
- The Adult Literacy and Life Skills Survey (ALL)
- The OECD Survey of Adult Skills. Also known as the Programme for the International Assessment of Adult Competencies (PIAAC)

In five of these studies (IAEP, PISA 2015, IALS, ALL, and PIAAC), ETS was directly involved in a leadership role and made significant methodological contributions. Two of the studies (TIMSS and PIRLS) have used ETS software directly under license with ETS and have received ETS scale validation services. These international assessments, including ETS's role and contributions, are described briefly below.

The existence of so many assessments brought about attempts to compare or link somewhat different tests. For example, comparing the IAEP test (Beaton and Gonzalez 1993) or linking the TIMSS test to NAEP tests might allow American students to be compared to students in foreign countries. ETS has carefully investigated the issues in linking and organized a special conference to address it. The conference produced a book outlining the problems and potential solutions (Dorans et al. 2007).

The IAEP assessments were conducted under the auspices of ETS and the UK's National Foundation for Educational Research, and funded by the National Science Foundation and NCES. In the middle of the 1980s there was concern about the start-up and reporting times of previously existing international assessments. In order to address these concerns, two assessments were conducted: IAEP1 in 1988 and IAEP2 in 1991. Archie Lapointe was the ETS director of these studies. Six countries were assessed in IAEP1. In IAEP2, students aged 9 and 13 from about 20 countries were tested in math, science, and geography. ETS applied the NAEP technology to these international assessments. These ventures showed that comprehensive assessments could be designed and completed quickly while maintaining rigorous standards. The results of the first IAEP are documented in a report titled *A World of Differences* (Lapointe et al. 1989). The IAEP methodologies are described in the *IAEP Technical Report* (1992).

The TIMSS assessments are conducted under the auspices of the International Association for the Evaluation of Educational Achievement (IEA). Conducted every 4 years since 1995, TIMSS assesses international trends in mathematics and science achievement at the fourth and eighth grades in more than 40 countries. For TIMSS, the ETS technology was adapted for the Rasch model by the Australian Council for Educational Research. The methodology used in these assessments was described in a TIMSS technical report (Martin and Kelly 1996).

The PIRLS assessments are also conducted under the auspices of the IEA. PIRLS is an assessment of reading comprehension that has been monitoring trends in student achievement at 5-year intervals in more than 50 countries around the world since 2001. PIRLS was described by Mullis et al. (2003).

The International Adult Literacy Survey (IALS), the world's first internationally comparative survey of adult skills, was administered in 22 countries in three waves of data collection between 1994 and 1998. The IALS study was developed by Statistics Canada and ETS in collaboration with participating national governments. The origins of the international adult literacy assessment program lie in the pioneering efforts employed in United States national studies that combined advances in large-scale assessment with household survey methodology. Among the national studies were the Young Adult Literacy Survey (Kirsch and Jungeblut 1986) undertaken by the NAEP program, and the National Adult Literacy Survey (described by Kirsch and ETS colleagues Norris, O'Reilly, Campbell, & Jenkins; Kirsch et al. 2000) conducted in 1992 by NCES.

ALL, designed and analyzed by ETS, continued to build on the foundation of IALS and earlier studies of adult literacy, and was conducted in 10 countries between 2003 and 2008 (Statistics Canada and OECD 2005).

The PIAAC study is an OECD Survey of Adult Skills conducted in 33 countries beginning in 2011. It measures the key cognitive and workplace skills needed for individuals to participate in society and for economies to prosper. The ETS Global Assessment Center, under the directorship of Irwin Kirsch, led the International Consortium and was responsible for the assessment's psychometric design, its analysis, and the development of cognitive assessment domains targeting skills in literacy, numeracy, and problem solving in technology-rich environments. ETS also coordinated development of the technology platform that brought the assessment to more than 160,000 adults, ages 16—65, in more than 30 language versions. The 2011 PIAAC survey broke new ground in international comparative assessment by being the first such instrument developed for computer-based delivery; the first to use multistage adaptive testing; the first to incorporate the use of computer-generated log file data in scoring and scaling; and the first to measure a set of reading components in more than 30 languages. The first PIAAC survey results were presented in an OECD publication (OECD 2013).

The PISA international study under the auspices of the OECD was launched in 1997. It aims to evaluate education systems worldwide every 3 years by assessing 15-year-olds' competencies in three key subjects: reading, mathematics, and science. To date, over 70 countries and economies have participated in PISA. For the sixth cycle of PISA in 2015, ETS is responsible for the design, delivery platform development, and analysis. To accomplish the new, complex assessment design, ETS Global continues to build on and expand the assessment methodologies it developed for PIAAC.

Kirsch et al. (Chap. 9, this volume) present a comprehensive history of Educational Testing Service's 25-year span of work in large-scale literacy assessments and resulting contributions to assessment methodology, innovative reporting, procedures, and policy information that "will lay the foundation for the new assessments yet to come."

In 2007, the Research and Development Division at ETS collaborated with the IEA Data Processing and Research Center to establish the IEA-ETS Research Institute (IERI). IERI publishes a SpringerOpen journal, *Large-Scale Assessments in Education*, which delivers state-of-the-art information on comparative international group score assessments. This IERI journal focuses on improving the science of large-scale assessments. A number of articles published in the IERI series present current research activities dealing with topics discussed in this paper, and also with issues surrounding the large-scale international assessments addressed here (TIMSS, PIRLS, PISA, IALS, ALL, and PIAAC).

In 2013, nine members of ETS's Research and Development division and two former ETSers contributed to a new handbook on international large-scale assessment (Rutkowski et al. 2014).

8.3.12.4 ETS Contributions to International Assessments

The ETS has also contributed to a number of international assessments in other ways, including the following:

- *GROUP Software*. GROUP software has been an important contribution of ETS to international assessments. This software gives many options for estimating the parameters of latent regression models, such as those used in national and international assessments. ETS offers licenses for the use of this software and consulting services as well. The software is described elsewhere in this paper and further described by Rogers et al. (2006).
- *International Data Explorer*. The NDE software has been adapted for international usage. The NDE allows a secondary researcher to create and manipulate tables from an assessment. ETS leveraged the NDE web-based technology infrastructure to produce the PIAAC Data Explorer (for international adult literacy surveys), as well as an International Data Explorer that reports on trends for PIRLS, TIMSS, and PISA data. The tools allow users to look up data according to survey, proficiency scale, country, and a variety of background variables, such as education level, demographics, language background, and labor force experiences. By selecting and organizing relevant information, stakeholders can use the large-scale data to answer questions of importance to them.
- *International linking*. Linking group assessments has taken on increased importance as new uses are proposed for large-scale assessment data. In addition to being linked to various state assessments, NAEP has been linked to TIMSS and PISA in order to estimate how well American students compare to students in other countries. In these cases, the tests being compared are designed to measure different—perhaps slightly different—student proficiencies. The question becomes whether or not the accuracy of a linking process is adequate for its proposed uses.

There is a wealth of literature on attempts at statistically linking national and international large-scale surveys to each other (Beaton and Gonzalez 1993; Johnson et al. 2003; Johnson and Siegendorf 1998; Pashley and Phillips 1993), as well as to state assessments (Braun and Qian 2007; McLaughlin 1998; Phillips 2007). Much of this work is based on concepts and methods of linking advocated by Mislevy (1992) and Linn (1993). In 2005, an ETS-sponsored conference focused on the general issue of score linking. The book that resulted from this conference (Dorans et al. 2007) examines the different types of linking both from theoretical and practical perspectives, and emphasizes the importance of both. It includes topics dealing with linking group assessments (such as NAEP and TIMSS). It also addresses mapping state or country standards to the NAEP scale.

There is an associated set of literature with arguments for and against the appropriateness of such mappings, and innovative attempts to circumvent some of the difficulties (Braun and Holland 1982; Linn and Kiplinger 1995; Thissen 2007; Wainer 1993). Past efforts to link large-scale assessments have met with varied levels of success. This called for continuing research to deal with problems such as linking instability related to differences in test content, format, difficulty, measurement precision, administration conditions, and valid use. Current linking studies draw on this research and experience to ameliorate linking problems. For example, the current 2011 NAEP-TIMSS linking study is intended to improve on previous attempts to link these two assessments by administering NAEP and TIMSS booklets at the same time under the same testing conditions, and using actual state TIMSS results in eight states to validate the predicted TIMSS average scores.

8.3.13 NAEP ETS Contributions

Large-scale group assessments lean heavily on the technology of other areas such as statistics, psychometrics, and computer science. ETS researchers have also contributed to the technology of these areas. This section describes a few innovations that are related to other areas as well as large-scale group assessments.

8.3.13.1 The FORTRAN IV Statistical System (F4STAT)

Although the development of F4STAT began in 1964, before ETS was involved in large-scale group assessments,[19] it quickly became the computation engine that made flexible, efficient data analysis possible. Statistical systems of the early 60s were quite limited and not generally available. Typically, they copied punch card systems that were used on earlier computers. Modern systems such as SAS, SPSS, and Stata were a long way off.

ETS had ordered an IBM 7040 computer for delivery in 1965, and it needed a new system that would handle the diverse needs of its research staff. For this reason, the organization decided to build its own statistical system, F4STAT (Beaton 1973b). Realizing that parameter-driven programs could not match the flexibility of available compilers, the decision was made to use the Fortran IV compiler as the driving force and then develop statistical modules as subroutines. Based on the statistical calculus operators defined by Beaton (1964), the F4STAT system was designed to be modular, general, and easily expandable as new analytic methods were conceived. Of note is that the Beaton operators are extensively cited and referenced throughout statistical computation literature (Dempster 1969; Milton and Nelder 1969), and that these operators or their variants are used in commercial statistical systems, such as SAS and SPSS (Goodnight 1979). Through incorporation of a modern integrated development environment (IDE), F4STAT continues to provide the computational foundation for ETS's large-scale assessment data analysis systems. This continual, technology-driven evolution is important for ETS researchers

[19] Albert Beaton, William Van Hassel, and John Barone implemented the early ETS F4STAT system. Ongoing development continued under Barone. Alfred Rogers is the current technical leader.

to respond to the ever increasing scope and complexity of large-scale and longitudinal surveys and assessments.

8.3.13.2 Fitting Robust Regressions Using Power Series

Many data analyses and, in particular large-scale group assessments, rely heavily on minimizing squared residuals, which overemphasizes the larger residuals. Extreme outliers may completely dominate an analysis. Robust regression methods have been developed to provide an alternative to least squares regression by detecting and minimizing the effect of deviant observations. The primary purpose of robust regression analysis is to fit a model that represents the information in the majority of the data. Outliers are identified and may be investigated separately.

As a result, the issue of fitting power series became an important issue at this time. Beaton and Tukey (1974) wrote a paper on this subject, which was awarded the Wilcoxon Award for the best paper in *Technometrics* in that year. The paper led to a method of computing regression analyses using least absolute value or minimax criteria instead of least squares. For more on this subject, see Holland and Welsch (1977), who reviewed a number of different computational approaches for robust linear regression and focused on iteratively reweighted least-squares (IRLS). Huber (1981, 1996) presented a well-organized overview of robust statistical methods.

8.3.13.3 Computational Error in Regression Analysis

An article by Longley (1967) brought about concern about the accuracy of regression programs. He found large discrepancies among the results of various regression programs. Although ETS software was not examined, the large differences were problematic for any data analyst. If regression programs were inconsistent, large-scale group studies would be suspect.

To investigate this problem, Beaton et al. (1976) looked carefully at the Longley data. The data were taken from economic reports and rounded to thousands, millions, or whatever depending on the variable. The various variables were highly collinear. To estimate the effect of rounding, they added a random uniform number to each datum in the Longley analysis. These random numbers had a mean of zero and a range of -.5 to +.5 after the last published digit. One thousand such data sets were produced, and each set would round to the published data.

The result was surprising. The effect of these random digits substantially affected the regression results more than the differences among various programs. In fact, the "highly accurate" results—computed by Longley to hundreds of places—were not even at the center of the distribution of the 1,000 regression results. The result was clear: increasing the precision of calculations with near-collinear data is not worth the effort, the "true" values are not calculable from the given data.

This finding points out that a greater source of inaccuracy may be the data themselves. Cases such as this, where slight variations in the original data cause large variations in the results, suggest further investigation is warranted before accepting the results. The cited ETS paper also suggests a ridge regression statistic to estimate the seriousness of collinearity problems.

8.3.13.4 Interpreting Least Squares

Regression analysis is an important tool for data analysis in most large- and small-scale studies. Generalizations from an analysis are based on assumptions about the population from which the data are sampled. In many cases, the assumptions are not met. For example, EOS had a complex sample and a 65% participation rate and therefore did not meet the assumptions for regression analysis. Small studies, such as those that take the data from an almanac, seldom meet the required assumptions. The purpose of this paper is to examine what can be stated without making any sampling assumptions.

Let us first describe what a typical regression analysis involves. Linear regression assumes a model such as $y = X\beta+\varepsilon$, where y is the phenomenon being studied, X represents explanatory variables, β is the set of parameters to be estimated, and ε is the residual. In practice, where N is the number of observations ($i = 1,2,\ldots,N$) and M ($j = 0,1,\ldots,M$) is the number of explanatory variables, y is an Nth order vector, X is an $N \times M$ matrix, β is an Mth order vector, and ε is an Nth order vector. The values $x_{i0} = 1$ and β_0 = the intercept. The values in y and X are assumed to be known. The values in ε are assumed to be independently distributed from a normal distribution with mean of 0 and variance of σ^2. Regression programs compute b, the least squares estimate of β, s^2 the estimate of σ^2, and e, the estimate of ε. Under the assumptions, regression creates a t-test for each regression coefficient in b, testing the hypotheses that $\beta_j = 0$. A two-tailed probability statistic p_j is computed to indicate the probability of obtaining a b_j if the true value is zero. A regression analysis often includes an F test that tests the hypothesis that all regression coefficients (excluding the intercept) are equal to zero.

The question addressed here is what we can say about the regression results if we do not assume that the error terms are randomly distributed. Here, we look at the regression analysis as a way of summarizing the relationship between the y and X variables. The regression coefficients are the summary. We expect a good summary to allow us to approximate the values of y using the X variables and their regression coefficients. The question then becomes: How well does the model fit?

Obviously, a good fit implies that the errors are small, near zero. Small errors should not have a substantial effect on the data summary, that is, the regression coefficients. The effect of the error can be evaluated by permuting the errors and then computing the regression coefficients using the permuted data. There are $N!$ ways to permute the errors. Paul Holland suggested flipping the signs of the errors. There are 2^N possible ways to flip the error signs. Altogether, there are $N!2^N$ possible signed permutations, which is a very large number. For example, 10 observations generate $3,628,800 \times 1,024 = 3,715,891,200$ possible signed permutations. We will denote each signed permutations as e_k (k = 1,2,..., $2^N N!$,), $y_k = X\beta + e_k$, and the corresponding regression coefficient as b_k with elements b_{jk}.

Fortunately, we do not need to compute these signed permutations to describe the model fit. Beaton (1981) has shown that the distribution of sign permuted regression coefficients rapidly approaches a normal distribution as the number of observations increases. The mean of the distribution is the original regression coefficient, and the standard deviation is approximately the same as the standard error in regression programs.

The model fit can be assessed from the p values computed in a regular regression analysis:

- The probability statistic p_j for an individual regression coefficient can be interpreted as the proportion of signed and permuted regression coefficients b_{jk} that are further away from b_j than the point where the b_{jk} have different signs.
- Since the distribution is symmetric, $.5p_j$ can be interpreted as the percentage of the b_{jk} that have different signs from b_j.
- The overall P statistic can be interpreted as the percentage of b_k that is as far from b as the point where all b_k have a different sign.
- Other fit criteria are possible, such as computing the number of b_{jk} that differ in the first decimal place.

In summary, the model fit is measured by comparing the sizes of the errors to their effect on the regression coefficients. The errors are not assumed to come from any outside randomization process. This interpretation is appropriate for any conforming data set. The ability to extrapolate to other similar data sets is lost by the failure to assume a randomization.

8.3.14 Impact on Policy—Publications Based on Large-Scale Assessment Findings

Messick (1986) described analytic techniques that provide the mechanisms for inspecting, transforming, and modeling large-scale assessment data with the goals of providing useful information, suggesting conclusions, and supporting decision making and policy research. In this publication, Messick eloquently espoused the enormous potential of large-scale educational assessment as effective policy research and examined critical features associated with transforming large-scale educational assessment into effective policy research. He stated that

> In policy research it is not sufficient simply to document the direction of change, which often may only signal the presence of a problem while offering little guidance for problem solution. One must also conceptualize and empirically evaluate the nature of the change and its contributing factors as a guide for rational decision making.

Among the critical features that he deemed necessary are the capacity to provide measures that are commensurable across time periods and demographic groups, correlational evidence to support construct interpretations, and multiple measures of diverse background and program factors to illuminate context effects and treatment or process differences. Combining these features with analytical methods and interpretative strategies that make provision for exploration of multiple perspectives can yield relevant, actionable policy alternatives. Messick noted that settling for less than full examination of plausible alternatives due to pressures of timeliness and limited funding can be, ironically, at the cost of timeliness.

With the above in mind, we refer the reader to the NCES and ETS websites to access the links to a considerable collection of large-scale assessment publications and data resources. Also, Coley, Goertz, and Wilder (Chap. 12, this volume) provide additional policy research insight.

Appendix: NAEP Estimation Procedures

The NAEP estimation procedures start with the assumption that the proficiency of a student in an assessment area can be estimated from a student's responses to the assessment items that the student received. The psychometric model is a latent regression consisting of four types of variables:

- Student proficiency
- Student item responses
- Conditioning variables
- Error variables

The true proficiency of a student is unobservable and thus unknown. The student item responses are known, since they are collected in an assessment. Also known are the conditioning variables that are collected for reporting (e.g., demographics) or may be otherwise considered related to student proficiency. The error variable is the difference between the actual student proficiency and its estimate from the psychometric model and is thus unknown.

The purpose of this appendix is to present the many ways in which ETS researchers have addressed the estimation problem and continue to look for more precise and efficient ways of using the model. Estimating the parameters of the model requires three steps:

1. Scaling
2. Conditioning
3. Variance estimation

Scaling processes the item-response statistics to develop estimates of student proficiency. Conditioning adjusts the proficiency estimates in order to improve their accuracy and reduce possible biases. Conditioning is an iterative process using the estimation–maximization (EM) algorithm (Dempster et al. 1977) that leads to maximum likelihood estimates. Variance estimation is the process by which the error in the parameter estimates is itself estimated. Both sampling and measurement error are examined.

The next section presents some background on the original application of this model. This is followed by separate sections on advances in scaling, conditioning, and variance estimation. Finally, a number of alternate models proposed by others are evaluated and discussed.

The presentation here is not intended to be highly technical. A thorough discussion of these topics is available in a section of the *Handbook of Statistics* titled "Marginal Estimation of Population Characteristics: Recent Developments and Future Directions" (von Davier et al. 2006).

The Early NAEP Estimation Process

NAEP procedures proposed by ETS were conceptually straightforward: the item responses are used to estimate student proficiency, and then the student estimates are summarized by gender, racial/ethnic groupings, and other factors of educational importance. The accuracy of the group statistics would be estimated using sampling weights and the jackknife method which would take into account the complex NAEP sample. The 3PL IRT model was to be used as described in Lord and Novick (1968).

This approach was first used in the 1983–1984 NAEP assessment of reading and writing proficiency. The proposed IRT methodology of that time was quite limited: it handled only multiple-choice items that could be scored either right or wrong. It also could not make any finite estimates for students who answered all items correctly or scored below the chance level. Since the writing assessment had graded-response questions, the standard IRT programs did not work, so the ARM was developed by Beaton and Johnson (1990). The ARM was later replaced by the PARSCALE program (Muraki and Bock 1997).

However, the straightforward approach to reading quickly ran into difficulties. The decision had been made to BIB spiral the reading and writing items, with the result that many students were assigned too few items to produce an acceptable estimate of their reading proficiency. Moreover, different racial/ethnic groupings had substantially different patterns of inestimable proficiencies, which would bias any results. Standard statistical methods did not offer any solution.

Fortunately, Mislevy had the insight that NAEP did not need individual student proficiency estimates; it needed only estimates of select populations and subpopulations. This led to the use of marginal maximum likelihood methods through the BILOG program (Mislevy and Bock 1982). The BILOG program could estimate group performance directly, but an alternative approach was taken in order to make the NAEP database useful to secondary researchers. BILOG did not develop acceptable individual proficiency estimates but did produce a posterior distribution for each student that indicated the likelihood of possible estimates. From these distributions, five plausible values were randomly selected. Using these plausible values made data analysis more cumbersome but produced a data set that could be used in most available statistical systems.

The adaptation and application of this latent regression model was used to produce the NAEP 1983–1984 Reading Report Card, which has served as a model for many subsequent reports. More details on the first application of the NAEP estimation procedures were described by Beaton (1987) and Mislevy et al. (1992).

Scaling

IRT is the basic component of NAEP scaling. As mentioned above, the IRT programs of the day were limited and needed to be generalized to address NAEP's future needs. There were a number of new applications, even in the early NAEP analyses:

- Vertical scales that linked students aged 9, 13, and 17.
- Across-year scaling to link the NAEP reading scales to the comparable assessments in the past.
- In 1986, subscales were introduced for the different subject areas. NAEP produced five subscales in mathematics. Overall mathematics proficiency was estimated using a composite of the subscales.
- In 1992, the generalized partial credit model was introduced to account for graded responses (polytomous items) such as those in the writing assessments (Muraki 1992; Muraki and Bock 1997).

Yamamoto and Mazzeo (1992) presented an overview of establishing the IRT-based common scale metric and illustrated the procedures used to perform these analyses for the 1990 NAEP mathematics assessment. Muraki et al. (2000) provided an overview of linking methods used in performance assessments, and discussed major issues and developments in linking performance assessments.

Conditioning

As mentioned, the NAEP reporting is focused on group scores. NAEP collected a large amount of demographic data, including student background information and school and teacher questionnaire data, which can be used to supplement the nonresponse due to BIB design and to improve the accuracy of group scores.

Mislevy (1984, 1985) has shown that maximum likelihood estimates of the parameters in the model can be obtained when the actual proficiencies are unknown using an EM algorithm.

The NAEP conditioning model employs both cognitive data and demographic data to construct a latent regression model. The implementation of the EM algorithm that is used in the estimation of the conditioning model leaves room for possible improvements in accuracy and efficiency. In particular, there is a complex multidimensional integral that must be calculated, and there are many ways in which this can be done, each method embodied by a computer program which has been carefully investigated for advantages and disadvantages. These programs have been generically labeled as GROUP programs. The programs that have been used or are currently in use are as follows:

- BGROUP (Sinharay and von Davier 2005). This program is a modification of BILOG (Mislevy and Bock 1982) and uses numerical quadrature and direct integration. This is typically used when there are one or two scales being analyzed
- MGROUP (Mislevy and Sheehan 1987) uses a Monte Carlo method to draw random normal estimates from posterior distributions as input to each estimation step.
- NGROUP (Allen et al. 1996; Mislevy 1985) uses Bayesian normal theory. The requirement of the assumption of a normal distribution results in little use of this method.
- CGROUP (Thomas 1993) uses a Laplace approximation for the posterior means and variance. This method is used when more than two scales are analyzed.
- DGROUP (Rogers et al. 2006) is the current operational program that brings together the BGROUP and CGROUP methods on a single platform. This platform is designed to allow inclusion of other methods as they are developed and tested.

To make these programs available in a single package, ETS researchers Ted Blew, Andreas Oranje, Matthias von Davier, and Alfred Rogers developed a single program called DESI that allows a user to try the different latent regression programs.

The end result of these programs is a set of plausible values for each student. These are random draws from each student's posterior distribution, which gives the likelihood of a student having a particular proficiency score. The plausible value methodology was developed by Mislevy (1991) based on the ideas of Little and Rubin (1987, 2002) on multiple imputation. These plausible values are not appropriate for individual proficiency scores or decision making. In their 2009 paper, "What Are Plausible Values and Why Are They Useful?," von Davier et al. described how plausible values are applied to ensure that the uncertainty associated with measures of skills in large scale surveys is properly taken into account. In 1988, NCME gave its Award for Technical Contribution to Educational Measurement to ETS researchers Robert Mislevy, Albert Beaton, Eugene Johnson, and Kathleen Sheehan for the development of plausible values methodology in the NAEP.

The student plausible values are merged with their sampling weights to compute population and subpopulation statistical estimates, such as the average student proficiency of a subpopulation.

It should be noted that the AM method (Cohen 1998) estimates population parameters directly and is a viable alternative to the plausible-value method that ETS has chosen. The AM approach has been studied in depth by Donoghue et al. (2006a).

These methods were subsequently evaluated for application in future large-scale assessments (Li and Oranje 2006; Sinharay et al. 2010; Sinharay and von Davier 2005; von Davier and Sinharay 2007, 2010). Their analysis of a real NAEP data set provided some evidence of a misfit of the NAEP model. However, the magnitude of the misfit was small, which means that the misfit probably had no practical significance. Research into alternative approaches and emerging methods is continuing.

Variance Estimation

Error variance has two components: sampling error and measurement error. These components are considered to be independent and are summed to estimate total error variance.

Sampling Error

The NAEP samples are obtained through a multistage probability sampling design. Because of the similarity of students within schools and of the effects of nonresponse, observations made of different students cannot be assumed to be independent of each other. To account for the unequal probabilities of selection and to allow for adjustments for nonresponse, each student is assigned separate sampling weights. If these weights are not applied in the computation of the statistics of interest, the resulting estimates can be biased. Because of the effects of a complex sample design, the true sampling variability is usually larger than a simple random sampling. More detailed information is available in reports by Johnson and Rust (1992, 1993), Johnson and King (1987), and Hsieh et al. (2009).

The sampling error is estimated by the jackknife method (Quenouille 1956; Tukey 1958). The basic idea is to divide a national or state population, such as in-school eighth graders, into primary sampling units (PSUs) that are reasonably similar in composition. Two schools are selected at random from each PSU. The sampling error is estimated by computing as many error estimates as there are PSUs. Each of these replicates consists of all PSU data except for one, in which one school is randomly removed from the estimate and the other is weighted doubly. The methodology for NAEP was described, for example, by E. G. Johnson and Rust (1992), and von Davier et al. (2006), and a possible extension was discussed by Hsieh et al. (2009).

The sampling design has evolved as NAEP's needs have increased. Certain ethnic groups are oversampled to ensure that reasonably accurate estimations and sampling weights are developed to ensure appropriately estimated national and state samples.

Also, a number of studies have been conducted about the estimation of standard errors for NAEP statistics. Particularly, an application of the Binder methodology (see also Cohen and Jiang 2001) was evaluated (Li and Oranje 2007) and a comparison with other methods was conducted (Oranje et al. 2009) showing that the Binder method under various conditions underperformed compared to sampling-based methods.

Finally, smaller studies were conducted on (a) the use of the coefficient of variation in NAEP (Oranje 2006b), which was discontinued as a result; (b) confidence intervals for NAEP (Oranje 2006a), which are now available in the NDE as a result; and (c) disclosure risk prevention (Oranje et al. 2007), which is currently a standard practice for NAEP.

Measurement Error

Measurement error is the difference between the estimated results and the "true" results that are not usually available. The plausible values represent the posterior distribution and can be used for estimating the amount of measurement error in statistical estimates such as a population mean or percentile. Five plausible values are computed for each student, and each is an estimate of the student's proficiency. If the five plausible values are close together, then the student is well measured; if the values differ substantially, the student is poorly measured. The variance of the plausible values over an entire population and subpopulation can be used to estimate the error variance. The general methodology was described by von Davier et al. (2009).

Researchers continue to explore alternative approaches to variance estimation for NAEP data. For example, Hsieh et al. (2009) explored a resampling-based approach to variance estimation that makes ability inferences based on replicate samples of the jackknife without using plausible values.

Alternative Psychometric Approaches

A number of modifications of the current NAEP methodology have been suggested in the literature. These evolved out of criticisms of (a) the complex nature of the NAEP model and (b) the approximations made at different stages of the NAEP estimation process. Several such suggestions are listed below:

- *Apply a group-specific variance term.* Thomas (2000) developed a version of the CGROUP program that allowed for a group-specific residual variance term instead of assuming a uniform term across all groups.
- *Apply seemingly unrelated regressions* (SUR; Greene 2002; Zellner 1962). Researchers von Davier and Yu (2003) explored this suggestion using a program called YGROUP and found that it generated slightly different results from CGROUP. Since YGROUP is faster, it may be used to produce better starting values for the CGROUP program.

- *Apply a stochastic EM method.* Researchers von Davier and Sinharay (2007) approximated the posterior expectation and variance of the examinees' proficiencies using importance sampling (e.g., Gelman et al. 2004). Their conclusion was that this method is a viable alternative to the MGROUP system but does not present any compelling reason for change.
- *Apply stochastic approximation.* A promising approach for estimation in the presence of high dimensional latent variables is stochastic approximation. Researchers von Davier and Sinharay (2010) applied this approach to the estimation of conditioning models and showed that the procedure can improve estimation in some cases.
- *Apply multilevel IRT using Markov chain Monte Carlo methods (MCMC).* M. S. Johnson and Jenkins (2004) suggested an MCMC estimation method (e.g., Gelman et al. 2004; Gilks et al. 1996) that can be adapted to combine the three steps (scaling, conditioning, and variance estimation) of the MGROUP program. This idea is similar to that proposed by Raudenbush and Bryk (2002). A maximum likelihood application of this model was implemented by Li et al. (2009) and extended to dealing with testlets by Wang et al. (2002).
- *Estimation using generalized least squares (GLS).* Researchers von Davier and Yon (2004) applied GLS methods to the conditioning model used in NAEP's MGROUP, employing an individual variance term derived from the IRT measurement model. This method eliminates some basic limitations of classical approaches to regression model estimation.
- *Other modifications.* Other important works on modification of the current NAEP methodology include those by Bock (2002) and Thomas (2002).

Possible Future Innovations

Random Effects Model

ETS developed and evaluated a random effects model for population characteristics estimation. This approach explicitly models between-school variability as a random effect to determine whether it is better aligned with the observed structure of NAEP data. It was determined that relatively small gains in estimation using this approach in NAEP were not sufficient to override the increase in computational complexity. However, this approach does appear to have potential for use in international assessments such as PISA and PIRLS.

Adaptive Numerical Quadrature

Use of adaptive numerical quadrature can improve estimation accuracy over using approximation methods in high-dimensional proficiency estimation. ETS researchers performed analytic studies (Antal and Oranje 2007; Haberman 2006) using adaptive quadrature to study the benefit of increased precision through numerical integration for multiple dimensions. Algorithmic development and resulting evaluation of gains in precision are ongoing, as are feasibility studies for possible operational deployment in large-scale assessment estimation processes.

Antal and Oranje (2007) posited that the Gauss-Hermite rule enhanced with Cholesky decomposition and normal approximation of the response likelihood is a fast, precise, and reliable alternative for the numerical integration in NAEP and in IRT in general.

Using Hierarchical Models

In addition, several studies have been conducted about the use of hierarchical models to estimate latent regression effects that ultimately lead to proficiency estimates for many student groups of interest. Early work based on MCMC (Johnson and Jenkins 2004) was extended into an MLE environment, and various studies were conducted to evaluate applications of this model to NAEP (Li et al. 2009).

The NAEP latent regression model has been studied to understand better some boundary conditions under which the model performs well or not so well (Moran and Dresher 2007). Research into different approaches to model selection has been initiated (e.g., Gladkova and Oranje 2007). This is an ongoing project.

References

Allen, N. L., Johnson, E. J., Mislevy, R. J., & Thomas, N. (1996). Scaling procedures. In N. L. Allen, D. L. Kline, & C. A. Zelenak (Eds.), *The NAEP 1994 technical report* (pp. 247–266). Washington, DC: National Center for Education Statistics.

Angoff, W. H. (1971). Scales, norms, and equivalent scores. In R. L. Thorndike (Ed.), *Educational measurement* (2nd ed., pp. 508–600). Washington, DC: American Council on Education.

Antal, T., & Oranje, A. (2007). *Adaptive numerical integration for item response theory* (Research Report No. RR-07-06). Princeton: Educational Testing Service. http://dx.doi.org/10.1002/j.2333-8504.2007.tb02048.x

Beall, G., & Ferris, J. (1971). *On discovering Youden rectangles with columns of treatments in cyclic order* (Research Bulletin No. RB-71-37). Princeton: Educational Testing Service. http://dx.doi.org/10.1002/j.2333-8504.1971.tb00611.x

Beaton, A. E. (1964). *The use of special matrix operators in statistical calculus* (Research Bulletin No. RB-64-51). Princeton: Educational Testing Service. http://dx.doi.org/10.1002/j.2333-8504.1964.tb00689.x

Beaton, Albert E. (1968). *Some considerations of technical problems in the Educational Opportunity Survey* (Research Memorandum No. RM-68-17). Princeton: Educational Testing Service.

Beaton, A. E. (1969). Scaling criterion of questionnaire items. *Socio–Economic Planning Sciences, 2*, 355–362. https://doi.org/10.1016/0038-0121(69)90030-5

Beaton, A. E. (1973a). *Commonality*. Retrieved from ERIC Database. (ED111829)

Beaton, A. E. (1973b). F4STAT statistical system. In W. J. Kennedy (Ed.), *Proceedings of the computer science and statistics: Seventh annual symposium of the interface* (pp. 279–282). Ames: Iowa State University Press.

Beaton, A. E. (1975). Ability scores. In F. T. Juster (Ed.), *Education, income, and human behavior* (pp. 427–430). New York: McGraw-Hill.

Beaton, A. E. (1981). *Interpreting least squares without sampling assumptions* (Research Report No. RR-81-38). Princeton: Educational Testing Service. http://dx.doi.org/10.1002/j.2333-8504.1981.tb01265.x

Beaton, A. E. (1987). *The NAEP 1983–84 technical report*. Washington, DC: National Center for Education Statistics.

Beaton, A. E. (2000). *Estimating the total population median.* Paper presented at the National Institute of Statistical Sciences workshop on NAEP inclusion strategies. Research Triangle Park: National Institute of Statistical Sciences.

Beaton, A. (2003). *A procedure for testing the fit of IRT models for special populations*. Unpublished manuscript.

Beaton, A. E., & Allen, N. L. (1992). Interpreting scales through scale anchoring. *Journal of Educational Statistics, 17*, 191–204. https://doi.org/10.2307/1165169

Beaton, A. E., & Chromy, J. R. (2007). *Partitioning NAEP trend data.* Palo Alto: American Institutes for Research.

Beaton, A. E., & Gonzalez, E. J. (1993). *Comparing the NAEP trial state assessment results with the IAEP international results. Report prepared for the National Academy of Education Panel on the NAEP Trial State Assessment.* Stanford: National Academy of Education.

Beaton, A. E., & Gonzalez, E. (1995). *NAEP primer*. Chestnut Hill: Boston College.

Beaton, A. E., & Johnson, E. G. (1990). The average response method of scaling. *Journal of Educational Statistics, 15*, 9–38. https://doi.org/10.2307/1164819

Beaton, A. E., & Tukey, J. W. (1974). The fitting of power series, meaning polynomials, illustrated on band–spectroscopic data. *Technometrics, 16*, 147–185. https://doi.org/10.1080/00401706.1974.10489171

Beaton, A.E., & Zwick, R. (1990). *The effect of changes in the national assessment: Disentangling the NAEP 1985–86 reading anomaly* (NAEP Report No. 17–TR–21). Princeton: Educational Testing Service.

Beaton, A. E., Rubin, D. B., & Barone, J. L. (1976). The acceptability of regression solutions: Another look at computational accuracy. *Journal of the American Statistical Association, 71*, 158–168. https://doi.org/10.1080/01621459.1976.10481507

Beaton, A. E., Hilton, T. L., & Schrader, W. B. (1977). *Changes in the verbal abilities of high school seniors, college entrants, and SAT candidates between 1960 and 1972* (Research Bulletin No. RB-77-22). Princeton: Educational Testing Service. http://dx.doi.org/10.1002/j.2333-8504.1977.tb01147.x

Beaton, A. E., Rogers, A. M., Gonzalez, E., Hanly, M. B., Kolstad, A., Rust, K. F., … Jia, Y. (2011). *The NAEP primer* (NCES Report No. 2011–463). Washington, DC: National Center for Education Statistics.

Bennett, R. E., Braswell, J., Oranje, A., Sandene, B., Kaplan, B., & Yan, F. (2008). Does it matter if I take my mathematics test on computer? A second empirical study of mode effects in NAEP. *Journal of Technology, Learning, and Assessment, 6*, 1–39.

Bock, R.D. (2002). *Issues and recommendations on NAEP data analysis.* Palo Alto: American Institutes for Research.

Bock, R. D., Gibbons, R., & Muraki, E. (1988). Full–information item factor analysis. *Applied Psychological Measurement, 12*, 261–280. https://doi.org/10.1177/014662168801200305

Bowles, S., & Levin, H. M. (1968). The determinants of scholastic achievement: An appraisal of some recent evidence. *Journal of Human Resources, 3*, 3–24.

Braun, H. I., & Holland, P. W. (1982). Observed–score test equating: A mathematical analysis of some ETS equating procedures. In P. W. Holland & D. B. Rubin (Eds.), *Test equating* (pp. 9–49). New York: Academic Press.

Braun, H. I., & Qian, J. (2007). An enhanced method for mapping state standards onto the NAEP scale. In N. J. Dorans, M. Pommerich, & P. W. Holland (Eds.), *Linking and aligning scores and scales* (pp. 313–338). New York: Springer. https://doi.org/10.1007/978-0-387-49771-6_17

Braun, H., Zhang, J., & Vezzu, S. (2008). *Evaluating the effectiveness of a full-population estimation method* (Research Report No. RR-08-18). Princeton: Educational Testing Service. https://doi.org/10.1002/j.2333-8504.2008.tb02104.x

Bridgeman, B., Blumenthal, J. B., & Andrews, S. R. (1981). *Parent child development center: Final evaluation report*. Unpublished manuscript.

Brown, L. D., Cai, T., & DasGupta, A. (2001). Interval estimation for a binomial proportion (with discussion). *Statistical Science, 16*, 101–133. https://doi.org/10.1214/ss/1009213286

Bryk, A. S., & Raudenbush, S. W. (1992). *Hierarchical linear models in social and behavioral research: Applications and data analysis methods*. Newbury Park: Sage.

Cain, G., & Watts, H. W. (1968). The controversy about the Coleman report: Comment. *The Journal of Human Resources, 3*, 389–392. https://doi.org/10.2307/145110

Carlson, J. E. (1993, April). *Dimensionality of NAEP instruments that incorporate polytomously-scored items*. Paper presented at the meeting of the American Educational Research Association, Atlanta, GA.

Carlson, J. E., & Jirele, T. (1992, April). *Dimensionality of 1990 NAEP mathematics data*. Paper presented at the meeting of the American Educational Research Association, San Francisco, CA.

Civil Rights Act, P.L. No. 88-352, 78 Stat. 241 (1964).

Cleary, T. A., Linn, R. L., & Rock, D. A. (1968). An exploratory study of programmed tests. *Educational and Psychological Measurement, 28*, 345–360. https://doi.org/10.1177/001316446802800212

Cohen, J. D. (1998). *AM online help content—Preview*. Washington, DC: American Institutes for Research.

Cohen, J., & Jiang, T. (2001). *Direct estimation of latent distributions for large-scale assessments with application to the National Assessment of Educational Progress (NAEP)*. Washington, DC: American Institutes for Research.

Coleman, J. S., Campbell, E. Q., Hobson, C. J., McPartland, J., Mood, A. M., Weinfeld, F. D., & York, R. L. (1966). *Equality of educational opportunity*. Washington, DC: U. S. Government Printing Office.

Dempster, A. P. (1969). *Elements of continuous multivariate analysis*. Reading: Addison–Wesley.

Dempster, A. P., Laird, N. M., & Rubin, D. B. (1977). Maximum likelihood from incomplete data via the EM algorithm (with discussion). *Journal of the Royal Statistical Society, Series B, 39*, 1–38.

Donoghue, J., Mazzeo, J., Li, D., & Johnson, M. (2006a). *Marginal estimation in NAEP: Current operational procedures and AM*. Unpublished manuscript.

Donoghue, J., McClellan, C. A., & Gladkova, L. (2006b). *Using rater effects models in NAEP*. Unpublished manuscript.

Dorans, N. J., Pommerich, M., & Holland, P. W. (Eds.). (2007). *Linking and aligning scores and scales*. New York: Springer.

Gamoran, A., & Long, D. A. (2006). *Equality of educational opportunity: A 40-year retrospective*. (WCER Working Paper No. 2006-9). Madison: University of Wisconsin–Madison, Wisconsin Center for Education Research.

Gelman, A., Carlin, J. B., Stern, H. S., & Rubin, D. B. (2004). *Bayesian data analysis*. Boca Raton: Chapman and Hall/CRC.

Gilks, W. R., Richardson, S., & Spiegelhalter, D. J. (Eds.). (1996). *Markov chain Monte Carlo in practice*. London: Chapman and Hall.

Gladkova, L., & Oranje, A. (2007, April). *Model selection for large scale assessments*. Paper presented at the meeting of the National Council of Measurement in Education, Chicago, IL.

Gladkova, L., Moran, R., Rogers, A., & Blew, T. (2005). Direct estimation software interactive (DESI) manual [Computer software manual]. Princeton: Educational Testing Service.

Goodnight, J. H. (1979). A tutorial on the SWEEP operator. *American Statistician, 33*, 149–158. https://doi.org/10.1080/00031305.1979.10482685
Greene, W. H. (2002). *Econometric analysis* (5th ed.). Upper Saddle River: Prentice Hall.
Haberman, S. J. (2006). *Adaptive quadrature for item response models* (Research Report No. RR-06-29). Princeton: Educational Testing Service. https://doi.org/10.1002/j.2333-8504.2006.tb02035.x
Hilton, T. L. (1992). *Using national data bases in educational research.* Hillsdale: Erlbaum.
Holland, P. W. (2000). Notes on Beaton's and McLaughlin's proposals. In L. V. Jones & I. Olkin, *NAEP inclusion strategies: The report of a workshop at the National Institute of Statistical Sciences.* Unpublished manuscript.
Holland, P. W., & Welsch, R. E. (1977). Robust regression using iteratively reweighted least squares. *Communications in Statistics – Theory and Methods, A6*, 813–827. https://doi.org/10.1080/03610927708827533
Hsieh, C., Xu, X., & von Davier, M. (2009). Variance estimation for NAEP data using a resampling–based approach: An application of cognitive diagnostic models. *IERI Monograph Series: Issues and methodologies in large scale assessments, 2*, 161–173.
Huber, P. J. (1981). *Robust statistics.* New York: Wiley. https://doi.org/10.1002/0471725250
Huber, P. J. (1996). *Robust statistical procedures* (2nd ed.). Philadelphia: Society for Industrial and Applied Mathematics. https://doi.org/10.1137/1.9781611970036
International Assessment of Educational Progress. (1992). *IAEP technical report.* Princeton: Educational Testing Service.
Johnson, M. S., & Jenkins, F. (2004). *A Bayesian hierarchical model for large–scale educational surveys: An application to the National Assessment of Educational Progress* (Research Report No. RR-04-38). Princeton: Educational Testing Service. https://doi.org/10.1002/j.2333-8504.2004.tb01965.x
Johnson, E. G., & King, B. F. (1987). Generalized variance functions for a complex sample survey. *Journal of Official Statistics, 3*, 235–250. https://doi.org/10.1002/j.2330-8516.1987.tb00210.x
Johnson, E. G., & Rust, K. F. (1992). Population inferences and variance estimation for NAEP data. *Journal of Educational Statistics, 17*, 175–190. https://doi.org/10.2307/1165168
Johnson, E. G., & Rust, K. F. (1993). Effective degrees of freedom for variance estimates from a complex sample survey. In *Proceedings of the Survey Research Methods Section, American Statistical Association* (pp. 863–866). Alexandria, VA: American Statistical Association.
Johnson, E. G., & Siegendorf, A. (1998). *Linking the National Assessment of Educational Progress (NAEP) and the Third International Mathematics and Science Study (TIMSS): Eighth–grade results* (NCES Report No. 98–500). Washington, DC: National Center for Education Statistics.
Johnson, E., Cohen, J., Chen, W. H., Jiang, T., & Zhang, Y. (2003). *2000 NAEP-1999 TIMSS linking report* (NCES Publication No. 2005–01). Washington, DC: National Center for Education Statistics.
Jones, L. V., & Olkin, I. (Eds.). (2004). *The Nation's Report Card: Evolutions and perspectives.* Bloomington: Phi Delta Kappa Educational Foundation.
Kirsch, I. S., & Jungeblut, A. (1986). *Literacy: Profiles of America's young adults* (NAEP Report No. 16-PL-01). Princeton: National Assessment of Educational Progress.
Kirsch, I., Yamamoto, K., Norris, N., Rock, D., Jungeblut, A., O'Reilly, P., … Baldi, S. (2000). *Technical report and data files user's manual For the 1992 National Adult Literacy Survey.* (NCES Report No. 2001457). U.S. Department of Education.
Lapointe, A. E., Mead, N. A., & Phillips, G. W. (1989). *A world of difference: An international assessment of mathematics and science.* Princeton: Educational Testing Service.
Li, D., & Oranje, A. (2007). *Estimation of standard errors of regression effects in latent regression models using Binder's linearization* (Research Report No. RR-07-09). Princeton: Educational Testing Service. https://doi.org/10.1002/j.2333-8504.2007.tb02051.x
Li, D., Oranje, A., & Jiang, Y. (2009). On the estimation of hierarchical latent regression models for large scale assessments. *Journal of Educational and Behavioral Statistics, 34*, 433–463. https://doi.org/10.3102/1076998609332757

Linn, R. L. (1993). Linking results of distinct assessments. *Applied Measurement in Education, 6*, 83–102. https://doi.org/10.1207/s15324818ame0601_5

Linn, R. L., & Kiplinger, V. L. (1995). Linking statewide tests to the National Assessment of Educational Progress: Stability of results. *Applied Measurement in Education, 8*, 135–155.

Little, R. J. A., & Rubin, D. B. (1987). *Statistical analysis with missing data*. New York: Wiley.

Little, R. J. A., & Rubin, D. B. (2002). *Statistical analysis with missing data* (2nd ed.). Hoboken: Wiley–Interscience. https://doi.org/10.1002/9781119013563

Longley, J. W. (1967). An appraisal of least-squares programs for the electronic computer from the point of view of the user. *Journal of the American Statistical Association, 62*, 819–841. https://doi.org/10.1080/01621459.1967.10500896

Lord, F. M. (1971). A theoretical study of two-stage testing. *Psychometrika, 36*, 227–242. https://doi.org/10.1007/BF02297844

Lord, F. M., & Novick, M. R. (1968). *Statistical theories of mental test scores*. Reading: Addison-Wesley.

Martin, M. O., & Kelly, D. L. (Eds.). (1996). *TIMSS technical report, Volume I: Design and development*. Chestnut Hill: Boston College.

Mayeske, G. W., & Beaton, A. E. (1975). *Special studies of our nation's students*. Washington, DC: U.S. Government Printing Office.

Mayeske, G. W., Cohen, W. M., Wisler, C. E., Okada, T., Beaton, A. E., Proshek, J. M., et al. (1972). *A study of our nation's schools*. Washington, DC: U.S. Government Printing Office.

Mayeske, G. W., Okada, T., & Beaton, A. E. (1973a). *A study of the attitude toward life of our nation's students*. Washington, DC: U.S. Government Printing Office.

Mayeske, G. W., Okada, T., Beaton, A. E., Cohen, W. M., & Wisler, C. E. (1973b). *A study of the achievement of our nation's students*. Washington, DC: U.S. Government Printing Office.

McLaughlin, D. H. (1998). *Study of the linkages of 1996 NAEP and state mathematics assessments in four states*. Washington, DC: National Center for Education Statistics.

McLaughlin, D. H. (2000). *Protecting state NAEP trends from changes in SD/LEP inclusion rates* (Report to the National Institute of Statistical Sciences). Palo Alto: American Institutes for Research.

McLaughlin, D. H. (2005). *Properties of NAEP full population estimates*. Palo Alto: American Institutes for Research.

Messick, S. (1986). *Large-scale educational assessment as policy research: Aspirations and limitations* (Research Report No. RR-86-27). Princeton: Educational Testing Service. https://doi.org/10.1002/j.2330-8516.1986.tb00182.x

Messick, S., Beaton, A. E., & Lord, F. (1983). *A new design for a new era*. Princeton: Educational Testing Service.

Milton, R. C., & Nelder, J. A. (Eds.). (1969). *Statistical computation*. Waltham: Academic Press.

Mislevy, R. J. (1984). Estimating latent distributions. *Psychometrika, 49*, 359–381. https://doi.org/10.1007/BF02306026

Mislevy, R. J. (1985). Estimation of latent group effects. *Journal of the American Statistical Association, 80*, 993–997. https://doi.org/10.1080/01621459.1985.10478215

Mislevy, R. J. (1991). Randomization-based inference about latent variables from complex samples. *Psychometrika, 56*, 177–196. https://doi.org/10.1007/BF02294457

Mislevy, R. J. (1992). *Linking educational assessments: Concepts, issues, methods and prospects*. Princeton: Educational Testing Service.

Mislevy, R. J., & Bock, R. D. (1982). *BILOG: Item analysis and test scoring with binary logistic models* [Computer program]. Chicago: Scientific Software.

Mislevy, R. J., & Sheehan, K. M. (1987). Marginal estimation procedures. In A. E. Beaton (Ed.), *Implementing the new design: The NAEP 1983–84 technical report* (No. 15–TR–20, pp. 293–360). Princeton: Educational Testing Service.

Mislevy, R., Johnson, E., & Muraki, E. (1992). Scaling procedures in NAEP. *Journal of Educational and Behavioral Statistics, 17*, 131–154. https://doi.org/10.3102/10769986017002131

Moran, R., & Dresher, A. (2007, April). *Results from NAEP marginal estimation research on multivariate scales*. Paper presented at the meeting of the National Council for Measurement in Education, Chicago, IL.

Mosteller, F., & Moynihan, D. P. (1972). A pathbreaking report: Further studies of the Coleman report. In F. Mosteller & D. P. Moynihan (Eds.), *On equality of educational opportunity* (pp. 3–68). New York: Vintage Books.

Mosteller, F., Fienberg, S. E., Hoaglin, D. C., & Tanur, J. M. (Eds.). (2010). *The pleasures of statistics: The autobiography of Frederick Mosteller*. New York: Springer.

Mullis, I. V. S., Martin, M. O., Gonzalez, E. J., & Kennedy, A. M. (2003). *PIRLS 2001 international report: IEA's study of reading literacy achievement in primary schools in 35 countries*. Chestnut Hill: International Study Center, Boston College.

Muraki, E. (1992). A generalized partial credit model: Application of an EM algorithm. *Applied Psychological Measurement, 16*, 159–176. https://doi.org/10.1177/014662169201600206

Muraki, E., & Bock, R. D. (1997). *PARSCALE: IRT item analysis and test scoring for rating scale data* [Computer software]. Chicago: Scientific Software.

Muraki, E., Hombo, C. M., & Lee, Y. W. (2000). Equating and linking of performance assessments. *Applied Psychological Measurement, 24*, 325–337. https://doi.org/10.1177/01466210022031787

National Assessment of Educational Progress. (1985.) *The reading report card: Progress toward excellence in our school: Trends in reading over four national assessments, 1971-1984* (NAEP Report No. 15-R-01). Princeton: Educational Testing Service.

National Commission on Excellence in Education. (1983). *A nation at risk: The imperative for educational reform*. Washington, DC: U. S. Government Printing Office.

Newton, R. G., & Spurrell, D. J. (1967a). A development of multiple regression for the analysis of routine data. *Applied Statistics, 16*, 51–64. https://doi.org/10.2307/2985237

Newton, R. G., & Spurrell, D. J. (1967b). Examples of the use of elements for clarifying regression analyses. *Applied Statistics, 16*, 165–172.

No Child Left Behind Act, P.L. 107-110, 115 Stat. § 1425 (2002).

Oranje, A. (2006a). *Confidence intervals for proportion estimates in complex samples* (Research Report No. RR-06-21). Princeton: Educational Testing Service. https://doi.org/10.1002/j.2333-8504.2006.tb02027.x

Oranje, A. (2006b). *Jackknife estimation of sampling variance of ratio estimators in complex samples: Bias and the coefficient of variation* (Research Report No. RR-06-19). Princeton: Educational Testing Service. https://doi.org/10.1002/j.2333-8504.2006.tb02025.x

Oranje, A., Freund, D., Lin, M.-J., & Tang, Y. (2007). *Disclosure risk in educational surveys: An application to the National Assessment of Educational Progress* (Research Report No. RR-07-24). Princeton: Educational Testing Service. https://doi.org/10.1002/j.2333-8504.2007.tb02066.x

Oranje, A., Li, D., & Kandathil, M. (2009). *Evaluation of methods to compute complex sample standard errors in latent regression models* (Research Report No. RR-09-49). Princeton: Educational Testing Service. https://doi.org/10.1002/j.2333-8504.2009.tb02206.x

Organisation for Economic Co-operation and Development. (2013). *OECD skills outlook 2013: First results from the survey of adult skills*. Paris: OECD Publishing.

Pashley, P. J., & Phillips, G. W. (1993). *Toward world-class standards: A research study linking international and national assessments*. Princeton: Educational Testing Service.

Pedhazur, E. J. (1997). *Multiple regression in behavioral research* (3rd ed.). Orlando: Harcourt Brace.

Phillips, G. (2007). *Chance favors the prepared mind: Mathematics and science indicators for comparing states and nations*. Washington, DC: American Institutes for Research.

Privacy Act, 5 U.S.C. § 552a (1974).

Qian, J. (1998). Estimation of the effective degree of freedom in t-type tests for complex data. *Proceedings of the Section on Survey Research Methods, American Statistical Association*, 704–708. Retrieved from http://www.amstat.org/sections/srms/Proceedings/

Qian, J., & Kaplan, B. (2001). Analysis of design effects for NAEP combined samples. *2001 Proceedings of the American Statistical Association, Survey Research Methods Section* [CD-ROM]. Alexandria: American Statistical Association.

Qian, J., Kaplan, B., & Weng, V. (2003) *Analysis of NAEP combined national and state samples* (Research Report No. RR-03-21). Princeton: Educational Testing Service.

Quenouille, M. H. (1956). Notes on bias in estimation. *Biometrika, 43*, 353–360. https://doi.org/10.1093/biomet/43.3-4.353

Raudenbush, S. W., & Bryk, A. S. (2002). *Hierarchical linear models: Applications and data analysis methods* (2nd ed.). Newbury Park: Sage.

Rijmen, F. (2011). Hierarchical factor item response theory models for PIRLS: Capturing clustering effects at multiple levels. *IERI Monograph Series: Issues and Methodologies in Large-Scale Assessment, 4*, 59–74.

Rijmen, F., Tuerlinckx, F., De Boeck, P., & Kuppens, P. (2003). A nonlinear mixed model framework for item response theory. *Psychological Methods, 8*, 185–205. https://doi.org/10.1037/1082-989X.8.2.185

Rock, D. A., Hilton, T., Pollack, J. M., Ekstrom, R., & Goertz, M. E. (1985). *Psychometric analysis of the NLS-72 and the high school and beyond test batteries* (NCES Report No. 85-218). Washington, DC: National Center for Education Statistics.

Rogers, A., Tang, C., Lin, M. J., & Kandathil, M. (2006). DGROUP [Computer software]. Princeton: Educational Testing Service.

Rosenbaum, P. (1984). Testing the conditional independence and monotonicity assumptions of item response theory. *Psychometrika, 49*, 425–435. https://doi.org/10.1007/BF02306030

Rubin, D. B. (1977). Formalizing subjective notions about the effect of nonrespondents in sample surveys. *Journal of the American Statistical Association, 72*, 538–543. https://doi.org/10.1080/01621459.1977.10480610

Rubin, D. B. (1987). *Multiple imputation for nonresponse in surveys*. New York: Wiley. https://doi.org/10.1002/9780470316696

Rutkowski, L., von Davier, M., & Rutkowski, D. (Eds.). (2014). *Handbook of international large-scale assessment: Background, technical issues, and methods of data analysis*. Boca Raton: CRC Press.

Satterthwaite, F. E. (1941). Synthesis of variance. *Psychometrika, 6*, 309–316. https://doi.org/10.1007/BF02288586

Sinharay, S., & von Davier, M. (2005). *Extension of the NAEP BGROUP program to higher dimensions* (Research Report No. RR-05-27). Princeton: Educational Testing Service. https://doi.org/10.1002/j.2333-8504.2005.tb02004.x

Sinharay, S., Guo, Z., von Davier, M., & Veldkamp, B. P. (2010). Assessing fit of latent regression models. *IERI Monograph Series, 3*, 35–55.

Statistics Canada & Organisation for Economic Co-operation and Development. (2005). *Learning a living: First results of the adult literacy and life skills survey*. Paris: OECD Publishing.

Thissen, D. (2007). Linking assessments based on aggregate reporting: Background and issues. In N. J. Dorans, M. Pommerich, & P. W. Holland (Eds.), *Linking and aligning scores and scales* (pp. 287–312). New York: Springer. https://doi.org/10.1007/978-0-387-49771-6_16

Thomas, N. (1993). Asymptotic corrections for multivariate posterior moments with factored likelihood functions. *Journal of Computational and Graphical Statistics, 2*, 309–322. https://doi.org/10.2307/1390648

Thomas, N. (2000). Assessing model sensitivity of imputation methods used in NAEP. *Journal of Educational and Behavioral Statistics, 25*, 351–371. https://doi.org/10.3102/10769986025004351

Thomas, N. (2002). The role of secondary covariates when estimating latent trait population distributions. *Psychometrika, 67*, 33–48. https://doi.org/10.1007/BF02294708

Thorndike, R. L., & Hagen, E. (1959). *Ten thousand careers*. New York: Wiley.

Tukey, J. W. (1958). Bias and confidence in not–quite large samples [abstract]. *The Annals of Mathematical Statistics, 29*, 614.

Viadero, D. (2006). Fresh look at Coleman data yields different conclusions. *Education Week, 25*(41), 21.

von Davier, M. (2003). *Comparing conditional and marginal direct estimation of subgroup distributions* (Research Report No. RR-03-02). Princeton: Educational Testing Service. https://doi.org/10.1002/j.2333-8504.2003.tb01894.x

von Davier, M. (2010). Hierarchical mixtures of diagnostic models. *Psychological Test and Assessment Modeling, 52*, 8–28.

von Davier, M., & Sinharay, S. (2007). An importance sampling EM algorithm for latent regression models. *Journal of Educational and Behavioral Statistics, 32*, 233–251. https://doi.org/10.3102/1076998607300422

von Davier, M., & Sinharay, S. (2010). Stochastic approximation for latent regression item response models. *Journal of Educational and Behavioral Statistics, 35*, 174–193. https://doi.org/10.3102/1076998609346970

von Davier, M., & Yon, H. (2004, April) *A conditioning model with relaxed assumptions*. Paper presented at the meeting of the National Council of Measurement in Education, San Diego, CA.

von Davier, M., & Yu, H. T. (2003, April). *Recovery of population characteristics from sparse matrix samples of simulated item responses*. Paper presented at the meeting of the American Educational Research Association, Chicago, IL.

von Davier, M., Sinharay, S., Oranje, A., & Beaton, A. E. (2006). Statistical procedures used in the National Assessment of Educational Progress (NAEP): Recent developments and future directions. In C. R. Rao & S. Sinharay (Eds.), *Handbook of statistics: Vol. 26. Psychometrics* (pp. 1039–1056). Amsterdam: Elsevier.

von Davier, M., Gonzalez, E., & Mislevy, R. J. (2009). What are plausible values and why are they useful? *IERI Monograph Series, 2*, 9–36.

Wainer, H. (1993). Measurement problems. *Journal of Educational Measurement, 30*, 1–21. https://doi.org/10.1111/j.1745-3984.1993.tb00419.x

Wang, X., Bradlow, E. T., & Wainer, H. (2002). A general Bayesian model for testlets: Theory and applications. *Applied Psychological Measurement, 26*, 109–128. https://doi.org/10.1177/0146621602026001007

Wingersky, M. S. (1983). LOGIST: A program for computing maximum likelihood procedures for logistic test models. In R. K. Hambleton (Ed.), *Applications of item response theory* (pp. 45–56). Vancouver: Educational Research Institute of British Columbia.

Wingersky, M. S., Barton, M.A., & Lord, F. M. (1982). LOGIST user's guide Logist 5, version 1.0 [Computer software manual]. Princeton: Educational Testing Service.

Wirtz, W. (Ed.). (1977). *On further examination: Report of the advisory panel on the scholastic aptitude test score decline* (Report No. 1977-07-01). New York: College Entrance Examination Board.

Wirtz, W., & Lapointe, A. (1982). Measuring the quality of education: A report on assessing educational progress. *Educational Measurement: Issues and Practice, 1*, 17–19, 23. https://doi.org/10.1111/j.1745-3992.1982.tb00673.x

Wood, R. L., Wingersky, M. S., & Lord, F. M. (1976). *LOGIST – A computer program for estimating examinee ability and item characteristic curve parameters* (Research Memorandum No. RM-76-06). Princeton: Educational Testing Service.

Yamamoto, K., & Mazzeo, J. (1992). Item response theory scale linking in NAEP. *Journal of Educational Statistics, 17*, 155–173. https://doi.org/10.2307/1165167

Zellner, A. (1962). An efficient method of estimating seemingly unrelated regression equations and tests for aggregation bias. *Journal of the American Statistical Association, 57*, 348–368. https://doi.org/10.1080/01621459.1962.10480664

Zwick, R. (1987a). Assessing the dimensionality of NAEP reading data. *Journal of Educational Measurement, 24*, 293–308. https://doi.org/10.1111/j.1745-3984.1987.tb00281.x

Zwick, R. (1987b). Some properties of the correlation matrix of dichotomous Guttman items. *Psychometrika, 52*, 515–520. https://doi.org/10.1007/BF02294816

Zwick, R. (1991). Effects of item order and context on estimation of NAEP reading proficiency. *Educational Measurement: Issues and Practice, 10*(3), 10–16. https://doi.org/10.1111/j.1745-3992.1991.tb00198.x

Chapter 9
Large-Scale Assessments of Adult Literacy

Irwin Kirsch, Mary Louise Lennon, Kentaro Yamamoto, and Matthias von Davier

Educational Testing Service's (ETS's) work in large-scale adult literacy assessments has been an ongoing and evolving effort, beginning in 1984 with the Young Adult Literacy Survey in the United States. This work has been designed to meet policy needs, both in the United States and internationally, based on the growing awareness of literacy as human capital. The impact of these assessments has grown as policy makers and other stakeholders have increasingly come to understand the critical role that foundational skills play in allowing individuals to maintain and enhance their ability to meet changing work conditions and societal demands. For example, findings from these surveys have provided a wealth of information about how the distribution of skills is related to social and economic outcomes. Of equal importance, the surveys and associated research activities have contributed to large-scale assessment methodology, the development of innovative item types and delivery systems, and methods for reporting survey data in ways that ensure its utility to a range of stakeholders and audiences.

The chronology of ETS's large-scale literacy assessments, as shown in Fig. 9.1, spans more than 30 years. ETS served as the lead contractor in the development of these innovative assessments, while the prime clients and users of the assessment outcomes were representatives of either governmental organizations such as the National Center for Education Statistics (NCES) and Statistics Canada, or trans-governmental entities such as the Organisation for Economic Co-operation and Development (OECD). These instruments have evolved from a single-language, paper-based assessment focusing on a U.S. population of 16- to 25-year-olds to an adaptive, computer-based assessment administered in almost 40 countries and close to 50 languages to adults through the age of 65. By design, the assessments have been linked at the item level, with sets of questions from previous assessments

This work was conducted while M. von Davier was employed with Educational Testing Service.

I. Kirsch (✉) • M.L. Lennon • K. Yamamoto • M. von Davier
Educational Testing Service, Princeton, NJ, USA
e-mail: ikirsch@ets.org

R.E. Bennett, M. von Davier (eds.), *Advancing Human Assessment*,
Methodology of Educational Measurement and Assessment,
DOI 10.1007/978-3-319-58689-2_9

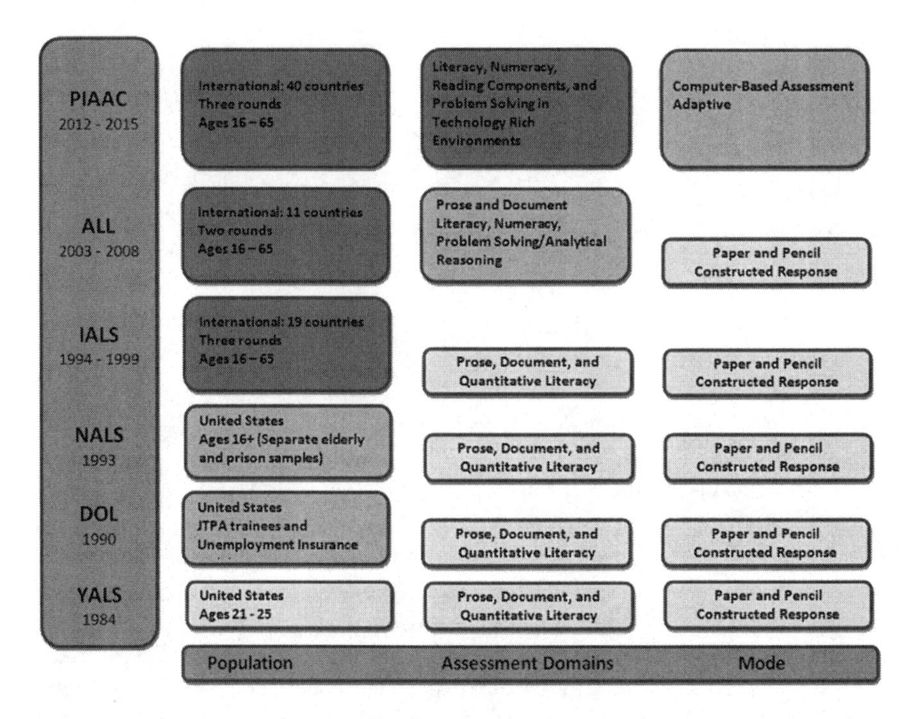

Fig. 9.1 ETS's large-scale literacy assessments. Note. *ALL* = Adult Literacy and Life Skills Survey (Statistics Canada, Organisation for Economic Co-operation and Development [OECD]), *DOL* = Department of Labor Survey, *JPTA* = Job Training Partnership Act, *IALS* = International Adult Literacy Survey (Statistics Canada, OECD), *PIAAC* = Programme for the International Assessment of Adult Competencies (OECD), *YALS* = Young Adult Literacy Survey (through the National Assessment of Educational Progress)

included in each new survey. This link has made it possible to look at changes in skill levels, as well as the distribution of those skills, over time. Each of the assessments has also expanded upon previous surveys. As Fig. 9.1 illustrates, the assessments have changed over the years in terms of who is assessed, what skills are assessed, and how those skills are assessed. The surveys have evolved to include larger and more diverse populations as well as new and expanded constructs. They have also evolved from a paper-and-pencil, open-ended response mode to an adaptive, computer-based assessment.

In many ways, as the latest survey in this 30-year history, the Programme for the International Assessment of Adult Competencies (PIAAC) represents the culmination of all that has been learned over several decades in terms of instrument design, translation and adaptation procedures, scoring, and the development of interpretive schemes. As the first computer-based assessment to be used in a large-scale household skills survey, the experience derived from developing and delivering PIAAC—including research focused on innovative item types, harvesting log files, and delivering an adaptive assessment—helped lay the foundation for new computer based large-scale assessments yet to come.

This paper describes the contributions of ETS to the evolution of large-scale adult literacy assessments in six key areas:

- Expanding the construct of literacy
- Developing a model for building construct-based assessments
- Expanding and implementing large-scale assessment methodology
- Linking real-life stimulus materials and innovative item types
- Developing extensive background questionnaires to link performance with experience and outcome variables
- Establishing innovative reporting procedures to better integrate research and survey data

9.1 Expanding the Construct of Literacy

Early work in the field of adult literacy defined literacy based on the attainment of certain grade level scores on standardized academic tests of reading achievement. Standards for proficiency increased over the decades with "functional literacy" being defined as performance at a fourth-grade reading level during World War II, eighth-grade level in the 1960s, and a 12th grade level by the early 1970s. This grade-level focus using instruments that consisted of school-based materials was followed by a competency-based approach that employed tests based on nonschool materials from adult contexts. Despite this improvement, these tests still viewed literacy along a single continuum, defining individuals as either *literate* or *functionally illiterate* based on where they performed along that continuum. The 1984 Young Adult Literacy Survey (YALS) was the first in a series of assessments that contributed to an increasingly broader understanding of what it means to be "literate" in complex modern societies. In YALS, the conceptualization of literacy was expanded to reflect the diversity of tasks that adults encounter at work, home, and school and in their communities. As has been the case for all of the large-scale literacy assessments, panels of experts were convened to help set the framework for this assessment. Their deliberations led to the adoption of the following definition of literacy: "using printed and written information to function in society, to achieve one's goals, and to develop one's knowledge and potential" (Kirsch and Jungeblut 1986, p. 3).

This definition both rejected an arbitrary standard for literacy, such as performing at a particular grade level on a test of reading, and implied that literacy comprises a set of complex information-processing skills that goes beyond decoding and comprehending text-based materials.

To better reflect this multi-faceted set of skills and abilities, performance in YALS was reported across three domains, defined as follows (Kirsch and Jungeblut 1986, p. 4):

- Prose literacy: the knowledge and skills needed to understand and use information from texts including editorials, news stories, poems, and the like
- Document literacy: the knowledge and skills required to locate and use information contained in job applications or payroll forms, bus schedules, maps, indexes, and so forth

- Quantitative literacy: the knowledge and skills required to apply arithmetic operations, either alone or sequentially, that are embedded in printed materials, such as in balancing a checkbook, figuring out a tip, completing an order form, or determining the amount of interest on a loan from an advertisement

Rather than attempt to categorize individuals, or groups of individuals, as literate or illiterate, YALS reported results for each of these three domains by characterizing the underlying information-processing skills required to complete tasks at various points along a 0–500-point reporting scale, with a mean of 305 and a standard deviation of about 50. This proficiency-based approach to reporting was seen as a more faithful representation of both the complex nature of literacy demands in society and the various types and levels of literacy demonstrated by young adults.

Subsequent research at ETS led to the definition of five levels within the 500-point scale. Analyses of the interaction between assessment materials and the tasks based on those materials defined points along the scale at which information-processing demands shifted. The resulting levels more clearly delineated the progression of skills required to complete tasks at different points on the literacy scales and helped characterize the skills and strategies underlying the prose, document, and quantitative literacy constructs. These five levels have been used to report results for all subsequent literacy surveys, and the results from each of those assessments have made it possible to further refine our understanding of the information-processing demands at each level as well as the characteristics of individuals performing along each level of the scale.[1]

With the 2003 Adult Literacy and Life Skills Survey (ALL), the quantitative literacy domain was broadened to reflect the evolving perspective of experts in the field. The new numeracy domain was defined as the ability to interpret, apply, and communicate numerical information. While quantitative literacy focused on quantitative information embedded in text and primarily required respondents to demonstrate computational skills, numeracy included a broader range of skills typical of many everyday and work tasks including sorting, measuring, estimating, conjecturing, and using models. This expanded domain allowed ALL to collect more information about how adults apply mathematical knowledge and skills to real-life situations. In addition, the ALL assessment included a problem-solving component that focused on analytical reasoning. This component collected information about the ability of adults to solve problems by clarifying the nature of a problem and developing and applying appropriate solution strategies. The inclusion of problem solving was seen as a way to improve measurement at the upper levels of the scales and to reflect a skill set of growing interest for adult populations.

Most recently, the concept of literacy was expanded again with the Programme for the International Assessment of Adult Competencies (PIAAC). As the first computer-based, large-scale adult literacy assessment, PIAAC reflected the changing nature of information, its role in society, and its impact on people's lives.

[1] See the appendix for a description of the information-processing demands associated with each of the five levels across the literacy domains.

The scope of the prose, document, and numeracy domains was broadened in PIAAC and the assessment incorporated two new domains, as follows:

- For the first time, this adult assessment addressed literacy in digital environments. As a computer-based assessment, PIAAC included tasks that required respondents to use electronic texts including web pages, e-mails, and discussion boards. These stimulus materials included hypertext and multiple screens of information and simulated real-life literacy demands presented by digital media.
- In PIAAC, the definition of numeracy was broadened again to include the ability to access, use, interpret, and communicate mathematical information and ideas in order to engage in and manage the mathematical demands of a range of situations in adult life. The inclusion of *engage* in the definition signaled that not only cognitive skills but also dispositional elements (i.e., beliefs and attitudes) are necessary to meet the demands of numeracy effectively in everyday life.
- PIAAC included the new domain of problem-solving in technology-rich environments (PS-TRE), the first attempt to assess this domain on a large scale and as a single dimension. PS-TRE was defined as:

  > using digital technology, communication tools and networks to acquire and evaluate information, communicate with others and perform practical tasks. The first PIAAC problem-solving survey focuses on the abilities to solve problems for personal, work and civic purposes by setting up appropriate goals and plans, and accessing and making use of information through computers and computer networks. (OECD 2012, p. 47)

  PS-TRE presented computer-based tasks designed to measure the ability to analyze various requirements of a task, define goals and plans, and monitor progress until the task purposes were achieved. Simulated web, e-mail and spreadsheet environments were created and respondents were required to use multiple, complex sources of information, in some cases across more than one environment, to complete the presented tasks. The focus of these tasks was not on computer skills per se, but rather on the cognitive skills required to access and make use of computer-based information to solve problems.
- Finally, PIAAC contained a reading components domain, which included measures of vocabulary knowledge, sentence processing, and passage comprehension. Adding this domain was an important evolution because it provided more information about the skills of individuals with low levels of literacy proficiency than had been available from previous international assessments. To have a full picture of literacy in any society, it is necessary to have more information about these individuals because they are at the greatest risk of negative social, economic, and labor market outcomes.

9.2 Developing a Model for Building Construct-Based Assessments

A key characteristic of the large-scale literacy assessments is that each was based on a framework that, following Messick's (1994) construct-centered approach, defined the construct to be measured, the performances or behaviors expected to reveal that construct, and the characteristics of assessment tasks to elicit those behaviors. In the course of developing these assessments, a model for the framework development process was created, tested, and refined. This six-part process, as shown in Fig. 9.2 and described in more detail below, provides a logical sequence of steps from clearly defining a particular skill area to developing specifications for item construction and providing a foundation for an empirically based interpretation of the assessment results. Through this process, the inferences and assumptions about what is to be measured and how the results will be interpreted and reported are explicitly described.

1. *Develop a general definition of the domain.* The first step in this model is to develop a working definition of the domain and the assumptions underlying it. It is this definition that sets the boundaries for what will and will not be measured in a given assessment.
2. *Organize the domain.* Once the definition is developed, it is important to think about the kinds of tasks that represent the skills and abilities included in that

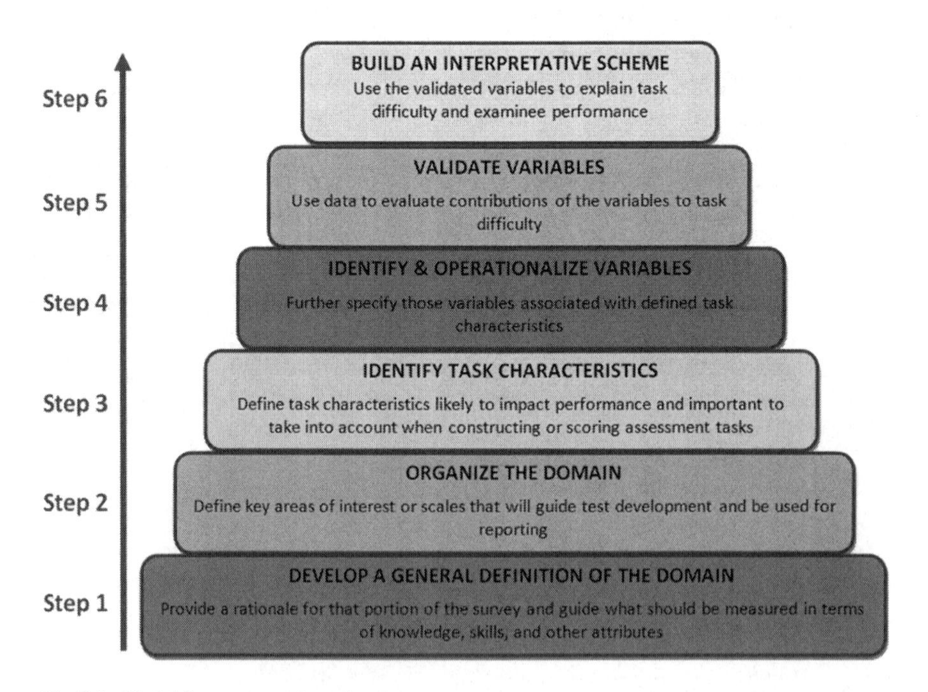

Fig. 9.2 Model for construct-based assessment

definition. Those tasks must then be categorized in relation to the construct definition to inform test design and result in meaningful score reporting. This step makes it possible to move beyond a laundry list of tasks or skills to a coherent representation of the domain that will permit policy makers and others to summarize and report information in more useful ways.

3. *Identify task characteristics*. Step 3 involves identifying a set of key characteristics, or task models, which will be used in constructing tasks for the assessment. These models may define characteristics of the stimulus materials to be used as well as characteristics of the tasks presented to examinees. Examples of key task characteristics that have been employed throughout the adult literacy assessments include contexts, material types, and information-processing demands.
4. *Identify and operationalize variables*. In order to use the task characteristics in designing the assessment and, later, in interpreting the results, the variables associated with each task characteristic need to be defined. These definitions are based on the existing literature and on experience with building and conducting other large-scale assessments. Defining the variables allows item developers to categorize the materials with which they are working, as well as the questions and directives they construct, so that these categories can be used in the reporting of the results. In the literacy assessments, for example, *context* has been defined to include home and family, health and safety, community and citizenship, consumer economics, work, leisure, and recreation; *materials* have been divided into continuous and noncontinuous texts with each of those categories being further specified; and *processes* have been identified in terms of type of match (focusing on the match between a question and text and including locating, integrating and generating strategies), type of information requested (ranging from concrete to abstract), and plausibility of distractors.[2]
5. *Validate variables*. In Step 5, research is conducted to validate the variables used to develop the assessment tasks. Statistical analyses determine which of the variables account for large percentages of the variance in the difficulty distribution of tasks and thereby contribute most towards understanding task difficulty and predicting performance. In the literacy assessments, this step provides empirical evidence that a set of underlying process variables represents the skills and strategies involved in accomplishing various kinds of literacy tasks.
6. *Build an interpretative scheme*. Finally in Step 6, an interpretative scheme is built that uses the validated variables to explain task difficulty and examinee performance. The definition of proficiency levels to explain performance along the literacy scales is an example of such an interpretative scheme. As previously explained, each scale in the literacy assessments has been divided into five progressive levels characterized by tasks of increasing complexity, as defined by the underlying information processing demands of the tasks. This scheme has been used to define what scores along a particular scale mean and to describe the survey results. Thus, it contributes to the construct validity of inferences based

[2] See Kirsch (2001) and Murray et al. (1997) for a more detailed description of the variables used in the IALS and subsequent assessments.

on scores from the measure (Messick 1989). Data from the surveys' background questionnaires have demonstrated consistent correlations between the literacy levels and social and economic outcomes, providing additional evidence for the validity of this particular scheme.

Advancing Messick's approach to construct-based assessment through the application of this framework development model has been one important contribution of the large-scale literacy surveys. This approach not only was used for each of these literacy assessments, but also has become an accepted practice in other assessment programs including the Organisation for Economic Co-operation and Development's (OECD's) Programme for International Student Achievement (PISA) and the United Nations Educational, Scientific, and Cultural Organization's (UNESCO's) Literacy Assessment and Monitoring Programme (LAMP).

Employing this model across the literacy assessments both informed the test development process and allowed ETS researchers to explore variables that explained differences in performance. Research based on data from the early adult literacy assessments led to an understanding of the relationship between the print materials that adults use in their everyday lives and the kinds of tasks they need to accomplish using such materials. Prior difficulty models for both assessments and learning materials tended to focus on the complexity of stimulus materials alone. ETS's research focused on both the linguistic features and the structures of prose and document materials, as well as a range of variables related to task demands.

Analyses of the linguistic features of stimulus materials first identified the important distinction between continuous and noncontinuous texts. Continuous texts (the prose materials used in the assessments) are composed of sentences that are typically organized into paragraphs. Noncontinuous texts (document materials) are more frequently organized in a matrix format, based on combinations of lists. Work by Mosenthal and Kirsch (1991) further identified a taxonomy of document structures that organized the vast range of matrix materials found in everyday life—television schedules, checkbook registers, restaurant menus, tables of interest rates, and so forth—into six structures: simple, combined, intersecting, and nested lists; and charts and graphs. In prose materials, analyses of the literacy data identified the impact of features such as the presence or absence of graphic organizers including headings, bullets, and bold or italicized print.

On the task side of the difficulty equation, these analyses also identified strategies required to match information in a question or directive with corresponding information in prose and document materials. These strategies—locate, cycle, integrate, and generate—in combination with text features, helped explain what made some tasks more or less difficult than others (Kirsch 2001). For example, locate tasks were defined as those that required respondents to match one or more features of information stated in the question to either identical or synonymous information in the stimulus. A locate task could be fairly simple if there was an exact match between the requested information in the question or directive and the wording in the stimulus and if the stimulus was relatively short, making the match easy to find.

IMPATIENS

Like many other cultured plants, impatiens plants have a long history behind them. One of the older varieties was sure to be found on grandmother's windowsill. Nowadays, the hybrids are used in many ways in the house and garden.

Origin: The ancestors of the impatiens, *Impatiens sultani* and *Impatiens holstii*, are probably still to be found in the mountain forests of tropical East Africa and on the islands off the coast, mainly Zanzibar. The cultivated European plant received the name *Impatiens walleriana*.

Appearance: It is a herbaceous bushy plant with a height of 30 to 40 cm. The thick, fleshy stems are branched and very juicy, which means, because of the tropical origin, that the plant is sensitive to cold. The light green or white speckled leaves are pointed, elliptical, and slightly indented on the edges. The smooth leaf surfaces and the stems indicate a great need of water.

Bloom: The flowers, which come in all shades of red, appear plentifully all year long, except for the darkest months. They grow from "suckers" (in the stem's "armpit").

Assortment: Some are compact and low-growing types, about 20 to 25 cm. high, suitable for growing in pots. A variety of hybrids can be grown in pots, window boxes, or flower beds. Older varieties with taller stems add dramatic colour to flower beds.

General care: In summer, a place in the shade without direct sunlight is best; in fall and spring, half-shade is best. When placed in a bright spot during winter, the plant requires temperatures of at least 20°C; in a darker spot, a temperature of 15°C will do. When the plant is exposed to temperatures of 12-14°C, it loses its leaves and won't bloom anymore. In wet ground, the stems will rot.

Watering: The warmer and lighter the plant's location, the more water it needs. Always use water without a lot of minerals. It is not known for sure whether or not the plant needs humid air. In any case, do not spray water directly onto the leaves, which causes stains.

Feeding: Feed weekly during the growing period from March to September.

Repotting: If necessary, repot in the spring or in the summer in light soil with humus (prepacked potting soil). It is better to throw the old plants away and start cultivating new ones.

Propagating: Slip or use seeds. Seeds will germinate in ten days.

Diseases: In summer, too much sun makes the plant woody. If the air is too dry, small white flies or aphids may appear.

Question 1: According to the article, what do the smooth leaf surfaces and the stems suggest about the plant?

____

____

Fig. 9.3 Sample prose task

As an example, see Fig. 9.3. Here there is an exact match between "the smooth leaf surfaces and the stems" in the question and in the last sentence in the second paragraph of the text.

Analyses showed that the difficulty of locate tasks increased when stimuli were longer and more complex, making the requested information more difficult to locate; or when there were distractors, or a number of plausible correct answers, within the text. Difficulty also increased when requested information did not exactly match the text in the stimulus, requiring respondents to locate synonymous information. By studying and defining the interaction between the task demands for locate, cycle, integrate, and generate tasks and features of various stimuli, the underlying information-processing skills could be more clearly understood. This research allowed for improved assessment design, increased interpretability of results, and development of derivative materials, including individual assessments[3] and instructional materials.[4]

In 1994, the literacy assessments moved from a national to an international focus. The primary goal of the international literacy assessments—International Adult Literacy Survey (IALS), ALL, and PIAAC—was to collect comparable international data that would provide a broader understanding of literacy across industrialized nations.

One challenge in meeting the goal of ensuring comparability across different national versions of the assessment was managing the translation process. Based on the construct knowledge gained from earlier assessments, it was clear that translators had to understand critical features of both the stimulus materials and the questions. Training materials and procedures were developed to help translators and project managers from participating countries reach this understanding. For example, the translation guidelines for the content shown in Fig. 9.3 specified the following:

- Translation must maintain literal match between the key phrase "the smooth leaf surfaces and the stems" in the question and in the last sentence in the second paragraph of the text.
- Translation must maintain a synonymous match between *suggest* in question and *indicate* in text.

Understanding task characteristics and the interaction between questions and stimulus materials allowed test developers to create precise translation guidelines to ensure that participating countries developed comparable versions of the assessment instruments. The success of these large-scale international efforts was in large part possible because of the construct knowledge gained from ETS research based on the results of earlier national assessments.

9.3 Expanding and Implementing Large-Scale Assessment Methodology

The primary purpose of the adult literacy large-scale assessments has been to describe the distribution of literacy skills in populations, as well as in subgroups within and across populations. The assessments have not targeted the production of

[3] These individual assessments include the Test of Applied Literacy Skills (TALS), a paper-and-pencil assessment with multiple forms; the *PDQ Profile*™ Series, an adaptive computer-based assessment of literacy proficiency; and the Health Activities Literacy Test, an adaptive computer-based assessment of literacy tasks focusing on health issues.

[4] Using information from this research, ETS developed P.D.Q. Building Skills for Using Print in the early 1990s. This multi-media, group-based system includes more than 100 h of instruction focusing on prose, document, and quantitative literacy, as well as workbooks and instructional support materials.

scores for individual test takers, but rather employed a set of specialized design principles and statistical tools that allow a reliable and valid description of skill distributions for policy makers and other stakeholders. To describe skills in a comparable manner in international contexts, the methodologies utilized needed to ensure that distributions were reported in terms of quantities that describe differences on scales across subgroups in meaningful ways for all participating entities.

The requirement to provide comparable estimates of skill distributions has been met by using the following methodological tools:

- Models that allow the derivation of comparable measures across populations and comparisons across literacy assessments
- Survey methodologies that provide representative samples of respondents
- Procedures to ensure scoring accuracy and to handle missing data
- Forward-looking designs that take advantage of context information in computer-based assessments

Taken together, these methodological tools facilitate the measurement goal of providing reliable, valid, and comparable estimates of skill distributions based on large-scale literacy assessments.

9.3.1 Models Allowing the Derivation of Comparable Measures and Comparisons Across Literacy Assessments

The goal of the literacy assessments discussed here has been to provide a description of skills across a broad range of ability, particularly given that the assessments target adults who have very different educational backgrounds and a wider range of life experiences than school-based populations. Thus the assessments have needed to include tasks that range from very easy to very challenging. To enable comparisons across a broad range of skill levels and tasks, the designs for all of the adult literacy assessments have used “incomplete block designs”. In such designs, each sampled individual takes a subset of the complete assessment. The method of choice for the derivation of comparable measures in incomplete block designs is based on measurement models that were developed for providing such measures in the analyses of test data (Lord 1980; Rasch 1960). These measurement models are now typically referred to as item response theory (IRT) models (Lord and Novick 1968).

IRT models are generally considered superior to simpler approaches based on sum scores, particularly in the way omitted responses and incomplete designs can be handled. Because IRT uses the full information contained in the set of responses, these models are particularly useful for assessment designs that utilize a variety of item types arranged in blocks that cannot be set up to be parallel forms of a test. Incomplete block designs do not allow the comparison of sum scores of aggregated responses because different blocks of items may vary in difficulty and even in the number of items. IRT models establish a comparable scale on which items from different blocks, and from respondents taking different sets of items, can be located, even in sparse incomplete designs. These models are powerful tools to evaluate whether the information provided for each individual item is comparable across populations of interest (see, for example, Yamamoto and Mazzeo 1992). In particular, the linking procedures typically used in IRT have been adapted, refined, and generalized for use in international assessments of adult literacy. More specifically, recent developments in IRT linking methods allow a more flexible approach to the alignment of scales that takes into account local deviations (Glas and Verhelst 1995; Yamamoto 1998; von Davier and von Davier 2007; Oliveri and von Davier 2011; Mazzeo and von Davier 2014; Glas and Jehangir 2014). The approach applied in IALS, ALL and PIAAC enables international assessments to be linked across a large number of common items while allowing for a small subset of items in each country to function somewhat differently to eliminate bias due to occasional item-by-country interactions. IRT has been the measurement method of choice not only for ETS's adult literacy assessments, but also for national and international assessments of school-age students such as the National Assessment of Educational Progress (NAEP), PISA, and Trends in International Mathematics and Science Study (TIMSS).

The integration of background information is a second important characteristic of the analytical methodologies used in the adult literacy assessments. Background data are used for at least two purposes in this context. First and foremost, they provide information about the relationship between demographic variables and skills. This makes it possible to investigate how the distribution of skills is associated with variables including educational attainment, gender, occupation, and immigration status of groups. These are among the variables needed to answer questions that are of interest to policy makers and other stakeholders, such as, "How are skills distributed in immigrant vs. nonimmigrant populations?" and "What is the relationship between literacy skills and measures of civic engagement such as voting?" In addition, background data provide auxiliary information that can be used to improve the precision of the skills measurement. This use of background data is particularly important because the available background data can help alleviate the effects of limited testing time for respondents by using the systematic differences between groups of respondents to strengthen the estimation of skills.[5]

While one of the main aims of ETS's large-scale literacy assessments has been to provide data on human capital at any given point in time, the extent to which skills change over time is also of fundamental interest. IRT models provide a powerful tool to link assessments over cycles conducted in different years. In much the same way that IRT allows linking of scales and provides comparable measures across blocks of different items within an assessment, and across countries, IRT can also be used to link different assessments over time. This link is only possible because significant efforts have been made across the literacy assessments to collect data in a manner that supports reusing sets of items over time while regularly renew-

[5]The interested reader is referred to Mislevy et al. (1992) for a description of this approach and to von Davier et al. (2006) for an overview and a description of recent improvements and extensions of the approach.

ing the item pool. The particular design principles applied ensure that new and previously used blocks of items are combined into test booklets in such a way that each assessment is also connected to multiple assessments over time. Because IRT estimation methods have been developed and extended to facilitate analyses of incomplete designs, these methods are particularly well suited to analyze multiple links across assessments. Statistical tools can be used to evaluate whether the items used repeatedly in multiple assessments are indeed comparable across assessments from different years and provide guidance as to which items to retain and which parts of the assessment have to be renewed by adding new task material.

9.3.2 Survey Methodologies That Provide Representative Samples of Respondents

The description of populations with respect to policy-relevant variables requires that members of the population of interest are observed with some positive probability. While it is not a requirement (or possibility) to assess every individual, a representative sample has to be drawn in order to provide descriptions of populations without bias. The adult literacy assessments have typically used methods common to household surveys, in which either a central registry of inhabitants or a list of addresses of dwellings/households of a country is used to randomly draw a representative random sample of respondents. This list is then used to select an individual at random, get in contact with those selected and ask the selected individual to participate in the survey. To account for unequal chances of being selected, the use of sampling weights is necessary. The importance of sampling and weighting for an accurate estimate of skill distributions is discussed in more detail in contributions summarizing analytic strategies involving sampling and weights for large-scale assessments by Rust (2014) and Rutkowski et al. (2010).

One particular use of these survey methodologies in large-scale assessments, and a contribution of ETS's adult assessments, is the projection of skill distributions based on expected changes in the population. The report, *America's Perfect Storm: Three Forces Changing Our Nation's Future* (Kirsch et al. 2007) shows how evidence regarding skill distributions in populations of interest can be projected to reflect changes in those populations, allowing a prediction of the increase or decline of human capital over time.

9.3.3 Procedures to Ensure Scoring Accuracy

One measurement issue that has been addressed in large-scale literacy assessments is the need to ensure that paper-and-pencil (as well as human-scored computer-based) tasks are scored accurately and reliably, both within and across countries participating in the international surveys. Many of the assessment tasks require respondents to provide short, written responses that typically range in length from single-word responses to short phrases or sentences. Some tasks ask for responses to be marked on the stimulus. On paper, respondents may be asked to circle or underline the correct answer whereas on the computer, respondents may be required to mark or highlight the response using the mouse or another input device. So while responses are typically quite short, scorers in all participating countries must follow a well-developed set of scoring rules to ensure consistent scoring. All of the adult literacy surveys prior to PIAAC were conducted as paper-and-pencil assessments, scored by national teams of trained scorers. While PIAAC is largely a computer-based assessment using automated scoring, a paper-and-pencil component has been retained, both to strengthen the link between modes and to provide an option for respondents without the requisite technical skills to complete the assessment on the computer. To ensure reliable and comparable data in all of the adult literacy surveys, it was critical that processes were developed to monitor the accuracy of human scoring for the short constructed responses in that mode within a country, across countries, and across assessments over time.

Without accurate, consistent and internationally comparable scoring of paper-and-pencil items, all subsequent psychometric analyses of these items would be severely jeopardized. For all of the large-scale adult literacy assessments, the essential activities associated with maintaining scoring consistency have been basically the same. Having items scored independently by two different scorers and then comparing the resulting scores has been the key required procedure for all participating countries. However, because the number of countries and number of languages has increased with each international assessment, the process has been refined over time. In IALS, the procedure used to ensure standardized scoring involved an exchange of booklets across countries with the same or similar languages. Country A and Country B thus would score their own booklets; then Country A would second score Country B's booklets and vice versa. In cases where a country could not be paired with another testing in the same language, the scorers within one country would be split into two independent groups, and booklets would be exchanged across groups for rescoring.

Beginning with ALL, the use of anchor booklets was introduced. This common set of booklets was prepared by test developers and distributed to all countries. Item responses in these booklets were based on actual responses collected in the field as well as responses that reflected key points on which scorers were trained. Because responses were provided in English, scoring teams in each country designated two bilingual scorers responsible for the double-scoring process. Anchor booklets were used in PIAAC as well. The new aspect introduced in PIAAC was the requirement that countries follow a specified design to ensure that each booklet was scored twice and that scorers functioned both as first and second scorer across all of the booklets. Figure 9.4 shows the PIAAC design for countries that employed three scorers. The completed booklets were divided up into 18 bundles of equal size. Bundle 0 was the set of anchor booklets to be scored by bilingual Scorers 1 and 2.

In an ideal world, the results of these double-scoring procedures would confirm that scoring accuracy was 100% and that scorers were perfectly consistent with each

	Bundle																		
Scorer	1	2	3	4	5	6	0	7	8	9	10	11	12	13	14	15	16	17	18
1	A	B		B	A		A	A	B		B	A		A	B		B	A	
2	B		A	A		B	B	B		A	A		B	B		A	A		B
3		A	B		B	A			A	B		B	A		A	B		B	A

Fig. 9.4 Double-scoring design for PIAAC. Cells marked with "A" represent the first scorer for each bundle

other. Although this level of consistency is never obtained due to random deviations, scoring accuracy in the adult literacy surveys tends to be around 96%.

When scoring discrepancies occur, experience has shown that they fall into two distinct classes. The first type of discrepancy reveals a consistent bias on the part of one scorer, for example when one scorer is consistently more lenient than others. Because countries are required to send rescoring data for analysis at set points during the scoring process, when this situation is found, problematic scorers must be retrained or, in some cases, dismissed.

The second type of discrepancy that can be revealed through analysis of the rescoring data is more challenging to address. This occurs when the scoring results reveal general inconsistencies between the scorers, with no pattern that can be attributed to one scorer or the other. This issue has been relatively rare in the adult literacy assessments. When it has occurred, it is generally the result of a problem with an item or an error in the scoring guides. One procedure for addressing this situation includes conducting a review of all inconsistently scored responses to determine if there is a systematic pattern and, if one is found, having those items rescored. Additionally, the scoring guides for such items can be revised to clarify any issue identified as causing inconsistent scoring. When a specific problem cannot be identified and resolved, model based adjustments such as assigning unique item parameters to account for this type of country-by-item deviation may be required for one or more countries to reflect this ambiguity in scoring.

9.3.4 Statistical Procedures for Handling Missing Data

A second key methodological issue developed through experience with the large-scale literacy assessments involves the treatment of missing data due to nonresponse. Missing responses reduce the amount of information available in the cognitive assessment and thus can limit the kinds of inferences that can be made about the distribution of skills in the population based on a given set of respondents. More specifically, the relationship between skills and key background characteristics is not measured well for respondents with a high proportion of item nonresponse. This issue has been addressed in the large-scale literacy assessments by estimating conditioning coefficients based on the performance of respondents with sufficient cognitive information and applying the parameters to those respondents for whom there is insufficient performance data. This solution allows stable estimation of the model and ensures that regression of performance data on background variables is based on cases that provide sufficiently accurate information.

The two most common but least desirable ways to treat missing cases are a) to ignore them and b) to assume all missing responses can be equated to incorrect responses. Ignoring missing responses is acceptable if one can assume that missing cases occur at random and that the remaining observed cases are representative of the target population. In this case, the result would be slightly larger standard errors due to reduced sample size, and the other estimates would remain unbiased. Randomly missing data rarely occur in real data collections, however, especially in surveys of performance. If the incidence of nonresponse varies for major subgroups of interest, or if the missing responses are related to the measurement objective— in this case, the measurement of literacy skills—then inferring the missing data from observed patterns results in biased estimates. If one can be sure that all missingness is due to a lack of skill, the treatment as incorrect is justified. This treatment may be appropriate in high-stakes assessments that are consequential for respondents. In surveys, however, the respondent will not be subjected to any consequences, so other reasons for missingness, such as a lack of motivation, may be present.

To address these issues, different approaches have been developed. In order to infer reasons for nonresponse, participants are classified into two groups based on standardized coding schemes used by interviewers to record reasons for nonparticipation: those who stop the assessment for literacy-related issues (e.g., reading difficulty, native language other than language of the assessment, learning disability) and those who stop for reasons unrelated to literacy (e.g., physical disability, refusal for unspecified reason). Special procedures are used to impute the proficiencies of individuals who complete fewer than the minimum number of tasks needed to estimate their proficiencies directly.

When individuals cite a literacy-related reason for not completing the cognitive items, this implies that they were unable to respond to the items. On the other hand, citing a reason unrelated to literacy implies nothing about a person's literacy proficiency. When an individual responds to fewer than five items per scale— the minimum number needed to directly estimate proficiencies—cases are treated as follows:

- If the individual cited a literacy-related reason for not completing the assessment, then all consecutively missing responses at the end of a block of items are scored as wrong.
- If the individual cited a reason unrelated to literacy, then all consecutively missing responses at the end of block are treated as not reached.

A respondent's proficiency is calculated from a posterior distribution that is the product of two functions: a conditional distribution of proficiency, given responses to the background questionnaire; and a likelihood function of proficiency, given responses to the cognitive items (see Murray et al. 1997, for more detail). By scoring missing responses as incorrect for individuals citing literacy-related reasons for stopping the assessment, the likelihood function is very peaked at the lower end of the scale—a result that is believed to accurately represent their proficiency.

Because PIAAC was a computer-based assessment, information was available to further refine the scoring rules for non-response. The treatment of item level missing data in paper-and-pencil assessments largely has to rely on the position of items. In order to define the reason for not responding as either volitional or being based on having never been exposed to (not reached) the items, the location of the 'last' item for which a response was observed is crucial. In computer-based assessments, non-response can be treated in a more sophisticated way by taking timing data and process information into account. While the problem of rapid guessing has been described in high-stakes assessment (Wise and DeMars 2005), the nature of literacy surveys does not compel respondents to guess, but rather to skip an item rapidly for some reasons that may be unrelated to skills, for example perceived time pressure or a lack of engagement. If an item was skipped in this way – a rapid move to the next item characterized by a very short overall time spent on the item (e.g., less than 5 s) and the minimal number of actions sufficient to 'skip' the item, PIAAC applied a coding of 'not reached/not administered' (OECD 2013; Weeks et al. 2014). If, however a respondent spent time on an item, or showed more than the minimum number of actions, a missing response would be assumed to be a volitional choice and counted as not correct.

9.3.5 *Forward-Looking Design for Using Context Information in Computer-Based Assessments*

The methodologies used in large-scale assessments are well developed, and variants of essentially these same methodologies are used in all major large-scale literacy assessments. While this repeated use implies that the current methodology is well suited for the analyses of assessments at hand, new challenges have arisen with the advent of PIAAC.

As a computer-based assessment, PIAAC presents two important advantages—and challenges—when compared to earlier paper-and-pencil assessments. First is the wealth of data that a computer can provide in terms of process information. Even seemingly simple information such as knowing precisely how much time a respondent spent on a particular item can reveal important data that were never available in the paper-and-pencil assessments. The use of such data to refine the treatment of non-response data, as described above, is one example of how this information can improve measurement. Second is the opportunity to design adaptive assessments that change the selection of items depending on a respondent's performance on previous sets of items. These differences result in both new sources of information about the performance of respondents and a change in the structure of the cognitive response data given that not all test takers respond to the same set of items.

Modern psychometric methodologies are available that can improve estimation in the face of such challenges. Such methods can draw upon process and navigation data to classify respondents (Lazarsfeld and Henry 1968) with respect to the typical paths they take through scenario-based tasks, such as the ones in PIAAC's problem-solving domain. Extensions of IRT models can reveal whether this or other types of classifications exist besides the skills that respondents apply (Mislevy and Verhelst 1990; Rost 1990; von Davier and Carstensen 2007; von Davier and Rost 1995; von Davier and Yamamoto 2004; Yamamoto 1989). Additional information such as response latency can be used to generate support variables that can be used for an in-depth analysis of the validity of responses. Rapid responders (DeMars and Wise 2010) who may not provide reliable response data can potentially be identified using this data. Nonresponse models (Glas and Pimentel 2008; Moustaki and Knott 2000; Rose et al. 2010) can be used to gain a deeper understanding of situations in which certain types of respondents tend not to provide any data on at least some of the items. Elaborate response-time models that integrate latency and accuracy (Klein Entink et al. 2009; Lee 2008) can be integrated with current large-scale assessment methodologies.

9.4 Linking Real-Life Stimulus Materials and Innovative Item Types

From the first adult literacy assessment onward, items have been based on everyday materials taken from various adult situations and contexts including the workplace, community, and home. In the 1993 National Adult Literacy Survey (NALS), for example, sets of open-ended questions required respondents to use a six-page newspaper that had been created from articles, editorials, and advertisements taken from real newspapers. In PIAAC, simulation tasks were based on content from real websites, advertisements, and e-mails. For each of the large-scale literacy assessments, original materials were used in their entirety, maintaining the range of wording, formatting, and presentation found in the source. The inclusion of real-life materials both increased the content validity of the assessments and improved respondent motivation, with participants commenting that the materials were both interesting and appropriate for adults.

Each of the large-scale literacy assessments also used open-ended items. Because they are not constrained by an artificial set of response options, these open-ended tasks allowed respondents to engage in activities that are similar to those they might perform if they encountered the materials in real life. In the paper-and-pencil literacy assessments, a number of different open-ended response types were employed. These included asking respondents to underline or circle information in the stimulus, copy or paraphrase information in the stimulus, generate a response, and complete a form.

With the move to computer-based tasks in PIAAC, new ways to collect responses were required. The design for PIAAC called for the continued use of open-ended response items, both to maintain the real-life simulation focus of the assessment and to maintain the psychometric link between PIAAC and prior surveys. While the paper-and-pencil surveys allowed respondents to compose answers ranging from a word or two to several sentences, the use of automated scoring for such responses was not possible, given that PIAAC was delivered in 33 languages. Instead, the response modes used for computer-based items in this assessment included highlighting, clicking, and typing numeric responses—all of which could be scored automatically. Throughout previous paper-and-pencil assessments, there had always been some subset of respondents who marked their responses on the stimulus rather than writing answers on the provided response lines. These had been considered valid answers, and scoring rubrics had been developed to train scorers on how such responses should be scored. Thus electronic marking of text by highlighting a phrase or sentence or clicking on a cell in a table fit within existing scoring schemes. Additionally, previous work on a derivative computer-based test for individuals, the PDQ Profile Series, had shown that item parameters for paper-and-pencil items adapted from IALS and ALL were not impacted when those items were presented on the computer and respondents were asked to highlight, click, or type a numeric response. PIAAC thus became the first test to employ these response modes on a large scale and in an international context.

Taking advantage of the computer-based context, PIAAC also introduced new types of simulation items. In reading literacy, items were included that required respondents to use scrolling and hyperlinks to locate text on a website or provide responses to an Internet poll. In the new problem-solving domain, tasks were situated in simulated web, e-mail, and spreadsheet environments that contained common functionality for these environments. Examples of these new simulation tasks included items that required respondents to access information in a series of e-mails and use that information to schedule meeting rooms via an online reservation system or to locate requested information in a complex spreadsheet where the spreadsheet environment included "find" and "sort" options that would facilitate the task.

In sum, by using real-life materials and open-ended simulation tasks, ETS's large-scale literacy assessments have sought to reflect and measure the range of literacy demands faced by adults in order to provide the most useful information to policy makers, researchers, and the public. Over time, the nature of the assessment materials and tasks has been expanded to reflect the changing nature of literacy as the role of technology has become increasingly prevalent and important in everyday life.

9.5 Developing Extensive Background Questionnaires to Link Performance With Experience and Outcome Variables

One important goal of the large-scale literacy assessments has been to relate skills to a variety of demographic characteristics and explanatory variables. Doing so has allowed ETS to investigate how performance is related to social and educational outcomes and thereby interpret the importance of skills in today's society. It has also enhanced our understanding of factors related to the observed distribution of literacy skills across populations and enabled comparisons with previous surveys.

For each of the literacy assessments, respondents completed a background questionnaire in addition to the survey's cognitive measures. The background questions were a significant component of each survey, taking up to one-third of the total survey time. In each survey, the questionnaire addressed the following broad issues:

- General language background
- Educational background and experience
- Labor force participation
- Literacy activities (types of materials read and frequency of use for various purposes)
- Political and social participation
- Demographic information

As explained earlier, information collected in the background questionnaires is used in the psychometric modeling to improve the precision of the skills measurement. Equally importantly, the background questionnaires provide an extensive database that has allowed ETS to explore questions such as the following: What is the relationship between literacy skills and the ability to benefit from employer-supported training and lifelong learning? How are educational attainment and literacy skills related? How do literacy skills contribute to health and well being? What factors may contribute to the acquisition and decline of skills across age cohorts? How are literacy skills related to voting and other indices of social participation? How do reading practices affect literacy skills?

The information collected via the background questionnaires has allowed researchers and other stakeholders to look beyond simple demographic information and examine connections between the skills being measured in the assessments and important personal and social outcomes. It has also led to a better understanding of factors that mediate the acquisition or decline of skills. At ETS, this work has provided the foundation for reports that foster policy debate on critical literacy issues. Relevant reports include Kirsch et al. (2007), Rudd et al. (2004) and Sum et al. (2002, 2004).

9.6 Establishing Innovative Reporting Procedures to Better Integrate Research and Survey Data

Reports for each of the large-scale surveys have gone beyond simply reporting distributions of scores on the assessment for each participating country. As noted above, using information from the background questionnaire has made it possible to link performance to a wide range of demographic variables. Other reporting innovations have been implemented to make the survey data more useful and understandable for policy makers, researchers, practitioners, and other stakeholders.

The PIAAC data, conjointly with IALS and ALL trend data, are available in the Data Explorer (http://piaacdataexplorer.oecd.org/ide/idepiaac/), an ETS-developed web-based analysis and reporting tool that allows users to query the PIAAC database and produce tabular and graphical summaries of the data. This tool has been designed for a wide range of potential users, including those with little or no statistical background. By selecting and organizing relevant information, stakeholders can use the large-scale data to address questions of importance to them.

In addition to linking performance and background variables, survey reports have also looked at the distribution of literacy skills and how performance is related to underlying information-processing skills. Reports have included item maps that present sample items in each domain, showing where these items performed on the literacy scale and discussing features of the stimuli and questions that impact difficulty. Such analyses have allowed stakeholders to understand how items represent the construct and thereby allow them to generalize beyond the pool of items in any one assessment. These reports were also designed to provide readers with a better understanding of the information-processing skills underlying performance. Such an understanding has important implications for intervention efforts.

9.7 Conclusion

During the 30 years over which the six large-scale adult literacy assessments have been conducted, literacy demands have increased in terms of the types and amounts of information adults need to manage their daily lives. The goal of the assessments has been to provide relevant information to the variety of stakeholders interested in the skills and knowledge adults have and the impact of those skills on both individuals and society in general. Meeting such goals in this ever-changing environment has required that ETS take a leading role in the following:

- Expanding the construct of literacy
- Developing a model for building construct-based assessments
- Expanding and implementing large-scale assessment methodology to ensure reliable, valid, and comparable measurement across countries and over time
- Taking an approach to test development that focuses on the use of real-life materials and response modes that better measure the kinds of tasks adults encounter in everyday life[6]
- Developing extensive background questionnaires that make it possible to link performance with experience and outcome variables, thereby allowing the survey data to address important policy questions
- Developing reporting procedures that better integrate survey data with research

These efforts have not just expanded knowledge of what adults know and can do; they have also made important contributions to understanding how to design, conduct, and report the results of large-scale international assessments.

---

[6] Sample PIAAC items are available at http://www.oecd.org/skills/piaac/samplequestionsandquestionnaire.htm.

Appendix: Description of the Five Levels for Prose, Document, and Numeracy Domains

	Prose	Document	Numeracy
Level 1 (0–225)	Most of the tasks in this level require the respondent to read a relatively short text to locate a single piece of information that is identical to or synonymous with the information given in the question or directive. If plausible but incorrect information is present in the text, it tends not to be located near the correct information.	Tasks in this level tend to require the respondent either to locate a piece of information based on a literal match or to enter information from personal knowledge onto a document. Little, if any, distracting information is present.	Tasks in this level require the respondent to show an understanding of basic numerical ideas by completing simple tasks in concrete, familiar contexts where the mathematical content is explicit with little text. Tasks consist of simple, one-step operations such as counting, sorting dates, performing simple arithmetic operations, or understanding common and simple percentages such as 50%.
Level 2 (226–275)	Some tasks in this level require respondents to locate a single piece of information in the text; however, several distractors or plausible but incorrect pieces of information may be present, or low-level inferences may be required. Other tasks require the respondent to integrate two or more pieces of information or to compare and contrast easily identifiable information based on a criterion provided in the question or directive.	Tasks in this level are more varied than those in level 1. Some require the respondents to match a single piece of information; however, several distractors may be present, or the match may require low-level inferences. Tasks in this level may also ask the respondent to cycle through information in a document or to integrate information from various parts of a document.	Tasks in this level are fairly simple and relate to identifying and understanding basic mathematical concepts embedded in a range of familiar contexts where the mathematical content is quite explicit and visual with few distractors. Tasks tend to include one-step or two-step processes and estimations involving whole numbers, interpreting benchmark percentages and fractions, interpreting simple graphical or spatial representations, and performing simple measurements.

(continued)

	Prose	Document	Numeracy
Level 3 (276–325)	Tasks in this level tend to require respondents to make literal or synonymous matches between the text and information given in the task, or to make matches that require low-level inferences. Other tasks ask respondents to integrate information from dense or lengthy text that contains no organizational aids such as headings. Respondents may also be asked to generate a response based on information that can be easily identified in the text. Distracting information is present but is not located near the correct information.	Some tasks in this level require the respondent to integrate multiple pieces of information from one or more documents. Others ask respondents to cycle through rather complex tables or graphs that contain information that is irrelevant or inappropriate to the task.	Tasks in this level require the respondent to demonstrate understanding of mathematical information represented in a range of different forms, such as in numbers, symbols, maps, graphs, texts, and drawings. Skills required involve number and spatial sense; knowledge of mathematical patterns and relationships; and the ability to interpret proportions, data, and statistics embedded in relatively simple texts where there may be distractors. Tasks commonly involve undertaking a number of processes to solve problems.
Level 4 (326–375)	These tasks require respondents to perform multiple-feature matches and to integrate or synthesize information from complex or lengthy passages. More complex inferences are needed to perform successfully. Conditional information is frequently present in tasks at this level and must be taken into consideration by the respondent.	Tasks in this level, like those at the previous levels, ask respondents to perform multiple-feature matches, cycle through documents, and integrate information; however, they require a greater degree of inference. Many of these tasks require respondents to provide numerous responses but do not designate how many responses are needed. Conditional information is also present in the document tasks at this level and must be taken into account by the respondent.	Tasks at this level require respondents to understand a broad range of mathematical information of a more abstract nature represented in diverse ways, including in texts of increasing complexity or in unfamiliar contexts. These tasks involve undertaking multiple steps to find solutions to problems and require more complex reasoning and interpretation skills, including comprehending and working with proportions and formulas or offering explanations for answers.

(continued)

	Prose	Document	Numeracy
Level 5 (376–500)	Some tasks in this level require the respondent to search for information in dense text that contains a number of plausible distractors. Others ask respondents to make high-level inferences or use specialized background knowledge. Some tasks ask respondents to contrast complex information.	Tasks in this level require the respondent to search through complex displays that contain multiple distractors, to make high-level text-based inferences, and to use specialized knowledge.	Tasks in this level require respondents to understand complex representations and abstract and formal mathematical and statistical ideas, possibly embedded in complex texts. Respondents may have to integrate multiple types of mathematical information, draw inferences, or generate mathematical justification for answers.

References

DeMars, C. E., & Wise, S. L. (2010). Can differential rapid-guessing behavior lead to differential item functioning? *International Journal of Testing, 10*(3), 207–229. https://doi.org/10.1080/15305058.2010.496347

Glas, C. A. W., & Jehangir, K. (2014). Modeling country specific differential item functioning. In L. Rutkowski, M. von Davier, & D. Rutkowski (Eds.), *Handbook of international large scale assessment* (pp. 97–116). New York: Chapman & Hall

Glas, C. A. W., & Pimentel, J. (2008). Modeling nonignorable missing data in speeded tests. *Educational and Psychological Measurement, 68*, 907–922. https://doi.org/10.1177/0013164408315262

Glas, C. A. W., & Verhelst, N. D. (1995). Testing the Rasch model. In G. H. Fischer & I. W. Molenaar (Eds.), *Rasch models: Foundations, recent developments, and applications* (pp. 69–96). New York: Springer.

Kirsch, I. S. (2001). *The international adult literacy survey (IALS): Understanding what was measured* (Research Report No. RR-01-25). Princeton: Educational Testing Service. http://dx.doi.org/10.1002/j.2333-8504.2001.tb01867.x

Kirsch, I. S., & Jungeblut, A. (1986). *Literacy: Profiles of America's young adults* (NAEP Report No. 16-PL-01). Princeton: Educational Testing Service.

Kirsch, I. S., Braun, H., Yamamoto, K., & Sum, A. (2007). *America's perfect storm: Three forces changing our nation's future* (Policy Information Report). Princeton: Educational Testing Service.

Klein Entink, R. H., van der Linden, W. J., & Fox, J.-P. (2009). A Box-Cox normal model for response times. *British Journal of Mathematical and Statistical Psychology, 62*, 621–640. https://doi.org/10.1348/000711008X374126

Lazarsfeld, P. F., & Henry, N. W. (1968). *Latent structure analysis*. Boston: Houghton Mifflin.

Lee, M. D. (2008). Three case studies in the Bayesian analysis of cognitive models. *Psychonomic Bulletin & Review, 15*, 1–15. https://doi.org/10.3758/PBR.15.1.1

Lord, F. M. (1980). *Applications of item response theory to practical testing problems*. Hillsdale: Erlbaum.

Lord, F. M., & Novick, M. R. (1968). *Statistical theories of mental test scores*. Reading: Addison-Wesley.

Mazzeo, J., & von Davier, M. (2014). Linking scales in international large-scale assessments. In L. Rutkowski, M. von Davier, & D. Rutkowski (Eds.), *Handbook of international large scale assessment* (pp. 229–258). New York: Chapman & Hall.

Messick, S. (1989). Validity. In R. Linn (Ed.), *Educational measurement* (3rd ed., pp. 13–103). New York: Macmillian.

Messick, S. (1994). The interplay of evidence and consequences in the validation of performance assessments. *Educational Researcher, 23*(1), 13–23. https://doi.org/10.3102/0013189X023002013

Mislevy, R. J., & Verhelst, N. (1990). Modeling item responses when different subjects employ different solution strategies. *Psychometrika, 55*, 195–215. https://doi.org/10.1007/BF02295283

Mislevy, R. J., Beaton, A. E., Kaplan, B., & Sheehan, K. M. (1992). Estimating population characteristics from sparse matrix samples of item responses. *Journal of Educational Measurement, 29*, 133–161. https://doi.org/10.1111/j.1745-3984.1992.tb00371.x

Mosenthal, P. B., & Kirsch, I. S. (1991). Toward an explanatory model of document process. *Discourse Processes, 14*, 147–180. https://doi.org/10.1080/01638539109544780

Moustaki, I., & Knott, M. (2000). Weighting for item non-response in attitude scales using latent variable models with covariates. *Journal of the Royal Statistical Society, Series A, 163*, 445–459. https://doi.org/10.1111/1467-985X.00177

Murray, T. S., Kirsch, I. S., & Jenkins, L. B. (Eds.). (1997). *Adult literacy in OECD countries: Technical report on the first international adult literacy survey*. Washington, DC: National Center for Education Statistics.

OECD. (2012). *Literacy, numeracy and problem solving in technology-rich environments: Framework for the OECD survey of adult skills*. Paris: OECD Publishing. http://dx.doi.org/10.1787/9789264128859-en

OECD. (2013). Technical report of the Survey of Adult Skills (PIAAC). Retrieved from https://www.oecd.org/skills/piaac/_Technical%20Report_17OCT13.pdf

Oliveri, M. E., & von Davier, M. (2011). Investigation of model fit and score scale comparability in international assessments. *Psychological Test and Assessment Modeling, 53*, 315–333.

Rasch, G. (1960). *Probabilistic models for some intelligence and attainment tests*. Copenhagen: Danish Institute for Educational Research.

Rose, N., von Davier, M., & Xu, X. (2010). *Modeling non-ignorable missing data with IRT* (Research Report No. RR-10-11). Princeton: Educational Testing Service. http://dx.doi.org/10.1002/j.2333-8504.2010.tb02218.x

Rost, J. (1990). Rasch models in latent classes: An integration of two approaches to item analysis. *Applied Psychological Measurement, 3*, 271–282. https://doi.org/10.1177/014662169001400305

Rudd, R., Kirsch, I., & Yamamoto, K. (2004). *Literacy and health in America* (Policy Information Report). Princeton: Educational Testing Service.

Rust, K. (2014). Sampling, weighting, and variance estimation in international large scale assessments. In L. Rutkowski, M. von Davier, & D. Rutkowski (Eds.), *Handbook of international large scale assessment* (pp. 117–154). New York: Chapman & Hall.

Rutkowski, L., Gonzalez, E., Joncas, M., & von Davier, M. (2010). International large-scale assessment data: Issues in secondary analysis and reporting. *Educational Researcher, 39*, 142–151. https://doi.org/10.3102/0013189X10363170

Sum, A., Kirsch, I. S., & Taggart, R. (2002). *The twin challenges of mediocrity and inequality: Literacy in the U.S. from an international perspective*. Princeton: Educational Testing Service.

Sum, A., Kirsch, I. S., & Yamamoto, K. (2004). *A human capital concern: The literacy proficiency of U.S. immigrants*. Princeton: Educational Testing Service.

von Davier, M., & Carstensen, C. (Eds.). (2007). *Multivariate and mixture distribution Rasch models*. New York: Springer. https://doi.org/10.1007/978-0-387-49839-3

von Davier, M., & Rost, J. (1995). Polytomous mixed Rasch models. In G. H. Fischer & I. W. Molenaar (Eds.), *Rasch models: Foundations, recent developments, and applications* (pp. 371–382). New York: Springer. https://doi.org/10.1007/978-1-4612-4230-7_20

von Davier, M., & von Davier, A. (2007). A unified approach to IRT scale linking and scale transformation. *Methodology, 3*, 115–124. https://doi.org/10.1027/1614-2241.3.3.115

von Davier, M., & Yamamoto, K. (2004, October). *A class of models for cognitive diagnosis*. Invited lecture at the ETS Spearman invitational conference, Philadelphia, PA.

von Davier, M., Sinharay, S., Oranje, A., & Beaton, A. (2006). Statistical procedures used in the National Assessment of Educational Progress (NAEP): Recent developments and future directions. In C. R. Rao & S. Sinharay (Eds.), *Handbook of statistics: Vol. 26. Psychometrics* (pp. 1039–1056). Amsterdam: Elsevier.

Weeks, J., von Davier, M., & Yamamoto, K. (2014). Design considerations for the Programme for International Student Assessment. In L. Rutkowski, M. von Davier, & D. Rutkowski (Eds.), *Handbook of international large scale assessment* (pp. 259–276). New York: Chapman & Hall.

Wise, S. L., & DeMars, C. E. (2005). Low examinee effort in low-stakes assessment: Problems and potential solutions. *Educational Assessment, 10*, 1–17.

Yamamoto, K. (1989). *HYBRID model of IRT and latent class model* (Research Report No. RR-89-41). Princeton: Educational Testing Service. http://dx.doi.org/10.1002/j.2333-8504.1982.tb01326.x

Yamamoto, K. (1998). Scaling and scale linking. In T. S. Murray, I. S. Kirsch, & L. B. Jenkins (Eds.), *Adult literacy in OECD countries: Technical report on the First International Adult Literacy Survey* (pp. 161–178). Washington, DC: National Center for Education Statistics.

Yamamoto, K., & Mazzeo, J. (1992). Item response theory scale linking in NAEP. *Journal of Educational Statistics, 17*, 155–173. https://doi.org/10.2307/1165167

Chapter 10
Modeling Change in Large-Scale Longitudinal Studies of Educational Growth: Four Decades of Contributions to the Assessment of Educational Growth

Donald A. Rock

ETS has had a long history of attempting to at least minimize, if not solve, many of the longstanding problems in measuring change (cf. Braun and Bridgeman 2005; Cronbach and Furby 1970; Rogosa 1995) in large-scale panel studies. Many of these contributions were made possible through the financial support of the Longitudinal Studies Branch of the U.S. Department of Education's National Center for Education Statistics (NCES). The combination of financial support from the Department of Education along with the content knowledge and quantitative skills of ETS staff over the years has led to a relatively comprehensive approach to measuring student growth. The present ETS model for measuring change argues for (a) the use of adaptive tests to minimize floor and ceiling effects, (b) a multiple-group Bayesian item response theory (IRT) approach to vertical scaling, which takes advantage of the adaptive test's potential to allow for differing ability priors both within and between longitudinal data waves, and (c) procedures for not only estimating how much an individual gains but also identifying where on the vertical scale the gain takes place. The latter concept argues that gains of equivalent size may well have quite different interpretations. The present model for change measurement was developed over a number of years as ETS's experience grew along with its involvement in the psychometric analyses of each succeeding NCES-sponsored national longitudinal study. These innovations in the measurement of change were not due solely to a small group of ETS staff members focusing on longitudinal studies, but also profited considerably from discussions and research solutions developed by the ETS NAEP group. The following historical summary recounts ETS's role in NCES's sequence of longitudinal studies and how each study contributed to the final model for measuring change.

D.A. Rock (✉)
Educational Testing Service, Princeton, NJ, USA
e-mail: donaldR706@aol.com

R.E. Bennett, M. von Davier (eds.), *Advancing Human Assessment*,
Methodology of Educational Measurement and Assessment,
DOI 10.1007/978-3-319-58689-2_10

For the purposes of this discussion, we will define large-scale longitudinal assessment of educational growth as data collections from national probability samples with repeated and direct measurements of cognitive skills. NCES funded these growth studies in order to develop longitudinal databases, which would have the potential to inform educational policy at the national level. In order to inform educational policy, the repeated waves of testing were supplemented with the collection of parent, teacher, and school process information. ETS has been or is currently involved in the following large-scale longitudinal assessments, ordered from the earliest to the most recent:

- The National Longitudinal Study of the High School Class of 1972 (NLS-72)
- High School and Beyond (HS&B 1980–1982), sophomore and senior cohorts
- The National Education Longitudinal Study of 1988 (NELS:88)
- The Early Childhood Longitudinal Studies (ECLS):
  - Early Childhood Longitudinal Study, Kindergarten Class of 1998–1999 (ECLS-K)
  - Early Childhood Longitudinal Study, Birth Cohort of 2001 (ECLS-B)
  - Early Childhood Longitudinal Study, Kindergarten Class of 2010–2011 (ECLS-K:2011)

We discuss the NLS-72 study briefly here, even though it is the only study in the list above that that does not meet one of the criteria we stated as part of our definition of large-scale, longitudinal assessment: Specifically, it does not include direct repeated cognitive measures across succeeding waves of data collection. While it was longitudinal with respect to educational attainment among post-high school participants, its shortcomings with respect to measuring change in developing cognitive skills led NCES to require the succeeding large-scale longitudinal studies to have direct repeated measures of cognitive skills. NCES and educational policy experts felt that the inclusion of repeated direct measures of cognitive skills would greatly strengthen the connection between school processes and cognitive growth. The reader should keep in mind that, while the notion of *value added* (Braun 2006) had not yet achieved its present currency, there was considerable concern about assessing the impact of selection effects on student outcomes independent of school and teaching practices. One way, or at least the first step in addressing this concern, was to measure *change* in cognitive skills during the school years. More specifically, it was hoped that measuring cognitive achievement at a relevant point in a student's development and again at a later date would help assess the impact on student growth of educational inputs and policies such as public versus private education, curriculum paths, tracking systems, busing of students across neighborhoods, and dropout rates.

As one progresses from the first to last of the above studies there was an evolutionary change in: (a) *what should be measured,* (b) *how it was measured,* and (c) *when it was measured.* The following summary of each of the studies will detail the evolution in both ETS's and NCES's thinking in each of these three dimensions, which in the end led to ETS's most recent thinking on measuring change in cognitive skills. Obviously, as the contracting agency, NCES and its policy advisors had the final say on what was measured and when it was measured. Although ETS's main role was to provide input on development, administration, and scoring of specific cognitive measures, psychometric findings from each succeeding large-scale longitudinal assessment informed decisions with respect to all three areas. While this paper records ETS's involvement in NCES longitudinal studies, we would be remiss not to mention our partners' roles in these studies. Typically, ETS had responsibility for the development of cognitive measures and psychometric and scaling analysis as a subcontractor to another organization that carried out the other survey activities. Specifically, ETS partnered with the Research Triangle Institute (RTI) on NLS-72 and ECLS-B, the National Opinion Research Center (NORC) on HS&B, NELS-88, and Phase I of ECLS-K, and Westat on ECLS-K Phases II-IV and ECLS-K:2011.

10.1 National Longitudinal Study of 1972 (NLS-72)

NCES has referred to NLS-72 as the "grandmother of the longitudinal studies" (National Center for Education Statistics [NCES] 2011, para. 1). When the NLS-72 request for proposals was initiated, principals at NCES were Dennis Carroll, who later became the director of longitudinal studies at NCES; and William Fetters and Kenneth Stabler, NCES project managers. NCES asked bidders responding to its NLS-72 request for proposals to submit plans and budgets for sample design, data collection, and the development and scoring of the instruments. Unlike succeeding longitudinal studies, NCES awarded a single organization (ETS) the contract including base-year sample design, data collection, and instrument design and scoring; NCES did not repeat this practice on succeeding longitudinal studies. In all future bidding on longitudinal study contracts, NCES welcomed, and in fact strongly preferred, that the prime contractor not undertake all the various components alone but instead assemble consortia of organizations with specific expertise in the various survey components. We would like to think that ETS's performance on this contract had little or no bearing on the change in contracting philosophy at NCES. It was, however, true that we did not have, at the time, in-house expertise in sampling design and operational experience in collecting data on a national probability sample.

At any rate, ETS had the winning bid under the direction of Tom Hilton of the Developmental Research division and Hack Rhett from the Program Direction area. Hilton's responsibilities included insuring the alignment of the cognitive measures, and to a lesser extent the other performance measures, with the long term goals of the study. Rhett's responsibilities were primarily in the operational areas and included overseeing the data collection, data quality, and scoring of the instruments.

The primary purpose of NLS-72 was to create a source of data that researchers could use to relate student achievement and educational experiences to postsecondary educational and occupational experiences. An earlier survey of educational policymakers and researchers suggested a need for student data on educational experiences that could be related to their post-secondary occupational/educational decisions and performance. Given time and budget constraints, it was decided that a battery of cognitive measures given in the spring of the senior year could provide a reasonable summary of a student's knowledge just prior to leaving high school. Limited information about school policies and processes were gathered from a school questionnaire, a student record document, a student questionnaire, and a counselor questionnaire. Unlike succeeding NCES longitudinal studies, NLS-72 provided only indirect measures of classroom practices and teacher qualifications since there was no teacher questionnaire. Indirect measures of teacher behavior were gathered from parts of the school and student questionnaire. The base-year student questionnaire included nine sections devoted to details about the student's plans and aspirations with respect to occupational/educational decisions, vocational training, financial resources, and plans for military service. This emphasis on post-secondary planning reflected the combined interest of the stakeholders and Dennis Carroll of NCES.

Five follow-ups were eventually carried out, documenting the educational attainment and occupational status (and, in some cases, performance) of individuals sampled from the high school class of 1972. In a publication released by NCES, NLS-72 is described as "probably the richest archive ever assembled on a single generation of Americans" (NCES 2011, para. 1). The publication goes on to say, "The history of the Class of 72 from its high school years through its early 30s is widely considered as the baseline against which the progress and achievements of subsequent cohorts will be measured" (NCES 2011, para 3). ETS was not directly involved in the five follow-up data collections. The primary contractor on the five follow-ups that tracked the post-graduate activities was the Research Triangle Institute (RTI); see, for example, Riccobono et al. (1981).

The NLS-1972 base-year national probability sample included 18,000 seniors in more than 1,000 public and nonpublic schools. In the larger schools, 18 students were randomly selected while in some smaller schools all students were assessed. Schools were selected from strata in such a way that there was an over-representation of minorities and disadvantaged students. The cognitive test battery included six measures: vocabulary, mathematics, reading, picture-number associations, letter groups, and mosaic comparisons. The battery was administered in a 69-min time period. Approximately 15,800 students completed the test battery. The reader should note that the battery included three nonverbal measures: picture-number associations (rote memory), letter groups (ability to apply general concepts), and mosaic comparisons (perceptual speed and accuracy). The inclusion of nonverbal measures seemed reasonable at the time since it was believed that: (a) the oversampled disadvantaged subpopulations could be hindered on the other language-loaded measures, and (b) a mixture of aptitude and achievement measures would give a more complete picture of the skills of students entering the workforce or post-high school education. It should be kept in mind that originally the primary goal of the NLS-72 battery was to enhance the prediction of career development choices and outcomes. The three aptitude measures were from the *Kit of Factor-Referenced Tests* developed by John French while at ETS (French 1964; Ekstrom et al. 1976). Subsequent NCES longitudinal studies dropped the more aptitude-based measures and focused more on repeated achievement measures. This latter approach was more appropriate for targeting school-related gains.

Part of ETS's contractual duties included scoring the base-year test battery. No new psychometric developments (e.g., item response theory) were used in the scoring; the reported scores on the achievement tests were simply number correct. Neither NCES nor the researchers who would use the public files could be expected to be familiar with IRT procedures under development at that time. Fred Lord's seminal book on applications of item response theory (Lord 1980) was yet to appear. As we will see later, the NLS-72 achievement tests were rescored using IRT procedures in order to put them on the same scale as comparable measures in the next NCES longitudinal study: High School and Beyond (Rock et al. 1985).

NLS-72 had lofty goals:

1. Provide a national picture of post-secondary career and educational decision making.
2. Show how these decisions related to student achievement and aptitude.
3. Contrast career decisions of subpopulations of interest.

However, as in the case of all comprehensive databases, it also raised many questions. It continued to fuel the public-versus-private-school debate that Coleman (1969), Coleman and Hoffer (1987), and subsequent school effects studies initiated. Once the comparable cognitive measures for high school seniors from three cohorts, NLS-72, HS&B first follow-up (1982), and NELS:88 second follow-up (1992), were placed on the same scale, generational trends in cognitive skills could be described and analyzed. Similarly, intergenerational gap studies typically began with NLS-72 and looked at trends in the gaps between groups defined by socioeconomic status, racial or ethnic identity, and gender groups and examined how they changed from 1972 to 1992 (Konstantopoulos 2006). Researchers analyzing NLS-72 data identified additional student and teacher information that would have been helpful in describing in-school and out-of-school processes that could be related to student outcomes. Based on the experience of having discovered these informational gaps in NLS-72, NCES called for an expanded student questionnaire and the addition of a parent questionnaire in the next NCES longitudinal study, High School and Beyond, in 1980–1982.

10.2 High School and Beyond (HS&B 1980–1982)

The NCES national education longitudinal survey called High School and Beyond (HS&B) was based on a national probability sample of 10th and 12th graders (often referred to in the literature as sophomores and seniors, respectively) in the same high schools during the spring of 1980. Two years later, in 1982, the students who were 10th graders in the initial survey were re-assessed as seniors. As in the NLS-72 survey, members of the 10th grade cohort (12th graders in 1982) were followed up in order to collect data on their post-secondary activities. The HS&B sample design was a two-stage stratified cluster design with oversampling of private and Catholic schools (Frankel et al. 1981). Thirty-six students were randomly selected from the 10th and 12th grade classes in each sampled school in 1980. HS&B was designed to serve diverse users and needs while attempting to collect data reasonably comparable to NLS-72. The oversampling of private and Catholic schools allowed for specific analysis by type of school. Although multi-level analysis (Raudenbush and Bryk 2002) had not yet been formally developed, the sample of 36 students in each class made this database particularly suitable for future multi-level school effectiveness studies. That is, having 36 students in each grade significantly enhanced the reliability of the within-school regressions used in multi-level analyses later on. The significant new contributions of HS&B as contrasted to NLS-72 were:

1. The repeated testing of cognitive skills for students in their 10th grade year and then again in their 12th grade year, allowing for the measurement of cognitive development. This emphasis on the measurement of change led to a move away from a more aptitude-related test battery to a more achievement-oriented battery in subsequent surveys.
2. The use of common items shared between NLS-72 and HS&B, making possible the introduction of IRT-based common item linking (Lord 1980) that allowed intergenerational contrasts between 12th graders in NLS-72 and 12th graders in HS&B-80 in mathematics and reading.
3. The expansion of the student questionnaire to cover many psychological and sociological concepts. In the past, NCES had considered such areas too risky and not sufficiently factual and/or sufficiently researched. This new material reflected the interests of the new outside advisory board consisting of many academicians along with support from Bill Fetters from NCES. It was also consistent with awarding the HS&B base-year contract to the National Opinion Research Center (NORC), which had extensive experience in measuring these areas.
4. The introduction of a parent questionnaire administered to a subsample of the HS&B sample. The inclusion of the parent questionnaire served as both a source of additional process variables as well as a check on the reliability of student self-reports.

The primary NCES players in HS&B were Dennis Carroll, then the head of the Longitudinal Studies Branch, William Fetters, Edith Huddleston, and Jeff Owings. Fetters prepared the original survey design. The principal players among the contractors were Steve Ingels at NORC who was the prime contractor for the base year and first follow-up study. Cognitive test development and psychometrics were ETS's responsibility, led by Don Rock and Tom Hilton. Tom Donlon played a major role in the selection of the cognitive test battery, and Judy Pollack carried out psychometric analyses with the advice and assistance of Fred Lord and Marilyn Wingersky.

The final selection of the HS&B test battery did not proceed as smoothly as hoped. ETS was given the contract to revise the NLS-72 battery. The charge was to replace some of the NLS-72 tests and items and add new items, yet make the HS&B scores comparable to those of the NLS-72 battery. ETS submitted a preliminary test plan that recommended that the letter groups, picture-number associations, and mosaic comparisons subtests be dropped from the battery. This decision was made because a survey of the users of the NLS-72 data tapes and the research literature suggested that these tests were little used. Donlon et al. suggested that science and a measure of career and occupational development be added to the HS&B 10th and 12th grade batteries. They also suggested adding a spatial relations measure to the 10th grade battery and abstract reasoning to the 12th grade battery. NCES accepted these recommendations; NORC field-tested these new measures. When the field test results were submitted to the National Planning Committee for HS&B, the committee challenged the design of the batteries (cf. Heyns and Hilton 1982). The committee recommended to NCES that:

> ...the draft batteries be altered substantially to allow for the measurement of school effects and cognitive change in a longitudinal framework. The concerns of the committee were twofold: First, conventional measures of basic cognitive skills are not designed to assess patterns of change over time, and there was strong feeling that the preliminary batteries would not be sufficiently sensitive to cognitive growth to allow analysis to detect differential effects among students. Second, the Committee recommended including items that would be valid measures of the skills or material a student might encounter in specific high school classes. (Rock et al. 1985, p. 27)

The batteries were then revised to make the HS&B 1980 12th grade tests a vehicle for measuring cross-sectional change from NLS-72 12th graders to HS&B 1980 12th graders. The HS&B 1980 12th grade test items were almost identical to those of NLS-72. The HS&B 1980 10th grade tests, however, were designed to be a baseline for the measurement of longitudinal change from the 10th grade to the 12th grade. The final HS&B 1980 10th grade test battery included vocabulary, reading, mathematics, science, writing, and civics education. With the possible exception of vocabulary, the final battery could be said to be more achievement-oriented than either the NLS-72 battery or the preliminary HS&B battery. The HS&B 1982 12th grade battery was identical to the HS&B-1980 10th grade battery. The purposes of the HS&B-1980 10th grade and 1982 12th grade test batteries were not just to predict post-secondary outcomes as in NLS-72, but also to measure school-related gains in achievement during the last 2 years of high school.

In 1983, NCES contracted with ETS to do a psychometric analysis of the test batteries for NLS-72 and both of the HS&B cohorts (1980 12th graders and 1980 10th graders who were 12th graders in 1982) to ensure the efficacy of:

1. Cross-sectional comparisons of NLS-72 12th graders with HS&B 12th graders.
2. The measurement of longitudinal change from the 10th grade year (HS&B 1980) to the 12th grade year (HS&B 1982).

This psychometric analysis was summarized in a comprehensive report (Rock et al. 1985) documenting the psychometric characteristics of all the cognitive measures as well as the change scores from the HS&B 1980 10th graders followed up in their 12th grade year.

ETS decided to use the three-parameter IRT model (Lord 1980) and the LOGIST computer program (Wood et al. 1976) to put all three administrations on the same scale based on common items spanning the three administrations. It is true that IRT was not necessarily required for the 10th grade to 12th grade gain-score analysis since these were identical tests. However, the crosswalk from NLS-72 12th graders to HS&B 1980 10th graders and then finally to HS&B 1982 12th graders became more problematic because of the presence of unique items, especially in the latter administration. There was one other change from NLS-72 to HS&B that argued for achieving comparability through IRT scaling, and that was the fact that NLS-72 12th graders marked an answer sheet while HS&B participants marked answers in the test booklet. As a result, HS&B test-takers attempted, on average, more items. This is not a serious problem operationally for IRT, which estimates scores based on items attempted and compensates for omitted items. Comparisons across cohorts were only done in reading and mathematics, which were present for all administrations. The IRT common crosswalk scale was carried out by pooling all test responses from all three administrations, with items not present for a particular administration treated as *not administered* for students in that particular cohort. Maximum likelihood estimates of number correct true scores were then computed for each individual.

For the longitudinal IRT scaling of the HS&B sophomore cohort tests, item parameters were calibrated separately for 10th graders and 12th graders and then transformed to the 12th grade scale. The HS&B 10th grade cohort science and writing tests were treated differently because of their shorter lengths. For the other tests, samples were used in estimating the pooled IRT parameters because the tests were sufficiently long to justify saving processing time and expense by selecting samples for item calibration. For the shorter science and writing tests, the whole sample was used.

With respect to the psychometric characteristics of the tests, it was found that:

1. The "sophomore tests were slightly more difficult than would be indicated by measurement theory" (Rock et al. 1985, p. 116). This was the compromise necessary because the same test was to be administered to 10th and 12th graders, and potential ceiling effects need to be minimized. Future longitudinal studies addressed this problem in different ways.
2. Confirmatory factor analysis (Joreskog and Sorbom 1996) suggested that the tests were measuring the same things with the same precision across racial/ethnic and gender groups.
3. Traditional estimates of reliability increased from the 10th grade to the 12th grade year in HS&B. Reliability estimates for IRT scores were not estimated. Reliability of IRT scores, however, would be estimated in subsequent longitudinal studies.
4. While the psychometric report argues that mathematics, reading, and science scores were sufficiently reliable for measuring individual change, they were borderline by today's criteria. Most of the subtests, with about 20 items each, had alpha coefficients between .70 and .80. The mathematics test, with 38 items, had

alpha coefficients close to .90 for the total group and most subgroups in both years, while the civics education subtest, with only 10 items, had reliabilities in the .50s, and was considered to be too low for estimating reliable individual change scores.

The HS&B experience taught us a number of lessons with respect to test development and methodological approaches to measuring change. These lessons led to significant changes in how students were tested in subsequent large-scale longitudinal studies. In HS&B, each student was administered six subject tests during a 69-min period, severely limiting the number of items that could be used, and thus the tests' reliabilities. Even so, there were those on the advisory committee who argued for subscores in mathematics and science. The amount of classroom time that schools would allow outside entities to use for testing purposes was shrinking while researchers and stakeholders on advisory committees increased their appetites for the number of things measured. NAEP's solution to this problem, which was just beginning to be implemented in the early 1980s, was to use sophisticated Bayesian algorithms to shrink individual scores towards their subgroup means, and then restrict reporting to summary statistics such as group means. The longitudinal studies approach has been to change the type of test administration in an attempt to provide individual scores that are sufficiently reliable that researchers can relate educational processes measured at the individual level with individual gain scores and/or gain trajectories. That is, ETS's longitudinal researchers' response to this problem was twofold: measure fewer things in a fixed amount of time, and develop procedures for measuring them more efficiently. ETS suggested that an adaptive test administration can help to increase efficiency by almost a factor of 2. That is, the IRT information function from an adaptive test can approximate that of a linear test twice as long. That is what ETS proposed for the next NCES longitudinal study.

ETS's longitudinal researchers also learned that maximum likelihood estimation (MLE) of item parameters and individual scores has certain limitations. Individuals with perfect or below-chance observed scores led to boundary condition problems, with the associated estimates of individual scores going to infinity. If we were to continue to use MLE estimation procedures, an adaptive test could help to minimize the occurrence of these problematic perfect and below-chance scores.

It is also the case that when the IRT procedures described in Lord (1980) first became popular, many applied researchers, policy stakeholders, members of advisory committees, and others got the impression that the weighted scoring in IRT would allow one to gather more reliable information in a shorter test. The fact was that solutions became very computationally unstable as the number of items became fewer in MLE estimation as used in the popular IRT program LOGIST (Wood et al. 1976). It was not until Bayesian IRT methods (Bock and Aiken 1981; Mislevy and Bock 1990) became available that stable solutions to IRT parameter estimation and scoring were possible for relatively short tests.

There is one other misconception that seems to be implicit, if not explicit, in thinking about IRT scoring—that is, the impression that IRT scores have the property of equal units along the score scale. This would be very desirable for the interpretation of gain scores. If this were the case, then a 2-point gain at the top of the test score scale would have a similar meaning with respect to progress as a 2-point gain at the bottom of the scale. This is the implicit assumption when gain scores from different parts of the test score scale are thrown in the same pool and correlated with process variables. For example, why would one expect a strong positive correlation between the number of advanced mathematics courses and this undifferentiated pool of mathematics gains? Gains at the lower end of the scale indicate progress in basic mathematics concepts while gains of an equivalent number of points at the top of the scale suggest progress in complex mathematical solutions. Pooling individual gains together and relating them to processes that only apply to gains at particular locations along the score scale is bound to fail and has little or nothing to do with the reliability of the gain scores. Policy makers who use longitudinal databases in an attempt to identify processes that lead to gains need to understand this basic measurement problem. Steps were taken in the next longitudinal study to develop measurement procedures to alleviate this concern.

10.3 The National Education Longitudinal Study of 1988 (NELS:88)

A shortcoming of the two longitudinal studies described above, NLS:72 and HS&B, is that they sampled students in their 10th or 12th-grade year of high school. As a result, at-risk students who dropped out of school before reaching their 10th or 12th-grade year were not included in the surveys. The National Education Longitudinal Study of 1988 (NELS:88) was designed to address this issue by sampling eighth graders in 1988 and then monitoring their transitions to later educational and occupational experiences. Students received a battery of tests in the eighth grade base year, and then again 2 and 4 years later when most sample members were in 10th and 12th grades. A subsample of dropouts was retained and followed up. Cognitive tests designed and scored by ETS were included in the first three rounds of data collection, in 1988, 1990, and 1992, as well as numerous questionnaires collecting data on experiences, attitudes, and goals from students, schools, teachers, and parents. Follow-ups conducted after the high school years as the students progressed to post-secondary education or entered the work force included questionnaires only, not cognitive tests. Transcripts collected from the students' high schools also became a part of this varied archive.

NELS:88 was sponsored by the Office of Educational Research and Improvement of the National Center for Education Statistics (NCES). NELS:88 was the third longitudinal study in the series of longitudinal studies supported by NCES and in which ETS longitudinal researchers participated. ETS's bidding strategy for the NELS:88 contract was to write a proposal for the test development, design of the testing procedure, and scoring and scaling of the cognitive tests. ETS's proposal was submitted as a subcontract with each of the competing prime bidders' proposals.

ETS continued to follow this bidding model for the next several longitudinal studies. Regardless of whom the prime contractor turned out to be, this strategy led to ETS furnishing considerable continuity, experience, and knowledge to the measurement of academic gain. The National Opinion Research Center (NORC) won the prime contract, and ETS was a subcontractor to NORC. Westat also was a subcontractor with responsibility for developing the teacher questionnaire. The contract monitors at NCES were Peggy Quinn and Jeff Owings, while Steven Ingels and Leslie Scott directed the NORC effort. Principals at ETS were Don Rock and Judy Pollack, aided by Trudy Conlon and Kalle Gerritz in test development. Kentaro Yamamoto at ETS also contributed very helpful advice in the psychometric area.

The primary purpose of the NELS:88 data collection was to provide policy-relevant information concerning the effectiveness of schools, curriculum paths, special programs, variations in curriculum content and exposure, and/or mode of delivery in bringing about educational growth (Rock et al. 1995; Scott et al. 1995). New policy-relevant information was available in NELS:88 with the addition of teacher questionnaires that could be directly connected with individual students. For the first time, a specific principal questionnaire was also included. Grades and course-taking history were collected in transcripts provided by the schools for a subset of students.

While the base-year (1988) sample consisted of 24,599 eighth graders, the first and second follow-up samples were smaller. As the base-year eighth graders moved on to high school, some high schools had a large number of sampled students, while others had only one or two. It would not have been cost effective to follow up on every student, which would have required going to thousands of high schools. Instead of simply setting a cutoff for retaining individual participants (e.g., only students in schools with at least ten sample members), individuals were followed up with varying probabilities depending on how they were clustered within schools. In this way, the representativeness of the sample could be maintained.

ETS test development under Trudy Conlon and Kalle Gerritz assembled an eighth-grade battery consisting of the achievement areas of reading comprehension, mathematics, science, and history/citizenship/geography. The battery was designed to measure school-related growth spanning a 4-year period during which most of the participants were in school. The construction of the NELS:88 eighth-grade battery was a delicate balancing act between several competing objectives—for example, general vs. specific knowledge and basic skills vs. higher-order thinking and problem solving. In the development of NELS:88 test items, efforts were made to take a middle road in the sense that our curriculum experts were instructed to select items that tapped the general knowledge that was found in most curricula but that typically did not require a great deal of isolated factual knowledge. The emphasis was to be on understanding concepts and measuring problem-solving skills (Rock and Pollack 1991; Ingels et al. 1993). However, it was thought necessary also to assess the basic operational skills (e.g., simple arithmetic and algebraic operations), which are the foundations for successfully carrying out the problem-solving tasks.

This concern with respect to developing tests that are sensitive to changes resulting from school related processes is particularly relevant to measuring change over relatively long periods of exposure to varied educational treatments. That is, the 2-year gaps between retesting coupled with a very heterogeneous student population were likely to coincide with considerable variability in course taking experiences. This fact, along with the constraints on testing time, made coverage of specific curriculum-related knowledge very difficult. Also, as indicated above, specificity in the knowledge being tapped by the cognitive tests could lead to distortions in the gain scores due to forgetting of specific details. The impact on gain scores due to forgetting should be minimized if the cognitive battery increasingly emphasizes general concepts and development of problem-solving abilities. This emphasis should increase as one goes to the tenth and twelfth grades. Students who take more high-level courses, regardless of the specific course content, are likely to increase their conceptual understanding as well as gain additional practice in problem-solving skills.

At best, any nationally representative longitudinal achievement testing program must attempt to balance testing-time burdens, the natural tensions between local curriculum emphasis and more general mastery objectives, and the psychometric constraints (in the case of NELS:88 in carrying out both vertical equating [year-to-year] and cross-sectional equating [form-to-form within year]). NELS:88, fortunately, did have the luxury of being able to gather cross-sectional pretest data on the item pools. Thus, we were able to take into consideration not only the general curriculum relevance but also whether or not the items demonstrated reasonable growth curves, in addition to meeting the usual item analysis requirements for item quality.

Additional test objectives included:

1. There should be little or no floor or ceiling effects. Tests should give every student the opportunity to demonstrate gain: some at the lower end of the scale and others making gains elsewhere on the scale. As part of the contract, ETS developed procedures for sorting out where the gain takes place.
2. The tests should be unspeeded.
3. Reliabilities should be high and the standard error of measurement should be invariant across ethnic and gender groups.
4. The comparable tests should have sufficient common items to provide crosswalks to HS&B tests.
5. The mathematics test should share common items with NAEP to provide a crosswalk to NAEP mathematics.
6. If psychometrically justified, the tests should provide subscale scores and/or proficiency levels, yet be sufficiently unidimensional as to be appropriate for IRT vertical scaling across grades.
7. The test battery should be administered within an hour and a half.

Obviously, certain compromises needed to be made, since some of the constraints are in conflict. In order to make the test reliable enough to support change-measurement within the time limits, adaptive testing had to be considered. It was decided that two new approaches would be introduced in the NELS:88 longitudinal study.

The first approach was the introduction of multi-stage adaptive testing (Cleary et al. 1968; Lord 1971) in Grade 10 and Grade 12. Theoretically, using adaptive tests would maximize reliability (i.e., maximize the expected IRT information function) across the ability distribution and do so with fewer items. Even more importantly, it would greatly minimize the potential for having floor and ceiling effects, the bane of all gain score estimations.

The second innovation was the identification of clusters of items identifying multiple proficiency levels marking a hierarchy of skill levels on the mathematics, reading comprehension, and science scales. These proficiency levels could be interpreted in much the same way as NAEP's proficiency levels, but they had an additional use in measuring gain: They could be used to pinpoint where on the scale the gain was taking place. Thus, one could tell not only *how much* a given student gained, but also *at what skill level* he or she was gaining. This would allow researchers and policymakers to select malleable factors that could influence gains at specific points (proficiency levels) on the scale. In short, this allowed them to match the educational process (e.g., taking a specific course), with the location on the scale where the maximum gain would be expected to be taking place.[1]

10.3.1 The Two-Stage Multilevel Testing in the NELS:88 Longitudinal Framework

The potentially large variation in student growth trajectories over a 4-year period argued for a longitudinal tailored testing approach to assessment. That is, to accurately assess a student's status both at a given point in time as well as over time, the individual tests must be capable of measuring across a broad range of ability or achievement. In the eighth-grade base year of NELS:88, all students received the same test battery, with tests designed to have broadband measurement properties. In the subsequent years, easier or more difficult reading and mathematics forms were selected according to students' performance in the previous years. A two-stage multilevel testing procedure was implemented that used the eighth-grade reading and mathematics test score results for each student to assign him or her to one of two forms in 10th-grade reading, and one of three forms in 10th grade mathematics, that varied in difficulty. If the student did very well (top 25%) on the eighth-grade

[1] The concept that score gains at different points on the scale should (a) be interpreted differently and (b) depending on that interpretation, be related to specific processes that affect that particular skill, has some intellectual forebears. For example, Cronbach and Snow (1977) described the frequent occurrence of aptitude-by-treatment interaction in educational pre-post test designs. We would argue that what they were observing was the fact that different treatments were necessary because they were looking for changes along different points on the aptitude scale. From an entirely different statistical perspective, Tukey, in a personal communication, once suggested that most if not all interactions can be reduced to nonsignificance by applying the appropriate transformations. That may be true operationally, but we might be throwing away the most important substantive findings.

mathematics test, he or she received the most difficult of the three mathematics forms in 10th grade; conversely, students scoring in the lowest 25% received the easiest form 2 years later. The remaining individuals received the middle form. With only two reading forms to choose from in the follow-up, the routing cut was made using the median of the eighth-grade scores. This branching procedure was repeated 2 years later, using 10th-grade performance to select the forms to be administered in 12th grade.

The 10th- and 12th-grade tests in reading and mathematics were designed to include sufficient linking items across grades, as well as across forms within grade, to allow for both cross-sectional and vertical scaling using IRT models. Considerable overlap between adjacent second-stage forms was desirable to minimize the loss of precision in case of any misassignment. If an individual were assigned to the most difficult second-stage form when he or she should have been assigned to the easiest form, then that student would not be well assessed, to say the least. Fortunately, we found no evidence for such two-level misclassifications. The science and history/citizenship/geography tests used the same relatively broad-ranged form for all students; linking items needed to be present only across grades.

To take advantage of this modest approach to paper-and-pencil adaptive testing, more recent developments in Bayesian IRT procedures (Mislevy and Bock 1990; Muraki and Bock 1991) were implemented in the first IRT analysis. The Bayesian procedures were able to take advantage of the fact that the adaptive procedure identified subpopulations, both within and across grades, who were characterized by different ability distributions. Both item parameters and posterior means were estimated for each individual at each point in time using a multiple-group version of PARSCALE (Muraki and Bock 1991), with updating of normal priors on ability distributions defined by grade and form within grade. PARSCALE does allow the shape of the priors to vary, but we have found that the smoothing that came from updating with normal ability priors leads to less jagged looking posterior ability distributions and does not over-fit items. It was our feeling that, often, lack of item fit was being absorbed in the shape of the ability distribution when the distribution was free to be any shape.

This procedure required the pooling of data as each wave was completed. This pooling often led to a certain amount of consternation at NCES, since item parameters and scores from the previous wave were updated as each new wave of data became available. In a sense, each wave of data remade history. However, this pooling procedure led to only very minor differences in the previous scores and tended to make the vertical scale more internally consistent. In most cases, it is best to use all available information in the estimation, and this use is particularly true in longitudinal analysis where each additional wave adds new supplementary information on item parameters and individual scores. The more typical approach fixes the linking item parameter values from the previous wave, but this procedure tends to underestimate the score variances in succeeding waves, contributing to the typical finding of a high negative correlation between initial status and gain.

It should be kept in mind that the multiple-group PARSCALE finds those item parameters that maximize the likelihood across all groups (in this case, forms):

seven in mathematics (one base-year form; three alternative forms in each follow-up), five in reading (two alternative forms per follow-up), and three each in science and history/citizenship/geography (one form per round). The version of the multiple-group PARSCALE used at that time only saved the subpopulation means and standard deviations and not the individual expected a posteriori (EAP) scores. The individual EAP scores, which are the means of their posterior distributions of the latent variable, were obtained from the NAEP B-group conditioning program, which uses the Gaussian quadrature procedure. This variation is virtually equivalent to conditioning (e.g., see Mislevy et al. 1992, as well as Barone and Beaton, Chap. 8, and Kirsch et al., Chap. 9, in this volume) on a set of dummy variables defining from which ability subpopulation an individual comes.

In summary, this procedure finds the item parameters that maximize the likelihood function across all groups (forms and grades) simultaneously. The items can be put on the same vertical scale because of the linking items that are common to different forms across years, or adjacent forms within year. Using the performance on the common items, the subgroup means can be located along the vertical scale. Individual ability scores are not estimated in the item parameter estimation step; only the subgroup means and variances are estimated. Next, NAEP's B-group program was used to estimate the individual ability scores as the mean of an individual's posterior distribution. (A detailed technical description of this procedure may be found in Rock et al. 1995). Checks on the goodness of fit of the IRT model to the observed data were then carried out.

Item traces were inspected to ensure a good fit throughout the ability range. More importantly, estimated proportions correct by item by grade were also estimated in order to ensure that the IRT model was both reproducing the item P-plus values and that there was no particular bias in favor of any particular grade. Since the item parameters were estimated using a model that maximizes the goodness-of-fit across the subpopulations, including grades, one would not expect much difference here. When the differences were summed across all items for each test, the maximum discrepancy between observed and estimated proportion correct for the whole test was .7 of a scale score point for Grade 12 mathematics, whose score scale had a range of 0 to 81. The IRT estimates tended to slightly underestimate the observed proportions. However, no systematic bias was found for any particular grade.

10.3.2 Criterion-Referenced Proficiency Levels

In addition to the normative interpretations in NELS:88 cognitive tests, the reading, mathematics, and science tests also provided criterion-referenced interpretations. The criterion-referenced interpretations were based on students demonstrating proficiencies on clusters of four items that mark ascending points on the test score scale. For example, there are three separate clusters consisting of four items each in reading comprehension that mark the low, middle, and high end of the reading scale. The items that make up these clusters exemplify the skills required to successfully answer the typical item located at these points along the scale. There were three levels in the reading comprehension test, five in the mathematics test, and three in the science test. Specific details of the skills involved in each of the levels may be found in Rock et al. (1995).

10.3.3 Criterion-Referenced Scores

There were two kinds of criterion-referenced proficiency scores reported in NELS:88 dichotomous scores and probability scores.

In the case of a dichotomous score, a 1 indicates mastery of the material in a given cluster of items marking a point on the scale, while a 0 implies nonmastery. A student was defined to be proficient at a given proficiency level if he or she got at least three out of four items correct that marked that level. Items were selected for a proficiency level if they shared similar cognitive processing demands and this cognitive demand similarity was reflected in similar item difficulties. Test developers were asked to build tests in which the more difficult items required all the skills of the easier items plus at least one additional higher level skill. Therefore, in the content-by-process test specifications, variation in item difficulty often coincided with variation in process. This logic leads to proficiency levels that are hierarchically ordered in the sense that mastery of the highest level among, for example, three levels implies that one would have also mastered the lower two levels. A student who mastered all three levels in reading had a proficiency score pattern of [1 1 1]. Similarly, a student who had only mastered the first two levels, but failed to answer at least three correct on the third level, had a proficiency score pattern of [1 1 0]. Dichotomous scores were not reported for students who omitted items that were critical to determining a proficiency level or who had reversals in their proficiency score pattern (a failed level followed by a passed level, such as 0 0 1). The vast majority of students did fit the hierarchical model; that is, they had no reversals.

Analyses using the dichotomous proficiency scores included descriptive statistics that showed the percentages of various subpopulations who demonstrated proficiencies at each of the hierarchical levels. They can also be used to examine patterns of change with respect to proficiency levels. An example of descriptive analysis using NELS:88 proficiency levels can be found in Rock et al. (1993).

The second kind of proficiency score is the probability of being proficient at each of the levels. These probabilities were computed using all of the information provided by students' responses on the whole test, not just the four-item clusters that marked the proficiency levels. After IRT calibration of item parameters and student ability estimates (thetas had been computed), additional *superitems* were defined marking each of the proficiency levels. These superitems were the dichotomous scores described above. Then, holding the thetas fixed, item parameters were calibrated for each of the superitems, just as if they were single items. Using these item parameters in conjunction with the students' thetas, probabilities of proficiency were computed for each proficiency level.

The advantages of the probability of being proficient at each of the levels over the dichotomous proficiencies are that (a) they are continuous scores and thus more powerful statistical methods may be applied, and (b) probabilities of being proficient at each of the levels can be computed for any individual who had a test score in a given grade, not only the students who answered enough items in a cluster. The latter advantage is true since the IRT model enables one to estimate how students would perform on those items that they were not given, for example, if the items were on a different form or not given in that grade.

The proficiency probabilities are particularly appropriate for relating specific processes to changes that occur at different points along the score scale. For example, one might wish to evaluate the impact of taking advanced mathematics courses on changes in mathematics achievement from Grade 10 to Grade 12. One approach to doing this evaluation would be to subtract every student's 10th-grade IRT-estimated number right from his or her 12th grade IRT-estimated number right and correlate this difference with the number of advanced mathematics courses taken between the 10th and 12th grades. The resulting correlation will be relatively low because lower achieving individuals taking no advanced mathematics courses are also gaining, *but probably at the low end of the test score scale.* Individuals who are taking advanced mathematics courses are making their greatest gains at the higher end of the test score scale. To be more concrete, let us say that the individuals who took none of the advanced math courses gained, on average, three points, all at the low end of the test score scale. Conversely, the individuals who took the advanced math courses gained three points, but virtually all of these individuals made their gains at the upper end of the test score scale. When the researcher correlates number of advanced courses with gains, the fact that, on average, the advanced math takers gained the same amount as those taking no advanced mathematics courses will lead to a very small or zero correlation between gain and specific processes (e.g., advanced math course taking). This low correlation has nothing to do with reliability of gain scores, but it has much to do with where on the test score scale the gains are taking place. Gains in the upper end of the test score distribution reflect increases in knowledge in advanced mathematical concepts and processes while gains at the lower end reflect gains in basic arithmetical concepts. In order to successfully relate specific processes to gains, one has to match the process of interest to where on the scale the gain is taking place.

The proficiency probabilities do this matching because they mark ascending places on the test score scale. If we wish to relate the number of advanced math courses taken to changes in mathematics proficiency, we should look at changes at the upper end of the test score distribution, not at the lower end, where students are making progress in more basic skills. There are five proficiency levels in mathematics, with Level 4 and Level 5 marking the two highest points along the test score scale. One would expect that taking advanced math courses would have its greatest impacts on changes in probabilities of being proficient at these highest two levels. Thus, one would simply subtract each individual's tenth grade probability of being

Table 10.1 Reliability of theta

	Baseyear	First follow-up	Second follow-up
Reading	.80	.86	.85
Math	.89	.93	.94
Science	.73	.81	.82
History/citizenship/geography	.84	.85	.85

proficient at, say, Level 4 from the corresponding probability of being proficient at Level 4 in 12th grade. Now, every individual has a continuous measure of change in mastery of advanced skills, not just a broadband change score. If we then correlate this change in Level 4 probabilities with the number of advanced mathematics courses taken, we will observe a substantial increase in the relationship between change and process (number of advanced mathematics courses taken) compared with change in the broad-band measure. We could do the same thing with the Level 5 probabilities as well. The main point here is that certain school processes, in particular course-taking patterns, target gains at different points along the test score distribution. It is necessary to match the type of school process we are evaluating with the location on the test score scale where the gains are likely to be taking place and then select the proper proficiency levels for appropriately evaluating that impact. For an example of the use of probability of proficiency scores to measure mathematics achievement gain in relation to program placement and course taking, see Chapter 4 of Scott et al. (1995).

10.3.4 Psychometric Properties of the Adaptive Tests Scores and the Proficiency Probabilities Developed in NELS:88

This section presents information on the reliability and validity of the adaptive test IRT (EAP) scores as well as empirical evidence of the usefulness of the criterion-referenced proficiency probabilities in measuring change. Table 10.1 presents the reliabilities of the thetas for the four tests. As expected, the introduction of the adaptive measures in Grades 10 and 12 lead to substantial increases in reliability. These IRT-based indices are computed as 1 minus the ratio of the average measurement error variance to the total variance.

The ETS longitudinal researchers moved from MLE estimation using LOGIST to multigroup PARSCALE and finally to NAEP's B-Group conditioning program for EAP estimates of theta and number-right true scores. The B-Group conditioning was based on ability priors associated with grade and test form. A systematic comparison was carried out among these competing scoring procedures. One of the reasons for introducing adaptive tests and Bayesian scoring procedures was to increase the accuracy of the measurement of gain by reducing floor and ceiling effects and thus enhance the relationships of test scores with relevant policy variables.

Table 10.2 Evaluation of alternative test scoring procedures for estimating gains in mathematics and their relationship with selected background/policy variables

Gains in theta metric	Any math last 2 years	Taking math now	Curriculum acad = 1; Gen/Voc = 0
Gain 8–10 LOG	0.07	0.06	0.06
Gain 8–10 STI	0.11	0.11	0.15
Gain 8–10 ST4	0.08	0.06	0.07
Gain 10–12 LOG	0.07	0.15	0.06
Gain 10–12 ST1	0.14	0.23	0.14
Gain10–12 ST4	0.10	0.18	0.06
Total gain LOG	0.12	0.18	0.11
Total gain ST1	0.19	0.26	0.22
Total gain ST4	0.14	0.18	0.10

Note. LOG = LOGIST, ST1 = NALS 1-step, ST4 = NAEP 4-step method

Table 10.3 Correlations between gains in proficiency at each mathematics level and mathematics course taking (no. of units), average grade, and precalculus course-taking

8th–12th grade gains in proficiency/ probabilities at each level in math	No. of units	Average grade	Precalculus Yes = 1; No = 0
Math level 1	−0.26	−0.28	−0.20
Math level 2	−0.01	−0.20	−0.20
Math level 3	0.22	0.05	−0.02
Math level 4	0.44	0.46	0.29
Math level 5	0.25	0.38	0.33

Table 10.2 presents a comparison of the relationships between MLE estimates and two Bayesian estimates with selected outside policy variables.

Inspection of Table 10.2 indicates that in the theta metric, the normal prior Bayesian procedure (ST1) shows stronger relationships between gains and course-taking than do the other two procedures. The differences in favor of ST1 are particularly strong where contrasts are being made between groups quite different in their mathematics preparation, for example, the relationship between being in the academic curriculum or taking math now and total gain.

When the correlations are based on the *number correct true score metric* (NCRT), the ST1 Bayesian approach still does as well or better than the other two approaches. The NCRT score metric is a nonlinear transformation of the theta scores, computed by adding the probabilities of a correct answer for all items in a selected item pool. Unlike the theta metric, the NCRT metric does not stretch out the tails of the score distribution. The stretching out at the tails has little impact on most analyses where group means are used. However, it can distort gain scores for individuals who are in or near the tails of the distribution. Gains in proficiency probabilities at each proficiency level and their respective correlations with selected process variables are shown in Table 10.3. The entries in Table 10.3 demonstrate the importance of relating specific processes with changes taking place at appropriate points along the score distribution.

Inspection of Table 10.3 indicates that gains between 8th and 12th grade in the probability of being proficient at Level 4 show a relatively high positive correlation with number of units of mathematics (.44) and with average grade in mathematics (.46). The changes in probability of mastery at each mathematics level shown in Table 10.3 are based on the ST1 scoring system.

When the dummy variable contrasting whether an individual took precalculus courses was correlated with gains in probabilities at the various proficiency levels, one observes negative correlations for demonstrated proficiencies at the two lower levels (simple operations and fractions and decimals) and higher positive correlations for Levels 4–5. That is, individuals with a score of 1 on the dummy variable, indicating they took precalculus courses, are making progressively greater gains in probabilities associated with mastery of Levels 4–5. As another example of the relation between scale region and educational process, students in the academic curriculum versus the general/vocational curriculum tend to have high positive correlations with changes in proficiency probabilities marking the high end of the scale. Conversely, students in the general/vocational curriculum tend to show positive correlations with gains in proficiency probabilities marking the low end of the scale. Other patterns of changes in lower proficiency levels and their relationship to appropriate process variables may be found in Rock et al. (1985).

10.3.5 Four New Approaches in Longitudinal Research

What did the ETS longitudinal studies group learn from NELS:88? Four new approaches were introduced in this longitudinal study. First, it was found that even a modest approach to adaptive testing improved measurement throughout the ability range and minimized floor and ceiling effects. Improved measurement led to significantly higher reliabilities as the testing moved from the 8th grade to more adaptive procedures in the 10th and 12th grades. Second, the introduction of the Bayesian IRT methodology with separate ability priors on subgroups of students taking different test forms, and/or in different grades, contributed to a more well-defined separation of subgroups both across and within grades. Third, on the advice of Kentaro Yamomoto, it became common practice in longitudinal research to pool and update item parameters and test scores as each succeeding wave of data was added. This pooling led to an internally consistent vertical scale across testing administrations. Last, we developed procedures that used criterion-referenced points to locate where on the vertical scale an individual was making his or her gains. As a result, the longitudinal researcher would have two pieces of information for each student: how much he or she gained in overall scale score points and where on the scale the gain took place. Changes in probabilities of proficiency at selected levels along the vertical scale could then be related to the appropriate policy variables that reflect learning at these levels.

While the above psychometric approaches contributed to improving longstanding problems in the measurement of change, there was still room for improvement.

For example, real-time two-stage adaptive testing would be a significant improvement over that used in the NELS:88 survey, where students' performance 2 years earlier was used to select test forms. Such an approach would promise a better fit of item difficulties to a student's ability level. This improvement would wait for the next NCES longitudinal study: The Early Childhood Longitudinal Study - Kindergarten Class of 1998–1999 (ECLS-K).

10.4 Early Childhood Longitudinal Study—Kindergarten Class of 1998–1999 (ECLS-K)

The Early Childhood Longitudinal Study, Kindergarten Class of 1998–1999 (ECLS-K) was sponsored by NCES and focused on children's school and home experiences beginning in fall kindergarten and continuing through 8th grade. Children were assessed in the fall and spring of kindergarten (1998–1999), the fall and spring of 1st grade (1999–2000), the spring of 3rd grade (2002), the spring of 5th grade (2004), and finally spring of 8th grade (2007). This was the first time that a national probability sample of kindergartners was followed up with repeated cognitive assessments throughout the critical early school years. ETS's longitudinal studies group continued the bidding strategy of writing the same psychometric proposal for inclusion in all the proposals of the prime contract bidders. NORC won the contract to develop instruments and conduct field tests prior to the kindergarten year; Westat was the winning bidder for the subsequent rounds, with ETS subcontracted to do the test development, scaling, and scoring. This study was by far the most complex as well as the largest undertaking to date with respect to the number and depth of the assessment instruments.

The spanning of so many grades with so many instruments during periods in which one would expect accelerated student growth complicated the vertical scaling. As a result, a number of subcontracts were also let reflecting the individual expertise required for the various instruments. Principals at NCES were Jeff Owings, the Longitudinal Studies Branch chief, with Jerry West, and later, Elvira Germino Hausken as project directors. The Westat effort was led by Karen Tourangeau, while NORC was represented by Tom Hoffer, who would be involved in student questionnaire construction, and Sally Atkins-Burnett and Sam Meisels from the University of Michigan led the development of indirect measures of socio-emotional and cognitive achievement. At ETS, Don Rock, Judy Pollack, and in the later rounds, Michelle Najarian, led the group responsible for developing and selecting test items and for scaling and scoring the direct measures of cognitive development. The test development endeavor benefited from the help and advice of the University of Michigan staff.

The ECLS-K base-year sample was a national probability sample of about 22,000 children who had entered kindergarten either full-day or part-day in fall 1998. About 800 public schools and 200 private schools were represented in the sample. Children in the kindergarten through fifth-grade rounds were assessed individually using computer-assisted interviewing methods, while group paper-and-pencil assessments were conducted in the eighth grade.[2] Children in the early grades (K-1) were assessed with socio-emotional and psychomotor instruments and ratings of cognitive development as well as direct cognitive assessments (Adkins-Burnett et al. 2000). The direct cognitive assessment in K-1 included a battery consisting of reading, mathematics, and general knowledge, all of which were to be completed in 75 min, on average, although the tests were not timed. In Grade 3, the general knowledge test was dropped and replaced with a science test. The original NCES plan was to assess children in fall and spring of their kindergarten year, fall and spring of their first-grade year, and in the spring only of each of their second- through fifth-grade years. Unfortunately, NCES budgetary constraints resulted in the second- and fourth-grade data collections being dropped completely; for similar reasons, data was collected from a reduced sample in fall of the first-grade year. At a later time, high school assessments were planned for 8th, 10th, and 12th grades, but again, due to budget constraints, only the 8th-grade survey was conducted.

Gaps of more than a year in a longitudinal study during a high-growth period can be problematic for vertical scaling. Dropping the second-grade data collection created a serious gap, particularly in reading. Very few children finish first-grade reading fluently; most are able to read with comprehension by the end of third grade. With no data collection bridging the gap between the early reading tasks of the first grade assessment and the much more advanced material in the third grade tests, the development of a vertical scale was at risk. As a result, a bridge study was conducted using a sample of about 1000 second graders; this study furnished the linking items to connect the first grade with the third grade and maintain the vertical scale's integrity. Subsequent gaps in data collection, from third to fifth grade and then to eighth grade were less serious because there was more overlap in the ability distributions.

While the changes referred to above did indeed complicate IRT scaling, one large difference between ECLS-K and the previous high school longitudinal studies was the relative uniformity of the curricula in the early grades. This standardization

[2]The individually administered test approach used in kindergarten through fifth grade had both supporters and critics among the experts. Most felt that individual administration would be advantageous because it would help maintain a high level of motivation in the children. In general, this was found to be true. In the kindergarten and first-grade rounds, however, some expressed a concern that the individual mode of administration may have contributed unwanted sources of variance to the children's performance in the direct cognitive measures. Unlike group administrations, which in theory are more easily standardized, variance attributable to individual administrators might affect children's scores. A multilevel analysis of fall-kindergarten and spring-first grade data found only a very small interviewer effect of about 1–3% of variance. A team leader effect could not be isolated, because it was almost completely confounded with primary sampling unit. Analysis of interviewer effect was not carried out for subsequent rounds of data for two reasons. First, the effect in kindergarten through first grade was about twice as large for the general knowledge assessment (which was not used beyond kindergarten) than for reading or mathematics. Second, the effect found was so small that it was inconsequential. Refer to Rock and Pollack (2002b) for more details on the analysis of interviewer effects.

holds reasonably well all the way through to the fifth grade. This curricular standardization facilitated consensus among clients, test developers, and outside advisors on the test specifications that would define the pools of test items that would be sensitive to changes in a child's development. However, there were some tensions with respect to item selection for measuring change across grades. While the curriculum experts emphasized the need for grade-appropriate items for children in a given grade, it is precisely the nongrade-appropriate items that also must be included in order to form links to the grade above and the grade below. Those items serve not only as linking items but also play an important role in minimizing floor and ceiling effects. Grade-appropriate items play a larger role in any cross-sectional assessment, but are not sufficient for an assessment in a particular grade as part of an ongoing longitudinal study.

Many of the psychometric approaches that were developed in the previous longitudinal studies, particularly in NELS:88, were applied in ECLS-K, with significant improvements. The primary example of this application was the introduction in ECLS-K of real-time, two-stage adaptive testing. That is, the cognitive tests in reading, mathematics, and general knowledge were individually administered in ECLS in Grades K–1. In each subject, the score on a short routing test determined the selection of an easier or more difficult second stage form. The reading and mathematics tests each had three second-stage forms of different difficulty; two forms were used for the general knowledge test. The same assessment package was used for the first four ECLS-K rounds, fall and spring kindergarten and fall and spring first grade. The reading and mathematics test forms were designed so that, in fall kindergarten, about 75% of the sample would be expected to be routed to the easiest of the three alternate forms; by spring of first grade, the intention was that about 75% of children would receive the hardest form. Assessments for the subsequent rounds were used in only one grade. The third- and fifth-grade tests were designed to route the middle half of the sample to the middle form, with the rest receiving the easiest or most difficult form. In the eighth grade, there were only two-second stage forms, each designed to be administered to half the sample. For the routing test, each item response was entered into a portable computer by the assessor. The computer would then score the routing test responses and based on the score select the appropriate second stage form to be administered.

As in NELS:88, multiple hierarchical proficiency levels were developed to mark critical developmental points along a child's learning curve in reading and mathematics. This development was easier to do in the early rounds of ECLS-K because of the relative standardization of the curriculum in the early grades along with the generally accepted pedagogical sequencing that was followed in early mathematics and reading. When the educational treatment follows a fairly standard pedagogical sequence (as in the early grades in school), we arguably have a situation that can be characterized by a common growth curve with children located at different points along that curve signifying different levels of development. Assuming a common growth curve, the job of the test developer and the psychometrician is to identify critical points along the growth curve that mark developmental milestones. Marking these points is the task of the proficiency levels.

10.4.1 Proficiency Levels and Scores in ECLS-K

Proficiency levels as defined in ECLS-K, as in NELS:88, provide a means for distinguishing status or gain in specific skills within a content area from the overall achievement measured by the IRT scale scores. Once again, clusters of four assessment questions having similar content and difficulty were located at several points along the score scale of the reading and mathematics assessments. Each cluster marked a learning milestone in reading or mathematics, agreed on by ECLS-K curriculum specialists. The sets of proficiency levels formed a hierarchical structure in the Piagetian sense in that the teaching sequence implied that one had to master the lower levels in the sequence before one could learn the material at the next higher level. This was the same basic procedure that was introduced in NELS:88.

Clusters of four items marking critical points on the vertical score scale provide a more reliable assessment of a particular proficiency level than do single items because of the possibility of guessing. It is very unlikely that a student who has not mastered a particular skill would be able to guess enough answers correctly to pass a four-item cluster. The proficiency levels were assumed to follow a Guttman model (Guttman 1950), that is, a student passing a particular skill level was expected to have mastered all lower levels; a failure at a given level should be consistent with nonmastery at higher levels. Only a very small percentage of students in ECLS-K had response patterns that did not follow the Guttman scaling model; that is, a failing score at a lower level followed by a pass on a more difficult item cluster. (For the first five rounds of data collection, fewer than 7% of reading response patterns and fewer than 5% of mathematics assessment results failed to follow the expected hierarchical pattern.) Divergent response patterns do not necessarily indicate a different learning sequence for these children. Because all of the proficiency level items were multiple choice, a number of these reversals simply may be due to children guessing as well as other random response errors.

Sections 4.2.2 and 4.3.2 of Najarian et al. (2009) described the ten reading and nine mathematics proficiency levels identified in the kindergarten through eighth-grade assessments. No proficiency scores were computed for the science assessment because the questions did not follow a hierarchical pattern. Two types of scores were reported with respect to the proficiency levels: a single indicator of highest level mastered, and a set of IRT-based probability scores, one for each proficiency level.

10.4.2 Highest Proficiency Level Mastered

As described above, mastery of a proficiency level was defined as answering correctly at least three of the four questions in a cluster. This definition results in a very low probability of guessing enough right answers to pass a cluster by chance. The probability varies depending on the guessing parameters (IRT *c* parameters) of the items in each cluster, but is generally less than 2%. At least two incorrect or "I don't know" responses indicated lack of mastery. Open-ended questions that were answered with an explicit "I don't know" response were treated as wrong, while omitted items were not counted. Since the ECLS-K direct cognitive child assessment was a two-stage design (where not all children were administered all items), and since more advanced assessment instruments were administered in third grade and beyond, children's data did not include all of the assessment items necessary to determine pass or fail for every proficiency level at each round of data collection. The missing information was not missing at random; it depended in part on children being routed to second-stage forms of varying difficulty within each assessment set and in part on different assessments being used for the different grades. In order to avoid bias due to the nonrandomness of the missing proficiency level scores, imputation procedures were undertaken to fill in the missing information.

Pass or fail for each proficiency level was based on actual counts of correct or incorrect responses, if they were present. If too few items were administered or answered to determine mastery of a level, a pass/fail score was imputed based on the remaining proficiency level scores only if they indicated a pattern that was unambiguous. That is, a fail might be inferred for a missing level if there were easier cluster(s) that had been failed and no higher cluster passed; or a pass might be assumed if harder cluster(s) were passed and no easier one failed. In the case of ambiguous patterns (e.g., pass, missing, fail for three consecutive levels, where the missing level could legitimately be either a pass or a fail), an additional imputation step was undertaken that relied on information from the child's performance in that round of data collection on all of the items answered within the domain that included the incomplete cluster. IRT-based estimates of the probability of a correct answer were computed for each missing assessment item and used to assign an imputed right or wrong score to the item. These imputed responses were then aggregated in the same manner as actual responses to determine mastery at each of the missing levels. Over all rounds of the study, the highest level scores were determined on the basis of item response data alone for about two-thirds of reading scores and 80% for mathematics; the rest utilized IRT-based probabilities for some or all of the missing items.

The need for imputation was greatest in the eighth-grade tests, as a result of the necessary placement of the proficiency level items on either the low or high second-stage form, based on their estimated difficulty levels. Scores were not imputed for missing levels for patterns that included a reversal (e.g., fail, blank, pass) because no resolution of the missing data could result in a consistent hierarchical pattern.

Scores in the public use data file represent the highest level of proficiency mastered by each child at each round of data collection, whether this determination was made by actual item responses, by imputation, or by a combination of methods. The highest proficiency level mastered implies that children demonstrated mastery of all lower levels and nonmastery of all higher levels. A zero score indicates nonmastery of the lowest proficiency level. Scores were excluded only if the actual or imputed mastery level data resulted in a reversal pattern as defined above. The highest proficiency level-mastered scores do not necessarily correspond to an interval scale, so in analyzing the data, they should be treated as ordinal.

10.4.3 Proficiency Probability Scores and Locus of Maximum Level of Learning Gains

Proficiency probability scores are reported for each of the proficiency levels described above, at each round of data collection. With respect to their use, these scores are essentially identical to those defined in NELS:88 above. They estimate the probability of mastery of each level and can take on any value from 0 to 1. As in NELS:88, the IRT model was employed to calculate the proficiency probability scores, which indicate the probability that a child would have passed a proficiency level, based on the child's whole set of item responses in the content domain. The item clusters were treated as single items for the purpose of IRT calibration, in order to estimate students' probabilities of mastery of each set of skills. The hierarchical nature of the skill sets justified the use of the IRT model in this way.

The proficiency probability scores can be averaged to produce estimates of mastery rates within population subgroups. These continuous measures can provide an accurate look at individuals' status and change over time. Gains in probability of mastery at each proficiency level allow researchers to study not only the amount of gain in total scale score points, but also where along the score scale different children are making their largest gains in achievement during a particular time interval. That is, when a child's difference in probabilities of mastery at each of the levels computed between adjacent testing sessions is largest, say at Level 3, we can then say the child's locus of maximum level of learning gains is in the skills defined at Level 3. Locus of maximum level of learning gains is not the same thing as highest proficiency level mastered. The latter score refers to the highest proficiency level in which the child got three out of four items correct. The locus of maximum level of learning gains could well be at the next higher proficiency level. At any rate, a student's school experiences at selected times can be related to improvements in specific skills. Additional details on the use of proficiency probabilities in ECLS-K can be found in Rock and Pollack (2002a) and Rock (2007a, b).

10.5 Conclusion

One might legitimately ask: What has been the impact of the above longitudinal studies on educational policy and research? Potential influences on policy were made possible by the implementation of extensive school, teacher, parent, and student process questionnaires and their relationships with student gains. While it is difficult to pinpoint specific impacts on policy, there is considerable evidence of the usefulness of the longitudinal databases for carrying out research on policy relevant questions. For example, NCES lists more than 1,000 publications and dissertations using the NELS:88 database. Similarly, the more recent ECLS-K study lists more than 350 publications and dissertations. As already noted, the availability of a wealth of process information gathered within a longitudinal framework is a useful first step in identifying potential causal relationships between educational processes and student performance.

In summary, the main innovations that were developed primarily in NELS:88 and improved upon in ECLS-K have become standard practices in the succeeding large-scale longitudinal studies initiated by NCES. These innovations are:

- *Real-time multistage adaptive testing* to match item difficulty to each student's ability level. Such matching of item difficulty and ability reduces testing time, as well as floor and ceiling effects, while improving accuracy of measurement.
- *The implementation of multiple-group Bayesian marginal maximum likelihood procedures for item parameter and EAP score estimation.* These procedures allow the estimation of item parameters that fit both within and across longitudinal data waves. In addition, the incorporation of ability priors for subpopulations defined by the adaptive testing procedure helps in minimizing floor and ceiling effects.
- *The pooling of succeeding longitudinal data waves to re-estimate item parameters and scores.* While this full-information approach has political drawbacks since it remakes history and is somewhat inconvenient for researchers, it helps to maintain the integrity of the vertical scale and yields more accurate estimates of the score variances associated with each wave.
- *The introduction of multiple proficiency levels that mark learning milestones in a child's development.* The concept of marking a scale with multiple proficiency points is not new, but their use within the IRT model to locate where an individual is making his/her maximum gains (locus of maximum level of learning gains) is a new contribution to measuring gains. Now the longitudinal data user has three pieces of information: how much each child gains; at what skill levels he/she is making those gains; and the highest level at which he/she has demonstrated mastery.
- The concept of *relating specific gains in proficiency levels to those process variables that can be logically expected to impact changes in the skill levels marked by these proficiency levels.*

References

Adkins-Burnett, S., Meisels, S. J., & Correnti, R. (2000). Analysis to develop the third grade indirect cognitive assessments and socioemotional measures. In *Early childhood longitudinal study, kindergarten class of 1998–99 (ECLS-K) spring 2000 field test report*. Rockville: Westat.

Bock, D., & Aiken, M. (1981). Marginal maximum likelihood estimation of item parameters, an application of an EM algorithm. *Psychometrika, 46*, 443–459. http://dx.doi.org/10.1002/j.2333-8504.1977.tb01147.x

Braun, H. (2006). *Using the value added modeling to evaluate teaching* (Policy Information Perspective). Princeton: Educational Testing Service.
Braun, H., & Bridgeman, B. (2005). *An introduction to the measurement of change problem* (Research Memorandum No. RM-05-01). Princeton: Educational Testing Service.
Cleary, T. A., Linn, R. L., & Rock, D. A. (1968). An exploratory study of programmed tests. *Educational and Psychological Measurement, 28*, 345–360. https://doi.org/10.1177/001316446802800212
Coleman, J. S. (1969). *Equality and achievement in education*. Boulder: Westview Press.
Coleman, J. S., & Hoffer, T. B. (1987). *Public and private schools: The impact of communities*. New York: Basic Books.
Cronbach, L. J., & Furby, L. (1970). How should we measure change—Or should we? *Psychological Bulletin, 74*, 68–80. https://doi.org/10.1037/h0029382
Cronbach, L., & Snow, R. (1977). *Aptitudes and instructional methods: A handbook for research on interactions*. New York: Irvington.
Ekstrom, R. B., French, J. W., & Harman, H. H. (with Dermen, D.). (1976). *Manual for kit of factor-referenced cognitive tests*. Princeton: Educational Testing Service.
Frankel, M. R., Kohnke, L., Buonanua, D., & Tourangeau, R. (1981). *HS&B base year sample design report*. Chicago: National Opinion Research Center.
French, J. W. (1964). *Experimental comparative prediction batteries: High school and college level*. Princeton: Educational Testing Service.
Guttman, L. (1950). The basis for scalogram analysis. In S. A. Stouffer (Ed.), *Studies in social psychology in world war II* (Vol. 4). Princeton: Princeton University Press.
Heyns, B., & Hilton, T. L. (1982). The cognitive tests for high school and beyond: An assessment. *Sociology of Education, 55*, 89–102. https://doi.org/10.2307/2112290
Ingels, S. J., Scott, L. A., Rock, D. A., Pollack, J. M., & Rasinski, K. A. (1993). *NELS-88 first follow-up final technical report*. Chicago: National Opinion Research Center.
Joreskog, K., & Sorbom, D. (1996). LISREL-8: Users reference guide [Computer software manual]. Chicago: Scientific Software.
Konstantopoulos, S. (2006). Trends of school effects on student achievement: Evidence from NLS:72, HSB:82, and NELS:92. *Teachers College Record, 108*, 2550–2581. https://doi.org/10.1111/j.1467-9620.2006.00796.x
Lord, F. M. (1971). A theoretical study of two-stage testing. *Psychometrika, 36*, 227–242. https://doi.org/10.1007/BF02297844
Lord, F. M. (1980). *Applications of item response theory to practical testing problems*. Hillsdale: Erlbaum.
Mislevy, R. J., & Bock, R. D. (1990). BILOG-3; Item analysis and test scoring with binary logistic models [Computer software]. Chicago: Scientific Software.
Mislevy, R. J., Johnson, E. G., & Muraki, E. (1992). Scaling procedures in NAEP. *Journal of Educational Statistics, 17*, 131–154. https://doi.org/10.2307/1165166
Muraki, E. J., & Bock, R. D. (1991). PARSCALE: Parameter scaling of rating data [Computer software]. Chicago: Scientific Software.
Najarian, M., Pollack, J. M., & Sorongon, A. G., (2009). *Early childhood longitudinal study, kindergarten class of 1998–99 (ECLS-K), Psychometric report for the eighth grade* (NCES Report No. 2009-002). Washington, DC: National Center for Education Statistics.
National Center for Education Statistics. (2011). National longitudinal study of 1972: Overview. Retrieved from http://nces.ed.gov/surveys/nls72/
Raudenbush, S. W., & Bryk, A. S. (2002). *Hierarchical linear models*. Thousand Oaks: Sage.
Riccobono, J., Henderson, L., Burkheimer, G., Place, C., & Levensohn, J. (1981). *National longitudinal study: Data file users manual*. Washington, DC: National Center for Education Statistics.
Rock, D. A. (2007a). *A note on gain scores and their interpretation in developmental models designed to measure change in the early school years* (Research Report No. RR-07-08). Princeton: Educational Testing Service. http://dx.doi.org/10.1002/j.2333-8504.2007.tb02050.x

Rock, D. A. (2007b). *Growth in reading performance during the first four years in school* (Research Report No. RR-07-39). Princeton: Educational Testing Service. http://dx.doi.org/10.1002/j.2333-8504.2007.tb02081.x
Rock, D. A., & Pollack, J. M. (1991). *The NELS-88 test battery*. Washington, DC: National Center for Education Statistics.
Rock, D. A., & Pollack, J. M. (2002a). *A model based approach to measuring cognitive growth in pre-reading and reading skills during the kindergarten year* (Research Report No. RR-02-18). Princeton: Educational Testing Service. http://dx.doi.org/10.1002/j.2333-8504.2002.tb01885.x
Rock, D. A., & Pollack, J. M. (2002b). *Early childhood longitudinal study–Kindergarten class of 1989–99 (ECLS-K), Psychometric report for kindergarten through the first grade* (Working Paper No. 2002–05). Washington, DC: National Center for Education Statistics.
Rock, D. A., Hilton, T., Pollack, J. M., Ekstrom, R., & Goertz, M. E. (1985). *Psychometric analysis of the NLS-72 and the High School and Beyond test batteries* (NCES Report No. 85-217). Washington, DC: National Center for Education Statistics.
Rock, D. A., Owings, J., & Lee, R. (1993). *Changes in math proficiency between 8th and 10th grades. Statistics in brief* (NCES Report No. 93-455). Washington, DC: National Center for Education Statistics.
Rock, D. A., Pollack, J. M., & Quinn, P. (1995). *Psychometric report of the NELS: 88 base year through second follow-up* (NCES Report No. 95-382). Washington, DC: National Center for Education Statistics.
Rogosa, D. R. (1995). Myths and methods: "Myths about longitudinal research" plus supplemental questions. In J. M. Gottman (Ed.), *The analysis of change* (pp. 3–46). Hillsdale: Erlbaum.
Scott, L. A., Rock, D. A., Pollack, J.M., & Ingels, S. J. (1995). *Two years later: Cognitive gains and school transitions of NELS: 88 eight graders. National education longitudinal study of 1998, statistical analysis report* (NCES Report No. 95-436). Washington, DC: National Center for Education Statistics.
Wood, R. L., Wingersky, M. S., & Lord, F. M. (1976). *LOGIST: A computer program for estimating ability and item characteristic curve parameters* (Research Memorandum No. RM-76-06). Princeton: Educational Testing Service.

Chapter 11
Evaluating Educational Programs

Samuel Ball

11.1 An Emerging Profession

Evaluating educational programs is an emerging profession, and Educational Testing Service (ETS) has played an active role in its development. The term *program evaluation* only came into wide use in the mid-1960s, when efforts at systematically assessing programs multiplied. The purpose of this kind of evaluation is to provide information to decision makers who have responsibility for existing or proposed educational programs. For instance, program evaluation may be used to help make decisions concerning whether to develop a program (*needs assessment*), how best to develop a program (*formative evaluation*), and whether to modify—or even continue—an existing program (*summative evaluation*).

Needs assessment is the process by which one identifies needs and decides upon priorities among them. *Formative evaluation* refers to the process involved when the evaluator helps the program developer—by pretesting program materials, for example. *Summative evaluation* is the evaluation of the program after it is in operation. Arguments are rife among program evaluators about what kinds of information should be provided in each of these forms of evaluation.

This chapter was written by Samuel Ball and originally published in 1979 by Educational Testing Service and later posthumously in 2011 as a research report in the ETS R&D Scientific and Policy Contributions Series. Ball was one of ETS's most active program evaluators for 10 years and directed several pacesetting studies including a large-scale evaluation of Sesame Street. The chapter documents the vigorous program of evaluation research conducted at ETS in the 1960s and 1970s, which helped lay the foundation for what was then a fledgling field. This work developed new viewpoints, techniques, and skills for systematically assessing educational programs and led to the creation of principles for program evaluation that still appear relevant today.

S. Ball (✉)
Educational Testing Service, Princeton, NJ, USA
e-mail: researchreports@ets.org

R.E. Bennett, M. von Davier (eds.), *Advancing Human Assessment*,
Methodology of Educational Measurement and Assessment,
DOI 10.1007/978-3-319-58689-2_11

In general, the ETS posture has been to try to obtain the best—that is, the most relevant, valid, and reliable—information that can be obtained within the constraints of cost and time and the needs of the various audiences for the evaluation. Sometimes, this means a tight experimental design with a national sample; at other times, the best information might be obtained through an intensive case study of a single institution. ETS has carried out both traditional and innovative evaluations of both traditional and innovative programs, and staff members also have cooperated with other institutions in planning or executing some aspects of evaluation studies. Along the way, the work by ETS has helped to develop new viewpoints, techniques, and skills.

11.2 The Range of ETS Program Evaluation Activities

Program evaluation calls for a wide range of skills, and evaluators come from a variety of disciplines: educational psychology, developmental psychology, psychometrics, sociology, statistics, anthropology, educational administration, and a host of subject matter areas. As program evaluation began to emerge as a professional concern, ETS changed, both structurally and functionally, to accommodate it. The structural changes were not exclusively tuned to the needs of conducting program evaluations. Rather, program evaluation, like the teaching of English in a well-run high school, became to some degree the concern of virtually all the professional staff. Thus, new research groups were added, and they augmented the organization's capability to conduct program evaluations.

The functional response was many-faceted. Two of the earliest evaluation studies conducted by ETS indicate the breadth of the range of interest. In 1965, collaborating with the Pennsylvania State Department of Education, Henry Dyer of ETS set out to establish a set of educational goals against which later the performance of the state's educational system could be evaluated (Dyer 1965a, b). A unique aspect of this endeavor was Dyer's insistence that the goal-setting process be opened up to strong participation by the state's citizens and not left solely to a professional or political elite. (In fact, ETS program evaluation has been marked by a strong emphasis, when at all appropriate, on obtaining community participation.)

The other early evaluation study in which ETS was involved was the now famous Coleman report (*Equality of Educational Opportunity*), issued in 1966 (Coleman et al. 1966). ETS staff, under the direction of Albert E. Beaton, had major responsibility for analysis of the massive data generated (see Beaton and Barone, Chap. 8, this volume). Until then, studies of the effectiveness of the nation's schools, especially with respect to programs' educational impact on minorities, had been small-scale. So the collection and analysis of data concerning tens of thousands of students and hundreds of schools and their communities were new experiences for ETS and for the profession of program evaluation.

In the intervening years, the Coleman report (Coleman et al. 1966) and the Pennsylvania Goals Study (Dyer 1965a, b) have become classics of their kind, and from these two auspicious early efforts, ETS has become a center of major program evaluation. Areas of focus include computer-aided instruction, aesthetics and creativity in education, educational television, educational programs for prison inmates, reading programs, camping programs, career education, bilingual education, higher education, preschool programs, special education, and drug programs. (For brief descriptions of ETS work in these areas, as well as for studies that developed relevant measures, see the appendix.) ETS also has evaluated programs relating to year-round schooling, English as a second language, desegregation, performance contracting, women's education, busing, Title I of the Elementary and Secondary Education Act (ESEA), accountability, and basic information systems.

One piece of work that must be mentioned is the *Encyclopedia of Educational Evaluation*, edited by Anderson et al. (1975). The encyclopedia contains articles by them and 36 other members of the ETS staff. Subtitled *Concepts and Techniques for Evaluating Education and Training Programs*, it contains 141 articles in all.

11.3 ETS Contributions to Program Evaluation

Given the innovativeness of many of the programs evaluated, the newness of the profession of program evaluation, and the level of expertise of the ETS staff who have directed these studies, it is not surprising that the evaluations themselves have been marked by innovations for the profession of program evaluation. At the same time, ETS has adopted several principles relative to each aspect of program evaluation. It will be useful to examine these innovations and principles in terms of the phases that a program evaluation usually attends to—goal setting, measurement selection, implementation in the field setting, analysis, and interpretation and presentation of evidence.

11.3.1 Making Goals Explicit

It would be a pleasure to report that virtually every educational program has a well-thought-through set of goals, but it is not so. It is, therefore, necessary at times for program evaluators to help verbalize and clarify the goals of a program to ensure that they are, at least, explicit. Further, the evaluator may even be given goal development as a primary task, as in the Pennsylvania Goals Study (Dyer 1965a, b). This need was seen again in a similar program, when Robert Feldmesser (1973) helped the New Jersey State Board of Education establish goals that underwrite conceptually that state's "thorough and efficient" education program.

Work by ETS staff indicates there are four important principles with respect to program goal development and explication. The first of these principles is as follows: What program developers say their program goals are may bear only a passing resemblance to what the program in fact seems to be doing.

This principle—the occasional surrealistic quality of program goals—has been noted on a number of occasions: For example, assessment instruments developed for a program evaluation on the basis of the stated goals sometimes do not seem at all sensitive to the actual curriculum. As a result, ETS program evaluators seek, whenever possible, to cooperate with program developers to help fashion the goals statement. The evaluators also will attempt to describe the program in operation and relate that description to the stated goals, as in the case of the 1971 evaluation of the second year of *Sesame Street* for Children's Television Workshop (Bogatz and Ball 1971). This comparison is an important part of the process and represents sometimes crucial information for decision makers concerned with developing or modifying a program.

The second principle is as follows: When program evaluators work cooperatively with developers in making program goals explicit, both the program and the evaluation seem to benefit.

The original *Sesame Street* evaluation (Ball and Bogatz, 1970) exemplified the usefulness of this cooperation. At the earliest planning sessions for the program, before it had a name and before it was fully funded, the developers, aided by ETS, hammered out the program goals. Thus, ETS was able to learn at the outset what the program developers had in mind, ensuring sufficient time to provide adequately developed measurement instruments. If the evaluation team had had to wait until the program itself was developed, there would not have been sufficient time to develop the instruments; more important, the evaluators might not have had sufficient understanding of the intended goals—thereby making sensible evaluation unlikely.

The third principle is as follows: There is often a great deal of empirical research to be conducted before program goals can be specified.

Sometimes, even before goals can be established or a program developed, it is necessary, through empirical research, to indicate that there is a need for the program. An illustration is provided by the research of Ruth Ekstrom and Marlaine Lockheed (1976) into the competencies gained by women through volunteer work and homemaking. The ETS researchers argued that it is desirable for women to resume their education if they wish to after years of absence. But what competencies have they picked up in the interim that might be worthy of academic credit? By identifying, surveying, and interviewing women who wished to return to formal education, Ekstrom and Lockheed established that many women had indeed learned valuable skills and knowledge. Colleges were alerted and some have begun to give credit where credit is due.

Similarly, when the federal government decided to make a concerted attack on the reading problem as it affects the total population, one area of concern was adult reading. But there was little knowledge about it. Was there an adult literacy problem? Could adults read with sufficient understanding such items as newspaper employment advertisements, shopping and movie advertisements, and bus schedules? And in investigating adult literacy, what characterized the reading tasks that should be taken into account? Murphy, in a 1973 study (Murphy 1973a), considered these factors: the *importance* of a task (the need to be able to read the material if only once a year as with income tax forms and instructions), the *intensity* of the task (a person who wants to work in the shipping department will have to read the shipping schedule each day), or the *extensivity* of the task (70% of the adult population read a newspaper but it can usually be ignored without gross problems arising). Murphy and other ETS researchers conducted surveys of reading habits and abilities, and this assessment of needs provided the government with information needed to decide on goals and develop appropriate programs.

Still a different kind of needs assessment was conducted by ETS researchers with respect to a school for learning disabled students in 1976 (Ball and Goldman 1976). The school catered to children aged 5–18 and had four separate programs and sites. ETS first served as a catalyst, helping the school's staff develop a listing of problems. Then ETS acted as an *amicus curiae*, drawing attention to those problems, making explicit and public what might have been unsaid for want of an appropriate forum. Solving these problems was the purpose of stating new institutional goals—goals that might never have been formally recognized if ETS had not worked with the school to make its needs explicit.

The fourth principle is as follows: The program evaluator should be conscious of and interested in the unintended outcomes of programs as well as the intended outcomes specified in the program's goal statement.

In program evaluation, the importance of looking for side effects, especially negative ones, has to be considered against the need to put a major effort into assessing progress toward intended outcomes. Often, in this phase of evaluation, the varying interests of evaluators, developers, and funders intersect—and professional, financial, and political considerations are all at odds. At such times, program evaluation becomes as much an art form as an exercise in social science.

A number of articles were written about this problem by Samuel J. Messick, ETS vice president for research (e.g., Messick 1970, 1975). His viewpoint—the importance of the medical model—has been illustrated in various ETS evaluation studies. His major thesis was that the medical model of program evaluation explicitly recognizes that "…prescriptions for treatment and the evaluation of their effectiveness should take into account not only reported symptoms but other characteristics of the organism and its ecology as well" (Messick 1975, p. 245). As Messick went on to point out, this characterization was a call for a systems analysis approach to program evaluation—dealing empirically with the interrelatedness of all the factors and monitoring all outcomes, not just the intended ones.

When, for example, ETS evaluated the first 2 years of *Sesame Street* (Ball and Bogatz 1970), there was obviously pressure to ascertain whether the intended goals of that show were being attained. It was nonetheless possible to look for some of the more likely unintended outcomes: whether the show had negative effects on heavy viewers going off to kindergarten, and whether the show was achieving impacts in attitudinal areas.

In summative evaluations, to study unintended outcomes is bound to cost more money than to ignore them. It is often difficult to secure increased funding for this purpose. For educational programs with potential national applications, however, ETS strongly supports this more comprehensive approach.

11.3.2 Measuring Program Impact

The letters *ETS* have become almost synonymous in some circles with standardized testing of student achievement. In its program evaluations, ETS naturally uses such tests as appropriate, but frequently the standardized tests are not appropriate measures. In some evaluations, ETS uses both standardized and domain-referenced tests. An example may be seen in *The Electric Company* evaluations (Ball et al. 1974). This televised series, which was intended to teach reading skills to first through fourth graders, was evaluated in some 600 classrooms. One question that was asked during the process concerned the interaction of the student's level of reading attainment and the effectiveness of viewing the series. Do good readers learn more from the series than poor readers? So standardized, norm-referenced reading tests were administered, and the students in each grade were divided into deciles on this basis, thereby yielding ten levels of reading attainment.

Data on the outcomes using the domain-referenced tests were subsequently analyzed for each decile ranking. Thus, ETS was able to specify for what level of reading attainment, in each grade, the series was working best. This kind of conclusion would not have been possible if a specially designed domain-referenced reading test with no external referent had been the only one used, nor if a standardized test, not sensitive to the program's impact, had been the only one used.

Without denying the usefulness of previously designed and developed measures, ETS evaluators have frequently preferred to develop or adapt instruments that would be specifically sensitive to the tasks at hand. Sometimes this measurement effort is carried out in anticipation of the needs of program evaluators for a particular instrument, and sometimes because a current program evaluation requires immediate instrumentation.

An example of the former is a study of doctoral programs by Mary Jo Clark et al. (1976). Existing instruments had been based on surveys in which practitioners in a given discipline were asked to rate the quality of doctoral programs in that discipline. Instead of this reputational survey approach, the ETS team developed an array of criteria (e.g., faculty quality, student body quality, resources, academic offerings, alumni performance), all open to objective assessment. This assessment tool can be used to assess changes in the quality of the doctoral programs offered by major universities.

Similarly, the development by ETS of the *Kit of Factor-Referenced Cognitive Tests* (Ekstrom et al. 1976) also provided a tool—one that could be used when evaluating the cognitive abilities of teachers or students if these structures were of interest in a particular evaluation. A clearly useful application was in the California study of teaching performance by Frederick McDonald and Patricia Elias (1976). Teachers with certain kinds of cognitive structures were seen to have differential impacts on student achievement. In the Donald A. Trismen study of an aesthetics program (Trismen 1968), the factor kit was used to see whether cognitive structures interacted with aesthetic judgments.

11.3.2.1 Developing Special Instruments

Examples of the development of specific instrumentation for ETS program evaluations are numerous. Virtually every program evaluation involves, at the very least, some adapting of existing instruments. For example, a questionnaire or interview may be adapted from ones developed for earlier studies. Typically, however, new instruments, including goal-specific tests, are prepared. Some ingenious examples, based on the 1966 work of E. J. Webb, D. F. Campbell, R. D. Schwartz, and L. Sechrest, were suggested by Anderson (1968) for evaluating museum programs, and the title of her article gives a flavor of the unobtrusive measures illustrated—"Noseprints on the Glass."

Another example of ingenuity is Trismen's use of 35 mm slides as stimuli in the assessment battery of the Education through Vision program (Trismen 1968). Each slide presented an art masterpiece, and the response options were four abstract designs varying in color. The instruction to the student was to pick the design that best illustrated the masterpiece's coloring.

11.3.2.2 Using Multiple Measures

When ETS evaluators have to assess a variable and the usual measures have rather high levels of error inherent in them, they usually resort to triangulation. That is, they use multiple measures of the same construct, knowing that each measure suffers from a specific weakness. Thus, in 1975, Donald E. Powers evaluated for the Philadelphia school system the impact of dual-audio television—a television show telecast at the same time as a designated FM radio station provided an appropriate educational commentary. One problem in measurement was assessing the amount of contact the student had with the dual-audio television treatment (Powers 1975a). Powers used home telephone interviews, student questionnaires, and very simple knowledge tests of the characters in the shows to assess whether students had in fact been exposed to the treatment. Each of these three measures has problems associated with it, but the combination provided a useful assessment index.

In some circumstances, ETS evaluators are able to develop measurement techniques that are an integral part of the treatment itself. This unobtrusiveness has clear benefits and is most readily attainable with computer-aided instructional (CAI) programs. Thus, for example, Donald L. Alderman, in the evaluation of TICCIT (a CAI program developed by the Mitre Corporation), obtained for each student such indices as the number of lessons passed, the time spent on line, the number of errors made, and the kinds of errors (Alderman 1978). And he did this simply by programming the computer to save this information over given periods of time.

11.3.3 Working in Field Settings

Measurement problems cannot be addressed satisfactorily if the setting in which the measures are to be administered is ignored. One of the clear lessons learned in ETS program evaluation studies is that measurement in field settings (home, school, community) poses different problems from measurement conducted in a laboratory.

Program evaluation, ether formative or summative, demands that its empirical elements usually be conducted in natural field settings rather than in more contrived settings, such as a laboratory. Nonetheless, the problems of working in field settings are rarely systematically discussed or researched. In an article in the *Encyclopedia of Educational Evaluation*, Bogatz (1975) detailed these major aspects:

- Obtaining permission to collect data at a site
- Selecting a field staff
- Training the staff
- Maintaining family/community support

Of course, all the aspects discussed by Bogatz interact with the measurement and design of the program evaluation. A great source of information concerning field operations is the ETS Head Start Longitudinal Study of Disadvantaged Children, directed by Virginia Shipman (1970). Although not primarily a program evaluation, it certainly has generated implications for early childhood programs. It was longitudinal, comprehensive in scope, and large in size, encompassing four sites and, initially, some 2000 preschoolers. It was clear from the outset that close community ties were essential if only for expediency—although, of course, more important ethical principles were involved. This close relationship with the communities in which the study was conducted involved using local residents as supervisors and testers, establishing local advisory committees, and thus ensuring free, two-way communication between the research team and the community.

The *Sesame Street* evaluation also adopted this approach (Ball and Bogatz 1970). In part because of time pressures and in part to ensure valid test results, the ETS evaluators especially developed the tests so that community members with minimal educational attainments could be trained quickly to administer them with proper skill.

11.3.3.1 Establishing Community Rapport

In evaluations of street academies by Ronald L. Flaugher (1971), and of education programs in prisons by Flaugher and Samuel Barnett (1972), it was argued that one of the most important elements in successful field relationships is the time an evaluator spends getting to know the interests and concerns of various groups, and lowering barriers of suspicion that frequently separate the educated evaluator and the less-educated program participants. This point may not seem particularly sophisticated or complex, but many program evaluations have floundered because of an evaluator's lack of regard for disadvantaged communities (Anderson 1970). Therefore, a firm principle underlying ETS program evaluation is to be concerned with the communities that provide the contexts for the programs being evaluated. Establishing two-way lines of communication with these communities and using community resources whenever possible help ensure a valid evaluation.

Even with the best possible community support, field settings cause problems for measurement. Raymond G. Wasdyke and Jerilee Grandy (1976) showed this idea to be true in an evaluation in which the field setting was literally that—a field setting. In studying the impact of a camping program on New York City grade school pupils, they recognized the need, common to most evaluations, to describe the treatment—in this case the camping experience. Therefore, ETS sent an observer to the campsite with the treatment groups. This person, who was herself skilled in camping, managed not to be an obtrusive participant by maintaining a relatively low profile.

Of course, the problems of the observer can be just as difficult in formal institutions as on the campground. In their 1974 evaluation of Open University materials, Hartnett and colleagues found, as have program evaluators in almost every situation, that there was some defensiveness in each of the institutions in which they worked (Hartnett et al. 1974). Both personal and professional contacts were used to allay suspicions. There also was emphasis on an evaluation design that took into account each institution's values. That is, part of the evaluation was specific to the institution, but some common elements across institutions were retained. This strategy underscored the evaluators' realization that each institution was different, but allowed ETS to study certain variables across all three participating institutions.

Breaking down the barriers in a field setting is one of the important elements of a successful evaluation, yet each situation demands somewhat different evaluator responses.

11.3.3.2 Involving Program Staff

Another way of ensuring that evaluation field staff are accepted by program staff is to make the program staff active participants in the evaluation process. While this integration is obviously a technique to be strongly recommended in formative evaluations, it can also be used in summative evaluations. In his evaluation of PLATO in junior colleges, Murphy (1977) could not afford to become the victim of a program developer's fear of an insensitive evaluator. He overcame this potential problem by enlisting the active participation of the junior college and program development staffs. One of Murphy's concerns was that there is no common course across colleges. Introduction to Psychology, for example, might be taught virtually everywhere, but the content can change remarkably, depending on such factors as who teaches the course, where it is taught, and what text is used. Murphy understood this variability and his evaluation of PLATO reflected his concern. It also necessitated considerable input and cooperation from program developers and college teachers working in concert—with Murphy acting as the conductor.

11.3.4 Analyzing the Data

After the principles and strategies used by program evaluators in their field operations are successful and data are obtained, there remains the important phase of data analysis. In practice, of course, the program evaluator thinks through the question of data analysis *before* entering the data collection phase. Plans for analysis help determine what measures to develop, what data to collect, and even, to some extent, how the field operation is to be conducted. Nonetheless, analysis plans drawn up early in the program evaluation cannot remain quite as immutable as the Mosaic Law. To illustrate the need for flexibility, it is useful to turn once again to the heuristic ETS evaluation of *Sesame Street*.

As initially planned, the design of the *Sesame Street* evaluation was a true experiment (Ball and Bogatz 1970). The analyses called for were multivariate analyses of covariance, using pretest scores as the covariate. At each site, a pool of eligible preschoolers was obtained by community census, and experimental and control groups were formed by random assignment from these pools. The evaluators were somewhat concerned that those designated to be the experimental (viewing) group might not view the show—it was a new show on public television, a loose network of TV stations not noted for high viewership. Some members of the *Sesame Street* national research advisory committee counseled ETS to consider paying the experimental group to view. The suggestion was resisted, however, because any efforts above mild and occasional verbal encouragement to view the show would compromise the results. If the experimental group members were paid, and if they then viewed extensively and outperformed the control group at posttest, would the improved performance be due to the viewing, the payment, or some interaction of payment and viewing? Of course, this nice argument proved to be not much more than an exercise in modern scholasticism. In fact, the problem lay not in the treatment group but in the uninformed and unencouraged-to-view control group. The members of that group, as indeed preschoolers with access to public television throughout the nation, were viewing the show with considerable frequency—and not much less than the experimental group. Thus, the planned analysis involving differences in posttest attainments between the two groups was dealt a mortal blow.

Fortunately, other analyses were available, of which the ETS-refined age cohorts design provided a rational basis. This design is presented in the relevant report (Ball and Bogatz 1970). The need here is not to describe the design and analysis but to emphasize a point made practically by the poet Robert Burns some time ago and repeated here more prosaically: The best laid plans of evaluators can "gang aft agley," too.

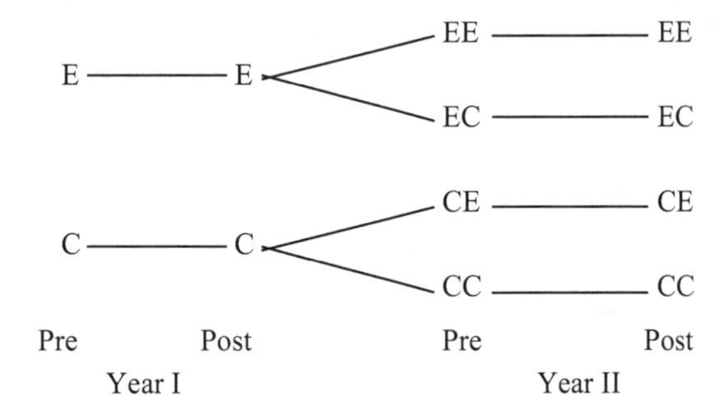

Fig. 11.1 The design for the new pool of classes. For Year II, EE represents children who were in E classrooms in Year I and again in Year II. That is, the first letter refers to status in Year I and the second to status in Year II

11.3.4.1 Clearing New Paths

Sometimes program evaluators find that the design and analysis they have in mind represent an untrodden path. This result is perhaps in part because many of the designs in the social sciences are built upon laboratory conditions and simply are not particularly relevant to what happens in educational institutions.

When ETS designed the summative evaluation of *The Electric Company*, it was able to set up a true experiment in the schools. Pairs of comparable classrooms within a school and within a grade were designated as the pool with which to work. One of each pair of classes was randomly assigned to view the series. Pretest scores were used as covariates on posttest scores, and in 1973 the first-year evaluation analysis was successfully carried out (Ball and Bogatz 1973). The evaluation was continued through a second year, however, and as is usual in schools, the classes did not remain intact.

From an initial 200 classes, the children had scattered through many more classrooms. Virtually none of the classes with subject children contained only experimental or only control children from the previous year. Donald B. Rubin, an ETS statistician, consulted with a variety of authorities and found that the design and analysis problem for the second year of the evaluation had not been addressed in previous work. To summarize the solution decided on, the new pool of classes was reassigned randomly to *E* (experimental) or *C* (control) conditions so that over the 2 years the design was portrayable as Fig. 11.1.

Further, the pretest scores of Year II were usable as new covariates when analyzing the results of the Year II posttest scores (Ball et al. 1974).

11.3.4.2 Tailoring to the Task

Unfortunately for those who prefer routine procedures, it has been shown across a wide range of ETS program evaluations that each design and analysis must be tailored to the occasion. Thus, Gary Marco (1972), as part of the statewide educational assessment in Michigan, evaluated ESEA Title I program performance. He assessed the amount of exposure students had to various clusters of Title I programs, and he included control schools in the analysis. He found that a regression-analysis model involving a correction for measurement error was an innovative approach that best fit his complex configuration of data.

Garlie Forehand, Marjorie Ragosta, and Donald A. Rock, in a national, correlational study of desegregation, obtained data on school characteristics and on student outcomes (Forehand et al. 1976). The purposes of the study included defining indicators of effective desegregation and discriminating between more and less effective school desegregation programs. The emphasis throughout the effort was on variables that were manipulable. That is, the idea was that evaluators would be able to suggest practical advice on what schools can do to achieve a productive desegregation program. Initial investigations allowed specification among the myriad variables of a hypothesized set of causal relationships, and the use of path analysis made possible estimation of the strength of hypothesized causal relationships. On the basis of the initial correlation matrices, the path analyses, and the observations made during the study, an important product—a nontechnical handbook for use in schools—was developed.

Another large-scale ETS evaluation effort was directed by Trismen et al. (1976). They studied compensatory reading programs, initially surveying more than 700 schools across the country. Over a 4-year period ending in 1976, this evaluation interspersed data analysis with new data collection efforts. One purpose was to find schools that provided exceptionally positive or negative program results. These schools were visited blind and observed by ETS staff. Whereas the Forehand evaluation analysis (Forehand et al. 1976) was geared to obtaining practical applications, the equally extensive evaluation analysis of Trismen's study was aimed at generating hypotheses to be tested in a series of smaller experiments.

As a further illustration of the complex interrelationship among evaluation purposes, design, analyses, and products, there is the 1977 evaluation of the use of PLATO in the elementary school by Spencer Swinton and Marianne Amarel (1978). They used a form of regression analysis—as did Forehand et al. (1976) and Trismen et al. (1976). But here the regression analyses were used differently in order to identify program effects unconfounded by teacher differences. In this regression analysis, teachers became fixed effects, and contrasts were fitted for each within-teacher pair (experimental versus control classroom teachers).

This design, in turn, provides a contrast to McDonald's (1977) evaluation of West New York programs to teach English as a second language to adults. In this instance, the regression analysis was directed toward showing which teaching method related most to gains in adult students' performance.

There is a school of thought within the evaluation profession that design and analysis in program evaluation can be made routine. At this point, the experience of ETS indicates that this would be unwise.

11.3.5 Interpreting the Results

Possibly the most important principle in program evaluation is that interpretations of the evaluation's meaning—the conclusions to be drawn—are often open to various nuances. Another problem is that the evidence on which the interpretations are based may be inconsistent. The initial premise of this chapter was that the role of program evaluation is to provide evidence for decision-makers. Thus, one could argue that differences in interpretation, and inconsistencies in the evidence, are simply problems for the decision-maker and not for the evaluator.

But consider, for example, an evaluation by Powers of a year-round program in a school district in Virginia (Powers 1974, 1975b). (The long vacation was staggered around the year so that schools remained open in the summer.) The evidence presented by Powers indicated that the year-round school program provided a better utilization of physical plant and that student performance was not negatively affected. The school board considered this evidence as well as other conflicting evidence provided by Powers that the parents' attitudes were decidedly negative. The board made up its mind, and (not surprisingly) scotched the program. Clearly, however, the decision was not up to Powers. His role was to collect the evidence and present it systematically.

11.3.5.1 Keeping the Process Open

In general, the ETS response to conflicting evidence or varieties of nuances in interpretation is to keep the evaluation process and its reporting as open as possible. In this way, the values of the evaluator, though necessarily present, are less likely to be a predominating influence on subsequent action.

Program evaluators do, at times, have the opportunity to influence decision-makers by showing them that there are kinds of evidence not typically considered. The Coleman Study, for example, showed at least some decision-makers that there is more to evaluating school programs than counting (or calculating) the numbers of books in libraries, the amount of classroom space per student, the student-teacher ratio, and the availability of audiovisual equipment (Coleman et al. 1966). Rather, the output of the schools in terms of student performance was shown to be generally superior as evidence of school program performance.

Through their work, evaluators are also able to educate decision makers to consider the important principle that educational treatments may have positive effects for some students and negative effects for others—that an interaction of treatment with student should be looked for. As pointed out in the discussion of unintended outcomes, a systems-analysis approach to program evaluation—dealing empirically with the interrelatedness of all the factors that may affect performance—is to be preferred. And this approach, as Messick emphasized, "properly takes into account those student-process-environment interactions that produce differential results" (Messick 1975, p. 246).

11.3.5.2 Selecting Appropriate Evidence

Finally, a consideration of the kinds of evidence and interpretations to be provided decision makers leads inexorably to the realization that different kinds of evidence are needed, depending on the decision-maker's problems and the availability of resources. The most scientific evidence involving objective data on student performance can be brilliantly interpreted by an evaluator, but it might also be an abomination to a decision maker who really needs to know whether teachers' attitudes are favorable.

ETS evaluations have provided a great variety of evidence. For a formative evaluation in Brevard County, Florida, Trismen (1970) provided evidence that students could make intelligent choices about courses. In the ungraded schools, students had considerable freedom of choice, but they and their counselors needed considerably more information than in traditional schools about the ingredients for success in each of the available courses. As another example, Gary Echternacht, George Temp, and Theodore Stolie helped state and local education authorities develop Title I reporting models that included evidence on impact, cost, and compliance with federal regulations (Echternacht et al. 1976). Forehand and McDonald (1972) had been working with New York City to develop an accountability model providing constructive kinds of evidence for the city's school system. On the other hand, as part of an evaluation team, Amarel provided, for a small experimental school in Chicago, judgmental data as well as reports and documents based on the school's own records and files (Amarel and The Evaluation Collective 1979). Finally, Michael Rosenfeld provided Montgomery Township, New Jersey, with student, teacher, and parent perceptions in his evaluation of the open classroom approach then being tried out (Rosenfeld 1973).

In short, just as tests are not valid or invalid (it is the ways tests are used that deserve such descriptions), so too, evidence is not good or bad until it is seen in relation to the purpose for which it is to be used, and in relation to its utility to decision-makers.

11.4 Postscript

For the most part, ETS's involvement in program evaluation has been at the practical level. Without an accompanying concern for the theoretical and professional issues, however, practical involvement would be irresponsible. ETS staff members have therefore seen the need to integrate and systematize knowledge about program evaluation. Thus, Anderson obtained a contract with the Office of Naval Research to draw together the accumulated knowledge of professionals from inside and outside ETS on the topic of program evaluation. A number of products followed. These products included a survey of practices in program evaluation (Ball and Anderson 1975a), and a codification of program evaluation principles and issues (Ball and

Anderson 1975b). Perhaps the most generally useful of the products is the aforementioned *Encyclopedia of Educational Evaluation* (Anderson et al. 1975).

From an uncoordinated, nonprescient beginning in the mid-1960s, ETS has acquired a great deal of experience in program evaluation. In one sense it remains uncoordinated because there is no specific “party line,” no dogma designed to ensure ritualized responses. It remains quite possible for different program evaluators at ETS to recommend differently designed evaluations for the same burgeoning or existing programs.

There is no sure knowledge where the profession of program evaluation is going. Perhaps, with zero-based budgeting, program evaluation will experience amazing growth over the next decade, growth that will dwarf its current status (which already dwarfs its status of a decade ago). Or perhaps there will be a revulsion against the use of social scientific techniques within the political, value-dominated arena of program development and justification. At ETS, the consensus is that continued growth is the more likely event. And with the staff’s variegated backgrounds and accumulating expertise, ETS hopes to continue making significant contributions to this emerging profession.

Appendix: Descriptions of ETS Evaluation and Some Related Studies in Some Key Categories

Aesthetics and Creativity in Education

For Bartlett Hayes III’s program of Education through Vision at Andover Academy, Donald A. Trismen developed a battery of evaluation instruments that assessed, inter alia, a variety of aesthetic judgments (Trismen 1968). Other ETS staff members working in this area have included Norman Frederiksen and William C. Ward, who have developed a variety of assessment techniques for tapping creativity and scientific creativity (Frederiksen and Ward 1975; Ward and Frederiksen 1977); Richard T. Murphy, who also has developed creativity-assessing techniques (Murphy 1973b, 1977); and Scarvia B. Anderson, who described a variety of ways to assess the effectiveness of aesthetic displays (Anderson 1968).

Bilingual Education

ETS staff have conducted and assisted in evaluations of numerous and varied programs of bilingual education. For example, Berkeley office staff (Reginald A. Corder, Patricia Elias, Patricia Wheeler) have evaluated programs in Calexico (Corder 1976a), Hacienda-La Puente (Elias and Wheeler 1972), and El Monte (Corder and Johnson 1972). For the Los Angeles office, J. Richard Harsh (1975) evaluated a bilingual program in Azusa, and Ivor Thomas (1970) evaluated one in Fountain Valley. Donald E. Hood (1974) of the Austin office evaluated the Dallas Bilingual Multicultural Program. These evaluations were variously formative and summative and covered bilingual programs that, in combination, served students from preschool (Fountain Valley) through 12th grade (Calexico).

Camping Programs

Those in charge of a school camping program in New York City felt that it was having unusual and positive effects on the students, especially in terms of motivation. ETS was asked to—and did—evaluate this program, using an innovative design and measurement procedures developed by Raymond G. Wasdyke and Jerilee Grandy (1976).

Career Education

In a decade of heavy federal emphasis on career education, ETS was involved in the evaluation of numerous programs in that field. For instance, Raymond G. Wasdyke (1977) helped the Newark, Delaware, school system determine whether its career education goals and programs were properly meshed. In Dallas, Donald Hood (1972) of the ETS regional staff assisted in developing goal specifications and reviewing evaluation test items for the Skyline Project, a performance contract calling for the training of high school students in 12 career clusters. Norman E. Freeberg (1970) developed a test battery to be used in evaluating the Neighborhood Youth Corps. Ivor Thomas (1973) of the Los Angeles office provided formative evaluation services for the Azusa Unified School District's 10th grade career training and performance program for disadvantaged students. Roy Hardy (1977) of the Atlanta office directed the third-party evaluation of Florida's Comprehensive Program of Vocational Education for Career Development, and Wasdyke (1976) evaluated the Maryland Career Information System. Reginald A. Corder, Jr. (1975) of the Berkeley office assisted in the evaluation of the California Career Education program and subsequently directed the evaluation of the Experience-Based Career Education Models of a number of regional education laboratories (Corder 1976b).

Computer-Aided Instruction

Three major computer-aided instruction programs developed for use in schools and colleges have been evaluated by ETS. The most ambitious is PLATO from the University of Illinois. Initially, the ETS evaluation was directed by Ernest Anastasio

(1972), but later the effort was divided between Richard T. Murphy, who focused on college-level programs in PLATO, and Spencer Swinton and Marianne Amarel (1978), who focused on elementary and secondary school programs. ETS also directed the evaluation of TICCIT, an instructional program for junior colleges that used small-computer technology; the study was conducted by Donald L. Alderman (1978). Marjorie Ragosta directed the evaluation of the first major in-school longitudinal demonstration of computer-aided instruction for low-income students (Holland et al. 1976).

Drug Programs

Robert F. Boldt (1975) served as a consultant on the National Academy of Science's study assessing the effectiveness of drug antagonists (less harmful drugs that will "fight" the impact of illegal drugs). Samuel Ball (1973) served on a National Academy of Science panel that designed, for the National Institutes of Health, a means of evaluating media drug information programs and spot advertisements.

Educational Television

ETS was responsible for the national summative evaluation of the ETV series *Sesame Street* for preschoolers (Ball and Bogatz 1970), and *The Electric Company* for students in Grades 1 through 4 (Ball and Bogatz 1973); the principal evaluators were Samuel Ball, Gerry Ann Bogatz, and Donald B. Rubin. Additionally, Ronald Flaugher and Joan Knapp (1972) evaluated the series *Bread and Butterflies* to clarify career choice; Jayjia Hsia (1976) evaluated a series on the teaching of English for high school students and a series on parenting for adults.

Higher Education

Much ETS research in higher education focuses on evaluating students or teachers, rather than programs, mirroring the fact that systematic program evaluation is not common at this level. ETS has made, however, at least two major forays in program evaluation in higher education. In their Open University study, Rodney T. Hartnett and associates joined with three American universities (Houston, Maryland, and Rutgers) to see if the British Open University's methods and materials were appropriate for American institutions Hartnett et al. 1974). Mary Jo Clark, Leonard L. Baird, and Hartnett conducted a study of means of assessing quality in doctoral programs (Clark et al. 1976). They established an array of criteria for use in obtaining more precise descriptions and evaluations of doctoral programs than the prevailing technique—reputational surveys—provides. P. R. Harvey (1974) also evaluated the National College of Education Bilingual Teacher Education project, while Protase Woodford, (1975) proposed a pilot project for oral proficiency interview tests of bilingual teachers and tentative determination of language proficiency criteria.

Preschool Programs

A number of preschool programs have been evaluated by ETS staff, including the ETV series *Sesame Street* (Ball and Bogatz 1970; Bogatz and Ball 1971). Irving Sigel (1976) conducted formative studies of developmental curriculum. Virginia Shipman (1974) helped the Bell Telephone Companies evaluate their day care centers, Samuel Ball, Brent Bridgeman, and Albert Beaton provided the U.S. Office of Child Development with a sophisticated design for the evaluation of Parent-Child Development Centers (Ball et al. 1976), and Ball and Kathryn Kazarow evaluated the To Reach a Child program (Ball and Kazarow 1974). Roy Hardy (1975) examined the development of CIRCO, a Spanish language test battery for preschool children.

Prison Programs

In New Jersey, ETS has been involved in the evaluation of educational programs for prisoners. Developed and administered by Mercer County Community College, the programs have been subject to ongoing study by Ronald L. Flaugher and Samuel Barnett (1972).

Reading Programs

ETS evaluators have been involved in a variety of ways in a variety of programs and proposed programs in reading. For example, in an extensive, national evaluation, Donald A. Trismen et al. (1976) studied the effectiveness of reading instruction in compensatory programs. At the same time, Donald E. Powers (1973) conducted a small study of the impact of a local reading program in Trenton, New Jersey. Ann M. Bussis, Edward A. Chittenden, and Marianne Amarel reported the results of their study of primary school teachers' perceptions of their own teaching behavior (Bussis et al. 1976). Earlier, Richard T. Murphy surveyed the reading competencies and needs of the adult population (Murphy 1973a).

Special Education

Samuel Ball and Karla Goldman (1976) conducted an evaluation of the largest private school for the learning disabled in New York City, and Carol Vale (1975) of the ETS office in Berkeley directed a national needs assessment concerning educational technology and special education. Paul Campbell (1976) directed a major study of an intervention program for learning disabled juvenile delinquents.

References

Alderman, D. L. (1978). *Evaluation of the TICCIT computer-assisted instructional system in the community college*. Princeton: Educational Testing Service.

Amarel, M., & The Evaluation Collective. (1979). *Reform, response, renegotiation: Transitions in a school-change project*. Unpublished manuscript.

Anastasio, E. J. (1972). *Evaluation of the PLATO and TICCIT computer-based instructional systems—A preliminary plan* (Program Report No. PR-72-19). Princeton: Educational Testing Service.

Anderson, S. B. (1968). Noseprints on the glass—Or how do we evaluate museum programs? In E. Larrabee (Ed.), *Museums and education* (pp. 115–126). Washington, DC: Smithsonian Institution Press.

Anderson, S. B. (1970). From textbooks to reality: Social researchers face the facts of life in the world of the disadvantaged. In J. Hellmuth (Ed.), *Disadvantaged child: Vol. 3. Compensatory education: A national debate*. New York: Brunner/Mazel.

Anderson, S. B., Ball, S., & Murphy, R. T. (Eds.). (1975). *Encyclopedia of educational evaluation: Concepts and techniques for evaluating education and training programs*. San Francisco: Jossey-Bass Publishers.

Ball, S. (1973, July). *Evaluation of drug information programs—Report of the panel on the impact of information on drug use and misuse, phase 2*. Washington, DC: National Research Council, National Academy of Sciences.

Ball, S., & Anderson, S. B. (1975a). *Practices in program evaluation: A survey and some case studies*. Princeton: Educational Testing Service.

Ball, S., & Anderson, S. B. (1975b). *Professional issues in the evaluation of education/training programs*. Princeton: Educational Testing Service.

Ball, S., & Bogatz, G. A. (1970). *The first year of Sesame Street: An evaluation* (Program Report No. PR-70-15). Princeton: Educational Testing Service.

Ball, S., & Bogatz, G. A. (1973). *Reading with television: An evaluation of the Electric Company* (Program Report No. PR-73-02). Princeton: Educational Testing Service.

Ball, S., & Goldman, K. S. (1976). *The Adams School An interim report*. Princeton: Educational Testing Service.

Ball, S., & Kazarow, K. M. (1974). *Evaluation of To Reach a Child*. Princeton: Educational Testing Service.

Ball, S., Bogatz, G. A., Kazarow, K. M., & Rubin, D. B. (1974). *Reading with television: A follow-up evaluation of The Electric Company* (Program Report No. PR-74-15). Princeton: Educational Testing Service.

Ball, S., Bridgeman, B., & Beaton, A. E. (1976). *A design for the evaluation of the parent-child development center replication project*. Princeton: Educational Testing Service.

Bogatz, G. A. (1975). Field operations. In S. B. Anderson, S. Ball, & R. T. Murphy (Eds.), *Encyclopedia of educational evaluation* (pp. 169–175). San Francisco: Jossey-Bass Publishers.

Bogatz, G. A., & Ball, S. (1971). *The second year of Sesame Street: A continuing evaluation* (Program Report No. PR-71-21). Princeton: Educational Testing Service.
Boldt, R. F. (with Gitomer, N.). (1975). *Editing and scaling of instrument packets for the clinical evaluation of narcotic antagonists* (Program Report No. PR-75-12). Princeton: Educational Testing Service.
Bussis, A. M., Chittenden, E. A., & Amarel, M. (1976). *Beyond surface curriculum. An interview study of teachers' understandings*. Boulder: Westview Press.
Campbell, P. B. (1976). *Psychoeducational diagnostic services for learning disabled youths* [Proposal submitted to Creighton Institute for Business Law and Social Research]. Princeton: Educational Testing Service.
Clark, M. J., Hartnett, R. Y., & Baird, L. L. (1976). *Assessing dimensions of quality in doctoral education* (Program Report No. PR-76-27). Princeton: Educational Testing Service.
Coleman, J. S., Campbell, E. Q., Hobson, C. J., McPartland, J., Mood, A. M., Weinfeld, F. D., & York, R. L. (1966). *Equality of educational opportunity*. Washington, DC: U.S. Government Printing Office.
Corder, R. A. (1975). *Final evaluation report of part C of the California career education program*. Berkeley: Educational Testing Service.
Corder, R. A. (1976a). *Calexico intercultural design. El Cid Title VII yearly final evaluation reports for grades 7–12 of program of bilingual education, 1970–1976*. Berkeley: Educational Testing Service.
Corder, R. A. (1976b). *External evaluator's final report on the experience-based career education program*. Berkeley: Educational Testing Service.
Corder, R. A., & Johnson, S. (1972). *Final evaluation report, 1971–1972, MANO A MANO*. Berkeley: Educational Testing Service.
Dyer, H. S. (1965a). *A plan for evaluating the quality of educational programs in Pennsylvania* (Vol. 1, pp 1–4, 10–12). Harrisburg: State Board of Education.
Dyer, H. S. (1965b). *A plan for evaluating the quality of educational programs in Pennsylvania* (Vol. 2, pp. 158–161). Harrisburg: State Board of Education.
Echternacht, G., Temp, G., & Storlie, T. (1976). *The operation of an ESEA Title I evaluation technical assistance center—Region* 2 [Proposal submitted to DHEW/O]. Princeton: Educational Testing Service.
Ekstrom, R. B., & Lockheed, M. (1976). Giving women college credit where credit is due. *Findings, 3*(3), 1–5.
Ekstrom, R. B., French, J., & Harman, H. (with Dermen, D.). (1976). *Kit of factor-referenced cognitive tests*. Princeton: Educational Testing Service.
Elias, P., & Wheeler, P. (1972). *Interim evaluation report: BUENO*. Berkeley: Educational Testing Service.
Feldmesser, R. A. (1973). *Educational goal indicators for New Jersey* (Program Report No. PR-73-01). Princeton: Educational Testing Service.
Flaugher, R. L. (1971). *Progress report on the activities of ETS for the postal academy program*. Unpublished manuscript, Educational Testing Service, Princeton.
Flaugher, R., & Barnett, S. (1972). *An evaluation of the prison educational network*. Unpublished manuscript, Educational Testing Service, Princeton.
Flaugher, R., & Knapp, J. (1972). *Report on evaluation activities of the Bread and Butterflies project*. Princeton: Educational Testing Service.
Forehand, G. A., & McDonald, F. J. (1972). *A design for an accountability system for the New York City school system*. Princeton: Educational Testing Service.
Forehand, G. A., Ragosta, M., & Rock, D. A. (1976). *Final report: Conditions and processes of effective school desegregation* (Program Report No. PR-76-23). Princeton: Educational Testing Service.
Frederiksen, N., & Ward, W. C. (1975). *Development of measures for the study of creativity* (Research Bulletin No. RB-75-18). Princeton: Educational Testing Service. http://dx.doi.org/10.1002/j.2333-8504.1975.tb01058.x
Freeberg, N. E. (1970). Assessment of disadvantaged adolescents: A different approach to research and evaluation measures. *Journal of Educational Psychology, 61*, 229–240. https://doi.org/10.1037/h0029243

Hardy, R. A. (1975). *CIRCO: The development of a Spanish language test battery for preschool children.* Paper presented at the Florida Educational Research Association, Tampa, FL.
Hardy, R. (1977). *Evaluation strategy for developmental projects in career education.* Tallahassee: Florida Department of Education, Division of Vocational, Technical, and Adult Education.
Harsh, J. R. (1975). *A bilingual/bicultural project. Azusa unified school district evaluation summary.* Los Angeles: Educational Testing Service.
Hartnett, R. T., Clark, M. J., Feldmesser, R. A., Gieber, M. L., & Soss, N. M. (1974). *The British Open University in the United States.* Princeton: Educational Testing Service.
Harvey, P. R. (1974). *National College of Education bilingual teacher education project.* Evanston: Educational Testing Service.
Holland, P. W., Jamison, D. T., & Ragosta, M. (1976). *Project report no. 1—Phase 1 final report research design.* Princeton: Educational Testing Service.
Hood, D. E. (1972). *Final audit report: Skyline career development center.* Austin: Educational Testing Service.
Hood, D. E. (1974). *Final audit report of the ESEA IV supplementary reading programs of the Dallas Independent School District. Bilingual education program.* Austin: Educational Testing Service.
Hsia, J. (1976). *Proposed formative evaluation of a WNET/13 pilot television program: The Speech Class* [Proposal submitted to educational broadcasting corporation]. Princeton: Educational Testing Service.
Marco, G. L. (1972). *Impact of Michigan 1970–71 grade 3 title I reading programs* (Program Report No. PR-72-05). Princeton: Educational Testing Service.
McDonald, F. J. (1977). *The effects of classroom interaction patterns and student characteristics on the acquisition of proficiency in English as a second language* (Program Report No. PR-77-05). Princeton: Educational Testing Service.
McDonald, F. J., & Elias, P. (1976). *Beginning teacher evaluation study, Phase 2. The effects of teaching performance on pupil learning* (Vol. 1, Program Report No. PR-76-06A). Princeton: Educational Testing Service.
Messick, S. (1970). The criterion problem in the evaluation of instruction: Assessing possible, not just intended outcomes. In M. Wittrock & D. Wiley (Eds.), *The evaluation of instruction: Issues and problems* (pp. 183–220). New York: Holt, Rinehart and Winston.
Messick, S. (1975). Medical model of evaluation. In S. B. Anderson, S. Ball, & R. T. Murphy (Eds.), *Encyclopedia of educational evaluation* (pp. 245–247). San Francisco: Jossey-Bass Publishers.
Murphy, R. T. (1973a). *Adult functional reading study* (Program Report No. PR-73-48). Princeton: Educational Testing Service.
Murphy, R. T. (1973b). *Investigation of a creativity dimension* (Research Bulletin No. RB-73-12). Princeton: Educational Testing Service. http://dx.doi.org/10.1002/j.2333-8504.1973.tb01027.x
Murphy, R. T. (1977). *Evaluation of the PLATO 4 computer-based education system: Community college component.* Princeton: Educational Testing Service.
Powers, D. E. (1973). *An evaluation of the new approach method* (Program Report No. PR-73-47). Princeton: Educational Testing Service.
Powers, D. E. (1974). *The Virginia Beach extended school year program and its effects on student achievement and attitudes—First year report* (Program Report No. PR-74-25). Princeton: Educational Testing Service.
Powers, D. E. (1975a). *Dual audio television: An evaluation of a six-month public broadcast* (Program Report No. PR-75-21). Princeton: Educational Testing Service.
Powers, D. E. (1975b). *The second year of year-round education in Virginia Beach: A follow-up evaluation* (Program Report No. PR-75-27). Princeton: Educational Testing Service.
Rosenfeld, M. (1973). *An evaluation of the Orchard Road School open space program* (Program Report No. PR-73-14). Princeton: Educational Testing Service.
Shipman, V. C. (1970). *Disadvantaged children and their first school experiences* (Vol. 1, Program Report No. PR-70-20). Princeton: Educational Testing Service.
Shipman, V. C. (1974). *Evaluation of an industry-sponsored child care center. An internal ETS report prepared for Bell Telephone Laboratories. Murray Hill, NJ.* Unpublished manuscript, Educational Testing Service, Princeton, NJ.

Sigel, I. E. (1976). *Developing representational competence in preschool children: A preschool educational program. In Basic needs, special needs: Implications for kindergarten programs. Selected papers from the New England Kindergarten Conference, Boston*. Cambridge, MA: The Lesley College Graduate School of Education.
Swinton, S., & Amarel, M. (1978). *The PLATO elementary demonstration: Educational outcome evaluation* (Program Report No. PR-78-11). Princeton: Educational Testing Service.
Thomas, I. J. (1970). *A bilingual and bicultural model early childhood education program. Fountain Valley School District title VII bilingual project*. Berkeley: Educational Testing Service.
Thomas, I. J. (1973). *Mathematics aid for disadvantaged students*. Los Angeles: Educational Testing Service.
Trismen, D. A. (1968). *Evaluation of the Education through Vision curriculum—Phase 1*. Princeton: Educational Testing Service.
Trismen, D. A. (with T. A. Barrows). (1970). *Brevard County project: Final report to the Brevard County (Florida) school system* (Program Report No. PR-70-06). Princeton: Educational Testing Service.
Trismen, D. A., Waller, M. I., & Wilder, G. (1976). *A descriptive and analytic study of compensatory reading programs* (Vols. 1 & 2, Program Report No. PR-76-03). Princeton: Educational Testing Service.
Vale, C. A. (1975). *National needs assessment of educational media and materials for the handicapped* [Proposal submitted to Office of Education]. Princeton: Educational Testing Service.
Ward, W. C., & Frederiksen, N. (1977). *A study of the predictive validity of the tests of scientific thinking* (Research Bulletin No. RB-77-06). Princeton: Educational Testing Service. http://dx.doi.org/10.1002/j.2333-8504.1977.tb01131.x
Wasdyke, R. G. (1976, August). *An evaluation of the Maryland Career Information System* [Oral report].
Wasdyke, R. G. (1977). *Year 3—Third party annual evaluation report: Career education instructional system project. Newark School District. Newark, Delaware*. Princeton: Educational Testing Service.
Wasdyke, R. G., & Grandy, J. (1976). *Field evaluation of Manhattan Community School District #2 environmental education program*. Princeton: Educational Testing Service.
Webb, E. J., Campbell, D. T., Schwartz, R. D., & Sechrest, L. (1966). *Unobtrusive measures: Nonreactive research in the social sciences*. Chicago: Rand McNally.
Woodford, P. E. (1975). *Pilot project for oral proficiency interview tests of bilingual teachers and tentative determination of language proficiency criteria* [Proposal submitted to Illinois State Department of Education]. Princeton: Educational Testing Service.

Chapter 12
Contributions to Education Policy Research

Richard J. Coley, Margaret E. Goertz, and Gita Z. Wilder

Since Educational Testing Service (ETS) was established in 1947, research has been a prominent gene in the organization's DNA. Nine days after its first meeting, the ETS Board of Trustees issued a statement on the new organization. "In view of the great need for research in all areas and the long-range importance of this work to the future development of sound educational programs, it is the hope of those who have brought the ETS into being that it may make fundamental contributions to the progress of education in the United States" (Nardi 1992, p. 22). Highlighting the important role of research, ETS's first president Henry Chauncey recalled, "We tried out all sorts of names. 'Educational Testing Service' has never been wholly satisfactory because it does leave out the research side" (Nardi 1992, p. 16).

As part of its nonprofit mission, ETS conducts and disseminates research to advance quality and equity in education. Education policy research at ETS was formally established with the founding of the Education Policy Research Institute (EPRI) some 40 years ago, and since then ETS research has focused on promoting equal educational opportunity for all individuals, including minority and educationally disadvantaged students, spanning infancy through adulthood. The major objectives of this work are to provide useful and accurate information on educational opportunity and educational outcomes to the public and to policy makers, to inform the debate on important education issues, and to promote equal educational opportunity for all.

R.J. Coley (✉)
Educational Testing Service, Princeton, NJ, USA
e-mail: richardjcoley@gmail.com

M.E. Goertz
University of Pennsylvania, Philadelphia, PA, USA

G.Z. Wilder
Princeton, NJ, USA

R.E. Bennett, M. von Davier (eds.), *Advancing Human Assessment*,
Methodology of Educational Measurement and Assessment,
DOI 10.1007/978-3-319-58689-2_12

The purpose of this chapter is to describe ETS's contribution to education policy research. The authors faced three main challenges in accomplishing this goal. First, we had to define what we mean by education policy research. We broadly defined this term to mean work serving to: define the nature of an educational problem that can be addressed by public or institutional policy (e.g., the achievement gap or unequal access to educational opportunities); identify the underlying causes of the problem; or examine the design, implementation, and impact of public or institutional policies or programs designed to address the problem (see, for example, AERA's *Handbook on Education Policy Research* by Sykes et al. 2009).

The second challenge was organizing the work that ETS has conducted. That research has covered three major areas, which were used to select and classify the work described in this chapter. While these areas do not capture the entire scope of ETS's education policy research, they provide important lenses through which to describe that work. The three major areas are:

- Analyzing, evaluating, and informing public policy in educational governance, including school finance; teacher policy; and federal, state, and local education policy.
- Examining differential access to educational opportunity in three areas of long-standing interest to ETS: the gender gap, advanced placement programs, and graduate education.
- Reporting on the educational outcomes of the U.S. population and describing the contexts for these outcomes and for the gaps in outcomes that exist among segments of the population.

The third challenge was selecting from the thousands of research studies that ETS staff have produced over more than half a century. An unfiltered search of ETS ReSEARCHER,[1] a database of publications by ETS staff members, produced nearly 9,000 publications. And while even this database is incomplete, its size is indicative of the scope of the organization's work in psychometrics, statistics, psychology, and education.

Over the past 40 years, the majority of ETS's education policy research was conducted under three organizational structures that operated at different times within the Research and Development division or its predecessors. EPRI was established at ETS in the early 1970s. Its work was expanded in the Education Policy Research division that existed during the 1980s and 1990s. In 1987, the ETS Board of Trustees established the Policy Information Center to inform the national debate on important education policy issues. Hundreds, if not thousands, of projects were conducted and reports produced within these organizational units. The Policy Information Center alone has produced more than 150 policy reports and other publications. These units and their work were heavily supported by internal funds, made possible by the organization's nonprofit status and mission. The organization's financial

[1]The ETS ReSEARCHER database (http://1340.sydneyplus.com/Authors/ETS_Authors/portal.aspx) is available to anyone interested in additional contributions made by the organization to education policy research and to research in measurement, psychology, statistics, and other areas.

commitment to education policy research has been, and continues to be, substantial.

Given this voluminous output, the authors applied the definition of education policy research and the areas described above to assemble what should be considered only a sample. That is, the work described here is reflective of this large body of work, but necessarily incomplete.

Many of ETS's other activities that are education-policy related and contribute to the field of education are not within the scope of this chapter. Some of this important work serves clearinghouse and collaboration functions. An important example includes the networking activities of the Policy Evaluation and Research Center, which collaborates with organizations such as the Children's Defense Fund and the National Urban League and its affiliates to convene a variety of stakeholders around issues related to the achievement gap. These conferences have focused on the particular challenges facing women and girls, the special circumstances of young Black males, issues related to the community college system, and the importance of family factors in students' success in school.

ETS has also had many long-standing relationships with important organizations such as Historically Black Colleges and Universities, the ASPIRA Association, and the Hispanic Association of Colleges and Universities. ETS researchers, in collaboration with the American Association of Community Colleges, examined a number of challenges faced by community colleges in effectively managing both their academic and vocational functions in the context of rapidly changing economic and demographic patterns and the rapid expansion of nondegreed, credentialing, and certification programs (Carnevale and Descrochers 2001). A second example is the Commission on Pathways through Graduate School and into Careers, led by the Council of Graduate Schools and ETS, which resulted in two important reports that identified the major enrollment, retention, and financial issues facing graduate education in the United States (Wendler et al. 2010, 2012).

ETS's policy research has had influence at several levels. It has played important roles in the development of government and institutional policy, in debates about how U.S. students are achieving and the context around student learning, in school and classroom practice, in assessing the status of the nation's human capital, in the shape of internal ETS programs and services, and in the lives of individuals that have been the focus of ETS's work.

In the next section, the first of three major areas, education policy and governance, is reviewed.

12.1 Education Policy and Governance

Over the years, ETS research in this area has covered school finance and governance, teacher policy, and monitoring education policy developments. Each of these areas will be briefly illustrated.

12.1.1 School Finance and Governance

In 1965, University of Chicago sociologist James Coleman led a team that produced the Coleman report, which shed light on unequal schooling conditions and educational opportunities in the United States (Coleman et al. 1966; see also Barone and Beaton, Chap. 8, this volume). At the same time, scholars began to examine how states' funding of elementary and secondary education contributed to these inequities and to raise questions about the constitutionality of these funding systems. ETS researchers played a major role in the subsequent school finance reform movement of the 1970s and 1980s. ETS undertook groundbreaking research on the design and effects of federal, state, and local finance systems—research that laid the foundation for challenges to the constitutionality of state school finance formulas, for the design of alternative funding formulas, and for the development of tools to assist policy makers and the public in its quest to create more equitable funding structures.

Joel Berke, the first director of EPRI, provided the statistical analyses relied upon by both majority and minority justices in the landmark U.S. Supreme Court decision in *Rodriquez vs. San Antonio.* When a closely divided Court ruled that school funding inequities did not violate the Equal Protection Clause of the 14th Amendment of the U.S. Constitution, school finance reformers turned to the education clauses of state constitutions and state courts for relief. Berke and his colleagues worked with attorneys, education groups, and commissions in several states to analyze the allocation of state and local education funds under existing formulas, to assess options for change, and to examine the effects of court-ordered reform systems. For example, a series of reports titled *Money and Education,* issued between 1978 and 1981, examined the implementation of New Jersey's Public School Education Act of 1975, which used a new formula designed to address the wealth-based disparities in education funding declared unconstitutional by the New Jersey Supreme Court (Goertz 1978, 1979, 1981). These reports, along with a follow-up study in the late 1980s, found that although the state increased its education funding, the law fell far short of equalizing expenditures between poor and wealthy communities. These analyses, along with expert testimony by ETS researcher Margaret Goertz, contributed to the New Jersey Supreme Court's 1990 decision in *Abbott v. Burke* to declare the law unconstitutional as applied to the state's poor urban school districts.

ETS staff also worked with policy makers to design new funding formulas in response to court-ordered change. For example, they assisted the New York City Board of Education and the United Federation of Teachers in designing formula adjustments that would address the special financial and educational needs of large urban school systems. The research culminated in *Politicians, Judges, and City Schools* (Berke et al. 1984), a book written to provide New York policy makers with reform options, as well as a better understanding of the political, economic, and social context for reform and of the trade-offs involved in developing a more equitable school finance system.

In addition to policy makers, ETS research has targeted the public. With support from the National Institute of Education and in collaboration with the American Federation of Teachers, ETS researchers sought to demystify the subject of school finance as a way of encouraging informed participation by educators and the general public in school finance debates. While describing school funding formulas in detail, *Plain Talk About School Finance* (Goertz and Moskowitz 1978) also showed that different school finance equalization formulas were mathematically equivalent. Therefore, the authors argued, the selection of a specific formula was secondary to value-laden political decisions about student and taxpayer equity goals for the system, as well as to how to define various components of the formulas (e.g., wealth, taxpayer effort, and student need) and establish the relationships among the components. Building on their analysis of the mathematical properties of school finance formulas, ETS researchers developed the School Finance Equalization Management System (SFEMS), the first generalizable computer software package for financial data analysis and school finance formula simulations (Educational Testing Service 1978a, b). With technical assistance and training from ETS staff, SFEMS was used by nearly a dozen state education agencies and urban school districts to build their capacity to analyze the equity of their state funding systems and to simulate and evaluate the results of different funding approaches.

The wave of legal and legislative struggles over school funding continued throughout the 1980s, and by 1985 more than 35 states had enacted new or revised education aid programs. ETS researchers took stock of this activity in light of the education reform movement that was taking shape in the early 1990s, calling for national standards and school restructuring. *The State of Inequality* (Barton et al. 1991) provided plain talk about school finance litigation and reform, as well as describing how differences in resources available to schools are related to disparities in educational programs and outcomes. The report detailed the disparity in education funding nationally and within states, reviewed data reported by teachers on the connection between instructional resources and student learning, and reviewed a new wave of court rulings on school funding.

School finance research such as that described above focused on disparities in the allocation of resources within states. ETS researchers, however, were among the first to explore disparities *within* school districts, a current focus of school funding debates and policy. In the early 1970s, ETS researcher Joan Baratz examined the implementation of the *Hobson v. Hansen* decision in Washington, DC, which called for the equalization of per-pupil expenditures for all teachers' salaries and benefits within the district. This remedy was designed to address disparities in spending and staffing between schools enrolling many Black and low-income students versus those enrolling many White and affluent students. Baratz (1975) found a significant reduction in the disparity in allocation of all professional staff among the schools as a result of funding equalization. Changes in resources generally involved exchanging highly paid classroom teachers for lower paid teachers, adding teachers in low-spending schools with high pupil/teacher ratios, and redistributing special subject teachers.

A decade later, ETS researchers conducted a congressionally mandated study of school districts' allocation of Title I resources (Goertz 1988). Because most prior research had focused on the distribution of federal funds to local school districts and the selection of schools and students for Title I services, federal policy makers were concerned about the wide range in per-pupil Title I expenditures across school districts and its impact on the delivery of services to students. The ETS study found that variation in program intensity reflected a series of district decisions about how to best meet the needs of students. These decisions concerned program design (e.g., staffing mixes, case loads, settings), type of program (e.g., prekindergarten, kindergarten, bilingual/English as a second language, basic skills replacement), availability and use of state compensatory education funds, and the extent to which allocation decisions reflected differences in student need across Title I schools.

As it is today, the proper organization of responsibility among federal, state, and local governments was a central issue in policy debates in the 1980s about how best to design programs for students with special educational needs. In July, 1981 a team led by ETS researchers began a congressionally mandated study of how federal and state governments interacted as they implemented major federal education programs and civil rights mandates. The study described how states responded to and were affected by federal education programs. Based on analyses of the laws, on case studies conducted in eight states, and interviews with more than 300 individuals at state and local levels, study results portrayed a robust, diverse, and interdependent federal/state governance system. Among the findings was the identification of three broad factors that appeared to explain states' differential treatment of federal programs—federal program signals, state political traditions and climate, and the management and programmatic priorities of state education agencies (Moore et al. 1983).

The topic of school finance was revisited in 2008 when ETS cosponsored a conference, "School Finance and the Achievement Gap: Funding Programs That Work," that explored the relationship between school finance and academic achievement, highlighted programs that successfully close gaps, and examined the costs and benefits of those programs. While much of the discussion was sobering, evidence supporting the cost effectiveness of prekindergarten programs as well as achievement gains made by students in a large urban school district offered evidence that achievement gaps can be narrowed—if the political will, and the money, can be found (Yaffe 2008).

12.1.2 Teacher Policy

While concern about the quality of the nation's teaching force can be traced back to the early twentieth century, during the past 30 years there has been a growing amount of evidence and recognition that teacher quality is a key factor in student achievement. From publication of *A Nation at Risk* in 1983 (National Commission on Excellence in Education [NCEE] 1983), to the National Education Summit in 1989, to the formation of the National Commission on Teaching and America's

Future in 1994, and the No Child Left Behind (NCLB) Act in 2001, teacher quality has remained squarely at the top of national and state education agendas. ETS policy research has responded to the central issues raised about teacher education and teacher quality at various junctures over this period.

12.1.2.1 Research on the Teacher Education Pipeline

Among the areas of education policy that drew significant attention from state policy makers in response to the perceived decline in the quality of the U.S. education system was a focus on improving the preparedness of individuals entering the teaching profession. In the early 1980s, these policies focused on screening program applicants with tests and minimum grade point averages, prescribing training and instruction for those aspiring to become teachers, and controlling access into the profession by requiring aspiring teachers to pass a licensing test or by evaluating a beginning teacher's classroom performance. While the level of state activity in this area was clear, little was known about the substance or impact of these policies. *The Impact of State Policy on Entrance Into the Teaching Profession* (Goertz et al. 1984) identified and described the policies used by states to regulate entrance into the teaching profession and collected information on the impact of these policies.

The study developed and described a *pipeline* model that identified the various points at which state policies can control the entry of individuals into the teaching profession and illustrated the relationships among these points. Next, the study collected information from all 50 states to identify the points of policy intervention and types of policies in effect in each state. In-depth case studies were also conducted in four states to provide details about the political environment and rationale behind the policies, the extent of coordination across policies, and the impact of the policies on teacher supply and equity. While the necessity of screens in the teacher supply pipeline was apparent, the study found that the approaches used by most states were inadequate to address the issues of equity, coordination, and accountability. For example, the study found that screening people out of teaching, rather than developing the talents of those who want to become teachers, is likely to reduce the socioeconomic and racial/ethnic diversity of the nation's teaching force at the very time that schools were becoming more diverse in the composition of their students. The study made recommendations to improve the quality of teachers coming into the profession while recognizing the importance of maintaining a sufficient supply of teachers to staff the nation's increasingly diverse classrooms.

Another movement that took hold during the 1980s in response to criticism directed at traditional teacher education programs was alternate routes to teaching. While these alternate routes took a variety of forms, The Holmes Group (a consortium of education deans from 28 prominent research universities) along with the American Association of Colleges for Teacher Education endorsed the idea of a 5-year teacher education program leading to a master's degree. The idea was that in addition to courses in pedagogy, teachers should have at least the equivalent of an undergraduate degree in the subject they intend to teach. Like the problem, this remedy was not entirely new. In an attempt to understand the likely impact of such an approach, ETS researchers set out to learn about the decades-old master of arts in teaching (MAT) programs, sponsored by the Ford Foundation in response to concerns about the quality of American education generated by the launching of Sputnik. These MAT programs sought to attract bright liberal arts graduates, prepare them for teaching by giving them graduate work in both their discipline and in pedagogy and by providing them with internships in participating school districts.

After searching the Ford Foundation's archives, the researchers put together profiles of the programs and surveyed nearly 1000 MAT program graduates from 1968 to 1969 to see what attracted them to the programs and to teaching, what were their careers paths, and what were their impressions of their preparation. Remarkably, 81% of the MAT program graduates responded to the survey. Among the results: Eighty-three percent entered teaching and one third who entered teaching were still teaching at the time of the survey. Among those who left teaching, the average time teaching was 5 years. Many of the former teachers pursued education careers outside of the classroom. The study, *A Look at the MAT Model of Teacher Education and Its Graduates: Lessons for Today*, concluded that the MAT model was a viable alternative to increase the supply and quality of the nation's teachers, although more modern programs should be designed to recognize the changing composition of the nation's school population (Coley and Thorpe 1985).

A related focus of ETS research during this period was on finding ways to increase the supply of minority teachers. Declining numbers of minority teachers can be attributed to the limited number of minority students entering and completing college, declining interest in education careers, and the policy screens identified in the study described earlier, including the teacher testing movement. *Characteristics of Minority NTE Test-Takers* (Coley and Goertz 1991) sought to inform interventions to increase minority representation in teaching by identifying the characteristics of minority students who met state certification requirements. The study was the first to collect information on candidates' demographic, socioeconomic, and educational background; education experience in college and graduate school; experiences in teacher education programs; career plans and teaching aspirations; and reasons for taking the certification test. The data analyses focused on determining whether successful and unsuccessful National Teachers Examination (NTE) candidates differed significantly on these background and educational characteristics. Four implications drawn were noteworthy. First, many of the minority candidates were the first generation in their families to attend college, and institutions must develop support programs geared to the academic and financial needs of these students. Second, in general, many low socioeconomic status (SES) students who succeeded in college passed the test. Colleges can and do make a difference for disadvantaged students. Third, recruiting and training policies should reflect the large number of minority students who take various routes into and out of teaching. Last, because only half of successful minority candidates planned to make teaching their career, changes to the structure of the teaching profession should be considered, and the professional environment of teaching should be improved to help retain these students.

A recent study by ETS researchers found that minorities remain underrepresented in the teaching profession and pool of prospective teachers (Nettles et al. 2011). The authors analyzed the performance of minority test takers who took ETS's *PRAXIS*® teacher-certification examinations for the first time between 2005 and 2009 and the relationship of performance with test takers' demographic, socioeconomic, and educational backgrounds, including undergraduate major and undergraduate grade point average (UGPA). They also interviewed students and faculty of teacher education programs at several minority-serving colleges and universities to identify challenges to, and initiatives for, preparing students to pass PRAXIS. The report revealed large score gaps between African American and White teacher candidates on selected *PRAXIS I*® and *PRAXIS II*® tests, gaps as large as those commonly observed on the SAT and *GRE*® tests. Selectivity of undergraduate institution, SES, UGPA, and being an education versus a noneducation major were consistently associated with PRAXIS I scores of African American candidates, particularly in mathematics. Recommendations included focusing on strengthening candidates' academic preparation for and achievement in college and providing students with the other skills and knowledge needed to pass PRAXIS.

ETS research has also informed the debate about how to improve teacher education by examining systems of teacher education and certification outside the United States. *Preparing Teachers Around the World* (Wang et al. 2003) compared teacher education in the United States with the systems in high-performing countries, systematically examining the kinds of policies and control mechanisms used to shape the quality of the teaching forces in countries that scored as well or better than the United States in international math and science assessments. The researchers surveyed the teaching policies of Australia, England, Hong Kong, Japan, Korea, the Netherlands, and Singapore. While no one way was identified that the best performing countries used to manage the teacher pipeline, by and large, they were able to control the quality of individuals who enter teacher education programs through more rigorous entry requirements and higher standards than exist in the United States. One of the most striking findings was that students in these countries are more likely to have teachers who have training in the subject matter they teach. And while much has been made in the United States about deregulating teacher education as a way to improve teacher quality, every high-performing country in the study employed significant regulatory controls on teaching, almost all more rigorous than what is found in the United States.

12.1.2.2 Research on the Academic Quality of the Teaching Force

ETS researchers have tracked the quality of the nation's teaching force in several studies. *How Teachers Compare: The Prose, Document, and Quantitative Literacy of America's Teachers* (Bruschi and Coley 1999) took advantage of the occupational data collected in the National Adult Literacy Survey (NALS) to provide a rare look at how the skill levels of teachers compare with other adults and with adults in other occupations. The results of this analysis were quite positive. America's teachers, on average, scored relatively highly on all three literacy scales and performed as well as other college-educated adults. In addition, the study found that teachers were a labor-market bargain, comparing favorably with other professionals in their literacy skills, yet earning less, dispelling some negative stereotypes that were gaining ground at the time.

In related work to determine whether the explosion of reform initiatives to increase teacher quality during the 1990s and early 2000s was accompanied by changes in the academic quality of prospective teachers, ETS research compared two cohorts of teachers (1994–1997 and 2002–2005) on licensure experiences and academic quality. *Teacher Quality in a Changing Policy Landscape: Improvements in the Teacher Pool* (Gitomer 2007) documented improvements in the academic characteristics of prospective teachers during the decade and cited reasons for those improvements. These reasons included greater accountability for teacher education programs, Highly Qualified Teacher provisions under the NCLB Act, increased requirements for entrance into teacher education programs, and higher teacher education program accreditation standards.

12.1.2.3 Research on Teaching and Student Learning

ETS policy research has also focused on trying to better understand the connection between teaching and classroom learning. ETS researchers have used the large-scale survey data available from the National Assessment of Educational Progress (NAEP) to provide insight into classroom practice and student achievement. *How Teaching Matters: Bringing the Classroom Back Into Discussions About Teacher Quality* (Wenglinsky 2000) attempted to identify which teacher classroom practices in eighth-grade mathematics and science were related to students' test scores. The research concluded that teachers should be encouraged to target higher-order thinking skills, conduct hands-on learning activities, and monitor student progress regularly. The report recommended that rich and sustained professional development that is supportive of these practices should be widely available.

ETS researchers conducted a similar analysis of NAEP data to identify teachers' instructional practices that were related to higher science scores and then examined the extent to which minority and disadvantaged students had access to these types of instruction. In addition to providing a rich description of the eighth-grade science classroom and its teachers, *Exploring What Works in Science Instruction: A Look at the Eighth-Grade Science Classroom* (Braun et al. 2009) found that two apparently effective practices—teachers doing science demonstrations and students discussing science in the news—were less likely to be used with minority students and might be useful in raising minority students' level of science achievement.

12.1.2.4 Research on Understanding Teacher Quality

Along with the recognition of the importance of teacher quality to student achievement have come a number of efforts to establish a quantitative basis for teacher evaluation. These efforts are typically referred to as value-added models (VAMs) and use student test scores to compare teachers. To inform the policy debate, ETS published a report on the topic. *Using Student Progress to Evaluate Teachers: A Primer on Value-Added Models* (Braun 2005) offered advice for policy makers seeking to understand both the potential and the technical limitations that are inherent in such models.

Also related to teacher evaluation, ETS partnered with several organizations as part of the National Comprehensive Center for Teacher Quality (NCCTQ) to produce reports aimed at improving the quality of teaching, especially in high-poverty, low-performing, and hard-to-staff schools. One effort by ETS researchers lays out an organizational framework for using evaluation results to target professional development opportunities for teachers, based on the belief that teacher accountability data can also be used to help teachers improve their practice (Goe et al. 2012). To help states and school districts construct high-quality teacher evaluation systems for employment and advancement, Goe and colleagues collaborated with NCCTQ partners to produce a practical guide for education policy makers on key areas to be addressed in developing and implementing new systems of teacher evaluation (Goe et al. 2011).

Work on teacher quality continues as ETS researchers grapple with policy makers' desire to hold teachers accountable for how much students learn. Studies that examine a range of potential measures of teaching quality, including classroom observation protocols, new measures of content knowledge for teaching, and measures based on student achievement, are ongoing. The studies investigate a wide range of approaches to measuring teaching quality, especially about which aspects of teaching and the context of teaching contribute to student learning and success.

12.1.3 Monitoring Education Policy Developments

Much of the Policy Information Center's work has focused on reporting on education policy developments and on analyzing the educational achievement and attainment of the U.S. population, as well as identifying and describing a range of factors that influence those outcomes. In monitoring and describing the changing education policy landscapes that evolved over the decades, the Center sought to anchor data on achievement and attainment to relevant educational reform movements. A sample of that work is provided next.

The decade of the 1980s that began with the publication of *A Nation at Risk* (NCEE 1983) witnessed extensive policy changes and initiatives led by governors and state legislatures, often with strong backing from business. *The Education Reform Decade* (Barton and Coley 1990) tracked changes at the state level between

1980 and 1990 in high school graduation requirements, student testing programs, and accountability systems, as well as sweeping changes in standards for teachers. Changes at the local level included stricter academic and conduct standards, more homework and longer school days, and higher pay for teachers. By the decade's end, 42 states had raised high school graduation requirements, 47 states had established statewide testing programs, and 39 states required passing a test to enter teacher education or begin teaching (Coley and Goertz 1990).

Against this backdrop often referred to as the *excellence movement*, the report provided a variety of data that could be used to judge whether progress was made. These data included changes in student achievement levels, several indicators of student effort, and success in retaining students in school. Data were also provided regarding progress toward increasing equality and decreasing gaps between minority and majority populations and between males and females. Some progress in closing the gaps in achievement, particularly between White and Black students, as well as modest progress in other areas, prompted this November 15, 1990, headline in *USA Today*: "Reforms Put Education on Right Track" (Kelly 1990). Then-ETS President Gregory R. Anrig noted at the press conference releasing the report, "The hallmark of the decade was a move toward greater equality rather than a move toward greater excellence" (Henry 1990, p. 1).

One of the more tangible outcomes of the education-reform decade was the near universal consensus that the high school curriculum should be strengthened. The National Commission on Excellence in Education recommended that all high school students should complete a core curriculum of 4 years of English; 3 years each of social studies, science, and mathematics; 2 years of a foreign language; and one-half year of computer science. Progress toward attaining this new standard was tracked by two ETS reports. *What Americans Study* (Goertz 1989) and *What Americans Study Revisited* (Coley 1994) reported steady progress in student course-taking between 1982 and 1990. While only 2% of high school students completed the core curriculum in 1982, the percentage rose to 19 in 1990. In addition, 40% of 1990 high school graduates completed the English, social studies, science, and mathematics requirements, up from 13% in 1982. The 1994 report also found that the level of mathematics course-taking increased in advanced sequences and decreased in remedial ones.

Along with changes in what students study, the explosion of state testing programs that occurred in the 1970s carried over and expanded in the 1980s with the excellence movement. Perhaps the most notable change was the growth of elementary and secondary school testing across the states. As the 1990s began, there were increasing calls to broaden educational assessment to include performance assessment, portfolios of students' work, and constructed-response for which students had to come up with an answer rather than fill in a bubble. By the 1992–1993 school year, only Iowa, Nebraska, New Hampshire, and Wyoming did not have a state testing program.

Testing in America's Schools (Barton and Coley 1994) documented the testing and assessment changes that were occurring across the country. The report used information from NAEP, a study from what was then the U.S. General Accounting

Office, and a survey of state testing directors conducted by the Council of Chief State School Officers to provide a profile of state testing programs in the early 1990s, as well as a view of classroom testing. The report noted that while the multiple-choice exam was still America's test of choice, the use of alternative methods was slowly growing, with many states using open-ended questions, individual performance assessments, and portfolios or learning records.

As the 1990s drew to a close, President Clinton and Vice President Al Gore called for connecting all of America's schools to the *information superhighway*, federal legislation was directing millions of dollars to school technology planning, and a National Education Summit of governors and business leaders pledged to help schools integrate technology into their teaching. Amid this activity and interest *Computers and Classrooms: The Status of Technology In U.S. Schools* (Coley et al. 1997) was published to meet a need for information on how technology is allocated among different groups of students, how computers are being used in schools, how teachers are being trained in its use, and what research shows about the effectiveness of technology. The report made headlines in *The Washington Post, USA Today, The Philadelphia Inquirer,* and *Education Week* for uncovering differences in computer use by race and gender. Among other findings were that poor and minority students had less access than other students to computers, multimedia technology, and the Internet.

While publications such as *Education Week* now take the lead in describing the policy landscape, there are occasions when ETS research fills a particular niche. Most recently, for example, information on pre-K assessment policies was collected and analyzed in *State Pre-K Assessment Policies: Issues and Status* (Ackerman and Coley 2012). In addition to information on each state's assessments, the report focused on reminding policy makers about the special issues that are involved in assessing young children and on sound assessment practices that respond to these challenges. In this area, ETS contributes by keeping track of important developments while at the same time providing leadership in disseminating tenets of proper test use.

12.2 Access to Educational Opportunities Along the Education Pipeline

ETS's mission has included broadening access to educational opportunities by groups other than the White middle-class population that had traditionally—and often disproportionately—enjoyed the benefits of those opportunities. Increasing access to graduate education, particularly for underrepresented groups, requires improving educational opportunities from early childhood through high school and college. Over the years, ETS researchers have studied differential access to quality education at all points along the educational pipeline. For example, ETS research on early childhood education has included seminal evaluations of the impact on traditionally underserved groups of such educational television programs as Sesame Street and The Electric Company (Ball and Bogatz 1970; Ball et al. 1974), and improving the quality of early childhood assessments (Ackerman and Coley 2012; Jones 2003). Other researchers have focused on minority students' access to mathematics and science in middle schools (see, for example, Clewell et al. 1992), and individual and school factors related to success in high school (see, for example, Ekstrom et al. 1988). ETS research on the access of underrepresented groups to higher education has also included evaluations of promising interventions, such as the Goldman Sachs Foundation's Developing High Potential Youth Program (Millett and Nettles 2009). These and other studies are too numerous to summarize in this chapter. Rather, we focus on contributions of ETS research in several areas of longstanding interest to the organization—gender differences, access to advanced placement courses in high school, and access to graduate education.

12.2.1 The Gender Gap

Much has been written about the gender gap. ETS has traditionally tracked the trajectories of scores on its own tests, and multiple reports have been dedicated to the topic. A 1989 issue of *ETS Policy Notes* examined male-female differences in NAEP results and in SAT and *PSAT/NMSQT*® scores (Coley 1989). An entire volume by Warren W. Willingham and Nancy Cole was devoted to the topic in the context of test fairness (Willingham and Cole 1997). And a 2001 report deconstructed male-female differences within racial/ethnic groups along with course-taking data, attempting to understand gender differences in educational achievement and opportunity across racial/ethnic groups (Coley 2001). The consensus from much of this work has been that the causes of the male-female achievement gap are many, varied, and complex.

In 1997, then-president of ETS Cole authored a report titled *The ETS Gender Study: How Males and Females Perform in Educational Settings* (Cole 1997). The report was based on 4 years of work by multiple researchers using data from more than 1500 data sets, many of them large and nationally representative. The collective studies used 400 different measures that cut across grades, academic subjects, and years and involved literally millions of students.

Although the study yielded many important and interesting findings, Cole chose to focus on several that were contrary to common expectations. Among them were the following:

- For many subjects, the differences between males and females are quite small, but there are some real differences in some subjects.
- The differences occur in both directions. In some areas, females outperform males, and in others the opposite is true.

- Dividing subjects by component skills produces a different picture of gender differences than those found for academic disciplines more generally.
- Gender differences increase over years in school. Among fourth-grade students, there are only minor differences in test performance on a range of school subjects. The differences grow as students progress in school and at different rates for different subjects.
- Gender differences are not easily explained by single variables such as course-taking or types of test. They are also reflected in differences in interests and out-of-school activities.

Cole concluded that "…while we can learn significant things from studying group behavior, these data remind us to look at each student as a unique individual and not stereotype anyone because of gender or other characteristics" (Cole 1997, p. 26).

Over the years, ETS researchers have sought to determine what factors contribute to the underrepresentation of women in the fields of science, technology, engineering, and mathematics (STEM), going back to elementary and secondary education. Marlaine E. Lockheed, for example, conducted studies of sex equity in classroom interactions (Lockheed 1984) and early research on girls' participation in mathematics and science and access to technology (Lockheed 1985; Lockheed et al. 1985). Building on this and related work, Clewell et al. (1992) identified what they determined were major barriers to participation by women and minorities in science and engineering: (a) negative attitudes toward mathematics and science; (b) lower performance levels than White males in mathematics and science courses, and on standardized tests; (c) limited exposure to extracurricular math- and science-related activities, along with failure to a participate in advanced math and science courses in high school; and (d) lack of information about or interest in math or science careers. Making a case for developing interventions aimed at the critical middle school years, they offered descriptions and case studies of ten intervention programs, then relatively recent phenomena, that the authors considered successful, along with a series of recommendations derived from the programs.

12.2.2 Access to Advanced Placement®

Providing high school students access to advanced coursework has long been considered an important means of preparing students for future success. This preparation is particularly important for minority students, who score, on average, lower than nonminority students. ETS researchers studied the characteristics of minority students with high SAT scores and found that these students tended to excel in advanced coursework in high school, including advanced placement courses (Bridgeman and Wendler 2005).

The College Board's *Advanced Placement Program®* (*AP®*) is a collaborative effort between secondary and postsecondary institutions that provides students opportunities to take freshman-level college courses while still in high school. The need for such opportunities is particularly acute for students from low-income families and students from racial/ethnic minorities. ETS researchers used a novel approach to examine data on AP program activity by merging AP-participation data from the College Board with a national database containing information on all U.S. high schools. By matching students with their high schools, the researchers were able to view AP program participation and performance in the context of high school characteristics, including such factors as school size, locale, and socioeconomic status. The unique view provided by *Access to Success: Patterns of Advanced Placement Participation in U.S. High Schools* (Handwerk et al 2008) was illuminating.

The report showed that while most students attended a high school at which the AP program was available, few students actually took an AP exam even after taking an AP course, and only a fraction of those who did take a test scored high enough to qualify for college credit or placement. In addition, patterns of participation for low-income and underrepresented minority students, and for students attending small, rural high schools, were particularly diminished.

The study concluded by identifying changes that could improve access to AP courses by schools and school districts. For more students to reap the benefits of AP program participation, the authors suggested that public schools make greater efforts to broaden their programs and to create a culture of academic rigor within their schools. The analyses demonstrated that students from underrepresented groups in particular were more likely to participate in the AP program in schools that offered more rigorous programs.

12.2.3 Access to and Participation in Graduate Education

In 1982, the then-called Minority Graduate Education Committee of the GRE Board took measures to address what it labeled "the severe underrepresentation of minority students in graduate education" (Baratz-Snowden et al. 1987, p. 3). In doing so, the Committee specified four critical stages in the graduate education process that a research agenda should address: preadmission, admission, enrollment, and retention/completion. The request resulted in a detailed research agenda and funded studies to address gaps in knowledge about the graduate education pipeline. Researchers were aided by a database, developed specifically for the purpose of studying talent flow, which contained responses from the GRE General Test background questionnaire for individuals taking the test between 1982 and 1993. This information included test takers' undergraduate majors, intended areas of graduate study, parents' education, undergraduate courses taken and grade-point averages, and whether test takers changed majors. Using this database, ETS researchers investigated the flow of minority students through the education pipeline from high school through graduate school (Brown 1987; Grandy 1995), the effects of financial aid on minority graduate school enrollment (Ekstrom et al. 1991; Nettles 1987; Wilder and Baydar

1990), and minority student persistence and attainment in graduate education (Clewell 1987; Nettles 1990; Thomas et al. 1992; Zwick 1991).

A comprehensive report issued in April 1994 summarized what had been learned about minority students in the upper portion of the education pipeline, such as their rates of completing high school, college, and graduate education; research findings that helped to explain the data; and suggestions for future research (Brown et al. 1994). This report concluded that the pipeline for Black and Hispanic students leading to completion of graduate and/or professional degrees grows narrower the higher the level. For example, while high school and college completion rates rose for African-American students, participation in undergraduate and graduate education differed markedly among minority groups, including in the types of institutions they attended and the fields of study they pursued. While the number of minority graduate students also grew, they remained a small proportion of total graduate enrollments, and even fewer minority students persisted to receive doctoral degrees. Minority graduate students were also heavily concentrated in the field of education and underrepresented in STEM fields.

Brown et al.'s (1994) synthesis also identified several factors that potentially explained the underrepresentation of minority students. These factors included a lack of minority recruitment programs at the graduate school level, a mismatch in academic interests between minority students and faculty, lack of financial aid, and unsupportive institutional climate. The level of undergraduate debt did not appear to affect enrollment in graduate school. The type of financial aid a graduate student received, however, did appear to affect both time to degree and integration into the academic life of a department. Minority students were more likely to receive grants and fellowships than hold the teaching and research assistantships that would give them access to mentoring and apprenticeship opportunities.

A qualitative study of minority students who did persist through doctoral study found that persisters came from low socioeconomic backgrounds, had been high achievers in high school, had supportive major advisers, were pursuing doctoral degrees to fulfill a desire for knowledge, and completed their doctoral study in spite of wanting to leave their programs to avoid experiencing failure (Clewell 1987). Institutional factors that supported persistence included institution-wide services for minority students beyond the level of the individual department, early identification of minority applicants, support services focused on these students' needs, and monitoring the effectiveness of such efforts.

Finally, ETS researchers conducted one of the largest surveys of American graduate students, collecting data from more than 9000 students in 21 of the nation's major doctorate-granting institutions and representing 11 fields of study. This decade-long project resulted in the publication of *Three Magic Letters: Getting to Ph.D.* (Nettles and Millett 2006). The authors' findings shed light on multiple factors that are critical to the progression of the doctoral degree, particularly adequate institutional funding and availability of engaged and accessible faculty mentors.

12.3 Reporting and Understanding Educational Outcomes

Enhancing educational opportunities for all individuals, particularly minority and educationally disadvantaged populations, requires an understanding of the educational achievement and attainment levels of the nation's population. Helping the public and policy makers to get a comprehensive view of the nation's educational achievement and attainment outcomes and how they differ across population groups has been a major focus of ETS's policy research, at both the elementary and the secondary education level and for the adult population. This section describes some of that work.

12.3.1 Elementary and Secondary Education

The achievement gap has deep roots in American society, and the nation's efforts to address it have a long history. Expectations for addressing the gap increased with the *Brown v. Board of Education* desegregation decision in 1954 and with the passage of the Elementary and Secondary Education Act of 1965 (ESEA), which focused on the inequality of school resources and sought to target more aid to disadvantaged children. The Civil Rights Act of 1964 sparked optimism for progress in education and in society at large. Reauthorizations of ESEA, such as the NCLB Act, required that achievement data from state assessments be disaggregated by population group to expose any uneven results, for which schools were to be held accountable.

In closing the achievement gap, there have been a few periods of progress. The ETS report, *The Black-White Achievement Gap: When Progress Stopped* (Barton and Coley 2010), documented the period starting from the 1970s until the late 1980s when the gap in NAEP reading and mathematics scores narrowed significantly and sought to understand what factors may have coincided with that narrowing. The report noted the irony that the very children born in the mid-1960s, when the landmark legislation was created, were the ones for whom progress slowed or stopped. While some of the progress is credited to changes in the education and income levels of minority families relative to White families, the reasons for most of the gap closure remain largely unexplained. The authors identified a number of factors that may have contributed to stalled progress, including the decline of minority communities and neighborhoods and stalled intergenerational mobility out of seriously disadvantaged neighborhoods. In dedicating the report to the late Senator Daniel Patrick Moynihan, the authors acknowledged his prescient warning on the deteriorating condition of low-income Black families nearly a half century ago.

Two other ETS reports helped increase understanding of how home, school, and environmental factors affected student achievement and contributed to the achievement gaps that exist across our population. When *Parsing the Achievement Gap: Baselines for Tracking Progress* (Barton 2003) and *Parsing the Achievement Gap II*

(Barton and Coley 2009) were released, they received considerable media attention and stimulated much debate about what actions to take.

The first report identified 14 life conditions and experiences that research had established as correlates of educational achievement and then gathered and examined data to determine whether these 14 factors differed across racial/ethnic or socioeconomic groups. For example, if research documents that low birth weight adversely affects a child's cognitive development, is there a greater incidence of this condition in minority or lower income populations? The 14 correlates comprised school-related factors, such as teacher experience, school safety, and curriculum rigor, as well as factors experienced before and outside of school, such as the number of parents in the home, television watching, and hunger and nutrition. The results were unambiguous—in all 14 correlates, there were gaps between minority and majority student populations. And for the 12 correlates where data were available, 11 also showed differences between low-income and higher income families.

The second report (Barton and Coley 2009) updated the first synthesis to see whether the gaps identified in the correlates narrowed, widened, or remained unchanged. In brief, the update concluded that while a few of the gaps in the correlates narrowed and a few widened, overall, the gaps identified in the first report remain unchanged. Both reports took care to emphasize that the correlates include school experiences as well as out-of-school experiences and cautioned that any effort to close the achievement gap would have to focus on both areas.

As the first decade of the 2000s was drawing to a close, ETS researchers made another effort to help policy makers, educators, and parents better understand that raising student achievement involves much more than improving what goes on in classrooms. Enhancing that understanding was critical given that a presidential election was on the horizon and that a debate in Congress was ongoing about the reauthorization of the NCLB Act. In *The Family: America's Smallest School* (Barton and Coley 2007), ETS researchers made the case that the family and home are where children begin learning long before they start school and where they spend much of their time after they enter school. The report took stock of the family's critical role as a child's first school, examining many facets of the home environment and experiences that foster children's cognitive and academic development. These facets included the number of parents in the home, family finances, early literacy activities, the availability of high-quality childcare, and parents' involvement with school.

12.3.2 The Literacy of the Nation's Adults

The education and skills of a nation's adult population represent the human capital that will allow it to compete in a changing labor market, both domestically and internationally (see Kirsch et al., Chap. 9, this volume). ETS's work in large-scale adult literacy assessments began in 1984 and continues today. One of these surveys, the NALS, provided a breakthrough in assessing the literacy of U.S. adults (Kirsch et al.

1990). While earlier studies tried to count the number of people unable to read or write in the nation, NALS profiled the literacy of adults based on their performance across a wide variety of tasks that reflects the types of materials and demands encountered in daily life, such as reading a bus schedule, filling out a job application, or balancing a checkbook. The definition of literacy used in NALS enabled researchers to profile the entire population in their use of printed and written information to function in society, to achieve one's goals, and to develop one's knowledge and potential. NALS rated adults' prose, document, and quantitative literacy in terms of five levels. In prose literacy, for example, someone scoring at Level 1 can read a short text to locate a single piece of information, while someone at Level 5 is able to make high-level inferences or use specialized background knowledge.

The NALS results indicated that nearly 40 million Americans were estimated to perform at Level 1 on all three scales, able only to perform simple routine tasks involving uncomplicated texts and documents. Another 50 million were estimated to be at Level 2, able to locate information in text, to make low-level inferences using printed materials, or to perform single-operation mathematics. Low literacy proficiency was not spread out uniformly among the population, however. Background information on demographics, education, labor market experiences, income, and activities such as voting, television watching, and reading habits that NALS collected from respondents enabled ETS researchers to connect individual characteristics with literacy skills.

The skills gaps revealed by NALS occurred at a time in our history when the rewards for literacy skills were growing, both in the United States and across the world. *Pathways to Labor Market Success: The Literacy Proficiency of U.S. Adults* (Sum et al. 2004) reviewed the literacy skills of the employed population in the United States and other countries and explored the links between the occupations, wages, and earnings of workers and their skills. Analyses revealed that low proficiency scores were associated with lower rates of labor-force participation and large gaps in earnings. Moreover, workers with higher skill levels were also more likely to participate in education and training, contributing to the gap between the haves and have-nots.

Literacy skills are not only connected with economic returns, but with other outcomes as well. Data show that these skills are associated with the likelihood of participating in lifelong learning, keeping abreast of social and political events, voting in national and local elections, and other important outcomes. *Literacy and Health in America* (Rudd et al. 2004) found that literacy was one of the major pathways linking education and health and that literacy skills may be a contributing factor to the disparities that have been observed in the quality of health care that individuals receive.

Results from NALS and from international literacy surveys conducted by ETS also provided a comparative perspective on the U.S. population. Despite its high ranking in the global economy, results from *The Twin Challenges of Mediocrity and Inequality: Literacy in the U.S. From an International Perspective* (Sum et al. 2002) found that the United States is average when the literacy skills of its adults are compared to those of adults in 20 other high-income countries, but is leads in the magnitude of the difference between those at the top and bottom of the skill distribution. These findings are supported by the results of school-age surveys such as NAEP, Trends in International Mathematics and Science Study (TIMSS), and Programme for International Student Assessment (PISA). It appears that other countries, recognizing the important role that human capital plays in social and economic development, have invested in the skills of their populations and have begun to catch up to the United States. All of this information was brought together with the release of *America's Perfect Storm: Three Forces Changing America's Future* (Kirsch et al. 2007).

America's Perfect Storm described three forces that are coming together to potentially create significant consequences: inadequate skill levels among large segments of the population, the continuing evolution of the economy and the changing nature of U.S. jobs, and a seismic shift in the demographic profile of the nation. As part of their analyses, given current skill levels and future demographic patterns, the authors estimated that the distribution of prose, document, and quantitative literacy in 2030 will shift in such a way that over the next 25 years or so the better educated individuals leaving the workforce will be replaced by those who, on average, have lower levels of education and skills. This downward proficiency shift will occur at a time when nearly half of the projected job growth will be concentrated in occupations requiring *higher* levels of education and skills. The authors argued that if our society's overall skill levels are not improved and if the existing gaps in achievement and attainment are not narrowed, these conditions will jeopardize American competitiveness and could ultimately threaten our democratic institutions.

12.4 Conclusion

ETS's nonprofit mission has supported a program of education policy research that has spanned nearly four decades. From studies that documented the promise of television as an educational tool, to analyses of state school finance systems that resulted in more equitable distribution of money for schools, to expanding the public's and policy makers' understanding of the achievement gap among America's students, ETS research has contributed a wealth of information on educational opportunity and educational outcomes to inform the policy debate in the United States. Of paramount importance has been a focus on enhancing educational opportunity for all individuals, especially for minority and disadvantaged groups.

The breadth and scope of this work have posed challenges to adequately summarizing it within a single chapter. The approach chosen by the authors was to produce a sampler organized around three broad themes selected to illustrate important areas of ETS's work. As such, this chapter is necessarily incomplete. At best, and in line with the authors' modest intentions, the chapter gives a flavor for the breadth and depth of the work undertaken since the establishment of a policy research unit at ETS in the early 1970s.

As the organization continues to contribute to the education policy debate, it is the hope and expectation of the authors that the work will continue to be of high quality, and relevant to the decision making needs of the public it serves.

References

Ackerman, D. J., & Coley, R. J. (2012). *State pre-K assessment policies: Issues and status* (Policy Information Report). Princeton: Educational Testing Service.

Ball, S., & Bogatz, G. A. (1970). *The first year of Sesame Street: An evaluation* (Program Report No. PR-70-15). Princeton: Educational Testing Service.

Ball, S., Bogatz, G. A., Kazarow, K. M., & Rubin, D. B. (1974). *Reading with television: A follow-up evaluation of the Electric Company* (Program Report No. PR-74-15). Princeton: Educational Testing Service.

Baratz, J. C. (1975). Court decisions and educational change: A case history of the DC public schools, 1954–1974. *Journal of Law and Education, 4*, 63–80.

Baratz-Snowden, J. C., Brown, S. V., Clewell, B. C., Nettles, M. T., & Wightman, L. (1987). *Research agenda for Graduate Record Examinations® Board Minority Graduate Education Project* (Minority Graduate Education Report No. MGE-94-01). Princeton: Educational Testing Service.

Barton, P. E. (2003). *Parsing the achievement gap: Baselines for tracking progress* (Policy Information Report). Princeton: Educational Testing Service.

Barton, P. E., & Coley, R. J. (1990). *The education reform decade* (Policy Information Report). Princeton: Educational Testing Service.

Barton, P. E., & Coley, R. J. (1994). *Testing in America's schools* (Policy Information Report). Princeton: Educational Testing Service.

Barton, P. E., & Coley, R. J. (2007). *The family: America's smallest school* (Policy Information Report). Princeton: Educational Testing Service.

Barton, P. E., & Coley, R. J. (2009). *Parsing the achievement gap II* (Policy Information Report). Princeton: Educational Testing Service.

Barton, P. E., & Coley, R. J. (2010). *The Black-White achievement gap: When progress stopped* (Policy Information Report). Princeton: Educational Testing Service.

Barton, P. E., Coley, R. J., & Goertz, M. E. (1991). *The state of inequality* (Policy Information Report). Princeton: Educational Testing Service.

Berke, J. S., Goertz, M. E., & Coley, R. J. (1984). *Politicians, judges, and city schools: Reforming school finance in New York*. New York: Russell Sage Foundation.

Braun, H. (2005). *Using student progress to evaluate teachers: A primer on value-added models* (Policy Information Perspective). Princeton: Educational Testing Service.

Braun, H., Coley, R., Jia, Y., & Trapani, C. (2009). *Exploring what works in science instruction: A look at the eighth-grade science classroom* (Policy Information Report). Princeton: Educational Testing Service.

Bridgeman, B., & Wendler, C. (2005). *Characteristics of minority students who excel on the SAT and in the classroom* (Policy Information Report). Princeton: Educational Testing Service.

Brown, S. V. (1987). *Minorities in the graduate education pipeline* (Minority Graduate Education Project Report No. MGE-87-02). Princeton: Educational Testing Service.

Brown, S. V., Clewell, B. E., Ekstrom, R. B., Goertz, M. E., & Powers, D. E. (1994). *Research agenda for the Graduate Record Examinations Board Minority Graduate Education Project: An update* (Minority Graduate Education Report No. MGE-94-01). Princeton: Educational Testing Service.

Bruschi, B. A., & Coley, R. J. (1999). *How teachers compare: The prose, document, and quantitative literacy of America's teachers* (Policy Information Report). Princeton: Educational Testing Service.
Carnevale, A., & Descrochers, D. M. (2001). *Help wanted...credentials required: Community colleges in the knowledge economy*. Princeton: Educational Testing Service.
Clewell, B. C., (1987). *Retention of Black and Hispanic doctoral students* (GRE Board Research Report No. 83-4R). Princeton: Educational Testing Service.
Clewell, B. C., Anderson, B. T., & Thorpe, M. E. (1992). *Breaking the barriers: Helping female and minority students succeed in mathematics and science*. San Francisco: Jossey-Bass.
Cole, N. (1997). *The ETS gender study: How males and females perform in educational settings*. Princeton: Educational Testing Service.
Coleman, J. S., Campbell, E. Q., Hobson, C. J., McPartland, J., Mood, A. M., Weinfeld, F. D., & York, R. L. (1966). *Equality of educational opportunity*. Washington, DC: U.S. Government Printing Office.
Coley, R. J. (1989, October). The gender gap in education: How early and how large? *ETS Policy Notes, 2*(1), 1–9.
Coley, R. J. (1994). *What Americans study revisited* (Policy Information Report). Princeton: Educational Testing Service.
Coley, R. J. (2001). *Differences in the gender gap: Comparisons across racial/ethnic groups in education and work* (Policy Information Report). Princeton: Educational Testing Service.
Coley, R. J., & Goertz, M. E. (1990). *Educational standards in the 50 states: 1990* (Research Report No. RR-90-15). Princeton: Educational Testing Service. http://dx.doi.org/10.1002/j.2333-8504.1990.tb01347.x
Coley, R. J., & Goertz, M. E. (1991). *Characteristics of minority NTE test-takers* (Research Report No. RR-91-04). Princeton: Educational Testing Service. http://dx.doi.org/10.1002/j.2333-8504.1991.tb01370.x
Coley, R. J., & Thorpe, M. E. (1985). *A look at the MAT model of teacher education and its graduates: Lessons for today*. Princeton: Educational Testing Service.
Coley, R. J., Cradler, J., & Engle, P. (1997). *Computers and classrooms: The status of technology in U.S. schools* (Policy Information Report). Princeton: Educational Testing Service.
Educational Testing Service. (1978a). *The School Finance Equalization Management System (SFEMS), Version 3.0: Volume I. The users' policy guide*. Princeton: Author.
Educational Testing Service. (1978b). *The School Finance Equalization Management System (SFEMS), Version 3.0: Volume II. The users' technical guide*. Princeton: Author.
Ekstrom, R. B., Goertz, M. E., & Rock, D. A. (1988). *Education and American youth: The impact of the high school experience*. London: Falmer Press.
Ekstrom, R. B., Goertz, M. E., Pollack, J. C., & Rock, D. A. (1991). *Undergraduate debt and participation in graduate education: The relationship between educational debt and graduate school aspirations, applications, and attendance among students with a pattern of full-time continuous postsecondary education* (GRE Board Research Report No. 86-05). Princeton: Educational Testing Service. http://dx.doi.org/10.1002/j.2333-8504.1982.tb01330.x
Gitomer, D. H. (2007). *Teacher quality in a changing policy landscape: Improvements in the teacher pool* (Policy Information Report). Princeton: Educational Testing Service.
Goe, L., Holdheide, L., & Miller, T. (2011). *A practical guide to designing comprehensive teacher evaluation systems*. Washington, DC: National Comprehensive Center for Teacher Quality.
Goe, L., Biggers, K., & Croft, A. (2012). *Linking teacher evaluation to professional development: Focusing on improving teaching and learning*. Washington, DC: National Comprehensive Center for Teacher Quality.
Goertz, M. E. (1978). *Money and education: Where did the 400 million dollars go? The impact of the New Jersey Public School Education Act of 1975*. Princeton: Educational Testing Service.
Goertz, M. E. (1979). *Money and education: How far have we come? Financing New Jersey education in 1979* (Research Report No. RR-79-04). Princeton: Educational Testing Service. http://dx.doi.org/10.1002/j.2333-8504.1979.tb01172.x

Goertz, M. E. (1981). *Money and education in New Jersey: The hard choices ahead.* Princeton: Educational Testing Service.
Goertz, M. E. (1988). *School districts' allocation of Chapter 1 resources* (Research Report No. RR-88-16). Princeton: Educational Testing Service. http://dx.doi.org/10.1002/j.2330-8516.1988.tb00272.x
Goertz, M. E. (1989). *What Americans study* (Policy Information Report). Princeton: Educational Testing Service.
Goertz, M. E., & Moskowitz, J. (1978). *Plain talk about school finance.* Washington, DC: National Institute of Education.
Goertz, M. E., Ekstrom, R., & Coley, R. (1984). *The impact of state policy on entrance into the teaching profession.* Princeton: Educational Testing Service.
Grandy, J. (1995). *Talent flow from undergraduate to graduate school: 1982–1993* (GRE Board Professional Report No. 92-02P). Princeton: Educational Testing Service.
Handwerk, P., Tognatta, N., Coley, R. J., & Gitomer, D. H. (2008). *Access to success: Patterns of Advanced Placement participation in U.S. high schools* (Policy Information Report). Princeton: Educational Testing Service.
Henry, T. (1990, November 15). *More equality is hallmark of 1980s education reform.* Retrieved from the AP Archive: http://www.apnewsarchive.com/1990/More-Equality-is-Hallmark-of-1980s-Education-Reform/id-0b990bf094e5830447272dca74d964cb
Jones, J. (2003). *Early literacy assessment systems: Essential elements* (Policy Information Perspective). Princeton: Educational Testing Service.
Kelly, D. (1990, November 15). *Reforms put education on right track.* USA TODAY, D1.
Kirsch, I. S., Jungeblut, A., Jenkins, L. B., & Kolstad, A. (1990). *Adult literacy in America: A first look at the results of the National Adult Literacy Survey.* Washington, DC: Government Printing Office.
Kirsch, I., Braun, H., Yamamoto, K., & Sum, A. (2007). *America's perfect storm: Three forces changing America's future* (Policy Information Report). Princeton: Educational Testing Service.
Lockheed, M. E. (1984). *A study of sex equity in classroom interaction: Final report* (Vol. 1). Princeton: Educational Testing Service.
Lockheed, M. E. (1985). Women, girls and computers: A first look at the evidence. *Sex Roles, 13*(3), 115–122. https://doi.org/10.1007/BF00287904
Lockheed, M. E., Thorpe, M., Brooks-Gunn, J., Casserly, P., & McAloon, A. (1985). *Sex and ethnic differences in middle school math, science and computer science: What do we know?* Princeton: Educational Testing Service.
Millett, C. M., & Nettles, M. T. (2009). *Developing high-potential youth program: A return on investment study for U.S. programs.* Princeton: Educational Testing Service.
Moore, M., Goertz, M. E., Hartle, T., Winslow, H., David, J., Sjogren, J., & Holland, R. P. (1983). *The interaction of federal and related state education programs* (Vol. I). Washington, DC: Educational Testing Service.
Nardi, W. (1992, Winter). *The origins of Educational Testing Service.* ETS Developments.
National Commission on Excellence in Education. (1983). *A nation at risk: The imperative for educational reform, a report to the Nation and the Secretary of Education, United States Department of Education.* Retrieved from the Department of Education website: http://www2.ed.gov/pubs/NatAtRisk/index.html
Nettles, M. T. (1987). *Financial aid and minority participation in graduate education* (Minority Graduate Education Project Report No. MGE-87-01). Princeton: Educational Testing Service.
Nettles, M. T. (1990). *Black, Hispanic, and White doctoral students: Before, during, and after enrolling in graduate school* (Minority Graduate Education Project Report No. MGE-90-01). Princeton: Educational Testing Service.
Nettles, M., & Millett, C. (2006). *Three magic letters: Getting to Ph.D.* Baltimore: Johns Hopkins University Press.

Nettles, M. T., Scatton, L. H., Steinberg, J. H., & Tyler, L. L. (2011). *Performance and passing rate differences of African American and White prospective teachers on PRAXIS examinations* (Research Report No. RR-11-08). Princeton: Educational Testing Service. http://dx.doi.org/10.1002/j.2333-8504.2011.tb02244.x

Rudd, R., Kirsch, I., & Yamamoto, K. (2004). *Literacy and health in America* (Policy Information Report). Princeton: Educational Testing Service.

Sum, A., Kirsch, I., & Taggart, R. (2002). *The twin challenges of mediocrity and inequality: Literacy in the U.S. from an international perspective* (Policy Information Report). Princeton: Educational Testing Service.

Sum, A., Kirsch, I., & Yamamoto, K. (2004). *Pathways to labor market success: The literacy proficiency of U.S. adults* (Policy Information Report). Princeton: Educational Testing Service.

Sykes, G., Schneider, B., & Plank, D. N. (Eds.). (2009). *Handbook on education policy research.* New York: American Educational Research Association/Routledge.

Thomas, G. E., Clewell, B. C., & Pearson Jr., W. (1992). *The role and activities of American graduate schools in recruitment, enrollment and retaining U.S. Black and Hispanic students.* Princeton: Educational Testing Service.

Wang, A. H., Coleman, A. B., Coley, R. J., & Phelps, R. P. (2003). *Preparing teachers around the world* (Policy Information Report). Princeton: Educational Testing Service.

Wendler, C., Bridgeman, B., Cline, F., Millett, C., Rock, J., Bell, N., & McAllister, P. (2010). *The path forward: The future of graduate education in the United States.* Princeton: Educational Testing Service.

Wendler, C., Bridgeman, B., Markle, R., Cline, F., Bell, N., McAllister, P., & Kent, J. (2012). *Pathways through graduate school and into careers.* Princeton: Educational Testing Service.

Wenglinsky, H. (2000). *How teaching matters: Bringing the classroom back into discussions of teacher quality* (Policy Information Report). Princeton: Educational Testing Service.

Wilder, G. Z., & Baydar, N. (1990). *Decisions about graduate education: A two-year longitudinal study of GRE-takers* (Interim Report to the Graduate Record Examinations Board). Princeton: Educational Testing Service.

Willingham, W. W., & Cole, N. S. (1997). *Gender and fair assessment.* Mahwah: Erlbaum.

Yaffe, D. (2008, Fall). School finance and the achievement gap: Funding programs that work. *ETS Policy Notes, 16*(3).

Zwick, R. (1991). *Differences in graduate school attainment: Patterns across academic programs and demographic groups* (Minority Graduate Education Project Report No. MGE-91-01). Princeton: Educational Testing Service.

Part III
ETS Contributions to Research in Scientific Psychology

Chapter 13
Research on Cognitive, Personality, and Social Psychology: I

Lawrence J. Stricker

Several months before ETS's founding in 1947, Henry Chauncey, its first president, described his vision of the research agenda:

> Research must be focused on objectives not on methods (they come at a later stage). Objectives would seem to be (1) advancement of test theory & statistical techniques (2) refinement of description & measurement of intellectual & personal qualities (3) development of tests for specific purposes (a) selection (b) guidance (c) measurement of achievement. (Chauncey 1947, p. 39)

By the early 1950s, research at ETS on intellectual and personal qualities was already proceeding. Cognitive factors were being investigated by John French (e.g., French 1951b), personality measurement by French, too (e.g., French 1952), interests by Donald Melville and Norman Frederiksen (e.g., Melville and Frederiksen 1952), social intelligence by Philip Nogee (e.g., Nogee 1950), and leadership by Henry Ricciuti (e.g., Ricciuti 1951). And a major study, by Frederiksen and William Schrader (1951), had been completed that examined the adjustment to college by some 10,000 veterans and nonveterans.

Over the years, ETS research on those qualities has evolved and broadened, addressing many of the core issues in cognitive, personality, and social psychology. The emphasis has continually shifted, and attention to different lines of inquiry has waxed and waned, reflecting changes in the Zeitgeist in psychology, the composition of the Research staff and its interests, and the availability of support, both external and from ETS. A prime illustration of these changes is the focus of research at ETS and in the field of psychology on level of aspiration in the 1950s, exemplified by the ETS studies of Douglas Schultz and Henry Ricciuti (e.g., Schultz and

---

This chapter was originally published in 2013 as a research report in the ETS R&D Scientific and Policy Contributions Series.

L.J. Stricker (✉)
Educational Testing Service, Princeton, NJ, USA
e-mail: lstricker@ets.org

R.E. Bennett, M. von Davier (eds.), *Advancing Human Assessment*,
Methodology of Educational Measurement and Assessment,
DOI 10.1007/978-3-319-58689-2_13

Ricciuti 1954), and on emotional intelligence 60 years later, represented by ETS investigations by Richard Roberts and his colleagues (e.g., Roberts et al. 2006).

What has been studied is so varied and so substantial that it defies easy encapsulation. Rather than attempt an encyclopedic account, a handful of topics that were the subjects of extensive and significant ETS research, very often in the forefront of psychology, will be discussed. In this chapter, the topics in cognitive psychology are the structure of abilities; in personality psychology, response styles, and social and emotional intelligence; and in social psychology, prosocial behavior and stereotype threat. Motivation is also covered. The companion chapter (Kogan, Chap. 14, this volume) discusses other topics in cognitive psychology (creativity), personality psychology (cognitive styles, kinesthetic after effects), and social psychology (risk taking).

13.1 The Structure of Abilities

Factor analysis has been the method of choice for mapping the ability domain almost from the very beginning of ability testing at the turn of the twentieth century. Early work, such as Spearman's (1904), focused on a single, general factor ("g"). But subsequent developments in factor analytic methods in the 1930s, mainly by Thurstone (1935), made possible the identification of multiple factors. This research was closely followed by Thurstone's (1938) landmark discovery of seven primary mental abilities. By the late 1940s, factor analyses of ability tests had proliferated, each analysis identifying several factors. However, it was unclear what factors were common across these studies and what were the best measures of the factors.

To bring some order to this field, ETS scientist John French (1951b) reviewed all the factor analyses of ability and achievement that had been conducted through the 1940s. He identified 59 different factors from 69 studies and listed tests that measured these factors. (About a quarter of the factors were found in a single study, and the same fraction did not involve abilities.)

This seminal work underscored the existence of a large number of factors, the importance of replicable factors, and the difficulty of assessing this replicability in the absence of common measures in different studies. It eventuated in a major ETS project led by French—with the long-term collaboration of Ruth Ekstrom and with the guidance and assistance of leading factor analysts and assessment experts across the country—that lasted almost two decades. Its objectives were both (a) substantive—to identify well-established ability factors and (b) methodological—to identify tests that define these factors and hence could be included in new studies as markers to aid in interpreting the factors that emerge. The project evolved over three stages.

At the first conference in 1951, organized by French, chaired by Thurstone, and attended by other factor analysts and assessment experts, French (1951a) reported that (a) 28 factors appeared to be reasonably well established, having been found in at least three different analyses; and (b) 29 factors were tentatively established, appearing with "reasonable clarity" (p. 8) in one or two analyses. (Several factors in each set were not defined by ability measures.) Committees were formed to verify the factors and identify the tests that defined them. Sixteen factors and three corresponding marker tests per factor were ultimately identified (French 1953, 1954). The 1954 Kit of Selected Tests for Reference Aptitude and Achievement Factors contained the tests selected to define the factors, including some commercially published tests (French 1954).

At a subsequent conference in 1958, plans were formulated to evaluate 46 replicable factors (including those already in the 1954 Kit) that were candidates for inclusion in a revised Kit and, as far as possible, develop new tests in place of the published tests to obviate the need for special permission for their use and to make possible a uniform format for all tests in the Kit (French 1958). Again, committees evaluated the factors and identified marker tests. The resulting 1963 Kit of Reference Tests for Cognitive Factors (French et al. 1963) had 24 factors, along with marker tests. Most of the tests were created for the 1963 Kit, but a handful were commercially published tests.

At the last conference, in 1971, plans were made for ETS staff to appraise existing factors and newly observed ones and to develop ETS tests for all factors (Harman 1975). The recent literature was reviewed and studies of 12 new factors were conducted to check on their viability (Ekstrom et al. 1979). The Kit of Factor-Referenced Cognitive

Table 13.1 Factors and sample marker tests in Kit of Factor-Referenced Cognitive Tests, 1976

Factor	Marker test
General reasoning	Necessary arithmetic operations
Induction	Letter sets
Logical reasoning	Nonsense syllogisms
Integrative processes	Following directions
Verbal comprehension	Vocabulary test 1
Number facility	Addition
Spatial orientation	Card rotations
Visualization	Paper folding
Spatial scanning	Maze tracing
Perceptual speed	Number comparison
Flexibility of closure	Hidden figures
Speed of closure	Gestalt completion
Verbal closure	Scrambled words
Memory span	Auditory number span
Associative memory	First and last names
Visual memory	Map memory
Figural Fluency	Ornamentation
Expressional Fluency	Arranging words
Word Fluency	Word beginnings
Associational Fluency	Opposites
Ideational Fluency	Thing categories
Flexibility of use	Different uses
Figural flexibility	Toothpicks

Note: Adapted from *Essentials of Psychological Testing* (5th ed.), by L. J. Cronbach, (1990), New York: Harper & Row

Tests, 1976 (Ekstrom et al. 1976) had 23 factors and 72 corresponding tests. The factors and sample marker tests appear in Table 13.1, as roughly grouped by Cronbach (1990).

Research and theory about ability factors has continued to advance in psychology since the work on the Kit ended in the 1970s, most notably Carroll's (1993) identification of 69 factors from a massive reanalysis of extant, factor-analytic studies through the mid-1980s, culminating in his three-stratum theory of cognitive abilities. Nonetheless, the Kit project has had a lasting impact on the field. The various Kits were, and are, widely used in research at ETS and elsewhere. The studies include not only factor analyses of large sets of tests that use a number from the Kit to define factors (e.g., Burton and Fogarty 2003), in keeping with its original purpose, but also many small-scale experiments and correlational investigations that simply use a few Kit tests to measure specific variables (e.g., Hegarty et al. 2000). It is noteworthy that versions of the Kit have been cited 2308 times through 2016, according to the *Social Science Citation Index*.

13.2 Response Styles

Response styles are

> … expressive consistencies in the behavior of respondents which are relatively enduring over time, with some degree of generality beyond a particular test performance to responses both in other tests and in non-test behavior, and usually reflected in assessment situations by consistencies in response to item characteristics other than specific content. (Jackson and Messick 1962a, p. 134)

Although a variety of response styles has been identified on tests, personality inventories, and other self-report measures, the best known and most extensively investigated are acquiescence and social desirability. Both have a long history in psychological assessment but were popularized in the 1950s by Cronbach's (1946, 1950) reviews of acquiescence and Edwards's (1957) research on social desirability. As originally defined, acquiescence is the tendency for an individual to respond *Yes, True,* etc. to test items, regardless of their content; social desirability is the tendency to give a socially desirable response to items on self-report measures, in particular.

ETS scientist Samuel Messick and his longtime collaborator at Pennsylvania State University and the University of Western Ontario, Douglas Jackson, in a seminal article in 1958 redirected this line of work by reconceptualizing response sets as response styles to emphasize that they represent consistent individual differences not limited to reactions to a particular test or other measure. Jackson and Messick underscored the impact of response styles on personality and self-report measures generally, throwing into doubt conventional interpretations of the measures based on their purported content:

> In the light of accumulating evidence it seems likely that the *major common factors in personality inventories of the true-false or agree-disagree type*, such as the MMPI and the California Personality Inventory, *are interpretable primarily in terms of style rather than specific item content.* (original italics; Jackson and Messick 1958, p. 247)

Messick, usually in collaboration with Jackson, carried out a program of research on response styles from the 1950s to the 1970s. The early work documented acquiescence on the California F scale, a measure of authoritarianism. But the bulk of the research focused on acquiescence and social desirability on the MMPI. In major studies (Jackson and Messick 1961, 1962b), the standard clinical and validity scales (separately scored for the true-keyed and false-keyed items) were factor analyzed in samples of college students, hospitalized mental patients, and prisoners. Two factors, identified as acquiescence and social desirability, and accounting for 72–76% of the common variance, were found in each analysis. The acquiescence factor was defined by an acquiescence measure and marked by positive loadings for the true-keyed scales and negative loadings for the false-keyed scales. The social desirability factor's loadings were closely related to the judged desirability of the scales.

A review by Fred Damarin and Messick (Damarin and Messick 1965; Messick 1967, 1991) of factor analytic studies by Cattell and his coworkers (e.g., Cattell et al. 1954; Cattell and Gruen 1955; Cattell and Scheier 1959) of response style measures and performance tests of personality that do not rely on self-reports, suggested two kinds of acquiescence: (a) uncritical agreement, a tendency to agree; and (b) impulsive acceptance, a tendency to accept many characteristics as descriptive of the self. In a subsequent factor analysis of true-keyed and false-keyed halves of original and reversed MMPI scales (items revised to reverse their meaning), two such acquiescence factors were found (Messick 1967).

The Damarin and Messick review (Damarin and Messick 1965; Messick 1991) also suggested that there are two kinds of socially desirable responding: (a) a partially deliberate bias in self-report and (b) a nondeliberate or autistic bias in self-regard. This two-factor theory of desirable responding was supported in later factor analytic research (Paulhus 1984).

The findings from this body of work led to the famous *response style controversy* (Wiggins 1973). The main critics were Rorer and Goldberg (1965a, b) and Block (1965). Rorer and Goldberg contended that acquiescence had a negligible influence on the MMPI, based largely on analyses of correlations between original and reversed versions of the scales. Block questioned the involvement of both acquiescence and social desirability response styles on the MMPI, based on his factor analyses of MMPI scales that had been balanced in their true-false keying to minimize acquiescence and his analyses of the correlations between a measure of the putative social desirability factor and the Edwards Social Desirability scale. These critics were rebutted by Messick (1967, 1991) and Jackson (1967). In recent years this controversy has reignited, focusing on whether response styles affect the criterion validity of personality measures (e.g., McGrath et al. 2010; Ones et al. 1996).

This work has had lasting legacies for both practice and research. Assessment specialists commonly recommend that self-report measures be balanced in keying (Hofstee et al. 1998; McCrae et al. 2001; Paulhus and Vazire 2007; Saucier and Goldberg 2002), and most recent personality inventories (Jackson Personality Inventory, NEO Personality Inventory, Personality Research Form) follow this practice. It is also widely recognized that social desirability response style is a potential threat to the validity of self-report measures and needs to be evaluated (American

Educational Research Association et al. 1999). Research on this response style continues, evolved from its conceptualization by Damarin and Messick (Damarin and Messick 1965; Messick 1991) and led by Paulhus (e.g., Paulhus 2002).

13.3 Prosocial Behavior

Active research on positive forms of social behavior began in psychology in the 1960s, galvanized at least in part by concerns about public apathy and indifference triggered by the famous Kitty Genovese murder (a New York City woman killed reportedly while 38 people watched from their apartments, making no efforts to intervene;[1] Latané and Darley 1970; Manning et al. 2007). This *prosocial behavior*, a term that ETS scientist David Rosenhan (Rosenhan and White 1967) and James Bryan (Bryan and Test 1967), an ETS visiting scholar and faculty member at Northwestern University, introduced into the social psychological literature to describe all manner of positive behavior (Wispé 1972), has many definitions. Perhaps the most useful is Rosenhan's (1972):

> ...while the bounds of prosocial behavior are not rigidly delineated, they include these behaviors where the emphasis is ...upon "concern for others." They include those acts of helpfulness, charitability, self-sacrifice, and courage where the possibility of reward from the recipient is presumed to be minimal or non-existent and where, on the face of it, the prosocial behavior is engaged in for its own end and for no apparent other. (p. 153)

Rosenhan and Bryan, working independently, were at the forefront of research on this topic in a short-lived but intensive program of research at ETS in the 1960s. The general thrust was the application of social learning theory to situations involving helping and donating, in line with the prevailing Zeitgeist. The research methods ran the gamut from surveys to field and laboratory experiments. And the participants included the general public, adults, college students, and children.

Rosenhan (1969, 1970) began by studying civil rights activists and financial supporters. They were extensively interviewed about their involvement in the civil rights movement, personal history, and ideology. The central finding was that fully committed activists had close affective ties with parents who were also fully committed to altruistic causes.

Rosenhan and White (1967) subsequently put this result to the test in the laboratory. Children who observed a model donate to charity and then donated in the model's presence were more likely to donate when they were alone, suggesting that both observation and rehearsal are needed to internalize norms for altruism. However, these effects occurred whether or not the children had positive or negative interactions with the model.

In a follow-up study, White (1972) found that children's observations of the model per se did not affect their subsequent donations; the donations were influ-

[1] Subsequent inquiries cast doubt on the number of witnesses and on whether any intervened (Manning et al. 2007).

enced by whether the children contributed in the model's presence. Hence, rehearsal, not observation, was needed to internalize altruistic norms. White also found that these effects persisted over time.

Bryan also carried out a mix of field studies and laboratory experiments. Bryan and Michael Davenport (Bryan and Davenport 1968), using data on contributions to *The New York Times* 100 Neediest Cases, evaluated how the reasons for being dependent on help were related to donations. Cases with psychological disturbances and moral transgressions received fewer donations, presumably because these characteristics reduce interpersonal attractiveness, specifically, likability; and cases with physical illnesses received more contributions.

Bryan and Test (1967) conducted several ingenious field experiments on the effects of modeling on donations and helping. Three experiments involved donations to Salvation Army street solicitors. More contributions were made after a model donated, and whether or not the solicitor acknowledged the donation (potentially reinforcing it). Furthermore, more White people contributed to White than Black solicitors when no modeling was involved, suggesting that interpersonal attraction—the donors' liking for the solicitors—is important. In the helping experiment, more motorists stopped to assist a woman with a disabled car after observing another woman with a disabled car being assisted.

Bryan and his coworkers also carried out several laboratory experiments about the effects of modeling on helping by college students and donations by children. In the helping study, by Test and Bryan (1969), the presence of a helping model (helping with arithmetic problems) increased subsequent helping when the student was alone, but whether the recipient of the helping was disabled and whether the participant had been offered help (setting the stage for reciprocal helping by the participant) did not affect helping.

In Bryan's first study of donations (Midlarsky and Bryan 1967), positive relationships with the donating model and the model's expression of pleasure when the child donated increased children's donations when they were alone. In a second study, by Bryan and Walbek (1970, Study 1), the presence of the donating model affected donations, but the model's exhortations to be generous or to be selfish in making donations did not.

Prosocial behavior has evolved since its beginnings in the 1960s into a major area of theoretical and empirical inquiry in social and developmental psychology, and sociology (e.g., see the review by Penner et al. 2005). The work has broadened over the years to include such issues as its biological and genetic causes, its development over the life span, and its dispositional determinants (demographic variables, motives, and personality traits). The focus has also shifted from the laboratory experiments on mundane tasks to investigations in real life that concern important social issues and problems (Krebs and Miller 1985), echoing Rosenhan's (1969, 1970) civil rights study at the very start of this line of research in psychology some 50 years ago.

13.4 Social and Emotional Intelligence

Social intelligence and its offshoot, emotional intelligence, have a long history in psychology, going back at least to Thorndike's famous *Harper's Monthly Magazine* article (Thorndike 1920) that described social intelligence as "the ability to understand and manage men and women, boys and girls—to act wisely in human relations" (p. 228). The focus of this continuing interest has varied over the years from accuracy in judging personality in the 1950s (see the review by Cline 1964); to skill in decoding nonverbal communication (see the review by Rosenthal et al. 1979) and understanding and coping with the behavior of others (Hendricks et al. 1969; O'Sullivan and Guilford 1975) in the 1970s; to understanding and dealing with emotions from the 1990s to the present. This latest phase, beginning with a seminal article by Salovey and Mayer (1990) on emotional intelligence and galvanized by Goleman's (1995) popularized book, *Emotional Intelligence: Why It Can Matter More Than IQ*, has engendered enormous interest in the psychological community and in the public.

ETS research on this general topic started in 1950 but until recently was scattered and modest, limited to scoring and validating situational judgment tests of social intelligence. These efforts included studies by Norman Cliff (1962), Philip Nogee (1950), and Lawrence Stricker and Donald Rock (1990). Substantial work on emotional intelligence at ETS by Roberts and his colleagues began more recently. They have conducted several studies on the construct validity of maximum-performance measures of emotional intelligence. Key findings are that the measures define several factors and relate moderately with cognitive ability tests, minimally with personality measures, and moderately with college grades (MacCann et al. 2010, 2011; MacCann and Roberts 2008; Roberts et al. 2006).

In a series of critiques, reviews, and syntheses of the extant research literature, Roberts and his colleagues have attempted to bring order to this chaotic and burgeoning field marked by a plethora of conceptions, "conceptual and theoretical incoherence" (Schulze et al. 2007, p. 200), and numerous measures of varying quality. These publications emphasize the importance of clear conceptualizations, adherence to conventional standards in constructing and validating measures, and the need to exploit existing measurement approaches (e.g., MacCann et al. 2008; Orchard et al. 2009; Roberts et al. 2005, 2008, 2010; Schulze et al. 2007).

More specifically, the papers make these major points:

1. In contrast to diffuse conceptions of emotional intelligence (e.g., Goleman 1995), it is reasonable to conceive of this phenomenon as consisting of four kinds of cognitive ability, in line with the view that emotional intelligence is a component of intelligence. This is the Mayer and Salovey (1997) four-branch model that posits these abilities: perceiving emotions, using emotions, understanding emotions, and managing emotions.
2. Given the ability conception of emotional intelligence, it follows that appropriate measures assess maximum performance, just like other ability tests. Self-report measures of emotional intelligence that appraise typical performance are inap-

propriate, though they are very widely used. It is illogical to expect that people lacking in emotional intelligence would be able to accurately report their level of emotional intelligence. And, empirically, these self-report measures have problematic patterns of relations with personality measures and ability tests: substantial with the former but minimal with the latter. In contrast, maximum performance measures have the expected pattern of correlations: minimal with personality measures and substantial with ability tests.
3. Maximum performance measures of emotional intelligence have unusual scoring and formats, unlike ability tests, that limit their validity. Scoring may be based on expert judgments or consensus judgments derived from test takers' responses. But the first may be flawed, and the second may disadvantage test takers with extremely high levels of emotional intelligence (their responses, though appropriate, diverge from those of most test takers). Standards-based scoring employed by ability tests obviates these problems. Unusual response formats include ratings (e.g., presence of emotion, effectiveness of actions) rather than multiple choice, as well as instructions to predict how the test taker would behave in some hypothetical situation rather than to identify what is the most effective behavior in the situation.
4. Only one maximum performance measure is widely used, the Mayer-Salovey-Caruso Emotional Intelligence Test (Mayer et al. 2002). Overreliance on a single measure to define this phenomenon is "a suboptimal state of affairs" (Orchard et al. 2009, p. 327). Other maximum performance methods, free of the measurement problems discussed, can also be used. They include implicit association tests to detect subtle biases (e.g., Greenwald et al. 1998), measures of ability to detect emotions in facial expressions (e.g., Ekman and Friesen 1978), inspection time tests to assess how quickly different emotions can be distinguished (e.g., Austin 2005), situational judgment tests (e.g., Chapin 1942), and affective forecasting of one's emotional state at a future point (e.g., Hsee and Hastie 2006).

It is too early to judge the impact of these recent efforts to redirect the field. Emotional intelligence continues to be a very active area of research in the psychological community (e.g., Mayer et al. 2008).

13.5 Stereotype Threat

Stereotype threat is a concern about fulfilling a negative stereotype regarding the ability of one's group when placed in a situation where this ability is being evaluated, such as when taking a cognitive test. These negative stereotypes exist about minorities, women, the working class, and the elderly. This concern has the potential for adversely affecting performance on the ability assessment (see Steele 1997). This phenomenon has clear implications for the validity of ability and achievement tests, whether used operationally or in research.

Stereotype threat research began with the seminal experiments by Steele and Aronson (1995). In one of the experiments (Study 2), for instance, they reported that the performance of Black research participants on a verbal ability test was lower when it was described as diagnostic of intellectual ability (priming stereotype threat) than when it was described as a laboratory task for solving verbal problems; in contrast, White participants' scores were unaffected.

Shortly after the Steele and Aronson (1995) work was reported, Walter McDonald, then director of the *Advanced Placement Program*® (*AP*®) examinations at ETS, commissioned Stricker to investigate the effects of stereotype threat on the AP examinations, arguing that ETS would be guilty of "educational malpractice" if the tests were being affected and ETS ignored it. This assignment eventuated in a program of research by ETS staff on the effects of stereotype threat and on the related question of possible changes that could be made in tests and test administration procedures.

The initial study with the AP Calculus examination and a follow-up study (Stricker and Ward 2004), with the Computerized Placement Tests (CPTs, now called the *ACCUPLACER*® test), a battery of basic skills tests covering reading, writing, and mathematics, were stimulated by a Steele and Aronson (1995, Study 4) finding. These investigators observed that the performance of Black research participants on a verbal ability test was depressed when asked about their ethnicity (making their ethnicity salient) prior to working on the test, while the performance of White participants was unchanged. The AP examinations and the CPTs, in common with other standardized tests, routinely ask examinees about their ethnicity and gender immediately before they take the tests, mirroring the Steele and Aronson experiment. The AP and CPTs studies, field experiments with actual test takers, altered the standard test administration procedures for some students by asking the demographic questions after the test and contrasted their performance with that of comparable students who were asked these questions at the outset of the standard test administration. The questions had little or no effect on the test performance of Black test takers or the others—Whites, Asians, women, and men—in either experiment. These findings were not without controversy (Danaher and Crandall 2008; Stricker and Ward 2008). The debate centered on whether the AP results implied that a substantial number of young women taking the test were adversely affected by stereotype threat.

Several subsequent investigations also looked at stereotype threat in field studies with actual test takers, all the studies motivated by the results of other laboratory experiments by academic researchers. Alyssa Walters et al. (2004) examined whether a match in gender or ethnicity between test takers and test-center proctors enhanced performance on the *GRE*® General Test. This study stemmed from the Marx and Roman (2002) finding that women performed better on a test of quantitative ability when the experimenter was a woman (a competent role model) while the experimenter's gender did not affect men's performance. Walters et al. reported that neither kind of match between test takers and their proctors was related to the test takers' scores for women, men, Blacks, Hispanics, or Whites.

Michael Walker and Brent Bridgeman (2008) investigated whether the stereotype threat that may affect women when they take the *SAT*® Mathematics section spills over to the Critical Reading section, though a reading test should not ordinarily be prone to stereotype threat for women (there are no negative stereotypes about their ability to read). The impetus for this study was the report by Beilock et al. (2007, Study 5) that the performance of women on a verbal task was lower when it followed a mathematics task explicitly primed to increase stereotype threat than when it followed the same task without such priming. Walker and Bridgeman compared the performance on a subsequent Critical Reading section for those who took the Mathematics section first with those who took the Critical Reading or Writing section first. Neither women's nor men's Critical Reading mean scores were lower when this section followed the Mathematics section than when it followed the other sections.

Stricker (2012) investigated changes in Black test takers' performance on the GRE General Test associated with Obama's 2008 presidential campaign. This study was modeled after one by Marx et al. (2009). In a field study motivated by the role-model effect in the Marx and Roman (2002) experiment—a competent woman experimenter enhanced women's test performance—Marx et al. observed that Black-White mean differences on a verbal ability test were reduced to nonsignificance at two points when Obama achieved concrete successes (after his nomination and after his election), though the differences were appreciable at other points. Stricker, using archival data for the GRE General Test's Verbal section, found that substantial Black-White differences persisted throughout the campaign and were virtually identical to the differences the year before the campaign.

The only ETS laboratory experiment thus far, by Lawrence Stricker and Isaac Bejar (2004), was a close replication of one by Spencer et al. (1999, Study 1). Spencer et al. found that women and men did not differ in their performance on an easy quantitative test, but they did differ on a hard one, consistent with the theoretical notion that stereotype threat is maximal when the test is difficult, at the limit of the test taker's ability. Stricker and Bejar used computer-adaptive versions of the GRE General Test, a standard version and one modified to produce a test that was easier but had comparable scores. Women's mean Quantitative scores, as well as their mean Verbal scores, did not differ on the easy and standard tests, and neither did the mean scores of the other participants: men, Blacks, and Whites.

In short, the ETS research to date has failed to find evidence of stereotype threat on operational tests in high-stakes settings, in common with work done elsewhere (Cullen et al. 2004, 2006). One explanation offered for this divergence from the results in other research studies is that motivation to perform well is heightened in a high-stakes setting, overriding any harmful effects of stereotype threat that might otherwise be found in the laboratory (Stricker and Ward 2004). The findings also suggest that changes in the test administration procedures or in the difficulty of the tests themselves are unlikely to ameliorate stereotype threat. In view of the limitations of field studies, the weight of laboratory evidence that document its robustness and potency, and its potential consequences for test validity (Stricker 2008), stereotype threat is a continuing concern at ETS.

13.6 Motivation

Motivation is at the center of psychological research, and its consequences for performance on tests, in school, and in other venues has been a long-standing subject for ETS investigations. Most of this research has focused on three related constructs: level of aspiration, need for achievement, and test anxiety. Level of aspiration, extensively studied by psychologists in the 1940s (e.g., see reviews by Lefcourt 1982; Powers 1986; Phares 1976), concerns the manner in which a person sets goals relative to that person's ability and past experience. Need for achievement, a very popular area of psychological research in the 1950s and 1960s (e.g., Atkinson 1957; McClelland et al. 1953), posits two kinds of motives in achievement-related situations: a motive to achieve success and a motive to avoid failure. Test anxiety is a manifestation of the latter. Research on test anxiety that focuses on its consequences for test performance has been a separate and active area of inquiry in psychology since the 1950s (e.g., see reviews by Spielberger and Vagg 1995; Zeidner 1998).

13.6.1 Test Anxiety and Test Performance

Several ETS studies have investigated the link between test anxiety and performance on ability and achievement tests. Two major studies by Donald Powers found moderate negative correlations between a test-anxiety measure and scores on the GRE General Test. In the first study (Powers 1986, 1988), when the independent contributions of the anxiety measure's Worry and Emotionality subscales were evaluated, only the Worry subscale was appreciably related to the test scores, suggesting that worrisome thoughts rather than physiological arousal affects test performance. The incidence of test anxiety was also reported. For example, 35% of test takers reported that they were tense and 36% that thoughts of doing poorly interfered with concentration on the test.

In the second study (Powers 2001), a comparison of the original, paper-based test and a newly introduced computer-adaptive version, a test-anxiety measure correlated similarly with the scores for the two versions. Furthermore, the mean level of test anxiety was slightly higher for the original version. These results indicate that the closer match between test-takers' ability and item difficulty provided by the computer-adaptive version did not markedly reduce test anxiety.

An ingenious experiment by French (1962) was designed to clarify the causal relationship between test anxiety and test performance. He manipulated test anxiety by administering sections of the SAT a few days before or after students took both the operational test and equivalent forms of these sections, telling the students that the results for the before and after sections would not be reported to colleges. The mean scores on these sections, which should *not* provoke test anxiety, were similar to those for sections administered with the SAT, which should provoke test anxiety, after adjusting for practice effects. The before and after sections and the sections administered with the SAT correlated similarly with high school grades. The results in toto suggest that test anxiety did not affect performance on the test or change what it measured.

Connections between test anxiety and other aspects of test-taking behavior have been uncovered in studies not principally concerned with test anxiety. Stricker and Bejar (2004), using standard and easy versions of a computer-adaptive GRE General Test in a laboratory experiment, found that the mean level for a test-anxiety measure was lower for the easy version. This effect interacted with ethnicity (but not gender): White participants were affected but Black participants were not.

Lawrence Stricker and Gita Wilder (2002) reported small positive correlations between a test anxiety measure and the extent of preparation for the Pre-Professional Skills Tests (tests of academic skills used for admission to teacher education programs and for teacher licensing).

Finally, Stricker et al. (2004) observed minimal or small negative correlations between a test-anxiety measure and attitudes about the *TOEFL®* test and about admissions tests in general in a survey of TOEFL test takers in three countries.

13.6.2 Test Anxiety/Defensiveness and Risk Taking and Creativity

Several ETS studies documented the relation between test anxiety, usually in combination with defensiveness, and both risk taking and creativity. Nathan Kogan and Michael Wallach (1967b), Kogan's long-time collaborator at Duke University, investigated this relation in the context of the risky-shift phenomenon (i.e., group discussion enhances the risk-taking level of the group relative to the members' initial level of risk taking; Kogan and Wallach 1967a). In their study, small groups were formed on the basis of participants' scores on test-anxiety and defensiveness measures. Risk taking was measured by responses to hypothetical life situations. The risky-shift effect was greater for the pure test-anxious groups (high on test anxiety, low on defensiveness) than for the pure defensiveness groups (high on defensiveness, low on test anxiety). This outcome was consistent with the hypothesis that test anxious groups, fearful of failure, diffuse responsibility to reduce the possibility of personal failure, and defensiveness groups, being guarded, interact insufficiently for the risky-shift to occur.

Henry Alker (1969) found that a composite measure of test anxiety and defensiveness correlated substantially with a risk-taking measure (based on performance on SAT Verbal items)—those with low anxiety and low defensiveness took greater risks. In contrast, a composite of the McClelland standard Thematic Apperception Test (TAT) measure of need for achievement and a test-anxiety measure correlated only moderately with the same risk-taking measure—those with high need for achievement and low anxiety took more risks. This finding suggested that the Kogan and Wallach (1964, 1967a) theoretical formulation of the determinants of risk taking (based on test anxiety and defensiveness) was superior to the Atkinson-McClelland (Atkinson 1957; McClelland et al. 1953) formulation (based on need for achievement and test anxiety).

Wallach and Kogan (1965) observed a sex difference in the relationships of test anxiety and defensiveness measures with creativity (indexed by a composite of several measures). For boys, defensiveness was related to creativity but test anxiety was not—the more defensive were less creative; for girls, neither variable was related to creativity. For both boys and girls, the pure defensiveness subgroup (high defensiveness and low test anxiety) were the least creative, consistent with the idea that defensive people's cognitive performance is impaired in unfamiliar or ambiguous contexts.

Stephen Klein et al. (1969), as part of a larger experiment, reported an unanticipated curvilinear, U-shaped relationship between a test-anxiety measure and two creativity measures: Participants in the midrange of test anxiety had the lowest creativity scores. Klein et al. speculated that the low anxious participants make many creative responses because they do not fear ridicule for the poor quality of their responses; the high anxious participants make many responses, even though the quality is poor, because they fear a low score on the test; and the middling anxious participants make few responses because their two fears cancel each other out.

13.6.3 Level of Aspiration or Need for Achievement and Academic Performance

Another stream of ETS research investigated the connection between level of aspiration and need for achievement on the one hand, and performance in academic and other settings on the other. The results were mixed. Schultz and Ricciuti (1954) found that level of aspiration measures, based on a general ability test, a code learning task, and regular course examinations, did not correlate with college grades.

A subsequent study by John Hills (1958) used a questionnaire measure of level of aspiration in several areas, TAT measures of need for achievement in the same areas, and McClelland's standard TAT measure of need for achievement to predict law-school criteria. The level of aspiration and need for achievement measures did not correlate with grades or social activities in law school, but one or more of the level of aspiration measures had small or moderately positive correlations with undergraduate social activities and law-school faculty ratings of professional promise.

A later investigation by Albert Myers (1965) reported that a questionnaire measure of achievement motivation had a substantial positive correlation with high school grades.

13.6.4 Overview

Currently, research on motivation outside of the testing arena is not an active area of inquiry at ETS, but work on test anxiety and test performance continues, particularly when new kinds of tests and delivery systems for them are introduced. The investigations of the connection between test anxiety and both risk taking and creativity, and the work on test anxiety on operational tests, are significant contributions to knowledge in this field.

13.7 Conclusion

The scope of the research conducted by ETS that is covered in this chapter is extraordinary. The topics range across cognitive, personality, and social psychology. The methods include not only correlational studies, but also laboratory and field experiments, interviews, and surveys. And the populations studied are children, adults, psychiatric patients, and the general public, as well as students.

The work represents basic research in psychology, sometimes far removed from either education or testing, much less the development of products. Prosocial behavior is a case in point.

The research on almost all of the topics discussed has had major impacts on the field of psychology, even the short-lived work on prosocial behavior. Although the effects of some of the newer work, such as that on emotional intelligence, are too recent to gauge, as this chapter shows, that work continues a long tradition of contributions to these three fields of psychology.

Acknowledgments Thanks are due to Rachel Adler and Jason Wagner for their assistance in retrieving reports, articles, and archival material, and to Randy Bennett, Jeremy Burrus, and Donald Powers for reviewing a draft of this chapter.

References

Alker, H. A. (1969). Rationality and achievement: A comparison of the Atkinson-McClelland and Kogan-Wallach formulations. *Journal of Personality, 37*, 207–224. https://doi.org/10.1111/j.1467-6494.1969.tb01741.x

American Educational Research Association, American Psychological Association, & National Council on Measurement in Education. (1999). *Standards for educational and psychological testing*. Washington, DC: American Educational Research Association.

Atkinson, J. W. (1957). Motivational determinants of risk-taking behavior. *Psychological Review, 64*, 359–372. https://doi.org/10.1037/h0043445

Austin, E. J. (2005). Emotional intelligence and emotional information processing. *Personality and Individual Differences, 19*, 403–414. https://doi.org/10.1016/j.paid.2005.01.017

Beilock, S. L., Rydell, R. J., & McConnell, A. R. (2007). Stereotype threat and working memory: Mechanisms, alleviation, and spillover. *Journal of Experimental Psychology: General, 136*, 256–276. https://doi.org/10.1037/0096-3445.136.2.256

Block, J. (1965). *The challenge of response sets—Unconfounding meaning, acquiescence, and social desirability in the MMPI*. New York: Appleton-Century-Crofts.

Bryan, J. H., & Davenport, M. (1968). *Donations to the needy: Correlates of financial contributions to the destitute* (Research Bulletin No. RB-68-01). Princeton: Educational Testing Service. http://dx.doi.org/10.1002/j.2333-8504.1968.tb00152.x

Bryan, J. H., & Test, M. A. (1967). Models and helping: Naturalistic studies in aiding behavior. *Journal of Personality and Social Psychology, 6*, 400–407. https://doi.org/10.1037/h0024826

Bryan, J. H., & Walbek, N. H. (1970). Preaching and practicing generosity: Children's actions and reactions. *Child Development, 41*, 329–353. https://doi.org/10.2307/1127035

Burton, L. J., & Fogarty, G. J. (2003). The factor structure of visual imagery and spatial abilities. *Intelligence, 31*, 289–318. https://doi.org/10.1016/S0160-2896(02)00139-3

Carroll, J. B. (1993). *Human cognitive abilities: A survey of factor-analytic studies*. New York: Cambridge University Press. https://doi.org/10.1017/CBO9780511571312

Cattell, R. B., & Gruen, W. (1955). The primary personality factors in 11-year-old children, by objective tests. *Journal of Personality, 23*, 460–478. https://doi.org/10.1111/j.1467-6494.1955.tb01169.x

Cattell, R. B., & Scheier, I. H. (1959). Extension of meaning of objective test personality factors: Especially into anxiety, neuroticism, questionnaire, and physical factors. *Journal of General Psychology, 61*, 287–315. https://doi.org/10.1080/00221309.1959.9710264

Cattell, R. B., Dubin, S. S., & Saunders, D. R. (1954). Verification of hypothesized factors in one hundred and fifteen objective personality test designs. *Psychometrika, 19*, 209–230. https://doi.org/10.1007/BF02289186

Chapin, F. S. (1942). Preliminary standardization of a social insight scale. *American Sociological Review, 7*, 214–228. https://doi.org/10.2307/2085176

Chauncey, H. (1947, July 13). [Notebook entry]. Henry Chauncey papers (Folder 1067). Carl. C. Brigham library, Educational Testing Service, Princeton, NJ.

Cliff, N. (1962). *Successful judgment in an interpersonal sensitivity task* (Research Bulletin No. RB-62-18). Princeton: Educational Testing Service. http://dx.doi.org/10.1002/j.2333-8504.1962.tb00296.x

Cline, V. B. (1964). Interpersonal perception. *Progress in Experimental Personality Research, 2*, 221–284.

Cronbach, L. J. (1946). Response sets and test validity. *Educational and Psychological Measurement, 6*, 475–494.

Cronbach, L. J. (1950). Further evidence of response sets and test design. *Educational and Psychological Measurement, 10*, 3–31. https://doi.org/10.1177/001316445001000101

Cronbach, L. J. (1990). *Essentials of psychological testing* (5th ed.). New York: Harper & Row.

Cullen, M. J., Hardison, C. M., & Sackett, P. R. (2004). Using SAT-grade and ability-job performance relationships to test predictions derived from stereotype threat theory. *Journal of Applied Psychology, 89*, 220–230. https://doi.org/10.1037/0021-9010.89.2.220

Cullen, M. J., Waters, S. D., & Sackett, P. R. (2006). Testing stereotype threat theory predictions for math-identified and non-math-identified students by gender. *Human Performance, 19*, 421–440. https://doi.org/10.1037/0021-9010.89.2.220

Damarin, F., & Messick, S. (1965). *Response styles and personality variables: A theoretical integration of multivariate research* (Research Bulletin No. RB-65-10). Princeton: Educational Testing Service. http://dx.doi.org/10.1002/j.2333-8504.1965.tb00967.x

Danaher, K., & Crandall, C. S. (2008). Stereotype threat in applied settings re-examined. *Journal of Applied Social Psychology, 38*, 1639–1655. https://doi.org/10.1111/j.1559-1816.2008.00362.x

Edwards, A. L. (1957). *The social desirability variable in personality assessment and research*. New York: Dryden.

Ekman, P., & Friesen, W. V. (1978). *Facial action coding system: A technique for the measurement of facial movement*. Palo Alto: Consulting Psychologists Press.
Ekstrom, R. B., French, J. W., & Harman, H. H. (with Dermen, D.). (1976). *Manual for Kit of Factor-Referenced Cognitive Tests, 1976*. Princeton: Educational Testing Service.
Ekstrom, R. B., French, J. W., & Harman, H. H. (1979). Cognitive factors: Their identification and replication. *Multivariate Behavioral Research Monographs*, No. 79-2.
Frederiksen, N., & Schrader, W. (1951). *Adjustment to college—A study of 10,000 veteran and non-veteran students in sixteen American colleges*. Princeton: Educational Testing Service.
French, J. W. (1951a). *Conference on factorial studies of aptitude and personality measures* (Research Memorandum No. RM-51-20). Princeton: Educational Testing Service.
French, J. W. (1951b). The description of aptitude and achievement factors in terms of rotated factors. *Psychometric Monographs*, No. 5.
French, J. W. (1952). *Validity of group Rorschach and Rosenzweig P. F. Study for adaptability at the U.S. coast guard academy* (Research Memorandum No. RM-52-02). Princeton: Educational Testing Service.
French, J. W. (1953). *Selected tests for reference factors—Draft of final report* (Research Memorandum No. RM-53-04). Princeton: Educational Testing Service.
French, J. W. (1954). *Manual for Kit of Selected Tests for Reference Aptitude and Achievement Factors*. Princeton: Educational Testing Service.
French, J. W. (1958). *Working plans for the reference test project* (Research Memorandum No. RM-58-10). Princeton: Educational Testing Service.
French, J. W. (1962). Effect of anxiety on verbal and mathematical examination scores. *Educational and Psychological Measurement, 22*, 553–564. https://doi.org/10.1177/001316446202200313
French, J. W., Ekstrom, R. B., & Price, L. A. (1963). *Manual for Kit of Reference Tests for Cognitive Factors*. Princeton: Educational Testing Service.
Goleman, D. (1995). *Emotional intelligence: Why it can matter more than IQ*. New York: Bantam.
Greenwald, A. G., McGhee, D. E., & Schwartz, J. L. K. (1998). Measuring individual differences in implicit cognition: The implicit association test. *Journal of Personality and Social Psychology, 74*, 1464–1480. https://doi.org/10.1037/0022-3514.74.6.1464
Harman, H. H. (1975). *Final report of research on assessing human abilities* (Project Report No. PR-75-20). Princeton: Educational Testing Service.
Hegarty, M., Shah, P., & Miyake, A. (2000). Constraints on using the dual-task methodology to specify the degree of central executive involvement in cognitive tasks. *Memory & Cognition, 28*, 373–385. https://doi.org/10.3758/BF03198553
Hendricks, M., Guilford, J. P., & Hoepfner, R. (1969). *Measuring creative social intelligence* (Psychological Laboratory Report No. 42). Los Angeles: University of Southern California.
Hills, J. R. (1958). Needs for achievement, aspirations, and college criteria. *Journal of Educational Psychology, 49*, 156–161. https://doi.org/10.1037/h0047283
Hofstee, W. K. B., ten Berge, J. M. F., & Hendriks, A. A. J. (1998). How to score questionnaires. *Personality and Individual Differences, 25*, 897–909. https://doi.org/10.1016/S0191-8869(98)00086-5
Hsee, C. K., & Hastie, R. (2006). Decision and experience: Why don't we choose what makes us happy? *Trends in Cognitive Science, 10*, 31–37. https://doi.org/10.1016/j.tics.2005.11.007
Jackson, D. N. (1967). Acquiescence response styles: Problems of identification and control. In I. A. Berg (Ed.), *Response set in personality assessment* (pp. 71–114). Chicago: Aldine.
Jackson, D. N., & Messick, S. (1958). Content and style in personality assessment. *Psychological Bulletin, 55*, 243–252. https://doi.org/10.1037/h0045996
Jackson, D. N., & Messick, S. (1961). Acquiescence and desirability as response determinants on the MMPI. *Educational and Psychological Measurement, 21*, 771–790. https://doi.org/10.1177/001316446102100402
Jackson, D. N., & Messick, S. (1962a). Response styles and the assessment of psychopathology. In S. Messick & J. Ross (Eds.), *Measurement in personality and cognition* (pp. 129–155). New York: Wiley.

Jackson, D. N., & Messick, S. (1962b). Response styles on the MMPI: Comparison of clinical and normal samples. *Journal of Abnormal and Social Psychology, 65*, 285–299. https://doi.org/10.1037/h0045340

Klein, S. P., Frederiksen, N., & Evans, F. R. (1969). Anxiety and learning to formulate hypotheses. *Journal of Educational Psychology, 60*, 465–475. https://doi.org/10.1037/h0028351

Kogan, N., & Wallach, M. A. (1964). *Risk taking: A study in cognition and personality*. New York: Holt, Rinehart and Winston.

Kogan, N., & Wallach, M. A. (1967a). Effects of physical separation of group members upon group risk taking. *Human Relations, 20*, 41–49. https://doi.org/10.1177/001872676702000104

Kogan, N., & Wallach, M. A. (1967b). Group risk taking as a function of members' anxiety and defensiveness levels. *Journal of Personality, 35*, 50–63. https://doi.org/10.1111/j.1467-6494.1967.tb01415.x

Krebs, D. L., & Miller, D. T. (1985). Altruism and aggression. In G. Lindzey & E. Aronson (Eds.), *Handbook of social psychology: Vol. 2. Special fields and applications* (3rd ed., pp. 1–71). New York, Random House.

Latané, B., & Darley, J. M. (1970). *The unresponsive bystander: Why doesn't he help?* New York: Appleton-Century-Crofts.

Lefcourt, H. M. (1982). *Locus of control: Current trends in theory and research* (2nd ed.). Hillsdale: Erlbaum.

MacCann, C., & Roberts, R. D. (2008). New paradigms for assessing emotional intelligence: Theory and data. *Emotions, 8*, 540–551. https://doi.org/10.1037/a0012746

MacCann, C., Schulze, R., Matthews, G., Zeidner, M., & Roberts, R. D. (2008). Emotional intelligence as pop science, misled science, and sound science: A review and critical synthesis of perspectives from the field of psychology. In N. C. Karafyllis & G. Ulshofer (Eds.), *Sexualized brains—Scientific modeling of emotional intelligence from a cultural perspective* (pp. 131–148). Cambridge: MIT Press.

MacCann, C., Wang, L., Matthews, G., & Roberts, R. D. (2010). Emotional intelligence and the eye of the beholder: Comparing self- and parent-rated situational judgments in adolescents. *Journal of Research in Personality, 44*, 673–676. https://doi.org/10.1016/j.jrp.2010.08.009

MacCann, C., Fogarty, G. J., Zeidner, M., & Roberts, R. D. (2011). Coping mediates the relationship between emotional intelligence (EI) and academic achievement. *Contemporary Educational Psychology, 36*, 60–70. https://doi.org/10.1016/j.cedpsych.2010.11.002

Manning, R., Levine, M., & Collins, A. (2007). The Kitty Genovese murder and the social psychology of helping behavior—The parable of the 38 witnesses. *American Psychologist, 68*, 555–562. https://doi.org/10.1037/0003-066X.62.6.555

Marx, D. M., & Roman, J. S. (2002). Female role models: Protecting women's math test performance. *Personality and Social Psychology Bulletin, 28*, 1183–1193.

Marx, D. M., Ko, S. J., & Friedman, R. A. (2009). The "Obama effect": How a salient role model reduces race-based differences. *Journal of Experimental Social Psychology, 45*, 953–956. https://doi.org/10.1016/j.jesp.2009.03.012

Mayer, J. D., & Salovey, P. (1997). What is emotional intelligence? In P. Salovey & D. Sluyter (Eds.), *Emotional development and emotional intelligence: Implications for educators* (pp. 3–31). New York: Basic Books.

Mayer, J. D., Salovey, P., & Caruso, D. R. (2002). *Mayer-Salovey-Caruso Emotional Intelligence Test (MSCEIT) user's manual*. Toronto: Multi-Health Systems.

Mayer, J. D., Roberts, R. D., & Barsade, S. G. (2008). Human abilities: Emotional intelligence. *Annual Review of Psychology, 59*, 507–536. https://doi.org/10.1146/annurev.psych.59.103006.093646

McClelland, D. C., Atkinson, J. W., Clark, R. A., & Lowell, E. L. (1953). *The achievement motive*. New York: Appleton-Century-Crofts. https://doi.org/10.1037/11144-000

McCrae, R. R., Herbst, J., & Costa Jr., P. T. (2001). Effects of acquiescence on personality factor structures. In R. Riemann, F. M. Spinath, & F. R. Ostendorf (Eds.), *Personality and temperament: Genetics, evolution, and structure* (pp. 216–231). Langerich: Pabst.

McGrath, R. E., Mitchell, M., Kim, B. H., & Hough, L. (2010). Evidence for response bias as a source of error variance in applied assessment. *Psychological Bulletin, 136*, 450–470. https://doi.org/10.1037/a0019216

Melville, S. D., & Frederiksen, N. (1952). Achievement of freshman engineering students and the Strong Vocational Interest Blank. *Journal of Applied Psychology, 36*, 169–173. https://doi.org/10.1037/h0059101

Messick, S. (1967). The psychology of acquiescence: An interpretation of research evidence. In I. A. Berg (Ed.), *Response set in personality assessment* (pp. 115–145). Chicago, IL: Aldine.

Messick, S. (1991). Psychology and methodology of response styles. In R. E. Snow & D. E. Wiley (Eds.), *Improving inquiry in social science—A volume in honor of Lee J. Cronbach* (pp. 161–200). Hillsdale: Erlbaum.

Midlarsky, E., & Bryan, J. H. (1967). Training charity in children. *Journal of Personality and Social Psychology, 5*, 408–415. https://doi.org/10.1037/h0024399

Myers, A. E. (1965). Risk taking and academic success and their relation to an objective measure of achievement motivation. *Educational and Psychological Measurement, 25*, 355–363. https://doi.org/10.1177/001316446502500206

Nogee, P. (1950). *A preliminary study of the "Social Situations Test"* (Research Memorandum No. RM-50-22). Princeton: Educational Testing Service.

O'Sullivan, M., & Guilford, J. P. (1975). Six factors of behavioral cognition: Understanding other people. *Journal of Educational Measurement, 12*, 255–271. https://doi.org/10.1111/j.1745-3984.1975.tb01027.x

Ones, D. S., Viswesvaran, C., & Reiss, A. D. (1996). Role of social desirability in personality testing for personnel selection: The red herring. *Journal of Applied Psychology, 81*, 660–679. https://doi.org/10.1037/0021-9010.81.6.660

Orchard, B., MacCann, C., Schulze, R., Matthews, G., Zeidner, M., & Roberts, R. D. (2009). New directions and alternative approaches to the measurement of emotional intelligence. In C. Stough, D. H. Saklofske, & J. D. A. Parker (Eds.), *Assessing human intelligence—Theory, research, and applications* (pp. 321–344). New York: Springer. https://doi.org/10.1007/978-0-387-88370-0_17

Paulhus, D. L. (1984). Two-component models of socially desirable responding. *Journal of Personality and Social Psychology, 46*, 598–609. https://doi.org/10.1037/0022-3514.46.3.598

Paulhus, D. L. (2002). Socially desirable responding: The evolution of a construct. In H. I. Braun, D. N. Jackson, & D. E. Wiley (Eds.), *The role of constructs in psychological and educational measurement* (pp. 49–69). Mahwah: Erlbaum.

Paulhus, D. L., & Vazire, S. (2007). The self-report method. In R. W. Robins, R. C. Fraley, & R. F. Krueger (Eds.), *Handbook of research methods in personality psychology* (pp. 224–239). New York: Guilford.

Penner, L. A., Dovidio, J. F., Pillavin, J. A., & Schroeder, D. A. (2005). Prosocial behavior: Multilevel perspectives. *Annual Review of Psychology, 56*, 365–392. https://doi.org/10.1146/annurev.psych.56.091103.070141

Phares, E. J. (1976). *Locus of control in personality*. Morristown: General Learning Press.

Powers, D. E. (1986). *Test anxiety and the GRE General Test* (GRE Board Professional Report No. 83-17P). Princeton: Educational Testing Service. http://dx.doi.org/10.1002/j.2330-8516.1986.tb00200.x

Powers, D. E. (1988). Incidence, correlates, and possible causes of test anxiety in graduate admissions testing. *Advances in Personality Assessment, 7*, 49–75.

Powers, D. E. (2001). Test anxiety and test performance: Comparing paper-based and computer-adaptive versions of the Graduate Record Examinations (GRE) General Test. *Journal of Educational Computing Research, 24*, 249–273. https://doi.org/10.2190/680W-66CR-QRP7-CL1F

Ricciuti, H. N. (1951). *A comparison of leadership ratings made and received by student raters* (Research Memorandum No. RM-51-04). Princeton: Educational Testing Service.
Roberts, R. D., Schulze, R., Zeidner, M., & Matthews, G. (2005). Understanding, measuring, and applying emotional intelligence: What have we learned? What have we missed? In R. Schulze & R. D. Roberts (Eds.), *Emotional intelligence—An international handbook* (pp. 311–341). Gottingen: Hogrefe & Huber.
Roberts, R. D., Schulze, R., O'Brien, K., McCann, C., Reid, J., & Maul, A. (2006). Exploring the validity of the Mayer-Salovey-Caruso Emotional Intelligence Test (MSCEIT) with established emotions measures. *Emotions, 6*, 663–669.
Roberts, R. D., Schulze, R., & MacCann, C. (2008). The measurement of emotional intelligence: A decade of progress? In G. Boyle, G. Matthews, & D. Saklofske (Eds.), *The Sage handbook of personality theory and assessment: Vol 2. Personality measurement and testing* (pp. 461–482).
Roberts, R. D., MacCann, C., Matthews, G., & Zeidner, M. (2010). Emotional intelligence: Toward a consensus of models and measures. *Social and Personality Psychology Compass, 4*, 821–840. https://doi.org/10.1111/j.1751-9004.2010.00277.x
Rorer, L. G., & Goldberg, L. R. (1965a). Acquiescence and the vanishing variance component. *Journal of Applied Psychology, 49*, 422–430. https://doi.org/10.1037/h0022754
Rorer, L. G., & Goldberg, L. R. (1965b). Acquiescence in the MMPI? *Educational and Psychological Measurement, 25*, 801–817. https://doi.org/10.1177/001316446502500311
Rosenhan, D. (1969). Some origins of concern for others. In P. H. Mussen, J. Langer, & M. V. Covington (Eds.), *Trends and issues in developmental psychology* (pp. 134–153). New York: Holt, Rinehart and Winston.
Rosenhan, D. (1970). The natural socialization of altruistic autonomy. In J. Macaulay & L. Berkowitz (Eds.), *Altruism and helping behavior—Social psychological studies of some antecedents and consequences* (pp. 251–268). New York: Academic Press.
Rosenhan, D. L. (1972). Learning theory and prosocial behavior. *Journal of Social Issues, 28*(3), 151–163. https://doi.org/10.1111/j.1540-4560.1972.tb00037.x
Rosenhan, D., & White, G. M. (1967). Observation and rehearsal as determinants of prosocial behavior. *Journal of Personality and Social Psychology, 5*, 424–431. https://doi.org/10.1037/h0024395
Rosenthal, R., Hall, J. A., DiMatteo, M. R., Rogers, P. L., & Archer, D. (1979). *Sensitivity to nonverbal communication—The PONS test.* Baltimore: Johns Hopkins University Press. https://doi.org/10.1016/b978-0-12-761350-5.50012-4
Salovey, P., & Mayer, J. D. (1990). Emotional intelligence. *Imagination, Cognition, and Personality, 9*, 185–211. https://doi.org/10.2190/DUGG-P24E-52WK-6CDG
Saucier, G., & Goldberg, L. R. (2002). Assessing the big five: Applications of 10 psychometric criteria to the development of marker scales. In B. de Raad & M. Perugini (Eds.), *Big five assessment* (pp. 29–58). Gottingen: Hogrefe & Huber.
Schultz, D. G., & Ricciuti, H. N. (1954). Level of aspiration measures and college achievement. *Journal of General Psychology, 51*, 267–275. https://doi.org/10.1080/00221309.1954.9920226
Schulze, R., Wilhelm, O., & Kyllonen, P. C. (2007). Approaches to the assessment of emotional intelligence. In G. Matthews, M. Zeidner, & R. D. Roberts (Eds.), *The science of emotional intelligence—Knowns and unknowns* (pp. 199–229). New York: Oxford University Press.
Spearman, C. (1904). "General intelligence" objectively determined and measured. *American Journal of Psychology, 15*, 201–293. https://doi.org/10.2307/1412107
Spencer, S. J., Steele, C. M., & Quinn, D. M. (1999). Stereotype threat and women's math performance. *Journal of Experimental Social Psychology, 35*, 4–28. https://doi.org/10.1006/jesp.1998.1373

Spielberger, C. D., & Vagg, P. R. (Eds.). (1995). *Test anxiety: Theory, assessment, and treatment.* Washington, DC: Taylor & Francis.

Steele, C. M. (1997). A threat in the air: How stereotypes shape intellectual identity and performance. *American Psychologist, 52*, 613–629. https://doi.org/10.1037/0003-066X.52.6.613

Steele, C. M., & Aronson, J. (1995). Stereotype threat and the intellectual test performance of African Americans. *Journal of Personality and Social Psychology, 69*, 797–811. https://doi.org/10.1037/0022-3514.69.5.797

Stricker, L. J. (2008). *The challenge of stereotype threat for the testing community* (Research Memorandum No. RM-08-12). Princeton: Educational Testing Service.

Stricker, L. J. (2012). *Testing: It's not just psychometrics* (Research Memorandum No. RM-12-07). Princeton: Educational Testing Service.

Stricker, L. J., & Bejar, I. (2004). Test difficulty and stereotype threat on the GRE General Test. *Journal of Applied Social Psychology, 34*, 563–597. https://doi.org/10.1111/j.1559-1816.2004.tb02561.x

Stricker, L. J., & Rock, D. A. (1990). Interpersonal competence, social intelligence, and general ability. *Personality and Individual Differences, 11*, 833–839. https://doi.org/10.1016/0191-8869(90)90193-U

Stricker, L. J., & Ward, W. C. (2004). Stereotype threat, inquiring about test takers' ethnicity and gender, and standardized test performance. *Journal of Applied Social Psychology, 34*, 665–693. https://doi.org/10.1111/j.1559-1816.2004.tb02564.x

Stricker, L. J., & Ward, W. C. (2008). Stereotype threat in applied settings re-examined: A reply. *Journal of Applied Social Psychology, 38*, 1656–1663. https://doi.org/10.1111/j.1559-1816.2008.00363.x

Stricker, L. J., & Wilder, G. Z. (2002). Why don't test takers prepare for the Pre-professional Skills Test? *Educational Assessment, 8*, 259–277. https://doi.org/10.1207/S15326977EA0803_03

Stricker, L. J., Wilder, G. Z., & Rock, D. A. (2004). Attitudes about the computer-based Test of English as a Foreign Language. *Computers in Human Behavior, 20*, 37–54. https://doi.org/10.1016/S0747-5632(03)00046-3

Test, M. A., & Bryan, J. H. (1969). The effects of dependency, models, and reciprocity upon subsequent helping behavior. *Journal of Social Psychology, 78*, 205–212. https://doi.org/10.1080/00224545.1969.9922357

Thorndike, E. L. (1920, January). Intelligence and its uses. *Harper's Monthly Magazine, 140*, 227–235.

Thurstone, L. L. (1935). *The vectors of mind—Multiple-factor analysis for the isolation of primary traits*. Chicago: University of Chicago Press. https://doi.org/10.1037/10018-000

Thurstone, L. L. (1938). Primary mental abilities. *Psychometric Monographs,* No. 1.

Walker, M. E., & Bridgeman, B. (2008). *Stereotype threat spillover and SAT scores* (College Board Report No. 2008–2). New York: College Board.

Wallach, M. A., & Kogan, N. (1965). *Modes of thinking in young children—A study of the creativity-intelligence distinction*. New York: Holt, Rinehart and Winston.

Walters, A. M., Lee, S., & Trapani, C. (2004) *Stereotype threat, the test-center environment, and performance on the GRE General Test* (GRE Board Research Report No. 01-03R). Princeton: Educational Testing Service. http://dx.doi.org/10.1002/j.2333-8504.2004.tb01964.x

White, G. M. (1972). Immediate and deferred effects of model observation and guided and unguided rehearsal on donating and stealing. *Journal of Personality and Social Psychology, 21*, 139–148. https://doi.org/10.1037/h0032308

Wiggins, J. S. (1973). *Personality and prediction: Principles of personality assessment*. Reading: Addison-Wesley.

Wispé, L. G. (1972). Positive forms of social behavior: An overview. *Journal of Social Issues, 28*(3), 1–19. https://doi.org/10.1111/j.1540-4560.1972.tb00029.x
Zeidner, M. (1998). *Test anxiety: The state of the art*. New York: Plenum.

Chapter 14
Research on Cognitive, Personality, and Social Psychology: II

Nathan Kogan

This is the second of two chapters describing research at Educational Testing Service (ETS) on cognitive, personality, and social psychology since its founding in 1947. The first chapter, Chap. 13 by Lawrence Stricker, also appears in this volume. Topics in these fields were selected for attention because they were the focus of extensive and significant ETS research. This chapter covers these topics: in cognitive psychology, creativity; in personality psychology, cognitive styles and kinesthetic aftereffect; and in social psychology, risk taking.

14.1 Creativity

Research on creativity thrived at ETS during the 1960s and 1970s. Three distinct strands of work can be distinguished. One of these strands was based largely on studies of children, with an emphasis on performance in the domain of divergent-thinking abilities. A second strand involved the construction of measures of scientific thinking and utilized samples of young adults. A third strand featured an emphasis on the products of creativity, mainly using young adult samples. The three strands were not entirely independent of each other as some studies explored possible links between the divergent-thinking and scientific-thinking domains or between ratings of products and characteristics of the individuals who produced them.

Lawrence J. Stricker edited a final draft of this chapter that Nathan Kogan completed before his death.

N. Kogan (✉)
Educational Testing Service, Princeton, NJ, USA
e-mail: researchreports@ets.org

R.E. Bennett, M. von Davier (eds.), *Advancing Human Assessment*,
Methodology of Educational Measurement and Assessment,
DOI 10.1007/978-3-319-58689-2_14

14.1.1 Divergent Thinking

We begin with studies in the divergent-thinking domain that employed children ranging across the preschool to primary-school years. The volume published by ETS scientist Nathan Kogan and Michael Wallach, his longtime collaborator at Duke University (Wallach and Kogan 1965a), set the tone for much of the research that followed. A major goal of that investigation was to bring clarity to the discriminant validity issue—whether divergent-thinking abilities could be statistically separated from the convergent thinking required by traditional tests of intellectual ability. In a paper by Thorndike (1966), evidence for such discriminant validity in investigations by Guilford and Christensen (1956) and by Getzels and Jackson (1962) was found to be lacking. A similar failure was reported by Wallach (1970) for the Torrance (1962) divergent-thinking test battery—the correlations within the divergent-thinking battery were of approximately the same magnitude as these tests' correlations with a convergent-thinking test. Accordingly, the time was ripe for another effort at psychometric separation of the divergent- and convergent-thinking domains.

Wallach and Kogan (1965a) made two fundamental changes in the research paradigms that had been used previously. They chose to purify the divergent-thinking domain by employing only ideational-fluency tasks and presented these tasks as games, thereby departing from the mode of administration typical of convergent-thinking tests. The rationale for these changes can be readily spelled out. Creativity in the real world involves the generation of new ideas, and this is what ideational-fluency tests attempt to capture. Of course, the latter represents a simple analogue of the former, but the actual relationship between them rests on empirical evidence (which is mixed at the present time). The choice of a game-like atmosphere was intended to reduce the test anxiety from which numerous test takers suffer when confronted with typical convergent-thinking tests.

The major outcome of these two modifications was the demonstration of both convergent validity of measures of the divergent- and convergent-thinking domains and the discrimination between them as reflected in near-zero correlations in a sample of fifth-grade children. As evidence has accumulated since the Wallach and Kogan study (Wallach and Kogan 1965a), the trend is toward a low positive correlation between measures of the two domains. For example, Silvia (2008), employing latent variable analysis, reanalyzed the Wallach and Kogan data and reported a significant correlation of .20, consistent with the predominant outcome of the majority of studies directed to the issue. It may well be a pseudo-issue at this point in time, reflecting the selectivity of the sample employed. As the range of IQ in a sample declines, its correlation with divergent-thinking measures should obviously decline as well. Thus, one would not expect to find divergent- and convergent-thinking tests correlated in a sample selected for giftedness.

The ideational-fluency tests developed by Wallach and Kogan (1965a) were scored for fluency and uniqueness. The two were highly correlated, consistent with what was expected by the principal theoretical conceptualization at the time—

Mednick's (1962) associative theory of creativity. In that theory, the associative process for divergent-thinking items initially favors common associates, and only with continued association would unique and original associations be likely to emerge. Accordingly, fluency represents the path through which originality was achieved. Individual differences in divergent-thinking performance are explained by the steepness-shallowness of the associative hierarchy. Low creatives exhibit a steep gradient in which strong common responses, upon their exhaustion, leave minimal response strength for the emergence of uncommon associates. High creatives, by contrast, demonstrate a shallow gradient in which response strength for common associates is weaker, allowing the person enough remaining power to begin emitting uncommon associates.

In a recent article, Silvia et al. (2008) ignored the Mednick (1962) formulation, criticized the scoring of divergent-thinking responses for uniqueness, concluded that scoring them for quality was psychometrically superior, and advocated that the administration of divergent-thinking tests urge test takers to be creative. A critical commentary on this work by Silvia and his associates appeared in the same issue of the journal (Kogan 2008). Particularly noteworthy is the indication that approximately 45 years after its publication, the issues raised in the Wallach and Kogan (1965a) volume remain in contention.

Beyond the topic of the creativity-intelligence (divergent vs. convergent thinking) distinction, the construct validity of divergent-thinking tests came under exploration. What psychological processes (beyond Mednick's 1962, response hierarchies) might account for individual differences in divergent-thinking performance? Pankove and Kogan (1968) suggested that tolerance for risk of error might contribute to superior divergent-thinking performance in elementary school children. A motor skill task (a shuffleboard game) allowed children to adjust their preferred risk levels by setting goal posts closer or further apart to make the task harder or easier, respectively. Children who challenged themselves by taking greater risks on the shuffleboard court (with motor skill statistically controlled) also generated higher scores on a divergent-thinking test, Alternate Uses.

In a provocative essay, Wallach (1971) offered the hypothesis that performance on divergent-thinking tests might be motivationally driven. In other words, test takers might vary in setting personal standards regarding an adequate number of responses. Some might stop well before their cognitive repertoire is exhausted, whereas others might continue to generate responses in a compulsive fashion. This hypothesis implies that the application of an incentive to continue for low-level responders should attenuate the range of fluency scores. Ward et al. (1972) tested this hypothesis in a sample of disadvantaged children by offering an incentive of a penny per response. The incentive increased the number of ideas relative to a control group but did not reduce the range of individual differences. Rather, the incentive added a constant to performance so that the original ordering of the children on the fluency dimension remained intact. In sum, the study bolstered the case for cognitive processes and repertoires underlying divergent-thinking performance and undermined the motivational claim that it is a simple matter of when one chooses to stop responding.

To designate divergent thinking as an indicator of creativity is credible only if divergent-thinking performance is predictive of a real-world criterion that expert judges would acknowledge to be relevant to creativity. This is the validity issue that has been examined in both its concurrent and long-term predictive forms. The concurrent validity of divergent-thinking performance has proven to be rather robust. Thus, third-grade and fourth-grade children's scores on the Wallach and Kogan (1965a) tasks correlated significantly with the originality and aesthetic quality of their art products, as evaluated by qualified judges (Wallbrown and Huelsman 1975). And college freshmen's scores on these tasks correlated significantly with their extracurricular attainments in leadership, art, writing, and sciences in their secondary-school years, whereas their *SAT*® scores did not (Wallach and Wing 1969). Efforts to predict future talented accomplishments from current divergent-thinking performance have yielded more equivocal outcomes. Kogan and Pankove (1972, 1974) failed to demonstrate predictive validity of fifth-grade Wallach and Kogan assessments against 10th-grade and 12th-grade accomplishments in extracurricular activities in the fields of art, writing, and science. On the other hand, Plucker's (1999) reanalysis of original data from the Torrance Tests of Creative Thinking (Torrance 1974) is suggestive of the predictive validity of that instrument.

The issue of predictive validity from childhood to adulthood continues to reverberate to the present day, with Kim (2011) insisting that the evidence is supportive for the Torrance tests while Baer (2011) notes that the adulthood creativity criterion employed is based exclusively on self-reports, hence rendering the claim for predictive validity highly suspect. Indeed, Baer extends his argument to the point of recommending that the Torrance creativity tests be abandoned.

Can the Mednick (1962) associative model of creativity be generalized to young children? Ward (1969b) offered an answer to this question by administering some of the Wallach and Kogan (1965a) tasks to seven- and eight-year old boys. The model was partially confirmed in the sense that the response rate (and the number of common responses) decreased over time while uniqueness increased over time. On the other hand, individual differences in divergent thinking did not seem to influence the steepness versus shallowness of the response gradients. Ward suggested that cognitive repertoires are not yet fully established in young children, while motivational factors (e.g., task persistence over time) that are not part of Mednick's theoretical model loom large.

Although Mednick's (1962) associative theory of creativity can explain individual differences in divergent-thinking performance, he chose to develop a creativity test—the Remote Associates Test (RAT; Mednick and Mednick 1962)—with a convergent-thinking structure. Items consist of verbal triads for which the test taker is required to find a word that is associatively linked to each of the three words in the triad. An example is "mouse, sharp, blue"; cheese is the answer. It is presumed that the correct answer to each item requires an associative verbal flow, with conceptual thinking of no value for problem solution. Working with children in the fourth-grade to sixth-grade, Ward (1975) administered the Wallach and Kogan (1965a) divergent-thinking tasks and alternate forms of the RAT, as well as IQ and achievement tests. Both forms of the RAT were substantially related to the IQ and achievement measures (*r*'s ranging from .50 to .64). Correlations of the RAT with the Wallach and Kogan tasks ranged from nonsignificant to marginally significant (*r*'s ranging from .19 to .34). These results demonstrated that the associative process is not similar across divergent- and convergent-thinking creativity measures and that the latter's strong relation to IQ and achievement indicates the RAT "represents an unusual approach to the measurement of general intellectual ability" (Ward 1975, p. 94), rather than being a creativity measure.

Among the different explanations for variation in divergent-thinking performance, breadth of attention deployment (Wallach 1970) has been considered important. This process has both an internal and external component, with the former reflecting the adaptive scanning of personal cognitive repertoires and the latter indicative of adaptive scanning of one's immediate environment. A demonstration of the latter can be found in Ward's (1969a) investigation of nursery school children's responses in cue-rich and cue-poor environments. Recognition and application of such cues enhanced divergent-thinking performance, as the cues were directly relevant to the divergent-thinking items presented to the child. Some of the cues in the cue-rich environment were highly salient; some were more subtle; and for some items, no cues were offered. Children were classified as more or less creative based on their pre-experimental divergent-thinking performance. Comparison of high and low creative children revealed no divergent-thinking performance difference with salient cues, and significant performance superiority for high creatives with subtle cues. The low-creative children performed worse in the cue-rich environment than in the cue-poor environment, suggesting that the cue-rich environment was distracting for them. Hence, children who performed well on divergent-thinking items in standard cue-poor conditions by virtue of internal scanning also took advantage of environmental cues for divergent-thinking items by virtue of adaptive external scanning.

To conclude the present section on children's divergent thinking, consider the issue of strategies children employed in responding to divergent-thinking tasks under test-like and game-like conditions (Kogan and Morgan 1969). In a verbal alternate-uses task, children (fifth graders) generated higher levels of fluency and uniqueness in a test-like than in a game-like condition. Yet, a spontaneous-flexibility test (number of response categories) showed no difference between the task contexts. Kogan and Morgan (1969) argued that a test-like condition stimulated a category-exhaustion strategy. Thus, when asked to list alternative uses for a knife, children seized upon some pivotal activity (such as cutting) and proceeded to exhaust exemplars that flow from it (e.g., cutting bread, butter, fruit). A child might eventually think of something to cut that is unique to the sample. Such a strategy is obviously antithetical to enhanced spontaneous-flexibility. The test-game difference did not emerge for the Wallach and Kogan (1965a) figural pattern-meanings task. This outcome may be attributed to the inability of a category-exhaustion strategy to work with a figural task where almost every response is likely to be a category in its own right. In sum, verbal and figural divergent-thinking tasks might elicit distinctive cognitive strategies in children that are moderated by the task context. Further discussion of the issue can be found in Kogan (1983).

14.1.2 Scientific Thinking

In an Office of Naval Research technical report, ETS scientist Norman Frederiksen (1959) described the development of the Formulating Hypotheses test that asks test takers to assume the role of a research investigator attempting to account for a set of results presented in tabular or figural form. An example of the latter is a graph demonstrating that "rate of death from infectious diseases has decreased markedly from 1900, while rate of death from diseases of old age has increased." Examples of possible explanations for the findings orient the test taker to the type of reasoning required by the test. Eight items were constructed, and in its initial version scoring simply involved a count of the number of hypotheses advanced. Subsequently, as a pool of item responses became available, each response could be classified as acceptable or not. The number of acceptable responses generated could then be treated as a quality score.

Publications making use of this test began with an article by Klein et al. (1969). In that study of a college undergraduate sample, Klein et al. explored the influence of feedback after each item relative to a control group with no feedback. The number of hypotheses offered with feedback increased significantly relative to the number for the control group, but no experimental-control difference was found for acceptable (higher quality) hypotheses. Further, no experimental-control difference was observed for Guilford's (1967) Consequences test, a measure of divergent production, indicating no transfer effects. An anxiety scale was also administered with the expectation that anxiety would enhance self-censorship on the items, which in turn would be mitigated in the feedback treatment as anxious participants become aware of the vast array of hypotheses available to them. No such effect was obtained. Klein et al. also examined the possibility that intermediate levels of anxiety would be associated with maximal scores on the test consistent with the U-shaped function of motivational arousal and performance described by Spence and Spence (1966). Surprisingly, this hypothesis also failed to be confirmed. In sum, this initial study by Klein et al. demonstrated the potential viability of the Formulating Hypotheses test as a measure of scientific thinking despite its failure to yield anticipated correlates.

A further advance in research on this test is displayed in a subsequent study by Frederiksen and Evans (1974). As in the previous investigation, this one featured an experimental-control contrast, but two treatments were now employed. Participants (college undergraduates) were exposed to either quantity or quality models. In the former, the feedback following each item consisted of a lengthy list of acceptable hypotheses (18 to 26); in the latter case, only the best hypotheses constituted the feedback (6 to 7 ideas). The control group did not receive any feedback. The Formulating Hypotheses test represented the dependent variable, and its scoring was expanded to include a rated quality-score and a measure of the average number of words per response. Highly significant effects of the treatments on performance were obtained. Relative to the control group, the quantity model increased the number of responses and decreased the average number of words per response; the quality model increased the rated quality of the responses and the average number of words per response but decreased the average number of responses. Of the two tests from the Kit of Reference Tests for Cognitive Factors (French et al. 1963) administered, Themes (ideational fluency) was significantly related to the number of responses and Advanced Vocabulary was significantly related to the rated quality of the responses. In their conclusion, Frederiksen and Evans expressed considerable doubt that the experimental treatments altered the participants' ability to formulate hypotheses. Rather, they maintained that the quantity and quality treatments simply changed participants' standards regarding a satisfactory performance.

Expansion of research on scientific thinking can be seen in the Frederiksen and Ward (1978) study, where measures extending beyond the Formulating Hypotheses test were developed. The general intent was to develop a set of measures that would have the potential to elicit creative scientific thinking while possessing psychometric acceptability. The authors sought to construct assessment devices in a middle ground between Guilford-type divergent-thinking tests (Guilford and Christensen 1956) and the global-creativity peer nominations of professional groups, typical of the work of MacKinnon (1962) and his collaborators. Leaning on the Flanagan (1949) study of critical incidents typical of scientists at work, Frederiksen and Ward attempted to develop instruments, called Tests of Scientific Thinking (TST), that would reflect problems that scientists often encounter in their work. The TST consisted of the Formulating Hypotheses test and three newly constructed tests: (a) Evaluating Proposals—test takers assume the role of an instructor and offer critical comments about proposals written by their students in a hypothetical science course; (b) Solving Methodological Problems—test takers offer solutions to a methodological problem encountered in planning a research study; and (c) Measuring Constructs—test takers suggest methods for eliciting relevant behavior for a specific psychological construct without resorting to ratings or self-reports. Scores tapped the quantity and quality of responses (statistical infrequency and ratings of especially high quality).

The TST was administered to students taking the *GRE*® Advanced Psychology Test. High levels of agreement prevailed among the four judges in scoring responses. However, the intercorrelations among the four tests varied considerably in magnitude, and Frederiksen and Ward (1978) concluded that there was "little evidence of generalized ability to produce ideas which are either numerous or good" (p. 11). It is to be expected, then, that factor analysis of the TST would yield multiple factors. A three-factor solution did in fact emerge, with Factor I reflecting the total number of responses and number of unusual responses, Factor II as a quality factor for Formulating Hypotheses and Measuring Constructs, and Factor III as a quality factor for Evaluating Proposals and Solving Methodological Problems. The Factor II tests were more divergent and imposed fewer constraints on the participants than did Factor III tests, which emphasized issues of design and analysis of experiments. The total number of responses and number of unusual responses cohering on Factor I parallels the findings with divergent-thinking tests where the number of unusual responses derives from the rate at which more obvious possibilities are exhausted. The factor analysis also makes clear that idea quality is unrelated to the number of proposed solutions.

Finally, Frederiksen and Ward (1978) inquired into the possible predictive validity of a composite of the four TSTs. A subgroup of the original sample, at the end of their first year in a graduate psychology-program, filled out a questionnaire with items inquiring into professional activities and accomplishments. Surprisingly, the scores for the number of responses from the TST composite yielded more significant relations with the questionnaire items than did the quality scores. Higher numbers of responses (mundane, unusual, and unusual high quality) were predictive of higher department quality, planning to work toward a Ph.D. rather than an M.A., generating more publications, engaging in collaborative research, and working with equipment. An inverse relation was found for enrollment in a program emphasizing the practice of psychology and for self-rated clinical ability. These outcomes strongly suggest that the TST may have value in forecasting the eventual productivity of a psychological scientist.

Two additional studies by ETS scientist Randy Bennett and his colleagues shed light on the validity of a computer-delivered Formulating Hypotheses test, which requires only general knowledge about the world, for graduate students from a variety of disciplines. Bennett and Rock (1995) used two four-item Formulating Hypotheses tests, one limiting the test takers' to seven-word responses and the other to 15-word responses. The tests were scored simply for the number of plausible, unduplicated hypotheses, based on the Frederiksen and Ward (1978) finding that the number of hypotheses is more highly related to criteria than their quality. A generalizability analysis showed high interjudge reliability. Generalizability coefficients for the mean ratings taken across judges and items were .93 for the seven-word version and .90 for the 15-word version. Three factors were identified in a confirmatory factor analysis of the two forms of the Formulating Hypotheses test and an ideational-fluency test (one item each from the Topics test of the Kit of Reference Tests for Cognitive Factors, French et al. 1963; the verbal form of the Torrance Tests of Creative Thinking, Torrance 1974; and two pattern-meaning tasks from the Wallach and Kogan 1965a, study). One factor was defined by the seven-word version, another by the 15-word version, and the third by the ideational-fluency test. The two formulating hypotheses factors correlated .90 with each other and .66 and .71 with the ideational-fluency factor. Bennett and Rock concluded that "the correlations between the formulating hypotheses factors …, though quite high, may not be sufficient to consider the item types equivalent" (p. 29).

Bennett and Rock (1995) also investigated the correlations of the two Formulating Hypotheses tests and the GRE General Test with two criterion measures: undergraduate grades and a questionnaire about extracurricular accomplishments in the college years (Stricker and Rock 2001), similar to the Baird (1979) and Skager et al. (1965) measures. The two tests had generally similar correlations with grades (r = .20 to .26 for the Formulating Hypotheses tests and .26 to .37 for the GRE General Test). The correlations were uniformly low between the tests and the six scales on the accomplishments questionnaire (Academic Achievement, Leadership, Practical Language [public speaking, journalism], Aesthetic Expression [creative writing, art, music, dramatics], Science, and Mechanical). Both Formulating Hypotheses tests correlated significantly with one of the scales: Aesthetic

Expression; and at least one of the GRE General Test sections correlated significantly with three scales: Aesthetic Expression, Academic Achievement, and Science.

The related issue of the Formulating Hypotheses test's incremental validity against these criteria was examined as well. The 15-word version of the test showed significant (but modest) incremental validity (vis-à-vis the GRE General Test) against grades (R^2 increased from .14 to .16). This version also demonstrated significant (but equally modest) incremental validity (vis-à-vis the GRE General Test and grades) against one of the six accomplishments scales: Aesthetic Expression (R^2 increased from .01 to .03). The seven-word version had no significant incremental validity against grades or accomplishments.

Enright et al. (1998), in a study to evaluate the potential of a Formulating Hypotheses test and experimental tests of reasoning for inclusion in the GRE General Test, replicated and extended the Bennett and Rock (1995) investigation of the Formulating Hypotheses tests. Enright et al. used the Bennett and Rock (1995) 15-word version of the test (renamed Generating Explanations), scored the same way. Four factors emerged in a confirmatory factor-analysis of the test with the GRE General Test's Verbal, Quantitative, and Analytical sections, and the three reasoning tests. The factors were Verbal, defined by the Verbal section, all the reasoning tests, and the logical-reasoning items from the Analytical section; Quantitative, defined only by the Quantitative section; Analytical, defined only by the analytical- reasoning items from the Analytical section; and Formulating Hypotheses, defined only by the Formulating Hypotheses test. The Formulating Hypotheses factor correlated .23 to .40 with the others.

Like Bennett and Rock (1995), Enright et al. (1998) examined the correlations of the Formulating Hypotheses test and the GRE General Test with undergraduate grades and accomplishments criteria. The Formulating Hypotheses test had lower correlations with grades ($r = .15$) than did the GRE General Test ($r = .22$ to .29). The two tests had consistently low correlations with the same accomplishments questionnaire (Stricker and Rock 2001) used by Bennett and Rock. The Formulating Hypotheses test correlated significantly with the Aesthetic Expression and Practical Language scales, and a single GRE General Test section correlated significantly with the Academic Achievement, Mechanical, and Science scales.

Enright et al. (1998) also looked into the incremental validity of the Formulating Hypotheses Test against these criteria. The test's incremental validity (vis-à-vis the GRE General Test) against grades was not significant for the total sample, but it was significant for the subsample of humanities and social-science majors (the increment was small, with R^2 increasing from .12 to .16). Enright et al. noted that the latter result is consistent with the test's demonstrated incremental validity for the total sample in the Bennett and Rock (1995) study, for over 60% of that sample were humanities and social-science majors. The test had no significant incremental validity (vis-à-vis the GRE General Test and grades) against an overall measure of accomplishments (pooling accomplishments across six areas), perhaps because of the latter's heterogeneity.

To sum up, the Bennett and Rock (1995) and Enright et al. (1998) investigations are remarkably consistent in demonstrating the distinctiveness of Formulating

Hypotheses tests from the GRE General Test and suggesting that the former can make a contribution in predicting important criteria.

In all of the TST research, a free-response format had been employed. The Formulating Hypotheses test lends itself to a machine-scorable version, and Ward et al. (1980) examined the equivalence of the two formats. In the machine-scorable version, nine possible hypotheses were provided, and the test taker was required to check those hypotheses that could account for the findings and to rank order them from best to worst. Comparable number and quality scores were derived from the two formats. The free-response/machine-scorable correlations ranged from .13 to .33 in a sample of undergraduate psychology majors, suggesting that the two versions were not alternate forms of the same test. When scores from the two versions were related to scores on the GRE Aptitude Test and the GRE Advanced Psychology Test, the correlations with the machine-scorable version were generally higher than those for the free-response version. Ward et al., in fact, suggested that the machine-scorable version offered little information beyond what is provided by the two GRE tests, whereas the free-response version did offer additional information. The obvious difference between the two versions is that the free-response requires test takers to produce solutions, whereas the machine-scorable merely calls for recognition of appropriate solutions. From the standpoint of ecological validity, it must be acknowledged that solutions to scientific problems rarely assume multiple-choice form. As Ward et al. point out, however, free-response tests are more difficult and time-consuming to develop and score, and yet are less reliable than multiple-choice tests of the same length.

14.1.3 Creative Products

Within a predictor-criterion framework, the previous two sections have focused on the former—individual differences in creative ability as reflected in performance on tests purportedly related to creativity on analogical or theoretical grounds. In some cases, various creativity criteria were available, making it possible to examine the concurrent or predictive validity of the creativity tests. Such research is informative about whether the creativity or scientific thinking label applied to the test is in fact warranted. In the present section, the focus is on the creative product itself. In some cases, investigators seek possible associations between the judged creativity of the product and the demographic or psychological characteristics of the individual who produced the product. In such instances, the predictor-to-criterion sequence is actually reversed.

Study of creative products can take two forms. The most direct form involves the evaluation of a concrete product for its creativity. A second and somewhat less direct form relies on test-takers' self-reports. The test taker is asked to describe his or her activities and accomplishments that may reflect different kinds of creative production, concrete or abstract, in such domains as science, literature, visual arts, and music. It is the test-taker's verbal description of a product that is evaluated for creativity rather than the product itself.

14.1.3.1 Concrete Products

A good example of the concrete product approach to creativity is a study by ETS scientists Skager et al. (1966a). They raised the issue of the extent of agreement among a group of 28 judges (24 artists and 4 nonartists) in their aesthetic-quality ratings of drawings produced by the 191 students in the sophomore class at the Rhode Island School of Design. The students had the common assignment of drawing a nature scene from a vantage point overlooking the city of Providence. The question was whether the level of agreement among the judges in their quality ratings would be so high as to leave little interjudge variance remaining to be explained. This did not prove to be the case, as a varimax rotation of a principal-axis factor analysis of the intercorrelations of the 28 judges across 191 drawings suggested that at least four points of view about quality were discernible. Different artist judges were located on the first three factors, and the nonartists fell on the fourth factor. Factor I clearly pointed to a contrast between judges who preferred more unconventional, humorous, and spontaneous drawings versus judges who favored more organized, static, and deliberate drawings. Factors II and III were not readily distinguished by drawing styles, but the nonartists of Factor IV clearly expressed a preference for drawings of a more deliberate, less spontaneous style. Skager et al. next turned to the characteristics of the students producing the drawings and whether these characteristics might relate to the location of the drawing on one of the four quality points of view. Correlations were reported between the points of view and the students' scores on a battery of cognitive tests as well as on measures of academic performance, cultural background, and socioeconomic status. Most of these correlations were quite low, but several were sufficiently intriguing to warrant additional study, notably, majoring in fine arts, cultural background, and socioeconomic status.

Further analysis of the Skager et al. (1966a) data is described in Klein and Skager (1967). Drawings with the highest positive and negative factor loadings on the first two factors extracted in the Skager et al. study (80 drawings in all) were selected with the aim of further clarifying the spontaneous-deliberate contrast cited earlier. Ten lay judges were given detailed definitions of spontaneity and deliberateness in drawing and were asked to classify the drawings by placing each of them in a spontaneous or deliberate pile. A three-dimensional (judges × viewpoint × high vs. low quality) contingency table was constructed, and its chi-square was partitioned to yield main and interaction effects. A highly significant viewpoint × quality interaction was found. High-quality drawings for both Factor I and Factor II viewpoints were more likely to be classified as spontaneous relative to low-quality drawings. However, the effect was much stronger for the Factor I viewpoint, thereby accounting for the significant interaction. In sum, for Factor I judges, spontaneity versus deliberateness was a key dimension in evaluating the quality of the drawings; Factor

II judges, on the other hand, were evidently basing their evaluations on a dimension relatively independent of spontaneity-deliberateness. Of further interest is the extent to which lay judges, although differing from art experts on what constitutes a good drawing, nevertheless can, with minimal instruction, virtually replicate the aesthetic judgments of art experts holding a particular viewpoint (Factor I). These findings point to the potential efficacy of art appreciation courses in teaching and learning about aesthetic quality.

In a third and final approach to the topic of judged aesthetic quality of drawings, Skager et al. (1966b) subjected their set of 191 drawings to multidimensional scaling (MDS). For this purpose, 26 judges were selected from the faculty of nine schools of design across the United States. Because the scaling procedure required similarity ratings in paired comparisons of an entire stimulus set, practicality required that the size of the set be reduced. Accordingly, 46 of the 191 drawings were selected, reflecting a broad range in aesthetic quality as determined in prior research, and the 46 was divided into two equally sized subsets. Three dimensions emerged from separate MDS analyses of the two subsets. When factor scores for these dimensions were correlated with the test scores and other measures (the same battery used in the Skager et al. 1966a, study), the corresponding correlations in the two analyses correlated .64, suggesting that the three dimensions emerging from the two analyses were reasonably comparable.

What was the nature of the three dimensions? Skager et al. (1966b) chose to answer this question by comparing the drawings with the highest and lowest scale values on each dimension. There is a subjective and impressionistic quality to this type of analysis, but the outcomes were quite informative nevertheless. Dimension I offered a contrast between relative simplification versus complexity of treatment. The contrast was suggestive of the simplicity-complexity dimension described by Barron (1953). Among the contrasts distinguishing Dimension II were little versus extensive use of detail and objects nearly obscured versus clearly delineated. There was no obvious label for these contrasts. Dimension III contrasted neatness versus carelessness and controlled versus chaotic execution. The correlations between the three dimensions, on the one hand, and the test scores and other measures, on the other, were low or inconsistent between the two analyses. Skager et al. noted that some of the contrasts in the drawing dimensions carried a personality connotation (e.g., impulsiveness vs. conscientiousness). Because no personality tests were administered to the students who produced the drawings, this interesting speculation about the basis for aesthetic preference could not be verified.

Finally, we consider two studies by Ward and Cox (1974) that explored creativity in a community sample. Listeners to a New York City radio station were invited to submit humorous and original small green objects to a Little Green Things contest, with a reward of $300 for the best entry and 10 consolation prizes of $20 each. In the first study, a total of 283 submitted objects were rated for originality on a seven-point scale by four judges. The rated originality of each object represented the average of the judges' ratings, yielding an interjudge reliability of .80. Some 58% of the objects were found things, 34% were made things, and 8% were verbal additions to found things. Names and addresses of the contestants made it possible to determine their gender and the census tract in which they resided. From the latter, estimates of family income and years of schooling could be derived. These demographic characteristics of the contestants could then be related to the originality ratings of their submissions. For all of the entries, these correlations were close to zero, but separating out made things yielded significant positive correlations of originality with estimates of family income and years of schooling. Unlike the case for verbal-symbolic forms of creativity assessment, where associations with socioeconomic status have generally not been found, a nonlaboratory context seemed to elevate the importance of this variable. Of course, this relationship occurred for made things—where some investment of effort was required. Why this should be so is unclear.

In a second study, using another set of objects, Ward and Cox (1974) attempted to uncover the dimensions possibly underlying the global originality rating of the objects. Judges were asked to rate the attractiveness, humor, complexity, infrequency, and effort involved in securing or making the object. A multiple *R* was computed indicating how much these five dimensions contributed to the object's originality rating. For found things, *R*s ranged from .25 to .73 (median = .53) with infrequency the strongest and humor the next strongest contributor; for made things, *R*s ranged from .47 to .71 (median = .64) with humor the strongest and amount of effort the next strongest contributor. It should be emphasized that the judges' evaluations were multidimensional, and for virtually every judge a combination of predictors accounted for more variance in the originality ratings than did any single predictor.

14.1.3.2 Reports of Products

A study by Skager et al. (1965) exemplifies the reporting of products approach. Using samples of college freshmen drawn from two institutions of higher learning (a technical institute and a state university), Skager et al. employed the Independent Activities Questionnaire, modeled after one devised by Holland (1961) and covering creative accomplishments outside of school during the secondary-school years. A sample item: "Have you ever won a prize or award for some type of original art work?" The number of these accomplishments served as a quantity score. Judges examined the participant's brief description of these activities with the goal of selecting the most significant achievement. These achievements were then given to a panel of judges to be rated on a 6-point scale to generate a quality score.

Quantity and quality scores were significantly correlated (*r*'s of .44 and .29). In a certain respect, this correlation is analogous to the significant correlations found between the fluency and uniqueness scores derived from ideational-fluency tests. The more divergent-thinking responses ventured or extracurricular activities undertaken by an individual, the more likely an original idea or high-quality accomplishment, respectively, will ensue. In neither sample did the quantity or quality scores relate to socioeconomic status, SAT Verbal, SAT Math, or high-school rank. The quantity score, however, did relate significantly in both samples with "an estimate from the student of the number of hours spent in discussing topics 'such as scientific issues, world affairs, art, literature, or drama' with adults living in the home" (Skager et al. 1965, p. 34). By combining the samples from the two institutions, the quality score began to show significant relationships with SAT Verbal and SAT Math. This result simply reflected the enhanced variance in SAT scores and is of greater methodological than substantive interest.

ETS scientists Baird and Knapp (1981) carried out a similar study with the Inventory of Documented Accomplishments (Baird 1979), devised for graduate school. The inventory, concerning extracurricular accomplishments in the college years, had four scales measuring the number of accomplishments in these areas: literary-expressive, artistic, scientific-technical, social service and organizational activity. It was administered to incoming, first-year graduate students in English, biology, and psychology departments. At the end of their first year, the students completed a follow-up questionnaire about their professional activities and accomplishments in graduate school. The four scales correlated significantly with almost all of these activities and accomplishments, though only one correlation exceeded .30 (r = .50 for the Scientific-Technical scale with working with equipment). Because the sample combined students from different fields, potentially distorting these correlations, the corresponding correlations within fields were explored. Most of the correlations were higher than those for the combined sample.

14.1.4 Overview

Creativity research has evolved since the heyday of ETS's efforts in the 1960s and 1970s, at the dawn of psychology's interest in this phenomenon. The first journal devoted to creativity, the *Journal of Creative Behavior*, was published in 1967, followed by others, notably the *Creativity Research Journal*; *Psychology of Aesthetics, Creativity, and the Arts;* and *Imagination, Creativity, and Personality.* Several handbooks have also appeared, beginning with the Glover et al. (1989) *Handbook of Creativity* (others are Kaufman and Sternberg 2006, 2010b; Sternberg 1999; Thomas and Chan 2013). The volume of publications has burgeoned from approximately 400 articles before 1962 (Taylor and Barron 1963) to more than 10,000 between 1999 and 2010 (Kaufman and Sternberg 2010a). And the research has broadened enormously, "a virtual explosion of topics, perspectives, and methodologies…." (Hennessey and Amabile 2010, p. 571).

Nonetheless, divergent-thinking tests, evaluations of products, and inventories of accomplishments, the focus of much of the ETS work, continue to be mainstays in appraising individual differences. Divergent-thinking tests remain controversial (Plucker and Makel 2010), as noted earlier. The evaluation of products, considered to be the gold standard (Plucker and Makel 2010), has been essentially codified by the wide use of the Consensus Assessment Technique (Amabile 1982), which neatly skirts the knotty problem of defining creativity by relying on expert judges' own implicit conceptions of it. And there now seems to be a consensus that inventories of accomplishments, which have proliferated (see Hovecar and Bachelor 1989;

Plucker and Makel 2010), are the most practical and effective assessment method (Hovecar and Bachelor 1989; Plucker 1990; Wallach 1976).

Creativity is not currently an active area of research at ETS, but its earlier work continues to have an influence on the field. According to the Social Science Citation Index, Wallach and Kogan's 1965 monograph, *Modes of Thinking in Young Children*, has been cited 769 times through 2014, making it a citation classic.

14.2 Cognitive Styles

Defined as individual differences in ways of organizing and processing information (or as individual variation in modes of perceiving, remembering, or thinking), cognitive styles represented a dominant feature of the ETS research landscape beginning in the late 1950s and extending well into the 1990s. The key players were Samuel Messick, Kogan, and Herman Witkin, along with his longtime collaborators, Donald Goodenough and Philip Oltman, and the best known style investigated was field dependence-independence (e.g., Witkin and Goodenough 1981). The impetus came from Messick, who had spent a postdoctoral year at the Menninger Foundation (then a center for cognitive-style research) before assuming the leadership of the personality research group at ETS. During his postdoctoral year at Menninger, Messick joined a group of researchers working within an ego-psychoanalytic tradition who sought to derive a set of cognitive constructs that mediated between motivational drives and situational requirements. These constructs—six in all—were assigned the label of cognitive-control principles and were assessed with diverse tasks in the domains of perception (field dependence-independence), attention (scanning), memory (leveling-sharpening), conceptualizing (conceptual differentiation), susceptibility to distraction and interference (constricted-flexible control), and tolerance for incongruent or unrealistic experience (Gardner et al. 1959). Messick's initial contribution to this effort explored links between these cognitive-control principles and traditional intellectual abilities (Gardner et al. 1960). This study initiated the examination of the style-ability contrast—whereas abilities almost always reflect maximal performance, styles generally tap typical performance.

The psychoanalytic origin of the cognitive-control principles accounts for the emphasis on links to drives and defenses in early theorizing, but later research and theory shifted to the study of the cognitive-control principles (relabeled cognitive styles) in their own right. Messick played a major role in this effort, launching a project, supported by the National Institute of Mental Health, focused on conceptual and measurement issues posed by the assessment of these new constructs. The project supported a series of empirical contributions as well as theoretical essays and scholarly reviews of the accumulating literature on the topic. In this effort, Messick was joined by Kogan, who collaborated on several of the empirical studies—conceptual differentiation (Messick and Kogan 1963), breadth of categorization and quantitative aptitude (Messick and Kogan 1965), and a MDS approach to cognitive complexity-simplicity (Messick and Kogan 1966). Other empirical work included studies of the influence of field dependence on memory by Messick and Damarin (1964) and Messick and Fritzky (1963). Scholarly reviews were published (Kagan and Kogan 1970; Kogan 1971) that enhanced the visibility of the construct of cognitive style within the broader psychological and educational community. Messick (1970) provided definitions for a total of nine cognitive styles, but this number expanded to 19 six years later (Messick 1976). It is evident that Messick's interest in cognitive styles at that latter point in time had moved well beyond his original psychoanalytic perspective to encompass cognitive styles generated by a diversity of conceptual traditions.

The reputation of ETS as a center for cognitive-style research was further reinforced by the 1973 arrival of Witkin, with Goodenough and Oltman. This team focused on field dependence-independence and its many ramifications. A field-independent person is described as able to separate a part from a whole in which it is embedded—the simple figure from the complex design in the Embedded Figures Test (EFT) and the rod from the tilted frame in the Rod and Frame Test (RFT). A field-dependent person is presumed to find it difficult to disembed part of a field from its embedding context. The Witkin team was exceptionally productive, generating empirical studies (e.g., Witkin et al. 1974, 1977a; Zoccolotti and Oltman 1978) and reviews (e.g., Goodenough 1976; Witkin and Berry 1975; Witkin and Goodenough 1977; Witkin et al. 1977b, 1979) that stamped Witkin as one of the foremost personality researchers of his era (Kogan 1980). His death in 1979 severely slowed the momentum of the field dependence-independence enterprise to the point where its future long-term viability was called into question. Nevertheless, further conceptual and methodological refinement of this construct continued in articles published by Messick (1984, 1987, 1994, 1996) and in empirical work and further conceptualizing by Goodenough and his colleagues (e.g., Goodenough 1981, 1986; Goodenough et al. 1987, 1991).

Kogan, who had by then departed for the New School for Social Research, continued to build upon his ETS experience and devoted several publications to field dependence-independence and other cognitive styles (Kogan 1976, 1983, 1994; Kogan and Saarni 1990). A conference bringing together the principal field dependence-independence theorists and researchers (domestic and foreign) was held at Clark University in 1989 and subsequently appeared as an edited book (Wapner and Demick 1991). Kogan and Block (1991) contributed a chapter to that volume on the personality and socialization aspects of field dependence-independence. That chapter served to resolve conceptual incongruities that arose when the Witkin team altered their original value-laden theory (Witkin et al. 1962) in a direction favoring a value-neutral formulation (Witkin and Goodenough 1981). The latter endowed field-dependent and field-independent individuals with distinctive sets of skills—analytic restructuring versus interpersonal, respectively. The extensive longitudinal research reported by Kogan and Block proved more consistent with the earlier formulation (Witkin et al. 1962) than with the more recent one (Witkin and Goodenough 1981).

Educational implications of cognitive styles were of particular interest at ETS, and ETS researchers made contributions along those lines. Working with the nine cognitive styles delineated by Messick (1970), a book chapter by Kogan (1971) pointed to much variation at that point in time in the degree to which empirical investigations based on those styles could be said to offer implications for education. Indeed, for some of the styles, no effort had been made to establish educational linkages, not surprising given that the origins of cognitive styles can be traced to laboratory-based research on personality and cognition. It took some years before the possibility of educational applications received any attention. By the time of a subsequent review by Kogan (1983), this dearth had been corrected, thanks in large part to the work of Witkin and his colleagues, and subsequently to Messick's (1984, 1987) persistent arguments for the importance of cognitive styles in accounting for educational processes and outcomes. Witkin and his colleagues considered the educational implications of field dependence-independence in a general survey of the field (Witkin et al. 1977b) and in an empirical study of the association between field dependence-independence and college students' fields of concentration (Witkin et al. 1977a). In the latter study, three broad categories of student majors were formed: (a) science; (b) social science, humanities, and arts; and (c) education. Field independence, assessed by EFT scores, was highest for science majors and lowest for education majors. Furthermore, students switching out of science were more field dependent than those who remained, whereas students switching out of education were more field independent than those who remained. An attempt to relate field dependence-independence to performance (i.e., grades) within each field yielded only marginal results. The findings clearly supported the relevance of field dependence-independence as an important educational issue.

Another topic of educational relevance is the matching hypothesis initially framed by Cronbach and Snow (1977) as a problem in aptitude-treatment interaction. The basic proposition of this interaction is that differences among learners (whether in aptitude, style, strategy, or noncognitive attributes) may imply that training agents or instructional methods can be varied to capitalize upon learners' strengths or to compensate for their weaknesses. A portion of this research inquired into cognitive styles as individual differences in the aptitude-treatment interaction framework, and the Witkin et al. (1977b) review showed inconsistent effects for field dependence-independence across several studies. There was some indication that style-matched teachers and students liked one another more than did mismatched pairs, but there was little evidence suggesting that matching led to improved learning outcomes. A more recent review by Davis (1991) of field dependence-independence studies of this kind again suggests a mixed picture of successes and failures. Messick (1994, 1996) has attributed many of these failures to the haphazard manner in which field dependence-independence has been assessed. Typically, the isolated use of the EFT to assess this cognitive style implies that only the cognitive restructuring or set-breaking component is represented in the field dependence-independence index to the exclusion of the component represented by the RFT, which Witkin and Goodenough (1981) described as visual versus vestibular sensitivity to perception of the upright. They, in fact, raised the possibility that the EFT and RFT may be tapping distinctive, although related, psychological processes. Multivariate studies of a diversity of spatial tasks have found that EFT and RFT load on separate factors (Linn and Kyllonen 1981; Vernon 1972). A discussion of the implications of these findings can be found in Kogan (1983).

14.2.1 Conclusion

There can be no doubt that the field dependence-independence construct has faded from view, but this in no way implies that the broader domain of cognitive styles has correspondingly declined. More recently, Kozhevnikov (2007) offered a review of the cognitive-style literature that envisions the future development of theoretical models incorporating neuroscience and research in other psychological fields. Thus, the style label has been attached to research on decision-making styles (e.g., Kirton 1989), learning styles (e.g., Kolb 1976), and thinking styles (e.g., Sternberg 1997). Possibly, the dominant approach at present is that of intellectual styles (Zhang and Sternberg 2006). Zhang and Sternberg view intellectual styles as a very broad concept that more or less incorporates all prior concepts characterizing stylistic variation among individuals. This view will undoubtedly be reinforced by the recent appearance of the *Handbook of Intellectual Styles* (Zhang et al. 2011). It is dubious, however, that so broad and diverse a field as stylistic variation among individuals would be prepared to accept such an overarching concept at the present stage in its history.

14.3 Risk Taking

Research on risk-taking behavior, conducted by Kogan and Wallach, was a major activity at ETS in the 1960s. Despite the importance and general interest in this topic, no review of research in the field had been published prior to their essay (Kogan and Wallach 1967c). In that review, they surveyed research on situational, personal, and group influences on risk-taking behavior. Also discussed were the assets and liabilities of mathematical models (e.g., Edwards 1961; Pruitt 1962) developed to account for economic decision making and gambling behavior. Simon (1957) rejected this rational economic view of the individual as maximizer in favor of the individual as satisfier—accepting a course of action as good enough. This latter perspective on decision making is more friendly to the possibility of systematic individual-variation in what constitutes good enough, and thus opened the door to a construct of risk taking.

14.3.1 Individuals

In the matter of situational influences, the distinction between chance tasks and skill tasks would seem critical, but the contrast breaks down when taking account of the degree of control individuals believe they can exert over decision outcomes. In the Kogan and Wallach (1964) research, a direct comparison between risk preferences under chance and skill conditions was undertaken. Participants gambled for money on games of chance (e.g., dice) and skill (e.g., shuffleboard), choosing the bets and resultant monetary payoffs (the games scored for maximizing gains, minimizing losses, and deviations from a 50-50 bet). There was no indication in the data of greater risk taking under skill conditions. Rather, there was a strategic preference for moderate risk taking (minimizing deviation from a 50-50 bet). By contrast, the chance condition yielded greater variance as some participants leaned toward risky strategies or alternatively toward cautious strategies.

Variation in risk-taking behavior can also be observed in an information-seeking context. The paradigm is one in which there is a desirable goal (e.g., a monetary prize for solving a problem) with informational cues helpful to problem solution offered at a price. To avail oneself of all the cues provided would reduce the prize to a negligible value. Hence, the risk element enters as the person attempts to optimize the amount of information requested. Venturing a solution early in the informational sequence increases the probability of an incorrect solution that would forfeit the prize. Such a strategy is indicative of a disposition toward risk taking. Irwin and Smith (1957) employed this information-seeking paradigm and observed that the number of cues requested was directly related to the value of the prize and inversely related to the monetary cost per cue. Kogan and Wallach (1964) employed information-seeking tasks in their risk-taking project.

The risk-taking measures described thus far were laboratory based and cast decisions in a gambling-type format with monetary incentives (while avoiding use of participants' own money). Most real-life decision making does not conform to the gambling paradigm, and accordingly, Kogan and Wallach (1964) constructed a series of choice dilemmas drawn from conceivable events in a variety of life domains. An abbreviated version of a scenario illustrates the idea: "Mr. A, an electrical engineer, had the choice of sticking with his present job at a modest, though adequate salary, or of moving on to another job offering more money but no long-term security." These scenarios (12 in all) constituted the Choice Dilemmas Questionnaire (CDQ). In each of these scenarios, the participant is asked to imagine advising the protagonist, who is faced with the choice between a highly desirable alternative with severe negative consequences for failure and a less desirable alternative where consequences for failure are considerably less severe. On a probability scale extending from 9 in 10 to 1 in 10, the participant is asked to select the minimum odds of success the protagonist should demand before opting for the highly desirable alternative. Descending the probability scale (toward 1 in 10) implies increasing preference for risk. (A 10 in 10 option is also provided for participants demanding complete certainty that the desirable alternative will be successful.) The

CDQ has also been claimed to measure the deterrence value of potential failure in the pursuit of desirable goals (Wallach and Kogan 1961). Its reliability has ranged from the mid-.50s to the mid-.80s.

Diverse tasks have been employed in the assessment of risk-taking dispositions. The basic question posed by Kogan and Wallach (1964) was whether participants demonstrate any consistency in their risk-taking tendencies across these tasks. The evidence derived from samples of undergraduate men and women pointed to quite limited generality, calling into question the possibility of risk-inclined versus prudent, cautious personalities. A comparable lack of cross-situational consistency had been observed earlier by Slovic (1962). Unlike Slovic, however, Kogan and Wallach chose to explore the role of potential moderators selected for their conceptual relevance to the risk-taking domain. The first moderator considered was test anxiety. Atkinson (1957) conceptualized test anxiety as fear of failure and offered a model in which fear-of-failure individuals would make exceedingly cautious or risky choices in a level-of-aspiration problem-solving paradigm. Cautious choices are obviously more likely to ensure success, and exceptionally risky choices offer a convenient rationalization for failure. Hence, test-anxious participants were expected to be sensitized to the success and failure potentialities of the risk-taking measures with the likely consequence of enhanced consistency in their choices. The second moderator under examination was defensiveness—also labeled need for approval, by Crowne and Marlowe (1964). Many of the tasks employed in the Kogan and Wallach research required a one-on-one interaction with an experimenter. Participants high in defensiveness were considered likely to engage in impression management—a desire to portray oneself consistently as a bold decision-maker willing to take risks, or as a cautious, prudent decision-maker seeking to avoid failure. Accordingly, enhanced cross-task consistency was anticipated for the highly defensive participants.

Both moderators proved effective in demonstrating the heightened intertask consistency of the high test-anxious and high-defensive participants relative to the participants low on both moderators. The latter subgroup's risk-taking preferences appeared to vary across tasks contingent on their stimulus properties, whereas the former, motivationally disturbed subgroups appeared to be governed by their inner motivational dispositions in tasks with a salient risk component. It should be emphasized that Kogan and Wallach (1964) were not the first investigators to discover the value of moderator analyses in the personality domain. Saunders (1956) had earlier reported enhanced predictability through the use of personality moderators. More recently, Paunonen and Jackson (1985) offered a multiple-regression model for moderator analyses as a path toward a more idiographic approach in personality research.

Of further interest in the Kogan and Wallach (1964) monograph is the evidence indicating an association between risk-taking indices and performance on the SAT Verbal section for undergraduate men. The relationship was moderated by test anxiety such that high test-anxious participants manifested an inverse association and low test-anxious participants a direct association between risk-taking level and SAT performance. In short, a disposition toward risk taking facilitated the low-anxious person (presumably enabling educated guessing) and hindered the anxiety-laden individual (presumably due to interference with cognitive processing). Hence, the penalty-for-guessing instructions for the SAT (retained by the College Board until recently) seemed to help some participants while hurting others.

Beyond the consistency of risk-taking dispositions in the motivationally-disturbed participants, Kogan and Wallach (1964) introduced the possibility of irrationality in the choices of those subgroups. After implementing their choices, participants were informed of their monetary winnings and offered the opportunity to make a final bet with those winnings on a single dice toss that could enhance those winnings up to six-fold if successful but with the risk of total loss if unsuccessful. The low anxious/low defensive participants exhibited the *protecting-one's-nest-egg* phenomenon in the sense of refusing to make a final bet or accepting less risk on the bet in proportion to the magnitude of their winnings. In the motivationally disturbed subgroups, on the other hand, the magnitude of winnings bore no relation to the risk level of the final bet. In other words, these subgroups maintained their consistently risky or cautious stance, essentially ignoring how much they had previously won. Further evidence for irrationality in the motivationally disturbed subgroups concerned post-decisional regret. Despite a frequent lack of success when playing their bets, participants in those subgroups expressed minimal regret about their original decisions unlike the low anxious/low defensive participants who wished they could alter original choices that failed to yield successful outcomes. In the sense that some participants ignored relevant situational properties whereas others took account of them, the issue of rationality-irrationality became germane.

The directions taken by risk-taking research subsequent to the Kogan and Wallach (1964, 1967c) contributions were summarized in the chapters of a book edited by Yates (1992). At that time, the issue of individual and situational influences on risk-taking preferences and behavior remained a focus of debate (Bromiley and Curley 1992). That risk taking continues as a hot topic is demonstrated in the research program undertaken by Figner and Weber (2011). They have introduced contrasts in the risk-taking domain that had received little attention earlier. For example, they distinguish between affective and deliberative risk-taking (also described as hot vs. cold risk-taking). Thus, a recreational context would be more likely to reflect the former, and a financial investment context the latter.

14.3.2 Small Groups

14.3.2.1 Intragroup Effects

It is often the case that major decisions are made, not by individuals acting alone, but by small groups of interacting individuals in an organizational context. Committees and panels are often formed to deal with problems arising in governmental, medical, and educational settings. Some of the decisions made by such groups entail risk assessments. The question then arises as to the nature of the relationship between the risk level of the individuals constituting the group and the risk level of the consensus they manage to achieve. Most of the research directed to this issue has employed groups of previously unacquainted individuals assembled solely for the purpose of the experiments. Hence, generalizability to longer-term groups of acquainted individuals remains an open question. Nevertheless, it would be surprising if the processes observed in minimally acquainted groups had no relevance for acquainted individuals in groups of some duration.

There are three possibilities when comparing individual risk preferences with a consensus reached through group discussion. The most obvious possibility is that the consensus approximates the average of the prior individual decisions. Such an outcome obviously minimizes the concessions required of the individual group members (in shifting to the mean), and hence would seem to be an outcome for which the members would derive the greatest satisfaction. A second possible outcome is a shift toward caution. There is evidence that groups encourage greater carefulness and deliberation in their judgments, members not wishing to appear foolhardy in venturing an extreme opinion. The third possibility is a shift toward greater risk taking. There is mixed evidence about this shift from brainstorming in organizational problem-solving groups (Thibaut and Kelley 1959), and the excesses observed in crowds have been described by Le Bon (1895/1960). Both of these situations would seem to have limited relevance for decision-making in small discussion groups.

This third possibility did emerge in an initial study of decision-making in small discussion groups (Wallach et al. 1962). College students made individual decisions on the CDQ items and were then constituted as small groups to reach a consensus and make individual post-discussion decisions. The significant risky shifts were observed in the all-male and all-female groups, and for both the consensus and post-consensus-individual decisions. Interpretation of the findings stressed a mechanism of diffusion of responsibility whereby a group member could endorse a riskier decision because responsibility for failure would be shared by all of the members of the group.

It could be argued, of course, that decision making on the CDQ is hypothetical—no concrete payoffs or potential losses are involved—and that feature could account for the consistent shift in the risky direction. A second study (Wallach et al. 1964) was designed to counter that argument. SAT items of varying difficulty levels (10% to 90% failure rate as indicated by item statistics) were selected from old tests, and monetary payoffs were attached proportional to difficulty level to generate a set of choices equal in expected value. College students individually made their choices about the difficulty level of the items they would be given and then were formed into small groups with the understanding they would be given the opportunity to earn the payoff if the item was answered correctly. A risky shift was observed (selecting more difficult, higher payoff items) irrespective of whether responsibility for answering the selected item was assigned to a particular group member or to the group as a whole. The monetary prize in each case for a successful solution was made available to each group member. Again, in a decision context quite different from the CDQ, group discussion to consensus yielded risky shifts that lent themselves to explanation in terms of diffusion of responsibility.

A partial replication of the foregoing study was carried out by Kogan and Carlson (1969). In addition to sampling college students, a sample of fourth graders and fifth graders was employed. Further, a condition of overt intragroup competition was added in which group members bid against one another to attempt more difficult items. Consistent with the Wallach et al. (1964) findings, risky shifts with group discussion to consensus were observed in the sample of college students. The competition condition did not yield risky shifts, and the outcomes for the elementary school sample were weaker and more ambiguous than those obtained for college students.

While the preceding two studies provided monetary payoffs contingent on problem solution, participants did not experience the prospect of losing their own money. To enhance the risk of genuinely aversive consequences, Bem et al. (1965) designed an experiment in which participants made choices that might lead to actual physical pain coupled with monetary loss. (In actuality, participants never endured these aversive effects, but they were unaware of this fact during the course of the experiment.) Participants were offered an opportunity to be in experiments that differed in the risks of aversive side effects from various forms of stimulation (e.g., olfactory, taste, movement). Monetary payoffs increased with the percentage of the population (10% to 90%) alleged to experience the aversive side effects. Again, discussion to consensus and private decisions following consensus demonstrated the risky-shift effect and hence provided additional evidence for a mechanism of responsibility diffusion.

With the indication that the risky-shift effect generalizes beyond the hypothetical decisions of the CDQ to such contexts as monetary gain and risk of painful side-effects, investigators returned to the CDQ to explore alternative interpretations for the effect with the knowledge that it is not unique to the CDQ. Thus, Wallach and Kogan (1965b) experimentally split apart the discussion and consensus components of the risky-shift effect. Discussion alone without the requirement of achieving a consensus generated risky shifts whose magnitude did not differ significantly from discussion with consensus. By contrast, the condition of consensus without discussion (a balloting procedure where group members were made aware of each other's decisions by the experimenter and cast as many ballots as necessary to achieve a consensus), yielded an averaging effect. It is thus apparent that actual verbal-interaction is essential for the risky shift to occur. The outcomes run contrary to Brown's (1965) interpretation that attributes the risky shift to the positive value of risk in our culture and the opportunity to learn in the discussion that other group members are willing to take greater risks than oneself. Hence, these members shift in a direction that yields the risky-shift effect. Yet, in the consensus-without-discussion condition in which group members became familiar with others' preferred risk levels, the outcome was an averaging rather than a risky-shift effect. Verbal interaction, on the other hand, not only allows information about others' preferences, but it also generates the cognitive and affective processes presumed necessary for responsibility diffusion to occur.

It could be contended that the balloting procedure omits the exchange of information that accompanies discussion, and it is the latter alone that might be sufficient to generate the risky-shift effect. A test of this hypothesis was carried out by Kogan and Wallach (1967d) who compared interacting and listening groups. Listeners were exposed to information about participants' risk preferences and to the pro and con arguments raised in the discussion as well. Both the interacting and listener groups manifested the risky-shift effect, but its magnitude was significantly smaller in the listening groups. Hence, the information-exchange hypothesis was not sufficient to account for the full strength of the effect. Even when group members were physically separated (visual cues removed) and communicated over an intercom system, the risky shift retained its full strength (Kogan and Wallach 1967a). Conceivably, the distinctiveness of individual voices and expressive styles allowed for the affective reactions presumed to underlie the mechanism of responsibility diffusion.

To what extent are group members aware that their consensus and individual post-consensus decisions are shifting toward greater risk-taking relative to their prior individual-decisions? Wallach et al. (1965) observed that group members' judgments were in the direction of shifts toward risk, but their estimates of the shifts significantly underestimated the actual shifts.

In a subsequent study, Wallach et al. (1968) inquired whether risk takers were more persuasive than their more cautious peers in group discussion. With risk-neutral material used for discussion, persuasiveness ratings were uncorrelated with risk-taking level for male participants and only weakly correlated for female participants. Overall, the results suggested that the risky shift could not be attributed to the greater persuasiveness of high risk takers. A different process seemed to be at work.

As indicated earlier, the paradigm employed in all of the previously cited research consisted of unacquainted individuals randomly assembled into small groups. Breaking with this paradigm, Kogan and Wallach (1967b) assembled homogeneous groups on the basis of participants' scores on test anxiety and defensiveness. Median splits generated four types of groups—high and/or low on the two dimensions. Both dimensions generated significant effects—test anxiety in the direction of a stronger risky shift and defensiveness in the direction of a weaker risky shift. These outcomes were consistent with a responsibility-diffusion interpretation. Test-anxious participants should be especially willing to diffuse responsibility so as to relieve the burden of possible failure. Defensive participants, by contrast, might be so guarded in relation to each other that the affective matrix essential for responsibility diffusion was hindered in its development.

In a related study, field dependence-independence served as the dimension for constructing homogeneous groups (Wallach et al. 1967). The magnitude of the risky shift was not significantly different between field-dependent and field-independent groups. There was a decision-time difference, with field-dependent groups arriving at a consensus significantly more quickly. The more time that was taken by field-dependent groups, the stronger the risky shift, whereas, the more time that was taken by field-independent groups, the weaker the risky shift. More time for field-dependent groups permitted affective bonds to develop, consistent with a process of responsibility diffusion. More time for field-independent groups, by contrast, entailed resistance to other group members' risk preferences and extended cognitively based analysis, a process likely to mitigate responsibility diffusion.

A slight change in the wording of instructions on the CDQ transforms it from a measure of risk taking into a measure of pessimism-optimism. On a probability scale ranging from 0 in 10 to 10 in 10, the test taker is asked to estimate the odds that the risky alternative would lead to a successful outcome if chosen. Descending the probability scale (toward 1 in 10) implies increasing pessimism. Contrary to the expectation that a risky shift would lead to a surge of optimism, the outcome was a significant shift toward pessimism (Lamm et al. 1970). The discussion generated a consensus probability more pessimistic than the prediscussion average of the participating group members. When estimating success/failure probabilities, the discussion focused on things that might go wrong and the best alternative for avoiding error. Hence, the pessimism outcomes can be viewed as a possible constraint on extremity in risky decision-making.

14.3.2.2 Intergroup Effects

With financial support from the Advanced Research Projects Agency of the US Defense Department, Kogan and his collaborators undertook a series of studies in France, Germany, and the United States that departed from the standard intragroup-paradigm by adding an intergroup component. Participants in small decision-making groups were informed that one or more of them would serve as delegates meeting with delegates from other groups with the intent of presenting and defending the decisions made in their parent groups. Such a design has real-world parallels in the form of local committees arriving at decisions, where a representative is expected to defend the decisions before a broader-based body of representatives from other localities.

In an initial exploratory study with a French university sample (Kogan and Doise 1969), 10 of the 12 CDQ items with slight modifications proved to be appropriate in the French cultural context and were accordingly translated into French. Discussion to consensus on the first five CDQ items was followed by an anticipated delegate condition for the latter five CDQ items. Three delegate conditions were employed in which the group members were told (a) the delegate would be selected by chance, (b) the delegate would be selected by the group, and (c) all group members would serve as delegates. The significant shift toward risk was observed in the initial five CDQ items, and the magnitude of the risky shift remained essentially at the same level for all three of the anticipated delegate conditions. It is evident, then, that the expectation of possibly serving as a delegate in the future does not influence the processes responsible for the risky-shift effect.

In subsequent studies, delegates were given the opportunity to negotiate with each other. In the Hermann and Kogan (1968) investigation with American undergraduate men, dyads pairing an upperclassman (seniors, juniors) with an underclassman (sophomores, freshmen) engaged in discussion to consensus on the CDQ items. The upperclassmen were designated as leaders, and the underclassmen as delegates. The risky shift prevailed at the dyadic level. Intergroup negotiation then followed among leaders and among delegates. The former manifested the risky shift, whereas the latter did not. This outcome is consistent with a responsibility-diffusion interpretation. Requiring delegates to report back to leaders would likely interfere with the affective processes presumed to underlie diffusion of responsibility. Leaders, by contrast, have less concern about reporting back to delegates. One cannot rule out loss-of-face motivation, however, and the magnitude of the risky shift in the leader groups was in fact weaker than that observed in the typical intragroup setting.

A follow-up to this study was carried out by Lamm and Kogan (1970) with a sample of German undergraduate men. As in the case of the French study (Kogan and Doise 1969), 10 of the 12 CDQ items (with slight modification) were considered appropriate in the German cultural context and were accordingly translated into German. Unlike the Hermann and Kogan (1968) study in which status was ascribed, the present investigation was based on achieved status. Participants in three-person groups designated a representative and an alternate, leaving a third individual designated as a nonrepresentative. Contrary to the Hermann and Kogan (1968) findings where leaders manifested the risky shift, the representative groups (presumed analogous to the leaders) failed to demonstrate the risky shift. On the other hand, the alternate and nonrepresentative groups did generate significant risky shifts. The argument here is that achieved, as opposed to ascribed, status enhanced loss-of-face motivation, making difficult the concessions and departures from prior intragroup decisions that are essential for risky shifts to occur. Having been assigned secondary status by the group, the alternates and nonrepresentatives were less susceptible to loss-of-face pressures and could negotiate more flexibly with their status peers.

In a third and final study of the delegation process, Kogan et al. (1972) assigned leader and subordinate roles on a random basis to German undergraduate men. The resultant dyads discussed the CDQ items to consensus (revealing the anticipated risky shift) and were assigned negotiating and observer roles in groups comprised exclusively of negotiators or observers. All four group types—leader-negotiators, subordinate-observers, subordinate-negotiators, and leader-observers—demonstrated the risky shift. However, the subordinate observers relative to their negotiating leaders preferred larger shifts toward risk. Evidently, loss-of-face motivation in the leaders in the presence of their subordinates served as a brake on willingness to shift from their initial-dyadic decisions. The nature of the arguments, however, convinced the observing subordinates of the merits of enhanced risk taking.

Two studies were conducted to examine preferred risk-levels when decisions are made for others. The first (Zaleska and Kogan 1971) utilized a sample of French undergraduate women selecting preferred probability and monetary stake levels in a series of equal-expected-value chance bets to be played for the monetary amounts involved. In addition to a control condition (self-choices on two occasions), three experimental conditions were employed: (a) individual choices for self and another, (b) individual and group choices for self, and (c) individual and group choices for others. The first condition generated cautious shifts, the second yielded risky shifts, and the third produced weakened risky-shifts. Evidently, making individual choices for another person enhances caution, but when such choices are made in a group, a significant risky shift ensues, though weaker than obtained in the standard intra-group condition.

The findings bear directly on competing interpretations of the risky-shift effect. The popular alternative to the responsibility-diffusion interpretation is the risk-as-value interpretation initially advanced by Brown (1965) and already described. As noted in the Zaleska and Kogan (1971) study, caution is a value for individuals making choices for others, yet when deciding for others as a group, the decisions shifted toward risk. Such an outcome is consistent with a responsibility-diffusion interpretation, but the lesser strength of the effect suggests that the value aspect exerts some influence. Hence, the two conflicting interpretations may not necessarily assume an either-or form. Rather, the psychological processes represented in the two interpretations may operate simultaneously, or one or the other may be more influential depending on the decision-making context.

Choices in the Zaleska and Kogan (1971) study were distinguished by reciprocity—individuals and groups choosing for unacquainted specific others were aware that those others would at the same time be choosing for them. A subsequent study by Teger and Kogan (1975), using the Zaleska and Kogan chance bets task, explored this reciprocity feature by contrasting gambling choices made under reciprocal versus nonreciprocal conditions in a sample of American undergraduate women. A significantly higher level of caution was observed in the reciprocal condition relative to the nonreciprocal condition. This difference was most pronounced for high-risk bets that could entail possible substantial loss for the reciprocating other. Hence, the enhanced caution with reciprocity was most likely intended to ensure at least a modest payoff for another who might benefit the self. Caution in such circumstances serves the function of guilt avoidance.

We might ask whether the research on group risk-taking represented a passing fad. The answer is no. The group risk-taking research led directly to the study of polarization in small groups—the tendency for group discussion on almost any attitudinal topic to move participants to adopt more extreme positions at either pole (e.g., Myers and Lamm 1976). This polarization work eventually led to the examination of the role of majorities and minorities in influencing group decisions (e.g., Moscovici and Doise 1994). In short, the dormant group-dynamics tradition in social psychology was invigorated. Reviewing the group risk-taking research 20 years after its surge, Davis et al. (1992) noted that the "decline of interest in investigating the parameters of group risk taking was unfortunate" (p. 170). They go on to note the many settings in which group decision-making takes place (e.g., parole boards, juries, tenure committees) and where the "conventional wisdom persists that group decisions are generally moderate rather than extreme, despite such contrary evidence as we have discussed above" (p. 170).

14.4 Kinesthetic Aftereffect

A phenomenon originally demonstrated by Köhler and Dinnerstein in 1947, the kinesthetic aftereffect captured the attention of psychologists for almost a half-century. Early interest in this phenomenon can be traced to experimental psychologists studying perception who sought to establish its parameters. In due course, individual differences in the kinesthetic aftereffect attracted personality psychologists who viewed it as a manifestation of the augmenter-reducer dimension, which distinguishes between people who reduce the subjective intensity to external stimuli and those who magnify it (Petrie 1967).

Consider the nature of the kinesthetic-aftereffect task. A blindfolded participant is handed (right hand) a wooden test block 2 inches in width and 6 inches in length. The participant then is requested to match the width of the test block on an adjustable wedge (30 inches long) located to the participant's left hand. This process constitutes the preinduction measurement. Next, the participant is handed an induction block 1/2 inch narrower or wider than the test block and asked to give it a back-and-forth rubbing. Then the participant returns to the test block, and the initial measurement is repeated. The preinduction versus postinduction difference in the width estimate constitutes the score.

Kinesthetic-aftereffect research at ETS was conducted by A. Harvey Baker and his colleagues. One question that they examined was the effect of experimental variations on the basic paradigm just described. Weintraub et al. (1973) had explored the contrast between a wider and narrower induction block (relative to the test block) on the magnitude and direction of the kinesthetic aftereffect. They also included a control condition eliminating the induction block that essentially reduced the score to zero. The kinesthetic aftereffect proved stronger with the wider induction block, probably the reason that subsequent research predominantly employed a wider induction block, too.

Taking issue with the absence of an appropriate control for the induction block in the kinesthetic-affereffect paradigm, Baker et al. (1986) included a condition in which the test and induction blocks were equal in size. Such a control permitted them to determine whether the unequal size of the two blocks was critical for the kinesthetic aftereffect. Both the induction > test and induction < test conditions generated a significant kinesthetic aftereffect. The induction = test condition also yielded a significant kinesthetic aftereffect, but it was not significantly different from the induction > test condition. On this basis, Baker et al. concluded that two processes rather than one are necessary to account for the kinesthetic-aftereffect phenomenon—induction (rubbing the induction block) and the size contrast. It should be noted that these findings were published as research on this phenomenon had begun to wane, and hence their influence was negligible.

Two additional questions investigated by Baker and his coworkers in other research were the kinesthetic aftereffect's reliability and personality correlates. A stumbling block in research on this phenomenon was the evidence of low test-retest reliability across a series of trials. Until the reliability issue could be resolved, the prospect for the kinesthetic aftereffect as an individual-differences construct remained dubious. Baker et al. (1976, 1978) maintained that test-retest reliability is inappropriate for the kinesthetic aftereffect. They noted that the kinesthetic aftereffect is subject to practice effects, such that the first preinduction-postinduction pairing changes the person to the extent that the second such pairing is no longer measuring the same phenomenon. In support of this argument, Baker et al. (1976, 1978) reviewed research based on a single-session versus a multiple-session kinesthetic aftereffect and reported that it was only the former that yielded significant validity coefficients with theoretically-relevant variables such as activity level and sensation seeking.

In another article on this topic, Mishara and Baker (1978) argued that internal-consistency reliability is most relevant for the kinesthetic aftereffect. Of the 10 samples studied, the first five employed the Petrie (1967) procedure in which a 45-minute rest period preceded kinesthetic-aftereffect measurements. Participants were not allowed to touch anything with their thumbs and forefingers during this period, and the experimenter used the time to administer questionnaires orally. The remaining five samples were tested with the Weintraub et al. (1973) procedure that did not employ the 45-minute rest period. For the samples using the Petrie procedure, the split-half reliabilities ranged from .92 to .97. For the samples tested with the Weintraub et al. procedure, the reliabilities ranged from .60 to .77. Mishara and Baker noted that the Weintraub et al. procedure employed fewer trials, but application of the Spearman-Brown correction to equate the number of trials in the Weintraub et al. procedure with the number in the Petrie procedure left the latter with substantially higher reliabilities. These results suggest that the 45-minute rest period may be critical to the full manifestation of the kinesthetic aftereffect, but a direct test of its causal role regarding differential reliabilities has not been undertaken.

Baker et al. (1976) continued the search for personality correlates of kinesthetic aftereffect begun by Petrie (1967). Inferences from her augmenter-reducer conception are that augmenters (their postinduction estimates smaller/narrower than their preinduction estimates) are overloaded with stimulation and hence motivated to avoid any more, whereas reducers (their postindduction estimated larger/wider than their preinduction estimates) are stimulus deprived and hence seek more stimulation. Supporting these inferences is empirical evidence (Petrie et al. 1958) indicating that reducers (relative to augmenters) are more tolerant of pain, whereas augmenters (relative to reducers) are more tolerant of sensory deprivation.

Baker et al. (1976), arguing that the first-session kinesthetic aftereffect was reliable and could potentially predict theoretically-relevant personality traits and behavioral dispositions, reanalyzed the earlier Weintraub et al. (1973) study. A 25-item scale was reduced to 18 items and an index was derived with positive scores reflecting the reducing end of the augmenting-reducing dimension. Some of the items in the index concerned responses to external stimulation (e.g., fear of an injection, lively parties, lengthy isolation). Other items concerned seeking or avoiding external stimulation (e.g., coffee and alcohol consumption, sports participation, smoking, friendship formation).

The kinesthetic-aftereffect scores for the first session were significantly related to the index ($r = -.36$, $p < .02$), as predicted, but the scores for the six subsequent sessions were not. Neither of the components of the kinesthetic-aftereffect score—preinduction and postinduction—were related to the index for any session. However, it is noteworthy that scores for subsequent sessions, made up from the preinduction score for the first session and the postinduction score for the subsequent session, were consistently related to the index. Baker et al. (1976) ended their article on a note of confidence, convinced that the kinesthetic-aftereffect task elicits personality differences in an augmenter-reducer dimension relevant to the manner in which external stimulation is sought and handled.

Nevertheless, in the very next year, an article by Herzog and Weintraub (1977) reported that an exact replication of the Baker et al. (1976) study found no link between the kinesthetic aftereffect and the personality behavior index. Herzog and Weintraub did, however, acknowledge the emergence of a reliable augmenter-reducer dimension. Disinclined to let the issue rest with so sharp a divergence from the Baker et al. study findings about the association between the kinesthetic aftereffect and the index, Herzog and Weintraub (1982) undertook a further replication. A slight procedural modification was introduced. Having failed to replicate the Baker et al. study with a wide inducing block, they chose to try a narrow inducing block for the first kinesthetic-aftereffect session and alternated the wide and narrow inducing blocks across subsequent sessions. Again, the results were negative, with the authors concluding that "we are unable to document any relationship between induction measures derived by the traditional kinesthetic-aftereffect procedure and questionnaire-derived personality measures" (Herzog and Weintraub 1982, p. 737).

Refusing to abandon the topic, Herzog et al. (1985) judged a final effort worthwhile if optimal procedures, identified in previous research, were applied. Accordingly, they employed the Petrie (1967) procedure (with the 45-minute initial rest period) that had previously generated exceptionally high reliabilities. They also selected the wide inducing block that had almost always been used whenever significant correlations with personality variables were obtained. In addition to the standard difference-score, Herzog et al. computed a residual change-score, "the deviation from the linear regression of post-induction scores on pre-induction scores" (p. 1342). In regard to the personality-behavior variables, a battery of measures was employed: a new 45-item questionnaire with two factor scales; the personality-behavior index used by Baker et al. (1976) and Herzog and Weintraub (1977, 1982); and several behavioral measures. Only those personality-behavior variables that had satisfactory internal-consistency reliability and at least two significant correlations with each other were retained for further analyses. All of these measurement and methodological precautions paid off in the demonstration that the kinesthetic aftereffect is indeed related to personality and behavior. Reducers (especially women) were significantly higher on the factor subscale Need for Sensory Stimulation, whose items have much in common with those on Zuckerman's (1994) sensation-seeking instruments. Consistent with earlier findings by Petrie et al. (1958) and Ryan and Foster (1967), reducers claimed to be more tolerant of cold temperatures and pain.

In sum, Herzog et al. (1985) have shown that the Petrie (1967) induction procedure generates reliable kinesthetic-aftereffect scores that correlate in the theoretically expected direction with reliable measures of personality and behavior. It is testimony to the importance of reliability when attempting to demonstrate the construct validity of a conceptually derived variable. However, a major disadvantage of the Petrie procedure must be acknowledged—an hour of individual administration—that is likely to limit the incentive of investigators to pursue further research with the procedure. It is hardly surprising then that research on the personality implications of the kinesthetic aftereffect essentially ended with the Herzog et al. investigation.

Virtually all of the research on the kinesthetic aftereffect-personality relationship has been interindividual (trait based). Baker et al. (1979) can be credited with one of the very few studies to explore intraindividual (state-based) variation. Baker et al. sought to determine whether the menstrual cycle influences kinesthetic-aftereffect. On the basis of evidence that maximal pain occurs at the beginning and end of the cycle, Baker et al. predicted greater kinesthetic aftereffect reduction (a larger aftereffect), "damping down of subjective intensity of incoming stimulation" (p. 236) at those points in the cycle and hence a curvilinear relationship between the kinesthetic aftereffect and locus in the menstrual cycle. Employing three samples of college-age women, quadratic-trend analysis yielded a significant curvilinear effect. The effect remained statistically significant when controlling for possible confounding variables—tiredness, oral contraceptive use, use of drugs or medication. Untested is the possibility of social-expectancy effects, participants at or near menses doing more poorly on the kinesthetic-aftereffect task simply because they believe women do poorly then. But, as Baker et al. observed, it is difficult to conceive of such effects in so unfamiliar a domain as the kinesthetic aftereffect.

Did personality research related to the kinesthetic aftereffect disappear from the psychological scene following the definitive Herzog et al. (1985) study? Not entirely, for the personality questionnaire used to validate the kinesthetic aftereffect became the primary instrument for assessing the augmenter-reducer dimension initially made operational in the kinesthetic-aftereffect laboratory tasks. A prime example of this change is represented by the Larsen and Zarate (1991) study. The 45-item questionnaire developed by Herzog et al. shifted from dependent to independent variable and was used to compare reducers' and augmenters' reactions to taking part in a boring and monotonous task. Compared to augmenters, reducers described the task as more aversive and were less likely to repeat it. Further, reducers, relative to augmenters, exhibited more novelty seeking and sensation seeking in their day-to-day activities.

Despite its promise, the augmenter-reducer construct seems to have vanished from the contemporary personality scene. Thus, it is absent from the index of the latest edition of the *Handbook of Personality* (John et al. 2008). The disappearance readily lends itself to speculation and a possible explanation. When there is a senior, prestigious psychologist advancing a construct whose predictions are highly similar to a construct advanced by younger psychologists of lesser reputation, the former's construct is likely to win out. Consider the theory of extraversion-introversion developed by Eysenck (e.g., Eysenck and Eysenck 1985). Under quiet and calm conditions, extraverts and introverts are presumed to be equally aroused. But when external stimulation becomes excessive—bright lights, loud noises, crowds—introverts choose to withdraw so as to return to what for them is optimal stimulation. Extraverts, by contrast, need that kind of excitement to arrive at what for them is optimal stimulation. It is readily apparent that the more recent introversion-extraversion construct is virtually indistinguishable from the earlier augmenter-reducer construct. Given the central role of the introversion-extraversion concept in personality-trait theory and the similarity in the two constructs' theoretical links with personality, it is no surprise that the augmenter-reducer construct has faded away.

14.5 Conclusion

The conclusions about ETS research in cognitive, personality, and social psychology in the companion chapter (Stricker, Chap. 13, this volume) apply equally to the work described here: the remarkable breadth of the research in terms of the span of topics addressed (kinesthetic aftereffect to risk taking), the scope of the methods used (experiments, correlational studies, multivariate analyses), and the range of populations studied (young children, college and graduate students in the United States and Europe, the general public); its major impact on the field of psychology; and its focus on basic research.

Another conclusion can also be drawn from this work: ETS was a major center for research in creativity, cognitive styles, and risk taking in the 1960s and 1970s, a likely product of the fortunate juxtaposition of a supportive institutional environment, ample internal and external funding, and a talented and dedicated research staff.

Acknowledgments Thanks are due to Rachel Adler, Irene Kostin, and Nick Telepak for their assistance in retrieving publications, reports, and archival material, and to Randy Bennett, Jeremy Burrus, James Kaufman, and Donald Powers for reviewing a draft of this chapter.

References

Amabile, T. M. (1982). Social psychology of creativity: A consensual assessment technique. *Journal of Personality and Social Psychology, 43*, 997–1013. https://doi.org/10.1037/0022-3514.43.5.997

Atkinson, J. W. (1957). Motivational determinants of risk-taking behavior. *Psychological Review, 64*, 359–372. https://doi.org/10.1037/h0043445

Baer, J. (2011). How divergent thinking tests mislead us: Are the Torrance tests still relevant in the 21st century? The Division 10 debate. *Psychology of Aesthetics, Creativity, and the Arts, 5*, 309–313. https://doi.org/10.1037/a0025210

Baird, L. L. (1979). *Development of an inventory of documented accomplishments for graduate admissions* (GRE Board Research Report No. 77-3R). Princeton: Educational Testing Service.

Baird, L. L., & Knapp, J. E. (1981). *The Inventory of Documented Accomplishments for Graduate Admissions: Results of a field trial study of its reliability, short-term correlates, and evaluation* (GRE Board Research Report No. 78-3R). Princeton: Educational Testing Service. https://doi.org/10.1002/j.2333-8504.1981.tb01253.x

Baker, A. H., Mishara, B. L., Kostin, I. W., & Parker, L. (1976). Kinesthetic aftereffect and personality: A case study of issues involved in construct validation. *Journal of Personality and Social Psychology, 34*, 1–13. https://doi.org/10.1037/0022-3514.34.1.1

Baker, A. H., Mishara, B. L., Parker, L., & Kostin, I. W. (1978). When "reliability" fails, must a measure be discarded?—The case of kinesthetic aftereffect. *Journal of Research in Personality, 12*, 262–273. https://doi.org/10.1016/0092-6566(78)90053-3

Baker, A. H., Mishara, B. L., Kostin, I. W., & Parker, L. (1979). Menstrual cycle affects kinesthetic aftereffect, an index of personality and perceptual style. *Journal of Personality and Social Psychology, 37*, 234–246. https://doi.org/10.1037/0022-3514.37.2.234

Baker, A. H., Mishara, B. L., & Kostin, I. W. (1986). Kinesthetic aftereffect: One phenomenon or two? *Perception and Psychophysics, 39*, 255–260. https://doi.org/10.3758/BF03204932

Barron, F. (1953). Complexity-simplicity as a personality dimension. *Journal of Abnormal and Social Psychology, 48*, 163–172. https://doi.org/10.1037/h0054907

Bem, D. J., Wallach, M. A., & Kogan, N. (1965). Group decision making under risk of aversive consequences. *Journal of Personality and Social Psychology, 1*, 453–460. https://doi.org/10.1037/h0021803

Bennett, R. E., & Rock, D. A. (1995). Generalizability, validity, and examinee perceptions of a computer-delivered formulating-hypotheses test. *Journal of Educational Measurement, 32*, 19–36. https://doi.org/10.1111/j.1745-3984.1995.tb00454.x

Bromiley, P., & Curley, S. P. (1992). Individual differences in risk taking. In J. F. Yates (Ed.), *Risk-taking behavior* (pp. 87–132). New York: Wiley.

Brown, R. (1965). *Social psychology*. New York: Free Press.

Cronbach, L. J., & Snow, R. E. (1977). *Aptitudes and instructional methods*. New York: Irvington.

Crowne, D. P., & Marlowe, D. (1964). *The approval motive*. New York: Wiley.

Davis, J. K. (1991). Educational implications of field dependence-independence. In S. Wapner & J. Demick (Eds.), *Field dependence-independence: Cognitive style across the life span* (pp. 149–175). Hillsdale: Erlbaum.

Davis, J. H., Kameda, T., & Stasson, M. F. (1992). Group risk taking: Selected topics. In J. F. Yates (Ed.), *Risk-taking behavior* (pp. 163–200). New York: Wiley.

Edwards, W. (1961). Behavioral decision theory. *Annual Review of Psychology, 12*, 473–498. https://doi.org/10.1146/annurev.ps.12.020161.002353

Enright, M. K., Rock, D. A., & Bennett, R. E. (1998). Improving measurement for graduate admissions. *Journal of Educational Measurement, 35*, 250–267. https://doi.org/10.1111/j.1745-3984.1998.tb00538.x

Eysenck, H. J., & Eysenck, M. W. (1985). *Personality and individual differences: A natural science approach*. New York: Plenum. https://doi.org/10.1007/978-1-4613-2413-3

Figner, B., & Weber, E. U. (2011). Who takes risks when and why: Determinants of risk taking. *Current Directions in Psychological Science, 20*, 211–216. https://doi.org/10.1177/0963721411415790

Flanagan, J. C. (1949). *Critical requirements for research personnel—A study of observed behaviors of personnel in research laboratories*. Pittsburgh: American Institute for Research.

Frederiksen, N. (1959). *Development of the test "Formulating Hypotheses": A progress report* (ONR Technical Report, Contract Nonr-2338[00]). Princeton: Educational Testing Service.

Frederiksen, N., & Evans, F. R. (1974). Effects of models of creative performance on ability to formulate hypotheses. *Journal of Educational Psychology, 66*, 67–82. https://doi.org/10.1037/h0035808

Frederiksen, N., & Ward, W. C. (1978). Measures for the study of creativity in scientific problem-solving. *Applied Psychological Measurement, 2*, 1–24. https://doi.org/10.1177/014662167800200101

French, J. W., Ekstrom, R. B., & Price, L. A. (1963). *Manual for Kit of Reference Tests for Cognitive Factors*. Princeton: Educational Testing Service.

Gardner, R. W., Holzman, P. S., Klein, G. S., Linton, H. B., & Spence, D. P. (1959). Cognitive control: A study of individual consistencies in cognitive behavior. *Psychological Issues, 1*(4, Whole No. 4).

Gardner, R. W., Jackson, D. N., & Messick, S. (1960). Personality organization in cognitive controls and intellectual abilities. *Psychological Issues, 2*(4, Whole No. 8). https://doi.org/10.1037/11215-000

Getzels, J. W., & Jackson, P. W. (1962). *Creativity and intelligence: Explorations with gifted students*. New York: Wiley.

Glover, J. A., Ronning, R. R., & Reynolds, C. R. (Eds.). (1989). *Handbook of creativity*. New York: Plenum. https://doi.org/10.1007/978-1-4757-5356-1

Goodenough, D. R. (1976). The role of individual differences in field dependence as a factor in learning and memory. *Psychological Bulletin, 83*, 675–694. https://doi.org/10.1037/0033-2909.83.4.675

Goodenough, D. R. (1981). An entry in the great frame-tilt judging contest. *Perceptual and Motor Skills, 52*, 43–46. https://doi.org/10.2466/pms.1981.52.1.43

Goodenough, D. R. (1986). History of the field dependence construct. In M. Bertini, L. Pizzamiglio, & S. Wapner (Eds.), *Field dependence in psychological theory, research, and application* (pp. 5–13). Hillsdale: Erlbaum.

Goodenough, D. R., Oltman, P. K., & Cox, P. W. (1987). The nature of individual differences in field dependence. *Journal of Research in Personality, 21*, 81–99. https://doi.org/10.1016/0092-6566(87)90028-6

Goodenough, D. R., Oltman, P. K., Snow, D., Cox, P. W., & Markowitz, D. (1991). Field dependence-independence and embedded figures performance. In S. Wapner & J. Demick (Eds.), *Field dependence-independence: Cognitive style across the life span* (pp. 131–148). Hillsdale: Erlbaum.

Guilford, J. P. (1967). *The nature of human intelligence*. New York: McGraw-Hill.

Guilford, J. P., & Christensen, P. R. (1956). *A factor-analytic study of verbal fluency* (Psychological Laboratory Report No. 17). Los Angeles: University of Southern California.

Hennessey, B. A., & Amabile, T. M. (2010). Creativity. *Annual Review of Psychology, 61*, 569–598. https://doi.org/10.1146/annurev.psych.093008.100416

Hermann, M. G., & Kogan, N. (1968). Negotiation in leader and delegate groups. *Journal of Conflict Resolution, 12*, 322–344. https://doi.org/10.1177/002200276801200304

Herzog, T. R., & Weintraub, D. J. (1977). Preserving the kinesthetic aftereffect: Alternating inducing blocks day by day. *American Journal of Psychology, 90*, 461–474. https://doi.org/10.2307/1421876

Herzog, T. R., & Weintraub, D. J. (1982). Roundup time at personality ranch: Branding the elusive augmenters and reducers. *Journal of Personality and Social Psychology, 42*, 729–737. https://doi.org/10.1037/0022-3514.42.4.729

Herzog, T. R., Williams, D. M., & Weintraub, D. J. (1985). Meanwhile, back at personality ranch: The augmenters and reducers ride again. *Journal of Personality and Social Psychology, 48*, 1342–1352. https://doi.org/10.1037/0022-3514.48.5.1342

Holland, J. L. (1961). Creative and academic performance among talented adolescents. *Journal of Educational Psychology, 52*, 136–147. https://doi.org/10.1037/h0044058

Hovecar, D., & Bachelor, P. (1989). A taxonomy and critique of measurements used in the study of creativity. In J. A. Glover, R. R. Ronning, & C. R. Reynolds (Eds.), *Handbook of creativity* (pp. 53–75). New York: Plenum.

Irwin, F. W., & Smith, W. A. S. (1957). Value, cost, and information as determiners of decision. *Journal of Experimental Psychology, 54*, 229–232. https://doi.org/10.1037/h0049137

John, O. P., Robins, R. W., & Pervin, L. A. (Eds.). (2008). *Handbook of personality: Theory and research* (3rd ed.). New York: Guilford Press.

Kagan, J., & Kogan, N. (1970). Individual variation in cognitive processes. In P. H. Mussen (Ed.), *Carmichael's manual of child psychology* (Vol. 1, 3rd ed., pp. 1273–1365). New York: Wiley.

Kaufman, J. C., & Sternberg, R. J. (Eds.). (2006). *The international handbook of creativity*. New York: Cambridge University Press. https://doi.org/10.1017/CBO9780511818240

Kaufman, J. C., & Sternberg, R. J. (2010a). Preface. In J. C. Kaufman & R. J. Sternberg (Eds.). *The Cambridge handbook of creativity* (pp. xiii–xv). New York: Cambridge University Press. https://doi.org/10.1017/CBO9780511763205.001

Kaufman, J. C., & Sternberg, R. J. (Eds.). (2010b). *The Cambridge handbook of creativity*. New York: Cambridge University Press.

Kim, K. H. (2011). The APA 2009 Division 10 debate: Are the Torrance Tests of Creative Thinking still relevant in the 21st century? *Psychology of Aesthetics, Creativity, and the Arts, 5*, 302–308. https://doi.org/10.1037/a0021917

Kirton, M. J. (Ed.). (1989). *Adaptors and innovators*. London: Routledge.

Klein, S. P., & Skager, R. W. (1967). "Spontaneity vs deliberateness" as a dimension of esthetic judgment. *Perceptual and Motor Skills, 25*, 161–168. https://doi.org/10.2466/pms.1967.25.1.161

Klein, S. P., Frederiksen, N., & Evans, F. R. (1969). Anxiety and learning to formulate hypotheses. *Journal of Educational Psychology, 60*, 465–475. https://doi.org/10.1037/h0028351

Kogan, N. (1971). Educational implications of cognitive styles. In G. Lesser (Ed.), *Psychology and educational practice* (pp. 242–292). Glencoe: Scott Foresman.

Kogan, N. (1976). *Cognitive styles in infancy and early childhood*. Hillsdale: Erlbaum.

Kogan, N. (1980). A style of life, a life of style [Review of the book *Cognitive styles in personal and cultural adaptation* by H. A. Witkin]. *Contemporary Psychology, 25*, 595–598. https://doi.org/10.1037/018043

Kogan, N. (1983). Stylistic variation in childhood and adolescence: Creativity, metaphor, and cognitive styles. In J. Flavell & E. Markman (Eds.), *Handbook of child psychology: Vol. 3. Cognitive development* (4th ed., pp. 630–706). New York: Wiley.

Kogan, N. (1994). Cognitive styles. In R. J. Sternberg (Ed.), *Encyclopedia of human intelligence* (pp. 266–273). New York: Macmillan.

Kogan, N. (2008). Commentary: Divergent-thinking research and the Zeitgeist. *Psychology of Aesthetics, Creativity, and the Arts, 2*, 10–102. https://doi.org/10.1037/1931-3896.2.2.100

Kogan, N., & Block, J. (1991). Field dependence-independence from early childhood through adolescence: Personality and socialization aspects. In S. Wapner & J. Demick (Eds.), *Field dependence-independence: Cognitive style across the life span* (pp. 177–207). Hillsdale: Erlbaum.

Kogan, N., & Carlson, J. (1969). Difficulty of problems attempted under conditions of competition and group consensus. *Journal of Educational Psychology, 60*, 155–167. https://doi.org/10.1037/h0027555

Kogan, N., & Doise, W. (1969). Effects of anticipated delegate status on level of risk taking in small decision-making groups. *Acta Psychologica, 29*, 228–243. https://doi.org/10.1016/0001-6918(69)90017-1

Kogan, N., & Morgan, F. T. (1969). Task and motivational influences on the assessment of creative and intellective ability in children. *Genetic Psychology Monographs, 80*, 91–127.

Kogan, N., & Pankove, E. (1972). Creative ability over a five-year span. *Child Development, 43*, 427–442. https://doi.org/10.2307/1127546

Kogan, N., & Pankove, E. (1974). Long-term predictive validity of divergent-thinking tests: Some negative evidence. *Journal of Educational Psychology, 66*, 802–810. https://doi.org/10.1037/h0021521

Kogan, N., & Saarni, C. (1990). Cognitive styles in children: Some evolving trends. In O. N. Saracho (Ed.), *Cognitive style and early education* (pp. 3–31). New York: Gordon and Breach.
Kogan, N., & Wallach, M. A. (1964). *Risk taking: A study in cognition and personality*. New York: Holt Rinehart and Winston.
Kogan, N., & Wallach, M. A. (1967a). Effects of physical separation of group members upon group risk taking. *Human Relations, 20*, 41–49. https://doi.org/10.1177/001872676702000104
Kogan, N., & Wallach, M. A. (1967b). Group risk taking as a function of members' anxiety and defensiveness levels. *Journal of Personality, 35*, 50–63. https://doi.org/10.1111/j.1467-6494.1967.tb01415.x
Kogan, N., & Wallach, M. A. (1967c). Risk taking as a function of the situation, the person, and the group. *New Directions in Psychology, 3*, 111–289.
Kogan, N., & Wallach, M. A. (1967d). Risky-shift phenomenon in small decision-making groups: A test of the information-exchange hypothesis. *Journal of Experimental Social Psychology, 3*, 75–84.
Kogan, N., Lamm, H., & Trommsdorff, G. (1972). Negotiation constraints in the risk-taking domain: Effects of being observed by partners of higher or lower status. *Journal of Personality and Social Psychology, 23*, 143–156. https://doi.org/10.1037/h0033035
Köhler, W., & Dinnerstein, D. (1947). Figural after-effects in kinesthesis. In *Miscellanea psychologica Albert Michotte* (pp. 196–220). Paris, France: Librairie Philosophique.
Kolb, D. A. (1976). *Learning Style Inventory: Technical manual*. Englewood Cliffs: Prentice Hall.
Kozhevnikov, M. (2007). Cognitive styles in the context of modern psychology: Toward an integrative framework of cognitive style. *Psychological Bulletin, 133*, 464–481. https://doi.org/10.1037/0033-2909.133.3.464
Lamm, H., & Kogan, N. (1970). Risk taking in the context of intergroup negotiation. *Journal of Experimental Social Psychology, 6*, 351–363. https://doi.org/10.1016/0022-1031(70)90069-7
Lamm, H., Trommsdorff, G., & Kogan, N. (1970). Pessimism-optimism and risk taking in individual and group contexts. *Journal of Personality and Social Psychology, 15*, 366–374. https://doi.org/10.1037/h0029602
Larsen, R. J., & Zarate, M. A. (1991). Extending reducer/augmenter theory into the emotion domain: The role of affect in regulating stimulation level. *Personality and Individual Differences, 12*, 713–723. https://doi.org/10.1016/0191-8869(91)90227-3
Le Bon, G. (1960). *The crowd: A study of the popular mind*. New York: Viking. (Original work published 1895).
Linn, M. C., & Kyllonen, P. (1981). The field dependence-independence construct: Some, one, or none. *Journal of Educational Psychology, 73*, 261–273. https://doi.org/10.1037/0022-0663.73.2.261
MacKinnon, D. W. (1962). The nature and nurture of creative talent. *American Psychologist, 17*, 484–495. https://doi.org/10.1037/h0046541
Mednick, S. A. (1962). The associative basis of the creative process. *Psychological Review, 69*, 220–232. https://doi.org/10.1037/h0048850
Mednick, S. A., & Mednick, M. T. (1962). *Examiner's manual, Remote Associates Test*. Rolling Meadows: Riverside.
Messick, S. (1970). The criterion problem in the evaluation of instruction: Assessing possible, not just intended, outcomes. In M. C. Wittrock & D. W. Wiley (Eds.), *The evaluation of instruction: Issues and problems* (pp. 183–202). New York: Holt Rinehart and Winston.
Messick, S. (1976). Personality consistencies in cognition and creativity. In S. Messick (Ed.), *Individuality in learning* (pp. 4–22). San Francisco: Jossey-Bass.
Messick, S. (1984). The nature of cognitive styles: Problems and promise in educational practice. *Educational Psychologist, 19*, 59–74. https://doi.org/10.1080/00461528409529283
Messick, S. (1987). Structural relationships across cognition, personality, and style. In R. E. Snow & M. J. Farr (Eds.), *Aptitude, learning, and instruction: Vol. 3. Conative and affective process analysis* (pp. 35–75). Hillsdale: Erlbaum.
Messick, S. (1994). The matter of style: Manifestations of personality in cognition, learning, and teaching. *Educational Psychologist, 29*, 121–136. https://doi.org/10.1207/s15326985ep2903_2

Messick, S. (1996). Bridging cognition and personality in education: The role of style in performance and development. *European Journal of Personality, 10*, 353–376. https://doi.org/10.1002/(SICI)1099-0984(199612)10:5<353::AID-PER268>3.0.CO;2-G

Messick, S., & Damarin, F. (1964). Cognitive styles and memory for faces. *Journal of Abnormal and Social Psychology, 69*, 313–318. https://doi.org/10.1037/h0044501

Messick, S., & Fritzky, F. J. (1963). Dimensions of analytic attitude in cognition and personality. *Journal of Personality, 31*, 346–370. https://doi.org/10.1111/j.1467-6494.1963.tb01304.x

Messick, S., & Kogan, N. (1963). Differentiation and compartmentalization in object-sorting measures of categorizing style. *Perceptual and Motor Skills, 16*, 47–51. https://doi.org/10.2466/pms.1963.16.1.47

Messick, S., & Kogan, N. (1965). Category width and quantitative aptitude. *Perceptual and Motor Skills, 20*, 493–497. https://doi.org/10.2466/pms.1965.20.2.493

Messick, S., & Kogan, N. (1966). Personality consistencies in judgment: Dimensions of role constructs. *Multivariate Behavioral Research, 1*, 165–175. https://doi.org/10.1207/s15327906mbr0102_3

Mishara, B. L., & Baker, A. H. (1978). Kinesthetic aftereffect scores are reliable. *Applied Psychological Measurement, 2*, 239–247. https://doi.org/10.1177/014662167800200206

Moscovici, S., & Doise, W. (1994). *Conflict and consensus*. Thousand Oaks: Sage.

Myers, D. G., & Lamm, H. (1976). The group polarization phenomenon. *Psychological Bulletin, 83*, 602–627. https://doi.org/10.1037/0033-2909.83.4.602

Pankove, E., & Kogan, N. (1968). Creative ability and risk taking in elementary school children. *Journal of Personality, 36*, 420–439. https://doi.org/10.1111/j.1467-6494.1968.tb01483.x

Paunonen, S. V., & Jackson, D. N. (1985). Idiographic measurement strategies for personality and prediction: Some unredeemed promissory notes. *Psychological Review, 92*, 486–511. https://doi.org/10.1037/0033-295X.92.4.486

Petrie, A. (1967). *Individuality in pain and suffering*. Chicago: University of Chicago Press.

Petrie, A., Collins, W., & Solomon, P. (1958). Pain sensitivity, sensory deprivation, and susceptibility to satiation. *Science, 128*, 1431–1433. https://doi.org/10.1126/science.128.3336.1431-a

Plucker, J. A. (1990). Reanalyses of student responses to creativity checklists: Evidence of content generality. *Journal of Creative Behavior, 33*, 126–137. https://doi.org/10.1002/j.2162-6057.1999.tb01042.x

Plucker, J. A. (1999). Is there proof in the pudding? Reanalyses of Torrance's (1958 to present) longitudinal data. *Creativity Research Journal, 12*, 103–114. https://doi.org/10.1207/s15326934crj1202_3

Plucker, J. A., & Makel, M. C. (2010). Assessment of creativity. In J. C. Kaufman & R. J. Sternberg (Eds.), *The Cambridge handbook of creativity* (pp. 48–79). New York: Cambridge University Press. https://doi.org/10.1017/CBO9780511763205.005

Pruitt, D. G. (1962). Pattern and level of risk in gambling decisions. *Psychological Review, 69*, 187–201. https://doi.org/10.1037/h0040607

Ryan, E. D., & Foster, R. (1967). Athletic participation and perceptual reduction and augmentation. *Journal of Personality and Social Psychology, 6*, 472–476. https://doi.org/10.1037/h0021226

Saunders, D. R. (1956). Moderator variables in prediction. *Educational and Psychological Measurement, 16*, 209–222. https://doi.org/10.1177/001316445601600205

Silvia, P. J. (2008). Creativity and intelligence revisited: A latent variable analysis of Wallach and Kogan (1965). *Creativity Research Journal, 20*, 34–39. https://doi.org/10.1080/10400410701841807

Silvia, P. J., Winterstein, B. P., Willse, J. T., Barona, C. M., Cram, J. T., Hess, K. I., et al. (2008). Assessing creativity with divergent-thinking tasks: Exploring the reliability and validity of new subjective scoring methods. *Psychology of Aesthetics, Creativity, and the Arts, 2*, 68–85. https://doi.org/10.1037/1931-3896.2.2.68

Simon, H. A. (1957). *Administrative behavior*. New York: Macmillan.

Skager, R. W., Schultz, C. B., & Klein, S. P. (1965). Quality and quantity of accomplishments as measures of creativity. *Journal of Educational Psychology, 56*, 31–39. https://doi.org/10.1037/h0021901

Skager, R. W., Schultz, C. B., & Klein, S. P. (1966a). Points of view about preference as tools in the analysis of creative products. *Perceptual and Motor Skills, 22*, 83–94. https://doi.org/10.2466/pms.1966.22.1.83

Skager, R. W., Schultz, C. B., & Klein, S. P. (1966b). The multidimensional scaling of a set of artistic drawings: Perceived structure and scale correlates. *Multivariate Behavioral Research, 1*, 425–436. https://doi.org/10.1207/s15327906mbr0104_2

Slovic, P. (1962). Convergent validation of risk taking measures. *Journal of Abnormal and Social Psychology, 65*, 68–71. https://doi.org/10.1037/h0048048

Spence, J. T., & Spence, K. W. (1966). The motivational components of manifest anxiety: Drive and drive stimuli. In C. D. Spielberger (Ed.), *Anxiety and behavior* (pp. 291–326). San Diego: Academic Press. https://doi.org/10.1016/B978-1-4832-3131-0.50017-2

Sternberg, R. J. (1997). *Thinking styles*. New York: Cambridge University Press. https://doi.org/10.1017/CBO9780511584152

Sternberg, R. J. (Ed.). (1999). *Handbook of creativity*. New York: Cambridge University Press.

Stricker, L. J., & Rock, D. A. (2001). Measuring accomplishments: Pseudoipsativity, quantity versus quality, and dimensionality. *Personality and Individual Differences, 31*, 103–115. https://doi.org/10.1016/S0191-8869(00)00114-8

Taylor, C., & Barron, F. (Eds.). (1963). *Scientific creativity*. New York: Wiley.

Teger, A. I., & Kogan, N. (1975). Decision making for others under reciprocal and non-reciprocal conditions. *British Journal of Social and Clinical Psychology, 14*, 215–222. https://doi.org/10.1111/j.2044-8260.1975.tb00174.x

Thibaut, J. W., & Kelley, H. H. (1959). *The social psychology of groups*. New York: Wiley.

Thomas, K., & Chan, J. (Eds.). (2013). *Handbook of research on creativity*. Cheltenham: Elgar.

Thorndike, R. L. (1966). Some methodological issues in the study of creativity. In A. Anastasi (Ed.), *Testing problems in perspective—Twenty-fifth anniversary volume of topical readings from the Invitational Conference on Testing Problems* (pp. 436–448). Washington, DC: American Council on Education.

Torrance, E. P. (1962). *Guiding creative talent*. Englewood Cliffs: Prentice-Hall. https://doi.org/10.1037/13134-000

Torrance, E. P. (1974). *Norms-technical manual, Torrance Tests of Creative Thinking*. Bensenville: Scholastic Testing Service.

Vernon, P. E. (1972). The distinctiveness of field independence. *Journal of Personality, 40*, 366–391. https://doi.org/10.1111/j.1467-6494.1972.tb00068.x

Wallach, M. A. (1970). Creativity. In P. H. Mussen (Ed.), *Carmichael's manual of child psychology* (Vol. 1, 3rd ed., pp. 1211–1272). New York: Wiley.

Wallach, M. A. (1971). *The intelligence/creativity distinction*. Morristown: General Learning Press.

Wallach, M. (1976). Tests tell us little about talent. *American Scientist, 64*, 57–63.

Wallach, M. A., & Kogan, N. (1961). Aspects of judgment and decision making: Interrelationships and changes with age. *Behavioral Science, 6*, 23–36. https://doi.org/10.1002/bs.3830060104

Wallach, M. A., & Kogan, N. (1965a). *Modes of thinking in young children—A study of the creativity-intelligence distinction*. New York: Holt Rinehart and Winston.

Wallach, M. A., & Kogan, N. (1965b). The roles of information, discussion, and consensus in group risk taking. *Journal of Experimental Social Psychology, 1*, 1–19. https://doi.org/10.1016/0022-1031(65)90034-X

Wallach, M. A., & Wing Jr., C. W. (1969). *The talented student: A validation of the creativity/intelligence distinction*. New York: Holt Rinehart and Winston.

Wallach, M. A., Kogan, N., & Bem, D. J. (1962). Group influence on individual risk taking. *Journal of Abnormal and Social Psychology, 65*, 75–86. https://doi.org/10.1037/h0044376

Wallach, M. A., Kogan, N., & Bem, D. J. (1964). Diffusion of responsibility and level of risk taking in groups. *Journal of Abnormal and Social Psychology, 68*, 263–274. https://doi.org/10.1037/h0042190

Wallach, M. A., Kogan, N., & Burt, R. B. (1965). Can group members recognize the effects of group discussion upon risk taking? *Journal of Experimental Social Psychology, 1*, 379–395. https://doi.org/10.1016/0022-1031(65)90016-8

Wallach, M. A., Kogan, N., & Burt, R. B. (1967). Group risk taking and field dependence-independence of group members. *Sociometry, 30*, 323–338.

Wallach, M. A., Kogan, N., & Burt, R. B. (1968). Are risk takers more persuasive than conservatives in group discussion? *Journal of Experimental Social Psychology, 4*, 76–88. https://doi.org/10.1016/0022-1031(68)90051-6

Wallbrown, F. H., & Huelsman Jr., C. B. (1975). The validity of the Wallach-Kogan creativity operations for the inner-city children in two areas of visual art. *Journal of Personality, 43*, 109–126. https://doi.org/10.1111/j.1467-6494.1975.tb00575.x

Wapner, S., & Demick, J. (Eds.). (1991). *Field dependence-independence: Cognitive style across the life span*. Hillsdale: Erlbaum.

Ward, W. C. (1969a). Creativity and environmental cues in nursery school children. *Developmental Psychology, 1*, 543–547. https://doi.org/10.1037/h0027977

Ward, W. C. (1969b). Rate and uniqueness in children's creative responding. *Child Development, 40*, 869–878. https://doi.org/10.2307/1127195

Ward, W. C. (1975). Convergent and divergent measurement of creativity in children. *Educational and Psychological Measurement, 35*, 87–95. https://doi.org/10.1177/001316447503500110

Ward, W. C., & Cox, P. W. (1974). A field study of nonverbal creativity. *Journal of Personality, 42*, 202–219. https://doi.org/10.1111/j.1467-6494.1974.tb00670.x

Ward, W. C., Kogan, N., & Pankove, E. (1972). Incentive effects in children's creativity. *Child Development, 43*, 669–676. https://doi.org/10.2307/1127565

Ward, W. C., Frederiksen, N., & Carlson, S. B. (1980). Construct validity of free-response and machine-scorable forms of a test. *Journal of Educational Measurement, 17*, 11–29. https://doi.org/10.1111/j.1745-3984.1980.tb00811.x

Weintraub, D. J., Green, G. S., & Herzog, T. R. (1973). Kinesthetic aftereffects day by day: Trends, task features, reliable individual differences. *American Journal of Psychology, 86*, 827–844. https://doi.org/10.2307/1422088

Witkin, H. A., & Berry, J. W. (1975). Psychological differentiation in cross-cultural perspective. *Journal of Cross-Cultural Psychology, 6*, 4–87. https://doi.org/10.1177/002202217500600102

Witkin, H. A., & Goodenough, D. R. (1977). Field dependence and interpersonal behavior. *Psychological Bulletin, 84*, 661–689. https://doi.org/10.1037/0033-2909.84.4.661

Witkin, H. A., & Goodenough, D. R. (1981). *Cognitive styles: Essence and origins*. New York: International Universities Press.

Witkin, H. A., Dyk, R. B., Faterson, H. F., Goodenough, D. R., & Karp, S. A. (1962). *Psychological differentiation*. New York: Wiley.

Witkin, H. A., Price-Williams, D., Bertini, M., Christiansen, B., Oltman, P. K., Ramirez, M., & van Meel, J. M. (1974). Social conformity and psychological differentiation. *International Journal of Psychology, 9*, 11–29. https://doi.org/10.1080/00207597408247089

Witkin, H. A., Moore, C. A., Oltman, P. K., Goodenough, D. R., Friedman, F., Owen, D. R., & Raskin, E. (1977a). Role of the field-dependent and field-independent cognitive styles in academic evolution: A longitudinal study. *Journal of Educational Psychology, 69*, 197–211. https://doi.org/10.1037/0022-0663.69.3.197

Witkin, H. A., Moore, C. A., Goodenough, D. R., & Cox, P. W. (1977b). Field-dependent and field-independent cognitive styles and their educational implications. *Review of Educational Research, 47*, 1–64. https://doi.org/10.3102/00346543047001001

Witkin, H. A., Goodenough, D. R., & Oltman, P. K. (1979). Psychological differentiation: Current status. *Journal of Personality and Social Psychology, 37*, 1127–1145. https://doi.org/10.1037/0022-3514.37.7.1127

Yates, J. F. (Ed.). (1992). *Risk-taking behavior*. New York: Wiley.
Zaleska, M., & Kogan, N. (1971). Level of risk selected by individuals and groups when deciding for self and for others. *Sociometry, 34*, 198–213. https://doi.org/10.2307/2786410
Zhang, L., & Sternberg, R. J. (2006). *The nature of intellectual styles*. Mahwah: Erlbaum.
Zhang, L., Sternberg, R. J., & Rayner, S. (Eds.). (2011). *Handbook of intellectual styles*. New York: Springer.
Zoccolotti, P., & Oltman, P. K. (1978). Field dependence and lateralization of verbal and configurational processing. *Cortex, 14*, 155–163. https://doi.org/10.1016/S0010-9452(78)80040-9
Zuckerman, M. (1994). *Behavioural expressions and biosocial bases of sensation seeking*. Cambridge: Cambridge University Press.

Chapter 15
Research on Developmental Psychology

Nathan Kogan, Lawrence J. Stricker, Michael Lewis, and Jeanne Brooks-Gunn

Developmental psychology was a major area of research at ETS from the late 1960s to the early 1990s, a natural extension of the work in cognitive, personality, and social psychology that had begun shortly after the organization's founding in 1947, consistent with Henry Chauncey's vision of investigating intellectual and personal qualities (see Stricker, Chap. 13, this volume). For a full understanding of these qualities, it is essential to know how they emerge and evolve. Hence the work in developmental psychology complemented the efforts already under way in other fields of psychology.

A great deal of the research in developmental psychology was conducted at ETS's Turnbull Hall in the Infant Laboratory, equipped with physiological recording equipment and observation rooms (e.g., Lewis 1974), and in a full-fledged Montessori school outfitted with video cameras (e.g., Copple et al. 1984). Hence, as Lewis (n.d.) recalled, the building "had sounds of infants crying and preschool children laughing" (p. 4). Other research was done in homes, schools, and hospitals, including a multisite longitudinal study of Head Start participants (e.g., Brooks-Gunn et al. 1989; Laosa 1984; Shipman 1972).

A handful of investigators directed most of the research, each carrying out a distinct program of extensive and influential work. This chapter covers research by Irving Sigel, on representational competence; Luis Laosa, on parental influences, migration, and measurement; Michael Lewis, on cognitive, personality, and social development of infants and young children; and Jeanne Brooks-Gunn, on cognitive,

N. Kogan • L.J. Stricker (✉)
Educational Testing Service, Princeton, NJ, USA
e-mail: lstricker@ets.org

M. Lewis
Rutgers Robert Wood Johnson Medical School, New Brunswick, NJ, USA

J. Brooks-Gunn
Columbia University, New York, NY, USA

R.E. Bennett, M. von Davier (eds.), *Advancing Human Assessment*,
Methodology of Educational Measurement and Assessment,
DOI 10.1007/978-3-319-58689-2_15

personality, and social development from infancy to adolescence. Other important research was conducted by Gordon Hale (e.g., Hale and Alderman 1978), on attention; Walter Emmerich (e.g., Emmerich 1968, 1982), on sex roles and personality development; and Nathan Kogan (e.g., Wallach and Kogan 1965) and William Ward (e.g., Ward 1968), on creativity. (The Kogan and Ward research is included in Kogan, Chap. 14, this volume.) In the present chapter, Kogan describes Sigel's research, and Stricker takes up Laosa's; Lewis and Brooks-Gunn discuss their own work.

15.1 Representational Competence and Psychological Distance

Representational competence was the focus of Sigel's research program. Roughly defined by Sigel and Saunders (1983), representational competence is the ability to transcend immediate stimulation and to remember relevant past events and project future possibilities. Also indicative of representational competence in preschoolers was the understanding of equivalence in symbol systems, whereby an object could be rendered three-dimensionally in pictorial form and in words.

The level of a child's representational competence was attributed in large part to parental beliefs and communicative behaviors and to family constellation (number of children and their birth order and spacing). Earlier research by Sigel and collaborators emphasized ethnicity and socioeconomic status (SES; see Kogan 1976). SES was retained in many of the ETS studies in addition to a contrast between typical children and those with communicative–language disabilities.

A conceptual model of the Sigel team's research approach is presented in a chapter by McGillicuddy-DeLisi et al. (1979): Mothers' and fathers' backgrounds determined their parental belief systems. Belief systems, in turn, influenced parental communication strategies, which then accounted for the child's level of cognitive development. It was a nonrecursive model, the child's developmental progress (relative to his or her age) feeding back to alter the parental belief systems. In terms of research design, then, parental background was the independent variable, parental belief systems and child-directed communicative behavior were mediating variables, and children's representational competence was the dependent variable. The full model was not implemented in every study, and other relevant variables were not included in the model. In most studies, family constellation (e.g., spacing and number of children), SES, the nature of the parent–child interaction task, the child's communicative status (with or without language disability), and the gender of the parent and child were shown to yield main or interaction effects on the child's representational competence.

In the view of Sigel and his associates, the critical component of parental teaching behavior was *distancing* (Sigel 1993). Parental teachings could reflect high- or low-level distancing. Thus, in a teaching context, asking the child to label an object was an example of low-level distancing, for the child's response was constrained to a single option with no higher-thinking processes invoked in the answer. By contrast, asking the child to consider possible uses of an object was an example of high-level distancing, for the child was forced to go beyond the overt stimulus properties of the object to adopt new perspectives toward it. In brief, the concept of distancing, as reflected in parental teaching behavior, referred to the degree of constraint versus openness that the parent imposed on the child. Sigel's principal hypothesis was that higher-level distancing in child-directed communication by an adult would be associated with greater representational competence for that child. Correspondingly, low-level distancing by an adult would inhibit the development of a child's representational competence.

An additional feature of Sigel's research program concerned the nature of the task in the parent–child interaction. Two tasks were selected of a distinctively different character. For the storytelling task, age-appropriate edited versions of children's books were used, with parents instructed to go through a story as they typically would do at home. The other task required paper folding, with the parent required to teach the child to make a boat or a plane.

15.1.1 Influence of Parental Beliefs and Behavior on Representational Competence

Having outlined the conceptual underpinning of Sigel's research program along with the nature of the variables selected and the research designs employed, we can now proceed to describe specific studies in greater detail. We begin with a study of 120 families in which the target child was 4 years of age (McGillicuddy-DeLisi 1982; Sigel 1982). Family variables included SES (middle vs. working class) and single-child versus three-child families. For the three-child families, there was variation in the age discrepancy between the first and second sibling (more than 3 years apart vs. less than 3 years apart), with the restriction that siblings be of the same sex. Each mother and father performed the storytelling and paper-folding tasks with their 4-year-old child. Proper controls were employed for order of task presentations. A total of 800 parent and child observations were coded by six raters with satisfactory interrater reliability.

The presentation of the research was divided into two parts, corresponding to the portion of the analytic model under investigation. In the first part (McGillicuddy-DeLisi 1982), the influence of the demographic variables, SES and family constellation, on parental beliefs was examined, and in turn the influence of parental beliefs for their prediction of overt parental behaviors in a teaching situation was explored. The second part, the child's representational competence, was treated separately in the Sigel (1982) chapter. Note that the assessment of beliefs was focused exclusively on the parents' views of how a preschool child acquired concepts and abilities, hence making such beliefs relevant to the parental strategies employed in facilitating the child's performance in a teaching context.

Parental beliefs were assessed in an interview based on 12 vignettes involving a 4-year-old and a mother or father. The interviewer asked the parent whether the child in the vignette had the necessary concepts or abilities to handle the problem being posed. Further inquiry focused more generally on parents' views of how children acquire concepts and abilities. Analysis of these data yielded 26 parental belief variables that were reliably scored by three coders. ANOVA was then employed to determine the influence of SES, family constellation, gender of child, and gender of parent on each of the 26 belief variables. Beliefs were found to vary more as a function of SES and family constellation than of gender of parent or child. More specifically, parents of three children had views of child development that differed substantially from those of single-child parents. For the parents of three children, development involved attributes more internal to the child (e.g., reference to self-regulation and impulsivity) as opposed to greater emphasis on external attributes (e.g., direct instruction) in single-child parents. The results as a whole constituted an intriguing mosaic, but they were post hoc in the absence of predictions derived from a theoretical framework. Of course, the exploratory nature of such research reflected the dearth at that time of theoretical development in the study of child-directed parental beliefs and behaviors.

Consider next the observed relationships between parental beliefs and teaching behaviors. Having shown that SES and family constellation influenced parental beliefs, the question of interest was whether such beliefs provided useful information about parents' teaching behaviors beyond what might be predicted from SES and family constellation. To answer the question, stepwise regressions were carried out with SES and family constellation entered into the analysis first, followed by the belief variables. Separate regressions—four in all—were conducted for mothers' and fathers' performance on the storytelling and paper-folding tasks, the dependent variables.

Demonstration of belief effects on teaching behaviors would require that multiple correlations show significant increments in magnitude when beliefs were entered into the regression analysis. Such increments were observed in all four regressions, indicating that parents' beliefs about their children's competencies were predictive of the way they went about teaching their children on selected tasks. Noteworthy is the evidence that the significant beliefs varied across the two tasks and that this variation was greater for mothers than for fathers. In other words, mothers appeared to be more sensitive to the properties of the task facing the child, whereas fathers appeared to have internalized a set of beliefs generally applied to different kinds of tasks. Mothers would seem to have a more differentiated view of their children's competencies and hence were more attuned to the nature of the task than were fathers.

Thus far, we have considered the relations among family demographics, parental beliefs, and teaching strategies. The missing link, the child's cognitive performance, was examined in the Sigel (1982) chapter, where it was specifically related to parental teaching behaviors. The child's responses to the storytelling and paper-folding tasks were considered (e.g., extent of engagement and problem solutions), as was the child's performance on tasks independent of parental instructions. These latter tasks included Piagetian conservation and imagery assessments and the Sigel Object Categorization Task (Sigel and Olmsted 1970). The major hypothesis was that the parents' uses of distancing strategies in their teaching behaviors would be associated with enhanced cognitive performances in their children—representational competence.

To address this hypothesis, stepwise regressions were analyzed. The results confirmed the basic hypothesis linking parental child-directed distancing to the child's representational competence. This general observation, however, conceals the specificity of the effects. Thus mothers and fathers employed different teaching strategies, and these strategies, in turn, varied across the storytelling and paper-folding tasks. Of special interest are those analyses in which the mothers' and fathers' teaching behaviors were entered into the same regression equation. Doing so in sequence often pointed to the complementarity of parental influences. In concrete terms, the multiple correlations sometimes demonstrated significant enhancements when both parents' teaching strategies entered into the analysis compared to the outcome for the parents considered separately. This result implied that the children could intellectually profit from the different, but complementary, teaching strategies of mothers and fathers.

15.1.2 Impact of a Communicative Disability

Sigel and McGillicuddy-DeLisi (1984) were able to recruit families who had a child with a communicative disability (CD), making it possible to compare such families with those where the child was not communicatively disabled (non-CD). It was possible to match the CD and non-CD children on SES, family size, gender, age, and birth order. Again, mothers' and fathers' distancing behaviors were examined in the task context of storytelling and paper folding.

In the case of the child's intellectual ability, assessed by the Wechsler Preschool and Primary School Scale of Intelligence (WPPSI; Wechsler 1949b), parental effects were largely confined to the CD sample. Low parental distancing strategies were tightly associated with lower WPPSI scores. Of course, we must allow for the possibility that the parent was adjusting his or her distancing level to the perceived cognitive ability of the child. In contrast, the child's representational competence, as defined by the assessments previously described in Sigel (1982), was linked with parental distancing behaviors in both CD and non-CD samples, with the magnitude of the relationship somewhat higher in the CD sample.

Of course, these associations could not address the causality question: The parent might be affecting the child or reacting to the child or, more likely, the influence was proceeding in both directions. Sigel and McGillicuddy-DeLisi (1984) argued that low-level distancing strategies by parents discouraged active thinking in the child; hence it was no surprise that such children did not perform well on representational tasks that required such thinking. They were optimistic about CD children, for high-level parental distancing seemed to encourage the kind of representational thinking that could partially compensate for their communicative disabilities (Sigel 1986).

15.1.3 Belief-Behavior Connection

Working with a subsample of the non-CD families described in the previous section, Sigel (1992) plunged into the controversial issue of the linkage between an individual's beliefs and actual behavior instantiating those beliefs. He also developed a measure of behavioral intentions—a possible mediator of the belief–behavior connection. Although the focus was naturally on parental beliefs and behaviors, similar work in social psychology on the belief and behavior connection (e.g., Ajzen and Fishbein 1977), where major advances in theory and research had occurred, was not considered.

Three categories of variables were involved: (a) parents' beliefs about how children acquired knowledge in four distinct domains (physical, social, moral, and self); (b) the strategies that parents claimed they would use to facilitate the children's acquisition of knowledge in those domains; and (c) the behavioral strategies employed by the parents in a teaching context with their children. The first two categories were assessed with a series of vignettes. Thus, in the vignette for the physical domain, the child asks the parent how to use a yardstick to measure the capacity of their bathtub. The parents' view about how children learn about measurement constituted the belief measure; the parents' statements about how they would help their child learn about measurement constituted the self-referent strategy measure. For the third category, the parents taught their child how to tie knots, and the strategies employed in doing so were observed. Note that the knots task involved different content than was used in the vignettes.

Parental beliefs regarding children's learning were categorized as emphasizing cognitive processing (e.g., children figuring out things on their own) or direct instruction (e.g., children learning from being told things by adults). Parental intended teaching strategies were classified as distancing, rational authoritative (e.g., parent gives reasons with commands), or direct authoritative (e.g., parent offers statement or rule without rationale). Parental behavioral strategies were scored for high-level versus low-level distancing.

The three variable classes—parental beliefs, parental intended teaching strategies, and parental behavioral strategies—were intercorrelated. Substantial relationships were observed between parental beliefs about learning (cognitive processing vs. direct instruction) and the strategies the parent intended to employ. As anticipated, cognitive processing was associated with distancing strategies, and direct instruction was linked to authoritative strategies. Of course, both the beliefs and self-referent strategies were derived from the same vignettes used in the parental interview, suggesting the likely influence of method variance on the correlational outcomes. When the foregoing variables were related to the parents' behavioral strategies in teaching the knots task, the magnitude of the correlations dropped precipitously, though the marginally significant correlations were in the predicted direction. Sigel (1992) attributed the correlational decline to variation across domains. Thus the belief–strategy linkages were not constant across physical, social, and moral problems. Aggregation across these domains could not be justified. Obviously, the shifting task content and context were also responsible for the absence of anticipated linkages. Conceivably, an analytic procedure in which parents' intended strategies were cast as mediators between their beliefs and their behavioral strategies would have yielded further enlightenment.

15.1.4 Collaborative Research

The large majority of Sigel's publications were either solely authored by him or coauthored with former or present members of his staff at ETS. A small number of papers, however, were coauthored with two other investigators, Anthony Pellegrini and Gene Brody, at the University of Georgia. These publications are of particular interest because they cast Sigel's research paradigm within a different theoretical framework, that of Vygotsky (1978), and they introduced a new independent variable into the paradigm, marital quality.

In the case of marital quality, Brody et al. (1986) raised the possibility that the quality of the marital relationship would influence mothers' and fathers' interactions with their elementary-school age children. More specifically, Brody et al., leaning on clinical reports, examined the assumption that marital distress would lead to compensatory behaviors by the parents when they interact with their children in a teaching context. Also under examination was the possibility that mothers and fathers would employ different teaching strategies when interacting with the children, with the nature of such differences possibly contingent on the levels of marital distress.

Again, storytelling and paper-folding tasks were used with the mothers and fathers. Level of marital distress was assessed by the Scale of Marriage Problems (Swenson and Fiore 1975), and a median split was used to divide the sample into distressed and nondistressed subgroups. Observation of parental teaching strategies and the child's responsiveness was accomplished with an event-recording procedure (Sigel et al. 1977) that yielded interrater reliability coefficients exceeding .75 for each of the eight behaviors coded. ANOVAs produced significant Marital Problems × Parent interactions for seven of the eight behavioral indices. Nondistressed mothers and fathers did not differ on any of the behavioral indices. By contrast, distressed mothers and fathers differed in their teaching strategies, the mothers' strategies being more effective: more questions, feedback, and suggestions and fewer attempts to take over the child's problem-solving efforts.

Fathers in the distressed group "behave in a more intrusive manner with their school-aged children, doing tasks for them rather than allowing them to discover their own solutions and displaying fewer positive emotions in response to their children's learning attempts" (p. 295). Mothers in distressed marriages, by contrast, responded with more effective teaching behaviors, inducing more responsive behavior from their children. Hence the hypothesis of compensatory maternal behaviors in a distressed marriage was supported. The psychological basis for such compensation, however, remained conjectural, with the strong likelihood that mothers were compensating for perceived less-than-satisfactory parenting by their husbands. Finally, Brody et al. (1986) offered the caveat that the outcomes could not be generalized to parents with more meager educational and economic resources than characterized the well-educated parents employed in their study.

In two additional studies (Pellegrini et al. 1985, 1986), the Sigel research paradigm was applied, but interpretation of the results leaned heavily on Vygotsky's (1978) theory of the zone of proximal development. Pellegrini et al. (1985) studied parents' book-reading behaviors with 4- and 5-year-old children. Families differed in whether their children were communicatively disabled. MANOVA was applied, with the parental interaction behavior as the dependent variable and age, CD vs. non-CD status, and parent (mother vs. father) as the independent variables. Only CD vs. non-CD status yielded a significant main effect. Parental behaviors were more directive and less demanding with CD children. Furthermore, stepwise regression analysis examined the link between the parental interaction variables and WPPSI verbal IQ. For non-CD children, high cognitive demand was significantly associated with higher IQ levels; for CD children, the strongest positive predictor of IQ was the less demanding strategy of verbal/emotional support.

In general, parents seemed to adjust the cognitive demands of their teaching strategies to the level of the children's communicative competences. In Vygotskyan terms, parents operated within the child's zone of proximal development. Other evidence indicated that parents engaged in scaffolding to enhance their children's cognitive–linguistic performances. Thus parents of non-CD children manifested more conversational turns in a presumed effort to elicit more language from their children. Similarly, more parental paraphrasing with non-CD children encouraged departures from the literal text, thereby fostering greater depth of interaction between parent and child. In sum, parental scaffolding of their children's task-oriented behavior activated the potential for children to advance toward more independent problem solving as outlined in Vygotsky's theory.

We turn, finally, to the second study (Pellegrini et al. 1986) influenced by Vygotsky's theory. The research paradigm was similar to studies previously described. Again, gender of parent, children's CD vs. non-CD status, and the tasks of book reading and paper folding constituted the independent variables, and the teaching strategies of the parents comprised the dependent variables. In addition, the extent of task engagement by the child was also examined. MANOVA was employed, and it yielded a significant main effect for the child's communicative status and for its interaction with the task variable. ANOVAs applied to the separate teaching variables indicated that (a) parents were more directive and less demanding with CD children than with non-CD children; (b) parents were more demanding, gave less emotional support, and asked fewer questions with the paper-folding task than with the book-reading task; and (c) communicative status and task variable interacted: A CD versus non-CD difference occurred only for the book-reading task, with parents of CD children asking more questions and making lower cognitive demands.

The teaching strategy measures were factor analyzed, and the resulting four orthogonal factors became the predictor variables in a regression analysis with children's rated task engagement as the criterion variable. For the paper-folding task, parents of both CD and non-CD children used high-demand strategies to keep their children engaged. For the book-reading task, parents of CD and non-CD children differed, with the CD parents using less demanding strategies and the non-CD parents using more demanding ones.

Pellegrini et al. (1986) had shown how ultimate problem-solving outcomes are of less significance than the processes by which such outcomes are achieved. Adult guidance is the key, with non-CD children requiring considerably less of it to remain engaged with the task than was the case for CD children. Hence the children's competence levels alert the parents to how demanding their teaching strategies should be. Pellegrini et al. further recommended the exploration of the sequence of parental teaching strategies, as parents found it necessary on occasion to switch from more demanding to less demanding strategies when the child encountered difficulty (see Wertsch et al. 1980). In sum, the findings strongly support the Vygotsky model of parents teaching children through the zone of proximal development and the adjustment of parental teaching consistent with the competence level of their children.

15.1.5 *Practice*

An important feature of Sigel's research program was linking research to practice (Renninger 2007). As Sigel (2006) noted,

> efforts to apply research to practice require acknowledging the inherent tensions of trying to validate theory and research in practical settings. They require stretching and/or adapting the root metaphors in which we have been trained so that collaborations between researchers and practitioners are the basis of research and any application of research to practice. (p. 1022)

The research on representational competence and psychological distance has had widespread impact, notably for early childhood education (Hyson et al. 2006) and cognitive behavior therapy (Beck 1967).

15.2 Parental Influences, Migration, and Measurement

Laosa's empirical work and his position papers spanned the psychological development of children, particularly Hispanics. His methodological contributions included test theory, especially as it relates to the assessment of minority children, and a standardized measure of parental teaching strategies. The major foci of Laosa's work to be considered here are parental influences on children's development, the consequences of migration for their adjustment and growth, and the measurement of their ability.

15.2.1 Parental Influences

Parental influence on children's intellectual development has been a topic of long-standing interest to developmental psychologists (e.g., Clarke-Stewart 1977). A particular concern in Laosa's work was Hispanic children, given the gap in their academic achievement. His early research concerned maternal teaching. Unlike much of the previous work in that area, Laosa made direct observations of the mothers teaching their children, instead of relying on mothers' self-reports about interactions with their children, and distinguished between two likely SES determinants of their teaching: education and occupation. In a study of Hispanic mother–child dyads (Laosa 1978), mother's education correlated positively with praising and asking questions during the teaching and correlated negatively with modeling (i.e., the mother working on the learning task herself while the child observes). However, mother's occupation did not correlate with any of the teaching variables, and neither did father's occupation. Laosa speculated that the education-linked differences in teaching strategies account for the relationship between mothers' education and their children's intellectual development found in other research (e.g., Bradley et al. 1977). Subsequently, Laosa (1980b) also suggested that the more highly educated mothers imitate how they were taught in school.

In a follow-up study of Hispanic and non-Hispanic White mother–child dyads (Laosa 1980b), the two groups differed on most of the teaching variables. Non-Hispanic White mothers praised and asked questions more, and Hispanic mothers modeled, gave visual cues, directed, and punished or restrained more. However, when mothers' education was statistically controlled, the differences between the groups disappeared; controlling for mothers' or fathers' occupation did not reduce the differences.

In a third study, with the Hispanic mother–child dyads (Laosa 1980a), mother's field independence, assessed by the Embedded Figure Test (Witkin et al. 1971) and WAIS Block Design (Wechsler 1955), correlated positively with mother's asking questions and praising, and correlated negatively with mother's modeling. The correlations were reduced, but their pattern was similar when mother's education was statistically controlled. Laosa suggested that asking questions and praising are self-discovery approaches to learning that reflect field independence, whereas modeling is a concrete approach that reflects field dependence; hence mothers were using strategies that foster their own cognitive style in their children. Mother's teaching strategies, in fact, correlated modestly but inconsistently with the children's field independence, as measured by the Children's Embedded Figures Test (Witkin et al. 1971), WISC Block Design (Wechsler 1949a), and Human Figure Drawing (Harris 1963), another measure of field independence. Most of the teaching strategies had scattered correlations with the Children's Embedded Figures Test and Block Design: positive correlations with asking questions and praising (field-independent strategies) and negative correlations with modeling, punishing or restraining, and giving visual cues (field-dependent strategies).

In Laosa's later research, a recurring topic was the impact of parents' education on their children's intellectual development; this line of work was presumably motivated by the influence of education in his maternal-teaching studies. Laosa (1982b) viewed parental education as impacting the parent–child interaction and presented a conceptual model of this interaction as the mediator between parent education and the child's development. He reported further analyses of the samples of Hispanic and non-Hispanic White mother–child dyads.

In one analysis, non-Hispanic White mothers and fathers read to their children more than did Hispanic parents. When parents' education was statistically controlled, the group difference disappeared, but controlling for parents' occupation did not reduce it. In addition, non-Hispanic mothers had higher *realistic* educational aspirations for their children ("*realistically*, how much education do you think your child will receive?"); this difference also disappeared when mothers' education was controlled but not when their occupation was controlled.

In another analysis, mother's education correlated positively in both the Hispanic and non-Hispanic White groups with mother's reading to the child, but father's education was uncorrelated with father's reading to the child in either group. Parent's occupation did not correlate with reading in the two groups. In both groups, mother's education also correlated positively with mother's educational aspirations for the child, but mother's occupation was uncorrelated.

Also, in an analysis of the Hispanic group, mother's education correlated positively with the child's ability to read or write before kindergarten, though father's education was uncorrelated. Parent's occupation was also uncorrelated with literacy. In addition, parent's education correlated positively with their use of English with the child; parent's occupation also correlated positively but weakly with English use.

Laosa argued that the set of findings, in total, suggests that the lower educational level of Hispanic parents produced a discontinuity between their children's home and school environments that adversely affected academic achievement.

He explored the consequences of these parental influences on the test performance of 3-year-olds in two studies. In the first study (Laosa 1982a), which targeted non-Hispanic White children, a path analysis was employed to assess the relationships, direct and indirect, between a host of family influences (e.g., mother's education and occupation, mother's reading to the child, nonparents in the household reading to the child, mother's teaching strategies) and performance on the Preschool Inventory (Caldwell 1970), a test of verbal, quantitative, and perceptual-motor skills for kindergarten children. A Mother's Socioeducational Values factor (defined by mother's education and occupation and mother's reading to child) was the strongest determinant of test performance. Less powerful determinants included nonparents in the household (probably older siblings) reading to the child and mother's use of modeling in teaching. Laosa highlighted two important and unanticipated findings:

the apparent influence of siblings and the substantial and positive influence of modeling, contrary to the conventional wisdom that verbal teaching strategies, such as asking questions, are superior to nonverbal ones, such as modeling.

In the second study (Laosa 1984) of Hispanic and non-Hispanic White children, the groups differed in their means on three of the five scales of the McCarthy Scales of Children's Abilities (McCarthy 1972): Verbal, Quantitative, and Memory. When a Sib Structure/Size factor (later-born child, many siblings) was statistically controlled, the group differences were unaffected. But when either a Language factor (mother uses English with child, child uses English with mother) or an SES factor (parents' education, father's occupation, household income) was controlled, the differences were reduced; when both factors were controlled, the differences were eliminated. The findings led Laosa to conclude that these early ethnic-group differences in ability were explainable by differences in SES and English-language usage.

15.2.2 Migration

In a series of white papers, Laosa reviewed and synthesized the extant research literature on the consequences of migration for children's adjustment and development, particularly Hispanic children, and laid out the salient issues (Laosa 1990, 1997, 1999). One theme was the need for—and the absence of—a developmental perspective in studying migration: "what develops, and *when, how,* and *why* it develops" (Laosa 1999, p. 370). The pioneering nature of this effort is underscored by the observation almost two decades later that migration is neglected by developmental psychology (Suárez-Orozco and Carhill 2008; Suárez-Orozco et al. 2008).

In a 1990 paper, Laosa proposed a multivariate, conceptual model that described the determinants of the adaptation of Hispanic immigrant children to the new society. Key features of the model were the inclusion of variables antedating immigration (e.g., sending community), moderator variables (e.g., receiving community), and mediating variables (e.g., child's perceptions and expectations) between the stresses of immigration and the outcomes.

In a complex, longitudinal survey of Puerto Rican migrants in New Jersey schools, Laosa (2001) found that the majority of the student body were Hispanic in 46% of the schools and were native speakers of Spanish in 31%. Additionally, the majority of the student body was eligible for free lunch in 77% of the schools and was from families on public assistance in 46%. Laosa concluded that the migrants faced considerable segregation by ethnicity or race as well as considerable isolation by language in high-poverty schools, factors with adverse consequences for the students' social and academic development.

15.2.3 Measurement

The measurement and evaluation of children's ability and achievement, particularly the unbiased assessment of minority children, has long been beset by controversies (see Laosa 1977; Oakland and Laosa 1977). These controversies were sparked in the 1960s and 1970s by the Coleman report (Coleman et al. 1966), which suggested that average differences in the academic performance of Black and White students are more affected by their home background than by their schools' resources, and by Jensen's (1969) review of research bearing on genetic and environmental influences on intelligence. He concluded that genetics is a stronger influence, which many observers interpreted as suggesting that the well-established disparity between Black and White children in their average scores on intelligence tests is largely genetic in origin. The upshot was widespread concerns that these tests are biased and calls for banning their use in schools. These arguments were reignited by *The Bell Curve* (Herrnstein and Murray 1994), which concluded that intelligence is mainly heritable. As Laosa (1996) noted, "thus, like a refractory strain of retrovirus, the issues tend to remain latent and from time to time resurge brusquely onto the fore of public consciousness" (p. 155).

In a 1977 paper, Laosa summarized the earlier controversies and other criticisms of testing and discussed alternatives to current testing practices that had been developed in response. The alternatives included constructing "culture-fair" tests "whose content is equally 'fair' or 'unfair' to different cultural groups" (p. 14), translating tests from English, using norms for subgroups, adjusting scores for test takers with deprived backgrounds, devising tests for subgroups (e.g., the BITCH, a vocabulary test based on Black culture; Williams 1972), using criterion-based interpretations of scores (i.e., how well a student achieves a specific objective) instead of norm-based interpretations (i.e., how well he or she does on the test relative to others), employing tests of specific abilities rather than global measures like IQ, and making observations of actual behavior. Laosa cautioned that these alternatives may also be problematic and would need to be carefully evaluated.

In a companion piece, Laosa, joined by Thomas Oakland of the University of Texas, Austin (Oakland and Laosa 1977), provided a comprehensive account of standards for minority-group testing that had been formulated by professional organizations, the government, and the courts. They argued for the need to consider these standards in testing minority-group children.

Laosa (1982c), in a subsequent paper on measurement issues in the evaluation of educational programs, specifically Head Start, delineated the concept of population validity and its applicability to program evaluation. Population validity deals with the question, "Do the results yielded by a given assessment technique have the same meaning when administered to persons of different sociocultural backgrounds?" (p. 512). Laosa discussed threats to population validity: familiarity (performance is influenced by familiarity with the task), communication, role relations (performance is influenced by the test taker's relationship with the tester), and situational (e.g., physical setting, people involved).

In another paper, Laosa (1991) explicated the links between population validity, cultural diversity, and professional ethics. As an illustration, he described a study by Bradley et al. (1989) of children in three ethnic groups, Black, Hispanic, and non-Hispanic White, matched on their HOME inventory (Caldwell and Bradley 1985) scores, a measure of the home environment. The HOME inventory scores correlated appreciably with performance on the Bayley Scales of Infant Development (Bayley 1969) and the Stanford–Binet Intelligence Test (Terman and Merrill 1960) for the Black and non-Hispanic White children but not for the Hispanic children. Laosa suggested that this finding highlights the importance of evaluating test results separately for different ethnic groups.

Laosa pointed out that population validity is a scientific concern in basic research and an ethical issue in applied work, given the inability to predict the results in different populations from the one studied. He also noted that when population differences are observed, two questions need to be answered. One, relevant to applied work, is, Which populations react differently? The other question, pertinent to scientific research, but rarely asked, is, Why do they differ?

In his last paper on measurement issues, Laosa (1996), responding to *The Bell Curve* controversy, made several general points about test bias. One was that bias reflects the absence of population validity. He noted that this view accords with the Cole and Moss (1989) definition of bias: "Bias is differential validity of a given interpretation of a test score for any definable, relevant group of test takers" (p. 205).

Another point was that the definition of predictive bias in the then current third edition of the *Standards for Educational and Psychological Testing* (American Educational Research Association, American Psychological Association, & National Council on Measurement in Education 1985) is insufficient. According to the *Standards*, predictive bias is absent if "the predictive relationship of two groups being compared can be adequately described by a common algorithm (e.g., regression line)" (p. 12). Laosa took up the argument that intelligence tests cannot be considered to be unbiased simply because their predictions are equally accurate for different races or social classes, noting Campbell's rejoinder that the same result would occur if the tests simply measured opportunity to learn (D. T. Campbell, personal communication, May 18, 1995).

The last point was that the consequences of test use also need to be considered (Messick 1989). Laosa cited Linn's (1989) example that requiring minimum high school grades and admissions test scores for college athletes to play during their freshman year can affect what courses minority athletes take in high school, whether they will attend college if they cannot play in their freshman year, and, ultimately, their education and employment.

15.3 Cognitive, Personality, and Social Development of Infants and Young Children

Lewis studied infant's cognitive attention and language ability, infants' and young children's physiological responses during attention, and infants' social and emotional development. He also formulated theories of development as well as theories about the integration of children's various competencies.

15.3.1 Social Development

Social development was a major interest, in particular, the mother–child interaction and the role this interaction played in the child's development. This work on social development revolved around four themes: (a) the mother–child relationship, (b) the growth of the child's social knowledge, (c) social cognition or the nature of the social world, and (d) the social network of children.

For example, in a 1979 volume (Lewis and Rosenblum 1979), *The Child and Its Family,* Lewis challenged the idea that the child's mother was the only important figure in the infant's early life and showed that fathers and siblings, as well as grandparents and teachers, were also key influences. And in a 1975 volume, on peer friendship in the opening years of life (Lewis and Rosenblum 1975), Lewis disputed the Piagetian idea that children could not form and maintain friendships before the age of 4 years. The finding that infants are attracted to and enjoy the company of other infants and young children, and that they can learn from them through observation and imitation, helped open the field of infant daycare (Goldman and Lewis 1976; Lewis 1982b; Lewis and Schaeffer 1981; Lewis et al. 1975). Because the infant's ability to form and maintain friends is important for the daycare context, where groups of infants are required to play together, this work also showed that the learning experience of young children and infants involved both direct and indirect interactions, such as modeling and imitation with their social world of peers, siblings, and teachers (Feiring and Lewis 1978; Lewis and Feiring 1982). This information also had an important consequence on hospital care; until this time, infants were kept far apart from each other in the belief that they could not appreciate or profit from the company of other children.

Another major theme of the research on social development involved infants' social knowledge. In a series of papers, Lewis was able to show that infants could discriminate among human faces (Lewis 1969); that they were learning about their gender (Lewis and Brooks 1975); that they were showing sex-role-appropriate behaviors (Feiring and Lewis 1979; Goldberg and Lewis 1969; Lewis 1975a); that they were learning about how people look, for example, showing surprise at the appearance of a midget—a child's height but an adult's face (Brooks and Lewis 1976); and that they were detecting the correspondence between particular faces and voices (McGurk and Lewis 1974). All of these results indicated that in the first

2 years, children were learning a great deal about their social worlds (Brooks-Gunn and Lewis 1981; Lewis 1981b; Lewis et al. 1971).

The most important aspect of this work on social development was the child's development of a sense of itself, something now called consciousness, which occurs in the second and third years of life. In the Lewis and Brooks-Gunn (1979a) book on self-recognition, *Social Cognition and the Acquisition of Self,* Lewis described his mirror self-recognition test, a technique that has now been used across the world. Results with this test revealed that between 15 and 24 months of age, typically developing children come to recognize themselves in mirrors. He subsequently showed that this ability, the development of the idea of "me," along with other cognitive abilities gives rise to the complex emotions of empathy, embarrassment, and envy as well as the later self-conscious emotions of shame, guilt, and pride (Lewis and Brooks 1975; Lewis and Brooks-Gunn 1981b; Lewis and Michalson 1982b; Lewis and Rosenblum 1974b).

These ideas, an outgrowth of the work on self-recognition, led to Lewis's interest in emotional development. They also resulted in a reinterpretation of the child's need for others. While children's attachment to their mothers was thought to be the most important relationship for the children, satisfying all of their needs, it became clear that others played an important role in children's social and emotional lives. His empirical work on fathers (Lewis and Weinraub 1974) and peers (Lewis et al. 1975) led to the formulation of a social network theory (Feiring and Lewis 1978; Lewis 1980; Lewis and Ban 1977; Lewis and Feiring 1979; Weinraub et al. 1977).

15.3.2 Emotional Development

Lewis's interest in social development and in consciousness led quite naturally to his research on emotional development, as already noted (Lewis 1973, 1977b, 1980; Lewis and Brooks 1974; Lewis et al. 1978; Lewis and Michalson 1982a, b; Lewis and Rosenblum 1978a, b). Two volumes framed this work on the development of emotional life (Lewis and Rosenblum 1974b, 1978b) and were the first published studies of emotional development. These early efforts were focused on the emotions of infants in the first year of life, including fear, anger, sadness, joy, and interest. To study emotional life, Lewis created experimental paradigms and devised a measurement system. So, for example, paradigms were developed for peer play (Lewis et al. 1975), social referencing (Feinman and Lewis 1983; Lewis and Feiring 1981), stranger approach (Lewis and Brooks-Gunn 1979a), mirror recognition (Lewis and Brooks-Gunn 1979a), and contingent learning (Freedle and Lewis 1970; Lewis and Starr 1979). A measurement system was created for observing infants' and young children's emotional behavior in a daycare situation that provided scales of emotional development (Lewis and Michalson 1983). These scales have been used by both American and Italian researchers interested in the effects of daycare on emotional life (Goldman and Lewis 1976).

15.3.3 Cognitive Development

Lewis's interests in development also extended to the study of infants' and children's cognitive development, including attentional processes, intelligence, and language development (Dodd and Lewis 1969; Freedle and Lewis 1977; Hale and Lewis 1979; Lewis 1971b, 1973, 1975b, 1976a, b, 1977a, 1978a, 1981a, 1982a; Lewis and Baldini 1979; Lewis and Baumel 1970; Lewis and Cherry 1977; Lewis and Freedle 1977; Lewis and Rosenblum 1977; Lewis et al. 1969a, 1971; McGurk and Lewis 1974).

Lewis first demonstrated that the Bayley Scales of Infant Development (Bayley 1969), which were—and still are—the most widely used test of infant intelligence, had no predictive ability up to 18 months of age (Lewis and McGurk 1973). In an effort to find an alternative, Lewis turned to research on infants' attentional ability, which he had begun at the Fels Research Institute, and developed it further at ETS. This work used a habituation–dishabituation paradigm where the infant was presented with the same visual stimulus repeatedly and then, after some time, presented with a variation of that stimulus. Infants show boredom to the repeated event, or habituation, and when the new event is presented, the infants show recovery of their interest, or dishabituation (Kagan and Lewis 1965; Lewis et al. 1967a, b). Infants' interest was measured both by observing their looking behavior and by assessing changes in their heart rate (Lewis 1974; Lewis et al. 1966a, b; Lewis and Spaulding 1967). He discovered that the infants' rate of habituation and degree of dishabituation were both related to their subsequent cognitive competence, in particular to their IQ. In fact, this test was more accurate than the Bayley in predicting subsequent IQ (Lewis and Brooks-Gunn 1981a, c; Lewis et al. 1969; Lewis and McGurk 1973).

This research on attentional processes convinced Lewis of the usefulness of physiological measures, such as heart rate changes, in augmenting behavior observation, work which he also began at the Fels Research Institute and continued and expanded at ETS (Lewis 1971a, b, 1974; Lewis et al. 1969, 1970, 1978; Lewis and Taft 1982; Lewis and Wilson 1970; Sontag et al. 1969; Steele and Lewis 1968).

15.3.4 Atypical Development

Lewis's research on normal development, especially on attentional processes as a marker of central nervous system functioning, led to an interest in atypical developmental processes and a program of research on children with disabilities (Brinker and Lewis 1982a, b; Brooks-Gunn and Lewis 1981, 1982a, b, c; Fox and Lewis 1982a, b; Lewis 1971c; Lewis and Fox 1980; Lewis and Rosenblum 1981; Lewis and Taft 1982; Lewis and Wehren 1982; Lewis and Zarin-Ackerman 1977; Thurman and Lewis 1979; Zarin-Ackerman et al. 1977). Perhaps of most importance was the development of an intervention strategy based on Lewis's work with typically developing children, the Learning to Learn Curriculum. Infants with disabilities were given home- and clinic-based interventions where their simple motor responses resulted in complex outcomes and where they had to learn to produce these outcomes, which served as operants—in effect, an applied-behavior-analysis intervention strategy (Brinker and Lewis 1982a, b; Lewis 1978a, b; Thurman and Lewis 1979).

15.3.5 Theories

Lewis formulated several influential theories about infant development. These included (a) a reconsideration of attachment theory (Weinraub and Lewis 1977) and (b) the infant as part of a social network (Weinraub et al. 1977). He also began work on a theory of emotional development (Lewis 1971b; Lewis and Michalson 1983).

15.3.6 The Origin of Behavior Series

Lewis and Leonard Rosenblum of SUNY Downstate Medical Center organized yearly conferences on important topics in child development for research scientists in both child and animal (primate) development to bring together biological, cultural, and educational points of view. These conferences resulted in a book series, *The Origins of Behavior* (later titled *Genesis of Behavior*), under their editorship, with seven highly cited volumes (Lewis and Rosenblum 1974a, b, 1975, 1977, 1978a, 1979, 1981). The initial volume, *The Effect of the Infant on the Caregiver* (Lewis and Rosenblum 1974a), was so influential that the term *caregiver* became the preferred term, replacing the old term *caretaker*. The book became the major reference on the interactive nature of social development—that the social development of the child involves an interaction between the mother's effect on the infant and the effect of the infant on the mother. It was translated into several languages, and 15 years after publication, a meeting sponsored by the National Institutes of Health reviewed the effects of this volume on the field.

15.4 Cognitive, Personality, and Social Development From Infancy to Adolescence

Brooks-Gunn's work encompassed research on the cognitive, personality, and social development of infants, toddlers, and adolescents, primarily within the framework of social-cognitive theory. Major foci were the acquisition of social knowledge in young children, reproductive processes in adolescence, and perinatal influences on children's development. These issues were attacked in laboratory experiments, other cross-sectional and longitudinal studies, and experimental interventions. (A fuller account appears in Brooks-Gunn 2013.)

15.4.1 *Social Knowledge in Infants and Toddlers*

In collaboration with Lewis, Brooks-Gunn carried out a series of studies on the development of early knowledge about the self and others in infancy and toddlerhood. They investigated how and when young children began to use social categories, such as gender, age, and relationship, to organize their world and to guide interactions (Brooks and Lewis 1976; Brooks-Gunn and Lewis 1979a, b, 1981) as well as the development of self-recognition as a specific aspect of social cognition (Lewis and Brooks-Gunn 1979b, c; Lewis et al. 1985). This developmental work was embedded in genetic epistemology theory as well as social-cognitive theory, with a strong focus on the idea that the self (or person) only develops in relation to others and that the self continues to evolve over time, as does the relation to others.

The studies demonstrated that social knowledge develops very early. Infants shown pictures of their parents, strange adults, and 5-year olds and asked, Who is that? were able to label their parents' pictures as mommy and poppy, labeling their fathers' pictures more accurately and earlier than their mothers' (Brooks-Gunn and Lewis 1979b). Shown pictures of their parents and strange adults, infants smiled more often and looked longer at their parents' pictures (Brooks-Gunn and Lewis 1981). And when infants were approached by strangers—5-year-old boys and girls, adult men and women, and a midget woman—the children discriminated among them on the basis of age and height, smiling and moving toward the children but frowning and moving away from the adults and, compared to the other adults, watching the midget more intently and averting their gaze less (Brooks and Lewis 1976).

15.4.2 *Reproductive Events*

15.4.2.1 Menstruation and Menarche

Brooks-Gunn's interest in the emergence of social cognition broadened to its role in the development of perceptions about reproductive events, at first menstruation and menarche. Her focus was on how social cognitions about menstruation and menarche emerge in adolescence and how males' and females' cognitions differ. Brooks-Gunn and Diane Ruble, then at Princeton University, began a research program on the salience and meaning of menarche and menstruation, especially in terms of definition of self and other in the context of these universal reproductive events

(Brooks-Gunn 1984, 1987; Brooks-Gunn and Ruble 1982a, b, 1983; Ruble and Brooks-Gunn 1979b). They found that menstruation was perceived as more physiologically and psychologically debilitating and more bothersome by men than by women (Ruble et al. 1982). In addition, their research debunked a number of myths about reproductive changes (Ruble and Brooks-Gunn 1979a), including the one that menarche is a normative crisis experienced very negatively by all girls. In fact, most girls reported mixed emotional reactions to menarche that were quite moderate. These reactions depended on the context the girls experienced: Those who were unprepared for menarche or reached it early reported more negative reactions as well as more symptoms (Ruble and Brooks-Gunn 1982).

15.4.2.2 Pubertal Processes

Brooks-Gunn's research further broadened to include pubertal processes. With Michelle Warren, a reproductive endocrinologist at Columbia University, she initiated a research program on pubertal processes and the transition from childhood to early adolescence. Brooks-Gunn and Warren conducted longitudinal studies of girls to chart their emotional experiences associated with pubertal changes and the socialization practices of families. The work included measurement of hormones to better understand pubertal changes and possible emotional reactions. The investigations followed girls who were likely to have delayed puberty because of exercise and food restriction (dancers training in national ballet companies as well as elite swimmers and gymnasts) and girls attending private schools—many of the girls were followed from middle school through college (Brooks-Gunn and Warren 1985, 1988a, b; Warren et al. 1986, 1991).

The private-school girls commonly compared their pubertal development and had no difficulty categorizing their classmates' development (Brooks-Gunn et al. 1986). Relatedly, the onset of breast development for these girls correlated positively with scores on measures of peer relationships, adjustment, and body image, but pubic hair was uncorrelated, suggesting that breast growth may be a visible sign of adulthood, conferring enhanced status (Brooks-Gunn and Warren 1988b).

The context in which the girls were situated influenced their reactions. In a context where delayed onset of puberty is valued (the dance world—most professional ballerinas are late maturers), dancers with delayed puberty had higher scores (relative to on-time dancers) on a body-image measure (Brooks-Gunn, Attie, Burrow, Rosso, & Warren, Brooks-Gunn et al. 1989; Brooks-Gunn and Warren 1985). (They also had lower scores on measures of psychopathology and bulimia; Brooks-Gunn and Warren 1985.) In contrast, in contexts where delayed onset is not valued (swimmers, private-school students/nonathletes), delayed and on-time girls did not differ in their body images (Brooks-Gunn, Attie et al., Brooks-Gunn et al. 1989; Brooks-Gunn and Warren 1985).

Two publications in this program, in particular, were very widely cited, according to the Social Science Citation Index: Attie and Brooks-Gunn (1989), on eating problems, and Brooks-Gunn et al. (1987), on measuring pubertal status, with 389 and 323 citations through 2015, respectively.

15.4.2.3 Adolescent Parenthood

Given Brooks-Gunn's research interest in menarche and other pubertal processes, it is not surprising that she moved on to research on pregnancy and parenthood, events that presage changes in self-definition as well as social comparisons with others. Brooks-Gunn joined Frank Furstenberg, a family sociologist at the University of Pennsylvania, in a 17-year follow-up of a group of teenage mothers who gave birth in Baltimore in the early 1970s (Furstenberg et al. 1987a, b, 1990). They charted the trajectories of these mothers as well as their children, who were about the age that their mothers had been when they gave birth to them. The interest was in both how well the mothers were doing and how the mothers' life course had influenced their children.

In brief, the teenage mothers differed widely in their life chances: About one third were on welfare and three quarters had jobs, usually full-time ones. Characteristics of the mothers' family of origin and of their own families (e.g., higher levels of education) and their attendance at a school for pregnant teenagers predicted the mothers' economic success.

The outcomes for their teenage children were "strikingly poor" (Brooks-Gunn 1996, p. 168). About one third were living with their biological father or stepfather. Half had repeated at least one grade in school, and most were sexually active. Maternal characteristics were linked to the children's behavior. Children of mothers who had not graduated from high school were 2.4 times as likely as other children, and children of unmarried mothers were 2.2 times as likely, to have repeated a grade. And children of unmarried mothers were 2.4 times as likely to have been stopped by the police, according to their mothers.

The Furstenberg et al. (1987b) monograph chronicling this study, *Adolescent Mothers in Later Life,* won the William J. Goode Book Award from the American Sociological Association's Sociology of the Family Section and is considered one of the classic longitudinal studies in developmental psychology.

Brooks-Gunn and Lindsay Chase-Lansdale, then at George Washington University, also began a study of low-income, Black multigenerational families (grandmother/grandmother figure–young mother–toddler) in Baltimore to investigate family relationships, via home visits (Chase-Lansdale et al. 1994). One issue was the parenting by the grandmother and mother, as observed separately in videotaped interactions of them aiding the child in working on a puzzle. The quality of parenting depended on whether they resided together and on the mother's age. Mothers' parenting was lower in quality when they lived with grandmothers. (Residing with the grandmothers and sharing child caring may be stressful for the mothers, interfering with their parenting.) Grandmothers' parenting was *higher* in quality when they lived with younger mothers than when they lived apart, but it was *lower* in quality when they lived with older mothers rather than apart. (Coresiding grandmothers may be more willing to help younger mothers, whom they view as needing assistance in parenting, than older mothers, whom they see as capable of parenting on their own.)

15.4.3 Perinatal Influences

Another line of research expanded beyond teenage parents to look at perinatal conditions, such as low birth weight and pregnancy behavior (e.g., smoking, no prenatal care), that influence parenting and children's development. Poor families and mothers with low education were often the focus of this research, given the differential rates of both early parenthood and adverse perinatal conditions as a function of social class.

In a joint venture between ETS, St. Luke's–Roosevelt Hospital, and Columbia University's College of Physicians and Surgeons, Brooks-Gunn studied low-birth-weight children and their parents, many from disadvantaged families because of the greater incidence of low-birth-weight children in these families. The work led to her thinking about how to ameliorate cognitive, emotional, and academic problems in these vulnerable children (Brooks-Gunn and Hearn 1982).

Brooks-Gunn joined Marie McCormick, a pediatrician then at the University of Pennsylvania, in a 9-year follow-up of low-birth-weight infants from previous multisite studies (Klebanov et al. 1994; McCormick et al. 1992). The focus was on very low birth weight infants, for more of them were surviving because of advances in neonatal intensive care.

At age 9, the low-birth-weight children did not differ from normal-birth-weight children on most aspects of classroom behavior, as reported by their teachers, but they had lower attention/ language skills and scholastic competence and higher daydreaming and hyperactivity; these differences were most pronounced for extremely low birth weight children. This pattern of differences resembles attention deficit disorder (Klebanov et al. 1994). The low-birth-weight children also had lower mean IQs and, at home, more behavioral problems, as reported by their mothers. The adverse health status of these children underscores the importance of efforts to reduce the incidence of premature births (McCormick et al. 1992).

15.4.4 Interventions With Vulnerable Children

15.4.4.1 Low-Birth-Weight Children

Brooks-Gunn and McCormick also collaborated on two other research programs involving interventions with biologically vulnerable children, the majority of whom were poor. One program focused on reducing the incidence of low-birth-weight deliveries by providing pregnant women with child-rearing and health information.

This program used a public health outreach model to locate pregnant women who were not enrolled in prenatal care; the intervention was located at Harlem Hospital. This effort was a logical extension of Brooks-Gunn's work on adolescent mothers in Baltimore (Brooks-Gunn et al. 1989; McCormick et al. 1987, 1989a, b).

The women in the program were very disadvantaged: One fifth were adolescents, three quarters were single, and half had not graduated from high school. The birth weight of their infants was unrelated to traditional risk factors: mother's demographic (e.g., education) and psychosocial characteristics (e.g., social support). This outcome suggests that low birth weight *in poor populations* is largely due to poverty per se. Birth weight *was* associated with the adequacy of prenatal care (Brooks-Gunn et al. 1988; McCormick et al. 1987).

The outreach program was extensive—four local people searching for eligible women over the course of a year, each making roughly 20 to 25 contacts daily—but recruited only 52 additional women, at a cost of about $850 each. The labor-intensive and expensive nature of this outreach effort indicates that more cost-effective alternatives are needed (Brooks-Gunn et al. 1988, 1989).

The other program involved the design and implementation of an early intervention for premature, low-birth-weight infants: enrollment in a child development education center and family support sessions. This program was initiated in eight sites and included almost 1000 children and their families; randomization was used to construct treatment and control groups. These children were followed through their 18th year of life, with the intervention from birth to 3 years of age being evaluated by Brooks-Gunn (Infant Health and Development Program 1990). The 3-year-olds in the intervention group had higher mean IQs and fewer maternally reported behavior problems, suggesting that early intervention may decrease low-birth-weight infants' risk of later developmental disability.

15.4.4.2 Head Start

Brooks-Gunn also carried out a notable evaluation of Head Start, based on data from an earlier longitudinal study conducted at ETS in the 1970s. The ETS–Head Start Longitudinal Study, directed by Shipman (1972, 1973), had canvassed poor school districts in three communities in an effort to identify and recruit for the study all children who were 3 ½ to 4 ½ years old, the Head Start population. The children were then assessed and information about their families was obtained. They were reassessed annually for the next 3 years. After the initial assessment, some children had entered Head Start, some had gone to other preschool programs, and some had not enrolled in any program. Clearly families chose whether to enroll their children in Head Start, some other program, or none at all (by processes that are difficult if not impossible to measure). But, by having the children's assessments and familial and demographic measures at age 3, it was possible to document and control statistically for initial differences among children and families in the three groups. Children's gains in ability in these groups could then be compared.

In several studies of two communities (Lee et al. 1988, 1990; Schnur et al. 1992), Brooks-Gunn and her collaborators investigated differences in the children's gains in the Head Start and other groups as well as preexisting group differences in the children's demographic and cognitive characteristics. Black children enrolled in Head Start made greater gains on a variety of cognitive tests than their Black peers in the other groups by the end of the program (Lee et al. 1988) and diminished gains after 2 years (Lee et al. 1990). (The gains for the small samples of White children did not differ between the Head Start and other groups in the initial study; these children were not included in the follow-up study.) These findings imply that Head Start may have some efficacy in improving participants' intellectual status. The Head Start children were the most disadvantaged (Schnur et al. 1992), seemingly allaying concerns that Head Start does not take the neediest children (Datta 1979).

15.5 Conclusion

As this review documents, ETS was a major center for basic and applied research in developmental psychology for decades. The number and quality of investigators (and their prodigious output) made for a developmental psychology program that rivaled the best in the academic community.

The research was wide ranging and influential, spanning the cognitive, personality, and social development of infants, children, and adolescents, with an emphasis on minority, working-class, and disabled individuals; addressing key theoretical, substantive, and methodological issues; using research methods that ran the gamut: laboratory and field experiments, correlational studies, surveys, and structured observations; and impacting theory, research, and practice across developmental psychology.

In common with ETS's research in cognitive, personality, and social psychology (Stricker, Chap. 13, and Kogan, Chap. 14, this volume), this achievement was probably attributable to the confluence of ample institutional and financial support, doubtless due to the vision of Chauncey, who saw the value of a broad program of psychological research.

Acknowledgments Thanks are due to Nick Telepak for retrieving publications and to Isaac Bejar, Randy Bennett, and Ann Renninger for reviewing a draft of this chapter.

References

Ajzen, I., & Fishbein, M. (1977). Attitude–behavior relations: A theoretical analysis and review of empirical research. *Psychological Bulletin, 84*, 888–918. https://doi.org/10.1037/0033-2909.84.5.888

American Educational Research Association, American Psychological Association, & National Council on Measurement in Education. (1985). *Standards for educational and psychological testing*. Washington, DC: American Psychological Association.

Attie, S., & Brooks-Gunn, J. (1989). Development of eating problems in adolescent girls: A longitudinal study. *Developmental Psychology, 25*, 70–79. https://doi.org/10.1037/0012-1649.25.1.70

Bayley, N. (1969). *Manual for the Bayley Scales of Infant Development*. New York: Psychological Corporation.

Beck, A. T. (1967). *Depression: Clinical, experimental, and theoretical aspects*. New York: Harper & Row.

Bradley, R. H., Caldwell, B. M., & Elardo, R. (1977). Home environment, social status, and mental test performance. *Journal of Educational Psychology, 69*, 697–701. https://doi.org/10.1037/0022-0663.69.6.697

Bradley, R. H., Caldwell, B. M., Rock, S. L., Ramey, C. T., Barnard, K. E., Gray, C., et al. (1989). Home environment and cognitive development in the first 3 years of life: A collaborative study involving six sites and three ethnic groups in North America. *Developmental Psychology, 25*, 217–235. https://doi.org/10.1037/0012-1649.25.2.217

Brinker, R. P., & Lewis, M. (1982a). Discovering the competent handicapped infant: A process approach to assessment and intervention. *Topics in Early Childhood Special Education, 2*(2), 1–16. https://doi.org/10.1177/027112148200200205

Brinker, R. P., & Lewis, M. (1982b). Making the world work with microcomputers: A learning prosthesis for handicapped infants. *Exceptional Children, 49*, 163–170. https://doi.org/10.1177/001440298204900210

Brody, G. H., Pellegrini, A. D., & Sigel, I. E. (1986). Marital quality and mother–child and father–child interactions with school-aged children. *Developmental Psychology, 22*, 291–296. https://doi.org/10.1037/0012-1649.22.3.291

Brooks, J., & Lewis, M. (1976). Infants' responses to strangers: Midget, adult, and child. *Child Development, 47*, 323–332. https://doi.org/10.2307/1128785

Brooks-Gunn, J. (1984). The psychological significance of different pubertal events in young girls. *Journal of Early Adolescence, 4*, 315–327. https://doi.org/10.1177/0272431684044003

Brooks-Gunn, J. (1987). Pubertal processes and girls' psychological adaptation. In R. Lerner & T. T. Foch (Eds.), *Biological–psychosocial interactions in early adolescence* (pp. 123–153). Hillsdale: Erlbaum.

Brooks-Gunn, J. (1996). Unexpected opportunities: Confessions of an eclectic developmentalist. In M. R. Merrens & G. C. Brannigan (Eds.), *The developmental psychologists—Research adventures across the life span* (pp. 153–171). New York: McGraw-Hill.

Brooks-Gunn, J. (2013). Person, time, and place—The life course of a developmental psychologist. In R. M. Lerner, A. C. Petersen, R. K. Silbereisen, & J. Brooks-Gunn (Eds.), *The developmental science of adolescence—History through autobiography* (pp. 32–44). New York: Psychology Press.

Brooks-Gunn, J., Attie, I., Burrow, C., Rosso, J. T., & Warren, M. P. (1989). The impact of puberty on body and eating concerns in athletic and nonathletics contexts. *Journal of Early Adolescence, 9*, 269–290. https://doi.org/10.1177/0272431689093006

Brooks-Gunn, J., & Hearn, R. P. (1982). Early intervention and developmental dysfunction: Implications for pediatrics. *Advances in Pediatrics, 29*, 497–527.

Brooks-Gunn, J., & Lewis, M. (1979a). The effects of age and sex on infants' playroom behavior. *Journal of Genetic Psychology, 134*, 99–105. https://doi.org/10.1080/00221325.1979.10533403

Brooks-Gunn, J., & Lewis, M. (1979b). "Why mama and papa?" The development of social labels. *Child Development, 50*, 1203–1206. https://doi.org/10.2307/1129349

Brooks-Gunn, J., & Lewis, M. (1981). Infant social perceptions: Responses to pictures of parents and strangers. *Developmental Psychology, 17*, 647–649. https://doi.org/10.1037/0012-1649.17.5.647

Brooks-Gunn, J., & Lewis, M. (1982a). Affective exchanges between normal and handicapped infants and their mothers. In T. Field & A. Fogel (Eds.), *Emotion and early interaction* (pp. 161–188). Hillsdale: Erlbaum.

Brooks-Gunn, J., & Lewis, M. (1982b). Development of play behavior in handicapped and normal infants. *Topics in Early Childhood Special Education, 2*(3), 14–27. https://doi.org/10.1177/027112148200200306

Brooks-Gunn, J., & Lewis, M. (1982c). Temperament and affective interaction in handicapped infants. *Journal of Early Intervention, 5*, 31–41. https://doi.org/10.1177/105381518200500105

Brooks-Gunn, J., McCormick, M. C., Gunn, R. W., Shorter, T., Wallace, C. Y., & Heagarty, M. C. (1989). Outreach as case finding—The process of locating low-income pregnant women. *Medical Care, 27,* 95–102. https://doi.org/10.1097/00005650-198902000-00001

Brooks-Gunn, J., & Ruble, D. N. (1982a). Developmental processes in the experience of menarche. In A. Baum & J. E. Singer (Eds.), *Hanbook of psychology and health: Vol. 2. Issues in child health and adolescent health* (pp. 117–147). Hillsdale: Erlbaum.

Brooks-Gunn, J., & Ruble, D. N. (1982b). The development of menstrual beliefs and behavior during early adolescence. *Child Development, 53*, 1567–1577. https://doi.org/10.2307/1130085

Brooks-Gunn, J., & Ruble, D. N. (1983). The experience of menarche from a developmental perspective. In J. Brooks-Gunn & A. C. Petersen (Eds.), *Girls at puberty—Biological and psychosocial perspectives* (pp. 155–177). New York: Plenum Press. https://doi.org/10.1007/978-1-4899-0354-9_8

Brooks-Gunn, J., & Warren, M. P. (1985). The effects of delayed menarche in different contexts: Dance and nondance students. *Journal of Youth and Adolescence, 141*, 285–300. https://doi.org/10.1007/BF02089235

Brooks-Gunn, J., & Warren, M. P. (1988a). Mother–daughter differences in menarcheal age in adolescent girls attending national dance company schools and non-dancers. *Annals of Human Biology, 15*, 35–43. https://doi.org/10.1080/03014468800009441

Brooks-Gunn, J., & Warren, M. P. (1988b). The psychological significance of secondary sexual characteristics in nine- to eleven-year-old girls. *Child Development, 59*, 1061–1069. https://doi.org/10.2307/1130272

Brooks-Gunn, J., Warren, M. P., Samelson, M., & Fox, R. (1986). Physical similarity to and disclosure of menarcheal status to friends: Effects of grade and pubertal status. *Journal of Early Adolescence, 6*, 3–34. https://doi.org/10.1177/0272431686061001

Brooks-Gunn, J., Warren, M. P., Rosso, J., & Gargiuolo, J. (1987). Validity of self-report measures of girls' pubertal status. *Child Development, 58*, 829–841. https://doi.org/10.2307/1130220

Brooks-Gunn, J., McCormick, M. C., & Heagarty, M. C. (1988). Preventing infant mortality and morbidity: Developmental perspectives. *American Journal of Orthopsychiatry, 58*, 288–296. https://doi.org/10.1111/j.1939-0025.1988.tb01590.x

Caldwell, B. M. (1970). *Handbook, Preschool Inventory* (Rev. ed.). Princeton: Educational Testing Service.

Caldwell, B., & Bradley, R. (1985). *Home observation for measurement of the environment*. Homewood: Dorsey.

Chase-Lansdale, P. L., Brooks-Gunn, J., & Zamsky, E. S. (1994). Young African-American multigenerational families in poverty: Quality of mothering and grandmothering. *Child Development, 65*, 373–393. https://doi.org/10.2307/1131390

Clarke-Stewart, A. (1977). *Child care in the family—A review of research and some propositions for policy*. New York: Academic Press.

Cole, N. S., & Moss, P. A. (1989). Bias in test use. In R. L. Linn (Ed.), *Educational measurement* (3rd ed., pp. 201–219). New York: Macmillan.

Coleman, J. S., Campbell, E. Q., Hobson, C. J., McPartland, J., Mood, A. M., Weinfeld, F. D., & York, R. L. (1966). *Equality of educational opportunity*. Washington, DC: U.S. Department of Health, Education, and Welfare, Office of Education.

Copple, C., Sigel, I. E., & Saunders, R. A. (1984). *Educating the young thinker—Classroom strategies for cognitive growth*. Hillsdale: Erlbaum.

Datta, L.-E. (1979). Another spring and other hopes: Some findings from national evaluations of Project Head Start. In E. Zigler & J. Valentine (Eds.), *Project Head Start—A legacy of the war on poverty* (pp. 405–432). New York: Free Press.

Dodd, C., & Lewis, M. (1969). The magnitude of the orienting response in children as a function of changes in color and contour. *Journal of Experimental Child Psychology, 8*, 296–305. https://doi.org/10.1016/0022-0965(69)90104-0

Emmerich, W. (1968). Personality development and the concept of structure. *Child Development, 39*, 671–690. https://doi.org/10.2307/1126978

Emmerich, W. (1982). Development of sex-differentiated preferences during late childhood and adolescence. *Developmental Psychology, 18*, 406–417. https://doi.org/10.1037/0012-1649.18.3.406

Feinman, S., & Lewis, M. (1983). Social referencing at ten months: A second-order effect on infants' responses to strangers. *Child Development, 50*, 848–853. https://doi.org/10.2307/1129892

Feiring, C., & Lewis, M. (1978). The child as a member of the family system. *Behavioral Science, 23*, 225–233. https://doi.org/10.1002/bs.3830230311

Feiring, C., & Lewis, M. (1979). Sex and age differences in young children's reactions to frustration: A further look at the Goldberg and Lewis subjects. *Child Development, 50*, 848–853. https://doi.org/10.2307/1128953

Fox, N., & Lewis, M. (1982a). Motor asymmetries in preterm infants: Effects of prematurity and illness. *Developmental Psychobiology, 15*, 19–23. https://doi.org/10.1002/dev.420150105

Fox, N., & Lewis, M. (1982b). Prematurity, illness, and experience as factors in preterm development. *Journal of Early Intervention, 6*, 60–71. https://doi.org/10.1177/105381518200600108

Freedle, R., & Lewis, M. (1970). On relating an infant's observation time of visual stimuli with choice-theory analysis. *Developmental Psychology, 2*, 129–133. https://doi.org/10.1037/h0028601

Freedle, R., & Lewis, M. (1977). Prelinguistic conversations. In M. Lewis & L. A. Rosenblum (Eds.), *The origins of behavior: Vol. 5. Interaction, conversation, and the development of language* (pp. 157–185). New York: Wiley.

Furstenberg Jr., F. F., Brooks-Gunn, J., & Morgan, S. P. (1987a). Adolescent mothers and their children in later life. *Family Planning Perspectives, 19*, 142–151. https://doi.org/10.2307/2135159

Furstenberg Jr., F. F., Brooks-Gunn, J., & Morgan, S. P. (1987b). *Adolescent mothers in later life*. New York: Cambridge University Press.

Furstenberg Jr., F. F., Levine, J. A., & Brooks-Gunn, J. (1990). The children of teenage mothers: Patterns of early childbearing in two generations. *Family Planning Perspectives, 22*, 54–61. https://doi.org/10.2307/2135509

Goldberg, S., & Lewis, M. (1969). Play behavior in the year-old infant: Early sex differences. *Child Development, 40*, 21–31. https://doi.org/10.2307/1127152

Goldman, K. S., & Lewis, M. (1976). *Child care and public policy: A case study*. Princeton: Educational Testing Service.

Hale, G. A., & Alderman, L. B. (1978). Children's selective attention and variation in amount of stimulus exposure. *Journal of Experimental Child Psychology, 25*, 320–327. https://doi.org/10.1016/0022-0965(78)90011-5

Hale, G. A., & Lewis, M. (Eds.). (1979). *Attention and cognitive development*. New York: Plenum Press. https://doi.org/10.1007/978-1-4613-2985-5

Harris, D. B. (1963). *Children's drawings as measures of intellectual maturity*. New York: Harcourt, Brace & World.

Herrnstein, R. J., & Murray, C. (1994). *The bell curve—Intelligence and class structure in American life*. New York: Free Press.

Hyson, M., Copple, C., & Jones, J. (2006). Early childhood development and education. In W. Damon & R. M. Lerner (Series Eds.), K. A. Renninger & I. E. Sigel (Vol. Eds.), *Handbook of child psychology: Vol. 4. Child psychology in practice* (6th ed., pp. 3–47). New York: Wiley.

Infant Health and Development Program. (1990). Enhancing the outcomes of low-birth-weight, premature infants: A multisite, randomized trial. *Journal of the American Medical Association, 263*, 3035–3042. https://doi.org/10.1001/jama.1990.03440220059030

Jensen, A. R. (1969). How much can we boost IQ and scholastic achievement? *Harvard Educational Review, 39*, 1–123. https://doi.org/10.17763/haer.39.1.l3u15956627424k7

Kagan, J., & Lewis, M. (1965). Studies of attention in the human infant. *Merrill-Palmer Quarterly, 11*, 95–127.

Klebanov, P. K., Brooks-Gunn, J., & McCormick, M. C. (1994). Classroom behavior of very low birth weight elementary school children. *Pediatrics, 94*, 700–708.

Kogan, N. (1976). *Cognitive styles in infancy and early childhood*. Hillsdale: Erlbaum.

Laosa, L. M. (1977). Nonbiased assessment of children's abilities: Historical antecedents and current issues. In T. Oakland (Ed.), *Psychological and educational assessment of minority children* (pp. 1–20). New York: Bruner/Mazel.

Laosa, L. M. (1978). Maternal teaching strategies in Chicano families of varied educational and socioeconomic levels. *Child Development, 49*, 1129–1135. https://doi.org/10.2307/1128752

Laosa, L. M. (1980a). Maternal teaching strategies and cognitive styles in Chicano families. *Journal of Educational Psychology, 72*, 45–54. https://doi.org/10.1037/0022-0663.72.1.45

Laosa, L. M. (1980b). Maternal teaching strategies in Chicano and Anglo-American families: The influence of culture and education on maternal behavior. *Child Development, 51*, 759–765. https://doi.org/10.2307/1129462

Laosa, L. M. (1982a). Families as facilitators of children's intellectual development at 3 years of age—A causal analysis. In L. M. Laosa & I. E. Sigel (Eds.), *Families as learning environments for children* (pp. 1–45). New York: Plenum Press. https://doi.org/10.1007/978-1-4684-4172-7_1

Laosa, L. M. (1982b). School, occupation, culture, and family: The impact of parental schooling on the parent–child relationship. *Journal of Educational Psychology, 74*, 791–827. https://doi.org/10.1037/0022-0663.74.6.791

Laosa, L. M. (1982c). The sociocultural context of evaluation. In B. Spodek (Ed.), *Handbook of research in early childhood education* (pp. 501–520). New York: Free Press.

Laosa, L. M. (1984). Ethnic, socioeconomic, and home language influences upon early performance on measures of abilities. *Journal of Educational Psychology, 76*, 1178–1198. https://doi.org/10.1037/0022-0663.76.6.1178

Laosa, L. M. (1990). Psychosocial stress, coping, and development of Hispanic immigrant children. In F. C. Serafica, A. I. Schwebel, R. K. Russell, P. D. Isaac, & L. B. Myers (Eds.), *Mental health of ethnic minorities* (pp. 38–65). New York: Praeger.

Laosa, L. M. (1991). The cultural context of construct validity and the ethics of generalizability. *Early Childhood Research Quarterly, 6*, 313–321. https://doi.org/10.1016/S0885-2006(05)80058-4

Laosa, L. M. (1996). Intelligence testing and social policy. *Journal of Applied Developmental Psychology, 17*, 155–173. https://doi.org/10.1016/S0193-3973(96)90023-4

Laosa, L. M. (1997). Research perspectives on constructs of change: Intercultural migration and developmental transitions. In A. Booth, A. C. Crouter, & N. Landale (Eds.), *Immigration and the family—Research and policy on U.S. Immigrants* (pp. 133–148). Mahwah: Erlbaum.

Laosa, L. M. (1999). Intercultural transitions in human development and education. *Journal of Applied Developmental Psychology, 20*, 355–406. https://doi.org/10.1016/S0193-3973(99)00023-4

Laosa, L. M. (2001). School segregation of children who migrate to the United States from Puerto Rico. *Education Policy Analysis Archives, 9*(1), 1–49. https://doi.org/10.14507/epaa.v9n1.2001

Lee, V. E., Brooks-Gunn, J., & Schnur, E. (1988). Does Head Start work? A 1-year follow-up comparison of disadvantaged children attending Head Start, no preschool, and other preschool programs. *Developmental Psychology, 24*, 210–222. https://doi.org/10.1037/0012-1649.24.2.210

Lee, V. E., Brooks-Gunn, J., Schnur, E., & Liaw, F.-R. (1990). Are Head Start effects sustained? A longitudinal follow-up comparison of disadvantaged children attending Head Start, no preschool, and other preschool programs. *Child Development, 61*, 495–507. https://doi.org/10.2307/1131110

Lewis, M. (1969). Infants' responses to facial stimuli during the first year of life. *Developmental Psychology, 1*, 75–86. https://doi.org/10.1037/h0026995

Lewis, M. (1971a). A developmental study of the cardiac response to stimulus onset and offset during the first year of life. *Psychophysiology, 8*, 689–698. https://doi.org/10.1111/j.1469-8986.1971.tb00506.x

Lewis, M. (1971b). Individual differences in the measurement of early cognitive growth. In J. Hellmuth (Ed.), *Studies in abnormalities: Vol. 2. Exceptional infant* (pp. 172–210). New York: Brunner/Mazel.

Lewis, M. (1971c). Infancy and early childhood in the urban environment: Problems for the 21st century. In European Cultural Foundation (Ed.), *Citizen and city in the year 2000* (pp. 167–172). Deventer: Kluwer.

Lewis, M. (1973). Infant intelligence tests: Their use and misuse. *Human Development, 16*, 108–118. https://doi.org/10.1159/000271270

Lewis, M. (1974). The cardiac response during infancy. In R. F. Thompson (Series Ed.), R. F. Thompson & M. M. Patterson (Vol. Eds.), *Methods in physiological psychology: Vol. 1, Part C. Bioelectric recording techniques: Receptor and effector processes* (pp. 201–229). New York: Academic Press.

Lewis, M. (1975a). Early sex differences in the human: Studies of socioemotional development. *Archives of Sexual Behavior, 4*, 329–335. https://doi.org/10.1007/BF01541719

Lewis, M. (1975b). The development of attention and perception in the infant and young child. In W. M. Cruickshank & D. P. Hallahan (Eds.), *Perceptual and learning disabilities in children* (Vol. 2, pp. 137–162). Syracuse: Syracuse University Press.

Lewis, M. (1976a). A theory of conversation. *Quarterly Newsletter of the Institute for Comparative Human Development, 1*(1), 5–7.

Lewis, M. (1976b). What do we mean when we say "infant intelligence scores"? A sociopolitical question. In M. Lewis (Ed.), *Origins of intelligence: Infancy and early childhood* (pp. 1–18). New York: Plenum Press. https://doi.org/10.1007/978-1-4684-6961-5_1

Lewis, M. (1977a). A new response to stimuli. *The Sciences, 17*(3), 18–19. https://doi.org/10.1002/j.2326-1951.1977.tb01524.x

Lewis, M. (1977b). Early socioemotional development and its relevance to curriculum. *Merrill-Palmer Quarterly, 23*, 279–286.

Lewis, M. (1978a). Attention and verbal labeling behavior in preschool children: A study in the measurement of internal representations. *Journal of Genetic Psychology, 133*, 191–202. https://doi.org/10.1080/00221325.1978.10533377

Lewis, M. (1978b). Situational analysis and the study of behavioral development. In L. A. Pervin & M. Lewis (Eds.), *Perspectives in interactional psychology* (pp. 49–66). New York: Plenum Press. https://doi.org/10.1007/978-1-4613-3997-7_3

Lewis, M. (1980). Developmental theories. In I. L. Kutash & L. B. Schlesinger (Eds.), *Handbook on stress and anxiety—Contemporary knowledge, theory, and treatment* (pp. 48–62). San Francisco: Jossey-Bass.

Lewis, M. (1981a). Attention as a measure of cognitive integrity. In M. Lewis & L. T. Taft (Eds.), *Developmental disabilities—Theory, assessment, and intervention* (pp. 185–212). Jamaica: Spectrum. https://doi.org/10.1007/978-94-011-6314-9_13

Lewis, M. (1981b). Self-knowledge: A social-cognitive perspective on gender identity and sex role development. In M. E. Lamb & L. R. Sherrod (Eds.), *Infant social cognition: Empirical and theoretical considerations* (pp. 395–414). Hillsdale: Erlbaum.

Lewis, M. (1982a). Play as whimsy. *Behavioral and Brain Sciences, 5*, 166. https://doi.org/10.1017/S0140525X00011067

Lewis, M. (1982b). The social network systems model—Toward a theory of social development. In T. M. Field, A. Huston, H. C. Quary, L. Troll, & G. E. Finley (Eds.), *Review of human development* (pp. 180–214). New York: Wiley.

Lewis, M. (n.d.). *My life at ETS.* Unpublished manuscript.

Lewis, M., & Baldini, N. (1979). Attentional processes and individual differences. In G. A. Hale & M. Lewis (Eds.), *Attention and cognitive development* (pp. 137–172). New York: Plenum Press. https://doi.org/10.1007/978-1-4613-2985-5_7

Lewis, M., & Ban, P. (1977). Variance and invariance in the mother–infant interaction: A cross-cultural study. In P. H. Leiderman, S. R. Tulkin, & A. Rosenfeld (Eds.), *Culture and infancy—Variations in the human experience* (pp. 329–355). New York: Academic Press.

Lewis, M., & Baumel, M. H. (1970). A study in the ordering of attention. *Perceptual and Motor Skills, 31*, 979–990. https://doi.org/10.2466/pms.1970.31.3.979

Lewis, M., & Brooks, J. (1974). Self, others and fear: Infants' reactions to people. In M. Lewis & L. A. Rosenblum (Eds.), *The origins of behavior: Vol. 2. The origins of fear* (pp. 195–227). New York: Wiley.

Lewis, M., & Brooks, J. (1975). Infants' social perception: A constructivist view. In L. B. Cohen & P. Salapatek (Eds.), *Perception of space, speech, and sound* (Infant perception: From sensation to cognition, Vol. 2, pp. 101–148). New York: Academic Press.

Lewis, M., & Brooks-Gunn, J. (1979a). *Social cognition and the acquisition of self.* New York: Plenum Press. https://doi.org/10.1007/978-1-4684-3566-5

Lewis, M., & Brooks-Gunn, J. (1979b). The search for the origins of self: Implications for social behavior and intervention. In L. Montada (Ed.), *Brennpunkte der entwicklungspsychologie [Foci of developmental psychology]* (pp. 157–172). Stuttgart: Kohlhammer.

Lewis, M., & Brooks-Gunn, J. (1979c). Toward a theory of social cognition: The development of self. *New Directions for Child Development, 4*, 1–20. https://doi.org/10.1002/cd.23219790403

Lewis, M., & Brooks-Gunn, J. (1981a). Attention and intelligence. *Intelligence, 5*, 231–238. https://doi.org/10.1016/S0160-2896(81)80010-4

Lewis, M., & Brooks-Gunn, J. (1981b). The self as social knowledge. In M. D. Lynch, A. A. Norem-Hebeisen, & K. J. Gergen (Eds.), *Self-concept: Advances in theory and research* (pp. 101–118). Cambridge, MA: Ballinger.

Lewis, M., & Brooks-Gunn, J. (1981c). Visual attention at three months as a predictor of cognitive functioning at two years of age. *Intelligence, 5*, 131–140. https://doi.org/10.1016/0160-2896(81)90003-9

Lewis, M., & Cherry, L. (1977). Social behavior and language acquisition. In M. Lewis & L. A. Rosenblum (Eds.), *The origins of behavior: Vol. 5. Interaction, conversation, and the development of language* (pp. 227–245). New York: Wiley.

Lewis, M., & Feiring, C. (1979). The child's social network: Social object, social functions, and their relationship. In M. Lewis & L. A. Rosenblum (Eds.), *Genesis of behavior: Vol. 2. The child and its family* (pp. 9–27). New York: Plenum Press.

Lewis, M., & Feiring, C. (1981). Direct and indirect interactions in social relationships. *Advances in Infancy Research, 1*, 129–161.

Lewis, M., & Feiring, C. (1982). Some American families at dinner. In L. Laosa & I. Sigel (Eds.), *Families as learning environments for children* (pp. 115–145). New York: Plenum Press. https://doi.org/10.1007/978-1-4684-4172-7_4

Lewis, M., & Fox, N. (1980). Predicting cognitive development from assessments in infancy. *Advances in Behavioral Pediatrics, 1*, 53–67.

Lewis, M., & Freedle, R. (1977). The mother and infant communication system: The effects of poverty. In H. McGurk (Ed.), *Ecological factors in human development* (pp. 205–215). Amsterdam: North-Holland.

Lewis, M., Goldberg, S., & Campbell, H. (1969a). A developmental study of information processing within the first three years of life: Response decrement to a redundant signal. *Monographs of the Society for Research in Child Development, 34*(9, Serial No. 133). https://doi.org/10.2307/1165696

Lewis, M., & McGurk, H. (1973). Evaluation of infant intelligence: Infant intelligence scores—True or false? *Science, 178*(4066), 1174–1177. https://doi.org/10.1126/science.178.4066.1174

Lewis, M., & Michalson, L. (1982a). The measurement of emotional state. In C. E. Izard & P. B. Read (Eds.), *Measuring emotions in infants and children* (Vol. 1, pp. 178–207). New York: Cambridge University Press.

Lewis, M., & Michalson, L. (1982b). The socialization of emotions. In T. Field & A. Fogel (Eds.), *Emotion and early interaction* (pp. 189–212). Hillsdale: Erlbaum.

Lewis, M., & Michalson, L. (1983). *Children's emotions and moods: Developmental theory and measurement.* New York: Plenum Press. https://doi.org/10.1007/978-1-4613-3620-4

Lewis, M., & Rosenblum, L. A. (Eds.). (1974a). *The origins of behavior: Vol. 1. The effect of the infant on its caregiver.* New York: Wiley.

Lewis, M., & Rosenblum, L. A. (Eds.). (1974b). *The origins of behavior: Vol. 2. The origins of fear*. New York: Wiley.
Lewis, M., & Rosenblum, L. A. (Eds.). (1975). *The origins of behavior: Vol. 4. Friendship and peer relations*. New York: Wiley.
Lewis, M., & Rosenblum, L. A. (Eds.). (1977). *The origins of behavior: Vol. 5. Interaction, conversation, and the development of language*. New York: Wiley.
Lewis, M., & Rosenblum, L. A. (Eds.). (1978a). *Genesis of behavior: Vol. 1. The development of affect*. New York: Plenum Press.
Lewis, M., & Rosenblum, L. A. (1978b). Introduction: Issues in affect development. In M. Lewis & L. A. Rosenblum (Eds.), *Genesis of behavior: Vol. 1. The development of affect* (pp. 1–10). New York: Plenum Press.
Lewis, M., & Rosenblum, L. A. (Eds.). (1979). *Genesis of behavior: Vol. 2. The child and its family*. New York: Plenum Press.
Lewis, M., & Rosenblum, L. A. (Eds.). (1981). *Genesis of behavior: Vol. 3. The uncommon child*. New York: Plenum Press.
Lewis, M., & Schaeffer, S. (1981). Peer behavior and mother–infant interaction in maltreated children. In M. Lewis & L. A. Rosenblum (Eds.), *Genesis of behavior: Vol. 3. The uncommon child* (pp. 193–223). New York: Plenum Press. https://doi.org/10.1007/978-1-4684-3773-7_9
Lewis, M., & Spaulding, S. J. (1967). Differential cardiac response to visual and auditory stimulation in the young child. *Psychophysiology, 3*, 229–237. https://doi.org/10.1111/j.1469-8986.1967.tb02700.x
Lewis, M., & Starr, M. D. (1979). Developmental continuity. In J. D. Osofsky (Ed.), *Handbook of infant development* (pp. 653–670). New York: Wiley.
Lewis, M., & Taft, L. T. (Eds.). (1982). *Developmental disabilities: Theory, assessment, and intervention*. Jamaica: Spectrum. https://doi.org/10.1007/978-94-011-6314-9
Lewis, M., & Wehren, A. (1982). The central tendency in study of the handicapped child. In D. Bricker (Ed.), *Intervention with at-risk and handicapped infants – From research to application* (pp. 77–89). Baltimore: University Park Press.
Lewis, M., & Weinraub, M. (1974). Sex of parent × sex of child: Socioemotional development. In R. C. Friedman, R. M. Richart, & R. L. Vande Wiele (Eds.), *Sex differences in behavior* (pp. 165–189). New York: Wiley.
Lewis, M., & Wilson, C. D. (1970). The cardiac response to a perceptual cognitive task in the young child. *Psychophysiology, 6*, 411–420. https://doi.org/10.1111/j.1469-8986.1970.tb01751.x
Lewis, M., & Zarin-Ackerman, J. (1977). Early infant development. In R. E. Behrman, J. M. Driscoll Jr., & A. E. Seeds (Eds.), *Neonatal-perinatal medicine—Diseases of the fetus and infant* (2nd ed., pp. 195–208). St. Louis: Mosby.
Lewis, M., Kagan, J., Campbell, H., & Kalafat, J. (1966a). The cardiac response as a correlate of attention in infants. *Child Development, 37*, 63–71. https://doi.org/10.2307/1126429
Lewis, M., Kagan, J., & Kalafat, J. (1966b). Patterns of fixation in the young infant. *Child Development, 37*, 331–341. https://doi.org/10.2307/1126807
Lewis, M., Bartels, B., Campbell, H., & Goldberg, S. (1967a). Individual differences in attention – The relation between infants' condition at birth and attention distribution within the first year. *American Journal of Diseases of Children, 113*, 461–465. https://doi.org/10.1001/archpedi.1967.02090190107010
Lewis, M., Goldberg, S., & Rausch, M. (1967b). Attention distribution as a function of novelty and familiarity. *Psychonomic Science, 7*, 227–228. https://doi.org/10.3758/BF03328553
Lewis, M., Dodd, C., & Harwitz, M. (1969b). Cardiac responsivity to tactile stimulation in waking and sleeping infants. *Perceptual and Motor Skills, 29*, 259–269. https://doi.org/10.2466/pms.1969.29.1.259
Lewis, M., Wilson, C. D., Ban, P., & Baumel, M. H. (1970). An exploratory study of resting cardiac rate and variability from the last trimester of prenatal life through the first year of postnatal life. *Child Development, 41*, 799–811. https://doi.org/10.2307/1127225
Lewis, M., Wilson, C. D., & Baumel, M. (1971). Attention distribution in the 24-month-old child: Variations in complexity and incongruity of the human form. *Child Development, 42*, 429–438.

Lewis, M., Young, G., Brooks, J., & Michalson, L. (1975). The beginning of friendship. In M. Lewis & L. Rosenblum (Eds.), *The origins of behavior: Vol. 4. Friendship and peer relations* (pp. 27–66). New York: Wiley.

Lewis, M., Brooks, J., & Haviland, J. (1978). Hearts and faces: A study in the measurement of emotion. In M. Lewis & L. A. Rosenblum (Eds.), *Genesis of behavior: Vol. 1. The development of affect* (pp. 77–123). New York: Plenum Press. https://doi.org/10.1007/978-1-4684-2616-8_4

Lewis, M., Brooks-Gunn, J., & Jaskir, J. (1985). Individual differences in visual self-recognition as a function of mother–infant attachment relationship. *Developmental Psychology, 21*, 1181–1187. https://doi.org/10.1037/0012-1649.21.6.1181

Linn, R. L. (1989). Current perspectives and future directions. In R. L. Linn (Ed.), *Educational measurement* (3rd ed., pp. 1–10). New York: Macmillan.

McCarthy, D. (1972). *Manual for the McCarthy Scales of Children's Abilities*. New York: Psychological Corporation.

McCormick, M. C., Brooks-Gunn, J., Shorter, T., Wallace, C. Y., Holmes, J. H., & Heagarty, M. C. (1987). The planning of pregnancy among low-income women in central Harlem. *American Journal of Obstetrics and Gynecology, 156*, 145–149. https://doi.org/10.1016/0002-9378(87)90226-2

McCormick, M. C., Brooks-Gunn, J., Shorter, T., Holmes, J. H., & Heagarty, M. C. (1989a). Factors associated with maternal rating of infant health in central Harlem. *Journal of Developmental & Behavioral Pediatrics, 10*, 139–144. https://doi.org/10.1097/00004703-198906000-00004

McCormick, M. C., Brooks-Gunn, J., Shorter, T., Holmes, J. H., Wallace, C. Y., & Heagarty, M. C. (1989b). Outreach as case finding: Its effect on enrollment in prenatal care. *Medical Care, 27*, 103–111. https://doi.org/10.1097/00005650-198902000-00002

McCormick, M. C., Brooks-Gunn, J., Workman-Daniels, K., Turner, J., & Peckham, G. J. (1992). The health and developmental status of very low-birth-weight children at school age. *JAMA, 267*, 2204–2208. https://doi.org/10.1001/jama.1992.03480160062035

McGillicuddy-DeLisi, A. V. (1982). The relationship between parents' beliefs about development and family constellation, socioeconomic status, and parents' teaching strategies. In L. M. Laosa & I. E. Sigel (Eds.), *Families as learning environments for children* (pp. 261–299). New York: Plenum Press. https://doi.org/10.1007/978-1-4684-4172-7_9

McGillicuddy-DeLisi, A. V., Sigel, I. E., & Johnson, J. E. (1979). The family as a system of mutual influences: Parental beliefs, distancing behaviors, and children's representational thinking. In M. Lewis & L. A. Rosenblum (Eds.), *Genesis of behavior: Vol. 2. The child and its family* (pp. 91–106). New York: Plenum Press.

McGurk, H., & Lewis, M. (1974). Space perception in early infancy: Perception within a common auditory-visual space? *Science, 186*(4164), 649–650. https://doi.org/10.1126/science.186.4164.649

Messick, S. (1989). Validity. In R. L. Linn (Ed.), *Educational measurement* (3rd ed., pp. 13–103). New York: Macmillan.

Oakland, T., & Laosa, L. M. (1977). Professional, legislative, and judicial influences on psychoeducational assessment practices in schools. In T. Oakland (Ed.), *Psychological and educational assessment of minority children* (pp. 21–51). New York: Bruner/Mazel.

Pellegrini, A. D., Brody, G. H., & Sigel, I. E. (1985). Parents' book-reading habits with their children. *Journal of Educational Psychology, 77*, 332–340. https://doi.org/10.1037/0022-0663.77.3.332

Pellegrini, A. D., Brody, G. H., McGillicuddy-DeLisi, A. V., & Sigel, I. E. (1986). The effects of children's communicative status and task on parents' teaching strategies. *Contemporary Educational Psychology, 11*, 240–252. https://doi.org/10.1016/0361-476X(86)90020-2

Renninger, K. A. (2007). Irving E. (Irv) Sigel (1921–2006). *American Psychologist, 62*, 321. https://doi.org/10.1037/0003-066X.62.4.321

Ruble, D. N., & Brooks-Gunn, J. (1979a). Menstrual myths. *Medical Aspects of Human Sexuality, 13*, 110–121.

Ruble, D. N., & Brooks-Gunn, J. (1979b). Menstrual symptoms: A social cognitive analysis. *Journal of Behavioral Medicine, 2*, 171–194. https://doi.org/10.1007/BF00846665

Ruble, D. N., & Brooks-Gunn, J. (1982). The experience of menarche. *Child Development, 53*, 1557–1566. https://doi.org/10.2307/1130084

Ruble, D. N., Boggiano, A. K., & Brooks-Gunn, J. (1982). Men's and women's evaluations of menstrual-related excuses. *Sex Roles, 8*, 625–638. https://doi.org/10.1007/BF00289896

Schnur, E., Brooks-Gunn, J., & Shipman, V. C. (1992). Who attends programs serving poor children? The case of Head Start attendees and non-attendees. *Journal of Applied Developmental Psychology, 13*, 405–421. https://doi.org/10.1016/0193-3973(92)90038-J

Shipman, V. C. (1972). Introduction. In V. C. Shipman (Ed.), *Disadvantaged children and their first school experiences: ETS–Head Start longitudinal study —Technical reports series* (Program Report No. 72-27, pp. 1–23). Princeton: Educational Testing Service.

Shipman, V. C. (1973). *Disadvantaged children and their first school experiences: ETS–Head Start longitudinal study—Interim report* (Program Report No. 73–35). Princeton: Educational Testing Service.

Sigel, I. E. (1982). The relationship between parental distancing strategies and the child's cognitive behavior. In L. M. Laosa & I. E. Sigel (Eds.), *Families as learning environments for children* (pp. 47–86). New York: Plenum Press. https://doi.org/10.1007/978-1-4684-4172-7_2

Sigel, I. E. (1986). Early social experience and the development of representational competence. *New Directions for Child Development, 1986*(32), 49–65. https://doi.org/10.1002/cd.23219863205

Sigel, I. E. (1992). The belief–behavior connection: A resolvable dilemma? In I. E. Sigel, A. V. McGillicuddy-DeLisi, & J. J. Goodnow (Eds.), *Parental belief systems: The psychological consequences for children* (2nd ed., pp. 433–456). Hillsdale: Erlbaum.

Sigel, I. E. (1993). The centrality of a distancing model for the development of representational competence. In R. R. Cocking & K. A. Renninger (Eds.), *The development and meaning of psychological distance* (pp. 141–158). Hillsdale: Erlbaum.

Sigel, I. E. (2006). Research to practice redefined. In W. Damon & R. M. Lerner (Series Eds.), K. A. Renninger & I. E. Sigel (Vol. Eds.), *Handbook of child psychology: Vol. 4. Child psychology in practice* (6th ed., pp. 1017–1023). New York: Wiley.

Sigel, I. E., & McGillicuddy-DeLisi, A. V. (1984). Parents as teachers of their children: A distancing behavior model. In A. D. Pellegrini & T. D. Yawkey (Eds.), *The development of oral and written language in social contexts* (pp. 71–92). Norwood: Ablex.

Sigel, I. E., & Olmsted, P. (1970). The modification of cognitive skills among lower-class Black children. In J. Hellsmuth (Ed.), *The disadvantaged child: Vol. 3. Compensatory education—A national debate* (pp. 330–338). New York: Bruner/Mazel.

Sigel, I. E., & Saunders, R. A. (1983). On becoming a thinker: An educational preschool program. *Early Child Development and Care, 12*, 39–65. https://doi.org/10.1080/0300443830120105

Sigel, I. E., Flaugher, J., Johnson, J. E., & Schauer, A. (1977). *Manual, Parent–Child Interaction Observation Schedule*. Princeton: Educational Testing Service.

Sontag, L. W., Steele, W. G., & Lewis, M. (1969). The fetal and maternal cardiac response to environmental stress. *Human Development, 12*, 1–9. https://doi.org/10.1159/000270674

Steele, W. G., & Lewis, M. (1968). A longitudinal study of the cardiac response during a problem-solving task and its relationship to general cognitive function. *Psychonomic Science, 11*, 275–276. https://doi.org/10.3758/BF03328189

Suárez-Orozco, C., & Carhill, A. (2008). Afterword: New directions in research with immigrant families and their children. *New Directions for Child and Adolescent Development, 121*, 87–104. https://doi.org/10.1002/cd.224

Suárez-Orozco, C., Suárez-Orozco, M. M., & Todorova, I. (2008). *Learning a new land: Immigrant students in American society*. Cambridge, MA: Harvard University Press.

Swenson, C. H., & Fiore, A. (1975). Scale of Marriage Problems. *1975 Annual Handbook for Group Facilitators, 7*, 72–89.

Terman, L. M., & Merrill, M. A. (1960). *Manual, Stanford Binet Intelligence Scale* (3rd ed.). Boston: Houghton Mifflin.

Thurman, S. K., & Lewis, M. (1979). Children's response to differences: Some possible implications for mainstreaming. *Exceptional Children, 45*, 468–470.
Vygotsky, L. (1978). *Mind in society – The development of higher psychological processes*. Cambridge, MA: Harvard University Press.
Wallach, M. A., & Kogan, N. (1965). *Modes of thinking in young children—A study of the creativity–intelligence distinction*. New York: Holt Rinehart and Winston.
Ward, W. C. (1968). Creativity in young children. *Child Development, 39*, 737–754. https://doi.org/10.2307/1126980
Warren, M. P., Brooks-Gunn, J., Hamilton, L. H., Warren, L. F., & Hamilton, W. G. (1986). Scoliosis and fractures in young ballet dancers – Relation to delayed menarche and secondary amenorrhea. *New England Journal of Medicine, 314*, 1348–1353. https://doi.org/10.1056/NEJM198605223142104
Warren, M. P., Brooks-Gunn, J., Fox, R. P., Lancelot, C., Newman, D., & Hamilton, W. G. (1991). Lack of bone accretion and amenorrhea: Evidence for a relative osteopenia in weight-bearing bones. *Journal of Clinical Endocrinology & Metabolism, 72*, 847–853. https://doi.org/10.1210/jcem-72-4-847
Wechsler, D. B. (1949a). *Wechsler Intelligence Scale for Children manual*. New York: Psychological Corporation.
Wechsler, D. (1949b). *Wechsler Preschool and Primary School Scale of Intelligence*. New York: Psychological Corporation.
Wechsler, D. B. (1955). *Manual for the Wechsler Adult Intelligence Scale*. New York: Psychological Corporation.
Weinraub, M., & Lewis, M. (1977). The determinants of children's responses to separation. *Monographs of the Society for Research in Child Development, 42*(4, Serial No. 172). https://doi.org/10.2307/1165949
Weinraub, M., Brooks, J., & Lewis, M. (1977). The social network: A reconsideration of the concept of attachment. *Human Development, 20*, 31–47. https://doi.org/10.1159/000271546
Wertsch, J. V., McNamee, G. D., McLane, J. B., & Budwig, N. A. (1980). The adult–child dyad as a problem solving system. *Child Development, 51*, 1215–1221. https://doi.org/10.2307/1129563
Williams, R. L. (1972). *Manual of directions, Black Intelligence Test of Cultural Homogeneity*. St. Louis: Author.
Witkin, H. A., Oltman, P. K., Raskin, E., & Karp, S. A. (1971). *A manual for the Embedded Figures Tests*. Palo Alto: Consulting Psychologists Press.
Zarin-Ackerman, J., Lewis, M., & Driscoll Jr., J. M. (1977). Language development in 2-year-old normal and risk infants. *Pediatrics, 59*, 982–986.

Part IV
ETS Contributions to Validity

Chapter 16
Research on Validity Theory and Practice at ETS

Michael Kane and Brent Bridgeman

Educational Testing Service (ETS) was founded with a dual mission: to provide high-quality testing programs that would enhance educational decisions and to improve the theory and practice of testing in education through research and development (Bennett 2005; Educational Testing Service 1992). Since its inception in 1947, ETS has consistently evaluated its testing programs to help ensure that they meet high standards of technical and operational quality, and where new theory and new methods were called for, ETS researchers made major contributions to the conceptual frameworks and methodology.

This chapter reviews ETS's contributions to validity theory and practice at various levels of generality, including overarching frameworks (Messick 1988, 1989), more targeted models for issues such as fairness, and particular analytic methodologies (e.g., reliability, equating, differential item functioning). The emphasis will be on contributions to the theory of validity and, secondarily, on the practice of validation rather than on specific methodologies.

16.1 Validity Theory

General conceptions of validity grew out of basic concerns about the accuracy of score meanings and the appropriateness of score uses (Kelley 1927), and they have necessarily evolved over time as test score uses have expanded, as proposed interpretations have been extended and refined, and as the methodology of testing has become more sophisticated.

M. Kane (✉) • B. Bridgeman
Educational Testing Service, Princeton, NJ, USA
e-mail: mkane@ets.org

R.E. Bennett, M. von Davier (eds.), *Advancing Human Assessment*,
Methodology of Educational Measurement and Assessment,
DOI 10.1007/978-3-319-58689-2_16

In the first edition of *Educational Measurement* (Lindquist 1951), which was released just after ETS was founded, Cureton began the chapter on validity by suggesting that "the essential question of test validity is how well a test does the job it is employed to do" (Cureton 1951, p. 621) and went on to say that

> validity has two aspects, which may be termed relevance and reliability. … To be valid—that is to serve its purpose adequately—a test must measure something with reasonably high reliability, and that something must be fairly closely related to the function it is used to measure. (p. 622)

In the late 1940s and early 1950s, tests tended to be employed to serve two kinds of purposes: providing an indication of the test taker's standing on some attribute (e.g., cognitive ability, personality traits, academic achievement) and predicting future performance in some context.

Given ETS's mission (Bennett, Chap. 1, this volume) and the then current conception of validity (Cureton 1951), it is not surprising that much of the early work on validity at ETS was applied rather than theoretical; it focused on the development of measures of traits thought to be relevant to academic success and on the use of these measures to predict future academic performance. For example, the second Research Bulletin published at ETS (i.e., Frederiksen 1948) focused on the prediction of first-year grades at a particular college.

This kind of applied research designed to support and evaluate particular testing programs continues to be an essential activity at ETS, but over the years, these applied research projects have also generated basic questions about the interpretations of test scores, the statistical methodology used in test development and evaluation, the scaling and equating of scores, the variables to be used in prediction, structural models relating current performance to future outcomes, and appropriate uses of test scores in various contexts and with various populations. In seeking answers to these questions, ETS researchers contributed to the theory and practice of educational measurement by developing general frameworks for validation and related methodological developments that support validation.

As noted earlier, at the time ETS was founded, the available validity models for testing programs emphasized score interpretations in terms of traits and the use of scores as predictors of future outcomes, but over the last seven decades, the concept of validity has expanded. The next section reviews ETS's contributions to the development and validation of trait interpretations, and the following section reviews ETS's contributions to models for the prediction of intended "criterion" outcomes. The fourth describes ETS's contributions to our conceptions and analyses of fairness in testing. The fifth section traces the development of Messick's comprehensive, unified model of construct validity, a particularly important contribution to the theory of validity. The sixth section describes ETS's development of argument-based approaches to validation. A seventh section, on validity research at ETS, focuses on the development of methods for the more effective interpretation and communication of test scores and for the control of extraneous variance. The penultimate section discusses fairness as a core validity concern. The last section provides some concluding comments.

This organization is basically thematic, with each section examining ETS's contributions to the development of aspects of validity theory, but it is also roughly chronological. The strands of the story (trait interpretations, prediction, construct interpretations, models for fairness, Messick's unified model of construct validity, models for the role of consequences of testing, and the development of better methods for encouraging clear interpretations and appropriate uses of test scores) overlap greatly, developed at different rates during different periods, and occasionally folded back on themselves, but there was also a gradual progression from simpler and more intuitive models for validity to more complex and comprehensive models, and the main sections in this chapter reflect this progression.

As noted, most of the early work on validity focused on trait interpretations and the prediction of desired outcomes. The construct validity model was proposed in the mid-1950s (Cronbach and Meehl 1955), but it took a while for this model to catch on. Fairness became a major research focus in the 1970s. In the 1970s and 1980s, Messick developed his unified framework for the construct validity of score interpretations and uses, and the argument-based approaches were developed at the turn of the century.

It might seem appropriate to begin this chapter by defining the term *validity*, but as in any area of inquiry (and perhaps more so than in many other areas of inquiry), the major developments in validity theory have involved changes in what the term means and how it is used. The definition of *validity* has been and continues to be a work in progress. Broadly speaking, validation has always involved an evaluation of the proposed interpretations and uses of test scores (Cronbach 1971; Kane 2006, 2013a; Messick 1989), but both the range of proposed interpretations and the evaluative criteria have gradually expanded.

16.2 Validity of Trait Interpretations

For most of its history from the late nineteenth century to the present, test theory has tended to focus on traits, which were defined in terms of dispositions to behave or perform in certain ways in response to certain kinds of stimuli or tasks, in certain kinds of contexts. Traits were assumed to be personal characteristics with some generality (e.g., over some domain of tasks, contexts, occasions). In the late 1940s and early 1950s, this kind of trait interpretation was being applied to abilities, skills, aptitudes, and various kinds of achievement as well as to psychological traits as such. Trait interpretations provided the framework for test development and, along with predictive inferences, for the interpretation of test scores (Gulliksen 1950a). As theory and methodology developed, *trait interpretations* tended to become more sophisticated in their conceptualizations and in the methods used to estimate the traits. As a result, trait interpretations have come to overlap with *construct interpretations* (which can have more theoretical interpretations), but in this section, we limit ourselves to basic trait interpretations, which involve dispositions to perform in some way in response to tasks of some kind.

Cureton (1951) summarized the theoretical framework for this kind of trait interpretation:

> When the item scores of a set of test-item performances correlate substantially and more or less uniformly with one another, the sum of the item scores (the summary score or test score) has been termed a quasi-measurement. It is a quasi-measurement of "whatever," in the reaction-systems of the individuals, is invoked in common by the test items as presented in the test situation. This "whatever" may be termed a "trait." The existence of the trait is demonstrated by the fact that the item scores possess some considerable degree of homogeneity; that is, they measure in some substantial degree the same thing. We term this "thing" the "trait." (pp. 647–648)

These traits can vary in their content (e.g., achievement in geography vs. anxiety), in their generality (e.g., mechanical aptitude vs. general intelligence), and in the extent to which they are context or population bound, but they share three characteristics (Campbell 1960; Cureton 1951). First, they are basically defined in terms of some relatively specific domain of performance or behavior (with some domains broader than others). Second, the performances or behaviors are assumed to reflect some characteristic of individuals, but the nature of this characteristic is not specified in any detail, and as a result, the interpretation of the trait relies heavily on the domain definition. Third, traits are assumed to be enduring characteristics of individuals, with some more changeable (e.g., achievement in some academic subject) than others (e.g., aptitudes, personality).

Note that the extent to which a trait is enduring is context dependent. Levels of achievement in an academic subject such as geography would be expected to increase while a student is studying the subject and then to remain stable or gradually decline thereafter. A personality trait such as conscientiousness is likely to be more enduring, but even the most stable traits can change over time.

An understanding of the trait (rudimentary as it might be) indicates the kinds of tasks or stimuli that could provide information about it. The test items are designed to reflect the trait, and to the extent possible nothing else, and differences in test scores are assumed to reflect mainly differences in level of the trait.

The general notion of a trait as a (somewhat) enduring characteristic of a person that is reflected in certain kinds of behavior in certain contexts is a basic building block of "folk psychology," and as such, it is ancient (e.g., Solomon was wise, and Caesar was said to be ambitious). As they have been developed to make sense of human behavior over the last century and a half, modern theories of psychology have made extensive use of a wide variety of traits (from introversion to mathematical aptitude) to explain human behavior. As Messick (1989) put it, "a trait is a relatively enduring characteristic of a person—an attribute, process, or disposition—which is consistently manifested to an appropriate degree when relevant, despite considerable variation in the range of settings and circumstances" (p. 15). Modern test theory grew out of efforts to characterize individuals in terms of traits, and essentially all psychometric theories (including classical test theory, generalizability theory, factor analysis, and item response theory) involve the estimation of traits of one kind or another.

From a psychological point of view, the notion of a trait suggests a persistent characteristic of a person that is prior to and independent of any testing program. The trait summarizes (and, in that sense, accounts for) performance or behavior. The trait is not synonymous with any statistical parameter, and it is reasonable to ask whether a parameter estimate based on a particular sample of behavior is an unbiased estimate of the trait of interest. Assuming that the estimate is unbiased, it is also reasonable to ask how precise the estimate is. An assessment of the trait may involve observing a limited range of performances or behaviors in a standardized context and format, but the trait is interpreted in terms of a tendency or disposition to behave in some way or an ability to perform some kinds of tasks in a range of test and non-test contexts. The trait interpretation therefore entails expectations that assessments of the trait using different methods should agree with each other, and assessments of different traits using common methods should not agree too closely (Campbell 1960; Campbell and Fiske 1959).

Traits have two complementary aspects. On one hand, a trait is thought of as an unobservable characteristic of a person, as some latent attribute or combination of such attributes of the person. However, when asked to say what is meant by a trait, the response tends to be in terms of some domain of observable behavior or performance. Thus traits are thought of as unobservable attributes and in terms of typical performance over some domain. Most of the work described in this section focuses on traits as dispositions to behave in certain ways. In a later section, we will focus more on traits as theoretical constructs that are related to domains of behavior or performance but that are defined in terms of their properties as underlying latent attributes or constructs.

16.2.1 ETS's Contributions to Validity Theory for Traits

Trait interpretations of test scores go back at least to the late nineteenth century and therefore predate both the use of the term *validity* and the creation of ETS. However ETS researchers made many contributions to theoretical frameworks and specific methodology for the validation of trait interpretations, including contributions to classical test theory (including reliability theory, standard errors, and confidence intervals), item response theory, equating, factor analysis, scaling, and methods for controlling trait-irrelevant variance. The remainder of this section concentrates on ETS's contributions to the development of these methodologies, all of which seek to control threats to validity.

ETS researchers have been involved in analyzing and measuring a wide variety of traits over ETS's history (Stricker, Chap. 13, this volume), including acquiescence (Messick 1965, 1967), authoritarian attitudes (Messick and Jackson 1958), emotional intelligence (Roberts et al. 2008), cognitive structure (Carroll 1974), response styles (Jackson and Messick 1961; Messick 1991), risk taking (Myers 1965), and social intelligence (Stricker and Rock 1990), as well as various kinds of aptitudes and achievement. ETS researchers have also made major contributions to the methodology for evaluating the assumptions inherent in trait interpretations and in ruling out factors that might interfere with the intended trait interpretations, particularly in classical test theory (Lord and Novick 1968), theory related to the sampling of target domains (Frederiksen 1984), and item response theory (Lord 1951, 1980).

16.2.2 Classical Test Theory and Reliability

Classical test theory (CTT) is based on trait interpretations, particularly on the notion of a trait score as the expected value over the domain of replications of a measurement procedure. The general notion is that the trait being measured remains invariant over replications of the testing procedure; the test scores may fluctuate to some extent over replications, but the value of the trait is invariant, and fluctuations in observed scores are treated as random errors of measurement. Gulliksen (1950b) used this notion as a starting point for his book, in which he summarized psychometric theory in the late 1940s but used the term *ability* instead of *trait*:

> It is assumed that the gross score has two components. One of these components (T) represents the actual ability of the person, a quantity that will be relatively stable from test to test as long as the tests are measuring the same thing. The other component (E) is an error. (p. 4)

Note that the true scores of CTT are expected values over replications of the testing procedure; they do not refer to an underlying, "real" value of the trait, which has been referred to as a platonic true score to differentiate it from the classical true score. Reliability coefficients were defined in terms of the ratio of true-score variance to observed-score variance, and the precision of the scores was evaluated in terms of the reliability or in terms of standard errors of measurement. Livingston (1972) extended the notion of reliability to cover the dependability of criterion-referenced decisions.

Evidence for the precision of test scores (e.g., standard errors, reliability) supports validity claims in at least three ways. First, some level of precision is necessary for scores to be valid for any interpretation; that is, if the trait estimates have low reliability (i.e., they fluctuate substantially over replications), the only legitimate interpretation of the scores is that they mostly represent error or "noise." Second, the magnitude of the standard error can be considered part of the interpretation of the scores. For example, to say that a test taker has an estimated score of 60 with a standard error of 2 is a much stronger claim than a statement that a test taker has an estimated score of 60 with a standard error of 20. Third, the relationships between the precision of test scores and the number and characteristics of the items in the test can be used to develop tests that are more reliable without sacrificing relevance, thereby improving validity.

Classical test theory was the state of the art in the late 1940s, and as ETS researchers developed and evaluated tests of various traits, they refined old methods and developed new methods within the context of the CTT model (Moses, Chaps. 2 and

3, this volume). The estimation of reliability and standard errors has been an ongoing issue of fundamental importance (Horst 1951; Jöreskog 1971; Keats 1957; Kristof 1962, 1970, 1974; Lord 1955; Novick and Lewis 1967; Tucker 1949). ETS's efforts to identify the implications of various levels of reliability began soon after its inception and have continued since (Angoff 1953; Haberman 2008; Horst 1950a, b; Kristof 1971; Livingston and Lewis 1995; Lord 1956, 1957, 1959).

An important early contribution of ETS researchers to the classical model was the development of conditional standard errors (Keats 1957; Lord 1955, 1956) and of associated confidence intervals around true-score estimates (Gulliksen 1950b; Lord and Novick 1968; Lord and Stocking 1976). Putting a confidence interval around a true-score estimate helps to define and limit the inferences that can be based on the estimate; for example, a decision to assign a test taker to one of two categories can be made without much reservation if a highly conservative confidence interval (e.g., 99%) for a test taker does not include the cutscore between the two categories (Livingston and Lewis 1995). Analyses of the reliability and correlations of subscores can also provide guidance on whether it would be meaningful to report the subscores separately (Haberman 2008).

Evaluations of the precision of test scores serve an important quality-control function, and they can help to ensure an adequate level of precision in the test scores generated by the testing program (Novick and Thayer 1969). Early research established the positive relationship between test length and reliability as well as the corresponding inverse relationship between test length and standard errors (Lord 1956, 1959). That research tradition also yielded methods for maximizing the reliability of composite measures (B.F. Green 1950).

One potentially large source of error in testing programs that employ multiple forms of a test (e.g., to promote security) is variability in content and statistical characteristics (particularly test difficulty) across different forms of the test, involving different samples of test items. Assuming that the scores from the different forms are to be interpreted and used interchangeably, it is clearly desirable that each test taker's score be more or less invariant across the forms, but this ideal is not likely to be met exactly, even if the forms are developed from the same specifications. Statistical equating methods are designed to minimize the impact of form differences by adjusting for differences in operating characteristics across the forms. ETS researchers have made major contributions to the theory and practice of equating (Angoff 1971; Holland 2007; Holland and Dorans 2006; Holland and Rubin 1982; Lord and Wingersky 1984; Petersen 2007; Petersen et al. 1989; A.A. von Davier 2011; A.A. von Davier et al. 2004). In the absence of equating, form-to-form differences can introduce substantial errors, and equating procedures can reduce this source of error.

On a more general level, ETS researchers have played major roles in developing the CTT model and in putting it on firm foundations (Lord 1965; Novick 1965). In 1968, Frederick Lord and Melvin Novick formalized and summarized most of what was known about the CTT model in their landmark book *Statistical Theories of Mental Test Scores*. They provided a very sophisticated statement of the classical test-theory model and extended it in many directions.

16.2.3 Adequate Sampling of the Trait

Adequate sampling of the trait domain requires a clear definition of the domain, and ETS researchers have devoted a lot of attention to developing a clear understanding of various traits and of the kinds of performances associated with these traits (Ebel 1962). For example, Dwyer et al. (2003) defined *quantitative reasoning* as "the ability to analyze quantitative information" (p. 13) and specified that its domain would be restricted to quantitative tasks that would be new to the student (i.e., would not require methods that the test takers had been taught). They suggested that quantitative reasoning includes six more specific capabilities: (a) understanding quantitative information presented in various formats, (b) interpreting and drawing inferences from quantitative information, (c) solving novel quantitative problems, (d) checking the reasonableness of the results, (e) communicating quantitative information, and (f) recognizing the limitations of quantitative methods. The quantitative reasoning trait interpretation assumes that the tasks do not require specific knowledge that is not familiar to all test takers and, therefore, any impact that such knowledge has on the scores would be considered irrelevant variance.

As noted earlier, ETS has devoted a lot of attention to developing assessments that reflect traits of interest as fully as possible (Lawrence and Shea 2011). Much of this effort has been devoted to more adequately sampling the domains associated with the trait, and thereby reducing the differences between the test content and format and the broader domain associated with the trait (Bejar and Braun 1999; Frederiksen 1984). For example, the "in basket" test (Frederiksen et al. 1957) was designed to evaluate how well managers could handle realistic versions of management tasks that required decision making, prioritizing, and delegating. Frederiksen (1959) also developed a test of creativity in which test takers were presented with descriptions of certain results and were asked to list as many hypotheses as they could to explain the results. Frederiksen had coauthored the chapter on performance assessment in the first edition of *Educational Measurement* (Ryans and Frederiksen 1951) and consistently argued for the importance of focusing assessment on the kinds of performance that are of ultimate interest, particularly in a landmark article, "The Real Test Bias: Influences of Testing on Teaching and Learning" (Frederiksen 1984). More recently, ETS researchers have been developing a performance-based program of Cognitively Based Assessment *of*, *for*, and *as* Learning (the *CBAL®* initiative) that elicits extended performances (Bennett 2010; Bennett and Gitomer 2009). For CBAL, and more generally for educational assessments, positive changes in the traits are the goals of instruction and assessment, and therefore the traits being assessed are not expected to remain the same over extended periods.

The evidence-centered design (ECD) approach to test development, which is discussed more fully later, is intended to promote adequate sampling of the trait (or construct) by defining the trait well enough up front to get a good understanding of the kinds of behaviors or performance that would provide the evidence needed to draw conclusions about the trait (Mislevy et al. 1999, 2002). To the extent that the testing program is carefully designed to reflect the trait of interest, it is more likely that the observed behaviors or performances will adequately achieve that end.

Based on early work by Lord (1961) on the estimation of norms by item sampling, matrix sampling approaches, in which different sets of test tasks are taken by different subsamples of test takers, have been developed to enhance the representativeness of the sampled test performances for the trait of interest (Mazzeo et al. 2006; Messick et al. 1983). Instead of drawing a single sample of tasks that are administered to all test takers, multiple samples of tasks are administered to different subsamples of test takers. This approach allows for a more extensive sampling of content in a given amount of testing time. In addition, because it loosens the time constraints on testing, the matrix sampling approach allows for the use of a wider range of test tasks, including performance tasks that require substantial time to complete. These matrix sampling designs have proven to be especially useful in large-scale monitoring programs like the National Assessment of Educational Progress (NAEP) and in various international testing programs (Beaton and Barone, Chap. 8, Kirsch et al. Chap. 9, this volume).

16.2.4 Factor Analysis

Although a test may be designed to reflect a particular trait, it is generally the case that the test scores will be influenced by many characteristics of the individuals taking the test (e.g., motivation, susceptibility to distractions, reading ability). To the extent that it is possible to control the impact of test-taker characteristics that are irrelevant to the trait of interest, it may be possible to interpret the assessment scores as relatively pure measures of that focal trait (French 1951a, b, 1954, 1963). More commonly, the assessment scores may also intentionally reflect a number of test-taker characteristics that, together, compose the trait. That is, broadly defined traits that are of practical interest may involve a number of more narrowly defined traits or factors that contribute to the test taker's performance. For example, as noted earlier, Dwyer et al. (2003) defined the performance domain for quantitative reasoning in terms of six capabilities, including understanding quantitative information, interpreting quantitative information, solving quantitative problems, and estimating and checking answers for reasonableness. In addition, most trait measures require ancillary abilities (e.g., the ability to read) that are needed for effective performance in the assessment context.

In interpreting test scores, it is generally helpful to develop an understanding of how different characteristics are related to each other. Factor analysis models have been widely used to quantify the contributions of different underlying characteristics, or "factors," to assessment scores, and ETS researchers have played a major role in the development of various factor-analytic methods (Moses, Chaps. 2 and 3, this volume), in part because of their interest in developing a variety of cognitive and noncognitive measures (French 1951a, b, 1954).

Basic versions of exploratory factor analysis were in general use when ETS was formed, but ETS researchers contributed to the development and refinement of more sophisticated versions of these methods (Browne 1968; B.F. Green 1952; Harman 1967; Lord and Novick 1968; Tucker 1955). Exploratory factor analysis makes it possible to represent the relationships (e.g., correlations or covariances) among observed scores on a set of assessments in terms of a statistical model describing the relationships among a relatively small number of underlying dimensions, or factors. The factor models decompose the observed total scores on the tests into a linear combination of factor scores, and they provide quantitative estimates of the relative importance of the different factors in terms of the variance explained by the factor.

By focusing on the traits as latent dimensions or factors or as some composite of more basic latent factors, and by embedding these factors within a web of statistical relationships, exploratory factor analysis provided a rudimentary version of the kind of nomological networks envisioned by Cronbach and Meehl (1955). The utility of exploratory analyses for explicating appropriate interpretations of test scores was enhanced by an extended research program at ETS to develop sets of reference measures that focused on particular basic factors (Ekstrom et al. 1979; French 1954; French et al. 1963). By including the reference tests with a more broadly defined trait measure, it would be possible to evaluate the factor structure of the broadly defined trait in terms of the reference factors.

As in other areas of theory development, the work done on factor analysis by ETS researchers tended to grow out of and be motivated by concerns about the need to build assessments that reflected certain traits and to evaluate how well the assessment actually reflected those traits. As a result, ETS's research on exploratory factor analysis has involved a very fruitful combination of applied empirical studies of score interpretations and sophisticated theoretical modeling (Browne 1968; French 1951a, b; Harman 1967; Lord and Novick 1968).

A major contribution to the theory and practice of validation that came out of research at ETS is confirmatory factor analysis (Jöreskog 1967, 1969; Jöreskog and Lawley 1967; Jöreskog and van Thillo 1972). As its name indicates, exploratory factor analysis does not propose strong constraints a priori; the analysis essentially partitions the observed-score variances by using statistical criteria to fit the model to the data. In a typical exploratory factor analysis, theorizing tends to occur after the analysis, as the resulting factor structure is used to suggest plausible interpretations for the factors. If reference factors are included in the analysis, they can help orient the interpretation.

In confirmatory factor analysis (CFA), a factor model is specified in advance by putting constraints on the factor structure, and the constrained model is fit to the data. The constraints imposed on the model are typically based on a priori theoretical assumptions, and the empirical data are used to check the hypotheses built into the models. As a result, CFAs can provide support for theory-based hypotheses or can result in refutations of some or all or the theoretical conjectures (Jöreskog 1969). This CFA model was extended as the basis for structural equation modeling (Jöreskog and van Thillo 1972). To the extent that the constraints incorporate theoretical assumptions, CFAs go beyond simple trait interpretations into theory-based construct interpretations.

CFA is very close in spirit and form to the nomological networks of Cronbach and Meehl (1955). In both cases, there are networks of hypothesized relationships between constructs (or latent variables), which are explicitly defined a priori and which may be extensive, and there are proposed measures of at least some of the constructs. Given specification of the network as a confirmatory factor model (and adequate data), the hypotheses inherent in the network can be checked by evaluating the fit of the model to the data. If the model fits, the substantive assumptions (about relationships between the constructs) in the model and the validity of the proposed measures of the constructs are both supported. If the model does not fit the data, either the substantive assumptions and/or the validity of the measures is likely to be questioned. As is the case in the classic formulation of the construct validity model (Cronbach and Meehl 1955), the substantive theory and the assessments are initially validated (or invalidated) holistically as a network of interrelated assumptions. If the constrained model fails to fit the data, the data can be examined to identify potential weaknesses in the network. In addition, the model fit can be compared to the fit of alternate models that make different (perhaps stronger or weaker) assumptions.

16.2.5 Latent Traits

Two major developments in test theory in the second half of the twentieth century (the construct validity model and latent trait theory) grew out of attempts to make the relationship between observed behaviors or performances and the relevant traits more explicit, and ETS researchers played major roles in both of these developments (see Carlson and von Davier, Chap. 5, this volume). Messick (1975, 1988, 1989) elaborated the construct validity model of Cronbach and Meehl (1955), which sought to explicate the relationships between traits and observed assessment performances through substantive theories that would relate trait scores to the constructs in a theory and to other trait scores attached to the theory. Item response theory (IRT) deployed measurement models to specify the relationships between test performances and postulated latent traits and to provide statistical estimates of these traits (Lord 1951). Messick's contributions to construct validity theory will be discussed in detail later in this chapter. In this section, we examine contributions to IRT and the implications of these developments for validity.

In their seminal work on test theory, Lord and Novick (1968) used trait language to distinguish true scores from errors:

> Let us suppose that we repeatedly administer a given test to a subject and thus obtain a measurement each day for a number of days. Further, let us assume that with respect to the particular *trait* the test is designed to measure, the person does not change from day to day and that successive measurements are unaffected by previous measurements. Changes in the environment or the *state* of the person typically result in some day-to-day variation in the measurements which are obtained. We may view this variation as the result of errors of measurement of the underlying trait characterizing the individual, or we may view it as a representation of a real change in this trait. (pp. 27–28)

In models for true scores, the true score captures the enduring component in the scores over repeated, independent testing, and the "random" fluctuations around this true score are relegated to error.

Lord and Novick (1968) also used the basic notion of a trait to introduce latent traits and item characteristic functions:

> Any theory of latent traits supposes that an individual's behavior can be accounted for, to a substantial degree, by defining certain human characteristics called *traits*, quantitatively estimating the individual's standing on each of these traits, and then using the numerical values obtained to predict or explain performance in relevant situations. (p. 358)

Within the context of the statistical model, the latent trait accounts for the test performances, real and possible, in conjunction with item or task parameters. The latent trait has model-specific meaning and a model-specific use; it captures the enduring contribution of the test taker's "ability" to the probability of success over repeated, independent performances on different tasks.

Latent trait models have provided a richer and in some ways firmer foundation for trait interpretations than offered by classical test theory. One motivation for the development of latent trait models (Lord 1951) was the realization that number-right scores and simple transformations of such scores would not generally yield the defining property of traits (i.e., invariance over measurement operations). The requirement that task performance data fit the model can also lead to a sharpening of the domain definition, and latent trait models can be helpful in controlling random errors by facilitating the development of test forms with optimal statistical properties and the equating of scores across different forms of a test.

A model-based trait interpretation depends on empirical evidence that the statistical model fits the data well enough; if it does, we can have confidence that the test scores reflect the trait conceived of as "whatever … is invoked in common by the test items" (Cureton 1951, p. 647). The application of a CTT or latent trait model to student responses to generate estimates of a true score or a latent trait does not in itself justify the interpretation of scores in terms of a construct that causes and explains the task performances, and it does not necessarily justify inferences to any nontest performance. A stronger interpretation in terms of a psychological trait that has implications beyond test scores requires additional evidence (Messick 1988, 1989). We turn to such construct interpretations later in this chapter.

16.2.6 Controlling Irrelevant Variance

As is the case in many areas of inquiry, a kind of negative reasoning can play an important role in validation of trait interpretations. Tests are generally developed to yield a particular score interpretation and often a particular use, and the test development efforts make a case for the interpretation and use (Mislevy et al. 2002). Once this initial positive case has been made, it can be evaluated by subjecting it to empirical challenge. We can have confidence in claims that have survived all serious challenges.

To the extent that an alternate proposal is as plausible, or more plausible, than a proposed trait interpretation, we cannot have much confidence in the intended interpretation. This notion, which is a fundamental methodological precept in science (Popper 1965), underlies, for example, multitrait–multimethod analyses (D. T. Campbell and Fiske 1959) and the assumption that reliability is a necessary condition for validity. As a result, to the extent that we can eliminate alternative interpretations of test scores, the proposed interpretation becomes more plausible, and if we can eliminate all plausible rivals for a proposed trait interpretation, we can accept that interpretation (at least for the time being).

In most assessment contexts, the question is not whether an assessment measures the trait or some alternate variable but rather the extent to which the assessment measures the trait of interest and is not overly influenced by sources of irrelevant variance. In their efforts to develop measures of various traits, ETS researchers have examined many potential sources of irrelevant variance, including anxiety (French 1962; Powers 1988, 2001), response styles (Damarin and Messick 1965), coaching (Messick 1981b, 1982a; Messick and Jungeblut 1981), and stereotype threat (Stricker 2008; Stricker and Bejar 2004; Stricker and Ward 2004). Messick (1975, 1989) made the evaluation of plausible sources of irrelevant variance a cornerstone of validation, and he made the evaluation of construct-irrelevant variance and construct underrepresentation central concerns in his unified model of validity.

It is, of course, desirable to neutralize potential sources of irrelevant variance before tests are administered operationally, and ETS has paid a lot of attention to the development and implementation of item analysis methodology, classical and IRT-based, designed to minimize irrelevant variance associated with systematic errors and random errors. ETS has played a particularly important role in the development of methods for the detection of differential item functioning (DIF), in which particular items operate inconsistently across groups of test takers while controlling for ability and thereby introduce systematic differences that may not reflect real differences in the trait of interest (Dorans, Chap. 7, this volume, 1989, 2004; Dorans and Holland 1993; Holland and Wainer 1993; Zieky 1993, 2011).

Trait interpretations continue to play a major role in the interpretation and validation of test scores (Mislevy 2009). As discussed earlier, trait interpretations are closely tied to domains of possible test performances, and these domains provide guidance for the development of assessment procedures that are likely to support their intended function. In addition, trait interpretations can be combined with substantive assumptions about the trait and the trait's relationships to other variables, thus going beyond the basic trait interpretation in terms of a domain of behaviors or performances to an interpretation of a theoretical construct (Messick 1989; Mislevy et al. 2002).

16.3 Validity of Score-Based Predictions

Between 1920 and 1950, test scores came to be used to predict future outcomes and to estimate concurrent criteria that were of practical interest but were not easily observed, and the validity of such criterion-based interpretations came to be evaluated mainly in terms of how well the test scores predicted the criterion (Angoff 1988; Cronbach 1971; Kane 2012; Messick 1988, 1989; Zwick 2006). In the first edition of *Educational Measurement*, which was written as ETS was being founded, Cureton (1951) associated validity with “the correlation between the actual test scores and the ‘true’ criterion score” (p. 623), which would be estimated by the correlation between the test scores and the criterion scores, with an adjustment for unreliability in the criterion.

The criterion variable of interest was assumed to have a definite value for each person, which was reflected by the criterion measure, and the test scores were to “predict” these values as accurately as possible (Gulliksen 1950b). Given this interpretation of the test scores as stand-ins for the true criterion measure, it was natural to evaluate validity in terms of the correlation between test scores and criterion scores:

> Reliability has been regarded as the correlation of a given test with a parallel form. Correspondingly, the validity of a test is the correlation of the test with some criterion. In this sense a test has a great many different “validities.” (Gulliksen 1950b, p. 88)

The criterion scores might be obtained at about the same time as the test scores (“concurrent validity”), or they might be a measure of future performance (e.g., on the job, in college), which was not available at the time of testing (“predictive validity”). If a good criterion were available, the criterion model could provide simple and elegant estimates of the extent to which scores could be used to estimate or predict criterion scores (Cureton 1951; Gulliksen 1950b; Lord and Novick 1968). For admissions, placement, and employment, the criterion model is still an essential source of validity evidence. In these applications, criterion-related inferences are core elements in the proposed interpretations and uses of the test scores. Once the criterion is specified and appropriate data are collected, a criterion-based validity coefficient can be estimated in a straightforward way.

As noted earlier, the criterion model was well developed and widely deployed by the late 1940s, when ETS was founded (Gulliksen 1950b). Work at ETS contributed to the further development of these models in two important ways: by improving the accuracy and generality of the statistical models and frameworks used to estimate various criteria (N. Burton and Wang 2005; Moses, Chaps. 2 and 3, this volume) and by embedding the criterion model in a more comprehensive analysis of the plausibility of the proposed interpretation and use of test scores (Messick 1981a, 1989). The criterion model can be implemented more or less mechanically once the criterion has been defined, but the specification of the criterion typically involves value judgments and a consideration of consequences (Messick 1989).

Much of the early research at ETS addressed the practical issues of developing testing programs and criterion-related validity evidence, but from the beginning, researchers were also tackling more general questions about the effective use of standardized tests in education. The criterion of interest was viewed as a measure of a trait, and the test was conceived of as a measure of another trait that was related to the criterion trait, as an aptitude is related to subsequent achievement. As discussed more fully in a later section, ETS researchers conducted extensive research on the factors that tend to have an impact on the correlations of predictors (particularly *SAT*® scores) with criteria (e.g., first-year college grades), which served as measures of academic achievement (Willingham et al. 1990).

In the 1940s and 1950s, there was a strong interest in measuring both cognitive and noncognitive traits (French 1948). One major outcome of this extensive research program was the finding that cognitive measures (test scores, grades) provided fairly accurate predictions of performance in institutions of higher education and that the wide range of noncognitive measures that were evaluated did not add much to the accuracy of the predictions (Willingham et al. 1990).

As noted by Zwick (2006), the validity of tests for selection has been judged largely in terms of how well the test scores can predict some later criterion of interest. This made sense in 1950, and it continues to make sense into the twenty-first century. The basic role of criterion-related validity evidence in evaluating the accuracy of such predictions continues to be important for the validity of any interpretation or use that relies on predictions of future performance (Kane 2013a), but these paradigm cases of prediction now tend to be evaluated in a broader theoretical context (Messick 1989) and from a broader set of perspectives (Dorans 2012; Holland 1994; Kane 2013b). In this broader context, the accuracy of predictions continues to be important, but concerns about fairness and utility are getting more attention than they got before the 1970s.

16.4 Validity and Fairness

Before the 1950s, the fairness of testing programs tended to be evaluated mainly in terms of equivalent or comparable treatment of test takers. This kind of procedural fairness was supported by standardizing test administration, materials, scoring, and conditions of observation, as a way of eliminating favoritism or bias; this approach is illustrated in the civil service testing programs, in licensure programs, and in standardized educational tests (Porter 2003). It is also the standard definition of fairness in sporting events and other competitions and is often discussed in terms of candidates competing on "a level playing field." Before the 1950s, this very basic notion of fairness in testing programs was evaluated mainly at the individual level; each test taker was to be treated in the same way, or if some adjustment were necessary (e.g., due to a candidate's disability or a logistical issue), as consistently as possible. In the 1950s, 1960s, and 1970s, the civil rights movement, legislation, and litigation raised a broader set of fairness issues, particularly issues of fair treatment of groups that had suffered discrimination in the past (Cole and Moss 1989; Willingham 1999).

With respect to the treatment of groups, concerns about fairness and equal opportunity prior to this period did exist but were far more narrowly defined. One of the goals of James Conant and others in promoting the use of the Scholastic Aptitude Test was to expand the pool of students admitted to major universities by giving all high school students an opportunity to be evaluated in terms of their aptitude and not just in terms of the schools they attended or the curriculum that they had experienced. As president of Harvard in the 1930s, Conant found that most of Harvard's students were drawn from a small set of elite prep schools and that the College Board examinations, as they then existed, evaluated mastery of prep school curricula (Bennett, Chap. 1, this volume): "For Conant, Harvard admission was being based largely on ability to pay. If a student could not afford to attend prep school, that student was not going to do well on the College Boards, and wasn't coming to Harvard" (p. 5). In 1947, when ETS was founded, standardized tests were seen as a potentially important tool for improving fairness in college admissions and other contexts, at least for students from diverse economic backgrounds. The broader issues of adverse impact and fairness as they related to members of ethnic, racial, and gender groups had not yet come into focus.

Those broader issues of racial, ethnic, and gender fairness and bias moved to center stage in the 1960s:

> Hard as it now may be to imagine, measurement specialists more or less discovered group-based test fairness as a major issue only some 30 years ago. Certainly, prior to that time, there was discussion of the cultural fairness of a test and its appropriateness for some examinees, but it was the Civil Rights movement in the 1960s that gave social identity and political dimension to the topic. That was the period when hard questions were first asked as to whether the egalitarian belief in testing was justified in the face of observed subgroup differences in test performance. The public and test specialists alike asked whether tests were inherently biased against some groups, particularly Black and Hispanic examinees. (Willingham 1999, p. 214)

As our conceptions of fairness and bias in testing expanded between the 1960s and the present, ETS played a major role in defining the broader notions of fairness and bias in testing. ETS researchers developed frameworks for evaluating fairness issues, and they developed and implemented methodology to control bias and promote fairness. These frameworks recognized the value of consistent treatment of individual test takers, but they focused on a more general conception of equitable treatment of individuals and groups (J. Campbell 1964; Anne Cleary 1968; Cleary and Hilton 1966; Cole 1973; Cole and Moss 1989; Dorans, Chap. 7, this volume; Frederiksen 1984; Linn 1973, 1975, 1976; Linn and Werts 1971; Messick 1975, 1980, 1989; Wild and Dwyer 1980; Willingham and Cole 1997; Xi 2010).

16.4.1 Fairness and Bias

Although the terms *fairness* and *bias* can be interpreted as covering roughly the same ground, with *fairness* being defined as the absence of bias, fairness often reflects a broader set of issues, including the larger issues of social equity. In contrast, *bias* may be given a narrower and more technical interpretation in terms of irrelevant factors that distort the interpretation of test scores:

> The word fairness suggests fairness that comes from impartiality, lacking in prejudice or favoritism. This implies that a fair test is comparable from person to person and group to group. Comparable in what respect? The most reasonable answer is validity, since validity is the raison d'etre of the entire assessment enterprise. (Willingham 1999, p. 220)

In its broadest uses, fairness tends to be viewed as an ethical and social issue concerned with "the justice and impartiality inherent in actions" (Willingham 1999, p. 221). Bias, conversely, is often employed as a technical concept, akin to the notion of bias in the estimation of a statistical parameter. For example, Cole and Moss (1989) defined bias as the "differential validity of a particular interpretation of a test score for any definable, relevant group of test takers" (p. 205).

Standardized testing programs are designed to treat all test takers in the same way (or if accommodations are needed, in comparable ways), thereby eliminating as many sources of irrelevant variance as possible. By definition, to the extent that testing materials or conditions are not standardized, they can vary from test taker to test taker and from one test administration to another, thereby introducing irrelevant variance, or bias, into test scores. Much of this irrelevant variance would be essentially random, but some of it would be systematic in the sense that some test scores (e.g., those from a test site with an especially lenient or especially severe proctor) would be consistently too high or too low. Standardization also tends to control some kinds of intentional favoritism or negative bias by mandating consistent treatment of all test takers. Test scores that consistently underestimate or overestimate the variable of interest for a subgroup for any reason are said to be biased, and standardization tends to control this kind of bias, whether it is inadvertent or intentional.

ETS and other testing organizations have developed systematic procedures designed to identify and eliminate any aspects of item content or presentation that might have an undue effect on the performance of some test takers: "According to the guidelines used at ETS, for example, the test 'must not contain language, symbols, words, phrases, or examples that are generally regarded as sexist, racist, or otherwise potentially offensive, inappropriate, or negative toward any group'" (Zwick 2006, p. 656). Nevertheless, over time, there was a growing realization that treating everyone in the same way does not necessarily ensure fairness or lack of bias. It is a good place to start (particularly as a way to control opportunities for favoritism, racism, and other forms of more or less overt bias), but it does not fully resolve the issue. As Turnbull (1951) and others recognized from mid-century, fairness depends on the appropriateness of the uses of test scores, and test scores that provide unbiased measures of a particular set of skills may not provide unbiased measures of a broader domain of skills needed in some context (e.g., in an occupation or in an educational program). In such cases, those test scores may not provide a fair basis for making decisions about test takers (Shimberg 1981, 1982, 1990).

Over the last 65 years or so, ETS researchers have been active in investigating questions about bias and fairness in testing, in defining issues of fairness and bias, and in developing approaches for minimizing bias and for enhancing fairness. Many of the issues are still not fully resolved, in part because questions of bias depend on the intended interpretation and because questions of fairness depend on values.

16.4.2 Adverse Impact and Differential Prediction

Unless we are willing to assume, a priori, that there are no differences between groups in the characteristic being measured, simple differences between groups in average scores or the percentages of test takers achieving some criterion score do not necessarily say anything about the fairness of test scores or of score uses. In 1971, the U.S. Supreme Court, in *Griggs v. Duke Power Co.*, struck down some employment practices at the Duke Power Company that had led to substantially different hiring rates between Black and White applicants, and in its decision, the Court relied on two concepts, adverse impact and business necessity, that have come to play an important role in discussions of possible bias in score-based selection programs. *Adverse impact* occurs if a protected group (defined by race, ethnicity, or gender, as specified in civil rights legislation) has a substantially lower rate of selection, certification, or promotion compared to the group with the highest rate. A testing program has *business necessity* if the scores are shown to be related to some important outcome (e.g., some measure of performance on the job). A testing program with adverse impact against one or more protected groups was required to demonstrate business necessity for the testing program; if there was no adverse impact, there was no requirement to establish business necessity. Employers and other organizations using test scores for selection would either have to develop selection programs that had little adverse impact or would have to demonstrate business necessity (Linn 1972). In *Griggs*, Duke Power's testing program was struck down because it had substantial adverse impact, and the company had made no attempt to investigate the relationship between test scores and performance on the job.

Although the terminology of *adverse impact* and *business necessity* was not in common use before *Griggs*, the notion that test scores can be considered fair if they reflect real differences in performance, even if they also suffer from adverse impact, was not new. Turnbull (1951) had pointed out the importance of evaluating fairness in terms of the proposed interpretation and use of the scores:

> That method is to define the criterion to which a test is intended to relate, and then to justify inter-group equality or inequality of test scores on the basis of its effect on prediction. It is necessarily true that an equality of test scores that would signify fairness of measurement for one criterion on which cultural groups performed alike would signify unfairness for another criterion on which group performance differed. Fairness, like its amoral brother, validity, resides not in tests or test scores but in the relation of test scores to criteria. (pp. 148–149)

Adverse impact does not necessarily say much about fairness, but it does act as a trigger that suggests that the relationship between test scores and appropriate criteria be evaluated (Dorans, Chap. 7, this volume; Linn 1973, 1975; Linn and Werts 1971; Messick 1989).

By 1971, when the Griggs decision was rendered, Cleary (1968) had already published her classic study of differential prediction, which was followed by a number of differential-prediction studies at ETS and elsewhere. The Cleary model stipulated that

> a test is biased for members of a subgroup of the population if, in the prediction of a criterion for which the test is designed, consistent nonzero errors of prediction are made for members of the subgroup. In other words, the test is biased if the criterion score predicted from the common regression line is consistently too high or too low for members of the subgroup. (p. 115)

The Cleary criterion is simple, clear, and direct; if the scores underpredict or overpredict the relevant criterion, the predictions can be considered biased. Note that although Cleary talked about the test being biased, her criterion applies to the predictions based on the scores and not directly to the test or test scores. In fact, the predictions can be biased in Cleary's sense without having bias in the test scores, and the predictions can be unbiased in Cleary's sense while having bias in the test scores (Zwick 2006). Nevertheless, assuming that the criterion measure is appropriate and unbiased (which can be a contentious assumption in many contexts; e.g., see Linn 1976; Wild and Dwyer 1980), the comparison of regressions made perfect sense as a way to evaluate predictive bias.

However, as a criterion for evaluating bias in the test scores, the comparison of regression lines is problematic for a number of reasons. Linn and Werts (1971) pointed out two basic statistical problems with the Cleary model; the comparisons of the regression lines can be severely distorted by errors of measurement in the independent variable (or variables) and by the omission of relevant predictor variables. Earlier, Lord (1967) had pointed to an ambiguity in the interpretation of differential-prediction analyses for groups with different means on the two measures, if the measures had less than perfect reliability or relevant predictors had been omitted.

In the 1970s, a concerted effort was made by many researchers to develop models of fairness that would make it possible to identify and remove (or at least ameliorate) group inequities in score-based decision procedures, and ETS researchers were heavily involved in these efforts (Linn 1973, 1984; Linn and Werts 1971; Myers 1975; Petersen and Novick 1976). These efforts raised substantive questions about what we might mean by fairness in selection, but by the early 1980s, interest in this line of research had declined for several reasons.

First, a major impetus for the development of these models was the belief in the late 1960s that at least part of the explanation for the observed disparities in test scores across groups was to be found in the properties of the test. The assumption was that cultural differences and differences in educational and social opportunities caused minority test takers to be less familiar with certain content and to be less adept at taking objective tests, and therefore the test scores were expected to underpredict performance in nontest settings (e.g., on the job, in various educational programs). Many of the fairness models were designed to adjust for inequities (defined in various ways) that were expected to result from the anticipated underprediction of performance. However, empirical results indicated that the test scores did not underpredict the scores of minority test takers, but rather overpredicted the performance of Black and Hispanic students on standard criteria, particularly first-year grade point average (GPA) in college (Cleary 1968; Young 2004; Zwick 2006). The test scores did underpredict the scores of women, but this difference was due in part to differences in courses taken (Wild and Dwyer 1980; Zwick 2006).

Second, Petersen and Novick (1976) pointed out some basic inconsistencies in the structures of the fairness models and suggested that it was necessary to explicitly incorporate assumptions about relative utilities of different outcomes for different test takers to resolve these discrepancies. However, it was not clear how to specify such utilities, and it was especially not clear how to get all interested stakeholders to agree on a specific set of such utilities.

As the technical difficulties mounted (Linn 1984; Linn and Werts 1971; Petersen and Novick 1976) and the original impetus for the development of the models (i.e., underprediction for minorities) turned out to be wrong (Cleary 1968; Linn 1984), interest in the models proposed to correct for underprediction faded.

An underlying concern in evaluating fairness was (and is) the acknowledged weaknesses in the criterion measures (Wild and Dwyer 1980). In addition to being less reliable than the tests being evaluated, and in representing proxy measures of success that are appealing in large part because of their ready availability, there is evidence that the criteria are, themselves, not free of bias (Wild and Dwyer 1980).

One major result of this extended research program is a clear realization that fairness and bias are very complex, multifaceted issues that cannot be easily reduced to a formal model of fairness or evaluated by straightforward statistical analyses (Cole and Moss 1989; Messick 1989; Wild and Dwyer 1980): "The institutions and professionals who sponsor and use tests have one view as to what is fair; examinees have another. They will not necessarily always agree, though both have a legitimate claim" (Willingham 1999, p. 224). Holland (1994) and Dorans (2012) suggested that analyses of test score fairness should go beyond the measurement perspective, which tends to focus on the elimination or reduction of construct-irrelevant variance (or measurement bias), to include the test taker's perspective, which tends to view tests as "contests," and Kane (2013b) has suggested adding an institutional perspective, which has a strong interest in eliminating any identifiable source of bias but also has an interest in reducing adverse impact, whether it is due to an identifiable source of bias or not.

16.4.3 Differential Item Functioning

ETS played a major role in the introduction of DIF methods as a way to promote fairness in testing programs (Dorans and Holland 1993; Holland and Thayer 1988). These methods identify test items that, after matching on an estimate of the attribute of interest, are differentially difficult or easy for a target group of test takers, as compared to some reference group. ETS pioneered the development of DIF methodology, including the development of the most widely used methods, as well as investigations of the statistical properties of these methods, matching variables, and sample sizes (Dorans, Chap. 7, this volume; Holland and Wainer 1993).

DIF analyses are designed to differentiate, across groups, between real differences in the construct being measured and sources of group-related construct-irrelevant variance. Different groups are not evaluated in terms of their differences in performance but rather in terms of differences in performance on each item, given the candidates' standings on the construct being measured, as indicated by the test taker's total score on the test (or some other relevant matching variable). DIF analyses provide an especially appealing way to address fairness issues, because the data required for DIF analyses (i.e., item responses and test scores) are readily available for most standardized testing programs and because DIF analyses provide a direct way to decrease construct-irrelevant differential impact (by avoiding the use of items with high DIF).

Zieky (2011) has provided a particularly interesting and informative analysis of the origins of DIF methodology. As noted earlier, from ETS's inception, its research staff had been concerned about fairness issues and had been actively investigating group differences in performance since the 1960s (Angoff and Ford 1973; Angoff and Sharon 1974; Cardall and Coffman 1964; Cleary 1968), but no fully adequate methodology for addressing group differences at the item level had been identified. The need to address the many obstacles facing the effective implementation of DIF was imposed on ETS researchers in the early 1980s:

> In 1984, ETS settled a lawsuit with the Golden Rule Insurance Company by agreeing to use raw differences in the percentages correct on an item in deciding on which items to include in a test to license insurance agents in Illinois; if two items were available that both met test specifications, the item with the smallest black-white difference in percentage correct was to be used; any difference in the percentages was treated as bias "even if it were caused by real and relevant differences between the groups in average knowledge of the tested subject." (Zieky 2011, p. 116)

The Golden Rule procedure was seen as causing limited harm in a minimum-competency licensing context but was seen as much more problematic in other contexts in which candidates would be ranked in terms of cognitive abilities or achievement, and concern grew that test quality would suffer if test developers were required to use only items "with the smallest raw differences in percent correct between Black and White test takers, regardless of the causes of these differences" (Zieky 2011, pp. 117–118):

> The goal was an empirical means of distinguishing between real group differences in the knowledge and skill measured by the test and unfair differences inadvertently caused by biased aspects of items. Test developers wanted help in ensuring that items were fair, but each method tried so far either had methodological difficulties or was too unwieldy to use on an operational basis with a wide variety of tests and several groups of test takers. The threat of legislation that would mandate use of the Golden Rule procedure for all tests further motivated ETS staff members to adopt a practical measure of DIF. (p. 118)

In response, ETS researchers (e.g., Dorans and Holland 1993; Holland and Thayer 1988) developed procedures that evaluated differential group performance, conditional on test takers' relative standing on the attribute of interest. The DIF methodology developed at ETS is now widely used in testing programs that aid in making high-stakes decisions throughout the world.

E. Burton and Burton (1993) found that the differences in scores across groups did not narrow substantially after the implementation of DIF analyses. Test items are routinely screened for sensitivity and other possible sources of differential functioning before administration, and relatively few items are flagged by the DIF statistics. As Zwick (2006) noted,

> even in the absence of evidence that it affects overall scores, … DIF screening is important as a precaution against the inclusion of unreasonable test content and as a source of information that can contribute to the construction of better tests in the future. (p. 668)

DIF screening addresses an issue that has to be confronted for psychometric and ethical reasons. That these checks on the quality of test items turn up relatively few cases of questionable item content is an indication that the item development and screening procedures are working as intended.

16.4.4 Identifying and Addressing Specific Threats to Fairness/ Validity

As illustrated in the two previous subsections, much of the research on fairness at ETS, and more generally in the measurement research community, has focused on the identification and estimation of differential impact and potential bias in prediction and selection, a global issue, and on DIF, which addresses particular group-specific item effects that can generate adverse impact or bias. However, some researchers have sought to address other potential threats to fairness and, therefore, to validity.

Xi (2010) pointed out that fairness is essential to validity and validity is essential to fairness. If we define validity in terms of the appropriateness of proposed interpretations and uses of scores, and fairness in terms of the appropriateness of proposed interpretations and uses of scores across groups, then fairness would be a necessary condition for validity; if we define fairness broadly in terms of social justice, then validity would be a necessary condition for fairness. Either way, the two concepts are closely related; as noted earlier, Turnbull referred to validity as the "amoral brother" of fairness (Dorans, Chap. 7, this volume; Turnbull 1951).

Xi (2010) combined fairness and validity in a common framework by evaluating fairness as comparable validity across groups within the population of interest. She proposed to identify and evaluate any fairness-based objections to proposed interpretations and uses of the test scores as a *fairness argument* that would focus on whether an interpretation is equally plausible for different groups and whether the decision rules are appropriate for the groups. Once the inferences and assumptions inherent in the proposed interpretation and use of the test scores have been specified, they can be evaluated in terms of whether they apply equally well to different groups. For example, it can be difficult to detect construct underrepresentation in a testing program by qualitatively evaluating how well the content of the test represents the content of a relevant domain, but empirical results indicating that there are substantial differences across groups in the relationship between performance on the test and more thorough measures of performance in the domain as a whole could raise serious questions about the representativeness of the test content. This argument-based approach can help to focus research on serious, specific threats to fairness/validity (Messick 1989).

Dorans and colleagues (Dorans, Chap. 7, this volume; Dorans and Holland 2000; Holland and Dorans 2006) have addressed threats to fairness/validity that can arise in scaling/equating test scores across different forms of a test:

> Scores on different forms or editions of a test that are supposed to be used interchangeably should be related to each other in the same way across different subpopulations. Score equity assessment (SEA) uses subpopulation invariance of linking functions across important subpopulations to assess the interchangeability of the scores. (Dorans, Chap. 7, this volume)

If the different forms of the test are measuring the same construct or combination of attributes in the different subpopulations, the equating function should not depend on the subpopulation on which it is estimated, and

> one way to demonstrate that two test forms are not equatable is to show that the equating functions used to link their scores are not invariant across different subpopulations of examinees. Lack of invariance in a linking function indicates that the differential difficulty of the two test forms is not consistent across different groups. (Dorans, Chap. 7, this volume)

SEA uses the invariance of the linking function across groups to evaluate consistency of the proposed interpretation of scores across groups and, thereby, to evaluate the validity of the proposed interpretation.

Mislevy et al. (2013) sought to develop systematic procedures for minimizing threats to fairness due to specific construct-irrelevant sources of variance in the assessment materials or procedures. To the extent that a threat to validity can be identified in advance or concurrently, the threat could be eliminated by suitably modifying the materials or procedures; for example, if it is found that English language learners have not had a chance to learn specific nontechnical vocabulary in a mathematics item, that vocabulary could be changed or the specific words could be defined. Mislevy et al. combined the general methodology of "universal design" with the ECD framework. In doing so, they made use of M. von Davier's (2008) general diagnostic model as a psychometric framework to identify specific requirements in test tasks. Willingham (1999) argued that test uses would be likely to be fairer across groups if "the implications of design alternatives are carefully examined at the outset" (p. 235) but recognized that this examination would be difficult to do "without much more knowledge of subgroup strengths and weaknesses… than is normally available" (p. 236). Mislevy et al. (2013) have been working to develop the kind of knowledge needed to build more fairness into testing procedures from the design stage.

16.5 Messick's Unified Model of Construct Validity

Samuel Messick spent essentially all of his professional life at ETS, and during his long and productive career, he made important contributions to many parts of test theory and to ETS testing programs, some of which were mentioned earlier. In this section, we focus on his central role in the development of the construct validity model and its transformation into a comprehensive, unified model of validity (Messick 1975, 1988, 1989). Messick's unified model pulled the divergent strands in validity theory into a coherent framework, based on a broad view of the meaning of test scores and the values and consequences associated with the scores, and in doing so, he gave the consequences of score use a prominent role.

Messick got his bachelor's degree in psychology and natural sciences from the University of Pennsylvania in 1951, and he earned his doctorate from Princeton University in 1954, while serving as an ETS Psychometric Fellow. His doctoral dissertation, "The Perception of Attitude Relationships: A Multidimensional Scaling Approach to the Structuring of Social Attitudes," reflected his dual interest in quantitative methods and in personality theory and social psychology. He completed postdoctoral fellowships at the University of Illinois, studying personality dynamics, and at the Menninger Foundation, where he did research on cognition and personality and received clinical training. He started as a full-time research psychologist at ETS in 1956, and he remained there until his death in 1998. Messick also served as a visiting lecturer at Princeton University on personality theory, abnormal psychology, and human factors between 1956 and 1958 and again in 1960–1961.

Messick completed his doctoral and postdoctoral work and started his career at ETS just as the initial version of construct validity was being developed (Cronbach and Meehl 1955). As noted, he came to ETS with a strong background in personality theory (e.g., see Messick 1956, 1972), where constructs play a major role, and a strong background in quantitative methods (e.g., see Gulliksen and Messick 1960; Messick and Abelson 1957; Schiffman and Messick 1963). Construct validity was originally proposed as a way to justify interpretations of test scores in terms of psychological constructs (Cronbach and Meehl 1955), and as such, it focused on psychological theory. Subsequently, Loevinger (1957) suggested that the construct model could provide a framework for all of validity, and Messick made this suggestion a reality. Between the late 1960s and the 1990s, he developed a broadly defined construct-based framework for the validation of test score interpretations and uses; his unified framework had its most complete statement in his validity chapter in the third edition of *Educational Measurement* (Messick 1989).

As Messick pursued his career, he maintained his dual interest in psychological theory and quantitative methods, applying this broad background to problems in educational and psychological measurement (Jackson and Messick 1965; Jackson et al. 1957; Messick and Frederiksen 1958; Messick and Jackson 1958; Messick and Ross 1962). He had close, long-term collaborations with a number of research psychologists (e.g., Douglas Jackson, Nathan Kogan, and Lawrence Stricker). His long-term collaboration with Douglas Jackson, whom he met while they were both postdoctoral fellows at the Menninger Foundation, and with whom he coauthored more than 25 papers and chapters (Jackson 2002), was particularly productive.

Messick's evolving understanding of constructs, their measurement, and their vicissitudes was, no doubt, strongly influenced by his background in social psychology and personality theory and by his ongoing collaborations with colleagues with strong substantive interest in traits and their roles in psychological theory. His work reflected an ongoing concern about how to differentiate between constructs (Jackson and Messick 1958; Stricker et al. 1969), between content and style (Jackson and Messick 1958; Messick 1962, 1991), and between constructs and potential sources of irrelevant variance (Messick 1962, 1964, 1981b; Messick and Jackson 1958).

Given his background and interests, it is not surprising that Messick became an "early adopter" of the construct validity model. Throughout his career, Messick tended to focus on two related questions: Is the test a good measure of the trait or construct of interest, and how can the test scores be appropriately used (Messick 1964, 1965, 1970, 1975, 1977, 1980, 1989, 1994a, b)? For measures of personality, he addressed the first of these questions in terms of "two critical properties for the evaluation of the purported personality measure … the measure's *reliability* and its *construct validity*" (Messick 1964, p. 111). Even in cases where the primary interest is in predicting behavior as a basis for decision making, and therefore, where it is necessary to develop evidence for adequate predictive accuracy, he emphasized the importance of evaluating the construct validity of the scores:

> Instead of talking about the reliability and construct validity (or even the empirical validity) of the *test* per se, it might be better to talk about the reliability and construct validity of the *responses* to the test, as summarized in a particular score, thereby emphasizing that these test properties are relative to the processes used by the subjects in responding. (Messick 1964, p. 112)

Messick also exhibited an abiding concern about ethical issues in research and practice throughout his career (Messick 1964, 1970, 1975, 1977, 1980, 1981b, 1988, 1989, 1998, 2000). In 1965, he examined some criticisms of psychological testing and discussed the possibilities for regulation and self-regulation for testing. He espoused "an 'ethics of responsibility,' in which pragmatic evaluations of the consequences of alternative actions form the basis for particular ethical decisions" (p. 140). Messick (1965) went on to suggest that policies based on values reflect and determine how we see the world, in addition to their intended regulatory effects, and he focused on "the value-laden nature of validity and fairness as psychometric concepts" (Messick 2000, p. 4) throughout his career. It is to this concern with meaning and values in measurement that we now turn.

16.5.1 Meaning and Values in Measurement

Messick was consistent in emphasizing ethical issues in testing, the importance of construct validity in evaluating meaning and ethical questions, and the need to consider consequences in evaluating test use: "But the ethical question of '*Should* these actions be taken?' cannot be answered by a simple appeal to empirical validity alone. The various social consequences of these actions must be contended with"

(Messick and Anderson 1970, p. 86). In 1975, Messick published a seminal paper that focused on meaning and values in educational measurement and explored the central role of construct-based analyses in analyzing meaning and in anticipating consequences. In doing so, he sketched many of the themes that he would subsequently develop in more detail. The paper (Messick 1975) was the published version of his presidential speech to Division 5 (Evaluation and Measurement) of the American Psychological Association. The title, "The Standard Problem: Meaning and Values in Measurement and Evaluation," indicates the intended breadth of the discussion and its main themes. As would be appropriate for such a speech, Messick focused on big issues in the field, and we will summarize five of these: (a) the central role of construct-based reasoning and analysis in validation, (b) the importance of ruling out alternate explanations, (c) the need to be precise about the intended interpretations, (d) the importance of consequences, and (e) the role of content-related evidence in validation.

First, Messick emphasized the central role of construct-based reasoning and analysis in validation. He started the paper by saying that any discussion of the meaning of a measure should center on construct validity as the "evidential basis" for inferring score meaning, and he associated construct validity with basic scientific practice:

> Construct validation is the process of marshalling evidence in the form of theoretically relevant empirical relations to support the inference that an observed response consistency has a particular meaning. The problem of developing evidence to support an inferential leap from an observed consistency to a construct that accounts for that consistency is a generic concern of all science. (Messick 1975, p. 955)

A central theme in the 1975 paper is the interplay between theory and data. Messick suggested that, in contrast to concurrent, predictive, and content-based approaches to validation, each of which focused on a specific question, construct validation involves hypothesis testing and "all of the philosophical and empirical means by which scientific theories are evaluated" (p. 956). He wrote, "The process of construct validation, then, links a particular measure to a more general theoretical construct, usually an attribute or process or trait, that itself may be embedded in a more comprehensive theoretical network" (Messick 1975, p. 955). Messick took construct validation to define validation in the social sciences but saw education as slow in adopting this view. A good part of Messick's (1975) exposition is devoted to suggestions for why education had not adopted the construct model more fully by the early 1970s and for why that field should expand its view of validation beyond simple content and predictive interpretations. He quoted Loevinger (1957) to the effect that, from a scientific point of view, construct validity is validity, but he went further, claiming that content and criterion analyses are not enough, even for applied decision making, and that "the meaning of the measure must also be pondered in order to evaluate responsibly the possible consequences of the proposed use" (Messick 1975, p. 956). Messick was not so much suggesting the adoption of a particular methodology but rather encouraging us to think deeply about meanings and consequences.

Second, Messick (1975) emphasized the importance of ruling out alternate explanations in evaluation and in validation. He suggested that it would be effective and efficient to

> direct attention from the outset to vulnerabilities in the theory by formulating *counterhypotheses*, or plausible alternative interpretations of the observed consistencies. If repeated challenges from a variety of plausible rival hypotheses can be systematically discounted, then the original interpretation becomes more firmly grounded. (p. 956)

Messick emphasized the role of convergent/divergent analyses in ruling out alternative explanations of test scores.

This emphasis on critically evaluating proposed interpretations by empirically checking their implications was at the heart of Cronbach and Meehl's (1955) formulation of construct validity, and it reflects Popper's (1965) view that conjecture and refutation define the basic methodology of science. Messick's insistence on the importance of this approach probably originates less in the kind of philosophy of science relied on by Cronbach and Meehl and more on his training as a psychologist and on his ongoing collaborations with psychologists, such as Jackson, Kogan, and Stricker. Messick had a strong background in measurement and scaling theory (Messick and Abelson 1957), and he maintained his interest in these areas and in the philosophy of science throughout his career (e.g., see Messick 1989, pp. 21–34). His writings, however, strongly suggested a tendency to start with a substantive problem in psychology and then to bring methodology and "philosophical conceits" (Messick 1989) to bear on the problem, rather than to start with a method and look for problems to which it can be applied. For example, Messick (1984, 1989; Messick and Kogan 1963) viewed cognitive styles as attributes of interest and not simply as sources of irrelevant variance.

Third, Messick (1975) recognized the need to be precise about the intended interpretations of the test scores. If the extent to which the test scores reflect the intended construct, rather than sources of irrelevant variance, is to be investigated, it is necessary to be clear about what is and is not being claimed in the construct interpretation, and a clear understanding of what is being claimed helps to identify plausible competing hypotheses. For example, in discussing the limitations of a simple content-based argument for the validity of a dictated spelling test, Messick pointed out that

> the inference of inability or incompetence from the absence of correct performance requires the elimination of a number of plausible rival hypotheses dealing with motivation, attention, deafness, and so forth. Thus, a report of failure to perform would be valid, but one of inability to perform would not necessarily be valid. The very use of the term *inability* invokes constructs of attribute and process, whereas a content-valid interpretation would stick to the outcomes. (p. 960)

To validate, or evaluate, the interpretation and use of the test scores, it is necessary to be clear about the meanings and values inherent in that interpretation and use.

Fourth, Messick (1975) gave substantial attention to values and consequences and suggested that, in considering any test use, two questions needed to be considered:

> First, is the test any good as a measure of the characteristic it is interpreted to assess? Second, should the test be used for the proposed purpose? The first question is a technical and scientific one and may be answered by appraising evidence bearing on the test's psychometric properties, especially construct validity. The second question is an ethical one, and its answer requires an evaluation of the potential consequences of the testing in terms of social values. We should be careful not to delude ourselves that answers to the first question are also sufficient answers to the second (except of course when a test's poor psychometric properties preclude its use). (p. 960)

Messick saw meaning and values as intertwined: "Just as values play an important role in measurement, where meaning is the central issue, so should meaning play an important role in evaluation, where values are the central issue" (p. 962). On one hand, the meanings assigned to scores reflect the intended uses of the scores in making claims about test takers and in making decisions. Therefore the meanings depend on the values inherent in these interpretations and uses. On the other hand, an analysis of the meaning of scores is fundamental to an evaluation of consequences because (a) the value of an outcome depends in part on how it is achieved (Messick 1970, 1975) and (b) an understanding of the meaning of scores and the processes associated with performance is needed to anticipate unintended consequences as well as intended effects of score uses.

Fifth, Messick (1975) recognized that content representativeness is an important issue in test development and score interpretation, but that, in itself, it cannot establish validity. For one, content coverage is a property of the test:

> The major problem here is that content validity in this restricted sense is focused upon test *forms* rather than test *scores*, upon *instruments* rather than *measurements*. Inferences in educational and psychological measurement are made from scores, … and scores are a function of subject responses. Any concept of validity of measurement must include reference to empirical consistency. (p. 960)

Messick suggested that Loevinger's (1957) substantive component of validity, defined as the extent to which the construct to be measured by the test can account for the properties of the items included in the test, "involves a confrontation between content representativeness and response consistency" (p. 961). The empirical analyses can result in the exclusion of some items because of perceived defects, or these analyses may suggest that the conception of the trait and the corresponding domain may need to be modified:

> These analyses offer evidence for the substantive component of construct validity to the extent that the resultant content of the test can be accounted for by the theory of the trait (along with collateral theories of test-taking behavior and method distortion). (p. 961)

Thus the substantive component goes beyond traditional notions of content validity to incorporate inferences and evidence on response consistency as well as on the extent to which the response patterns are consistent with our understanding of the corresponding construct.

16.5.2 A Unified but Faceted Framework for Validity

Over the following decade, Messick developed his unified, construct-based conception of validity in several directions. In the third edition of *Educational Measurement* (Messick 1989), he proposed a very broad and open framework for validity as scientific inquiry. The framework allows for different interpretations at different levels of abstraction and generality, and it encourages the use of multiple modes of inquiry. It also incorporates values and consequences. Given the many uses of testing in our society and the many interpretations entailed by these uses, Messick's unified model inevitably became complicated, but he wanted to get beyond the narrow views of validation in terms of content-, criterion-, and construct-related evidence:

> What is needed is a way of cutting and combining validity evidence that forestalls undue reliance on selected forms of evidence, that highlights the important though subsidiary role of specific content- and criterion-related evidence in support of construct validity in testing applications, and that formally brings consideration of value implications and social consequences into the validity framework. (Messick 1989, p. 20)

Messick organized his discussion of the roles of different kinds of evidence in validation in a 2 × 2 table (see Fig. 16.1) that he had introduced a decade earlier (Messick 1980). The table has four cells, defined in terms of the function of testing (interpretation or use) and the justification for testing (evidence or consequences):

> The *evidential basis of test interpretation* is construct validity. The *evidential basis of test use* is also construct validity, but as buttressed by evidence for the relevance of the test to the specific applied purpose and for the utility of the testing in the applied setting. The *consequential basis of test interpretation* is the appraisal of the value implications of the construct label, of the theory underlying test interpretation, and of the ideology in which the theory is embedded.... Finally, the *consequential basis of test use* is the appraisal of both potential and actual social consequences of the applied testing. (Messick 1989, p. 20, emphasis added)

Messick acknowledged that these distinctions were "interlocking and overlapping" (p. 20) and therefore potentially "fuzzy" (p. 20), but he found the distinctions and resulting fourfold classification to be helpful in structuring his description of the unified model of construct validity.

	Test Interpretation	**Test Use**
Evidential Basis	**Construct Validity**	**Construct Validity + Relevance/Utility**
Consequential Basis	**Value Implications**	**Social Consequences**

Fig. 16.1 Messick's facets of validity. From *Test Validity and the Ethics of Assessment* (p. 30, Research Report No. RR-79-10), by S. Messick, 1979, Princeton, NJ: Educational Testing Service. Copyright 1979 by Educational Testing Service. Reprinted with permission

16.5.3 The Evidential Basis of Test Score Interpretations

Messick (1989) began his discussion of the evidential basis of score interpretation by focusing on construct validity: "Construct validity, in essence, comprises the evidence and rationales supporting the trustworthiness of score interpretation in terms of explanatory concepts that account for both test performance and relationships with other variables" (p. 34). Messick saw convergent and discriminant evidence as "overarching concerns" in discounting construct-irrelevant variance and construct underrepresentation. *Construct-irrelevant variance* occurs to the extent that test score variance includes "excess reliable variance that is irrelevant to the interpreted construct" (p. 34). *Construct underrepresentation* occurs to the extent that "the test is too narrow and fails to include important dimensions or facets of the construct" (p. 34).

Messick (1989) sought to establish the "trustworthiness" of the proposed interpretation by ruling out the major threats to this interpretation. The basic idea is to develop a construct interpretation and then check on plausible threats to this interpretation. To the extent that the interpretation survives all serious challenges (i.e., the potential sources of construct-irrelevant variance and construct underrepresentation), it can be considered trustworthy. Messick was proposing that strong interpretations (i.e., in terms of constructs) be adopted, but he also displayed a recognition of the essential limits of various methods of inquiry. This recognition is the essence of the constructive-realist view he espoused; our constructed interpretations are ambitious, but they are constructed by us, and therefore they are fallible. As he concluded,

> validation in essence is scientific inquiry into score meaning—nothing more, but also nothing less. All of the existing techniques of scientific inquiry, as well as those newly emerging, are fair game for developing convergent and discriminant arguments to buttress the construct interpretation of test scores. (p. 56)

That is, rather than specify particular rules or guidelines for conducting construct validations, he suggested broad scientific inquiry that could provide support for and illuminate the limitations of proposed interpretations and uses of test scores.

Messick suggested that construct-irrelevant variance and construct underrepresentation should be considered serious when they interfere with intended interpretations and uses of scores to a substantial degree. The notion of "substantial" in this context is judgmental and depends on values, but the judgments are to be guided by the intended uses of the scores. This is one way in which interpretations and meanings are not value neutral.

16.5.4 The Evidential Basis of Test Score Use

According to Messick (1989), construct validity provides support for test uses. However, the justification of test use also requires evidence that the test is appropriate for a particular applied purpose in a specific applied setting: "The construct validity of score interpretation undergirds *all* score-based inferences, not just those related to interpretive meaningfulness but also the content- and criterion-related inferences specific to applied decisions and actions based on test scores" (pp. 63–64). Messick rejected simple notions of content validity in terms of domain representativeness in favor of an analysis of the constructs associated with the performance domain. "By making construct theories of the performance domain and of its key attributes more explicit, however, test construction and validation become more rational, and the supportive evidence sought becomes more attuned to the inferences made" (p. 64). Similarly, Messick rejected the simple model of predictive validity in terms of a purely statistical relationship between test scores and criterion scores in favor of a construct-based approach that focuses on hypotheses about relationships between predictor constructs and criterion constructs: "There is simply no good way to judge the appropriateness, relevance, and usefulness of predictive inferences in the absence of evidence as to what the predictor and criterion scores mean" (p. 64). In predictive contexts, it is the relationship between the characteristics of test takers and their future performances that is of interest. The observed relationship between predictor scores and criterion scores provides evidence relevant to this hypothetical relationship, but it does not exhaust the meaning of that relationship.

In elaborating on the evidential basis of test use, Messick (1989) discussed a number of particular kinds of score uses (e.g., employment, selection, licensure), and a number of issues that would need to be addressed (e.g., curriculum, instructional, or job relevance or representativeness; test–criterion relationships; the utility of criteria; and utility and fairness in decision making), rather than relying on what he called ad hoc targets. He kept the focus on construct validation and suggested that "one should strive to maximize the meaningfulness of score interpretation and to minimize construct-irrelevant test variance. The resulting construct-valid scores then provide empirical components for rationally defensible prediction systems and rational components for empirically informed decision making" (p. 65). Messick (1989) was quite consistent in insisting on the primacy of construct interpretations in validity, even in those areas where empirical methods had tended to predominate. He saw the construct theory of domain performance as the basis for developing both the criterion and the predictor. Constructs provided the structure for validation and the glue that held it all together.

16.5.5 The Consequential Basis of Test Score Interpretation

Messick (1989) saw the consequential basis of test score interpretation as involving an analysis of the value implications associated with the construct label, with the construct theory, and with the general conceptual frameworks, or ideologies, surrounding the theory. In doing so, he echoed his earlier emphasis (Messick 1980) on the role of values in validity:

> Constructs are broader conceptual categories than are test behaviors and they carry with them into score interpretation a variety of value connotations stemming from at least three major sources: the evaluative overtones of the construct labels themselves; the value connotations of the broader theories or nomological networks in which constructs are embedded; and the value implications of still broader ideologies about the nature of humankind, society, and science that color our manner of perceiving and proceeding. (Messick 1989, p. 59)

Neither constructs nor the tests developed to estimate constructs are dictated by the data, as such. We make decisions about the kinds of attributes that are of interest to us, and these choices are based on the values inherent in our views.

Messick (1989) saw values as pervading and shaping the interpretation and use of test scores and therefore saw the evaluation of value implications as an integral part of validation:

> In sum, the aim of this discussion of the consequential basis of test interpretation was to raise consciousness about the pervasive consequences of value-laden terms (which in any event cannot be avoided in either social action or social science) and about the need to take both substantive aspects and value aspects of score meaning into account in test validation. (p. 63)

Under a constructive-realist model, researchers have to decide how to carve up and interpret observable phenomena, and they should be clear about the values that shape these choices.

16.5.6 The Consequential Basis of Test Score Use

The last cell (bottom right) of Messick's progressive matrix addresses the social consequences of testing as an "integral part of validity" (Messick 1989, p. 84). The validity of a testing program is to be evaluated in terms of how well the program achieves its intended function or purpose without undue negative consequences:

> Judging validity in terms of whether a test does the job it is employed to do … that is, whether it serves its intended function or purpose—requires evaluation of the intended or unintended social consequences of test interpretation and use. The appropriateness of the intended testing purpose and the possible occurrence of unintended outcomes and side effects are the major issues. (pp. 84–85)

The central question is whether the testing program achieves its goals well enough and at a low enough cost (in terms of negative consequences, anticipated and unanticipated) that it should be used.

Messick's (1989) discussion of the consequences of testing comes right after an extended discussion of criterion-related evidence and analyses of utility, in terms of a specific criterion in selection, and it emphasizes that such utility analyses are important, but they are not enough. The evaluation, or validation, of a test score use requires an evaluation of all major consequences of the testing program and not simply evidence that a particular criterion is being estimated and optimized:

> Even if adverse testing consequences derive from valid test interpretation and use, the appraisal of the functional worth of the testing in pursuit of the intended ends should take into account all of the ends, both intended and unintended, that are advanced by the testing application, including not only individual and institutional effects but societal or systemic effects as well. Thus, although appraisal of intended ends of testing is a matter of social policy, it is not only a matter of policy formation but also of policy evaluation that weighs all of the outcomes and side effects of policy implementation by means of test scores. Such evaluation of the consequences and side effects of testing is a key aspect of the validation of test use. (p. 85)

Messick used the term *functional worth* to refer to the extent that a testing program achieves its intended goals and is relatively free of unintended negative consequences. He seems to contrast this concept with *test validity*, which focuses on the plausibility of the proposed interpretation of the test scores. The approach is unified, but the analysis in terms of the progressive matrix is structured, complex, and nuanced.

Messick (1989) made several points about the relationship between validity and functional worth. First, to the extent that consequences are relevant to the evaluation of a testing program (in terms of either validity or functional worth), both intended and unintended consequences are to be considered. Second, consequences are relevant to the evaluation of test validity if they result from construct-irrelevant characteristics of the testing program. Third, if the unintended consequences cannot be traced to construct-irrelevant aspects of the testing program, the evaluation of consequences, intended and unintended, becomes relevant to the functional worth of the testing program, which is in Messick's progressive matrix "an aspect of the validation of test use" (p. 85). Messick's main concern in his discussion of functional worth was to emphasize that in evaluating such worth, it is necessary to evaluate unintended negative consequences as well as intended, criterion outcomes so as to further inform judgments about test use.

Construct meaning entered Messick's (1989) discussion of the consequential basis of test use in large part as a framework for identifying unintended consequences that merit further study:

> But once again, the construct interpretation of the test scores plays a facilitating role. Just as the construct meaning of the scores afforded a rational basis for hypothesizing predictive relationships to criteria, construct meaning provides a rational basis for hypothesizing potential testing outcomes and for anticipating possible side effects. That is, the construct theory, by articulating links between processes and outcomes, provides clues to possible effects. Thus, evidence of construct meaning is not only essential for evaluating the import of testing consequences, it also helps determine where to look for testing consequences. (pp. 85–86)

Messick's unified framework for validity encourages us to think broadly and deeply, in this case in evaluating unintended consequences. He encouraged the use of multiple value perspectives in identifying and evaluating consequences. The unified framework for validity incorporates evaluations of the extent to which test scores reflect the construct of interest (employing a range of empirical and conceptual methods) and an evaluation of the appropriateness of the construct measures for the use at hand (employing a range of values and criteria), but ultimately, questions about how and where tests are used are policy issues.

Messick (1989) summarized the evidential and consequential bases of score interpretation and use in terms of the four cells in his progressive matrix:

> The process of construct interpretation inevitably places test scores both in a theoretical context of implied relationships to other constructs and in a value context of implied relationships to good and bad valuations, for example, of the desirability or undesirability of attributes and behaviors. Empirical appraisals of the former substantive relationships contribute to an *evidential basis of test interpretation*, that is, to construct validity. Judgmental appraisals of the latter value implications provide a *consequential basis of test interpretation*.
>
> The process of test use inevitably places test scores both in a theoretical context of implied relevance and utility and in a value context of implied means and ends. Empirical appraisals of the former issues of relevance and utility, along with construct validity contribute to an *evidential basis for test use*. Judgmental appraisals of the ends a proposed test use might lead to, that is, of the potential consequences of a proposed use and of the actual consequences of applied testing, provide a *consequential basis for test use*. (p. 89)

The four aspects of the unified, construct-based approach to validation provide a comprehensive framework for validation, but it is a framework intended to encourage and guide conversation and investigation. It was not intended as an algorithm or a checklist for validation.

Messick's (1989) chapter is sometimes criticized for being long and hard to read, and it is in places, but this perception should not be so surprising, because he was laying out a broad framework for validation; making the case for his proposal; putting it in historical context; and, to some extent, responding to earlier, current, and imagined future critics—not a straightforward task. When asked about the intended audience for his proposed framework, he replied, "Lee Cronbach" (M. Zieky, personal communication, May 20, 2014). As is true in most areas of scientific endeavor, theory development is an ongoing dialogue between conjectures and data, between abstract principles and applications, and between scholars with evolving points of view.

16.5.7 Validity as a Matter of Consequences

In one of his last papers, Messick (1998) revisited the philosophical conceits of his 1989 chapter, and in doing so, he reiterated the importance of values and consequences for validity:

> What needs to be valid are the inferences made about score meaning, namely, the score interpretation and its action implications for test use. Because value implications both derive from and contribute to score meaning, different value perspectives may lead to different score implications and hence to different validities of interpretation and use for the same scores. (p. 37)

Messick saw construct underrepresentation and construct-irrelevant variance as serious threats to validity in all cases, but he saw them as especially serious if they led to adverse consequences:

> All educational and psychological tests underrepresent their intended construct to some degree and all contain sources of irrelevant variance. The details of this underrepresentation and irrelevancy are typically unknown to the test maker or are minimized in test interpretation and use because they are deemed to be inconsequential. If noteworthy adverse consequences occur that are traceable to these two major sources of invalidity, however, then both score meaning and intended uses need to be modified to accommodate these findings. (p. 42)

And he continued, "This is precisely why unanticipated consequences constitute an important form of validity evidence. Unanticipated consequences signal that we may have been incomplete or off-target in test development and, hence, in test interpretation and use" (p. 43). Levels of construct underrepresentation and construct-irrelevant variance that would otherwise be acceptable would become unacceptable if it were shown that they had serious negative consequences.

16.5.8 The Central Messages

Messick's (1975, 1980, 1981a, 1988, 1989, 1995) treatment of validity is quite thorough and complex, but he consistently emphasizes a few basic conclusions.

First, validity is a unified concept. It is "an integrated evaluative judgment" of the degree to which evidence and rationales support the inferences and actions based on test scores. We do not have "kinds" of validity for different score interpretations or uses.

Second, all validity is construct validity. Construct validity provides the framework for the unified model of validity because it subsumes both the content and criterion models and reflects the general practice of science in which observation is guided by theory.

Third, validation is scientific inquiry. It is not a checklist or procedure but rather a search for the meaning and justification of score interpretations and uses. The meaning of the scores is always important, even in applied settings, because meaning guides both score interpretation and score use. Similarly, values guide the construction of meaning and the goals of test score use.

Fourth, validity and science are value laden. Construct labels, theories, and supporting conceptual frameworks involve values, either explicitly or implicitly, and it is good to be clear about the underlying assumptions. It is better to be explicit than implicit about our values.

Fifth, Messick maintained that validity involves the appraisal of social consequences of score uses. Evaluating whether a test is doing what it was intended to do necessarily involves an evaluation of intended and unintended consequences.

There were two general concerns that animated Messick's work on validity theory over his career, both of which were evident from his earliest work to his last papers. One was his abiding interest in psychological theory and in being clear and explicit about the theoretical and pragmatic assumptions being made. Like Cronbach, he was convinced that we cannot do without theory and, more specifically, theoretical constructs, and rather than ignoring substantive, theoretical assumptions, he worked to understand the connections between theories, constructs, and testing.

The second was his abiding interest in values, ethics, and consequences, which was evident in his writing from the 1960s (Messick 1965) to the end of his career (Messick 1998). He recognized that values influence what we look at and what we see and that if we try to exclude values from our testing programs, we will tend to make the values implicit and unexamined. So he saw a role for values in evaluating the validity of both the interpretations of test scores and the uses of those scores. He did not advocate that the measurement community should try to impose any particular set of values, but he was emphatic and consistent in emphasizing that we should recognize and make public the value implications inherent in score interpretations and uses.

16.6 Argument-Based Approaches to Validation

Over a period of about 25 years, from the early 1960s to 1989, Messick developed a broad construct-based framework for validation that incorporated concerns about score interpretations and uses, meaning and values, scientific reasoning and ethics, and the interactions among these different components. As a result, the framework was quite complex and difficult to employ in applied settings.

Since the early 1990s, researchers have developed several related approaches to validation (Kane 1992, 2006, 2013a; Mislevy 2006, 2009; Mislevy et al. 1999; Mislevy et al. 2003b; Shepard 1993) that have sought to streamline models of validity and to add some more explicit guidelines for validation by stating the intended interpretation and use of the scores in the form of an argument. The argument would provide an explicit statement of the claims inherent in the proposed interpretation and use of the scores (Cronbach 1988).

By explicitly stating the intended uses of test scores and the score interpretations supporting these uses, these argument-based approaches seek to identify the kinds of evidence needed to evaluate the proposed interpretation and use of the test scores and thereby to specify necessary and sufficient conditions for validation.

Kane (1992, 2006) suggested that the proposed interpretation and use of test scores could be specified in terms of an interpretive argument. After coming to ETS, he extended the argument-based framework to focus on an interpretation/use argument (IUA), a network of inferences and supporting assumptions leading from a test taker's observed performances on test tasks or items to the interpretive claims and decisions based on the test scores (Kane 2013a). Some of the inferences in the IUA would be statistical (e.g., generalization from an observed score to a universe score or latent variable, or a prediction of future performance); other inferences would rely on expert judgment (e.g., scoring, extrapolations from the testing context to nontest contexts); and many of the inferences might be evaluated in terms of several kinds of evidence.

Most of the inferences in the IUA would be presumptive in the sense that the inference would establish a presumption in favor of its conclusion, or claim, but it would not prove the conclusion or claim. The inference could include qualitative qualifiers (involving words such as "usually") or quantitative qualifiers (e.g., standard errors or confidence intervals), as well as conditions under which the inference would not apply. The IUA is intended to represent the claims being made in interpreting and using scores and is not limited to any particular kind of claim.

The IUAs for most interpretations and uses would involve a chain of linked inferences leading from the test performances to claims based on these performances; the conclusion of one inference would provide the starting point, or datum, for subsequent inferences. The IUA is intended to provide a fairly detailed specification of the reasoning inherent in the proposed interpretation and uses of the test scores. Assuming that the IUA is coherent, in the sense that it hangs together, and complete, in the sense that it fully represents the proposed interpretation and use of the scores, it provides a clear framework for validation. The inferences and supporting assumptions in the IUA can be evaluated using evidence relevant to their plausibility. If all of the inferences and assumptions hold up under critical evaluation (conceptual and empirical), the interpretation and use can be accepted as plausible, or valid; if any of the inferences or assumptions fail to hold up under critical evaluation, the proposed interpretation and use of the scores would not be considered valid.

An argument-based approach provides a validation framework that gives less attention to philosophical foundations and general concerns about the relationship between meaning and values than did Messick's unified, construct-based validation framework, and more attention to the specific IUA under consideration. In doing so, an argument-based approach can provide necessary and sufficient conditions for validity in terms of the plausibility of the inferences and assumptions in the IUA. The validity argument is contingent on the specific interpretation and use outlined in the IUA; it is the proposed interpretation and uses that are validated and not the test or the test scores.

The argument-based approach recognizes the importance of philosophical foundations and of the relationship between meaning and values, but it focuses on how these issues play out in the context of particular testing programs with a particular interpretation and use proposed for the test scores. The conclusions of such argument-based analyses depend on the characteristics of the testing program and the proposed interpretation and uses of the scores; the claims being based on the test scores are specified and the validation effort is limited to evaluating these claims.

Chapelle et al. (2008, 2010) used the argument-based approach to analyze the validity of the *TOEFL*® test in some detail and, in doing so, provided insight into the meaning of the scores as well as their empirical characteristics and value implications. In this work, it is clear how the emphasis in the original conception of construct validity (Cronbach and Meehl 1955) on the need for a program of validation research rather than a single study and Messick's emphasis on the need to rule out threats to validity (e.g., construct-irrelevant variance and construct underrepresentation) play out in an argument-based approach to validation.

Mislevy (1993, 1994, 1996, 2007) focused on the role of evidence in validation, particularly in terms of model-based reasoning from observed performances to more general claims about students and other test takers. Mislevy et al. (1999, 2002, 2003a, b) developed an ECD framework that employs argument-based reasoning. ECD starts with an analysis of the attributes, or constructs, of interest and the social and cognitive contexts in which they function and then designs the assessment to generate the kinds and amounts of evidence needed to draw the intended inferences. The ECD framework involves several stages of analysis (Mislevy and Haertel 2006; Mislevy et al. 1999, 2002, 2003a). The first stage, *domain analysis*, concentrates on building substantive understanding of the performance domain of interest, including theoretical conceptions and empirical research on student learning and performance, and the kinds of situations in which the performances are likely to occur. The goal of this first stage is to develop an understanding of how individuals interact with tasks and contexts in the domain.

At the second stage, *domain modeling*, the relationships between student characteristics, task characteristics, and situational variables are specified (Mislevy et al. 2003a, b). The structure of the assessment to be developed begins to take shape, as the kinds of evidence that would be relevant to the goals of the assessment are identified.

The third stage involves the development of a *conceptual assessment framework* that specifies the operational components of the test and the relationships among these components, including a student model, task models, and evidence models. The student model provides an abstract account of the student in terms of ability parameters (e.g., in an IRT model). Task models posit schemas for collecting data that can be used to estimate the student parameters and guidelines for task development. The evidence model describes how student performances are to be evaluated, or scored, and how estimates of student parameters can be made or updated. With this machinery in place, student performances on a sample of relevant tasks can be used to draw probabilistic inferences about student characteristics.

The two dominant threads in these argument-based approaches to validation are the requirement that the claims to be made about test takers (i.e., the proposed interpretation and use of the scores) be specified in advance, and then justified, and that inferences about specific test takers be supported by warrants or models that have been validated, using empirical evidence and theoretical rationales. The argument-based approaches are consistent with Messick's unified framework, but they tend to focus more on specific methodologies for the validation of proposed interpretations and uses than did the unified framework.

16.7 Applied Validity Research at ETS

In addition to the contributions to validity theory described above, ETS research has addressed numerous practical issues in documenting the validity of various score uses and interpretations and in identifying the threats to the validity of ETS tests. Relatively straightforward predictive validity studies were conducted at ETS from its earliest days, but ETS research also has addressed problems in broadening both the predictor and criterion spaces and in finding better ways of expressing the results of predictive validity studies. Samuel Messick's seminal chapter in the third edition of *Educational Measurement* (Messick 1989) focused attention on the importance of identifying factors contributing to construct-irrelevant variance and identifying instances of construct underrepresentation, and numerous ETS studies have focused on both of these problems.

16.7.1 Predictive Validity

Consistent with the fundamental claim that tests such as the SAT test were useful because they could predict academic performance, predictive validity studies were common throughout the history of ETS. As noted earlier, the second Research Bulletin published by ETS (RB-48-02) was a predictive study titled *The Prediction of First Term Grades at Hamilton College* (Frederiksen 1948). The abstract noted, "It was found that the best single predictor of first term average grade was rank in secondary school ($r = .57$). The combination of SAT scores with school rank was found to improve the prediction considerably ($R = .67$)." By 1949, enough predictive validity studies had been completed that results of 17 such studies could be summarized by Allen (1949). This kind of study was frequently repeated over the years, but even in the very earliest days there was considerable attention to a more nuanced view of predictive validity from both the perspective of potential predictors and potential criteria. As noted, the Frederiksen study cited earlier was the *second* Research Bulletin published by ETS, but the *first* study published (College Board 1948) examined the relationship of entrance test scores at the U.S. Coast Guard Academy to outcome variables that included both course grades and nonacademic ratings. On the predictor side, the study proposed that "a cadet's standing at the Academy be based on composite scores based on three desirable traits: athletic ability, adaptability, and academic ability." A follow-up study (French 1948) included intercorrelations of 76 measures that included academic and nonacademic tests as predictors and included grades and personality ratings as criteria. The conclusions supported the use of the academic entrance tests but noted that the nonacademic tests in that particular battery did not correlate with either grades or personality ratings.

Although there were a number of studies focusing on the prediction of first-year grades in the 1950s (e.g., Abelson 1952; Frederiksen et al. 1950a, b; Mollenkopf

1951; Schultz 1952), a number of studies went beyond that limited criterion. For example, Johnson and Olsen (1952) compared 1-year and 3-year predictive validities of the Law School Admissions Test in predicting grades. Mollenkopf (1950) studied the ability of aptitude and achievement tests to predict both first- and second-year grades at the U.S. Naval Postgraduate School. Although the second-year validities were described as "fairly satisfactory," they were substantially lower than the Year 1 correlations. This difference was attributed to a number of factors, including differences in the first- and second-year curricula, lower reliability of second-year grades, and selective dropout. Besides looking beyond the first year, these early studies also considered other criteria. French (1957), in a study of 12th-grade students at 42 secondary schools, related SAT scores and scores on the Tests of Developed Ability (TDA) to criteria that included high school grades but also included students' self-reports of their experiences and interests and estimations of their own abilities. In addition, teachers nominated students who they believed exhibited outstanding ability. The study concluded not only that the TDA predicted grades in physics, chemistry, biology, and mathematics but that, more so than the SAT, it was associated with self-reported scientific interests and experiences.

From the 1960s through the 1980s, ETS conducted a number of SAT validity studies that focused on routine predictions of the freshman grade point average (FGPA) with data provided from colleges using the College Board/ETS Validity Study Service as summarized by Ramist and Weiss (1990). In 1994, Ramist et al. (1994) produced a groundbreaking SAT validity study that introduced a number of innovations not found in prior work. First, the study focused on course grades, rather than FGPA, as the criterion. Because some courses are graded much more strictly than others, when grades from these courses are combined without adjustment in the FGPA, the ability of the SAT to predict freshman performance is underestimated. Several different ways of making the adjustment were described and demonstrated. Second, the study corrected for the range restriction in the predictors caused by absence of data for the low-scoring students not admitted to college. (Although the range restriction formulas were not new, they had not typically been employed in multicollege SAT validity studies.) Third, the authors adjusted course grades for unreliability. Fourth, they provided analyses separately for a number of subgroups defined by gender, ethnicity, best language, college selectivity, and college size. When adjustments were made for multivariate range restriction in the predictors, grading harshness/leniency for specific courses, and criterion unreliability, the correlation of the SAT with the adjusted grades was .64, and the multiple correlation of SAT and high school record with college grades was .75.

Subsequent SAT validity studies incorporated a number of these methods and provided new alternatives. Bridgeman et al. (2000), for example, used the course difficulty adjustments from the 1994 study but noted that the adjustments could be quite labor intensive for colleges trying to conduct their own validity studies. They showed that simply dividing students into two categories based on intended major (math/science [where courses tend to be severely graded] vs. other) recovered many of the predictive benefits of the complex course difficulty adjustments. In a variation on this theme, a later study by Bridgeman et al. (2008c) provided correlations separately for courses in four categories (English, social science, education, and science/math/engineering) and focused on cumulative grades over an entire college career, not just the first year. This study also showed that, contrary to the belief that the SAT predicts only FGPA, predictions of cumulative GPA over 4 or 5 years are similar to FGPA predictions.

16.7.2 Beyond Correlations

From the 1950s through the early 2000s, the predictive validity studies for the major admissions testing programs (e.g., SAT, the *GRE*® test, GMAT) tended to rely on correlations to characterize the relationship between test scores and grades. Test critics would often focus on unadjusted correlations (typically around .30). Squaring this number to get "variance accounted for," the critics would suggest that a test that explained less than 10% of the variance in grades must be of very little practical value (e.g., Fairtest 2003). To counter this perception, Bridgeman and colleagues started supplementing correlational results by showing the percentage of students who would succeed in college at various score levels (e.g., Bridgeman et al. 2008a, b, c; Cho and Bridgeman 2012). For example, in one study, 12,529 students at moderately selective colleges who had high school GPAs of at least 3.7 were divided into groups based on their combined Verbal and Mathematics SAT scores (Bridgeman et al. 2008a). Although college success can be defined in many different ways, this study defined success relatively rigorously as achieving a GPA of 3.5 or higher at the end of the college career. For students with total SAT scores (verbal + mathematics) of 1000 or lower, only 22% had achieved this level of success, whereas 73% of students in the 1410–1600 score category had finished college with a 3.5 or higher. Although SAT scores explained only about 12% of the variance in the overall group (which may seem small), the difference between 22% and 73% is substantial. This general approach to meaningful presentation of predictive validity results was certainly not new; rather, it is an approach that must be periodically rediscovered. As Ben Schrader noted in 1965,

> during the past 60 years, correlation and regression have come to occupy a central position in measurement and research.… Psychologists and educational researchers use these methods with confidence based on familiarity. Many persons concerned with research and testing, however, find results expressed in these terms difficult or impossible to interpret, and prefer to have results expressed in more concrete form. (p. 29)

He then went on to describe a method using expectancy tables that showed how standing on the predictor, in terms of fifths, related to standing on the criterion, also in terms of fifths. He used scores on the Law School Admission Test as the predictor and law school grades as the criterion. Even the 1965 interest in expectancy tables was itself a rediscovery of their explanatory value. In their study titled "Prediction of First Semester Grades at Kenyon College, 1948–1949," Frederiksen et al. (1950a) included an expectancy table that showed the chances in 100 that a student would

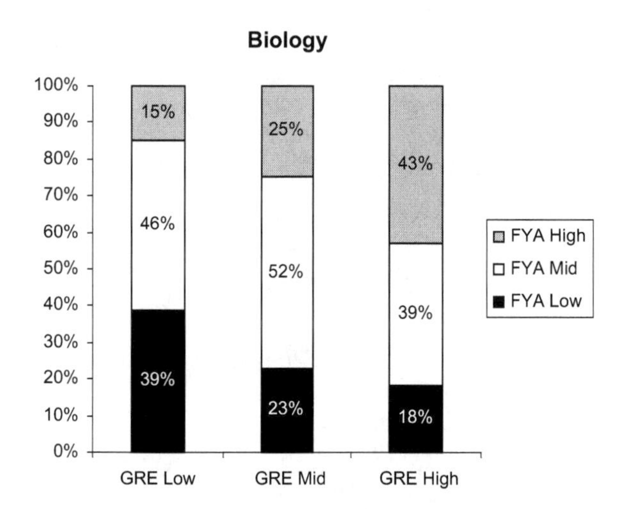

Fig. 16.2 Percentage of biology graduate students in GRE quartile categories whose graduate grade point averages were in the bottom quartile, mid-50%, or top quartile of the students in their departments across 24 programs (Adapted from Bridgeman et al. 2008a. Copyright 2008 by Educational Testing Service. Used with permission)

earn an average of at least a specified letter grade given a predicted grade based on a combination of high school rank and SAT scores. For example, for a predicted grade of B, the chance in 100 of getting at least a C+ was 88, at least a B was 50, and at least an A− was 12.

Despite the appeal of the expectancy table approach, it lay dormant until modest graphical extensions of Schrader's ideas were again introduced in 2008 and beyond. An example of this more graphical approach is in Fig. 16.2 (Bridgeman et al. 2008a, p. 10). Within each of 24 graduate biology programs, students were divided into quartiles based on graduate grades and into quartiles based on combined GRE verbal and quantitative scores. These results were then aggregated across the 24 programs and graphed. The graph shows that almost three times as many students with top quartile GRE scores were in the high-GPA category (top quartile) compared to the number of high-GPA students in the bottom GRE quartile.

The same report also used a graphical approach to show a kind of incremental validity information. Specifically, the bottom and top quartiles in each department were defined in terms of both undergraduate grade point average (UGPA) and GRE scores. Then, within the bottom UGPA quartile, students with top or bottom GRE scores could be compared (and similarly for the top UGPA quartile). Because graduate grades tend to be high, success was defined as achieving a 4.0 grade average. Figure 16.3 indicates that, even within a UGPA quartile, GRE scores matter for identifying highly successful students (i.e., the percentage achieving a 4.0 average).

16.7.3 Construct-Irrelevant Variance

The construct-irrelevant factors that can influence test scores are almost limitless. A comprehensive review of all ETS studies related to construct-irrelevant variance would well exceed the space limitations in this document; rather, a sampling of

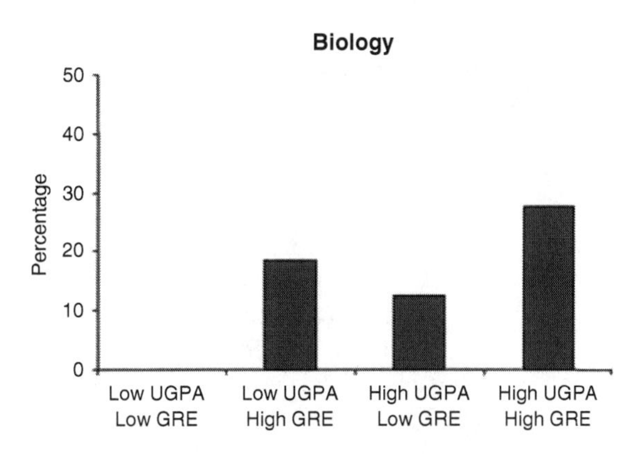

Fig. 16.3 Percentage of students in graduate biology departments earning a 4.0 grade point average by undergraduate grade point average and GRE high and low quartiles (Adapted from Bridgeman et al. 2008a Copyright 2008 by Educational Testing Service. Used with permission)

studies that explore various aspects of construct-irrelevant variance is presented. Research on one source of irrelevant variance, coaching, is described in a separate chapter by Donald Powers (Chap. 17, this volume).

16.7.3.1 Fatigue Effects

The potential for test-taker fatigue to interfere with test scores was already a concern in 1948, as suggested by the title of ETS Research Memorandum No. 48-02 by Tucker (1948), *Memorandum Concerning Study of Effects of Fatigue on Afternoon Achievement Test Scores Due to Scholastic Aptitude Test Being Taken in the Morning*. A literature review on the effects of fatigue on test scores completed in 1966 reached three conclusions:

> 1) Sufficient evidence exists in the literature to discount any likelihood of physiological consequences to the development of fatigue during a candidate's taking the College Board SAT or Achievement Tests; 2) the decline in feeling-tone experienced by an individual is often symptomatic of developing fatigue, but this decline does not necessarily indicate a decline in the quantity or quality of work output; and 3) the amount of fatigue that develops as a result of mental work is related to the individual's conception of, and attitude and motivation toward, the task being performed. (Wohlhueter 1966, Abstract)

A more recent experimental study conducted when the SAT was lengthened by the addition of the writing section reached a similar conclusion: "Results indicated that while the extended testing time for the new SAT may cause test takers to feel fatigued, fatigue did not affect test taker performance" (Liu et al. 2004, Abstract).

16.7.3.2 Time Limits

If a test is designed to assess speed of responding, then time limits merely enforce construct-relevant variance. But if the time limit is imposed primarily for administrative convenience, then a strict time limit might not be construct relevant. On one hand, an early study on the influence of timing on Cooperative Reading Test scores suggested no significant changes in means or standard deviations with extended time (Frederiksen 1951). On the other hand, Lord (1953, Abstract) concluded that "unspeeded (power) tests are more valid" based on a study of 649 students at one institution. Evans (1980) created four SAT-like test forms that were administered in one of three speededness conditions: normal, speeded, and unspeeded. Degree of speededness affected scores but did not interact with gender or ethnicity. The technical handbook for the SAT by Donlon (1984) indicated that the speed with which students can answer the questions should play only a very minor role in determining scores. A study of the impact of extending the amount of time allowed per item on the SAT concluded that there were some effects of extended time (1.5 times regular time); average gains for the verbal score were less than 10 points on the 200–800 scale and about 30 points for the mathematics scores (Bridgeman et al. 2004b). But these effects varied considerably depending on the ability level of the test taker. Somewhat surprisingly, for students with SAT scores of 400 or lower, extra time had absolutely no impact on scores. Effects did not interact with either gender or ethnicity. Extended time on the GRE was similarly of only minimal benefit with an average increase of 7 points for both verbal and quantitative scores on the 200–800 scale when the time limit was extended to 1.5 times standard time (Bridgeman et al. 2004a).

When new tests are created or existing tests are modified, appropriate time limits must be set. A special timing study was conducted when new item types were to be introduced to the SAT to provide an estimate of the approximate amount of time required to answer new and existing item types (Bridgeman and Cahalan 2007). The study used three approaches to estimate the amount of time needed to answer questions of different types and difficulties: (a) Item times were automatically recorded from a computer-adaptive version of the SAT, (b) students were observed from behind a one-way mirror in a lab setting as they answered SAT questions under strict time limits and the amount of time taken for each question was recorded, and (c) high school students recorded the amount of time taken for test subsections that were composed of items of a single type. The study found that the rules of thumb used by test developers were generally accurate in rank ordering the item types from least to most time consuming but that the time needed for each question was higher than assumed by test developers.

Setting appropriate time limits that do not introduce construct-irrelevant variance is an especially daunting challenge for evaluating students with disabilities, as extended time is the most common accommodation for these students. Evaluating the appropriateness of extended time limits for students with disabilities has been the subject of several research reports (e.g., Cahalan et al. 2006; Packer 1987; Ragosta and Wendler 1992) as well as receiving considerable attention in the book *Testing Handicapped People* (Willingham et al. 1988).

Setting appropriate time limits on a computer-adaptive test (CAT) in which different students respond to different items can be especially problematic. Bridgeman and Cline (2000) showed that when the GRE was administered as a CAT, items at the same difficulty level and meeting the same general content specifications could vary greatly in the time needed to answer them. For example, a question assessing the ability to add numbers with negative exponents could be answered very quickly while a question at the same difficulty level that required the solution of a pair of simultaneous equations would require much more time even for very able students. Test takers who by chance received questions that could be answered quickly would then have an advantage on a test with relatively strict time limits. Furthermore, running out of time on a CAT and guessing to avoid the penalty for an incomplete test can have a substantial impact on the test score because the CAT scoring algorithm assumed that an incorrect answer reflected a lack of ability and not an unlucky guess (Bridgeman and Cline 2004). A string of unlucky guesses at the end of the GRE CAT (because the test taker ran out of time and had to randomly respond) could lower the estimated score by more than 100 points (on a 200–800 scale) compared to the estimated score when the guessing began.

16.7.3.3 Guessing

Guessing can be a source of construct-irrelevant variance because noise is added to measurement precision when test takers answer correctly by guessing but actually know nothing about the answer (Wendler and Walker 2006). Corrections for guessing often referred to as formula scoring attempt to limit this irrelevant variance by applying a penalty for incorrect answers so that answering incorrectly has more negative consequences than merely leaving a question blank. For example, with the five-option multiple-choice questions on the SAT (prior to 2016), a test taker received 1 point for a correct answer and 0 points for an omitted answer, and one-fourth of a point was subtracted for each incorrect answer. (The revised SAT introduced in 2016 no longer has a correction for guessing.) By the time ETS was founded, there were already more than 20 years of research on the wisdom and effects of guessing corrections. Freeman (1952) surveyed this research and observed,

> At the outset, it may be stated that the evidence is not conclusive. While much that is significant has been written about the theoretical need to correct for guessing, and about the psychological and instructional value of such a correction, the somewhat atomistic, or at least uncoordinated, research that has been done during the last 25 years fails to provide an answer that can be generalized widely. (p. 1)

More than 60 years later, research is still somewhat contradictory and a definitive answer is still illusive. Lord (1974) argued that under certain assumptions, formula scoring is "clearly superior" to number-right scoring, though it remains unclear how often those assumptions are actually met. Angoff (1987) conducted an experimental study with different guessing instructions for SAT Verbal items and concluded, "Formula scoring is not disadvantageous to students who are less willing to guess and attempt items when they are not sure of the correct answer" (abstract). Conversely, some individuals and population subgroups may differ in their willingness to guess so that conclusions based on averages in the population as a whole may not be valid for all people. Rivera and Schmitt (1988), for example, noted a difference in willingness to guess on the part of Hispanic test takers, especially

Mexican Americans. Beginning in the 1981–1982 test year, the GRE General Test dropped formula scoring and became a rights-only scored test, but the GRE Subject Tests retained formula scoring. In 2011, the *Advanced Placement*® (*AP*®) test program dropped formula scoring and the penalty for incorrect answers. At the end of 2014, the SAT was still using formula scoring, but the announcement had already been made that the revised SAT would use rights-only scoring.

16.7.3.4 Scoring Errors

Any mistakes made in scoring a test will contribute to irrelevant variance. Although the accuracy of machine scoring of multiple-choice questions is now almost taken for granted, early in the history of ETS, there were some concerns with the quality of the scores produced by the scanner. Note that the formula scoring policy put special demands on the scoring machine because omitted answers and incorrect answers were treated differently. The machine needed to determine if a light mark was likely caused by an incomplete erasure (indicating intent to omit) or if the relatively light mark was indeed the intended answer. The importance of the problem may be gauged by the status of the authors, Fan, Lord, and Tucker, who devised "a system for reducing the number of errors in machine-scoring of multiple-choice answer sheets" (Fan et al. 1950, Abstract). Measuring and reducing rater-related scoring errors on essays and other constructed responses were also of very early concern. A study of the reading reliability of the College Board English Composition test was completed in 1948 (Aronson 1948; ETS 1948). In the following years, controlling irrelevant variance introduced by raters of constructed responses (whether human or machine) was the subject of a great deal of research, which is discussed in another chapter (Bejar, Chap. 18, this volume).

16.7.4 Construct Underrepresentation

Whereas construct-irrelevant variance describes factors that should not contribute to test scores, but do, construct underrepresentation is the opposite—failing to include factors in the assessment that should contribute to the measurement of a particular construct. If the purpose of a test or battery of tests is to assess the likelihood of success in college (i.e., the construct of interest), failure to measure the noncognitive skills that contribute to such success could be considered a case of construct underrepresentation. As noted, from the earliest days of ETS, there was interest in assessing more than just verbal and quantitative skills. In 1948, the organization's first president, Chauncey, called for a "Census of Abilities" that would assess attributes that went beyond just verbal and quantitative skills to include "personal qualities, … drive (energy), motivation (focus of energy), conscientiousness, … ability to get along with others" (Lemann 1995, p. 84). From 1959 to 1967, ETS had a personality research group headed by Samuel Messick. The story of personality research at ETS is described in two other chapters (Kogan, Chap. 14, this volume; Stricker, Chap. 13, this volume).

Despite the apparent value of broadening the college-readiness construct beyond verbal and quantitative skills, the potential of such additional measures as a part of operational testing programs needed to be rediscovered from time to time. Frederiksen and Ward (1978) described a set of tests of scientific thinking that were developed as potential criterion measures, though they could also be thought of as additional predictors. The tests assessed both quality and quantity of ideas in formulating hypotheses and solving methodological problems. In a longitudinal study of 3,500 candidates for admission to graduate programs in psychology, scores were found to be related to self-appraisals of professional skills, professional accomplishments in collaborating in research, designing research apparatus, and publishing scientific papers. In a groundbreaking article in the *American Psychologist*, Norman Frederiksen (1984) expanded the argument for a broader conception of the kinds of skills that should be assessed. In the article, titled "The Real Test Bias: Influences of Testing on Teaching and Learning," Frederiksen argued that

> there is evidence that tests influence teacher and student performance and that multiple-choice tests tend not to measure the more complex cognitive abilities. The more economical multiple-choice tests have nearly driven out other testing procedures that might be used in school evaluation. (Abstract)

Another article, published in the same year, emphasized the critical role of social intelligence (Carlson et al. 1984). The importance of assessing personal qualities in addition to academic ability for predicting success in college was further advanced in a multiyear, multicampus study that was the subject of two books (Willingham 1985; Willingham and Breland 1982). This study indicated the importance of expanding both the predictor and criterion spaces. The study found that if the only criterion of interest is academic grades, SAT scores and high school grades appear to be the best available predictors, but, if criteria such as leadership in school activities or artistic accomplishment are of interest, the best predictors are previous successes in those areas.

Baird (1979) proposed a measure of documented accomplishments to provide additional evidence for graduate admissions decisions. In contrast to a simple listing of accomplishments, documented accomplishments require candidates to provide verifiable evidence for their claimed accomplishments. The biographical inventory developed in earlier stages was evaluated in 26 graduate departments that represented the fields of English, biology, and psychology. Responses to the inventory were generally not related to graduate grades, but a number of inventory responses reflecting preadmission accomplishments were significantly related to accomplishments in graduate school (Baird and Knapp 1981). Lawrence Stricker and colleagues further refined measures of documented accomplishments (Stricker et al. 2001).

Moving into the twenty-first century, there was rapidly increasing interest in noncognitive assessments (Kyllonen 2005), and a group was established at ETS to deal specifically with these new constructs (or to revisit older noncognitive constructs that in earlier years had failed to gain traction in operational testing programs). The label "noncognitive" is not really descriptive and was a catch-all that included any assessment that went beyond the verbal, quantitative, writing, and subject matter skills and knowledge that formed the backbone of most testing programs at ETS. Key noncognitive attributes include persistence, dependability, motivation, and teamwork. One measure that was incorporated into an operational program was the *ETS*® Personal Potential Index (*ETS*® PPI) service, which was a standardized rating system in which individuals who were familiar with candidates for graduate school, such as teachers or advisors, could rate core personal attributes: knowledge and creativity, resilience, communication skills, planning and organization, teamwork, and ethics and integrity. All students who registered to take the GRE were given free access to the PPI and a study was reported that demonstrated how the diversity of graduate classes could be improved by making the PPI part of the selection criteria (Klieger et al. 2013). Despite its potential value, the vast majority of graduate schools were reluctant to require the PPI, at least in part because they were afraid of putting in place any additional requirements that they thought might discourage applicants, especially if their competition did not have a similar requirement. Because of this very low usage, ETS determined that the resources needed to support this program could be better used elsewhere and, in 2015, announced the end of the PPI as part of the GRE program. This announcement certainly did not signal an end to interest in noncognitive assessments. A noncognitive assessment, the *SuccessNavigator*® assessment, which was designed to assist colleges in making course placement decisions, was in use at more than 150 colleges and universities in 2015. An ongoing research program provided evidence related to placement validity claims, reliability, and fairness of the measure's scores and placement recommendations (e.g., Markle et al. 2013; Rikoon et al. 2014).

The extent to which writing skills are an important part of the construct of readiness for college or graduate school also has been of interest for many years. Although a multiple-choice measure of English writing conventions, the Test of Standard Written English, was administered along with the SAT starting in 1977, it was seen more as an aid to placement into English classes than as part of the battery intended for admissions decisions. Rather than the 200–800 scale used for Verbal and Mathematics tests, it had a truncated scale running from 20 to 60. By 2005, the importance of writing skills to college preparedness was recognized by inclusion of a writing score based on both essay and multiple-choice questions and reported on the same 200–800 scale as Verbal and Mathematics. Starting in the mid-1990s, separately scored essay-based writing sections became a key feature of high-stakes admissions tests at ETS, starting with the GMAT, then moving on to the GRE and the *TOEFL iBT*® test. A major reason for the introduction of TOEFL iBT in 2005 was to broaden the academic English construct assessed (i.e., reduce the construct underrepresentation) by adding sections on speaking and writing skills. By 2006, the *TOEIC®* tests, which are designed to evaluate English proficiency in the workplace, were also offering an essay section.

The importance of writing in providing adequate construct representation was made clear for AP tests by the discovery of nonequivalent gender differences on the multiple-choice and constructed-response sections of many AP tests (Mazzeo et al. 1993). That finding meant that a different gender mix of students would be granted AP credit depending on which item type was given more weight, including if only one question type was used. Bridgeman and Lewis (1994) noted that men scored substantially higher than women (by about half of a standard deviation) on multiple-choice portions of AP history examinations but that women and men scored almost the same on the essays and that women tended to get slightly higher grades in their college history courses. Furthermore, the composite of the multiple-choice and essay sections provided better prediction of college history grades than either section by itself for both genders. Thus, if the construct were underrepresented by a failure to include the essay section, not only would correlations have been lower but substantially fewer women would have been granted AP credit. Bridgeman and McHale (1998) performed a similar analysis for the GMAT, demonstrating that the addition of the essay would create more opportunities for women.

16.8 Fairness as a Core Concern in Validity

Fairness is a thread that has run consistently through this chapter because, as Turnbull (1951) and others have noted, the concepts of fairness and validity are very closely related. Also noted at a number of points in this chapter, ETS has been deeply concerned about issues of fairness and consequences for test takers as individuals throughout its existence, and these concerns have permeated its operational policies and its research program (Bennett, Chap. 1, this volume; Messick 1975, 1989, 1994a, 1998, 2000; Turnbull 1949, 1951). However, with few exceptions, measurement professionals paid little attention to fairness across groups until the 1960s (D.R. Green 1982), when this topic became a widespread concern among test developers and many test publishers instituted fairness reviews and empirical analyses to promote item and test fairness (Zieky 2006).

Messick's (1989) fourfold analysis of the evidential and consequential bases of test score interpretations and uses gave a lot of attention to evaluations of the fairness and overall effectiveness of testing programs in achieving intended outcomes and in minimizing unintended negative consequences. As indicated earlier, ETS researchers have played a major role in developing statistical models and methodology for identifying and controlling likely sources of construct-irrelevant variance and construct underrepresentation and thereby promoting fairness and reducing bias. In doing so, they have tried to clarify how the evaluation of consequences fits into a more general validation framework.

Frederiksen (1984, 1986) made the case that objective (multiple-choice) formats tended to measure a subset of the skills important for success in various contexts but that reliance on that format could have a negative, distorting effect on instruction. He recalled that, while conducting validity studies during the Second World War, he was surprised that reading comprehension tests and other verbal tests were the best predictors of grades in gunner's mate school. When he later visited the school, he found that the instruction was mostly lecture–demonstration based on the content of manuals, and the end-of-course tests were based on the lectures and manuals. Frederiksen's group introduced performance tests that required students to service real guns, and grades on the end-of-course tests declined sharply. As a result, the students began assembling and disassembling guns, and the instructors "moved out the classroom chairs and lecture podium and brought in more guns and gunmounts" (Frederiksen 1984, p. 201). Scores on the new performance tests improved. In addition, mechanical aptitude and knowledge became the best predictors of grades:

> No attempt was made to change the curriculum or teacher behavior. The dramatic changes in achievement came about solely through a change in the tests. The moral is clear: It is possible to influence teaching and learning by changing the tests of achievement. (p. 201)

Testing programs can have dramatic systemic consequences, positive or negative.

Negative consequences count against a decision rule (e.g., the use of a cut score), but they can be offset by positive consequences. A program can have substantial negative consequences and still be acceptable, if the benefits outweigh those costs. Negative consequences that are not offset by positive consequences tend to render a decision rule unacceptable (at least for stakeholders who are concerned about these consequences).

In reviewing a National Academy of Sciences report on ability testing (Wigdor and Garner 1982), Messick (1982b) suggested that the report was dispassionate and wise but that it "evinces a pervasive institutional bias" (p. 9) by focusing on common analytic models for selection and classification, which emphasize the intended outcomes of the decision rule:

> Consider that, for the most part, the utility of a test for selection is appraised statistically in terms of the correlation coefficient between the test and the criterion … but this correlation is directly proportional to the obtained gains over random selection in the criterion performance of the selected group…. Our traditional statistics tend to focus on the accepted group and on minimizing the number of poor performers who are accepted, with little or no attention to the rejected group or those rejected individuals who would have performed adequately if given the chance. (p. 10)

Messick went on to suggest that "by giving primacy to productivity and efficiency, the Committee simultaneously downplays the significance of other important goals in education and the workplace" (p. 11). It is certainly appropriate to evaluate a decision rule in terms of the extent to which it achieves the goals of the program, but it is also important to attend to unintended effects that have potentially serious consequences.

Holland (1994) and Dorans (2012) have pointed out that that different stakeholders (test developers, test users, test takers) can have very different but legitimate perspectives on testing programs and on the criteria to be used in evaluating the programs. For some purposes and in some contexts, it is appropriate to think of testing programs primarily as measurement procedures designed to produce accurate and precise estimates of some variable of interest; within this *measurement perspective* (Dorans 2012; Holland 1994), the focus is on controlling potential sources of random error and potential sources of bias (e.g., construct-irrelevant score variance, construct underrepresentation, method effects). However, in any applied context, additional considerations are relevant. For example, test takers often view testing programs as contests in which they are competing for some desired outcome, and whether they achieve their goal or not, they want the process to be fair; Holland (1994) and Dorans (2012) referred to this alternate, and legitimate, point of view as the *contest perspective*.

A *pragmatic perspective* (Kane 2013b) focuses on how well the program, as implemented, achieves its goals and avoids unintended negative effects. The pragmatic perspective is particularly salient for testing programs that serve as the bases for high-stakes decisions in public contexts. To the extent that testing programs play important roles in the public arena, their claims need to be justified. The pragmatic perspective is particularly concerned about fairness but also values objectivity (defined as the absence of subjectivity or preference) as a core concern; decision makers want testing procedures to be clearly relevant, fair, and practical. In general, it is important to evaluate how well testing programs work in practice, in the contexts in which they are operating (e.g., as the basis for decisions in employment, in academic selection, in placement, in licensure and certification). Testing programs can have strong effects on individuals and institutions, both positive and negative (Frederiksen 1984). The pragmatic perspective suggests identifying those effects and explicitly weighing them against one another in considering the value, or functional worth, of a testing program.

16.9 Concluding Remarks

ETS has been heavily involved in the development of validity theory, the creation of models for validation, and the practice of validation since the organization's creation. All of the work involved in designing and developing tests, score scales, and the materials and procedures involved in reporting and interpreting scores contributes to the soundness and plausibility of the results. Similarly, all of the research conducted on how testing programs function, on how test scores are used, and on the impact of such uses on test takers and institutions contributes to the evaluation of the functional worth of programs.

This chapter has focused on the development of validity theory, but the theory developed out of a need to evaluate testing programs in appropriate ways, and therefore it has been based on the practice of assessment. At ETS, most theoretical innovations have come out of perceived needs to solve practical problems, for which the then current theory was inadequate or unwieldy. The resulting theoretical frameworks may be abstract and complex, but they were suggested by practical problems and were developed to improve practice.

This chapter has been organized to reflect a number of major developments in the history of validity theory and practice. The validity issues and validation models were developed during different periods, but the fact that a new issue or model appeared did not generally lead to a loss of interest in the older topics and models. The issues of fairness and bias in selection and admissions were topics of interest in the early days of ETS; their conceptualization and work on them were greatly expanded in the 1960s and 1970s, and they continue to be areas of considerable emphasis today. Although the focus has shifted and the level of attention given to different topics has varied over time, the old questions have neither died nor faded away; rather, they have evolved into more general and sophisticated analyses of the issues of meaning and values that test developers and users have been grappling with for longer than a century.

Messick shaped validity theory in the last quarter of the twentieth century; therefore this chapter on ETS's contributions has given a lot of attention to his views, which are particularly comprehensive and complex. His unified, construct-based framework assumes that "validation in essence is scientific inquiry into score meaning—nothing more, but also nothing less" (Messick 1989, p. 56) and that "judging validity in terms of whether a test does the job it is employed to do … requires evaluation of the intended or unintended social consequences of test interpretation and use" (pp. 84–85). Much of the work on validity theory at the beginning of the twenty-first century can be interpreted as attempts to build on Messick's unified, construct-based framework, making it easier to apply in a straightforward way so that tests can be interpreted and used to help achieve the goals of individuals, education, and society.

Acknowledgments The authors wish to thank Randy Bennett, Cathy Wendler, and James Carlson for comments and suggestions on earlier drafts of the chapter.

References

Abelson, R. P. (1952). Sex differences in predictability of college grades. *Educational and Psychological Measurement, 12*, 638–644. https://doi.org/10.1177/001316445201200410

Allen, C. D. (1949). *Summary of validity studies of the College Entrance Examination Board tests in current use* (Research Bulletin No. RB-49-09). Princeton: Educational Testing Service. https://doi.org/10.1002/j.2333-8504.1949.tb00872.x

Angoff, W. H. (1953). Test reliability and effective test length. *Psychometrika, 18*, 1–14. https://doi.org/10.1007/BF02289023

Angoff, W. H. (1971). Scales, norms, and equivalent scores. In R. L. Thorndike (Ed.), *Educational measurement* (2nd ed., pp. 508–600). Washington, DC: American Council on Education.

Angoff, W. H. (1987). *Does guessing really help?* (Research Report No. RR-87-16). Princeton: Educational Testing Service. https://doi.org/10.1002/j.2330-8516.1987.tb00220.x

Angoff, W. H. (1988). Validity: An evolving concept. In H. Wainer & H. Braun (Eds.), *Test validity* (pp. 9–13). Hillsdale: Erlbaum.

Angoff, W. H., & Ford, S. (1973). Item-race interaction on a test of scholastic aptitude. *Journal of Educational Measurement, 10*, 95–106. https://doi.org/10.1111/j.1745-3984.1973.tb00787.x

Angoff, W. H., & Sharon, A. (1974). The evaluation of differences in test performance of two or more groups. *Educational and Psychological Measurement, 34*, 807–816. https://doi.org/10.1177/001316447403400408

Aronson, J. E. R. (1948). *April 1948 English composition reading-reliability study* (Research Bulletin No. RB-48-10). Princeton: Educational Testing Service. https://doi.org/10.1002/j.2333-8504.1948.tb00912.x

Baird, L. L. (1979). *Development of an inventory of documented accomplishments for graduate admissions* (GRE Board Research Report No. 77-03R). Princeton: Educational Testing Service.

Baird, L. L., & Knapp, J. E. (1981). *The inventory of documented accomplishments for graduate admissions: Results of a field trial study of its reliability, short-term correlates, and evaluation* (GRE Board Research Report No. 78-03R). Princeton: Educational Testing Service. https://doi.org/10.1002/j.2333-8504.1981.tb01253.x

Bejar, I. I., & Braun, H. I. (1999). *Architectural simulations: From research to implementation: Final report to the National Council on Architectural Registration* (Research Memorandum No. RM-99-02). Princeton: Educational Testing Service.

Bennett, R. E. (2010). Cognitively based assessment of, for, and as learning (CBAL): A preliminary theory of action for summative and formative assessment. *Measurement: Interdisciplinary Research and Perspectives, 8*(2–3), 70–91. https://doi.org/10.1080/15366367.2010.508686

Bennett, R. E., & Gitomer, D. H. (2009). Transforming K-12 assessment: Integrating accountability testing, formative assessment, and professional support. In C. Wyatt-Smith & J. Cumming (Eds.), *Educational assessment in the 21st century* (pp. 43–61). New York: Springer. https://doi.org/10.1007/978-1-4020-9964-9_3

Bridgeman, B., & Cahalan, C. (2007). *Time requirements for the different item types proposed for use in the revised SAT* (Research Report No. RR-07-35). Princeton: Educational Testing Service. https://doi.org/10.1002/j.2333-8504.2007.tb02077.x

Bridgeman, B., & Cline, F. (2000). *Variations in mean response times for questions on the computer-adaptive GRE General Test: Implications for fair assessment* (GRE Board Report No. 96-20P). Princeton: Educational Testing Service. https://doi.org/10.1002/j.2333-8504.2000.tb01830.x

Bridgeman, B., & Cline, F. (2004). Effects of differentially time-consuming tests on computer-adaptive test scores. *Journal of Educational Measurement, 41*, 137–148. https://doi.org/10.1111/j.1745-3984.2004.tb01111.x

Bridgeman, B., & Lewis, C. (1994). The relationship of essay and multiple-choice scores with grades in college courses. *Journal of Educational Measurement, 31*, 37–50. https://doi.org/10.1111/j.1745-3984.1994.tb00433.x

Bridgeman, B., & McHale, F. J. (1998). Potential impact of the addition of a writing assessment on admissions decisions. *Research in Higher Education, 39*, 663–677. https://doi.org/10.1023/A:1018709924672

Bridgeman, B., McCamley-Jenkins, L., & Ervin, N. S. (2000). *Predictions of freshman grade-point average from the revised and recentered SAT I: Reasoning Test* (College Board Report No. 2000-01). Princeton: Educational Testing Service. https://doi.org/10.1002/j.2333-8504.2000.tb01824.x

Bridgeman, B., Cline, F., & Hessinger, J. (2004a). Effect of extra time on verbal and quantitative GRE scores. *Applied Measurement in Education, 17*, 25–37. https://doi.org/10.1207/s15324818ame1701_2

Bridgeman, B., Trapani, C., & Curley, E. (2004b). Impact of fewer questions per section on SAT I scores. *Journal of Educational Measurement, 41*, 291–310. https://doi.org/10.1111/j.1745-3984.2004.tb01167.x

Bridgeman, B., Burton, N., & Cline, F. (2008a). *Understanding what the numbers mean: A straightforward approach to GRE predictive validity* (GRE Board Research Report No. 04-03). Princeton: Educational Testing Service. https://doi.org/10.1002/j.2333-8504.2008.tb02132.x

Bridgeman, B., Burton, N., & Pollack, J. (2008b). Predicting grades in college courses: A comparison of multiple regression and percent succeeding approaches. *Journal of College Admission, 199*, 19–25.

Bridgeman, B., Pollack, J., & Burton, N. (2008c). *Predicting grades in different types of college courses* (Research Report No. RR-08-06). Princeton: Educational Testing Service. https://doi.org/10.1002/j.2333-8504.2008.tb02092.x

Browne, M. W. (1968). A comparison of factor analytic techniques. *Psychometrika, 33*, 267–334. https://doi.org/10.1007/BF02289327

Burton, E., & Burton, N. (1993). The effect of item screening on test scores and test characteristics. In P. Holland & H. Wainer (Eds.), *Differential item functioning* (pp. 321–336). Hillsdale: Erlbaum.

Burton, N., & Wang, M. (2005). *Predicting long-term success in graduate school: A collaborative validity study* (Research Report No. RR-05-03). Princeton: Educational Testing Service. https://doi.org/10.1002/j.2333-8504.2005.tb01980.x

Cahalan, C., King, T. C., Cline, F., & Bridgeman, B. (2006). *Observational timing study on the SAT Reasoning Test for test-takers with learning disabilities and/or ADHD* (Research Report No. RR-06-23). Princeton: Educational Testing Service. https://doi.org/10.1002/j.2333-8504.2006.tb02029.x

Campbell, D. T. (1960). Recommendations for APA test standards regarding construct, trait, or discriminant validity. *American Psychologist, 15*, 546–553. https://doi.org/10.1037/h0048255

Campbell, J. (1964). *Testing of culturally different groups* (Research Bulletin No. RB-64-34). Princeton: Educational Testing Service. https://doi.org/10.1002/j.2333-8504.1964.tb00506.x

Campbell, D. T., & Fiske, D. W. (1959). Convergent and discriminant validation by the multitrait-multimethod matrix. *Psychological Bulletin, 56*, 81–105. https://doi.org/10.1037/h0046016

Cardall, C., & Coffman, W. (1964). *A method for comparing the performance of different groups on the items in a test* (Research Bulletin No. RB-64-61). Princeton: Educational Testing Service.

Carlson, S. B., Ward, W. C., & Frederiksen, N. O. (1984). The place of social intelligence in a taxonomy of cognitive abilities. *Intelligence, 8*, 315–337. https://doi.org/10.1016/0160-2896(84)90015-1

Carroll, J. B. (1974). *Psychometric tests as cognitive tasks: A new structure of intellect* (Research Bulletin No. RB-74-16). Princeton: Educational Testing Service.

Chapelle, C. A., Enright, M. K., & Jamieson, J. (Eds.). (2008). *Building a validity argument for the Test of English as a Foreign Language*. New York: Routledge.

Chapelle, C. A., Enright, M. K., & Jamieson, J. (2010). Does an argument-based approach to validity make a difference? *Educational Measurement: Issues and Practice, 29*, 3–13. https://doi.org/10.1111/j.1745-3992.2009.00165.x

Cho, Y., & Bridgeman, B. (2012). Relationship of *TOEFL iBT* Scores to academic performance: Some evidence from American universities. *Language Testing, 29*, 421–442. https://doi.org/10.1177/0265532211430368

Cleary, A. (1968). Test bias: Prediction of grades of Negro and White students in integrated colleges. *Journal of Educational Measurement, 5*, 115–124.

Cleary, A., & Hilton, T. (1966). *An investigation of item bias* (Research Bulletin No. RB-66-17). Princeton: Educational Testing Service. https://doi.org/10.1002/j.2333-8504.1966.tb00355.x

Cole, N. (1973). Bias in selection. *Journal of Educational Measurement, 5*, 237–255. https://doi.org/10.1111/j.1745-3984.1973.tb00802.x

Cole, N. S., & Moss, P. A. (1989). Bias in test use. In R. L. Linn (Ed.), *Educational measurement* (3rd ed., pp. 201–219). New York: Macmillan.

College Board. (1948). *Report on the study adaptability ratings of cadets in the class of 1947 at the U.S. Coast Guard Academy* (Research Bulletin No. RB-48-01A). Princeton: Educational Testing Service. https://doi.org/10.1002/j.2333-8504.1948.tb00001.x

Cronbach, L. J. (1971). Test validation. In R. L. Thorndike (Ed.), *Educational measurement* (2nd ed., pp. 443–507). Washington, DC: American Council on Education.

Cronbach, L. J. (1988). Five perspectives on validity argument. In H. Wainer & H. Braun (Eds.), *Test validity* (pp. 3–17). Hillsdale: Erlbaum.

Cronbach, L. J., & Meehl, P. E. (1955). Construct validity in psychological tests. *Psychological Bulletin, 52*, 281–302. https://doi.org/10.1037/h0040957

Cureton, E. E. (1951). Validity. In E. F. Lindquist (Ed.), *Educational measurement* (pp. 621–694). Washington, DC: American Council on Education.

Damarin, F., & Messick, S. (1965). *Response styles and personality variables: A theoretical integration of multivariate research* (Research Bulletin No. RB-65-10). Princeton: Educational Testing Service. https://doi.org/10.1002/j.2333-8504.1965.tb00967.x

Donlon, T. F. (Ed.). (1984). *The College Board technical handbook for the Scholastic Aptitude Test and Achievement Tests*. New York: College Board.

Dorans, N. J. (1989). Two new approaches to assessing differential item functioning: Standardization and the Mantel–Haenszel method. *Applied Measurement in Education, 2*, 217–233. https://doi.org/10.1207/s15324818ame0203_3

Dorans, N. J. (2004). Using population invariance to assess test score equity. *Journal of Educational Measurement, 41*, 43–68. https://doi.org/10.1111/j.1745-3984.2004.tb01158.x

Dorans, N. J. (2012). The contestant perspective on taking tests: Emanations from the statue within. *Educational Measurement: Issues and Practice, 31*(4), 20–37. https://doi.org/10.1111/j.1745-3992.2012.00250.x

Dorans, N. J., & Holland, P. W. (1993). DIF detection and description: Mantel–Haenszel and standardization. In P. W. Holland & H. Wainer (Eds.), *Differential item functioning* (pp. 35–66). Hillsdale: Erlbaum. https://doi.org/10.1111/j.1745-3984.2000.tb01088.x

Dorans, N. J., & Holland, P. W. (2000). Population invariance and the equatability of tests: Basic theory and the linear case. *Journal of Educational Measurement, 37*, 281–306. https://doi.org/10.1111/j.1745-3984.2000.tb01088.x

Dwyer, C. A., Gallagher, A., Levin, J., & Morley, M. E. (2003). *What is quantitative reasoning? Defining the construct for assessment purposes* (Research Report No. RR-03-30). Princeton: Educational Testing Service. https://doi.org/10.1002/j.2333-8504.2003.tb01922.x

Ebel, R. (1962). Content standard test scores. *Educational and Psychological Measurement, 22*, 15–25. https://doi.org/10.1177/001316446202200103

Educational Testing Service. (1948). *The study of reading reliability of the College Board English composition tests of April and June 1947* (Research Bulletin No. RB-48-07). Princeton: Author. https://doi.org/10.1002/j.2333-8504.1948.tb00005.x

Educational Testing Service. (1992). *The origins of Educational Testing Service*. Princeton: Author.

Ekstrom, R. B., French, J. W., & Harman, H. H. (1979). Cognitive factors: Their identification and replication. *Multivariate Behavioral Research Monographs, 79*(2), 3–84.

Evans, F. R. (1980). *A study of the relationship among speed and power, aptitude test scores, and ethnic identity* (Research Report No. RR-80-22). Princeton: Educational Testing Service. https://doi.org/10.1002/j.2333-8504.1980.tb01219.x

Fairtest. (2003). *SAT I: A faulty instrument for predicting college success*. Retrieved from http://fairtest.org/facts/satvalidity.html

Fan, C. T., Lord, F. M., & Tucker, L. R. (1950). *A score checking machine* (Research Memorandum No. RM-50-08). Princeton: Educational Testing Service.

Frederiksen, N. O. (1948). *The prediction of first term grades at Hamilton College* (Research Bulletin No. RB-48-02). Princeton: Educational Testing Service. https://doi.org/10.1002/j.2333-8504.1948.tb00867.x

Frederiksen, N. (1951). The influence of timing and instructions on Cooperative Reading Test scores. *Educational and Psychological Measurement, 12*, 598–607. https://doi.org/10.1002/j.2333-8504.1951.tb00022.x

Frederiksen, N. (1959). *Development of the test "Formulating Hypotheses": A progress report* (Office of Naval Research Technical Report No. NR-2338[00]). Princeton: Educational Testing Service.

Frederiksen, N. (1984). The real test bias: Influences of testing on teaching and learning. *American Psychologist, 39*, 193–202. https://doi.org/10.1037/0003-066X.39.3.193

Frederiksen, N. (1986). Toward a broader conception of human intelligence. *American Psychologist, 41*, 445–452. https://doi.org/10.1037/0003-066X.41.4.445

Frederiksen, N., & Ward, W. (1978). Measures for the study of creativity in scientific problem-solving. *Applied Psychological Measurement, 2*, 1–24. https://doi.org/10.1177/014662167800200101

Frederiksen, N., Olsen, M., & Schrader, W. (1950a). *Prediction of first-semester grades at Kenyon College, 1948–1949* (Research Bulletin No. RB-50-49). Princeton: Educational Testing Service. https://doi.org/10.1002/j.2333-8504.1950.tb00879.x

Frederiksen, N., Schrader, W., Olsen, M., & Wicoff, E. (1950b). *Prediction of freshman grades at the University of Rochester* (Research Bulletin No. RB-50-44). Princeton: Educational Testing Service. http://dx.doi.org/10.1002/j.2333-8504.1950.tb00481.x

Frederiksen, N., Saunders, D. R., & Wand, B. (1957). The in-basket test. *Psychological Monographs, 71*(9), 1–28. https://doi.org/10.1037/h0093706

Freeman, P. M. (1952). *Survey of studies on correction for guessing and guessing instructions* (Research Memorandum No. RM-52-04). Princeton: Educational Testing Service.

French, J. W. (1948). *Report on the study intercorrelations of entrance tests and grades, class of 1947 at the U.S. Coast Guard Academy* (Research Bulletin No. RB-48-03). Princeton: Educational Testing Service. https://doi.org/10.1002/j.2333-8504.1948.tb00002.x

French, J. W. (1951a). *Conference on factorial studies of aptitude and personality measures* (Research Memorandum No. RM-51-20). Princeton: Educational Testing Service.

French, J. W. (1951b). *The description of aptitude and achievement factors in terms of rotated factors*. (Psychometric Monograph No. 5). Richmond: Psychometric Society.

French, J. W. (1954). *Manual for Kit of Selected Tests for Reference Aptitude and Achievement Factors*. Princeton: Educational Testing Service.

French, J. W. (1957). *The relation of ratings and experience variables to Tests of Developed Abilities profiles* (Research Bulletin No. RB-57-14). Princeton: Educational Testing Service. https://doi.org/10.1002/j.2333-8504.1957.tb00934.x

French, J. W. (1962). Effect of anxiety on verbal and mathematical examination scores. *Educational and Psychological Measurement, 22*, 553–564. https://doi.org/10.1177/001316446202200313

French, J. W., Ekstrom, R. B., & Price, L. A. (1963). *Manual for Kit of Reference Tests for Cognitive Factors*. Princeton: Educational Testing Service.

Green, B. F., Jr. (1950). A note on the calculation of weights for maximum battery reliability. *Psychometrika, 15*, 57–61. https://doi.org/10.1007/BF02289178

Green, B. F., Jr. (1952). The orthogonal approximation of an oblique structure in factor analysis. *Psychometrika, 17*(4), 429–440. https://doi.org/10.1007/BF02288918

Green, D. R. (1982). Methods used by test publishers to "debias" standardized tests. In R. A. Berk (Ed.), *Handbook of methods for detecting test bias* (pp. 278–313). Baltimore: The Johns Hopkins University Press.

Griggs v. Duke Power Co., 401 U.S. 424 (1971).

Gulliksen, H. (1950a). *Intrinsic versus correlational validity* (Research Bulletin No. RB-50-37). Princeton: Educational Testing Service. https://doi.org/10.1002/j.2333-8504.1950.tb00477.x

Gulliksen, H. (1950b). *Theory of mental tests*. New York: Wiley. https://doi.org/10.1037/13240-000

Gulliksen, H., & Messick, S. (1960). *Psychological scaling: Theory and applications*. New York: Wiley.

Haberman, S. J. (2008). When can subscores have value? *Journal of Educational and Behavioral Statistics, 33*, 204–229. https://doi.org/10.3102/1076998607302636

Harman, H. H. (1967). *Modern factor analysis* (3rd ed.). Chicago: University of Chicago Press.

Holland, P. W. (1994). Measurements or contests? Comment on Zwick, Bond, and Allen/Donogue. In *Proceedings of the Social Statistics Section of the American Statistical Association* (pp. 27–29). Alexandria: American Statistical Association.

Holland, P. W. (2007). A framework and history for score linking. In N. J. Dorans, M. Pommerich, & P. W. Holland (Eds.), *Linking and aligning scores and scales* (pp. 5–30). New York: Springer. https://doi.org/10.1007/978-0-387-49771-6_2

Holland, P. W., & Dorans, N. J. (2006). Linking and equating. In R. L. Brennan (Ed.), *Educational measurement* (4th ed., pp. 187–220). Westport: American Council on Education and Praeger.
Holland, P. W., & Rubin, D. B. (Eds.). (1982). *Test equating*. New York: Academic Press.
Holland, P. W., & Thayer, D. T. (1988). Differential item performance and the Mantel–Haenszel procedure. In H. Wainer & H. I. Braun (Eds.), *Test validity* (pp. 129–145). Hillsdale: Erlbaum.
Holland, P. W., & Wainer, H. (Eds.). (1993). *Differential item functioning*. Hillsdale: Erlbaum.
Horst, P. (1950a). *Optimal test length for maximum battery validity* (Research Bulletin No. RB-50-36). Princeton: Educational Testing Service. https://doi.org/10.1002/j.2333-8504.1950.tb00476.x
Horst, P. (1950b). *The relationship between the validity of a single test and its contribution to the predictive efficiency of a test battery* (Research Bulletin No. RB-50-32). Princeton: Educational Testing Service. https://doi.org/10.1002/j.2333-8504.1950.tb00472.x
Horst, P. (1951). Estimating total test reliability from parts of unequal length. *Educational and Psychological Measurement, 11*, 368–371. https://doi.org/10.1177/001316445101100306
Jackson, D. (2002). The constructs in people's heads. In H. I. Braun, D. N. Jackson, D. E. Wiley, & S. Messick (Eds.), *The role of constructs in psychological and educational measurement* (pp. 3–17). Mahwah: Erlbaum.
Jackson, D., & Messick, S. (1958). Content and style in personality assessment. *Psychological Bulletin, 55*, 243–252. https://doi.org/10.1037/h0045996
Jackson, D. N., & Messick, S. (1961). Acquiescence and desirability as response determinants on the MMPI. *Educational and Psychological Measurement, 21*, 771–790. https://doi.org/10.1177/001316446102100402
Jackson, D. N., & Messick, S. (1965). The person, the product, and the response: Conceptual problems in the assessment of creativity. *Journal of Personality, 33*, 309–329. https://doi.org/10.1111/j.1467-6494.1965.tb01389.x
Jackson, D., Messick, S., & Solley, C. (1957). A multidimensional scaling approach to the perception of personality. *Journal of Psychology, 22*, 311–318. https://doi.org/10.1080/00223980.1957.9713088
Johnson, A. P., & Olsen, M. A. (1952). *Comparative three-year and one-year validities of the Law School Admissions Test at two law schools* (Law School Admissions Council Report No. 52-02). Princeton: Educational Testing Service.
Jöreskog, K. G. (1967). Some contributions to maximum likelihood factor analysis. *Psychometrika, 32*, 443–482. https://doi.org/10.1007/BF02289658
Jöreskog, K. G. (1969). A general approach to confirmatory maximum likelihood factor analysis. *Psychometrika, 34*, 183–202. https://doi.org/10.1007/BF02289343
Jöreskog, K. G. (1971). Statistical analysis of sets of congeneric tests. *Psychometrika, 36*, 109–133.
Jöreskog, K. G., & Lawley, D. N. (1967). *New methods in maximum likelihood factor analysis* (Research Bulletin No. RB-67-49). Princeton: Educational Testing Service. https://doi.org/10.1002/j.2333-8504.1967.tb00703.x
Jöreskog, K. G., & van Thillo, M. (1972). *LISREL: A general computer program for estimating a linear structural equation system involving multiple indicators of unmeasured variables* (Research Bulletin No. RB-72-56). Princeton: Educational Testing Service. https://doi.org/10.1002/j.2333-8504.1972.tb00827.x
Kane, M. (1992). An argument-based approach to validation. *Psychological Bulletin, 112*, 527–535. https://doi.org/10.1037/0033-2909.112.3.527
Kane, M. (2006). Validation. In R. Brennan (Ed.), *Educational measurement* (4th ed., pp. 17–64). Westport: American Council on Education and Praeger.
Kane, M. (2012). Validating score interpretations and uses: Messick Lecture, Language Testing Research Colloquium, Cambridge, April 2010. *Language Testing, 29*, 3–17. https://doi.org/10.1177/0265532211417210
Kane, M. (2013a). Validating the interpretations and uses of test scores. *Journal of Educational Measurement, 50*, 1–73. https://doi.org/10.1111/jedm.12000

Kane, M. (2013b). Validity and fairness in the testing of individuals. In M. Chatterji (Ed.), *Validity and test use* (pp. 17–53). Bingley: Emerald.

Keats, J. A. (1957). Estimation of error variances of test scores. *Psychometrika, 22*, 29–41. https://doi.org/10.1007/BF02289207

Kelley, T. (1927). *Interpretation of educational measurements*. Yonkers: World Book.

Klieger, D. M., Holtzman, S., & Ezzo, C. (2013, April). *The promise of non-cognitive assessment in graduate and professional school admissions*. Paper presented at the annual meeting of the American Educational Research Association, San Francisco, CA.

Kristof, W. (1962). *Statistical inferences about the error variance* (Research Bulletin No. RB-62-21). Princeton: Educational Testing Service. https://doi.org/10.1002/j.2333-8504.1962.tb00299.x

Kristof, W. (1970). On the sampling theory of reliability estimation. *Journal of Mathematical Psychology, 7*, 371–377. https://doi.org/10.1016/0022-2496(70)90054-4

Kristof, W. (1971). On the theory of a set of tests which differ only in length. *Psychometrika, 36*, 207–225. https://doi.org/10.1007/BF02297843

Kristof, W. (1974). Estimation of reliability and true score variance from a split of a test into three arbitrary parts. *Psychometrika, 39*, 491–499.

Kyllonen, P. (2005, September). The case for noncognitive assessments. *ETS R&D Connections*, pp. 1–7.

Lawrence, I., & Shea, E. (2011, April). *A brief history of Educational Testing Service as a scientific organization*. Paper presented at the annual meeting of the National Council of Measurement in Education, New Orleans, LA.

Lemann, N. (1995). The great sorting. *Atlantic Monthly, 276*(3), 84–100.

Lindquist, E. F. (Ed.). (1951). *Educational measurement*. Washington, DC: American Council on Education.

Linn, R. L. (1972). *Some implications of the Griggs decision for test makers and users* (Research Memorandum No. RM-72-13). Princeton: Educational Testing Service.

Linn, R. L. (1973). Fair test use in selection. *Review of Educational Research, 43*, 139–161. https://doi.org/10.3102/00346543043002139

Linn, R. L. (1975). *Test bias and the prediction of grades in law school* (Research Report No. 75-01). Newtown: Law School Admissions Council.

Linn, R. L. (1976). In search of fair selection procedures. *Journal of Educational Measurement, 13*, 53–58. https://doi.org/10.1111/j.1745-3984.1976.tb00181.x

Linn, R. L. (1984). Selection bias: Multiple meanings. *Journal of Educational Measurement, 21*, 33–47. https://doi.org/10.1111/j.1745-3984.1984.tb00219.x

Linn, R. L., & Werts, C. (1971). Considerations for studies of test bias. *Journal of Educational Measurement, 8*, 1–4. https://doi.org/10.1111/j.1745-3984.1971.tb00898.x

Liu, J., Allsbach, J. R., Feigenbaum, M., Oh, H.-J., & Burton, N. W. (2004). *A study of fatigue effects from the New SAT* (Research Report No. RR-04-46). Princeton: Educational Testing Service. https://doi.org/10.1002/j.2333-8504.2004.tb01973.x

Livingston, S. (1972). Criterion-referenced applications of classical test theory. *Journal of Educational Measurement, 9*, 13–26. https://doi.org/10.1111/j.1745-3984.1972.tb00756.x

Livingston, S. A., & Lewis, C. (1995). Estimating the consistency and accuracy of classifications based on test scores. *Journal of Educational Measurement, 32*, 179–197. https://doi.org/10.1111/j.1745-3984.1995.tb00462.x

Loevinger, J. (1957). Objective tests as instruments of psychological theory. *Psychological Reports, Monograph Supplement, 3*, 635–694.

Lord, F. M. (1951). *A theory of test scores and their relation to the trait measured* (Research Bulletin No. RB-51-13). Princeton: Educational Testing Service. https://doi.org/10.1002/j.2333-8504.1951.tb00922.x

Lord, F. M. (1953). *Speeded tests and power tests—An empirical study of validities* (Research Bulletin No. RB-53-12). Princeton: Educational Testing Service. https://doi.org/10.1002/j.2333-8504.1953.tb00228.x

Lord, F. M. (1955). *Estimating test reliability* (Research Bulletin No. RB-55-07). Princeton: Educational Testing Service. https://doi.org/10.1002/j.2333-8504.1955.tb00054.x
Lord, F. M. (1956). *Do tests of the same length have the same standard error of measurement?* (Research Bulletin No. RB-56-07). Princeton: Educational Testing Service. https://doi.org/10.1002/j.2333-8504.1956.tb00063.x
Lord, F. M. (1957). *Inferring the shape of the frequency distribution of true scores and of errors of measurement* (Research Bulletin No. RB-57-09). Princeton: Educational Testing Service. https://doi.org/10.1002/j.2333-8504.1957.tb00076.x
Lord, F. M. (1959). Tests of the same length do have the same standard error of measurement. *Educational and Psychological Measurement, 19*, 233–239. https://doi.org/10.1177/001316445901900208
Lord, F. M. (1961). *Estimating norms by item sampling* (Research Bulletin No. RB-61-02). Princeton: Educational Testing Service. https://doi.org/10.1002/j.2333-8504.1961.tb00103.x
Lord, F. M. (1965). A strong true score theory with applications. *Psychometrika, 30*, 239–270. https://doi.org/10.1007/BF02289490
Lord, F. M. (1967). A paradox in the interpretation of group comparisons. *Psychological Bulletin, 68*, 304–305. https://doi.org/10.1037/h0025105
Lord, F. M. (1974). *Formula scoring and number-right scoring* (Research Memorandum No. RM-52-04). Princeton: Educational Testing Service.
Lord, F. M. (1980). *Applications of item response theory to practical testing problems*. Hillsdale: Erlbaum.
Lord, F. M., & Novick, M. R. (1968). *Statistical theories of mental test scores*. Reading: Addison-Wesley.
Lord, F. M., & Stocking, M. (1976). An interval estimate for making statistical inferences about true score. *Psychometrika, 41*, 79–87. https://doi.org/10.1007/BF02291699
Lord, F. M., & Wingersky, M. S. (1984). Comparison of IRT true-score and equipercentile observed-score "equatings". *Applied Psychological Measurement, 8*, 453–461. https://doi.org/10.1177/014662168400800409
Markle, R., Olivera-Aguilar, M., Jackson, R., Noeth, R., & Robbins, S. (2013). *Examining evidence of reliability, validity, and fairness for the Success Navigator Assessment* (Research Report No. RR-13-12). Princeton: Educational Testing Service. https://doi.org/10.1002/j.2333-8504.2013.tb02319.x
Mazzeo, J., Schmitt, A. P., & Bleistein, C. A. (1993). *Sex-related performance differences on constructed-response and multiple-choice sections of the Advanced Placement Examinations* (College Board Report No. 92-07). Princeton: Educational Testing Service. https://doi.org/10.1002/j.2333-8504.1993.tb01516.x
Mazzeo, J., Lazer, S., & Zieky, M. J. (2006). Monitoring educational progress with group-score assessment. In R. L. Brennan (Ed.), *Educational measurement* (4th ed., pp. 681–689). Westport: American Council on Education and Praeger.
Messick, S. (1956). The perception of social attitudes. *Journal of Abnormal and Social Psychology, 52*, 57–66. https://doi.org/10.1037/h0038586
Messick, S. (1962). Response style and content measures from personality inventories. *Educational and Psychological Measurement, 22*, 41–56. https://doi.org/10.1177/001316446202200106
Messick, S. (1964). Personality measurement and college performance. In A. G. Wesman (Ed.), *Proceedings of the 1963 invitational conference on testing problems* (pp. 110–129). Princeton: Educational Testing Service.
Messick, S. (1965). Personality measurement and the ethics of assessment. *American Psychologist, 20*, 136–142. https://doi.org/10.1037/h0021712
Messick, S. (1967). The psychology of acquiescence: An interpretation of research evidence. In I. A. Berg (Ed.), *Response set in personality assessment* (pp. 115–145). Chicago: Aldine.
Messick, S. (1970). The criterion problem in the evaluation of instruction: Assessing possible, not just intended outcomes. In M. Wittrock & D. Wiley (Eds.), *The evaluation of instruction: Issues and problems* (pp. 183–202). New York: Holt, Rinehart and Winston.
Messick, S. (1972). Beyond structure: In search of functional models of psychological process. *Psychometrika, 37*, 357–375. https://doi.org/10.1007/BF02291215

Messick, S. (1975). The standard problem: Meaning and values in measurement and evaluation. *American Psychologist, 30*, 955–966. https://doi.org/10.1037/0003-066X.30.10.955
Messick, S. (1977). Values and purposes in the uses of tests. In *Speaking out: The use of tests in the policy arena* (pp. 20–26). Princeton: Educational Testing Service.
Messick, S. (1979). *Test validity and the ethics of assessment*. (Research Report No. RR-79-10). Princeton: Educational Testing Service.
Messick, S. (1980). Test validity and the ethics of assessment. *American Psychologist, 35*, 1012–1027. https://doi.org/10.1037/0003-066X.35.11.1012
Messick, S. (1981a). Constructs and their vicissitudes in educational and psychological measurement. *Psychological Bulletin, 89*, 575–588. https://doi.org/10.1037/0033-2909.89.3.575
Messick, S. (1981b). The controversy over coaching: Issues of effectiveness and equity. *New Directions for Testing and Measurement, 11*, 21–53.
Messick, S. (1982a). Issues of effectiveness and equity in the coaching controversy: Implications for educational and testing policy. *Educational Psychologist, 17*, 69–91. https://doi.org/10.1080/00461528209529246
Messick, S. (1982b). The values of ability testing: Implications of multiple perspectives about criteria and standards. *Educational Measurement: Issues and Practice, 1*, 9–12. https://doi.org/10.1111/j.1745-3992.1982.tb00660.x
Messick, S. (1984). The nature of cognitive styles: Problems and promise in educational practice. *Educational Psychologist, 19*, 59–74. https://doi.org/10.1080/00461528409529283
Messick, S. (1988). The once and future issues of validity: Assessing the meaning and consequences of measurement. In H. Wainer & H. Braun (Eds.), *Test validity* (pp. 33–45). Hillsdale: Erlbaum.
Messick, S. (1989). Validity. In R. L. Linn (Ed.), *Educational measurement* (3rd ed., pp. 13–103). New York: Macmillan.
Messick, S. (1991). Psychology and methodology of response styles. In R. E. Snow & D. E. Wiley (Eds.), *Improving inquiry in social science—A volume in honor of Lee J. Cronbach* (pp. 161–200). Hillsdale: Erlbaum.
Messick, S. (1994a). The interplay of evidence and consequences in the validation of performance assessments. *Educational Researcher, 23*, 13–23. https://doi.org/10.3102/0013189X023002013
Messick, S. (1994b). *Standards-based score interpretation: Establishing valid grounds for valid inferences* (Research Report No. RR-94-57). Princeton: Educational Testing Service. https://doi.org/10.1002/j.2333-8504.1994.tb01630.x
Messick, S. (1995). Standards of validity and the validity of standards in performance assessment. *Educational Measurement: Issues and Practice, 14*, 5–8. https://doi.org/10.1111/j.1745-3992.1995.tb00881.x
Messick, S. (1998). Test validity: A matter of consequences. *Social Indicators Research, 45*, 35–44. https://doi.org/10.1023/A:1006964925094
Messick, S. (2000). Consequences of test interpretation and use: The fusion of validity and values in psychological assessment. In R. Goffin & E. Helmes (Eds.), *Problems and solutions in human assessment: Honoring Douglas N. Jackson at seventy* (pp. 3–20). Boston: Kluwer. https://doi.org/10.1007/978-1-4615-4397-8_1
Messick, S., & Abelson, R. (1957). Research tools: Scaling and measurement theory. *Review of Educational Research, 27*, 487–497. https://doi.org/10.2307/1169167
Messick, S., & Anderson, S. (1970). Educational testing, individual development, and social responsibility. *The Counseling Psychologist, 2*, 80–88. https://doi.org/10.1177/001100007000200215
Messick, S., & Frederiksen, N. (1958). Ability, acquiescence, and "authoritarianism." *Psychological Reports, 4*, 687–697. https://doi.org/10.2466/pr0.1958.4.3.687
Messick, S., & Jackson, D. (1958). The measurement of authoritarian attitudes. *Educational and Psychological Measurement, 18*, 241–253. https://doi.org/10.1177/001316445801800202
Messick, S., & Jungeblut, A. (1981). Time and method in coaching for the SAT. *Psychological Bulletin, 89*, 191–216. https://doi.org/10.1037/0033-2909.89.2.191

Messick, S., & Kogan, N. (1963). Differentiation and compartmentalization in object-sorting measures of categorizing style. *Perceptual and Motor Skills, 16*, 47–51. https://doi.org/10.2466/pms.1963.16.1.47
Messick, S., & Ross, J. (Eds.). (1962). *Measurement in personality and cognition*. New York: Wiley.
Messick, S., Beaton, A., & Lord, F. (1983). *National Assessment of Educational Progress reconsidered: A new design for a new era* (NAEP Report No. 83-1). Princeton: Educational Testing Service.
Mislevy, R. (1993). Foundations of a new test theory. In N. Frederiksen, R. J. Mislevy, & I. I. Bejar (Eds.), *Test theory for a new generation of tests* (pp. 19–39). Hillsdale: Erlbaum.
Mislevy, R. (1994). Evidence and inference in educational assessment. *Psychometrika, 59*, 439–483. https://doi.org/10.1007/BF02294388
Mislevy, R. (1996). Test theory reconceived. *Journal of Educational Measurement, 33*, 379–416. https://doi.org/10.1111/j.1745-3984.1996.tb00498.x
Mislevy, R. J. (2006). Cognitive psychology and educational assessment. In R. L. Brennan (Ed.), *Educational measurement* (4th ed., pp. 257–306). Westport: American Council on Education and Praeger.
Mislevy, R. (2007). Validity by design. *Educational Researcher, 36*, 463–469. https://doi.org/10.3102/0013189X07311660
Mislevy, R. (2009). Validity from the perspective of model-based reasoning. In R. W. Lissitz (Ed.), *The concept of validity* (pp. 83–108). Charlotte: Information Age.
Mislevy, R. J., & Haertel, G. D. (2006). Implications of evidence-centered design for educational testing. *Educational Measurement: Issues and Practice, 25*(4), 6–20. https://doi.org/10.1111/j.1745-3992.2006.00075.x
Mislevy, R. J., Steinberg, L. S., & Almond, R. G. (1999). *Evidence-centered assessment design*. Princeton: Educational Testing Service.
Mislevy, R., Steinberg, L., & Almond, R. (2002). On the roles of task model variables in assessment design. In S. Irvine & P. Kyllonen (Eds.), *Generating items for cognitive tests: Theory and practice* (pp. 97–128). Mahwah: Erlbaum.
Mislevy, R. J., Almond, R. G., & Lukas, J. F. (2003a). *A brief introduction to evidence-centered design* (Research Report No. RR-03-16). Princeton: Educational Testing Service. https://doi.org/10.1002/j.2333-8504.2003.tb01908.x
Mislevy, R. J., Steinberg, L. S., & Almond, R. G. (2003b). On the structure of educational assessments. *Measurement: Interdisciplinary Research and Perspectives, 1*, 3–67. https://doi.org/10.1207/S15366359MEA0101_02
Mislevy, R., Haertel, G., Cheng, B., Ructtinger, L., DeBarger, A., Murray, E., et al. (2013). A "conditional" sense of fairness in assessment. *Educational Research and Evaluation, 19*, 121–140. https://doi.org/10.1080/13803611.2013.767614
Mollenkopf, W. G. (1950). Predicted differences and differences between predictions. *Psychometrika, 15*, 259–269. https://doi.org/10.1002/j.2333-8504.1950.tb00018.x
Mollenkopf, W. G. (1951). *Effectiveness of Naval Academy departmental standings for predicting academic success at the U.S. Naval Postgraduate School* (Research Bulletin No. RB-51-22). Princeton: Educational Testing Service. https://doi.org/10.1002/j.2333-8504.1951.tb00221.x
Myers, A. E. (1965). Risk taking and academic success and their relation to an objective measure of achievement motivation. *Educational and Psychological Measurement, 25*, 355–363. https://doi.org/10.1177/001316446502500206
Myers, C. T. (1975). *Test fairness: A comment on fairness in statistical analysis* (Research Bulletin No. RB-75-12). Princeton: Educational Testing Service. https://doi.org/10.1002/j.2333-8504.1975.tb01051.x
Novick, M. R. (1965). *The axioms and principal results of classical test theory* (Research Report No. RR-65-02). Princeton: Educational Testing Service. https://doi.org/10.1002/j.2333-8504.1965.tb00132.x

Novick, M. R., & Lewis, C. (1967). Coefficient alpha and the reliability of composite measurements. *Psychometrika, 32*, 1–13. https://doi.org/10.1007/BF02289400

Novick, M. R., & Thayer, D. T. (1969). *Some applications of procedures for allocating testing time* (Research Bulletin No. RB-69-01). Princeton: Educational Testing Service. https://doi.org/10.1002/j.2333-8504.1969.tb00161.x

Packer, J. (1987). *SAT testing time for students with disabilities* (Research Report No. RR-87-37). Princeton: Educational Testing Service. https://doi.org/10.1002/j.2330-8516.1987.tb00241.x

Petersen, N. S. (2007). Equating: Best practices and challenges to best practices. In N. J. Dorans, M. Pommerich, & P. W. Holland (Eds.), *Linking and aligning scores and scales* (pp. 59–72). New York: Springer. https://doi.org/10.1007/978-0-387-49771-6_4

Petersen, N., & Novick, M. (1976). An evaluation of some models for culture-fair selection. *Journal of Educational Measurement, 13*, 3–29. https://doi.org/10.1111/j.1745-3984.1976.tb00178.x

Petersen, N. S., Kolen, M. J., & Hoover, H. D. (1989). Scaling, norming, and equating. In R. L. Linn (Ed.), *Educational measurement* (3rd ed., pp. 221–262). New York: Macmillan.

Popper, K. (1965). *Conjectures and refutations*. New York: Harper and Row.

Porter, T. (2003). Measurement, objectivity, and trust. *Measurement: Interdisciplinary Research and Perspectives, 1*, 241–255. https://doi.org/10.1207/S15366359MEA0104_1

Powers, D. E. (1988). Incidence, correlates, and possible causes of test anxiety in graduate admissions testing. *Advances in Personality Assessment, 7*, 49–75.

Powers, D. E. (2001). Test anxiety and test performance: Comparing paper-based and computer-adaptive versions of the *Graduate Record Examinations*® (GRE) General Test. *Journal of Educational Computing Research, 24*, 249–273. https://doi.org/10.2190/680W-66CR-QRP7-CL1F

Ragosta, M., & Wendler, C. (1992). *Eligibility issues and comparable time limits for disabled and nondisabled SAT examinees* (Research Report No. RR-92-35). Princeton: Educational Testing Service. https://doi.org/10.1002/j.2333-8504.1992.tb01466.x

Ramist, L., & Weiss, G. E. (1990). The predictive validity of the SAT, 1964 to 1988. In W. W. Willingham, C. Lewis, R. L. Morgan, & L. Ramist (Eds.), *Predicting college grades: An analysis of institutional trends over two decades* (pp. 117–140). Princeton: Educational Testing Service.

Ramist, L., Lewis, C., & McCamley-Jenkins, L. (1994). *Student group differences in predicting college grades: Sex, language, and ethnic groups* (Research Report No. RR-94-27). Princeton: Educational Testing Service. https://doi.org/10.1002/j.2333-8504.1994.tb01600.x

Rikoon, S., Liebtag, T., Olivera-Aguilar, M., Robbins, S., & Jackson, T. (2014). *A pilot study of holistic assessment and course placement in community college samples: Findings and recommendations* (Research Memorandum No. RM-14-10). Princeton: Educational Testing Service.

Rivera, C., & Schmitt, A. P. (1988). *A comparison of Hispanic and White non-Hispanic students' omit patterns on the Scholastic Aptitude Test* (Research Report No. RR-88-44). Princeton: Educational Testing Service. https://doi.org/10.1002/j.2330-8516.1988.tb00300.x

Roberts, R. D., Schulze, R., & MacCann, C. (2008). The measurement of emotional intelligence: A decade of progress? In G. Boyle, G. Matthews, & D. Saklofske (Eds.), *The Sage handbook of personality theory and assessment: Vol. 2. Personality measurement and testing* (pp. 461–482). London: Sage. https://doi.org/10.4135/9781849200479.n22

Ryans, D. G., & Frederiksen, N. (1951). Performance tests of educational achievement. In E. F. Lindquist (Ed.), *Educational measurement* (pp. 455–494). Washington, DC: American Council on Education.

Schiffman, H., & Messick, S. (1963). Scaling and measurement theory. *Review of Educational Research, 33*, 533–542. https://doi.org/10.2307/1169654

Schrader, W. B. (1965). A taxonomy of expectancy tables. *Journal of Educational Measurement, 2*, 29–35. https://doi.org/10.1111/j.1745-3984.1965.tb00388.x

Schultz, D. G. (1952). Item validity and response change under two different testing conditions. *Journal of Educational Psychology, 45*, 36–43. https://doi.org/10.1037/h0059845

Shepard, L. (1993). Evaluating test validity. In L. Darling-Hammond (Ed.), *Review of research in education* (pp. 405–450). Washington, DC: American Educational Research Association. https://doi.org/10.2307/1167347

Shimberg, B. (1981). Testing for licensure and certification. *American Psychologist, 36*, 1138–1146. https://doi.org/10.1037/0003-066X.36.10.1138

Shimberg, B. (1982). *Occupational licensing: A public perspective* (Center for Occupational and Professional Assessment Report). Princeton: Educational Testing Service.

Shimberg, B. (1990). Social considerations in the validation of licensing and certification exams. *Educational Measurement: Issues and Practice, 9*, 11–14. https://doi.org/10.1111/j.1745-3992.1990.tb00386.x

Stricker, L. J. (2008). *The challenge of stereotype threat for the testing community* (Research Memorandum No. RM-08-12). Princeton: Educational Testing Service.

Stricker, L. J., & Bejar, I. (2004). Test difficulty and stereotype threat on the GRE General Test. *Journal of Applied Social Psychology, 34*, 563–597. https://doi.org/10.1111/j.1559-1816.2004.tb02561.x

Stricker, L. J., & Rock, D. A. (1990). Interpersonal competence, social intelligence, and general ability. *Personality and Individual Differences, 11*, 833–839. https://doi.org/10.1016/0191-8869(90)90193-U

Stricker, L. J., & Ward, W. C. (2004). Stereotype threat, inquiring about test takers' ethnicity and gender, and standardized test performance. *Journal of Applied Social Psychology, 34*, 665–693. https://doi.org/10.1111/j.1559-1816.2004.tb02564.x

Stricker, L., Messick, S., & Jackson, D. (1969). *Conformity, anticonformity, and independence: Their dimensionality and generality* (Research Bulletin No. RB-69-17). Princeton: Educational Testing Service. https://doi.org/10.1002/j.2333-8504.1969.tb01006.x

Stricker, L. J., Rock, D. A., & Bennett, R. E. (2001). Sex and ethnic-group differences on accomplishments measures. *Applied Measurement in Education, 14*, 205–218. https://doi.org/10.1207/S15324818AME1403_1

Tucker, L. R. (1948). *Memorandum concerning study of effects of fatigue on afternoon achievement test scores due to scholastic aptitude test being taken in the morning* (Research Memorandum No. RM-48-02). Princeton: Educational Testing Service.

Tucker, L. R. (1949). A note on the estimation of test reliability by the Kuder–Richardson formula (20). *Psychometrika, 14*, 117–119. https://doi.org/10.1007/BF02289147

Tucker, L. R. (1955). The objective definition of simple structure in linear factor analysis. *Psychometrika, 20*, 209–225. https://doi.org/10.1007/BF02289018

Turnbull, W. (1949). Influence of cultural background on predictive test scores. In *Proceedings of the ETS invitational conference on testing problems* (pp. 29–34). Princeton: Educational Testing Service.

Turnbull, W. (1951). Socio-economic status and predictive test scores. *Canadian Journal of Psychology, 5*, 145–149. https://doi.org/10.1037/h0083546

von Davier, A. A. (2011). *Statistical models for test equating, scaling, and linking*. New York: Springer. https://doi.org/10.1007/978-0-387-98138-3

von Davier, A. A., Holland, P. W., & Thayer, D. T. (2004). *The kernel method of test equating*. New York: Springer. https://doi.org/10.1007/b97446

von Davier, M. (2008). A general diagnostic model applied to language testing data. *British Journal of Mathematical and Statistical Psychology, 61*, 287–307. https://doi.org/10.1348/000711007X193957

Wendler, C., & Walker, M. E. (2006). Practical issues in designing and maintaining multiple test forms for large-scale programs. In S. Downing & T. Haladyna (Eds.), *Handbook of test development* (pp. 445–467). Mahwah: Erlbaum.

Wigdor, A., & Garner, W. (1982). *Ability testing: Uses, consequences, and controversies*. Washington, DC: National Academy Press.

Wild, C., & Dwyer, C. A. (1980). Sex bias in selection. In L. van der Kamp, W. Langerat, & D. de Gruijter (Eds.), *Psychometrics for educational debates* (pp. 153–168). New York: Wiley.

Willingham, W. W. (1985). *Success in college: The role of personal qualities and academic ability*. New York: College Entrance Examination Board.
Willingham, W. W. (1999). A systematic view of test fairness. In S. Messick (Ed.), *Assessment in higher education: Issues of access, quality, student development, and public policy* (pp. 213–242). Mahwah: Erlbaum.
Willingham, W. W., & Breland, H. M. (1982). *Personal qualities and college admissions*. New York: College Entrance Examination Board.
Willingham, W. W., & Cole, N. S. (1997). *Gender and fair assessment*. Mahwah: Erlbaum.
Willingham, W. W., Bennett, R. E., Braun, H. I., Rock, D. A., & Powers, D. A. (Eds.). (1988). *Testing handicapped people*. Needham Heights: Allyn and Bacon.
Willingham, W., Lewis, C., Morgan, R., & Ramist, L. (1990). *Predicting college grades: An analysis of institutional trends over two decades*. Princeton: Educational Testing Service.
Wohlhueter, J. F. (1966). *Fatigue in testing and other mental tasks: A literature survey* (Research Memorandum No. RM-66-06). Princeton: Educational Testing Service.
Xi, X. (2010). How do we go about investigating test fairness? *Language Testing, 27*, 147–170. https://doi.org/10.1177/0265532209349465
Young, J. W. (2004). Differential validity and prediction: Race and sex differences in college admissions testing. In R. Zwick (Ed.), *Rethinking the SAT* (pp. 289–301). New York: Routledge Falmer.
Zieky, M. (1993). Practical questions in the use of DIF statistics in test development. In P. Holland & H. Wainer (Eds.), *Differential item functioning* (pp. 337–347). Mahwah: Erlbaum.
Zieky, M. (2006). Fairness review. In S. Downing & T. Haladyna (Eds.), *Handbook of test development* (pp. 359–376). Hillsdale: Erlbaum.
Zieky, M. (2011). The origins of procedures for using differential item functioning statistics at Educational Testing Service. In N. Dorans & S. Sinharay (Eds.), *Looking back: Proceedings of a conference in honor of Paul Holland* (pp. 115–127). New York: Springer. https://doi.org/10.1007/978-1-4419-9389-2_7
Zwick, R. (2006). Higher education admissions testing. In R. L. Brennan (Ed.), *Educational measurement* (4th ed., pp. 647–679). Westport: American Council on Education and Praeger.

Chapter 17
Understanding the Impact of Special Preparation for Admissions Tests

Donald E. Powers

By examining unique developments and singular advancements, it is possible to sort the history of educational and psychological testing into a number of distinct phases. One topic that seems to permeate all stages, however, is the question of how best to *prepare* for such tests. This chapter documents some of Educational Testing Service's (ETS's) contributions to understanding the role of test preparation in the testing process. These contributions include (a) analyzing key features of test preparation, (b) understanding the effects of various sorts of preparation on test performance, and (c) devising tests that will yield meaningful scores in the face of both legitimate as well as questionable attempts to improve test-taker performance. The chapter begins with a definition of special test preparation and then elaborates on its significance. Next, it examines the nature of interest in the topic. Finally, it explores ETS Research and Development (R&D) contributions to explicating the issues associated with special test preparation.

17.1 Definitions

The first issue that one encounters when discussing test preparation is terminology. This terminology applies both to the tests that are involved and to the kinds of preparation that are directed at test takers. Most of the research described below pertains to several tests that are designed to measure academic abilities (e.g., verbal and quantitative reasoning abilities) that develop relatively slowly over a significant

This chapter was originally published in 2012 by Educational Testing Service as a research report in the ETS R&D Scientific and Policy Contributions Series.

D.E. Powers (✉)
Educational Testing Service, Princeton, NJ, USA
e-mail: dpowers@ets.org

R.E. Bennett, M. von Davier (eds.), *Advancing Human Assessment*,
Methodology of Educational Measurement and Assessment,
DOI 10.1007/978-3-319-58689-2_17

period of time. This improvement occurs as a result of both formal schooling as well as other less formal experiences outside of school. Thus, to varying degrees, all students who take these kinds of tests receive highly relevant (but certainly differentially effective) preparation that should improve the skills and abilities being tested.

With respect to preparation, we have chosen here to use the word *special* to refer to a particular category of test preparation that focuses on readying test takers for a specific test. This special preparation may be of different sorts. For example, test familiarization is designed to ensure that prospective test takers are well versed in the general skills required for test taking and to help them gain familiarity with the procedures that are required to take a particular test. This type of preparation may entail, for instance, exposing test takers to the kinds of item formats they will encounter, making certain that they know when to guess, and helping them learn to apportion their time appropriately. Special preparation of this sort is generally regarded as desirable, as it presumably enables individuals to master the mechanics of test taking, thereby freeing them to focus on, and accurately demonstrate, the skills and abilities that are being assessed.

Coaching, on the other hand, has had a decidedly more negative connotation insofar as it is typically associated with short-term efforts aimed at teaching test-taking strategies or "tricks" to enable test takers to "beat the test;" that is, to take advantage of flaws in the test or in the testing system (e.g., never choose a particular answer choice if a question has these characteristics…). As Messick (1982) has noted, however, the term *coaching* has often been used in a variety of ways. At one extreme, it may signify short-term cramming and practice on sample item types, while on the other it may denote long-term instruction designed to develop the skills and abilities that are being tested. In practice, the distinctions among (a) relevant instruction, (b) test familiarization, and (c) coaching are sometimes fuzzy, as many programs contain elements of each type of preparation.

17.1.1 Significance of Special Test Preparation

Messick (1982) noted three ways in which special preparation may improve test scores. Each of these ways has a very different implication for score use. First, like real instruction, some types of special test preparation may genuinely improve the skills and abilities being tested, thereby resulting in higher test scores also. This outcome should have no detrimental effect on the validity of scores.

Second, some special test preparation (or familiarization) may enhance general test-taking skills and reduce test anxiety, thereby increasing test scores that may otherwise have been inaccurately low indicators of test takers' true abilities. Insofar as this kind of preparation reduces or eliminates unwanted sources of test difficulty, it should serve only to improve score validity.

The third possibility is that if it entails the teaching of test-taking tricks or other such strategies, special test preparation may increase test scores without necessarily improving the underlying abilities that are being assessed. A likely result is inaccurately high test scores and diminished score validity.

Finally, along with score validity, equity is often at issue in special test preparation, as typically not all students have equal opportunity to benefit in the ways described above. If special preparation is effective, its benefits may accrue only to those who can afford it.

17.1.2 Interest in Special Test Preparation

At first blush, the issue of special test preparation might seem to be of interest mainly to a relatively small group of test developers and psychometricians. Historically, however, attention to this topic has been considerably more widespread. Naturally, test takers (and for some tests, their parents) are concerned with ensuring that they are well prepared to take any tests that have high-stakes consequences. However, other identifiable groups have also shown considerable interest in the topic.

For instance, concern is clearly evident in the professional community. The current version of the *Standards for Educational and Psychological Testing* (American Educational Research Association, American Psychological Association, and National Council on Measurement in Education 2014) suggests a need to establish the degree to which a test is susceptible to improvement from special test preparation (Standard 1.7: "If test performance, or a decision made therefrom, is claimed to be essentially unaffected by practice and coaching, then the propensity for test performance to change with these forms of instruction should be documented," p. 24). In addition, a previous edition of *Educational Measurement* (Linn 1989), perhaps the most authoritative work on educational testing, devoted an entire chapter to special test preparation (Bond 1989).

General public interest is apparent also, as coaching has been the subject of numerous articles in the popular media (e.g., "ETS and the Coaching Cover Up," Levy 1979). One study of the effects of coaching (Powers and Rock 1999) was even a topic of discussion on a prominent national television show when the host of the *Today Show*, Matt Lauer, interviewed College Board Vice President Wayne Camara.

Besides being of general interest to the public, ETS coaching studies have also had a major impact on testing policy and practice. For example, in the early 1980s a previously offered section of the *GRE*® General Test (the analytical ability measure) was changed radically on the basis of the results of a GRE Board-sponsored test preparation study (Powers and Swinton 1984).

As a final indication of the widespread interest in the topic, in the late 1970s the U.S. Federal Trade Commission (FTC) became so troubled by the possibly misleading advertising of commercial coaching companies that it launched a major national investigation of the efficacy of such programs (Federal Trade Commission 1978, 1979). As described below, ETS contributed in several ways to this effort.

17.2 Studying the Effects of Special Test Preparation

What follows is an account of several key ETS contributions to understanding the role and effects of special test preparation. The account is organized within each of the two major testing programs on which special test preparation research has concentrated, the *SAT*® test and the GRE General Test.

17.2.1 The SAT

17.2.1.1 The College Board Position

The *effectiveness* of special test preparation has long been a contentious issue. Perhaps a reasonable place to begin the discussion is with the publication of the College Board's stance on coaching, as proclaimed by the Board's trustees in *Effects of Coaching on Scholastic Aptitude Test Scores* (College Entrance Examination Board 1965). This booklet summarized the (then) relatively few, mostly ETS-sponsored studies of coaching for the SAT (e.g., Dyer 1953, and French and Dear 1959) and concluded, "... the magnitude of gains resulting from coaching vary slightly, but they are always small ..." (p. 4), the average gain being fewer than 10 points on the 200-800 SAT scale.

17.2.1.2 Early Studies

The first significant challenge to the Board's stance seems to have come with the completion of a study by ETS researchers Evans and Pike (1973), who demonstrated that two SAT quantitative item types being considered for inclusion in the SAT were susceptible to improvement through special preparation—in particular, to the Saturday morning test preparation classes that the researchers designed for implementation over a 7-week period. The researchers' best estimate of effects was about 25 points on the 200–800 SAT Math (SAT-M) scale.

Besides the significant program of instruction that Evans and Pike developed, another particularly noteworthy aspect of this effort was the researchers' ability to implement a true experimental design. Students were randomly assigned to either (a) one of three treatment groups, each of which focused specifically on a different item type, or (b) a comparison condition that involved only more general test-taking skills. Previously, virtually no such studies had successfully carried out a true experiment.

At least partly because of the Evans and Pike (1973) study, interest also increased in the effects of special preparation for the verbal section of the SAT. The College Board subsequently funded ETS researchers to study the effectiveness of special secondary school programs geared to improving SAT Verbal (SAT-V) scores (Alderman and Powers 1980). A contribution here was that instead of relying on strictly observational methods or quasi-experimental designs, the investigators were able, through careful collaboration with a set of secondary schools, to exert a reasonably strong degree of experimental control over *existing* special preparation programs, assigning students randomly to treatment or control groups. This task was accomplished, for example, by taking advantage of demand for preparation that, in some cases, exceeded the schools' ability to offer it. In other cases, it was possible to simply delay preparation for randomly selected students. The results suggested that secondary school programs can affect SAT-V scores, albeit modestly, increasing them by about 4–16 points on the 200–800 SAT-V scale.

17.2.1.3 Test Familiarization

About the same time, the College Board, realizing the need to ensure that all test takers were familiar with the SAT, developed a much more extensive information bulletin than had been available previously. The new booklet, called *Taking the SAT*, contained extensive information about the test and about test-taking strategies, a review of math concepts, and a full-length practice SAT. Much to its credit, the Board was interested not only in offering the more extensive preparation material, but also in learning about its impact, and so it commissioned a study to assess the booklet's effects on both test-taking behavior and test scores (Powers and Alderman 1983). The study was an experiment in which a randomly selected group of SAT registrants received a prepublication version of the new booklet. Subsequently, their test performance was compared with that of an equivalent randomly selected group of test takers who had not received the booklet. (Only high school juniors were included in the study, partly to ensure that, should the booklet prove effective in increasing scores, all students in the cohort would have the opportunity to benefit from it before they graduated.)

The results showed increases in knowledge of appropriate test-taking behavior (e.g., when to guess), decreased anxiety, and increased confidence. There were no statistically significant effects on SAT-V scores but a small, significant effect on SAT-M scores of about 8 points.

17.2.1.4 Federal Interest

Perhaps the single most significant factor in the rising interest in coaching and test preparation was the involvement of the U.S. Federal Trade Commission (FTC). The FTC became increasingly concerned about the veracity of claims being made by commercial coaching companies, which promised to increase SAT takers' scores by hundreds of points. The issue became so important that the FTC eventually undertook its own study to investigate the effectiveness of commercial coaching programs.

Both ETS and several of the major commercial coaching companies cooperated with the FTC investigation. ETS provided students' SAT scores, and the coaching companies provided information about students' enrollment in their programs. FTC researchers analyzed the data and eventually issued a report, finding the effects of commercial coaching for the SAT to be statistically significant—in the range of 20–30 points for both SAT-V and SAT-M at the most effective of the coaching schools that were studied (Federal Trade Commission 1978, 1979; Sesnowitz et al. 1982). Needless to say, the study attracted considerable attention.

ETS responded to the FTC's findings as follows. Samuel Messick, then Vice President for Research at ETS, assembled a team of researchers to take a critical look at the methods the FTC had used and the conclusions it had reached. Messick and his team critiqued the FTC's methodology and, in order to address some serious flaws in the FTC analyses, reanalyzed the data. Various methods were employed to correct mainly for test taker self-selection in attending coaching programs.

Messick's contribution was released as a monograph titled, "The Effectiveness of Coaching for the SAT: Review and Reanalysis of Research from the Fifties to the FTC" (Messick 1980). In the book, Messick summarized and critiqued previous research on coaching, and several ETS researchers offered their critiques of the FTC study. Most importantly, the researchers conducted several reanalyses of the data obtained from the FTC. For example, ETS consultant Thomas Stroud reanalyzed the data, controlling for a variety of background variables, and found results similar to those reported by the FTC. In addition, by considering *PSAT/NMSQT*® scores, as well as pre- and postcoaching SAT scores, ETS researcher Don Rock was able to apply a differential growth model to the FTC data. His analysis showed that, at least for SAT-V scores, some of the difference between the posttest SAT scores of coached and uncoached test takers could be attributed, not to any specific effect of coaching, but rather to the faster growth expected of coached students. (The differential growth rate of coached and uncoached students was determined from PSAT/NMSQT to SAT score changes *before* students were coached.) The results of the various ETS analyses differed somewhat, but in total they revealed that only one of the three coaching schools had a significant impact on SAT scores—about 12–18 points on the SAT-V scale and about 20–30 points on the SAT-M scale.

One of the main lessons from the critique and reanalysis of the FTC study was stated by Messick (1980) in the preface to the report. Messick wrote that the issue of the effectiveness of coaching for the SAT is much more complicated than the simplistic question of whether coaching works or not. Coaching in and of itself is not automatically to be either rejected or encouraged. Rather, it matters what materials and practices are involved, at what cost in student time and resources, and with what effect on student skills, attitudes, and test scores (p. v).

Messick's (1980) insight was that complex issues, like the coaching controversy, are rarely ever usefully framed as either/or, yes/no questions. Rather, those questions turn out to involve degrees and multiple factors that need to be appreciated and sorted out. As a consequence, the answer to most questions is usually not a simple "yes" or "no," but more often a sometimes frustrating, "it depends." The task of researchers, then, is usually to determine, as best they can, the factors on which the effects depend.

17.2.1.5 Extending Lessons Learned

Messick followed through with this theme by analyzing the relationship of test preparation effects to the duration or length of test preparation programs. He published these results in the form of a meta-analysis (Messick and Jungeblut 1981), in which the authors noted "definite regularities" (p. 191) between SAT coaching effects and the amount of student contact time in coaching programs. On this basis, Messick and Jungeblut concluded that the size of the effects being claimed by coaching companies could probably be obtained only with programs that were tantamount to full-time schooling.

Powers (1986) followed Messick and Jungeblut's (1981) lead by reviewing a variety of other features of test preparation and coaching programs, and relating these features to the size of coaching effects. The advance here was that instead of focusing on the features of coaching programs, Powers analyzed the characteristics of the *item types* that comprised a variety of tests—for instance, how complex their directions were, whether they were administered under timed or untimed conditions, and what kinds of formats they employed. The results suggested that some features of test items (e.g., the complexity of directions) did render them more susceptible to improvement through coaching and practice than did others.

Several of the studies that Powers reviewed were so-called within-test practice studies, which were conducted by ETS statistical analysts (e.g., Faggen and McPeek 1981; Swinton et al. 1983; Wightman 1981). This innovative method involved trying out new test item types in early and later sections of the same test form. Then, differences in performance were compared for these early and later administered items. For some item types, it was routinely noticed that examinees performed better on new items that appeared later in the test, after earlier appearances of items of that type. A large within-test practice effect was viewed as a sufficient condition to disqualify a proposed new item type from eventual operational use. The rationale was the following: If an item type exhibited susceptibility to simple practice *within* a single test session, surely it would be *at least* as susceptible to more intensive coaching efforts.

17.2.1.6 Studying the 1994 Revision to the SAT

In 1994, a revision of the SAT was introduced. Many of the changes suggested that the revision should be even less susceptible to coaching than the earlier version. However, claims being made by coaching companies did not subside. For example, the January, 8, 1995, issue of the Philadelphia *Inquirer* proclaimed "New SAT proves more coachable than old." At least partly in response to such announcements, the College Board sponsored research to examine the effects of commercial coaching on SAT scores. Powers and Rock (1999) surveyed SAT takers about their test preparation activities, identifying a subset of test takers who had attended commercial coaching programs. Although the study was observational in nature, the researchers obtained a wide variety of background information on test takers and used this information to control statistically for self-selection effects. This approach was necessary, as it was widely acknowledged that coached and uncoached students differ on numerous factors that are also related to SAT scores. One of the differences noted by Powers and Rock, and controlled in their analysis, was that coached test takers were more likely than their uncoached counterparts to have engaged in a variety of *other* test preparation activities (e.g., self-study of various sorts), which may also have affected SAT scores. Several alternative analyses were employed to control for self-selection effects, and although each of the analyses produced slightly different estimates, all of them suggested that the effects of coaching were far less than was being alleged by coaching enterprises—perhaps only a quarter as large as claimed.

The alternative analyses yielded coaching effect estimates of 6–12 for SAT-V and 13–26 points for SAT-M. When analyses were undertaken separately for major coaching companies, the results revealed SAT-V effects of 12–19 points for one company and 5–14 points for another. The effects for SAT-M were 5–17 and 31–38, respectively, suggesting that the two programs were differentially effective for the two portions of the SAT.

The results of the study were featured in a New York *Times* article (Bronner 1998). The article quoted Professor Betsy Jane Becker, who had reviewed numerous SAT coaching studies (Becker 1990), as saying that the study was "perhaps the finest piece of coaching research yet published" (p. A23). This assessment may of course reflect either a regard for the high quality of the study or, on the other hand, concern about the limitations of previous ones.

17.2.2 The GRE General Test

Although the SAT program has been a major focus of test preparation and coaching studies, the GRE Board has also sponsored a number of significant efforts by ETS researchers. For instance, the GRE program revised its General Test in the late 1970s, introducing an analytical ability measure to complement the long-offered verbal and quantitative reasoning measures (Powers and Swinton 1981). Concurrently, the GRE Board sponsored several studies to examine the susceptibility of the new measure to coaching and other forms of special test preparation. Swinton and Powers (1983) designed a brief course to prepare students for the new analytical section of the GRE General Test and offered it to a small group of volunteer GRE test takers at a local university. Controlling for important pre-existing differences between groups, they compared the postcourse GRE performance of these specially prepared individuals with that of all other GRE test takers at the same university. They found that the specially prepared group did much better on the analytical section (by about 66 points on the 200–800 scale) than did the larger comparison group, even after controlling for differences in the GRE verbal and quantitative scores of the two groups.

Powers and Swinton (1984) subsequently packaged the course and used it in an experimental study in which a randomly selected sample of GRE test takers received the course materials by mail. A comparison of the test scores of the prepared sample with those of a randomly selected equivalent sample of nonprepared GRE test takers revealed score improvements that were nearly as large (about 53 points with about 4 hours of self-preparation) as those observed in the face-to-face classroom preparation. A major implication of this latter study was that test preparation designed for self-study by test takers themselves was a viable alternative to more expensive, formal face-to-face interventions. The ramifications for fairness and equity were obvious. However, although the researchers were relatively sanguine about the prospects for ensuring that all examinees could be well prepared for the "coachable" item types on the GRE, the GRE Board took a conservative stance, deciding instead to remove the two most susceptible item types from the analytical ability measure.

Data collected in the studies of the GRE analytical measure were also used to gauge the effectiveness of formal commercial coaching for the verbal and quantitative sections (Powers 1985a). That is, since the analytical measure had been shown to be coachable, it could serve as a baseline against which to judge the coachability of the other test sections.

For this analysis, Powers identified test takers who had attended formal coaching programs for any or all of the GRE test sections. For the analytical ability section, the analysis revealed a strong relationship between the effect of coaching and its duration (in terms of hours devoted to instruction). However, applying the same methodology to the verbal and quantitative sections revealed little if any such relationship, contrary to claims being made by commercial coaching firms. Increasing the duration of preparation for the verbal and quantitative GRE measures was not associated with commensurate increases in scores for these two measures.

17.2.2.1 Effects on Relationships of Test Scores with Other Measures

While Messick (1982) provided an insightful *logical* analysis of the ways in which special test preparation may impact validity, there appears to have been little *empirical* research to demonstrate how such practices may affect, for example, the relationship of test scores to other relevant measures. An exception is a study by Powers (1985b), who examined the relationship of GRE analytical ability scores, obtained under ten different randomly assigned test preparation conditions, to indicators of academic performance. Each of the various test preparation conditions was designed, mainly, to help test takers become familiar with each of several novel analytical ability item types. The results suggested that the more time test takers devoted to using the test preparation materials, the stronger the relationship was between academic performance and scores on the GRE analytical ability measure. Specifically, over the ten treatment groups, the correlation between (a) GRE analytical ability score and (b) undergraduate grade point average in the final 2 years of undergraduate study increased according to mean time devoted to preparing for the analytical measure ($r = .70, p < .05$). In addition, correlations of GRE analytical ability scores with GRE verbal and quantitative scores were not significantly related to amounts of test preparation. Thus, both the convergent and (possibly) the discriminant aspects of construct validity of test scores may have been enhanced.

17.3 Summary

ETS has made several contributions to understanding the effects of special test preparation and coaching on (a) test-taking behavior, (b) test performance, and (c) test validity. First, ETS researchers have brought more methodological rigor to the field by demonstrating the feasibility of conducting experimental studies of the effects of test preparation. Rigor has also been increased by introducing more sophisticated methods for controlling self-selection bias in nonexperimental studies.

Moreover, ETS researchers have evaluated the effects of a variety of different types of test preparation: formal commercial coaching, school-offered test preparation programs, and test sponsor-provided test familiarization. With respect to the last type, a significant portion of the ETS-conducted research has focused on making certain that *all* test takers are well prepared, not just those who can afford extensive coaching. Along these lines, researchers have evaluated the effects of test familiarization and other means of test preparation that can be offered, usually remotely for independent study, to *all* test takers. Both secondary and postsecondary student populations have been studied.

Thanks largely to Messick (1980, 1981, 1982), the question of the effectiveness of coaching and test preparation has been reformulated—that is, extended beyond the search for a yes/no answer to the oversimplified question "Does coaching work?" Partly as a result, researchers now seem more inclined to examine the components of test preparation programs in order to ascertain the particular features that are implicated in its effectiveness.

ETS researchers have also stressed that every test is typically composed of a variety of item types and that some of these item types may be more susceptible to coaching and practice than others. In this vein, they have determined some of the features of test item types that seem to render them more or less susceptible. As a consequence, there is now a greater realization that it is insufficient to simply consider the coachability of a test as a whole, but rather it is necessary to consider the characteristics of the various item types that comprise it.

In addition, at least in the view of the scientific community, if not among the general public, a more accurate estimate of the true value of commercial coaching programs now exists. Consumers have information to make more informed choices about whether to seek commercial coaching, for instance. The true effect of coaching on test performance seems neither as negligible as some have claimed nor as large as has been advertised by the purveyors of coaching services.

Most of the studies of coaching and test preparation have focused on the extent to which these practices cause spurious test score improvement. However, although relatively rare, ETS researchers have also examined, in both a logical and an empirical manner, the effects of test preparation and coaching on the empirical relationships of test scores to other indicators of developed ability.

Finally, ETS research on test preparation has been more than an academic exercise. It has resulted in significant—even dramatic—modifications to several tests that ETS offers. These changes are perhaps the clearest example of the impact of ETS's research on test preparation. However, there have, arguably, been more subtle effects as well. Now, when new assessments are being developed, the potential coachability of proposed new test item types is likely to be a factor in decisions about the final composition of a test. Considerations about test preparation figure into the *design* of tests, well before these tests are ever administered to test takers.

References

Alderman, D. L., & Powers, D. E. (1980). The effects of special preparation on SAT verbal scores. *American Educational Research Journal, 17*, 239–251. https://doi.org/10.3102/00028312017002239

American Educational Research Association, American Psychological Association, & National Council on Measurement in Education. (2014). *Standards for educational and psychological testing*. Washington, DC: American Educational Research Association.

Becker, B. J. (1990). Coaching for the Scholastic Aptitude Test: Further synthesis and appraisal. *Review of Educational Research, 60*, 373–417. https://doi.org/10.3102/00346543060003373

Bond, L. (1989). The effects of special preparation on measures of scholastic ability. In R. L. Linn (Ed.), *Educational measurement* (3rd ed., pp. 429–444). New York: Macmillan.

Bronner, E. (1998, November 24). Study casts doubt on the benefits of S.A.T.-coaching courses. *The New York Times National*, p. A23.

College Entrance Examination Board. (1965). *Effects of coaching on Scholastic Aptitude Test scores*. New York: Author.

Dyer, H. S. (1953). Does coaching help? *College Board Review, 19*, 331–335.

Evans, F. R., & Pike, L. W. (1973). The effects of instruction for three mathematics item formats. *Journal of Educational Measurement, 10*, 257–272. https://doi.org/10.1111/j.1745-3984.1973.tb00803.x

Faggen, J., & McPeek, M. (1981, April). *Practice effects for four different item types*. Paper presented at the annual meeting of the National Council on Measurement in Education, Los Angeles.

Federal Trade Commission, Boston Regional Office. (1978, September). *Staff memorandum of the Boston Regional office of the Federal Trade Commission: The effects of coaching on standardized admission examinations*. Boston: Author.

Federal Trade Commission, Bureau of Consumer Protection. (1979). *Effects of coaching on standardized admission examinations: Revised statistical analyses of data gathered by Boston Regional office, Federal Trade Commission*. Washington, DC: Author.

French, J. W., & Dear, R. E. (1959). Effect of coaching on an aptitude test. *Educational and Psychological Measurement, 19*, 319–330. https://doi.org/10.1177/001316445901900304

Levy, S. (1979, March). ETS and the coaching cover-up. *New Jersey Monthly, 3*(4), 50–54, 82–89.

Linn, R. L. (Ed.). (1989). *Educational measurement* (3rd ed.). New York: Macmillan.

Messick, S. (1980). *The effectiveness of coaching for the SAT: Review and reanalysis of research from the fifties to the FTC*. Princeton: Educational Testing Service.

Messick, S. (1981). The controversy over coaching: Issues of effectiveness and equity. In B. F. Green (Ed.), *Issues in testing: Coaching, disclosure, and ethnic bias*. San Francisco: Jossey-Bass.

Messick, S. (1982). Issues of effectiveness and equity in the coaching controversy: Implications for educational testing and practice. *Educational Psychologist, 17*, 67–91. https://doi.org/10.1080/00461528209529246

Messick, S., & Jungeblut, A. (1981). Time and method in coaching for the SAT. *Psychological Bulletin, 89*, 191–216. https://doi.org/10.1037/0033-2909.89.2.191

Powers, D. E. (1985a). Effects of coaching on GRE Aptitude Test scores. *Journal of Educational Measurement, 22*, 121–136. https://doi.org/10.1111/j.1745-3984.1985.tb01052.x

Powers, D. E. (1985b). Effects of test preparation on the validity of a graduate admissions test. *Applied Psychological Measurement, 9*, 179–190. https://doi.org/10.1177/014662168500900206

Powers, D. E. (1986). Relations of test item characteristics to test preparation/test practice effects: A quantitative summary. *Psychological Bulletin, 100*, 67–77. https://doi.org/10.1037/0033-2909.100.1.67

Powers, D. E., & Alderman, D. L. (1983). Effects of test familiarization on SAT performance. *Journal of Educational Measurement, 20*, 71–79. https://doi.org/10.1111/j.1745-3984.1983.tb00191.x

Powers, D. E., & Rock, D. A. (1999). Effects of coaching on SAT I: Reasoning scores. *Journal of Educational Measurement, 36*, 93–118. https://doi.org/10.1111/j.1745-3984.1999.tb00549.x

Powers, D. E., & Swinton, S. S. (1981). Extending the measurement of graduate admission abilities beyond the verbal and quantitative domains. *Applied Psychological Measurement, 5*, 141–158. https://doi.org/10.1177/014662168100500201

Powers, D. E., & Swinton, S. S. (1984). Effects of self-study for coachable test item types. *Journal of Educational Psychology, 76*(2), 266–278. https://doi.org/10.1037/0022-0663.76.2.266

Sesnowitz, M., Bernhardt, K. L., & Knain, D. M. (1982). An analysis of the impact of commercial test preparation courses on SAT scores. *American Educational Research Journal, 19*, 429–441. https://doi.org/10.3102/00028312019003429

Swinton, S. S., & Powers, D. E. (1983). A study of the effects of special preparation on GRE analytical scores and item types. *Journal of Educational Psychology, 75*, 104–115.

Swinton, S. S., Wild, C. L., & Wallmark, M. M. (1983). *Investigating practice effects on item types in the graduate record examinations aptitude test* (Research Report No. RR-82-56). Princeton: Educational Testing Service.

Wightman, L. E. (1981, April). *GMAT within-test practice effects studies*. Paper presented at the annual meeting of the National Council of Measurement in Education, Los Angeles.

Chapter 18
A Historical Survey of Research Regarding Constructed-Response Formats

Isaac I. Bejar

This chapter chronicles ETS research and development contributions related to the use of constructed-response item formats.[1] The use of constructed responses in testing dates back to imperial China, where tests were used in the selection of civil servants. However, in the United States, the multiple-choice format became dominant during the twentieth century, following its invention and use by the *SAT*® examinations created by the College Board in 1926. When ETS was created in 1947, post-secondary admissions testing was largely based on tests consisting of multiple-choice items. However, from the start, there were two camps at ETS: those who believed that multiple-choice tests were sufficiently adequate for the purpose of assessing "verbal" skills and those who believed that "direct" forms of assessment requiring written responses had a role to play. For constructed-response formats to regain a foothold in American education several hurdles would need to be overcome. Research at ETS was instrumental in overcoming those hurdles.

The first hurdle was that of reliability, specifically the perennial issue of low interrater agreement, which plagued the acceptance of constructed-response formats for most of the twentieth century. The second hurdle was broadening the conception of validity to encompass more than predictive considerations, a process that began with the introduction of construct validity by Cronbach and Meehl (1955). Samuel Messick at ETS played a crucial role in this process by making construct validity relevant to educational tests. An inflexion point in the process of reincorporating constructed-response formats more widely in educational tests was marked

[1] Constructed responses to a prompt or question can range in scope and complexity. Perhaps the most common constructed response is the written essay. However, short written responses to questions are also considered to be constructed, as are spoken answers in response to a prompt, mathematical responses (equations, plotted functions, etc.), computer programs, and graphical responses such as architectural designs.

I.I. Bejar (✉)
Educational Testing Service, Princeton, NJ, USA
e-mail: ibejar@ets.org

R.E. Bennett, M. von Davier (eds.), *Advancing Human Assessment*, Methodology of Educational Measurement and Assessment,
DOI 10.1007/978-3-319-58689-2_18

by the publication of *Construction Versus Choice in Cognitive Measurement* (Bennett and Ward 1993), following the indictment of the multiple-choice format by Norm Frederiksen (1984) regarding the format's potentially pernicious influence on education. The chapters in the book made it clear that the choice of format (multiple choice vs. constructed response) includes considerations of validity broadly conceived. Even when there was growing concern about the almost exclusive reliance on the multiple-choice format, there was much more work to be done to facilitate the operational use of constructed-response items since over the preceding decades the profession had come to rely on the multiple-choice format. That work continues to this day at ETS and elsewhere.

Clearly there is more than one way to convey the scope of research and development at ETS to support constructed-response formats. The chapter proceeds largely chronologically in several sections. The first section focuses on the ETS contributions to scoring reliability, roughly through the late 1980's. The next section considers the evolution of validity toward a unitary conception and focuses on the critical contributions by Samuel Messick with implications for the debate around constructed-response formats.

The third section argues that the interest in technology for testing purposes at ETS from early on probably accelerated the eventual incorporation of writing assessment into several ETS admissions tests. That section reviews work related to computer-mediated scoring, task design in several domains, and the formulation of an assessment design framework especially well-suited for constructed-response tasks, evidence-centered design (ECD).

The fourth section describes ETS's involvement in school-based testing, including *Advanced Placement®* (*AP®*), the National Assessment of Educational Progress (NAEP), and the *CBAL®* initiative. A fifth section briefly discusses validity and psychometric research related to constructed-response formats. The chapter closes with some reflections on six decades of research.

18.1 Reliability

The acceptance of the multiple-choice format, after its introduction in the 1926 SAT, together with the growing importance of reliability as a critical attribute of the scores produced by a test, seems to have contributed to the decline of widely used constructed-response forms of assessment in the United States. However, research at ETS was instrumental in helping to return those formats to the assessment of writing in high-stakes contexts. In this section, some of that research is described. Specifically, among the most important ETS contributions are

1. developing holistic scoring
2. advancing the understanding of rater cognition
3. conducting psychometric research in support of constructed responses

Reliability (Haertel 2006) refers to the level of certainty associated with scores from a given test administered to a specific sample and is quantified as a reliability or generalizability coefficient or as a standard error. However, the first sense of *reliability* that comes to mind in the context of constructed responses is that of *interrater reliability,* or agreement. Unlike responses to multiple-choice items, constructed responses need to be scored by a process (cf. Baldwin et al. 2005) that involves human judgment or, more recently (Williamson et al. 2006), by an automated process that is guided by human judgment. Those human judgments can be more or less fallible and give rise to concerns regarding the replicability of the assigned score by an independent scorer. Clearly, a low level of interrater agreement raises questions about the meaning of scores.

The quantification of interrater disagreement begins with the work of the statistician F. Y. Edgeworth (as cited by Mariano 2002). As Edgeworth noted,

> let a number of equally competent critics independently assign a mark to the (work) … even supposing that the examiners have agreed beforehand as to … the scale of excellence to be adopted … there will occur a certain divergence between the verdicts of competent examiners. (p. 2)

Edgeworth also realized that individual differences among readers could be the source of those errors by noting, for example, that some raters could be more or less severe than others, thus providing the first example of theorizing about rater cognition, a topic to which we will return later in the chapter. Edgeworth (1890) noted,

> Suppose that a candidate obtains 95 at such an examination, it is reasonably certain that he deserves his honours. Still there is an appreciable probability that his real mark, as determined by a jury of competent examiners (marking independently and taking the average of those marks) is just below 80; and that he is pushed up into the honour class by the accident of having a *lenient examiner*. Conversely, his real mark might be just above 80; and yet by accident he might be compelled without honour to take a lower place as low as 63. (emphasis added, p. 470)

The lack of interrater agreement would plague attempts to reincorporate constructed responses into post-secondary admissions testing once multiple-choice items began to supplant them. An approach was needed to solve the interrater reliability problem. A key player in that effort was none other than Carl Brigham (1890–1943), who was the chief architect behind the SAT, which included only multiple-choice items.[2] Brigham was an atypical test developer and psychometrician in that he viewed the administration of a test as an opportunity to experiment and further learn about students' cognition. And experiment he did. He developed an "experimental section" (N. Elliot 2005, p. 75) that would contain item types that were not being used operationally, for example. Importantly, he was keenly interested in the possibility of incorporating more "direct" measures of writing (Valentine 1987, p. 44). However, from the perspective of the College Board, the sponsor of the test, by the 1930s, the SAT was generating significant income, and the Board seemed to have set some limits on the degree of experimentation. According to Hubin (1988),

[2] For a historical account of how Brigham came to lead the development of the SAT, see Hubin (1988).

> the growth of the Scholastic Aptitude Test in the thirties, although quite modest by standards of the next decade, contrasted sharply with a constant decline in applicants for the traditional Board essay examinations. Board members saw the SAT's growth as evidence of its success and increasingly equated such success with the Board's very existence. The Board's perception decreased Brigham's latitude to experiment with the instrument. (p. 241)

Nevertheless, Brigham and his associates continued to experiment with direct measures of writing. As suggested by the following excerpt (Jones and Brown 1935), there appeared to be progress in solving the rater agreement challenge:

> Stalnaker and Stalnaker … present evidence to show that the essay-type test can be scored with rather high reliability if certain rules are followed in formulating questions and in scoring. Brigham … has made an analysis of the procedures used by readers of the English examinations of the College Entrance Examination Board, and believes that the major sources of errors in marking have been identified. A new method of grading is being tried which, he thinks, will lead to greatly increased reliability. (p. 489)

There were others involved in the improvement of the scoring of constructed responses. For example, Anderson and Traxler (1940) argued that[3]

> by carefully formulating the test material and training the readers, it is possible to obtain highly reliable readings of essay examinations. Not only is the reliability high for the total score, but it is also fairly high for most of the eight aspects of English usage that were included in this study. The reliability is higher for some of those aspects that are usually regarded as fairly intangible than for the aspects that one would expect to be objective and tangible. The test makes fair, though by no means perfect, discrimination among the various years of the secondary school in the ability of the pupils to write a composition based on notes supplied to them. The results of the study are not offered as conclusive, but it is believed that, when they are considered along with the results of earlier studies, they suggest that it is highly desirable for schools to experiment with essay-test procedures as means for supplementing the results of objective tests of English usage in a comprehensive program of evaluation in English expression. (p. 530)

Despite these positive results, further resistance to constructed responses was to emerge. Besides reliability concerns, costs and efficiency also were part of the equation. For example, we can infer from the preceding quotation that the scoring being discussed is "analytic" and would require multiple ratings of the same response. At the same time, machine scoring of multiple-choice responses was rapidly becoming a reality[4] (Hubin 1988, p. 296). The potential efficiencies of machine scoring contrasted sharply with the inefficiencies and logistics of human scoring. In fact, the manpower shortages during World War II led the College Board to suspend examinations relying on essays (Hubin 1988, p. 297).

---

[3] In this passage, *reliability* refers to interrater agreement.

[4] Du Bois (1970, p. 119), citing Downey (1965), notes that the scoring machine was invented in 1934 by Reynold B. Johnson, inspired by Ben D. Wood's vision of large-scale testing. According to Du Bois, these scoring machines greatly reduced the cost and "accelerated the trend toward more or less complete reliance on objective tests, especially the multiple-choice item." Of course, testing volume increased over the decades and motivated significant innovations. E. F. Lindquist at the University of Iowa invented the first successful optical scanner in 1962 (U.S. Patent 3,050,248) that was capable of processing larger numbers of answer sheets than the prior electrical mark sense scanner Johnson invented.

However, the end of the war did not help. Almost 10 years on, a study published by ETS (Huddleston 1954) concluded that

> the investigation points to the conclusion that in the light of present knowledge, measurable "ability to write" is no more than verbal ability. It has been impossible to demonstrate by the techniques of this study that essay questions, objective questions, or paragraph-revision exercises contain any factor other than verbal; furthermore, these types of questions measure writing ability less well than does a typical verbal test. The high degree of success of the verbal test is, however, a significant outcome.
>
> The results are discouraging to those who would like to develop reliable and valid essay examinations in English composition—a hope that is now more than half a century old. Improvement in such essay tests has been possible up to a certain point, but professional workers have long since reached what appears to be a stone wall blocking future progress. New basic knowledge of human capacities will have to be unearthed before better tests can be made or more satisfactory criteria developed. To this end the Educational Testing Service has proposed, pending availability of appropriate funds, a comprehensive factor study in which many types of exercises both new and traditional are combined with tests of many established factors in an attempt to discover the fundamental nature of writing ability. The present writer would like to endorse such a study as the only auspicious means of adding to our knowledge in this field. Even then, it appears unlikely that significant progress can be made without further explorations in the area of personality measurement.[5] (pp. 204–205)

In light of the limited conception of both "verbal ability" and "writing ability" at the time, Huddleston's conclusions appear, in retrospect, to be unnecessarily strong and overreaching. The evolving conception of "verbal ability" continues to this day, and it is only recently that even basic skills, like vocabulary knowledge, have become better understood (Nagy and Scott 2000); it was not by any means settled in the early 1950s. Importantly, readily available research at the time was clearly pointing to a more nuanced understanding of writing ability. Specifically, the importance of the role of "fluency" in writing was beginning to emerge (C. W. Taylor 1947) well within the psychometric camp. Today, the assessment of writing is informed by a view of writing as a "complex integrated skill" (Deane et al. 2008; Sparks et al. 2014) with fluency as a key subskill.

By today's standards, the scope of the concept of *reliability* was not fully developed in the 1930s and 1940s in the sense of understanding the *components* of unreliability. The conception of reliability emerged from Spearman's work (see Stanley 1971, pp. 370–372) and was focused on *test-score* reliability. If the assignment of a score from each component (item) is error free, because it is scored objectively, then the scoring does not contribute error to the total score, and in that case score reliability is a function of the number of items and their intercorrelations. In the case of constructed responses, the scoring is not error free since the scorer renders a judgment, which is a fallible process.[6] Moreover, because items that require constructed responses require more time, typically, fewer of them can be administered which, other things

---

[5] The mysterious reference to "personality measurement" appears to be reference to the thinking that personality measurement would be the next frontier in admissions testing. In fact, ETS, specifically Henry Chauncey, was interested in personality measurement (see Lemann 1999, p. 91).

[6] Fallibility is relative; even the scoring of multiple-choice items is not 100% error free, at least not without many preventive measures to make it so. For a discussion, see Baker (1971).

being equal, reduces score reliability. The estimation of error components associated with ratings would develop later (Ebel 1951; Finlayson 1951), as would the interplay among those components (Coffman 1971), culminating in the formulation of generalizability theory (Cronbach et al. 1972).[7] Coffman (1971), citing multiple sources, summarized the state of knowledge on interreader agreement as follows:

> The accumulated evidence leads, however, to three inescapable conclusions: a) different raters tend to assign different grades to the same paper; b) a single rater tends to assign different grades to the same paper on different occasions; and c) the differences tend to increase as the essay question permits greater freedom of response. (p. 277)

Clearly, this was a state of affairs not much different than what Edgeworth had observed 80 years earlier.

18.1.1 The Emergence of a Solution

The Huddleston (1954) perspective could have prevailed at ETS and delayed the wider use of constructed responses, specifically in writing.[8] Instead, from its inception ETS research paved the way for a solution to reducing interrater disagreement. First, a groundbreaking investigation at ETS (funded by the Carnegie Corporation) established that raters operated with different implied scoring criteria (Diederich et al. 1961). The investigation was motivated by the study to which Huddleston refers in the preceding quotation. That latter study did not yield satisfactory results, and a different approach was suggested: "It was agreed that further progress in grading essays must wait upon a factor analysis of judgments of a diverse group of competent readers in an unstructured situation, where each could grade as he liked" (Diederich et al. 1961, p. 3). The motivating hypothesis was that different readers belong to different "schools of thought" that would presumably value qualities of writing differently. The methodology that made it possible to identify types of readers was first suggested by Torgerson and Green (1952) at ETS. To identify the schools of thought, 53 "distinguished readers" were asked to rate and annotate 300 papers without being given standards or criteria for rating. The factors identified from the interrater correlations consisted of groupings of raters (e.g., raters that loaded highly on a specific factor). What school of thought was represented by a given factor would not be immediately obvious without knowing the specifics of the reasoning underlying a rater's judgment. The reasoning of the readers was captured by means of the annotations each judge had been asked to make, which then had to be coded and classified.[9] The results showed that agreement among readers was

[7] Brennan (2001, p. 3) credits Burt in 1936 and Lindquist in 1953 with anticipating the essence of univariate generalizability theory.

[8] See Diederich (1957) for a candid description of the state of affairs with respect to using essays in admissions testing.

[9] A very laborious process carried out by Sydell Carlton, an employee at ETS until 2017. She recalls (personal communication, July 19, 2010) that no one could initially interpret the factors.

poor and that the nature of the schools of thought was that they valued different aspects of writing. However, the two most sharply defined groups were those that valued "ideas" or that valued "mechanics."

The Diederich et al. (1961) study showed that judges, when left to their own analytical devices, will resort to particular, if not idiosyncratic, evaluative schemes and that such particularities could well explain the perennial lack of adequate interrater agreement. Important as that finding was, it still did not formulate a solution to the problem of lack of interrater agreement. That solution took a few more years, also leading to a milestone in testing by means of constructed responses. The study was carried out at ETS and led by Fred I. Godshalk. The study was ambitious and included five 20-minute essays, six objective tests, and two interlinear exercises, administered to 646 12th graders over a period of several weeks. Importantly, the scoring of the essays was *holistic*. They defined the scoring procedure of the essays as follows (Godshalk et al. 1966):

> The readers were asked to make global or holistic, not analytical, judgments of each paper, *reading rapidly for a total impression*. There were only three ratings: a score of "3" for a superior paper, "2" for an average paper, and "1" for an inferior paper. The readers were told to judge each paper on its merits without regard to other papers on the same topic; that is, they were not to be concerned with any ideas of a normal distribution of the three scores. They were advised that scores of "3" were possible and that the "safe" procedure of awarding almost all "2s" was to be avoided. Standards for the ratings were established in two ways: by furnishing each reader with copies of the sample essays for inspection and discussion, and by explaining the conditions of administration and the nature of the testing population; and by having all readers score reproduced sets of carefully selected sample answers to all five questions and to report the results. The scores were then tabulated and announced. No effort was made to identify any reader whose standards were out of line, because that fact would be known to him and would be assumed to have a corrective effect. The procedure was repeated several times during the first two days of scoring to assist readers in maintaining standards. (p. 10, emphasis added)

Perhaps the critical aspect of the directions was to "to make global or holistic, not analytical, judgments" and the use of what is known today (Baldwin et al. 2005) as benchmark or range finding papers to illustrate the criteria. The authors describe the procedure in the preceding quotation and do not provide a theoretical rationale. They were, of course, aware of the earlier Diederich study, and it could have influenced the conception of the holistic scoring instructions. That is, stressing that the scoring was to be holistic and not analytical could have been seen as way to prevent the schools of thought from entering the scoring process and to make the scoring process that much faster.[10]

After she classified a few of the annotations, she formulated a coding scheme that could be used to systematically annotate the rest of essays. The actual coding of more than 10,000 papers was hired out. By examining the annotation of readers that loaded highly on one factor or another, it became possible to interpret the factors as schools of thought.

[10]Although Godshalk et al. (1966) are associated with making *holistic* scoring a widely accepted approach, the term "wholistic" was used first at ETS by Ann F. Coward with the same meaning. In a project published as a brief internal report (Coward 1950) that was subsequently published (Coward 1952), she compared "wholistic," which corresponded with what later was called *holistic*,

Outside of ETS, the development of holistic scoring was well received by teachers of English (White 1984) and characterized as "undoubtedly one of the biggest breakthroughs in writing assessment" (Huot 1990, p. 201). Interestingly, other concurrent work in psychology, although relevant in retrospect, was not considered at the time as related to scoring of essays. For example, N. Elliot (2005) postulated the relevance of Gestalt psychology to a possible adoption of holistic scoring, although there is no such evidence in the Godshalk et al. (1966) report. Another line of research that was relevant was models of judgment, such as the lens model proposed by Egon Brunswik (Brunswik 1952; Hammond et al. 1964; Tucker 1964).[11] The lens model, although intended as a perceptual model, has been used primarily in decision-making (Hammond and Stewart 2001). According to the model, the perceiver or decision maker decomposes an object into its attributes and weighs those attributes in arriving at a judgment. The model is clearly applicable in modeling raters (Bejar et al. 2006). Similarly, a theory of personality of the same period, George Kelly's personal construct theory, included a method for eliciting "personal constructs" by means of analysis of sets of important others.[12] The method, called the repertory grid technique, was later found useful for modeling idiographic or reader-specific rating behavior (Bejar et al. 2006; Suto and Nadas 2009).

One additional area of relevant research was the work on clinical judgment. Meehl's (1954) influential monograph concluded that actuarial methods were superior to clinical judgment in predicting clinical outcomes. One reason given for the superiority of actuarial methods, often implemented as a regression equation or even the sum of unweighted variables (Dawes and Corrigan 1974), is that the actuarial method is provided with variables from which to arrive at a judgment. By contrast, the clinician first needs to figure out the variables that are involved, the rubric, so to speak, and determine the value of the variables to arrive at a judgment. As Meehl stressed, the clinician has limited mental resources to carry out the task. Under such conditions, it is not unreasonable for the clinician to perform inconsistently relative to actuarial methods. The overall and quick impression called for by the holistic instructions could have the effect of reducing the cognitive load demanded by a very detailed analysis. Because of this load, such an analysis is likely to play upon the differences that might exist among readers with respect to background and capacity to carry out the task.

There was such relief once the holistic method had been found to help to improve interrater agreement that no one seems to have noted that the idea of holistic scoring is quite counterintuitive. How can a quick impression substitute for a deliberate and

---

and "atomistic" approaches to scoring. No great differences between the two methods were reported, nor was any rationale proposed for the "wholistic" method. There was also experimentation in the UK on impressionistic scoring in the early 1960s (N. Elliot, personal communication, May 15, 2015)

[11] Ledyard Tucker, the eminent ETS psychometrician, had been a reviewer of the Hammond et al. (1964) paper. His review so influenced the Hammond et al. paper that Hammond suggested to the *Psychological Review* editors that Tucker's formulation of the lens model appear as an independent paper (Hammond, personal communication, March 29, 2010).

[12] George Kelly's work was well known at ETS (Messick and Kogan 1966).

extensive analysis of a constructed response by a subject matter *expert*? Research on decision making suggests, in fact, that experts operate in a holistic sort of fashion and that it is a sign of their expertise to do so. Becoming an expert in any domain involves developing "fast and frugal heuristics" (Gigerenzer and Goldstein 1996) that can be applied to arrive at accurate judgments quickly.

Eventually, questions would be raised about holistic scoring, however. As Cumming et al. (2002) noted,

> holistic rating scales can conflate many of the complex traits and variables that human judges of students' written compositions perceive (such as fine points of discourse coherence, grammar, lexical usage, or presentation of ideas) into a few simple scale points, rendering the meaning or significance of the judges' assessments in a form that many feel is either superficial or difficult to interpret. (p. 68)

That is, there is a price for the increased interreader agreement made possible by holistic scoring, namely, that we cannot necessarily document the mental process that scorers are using to arrive at a score. In the absence of that documentation, strictly speaking, we cannot be sure by what means scores are being assigned and whether those means are appropriate until evidence is presented.

Concerns such as these have given rise to research on rater cognition (Bejar 2012). The Diederich et al. (1961) study at ETS started the research tradition by attempting to understand the basis of lack of agreement among scorers (see also Myers et al. 1966). The range of the literature, a portion of it carried out at ETS, is vast and aims, in general, to unpack what goes on in the minds of the raters as they score (Bejar et al. 2006; Crisp 2010; Elbow and Yancey 1994; Huot and Neal 2006; Lumley 2002; Norton 1990; Pula and Huot 1993; Vaughan 1991), the effect of a rater's background (Myford and Mislevy 1995; Shohamy et al. 1992), rater strategies (Wong and Kwong 2007), and methods to elicit raters' personal criteria (Bejar et al. 2006; Heller et al. 1998). Descriptions of the qualifications of raters have also been proposed (Powers et al. 1998; Suto et al. 2009). In addition, the nature of scoring expertise has been studied (Wolfe 1997; Wolfe et al. 1998). Methods to capture and monitor rater effects during scoring as a function of rater characteristics are similarly relevant (Myford et al. 1995; Myford and Mislevy 1995; Patz et al. 2002). Experimental approaches to modeling rater cognition have also emerged (Freedman and Calfee 1983), where the interest is on systematic study of different factors that could affect the scoring process. The effectiveness of different approaches to the training of readers (Wolfe et al. 2010) and the qualifying of raters (Powers et al. 1998) has also been studied. In short, the Diederich et al. study was the first in a long line of research concerned with better understanding and improving the processes in which raters engage.

A second concern regarding holistic scoring is the nature of the inferences that can be drawn from scores. Current rubrics described as holistic, such as those used for scoring the *GRE*® analytical writing assessment, are very detailed, unlike the early rubrics. That is, holistic scoring has evolved from its inception, although quietly. Early holistic scoring had as a goal the *ranking* of students' responses.

Holistic scoring emerged in the context of admissions testing, which means in a norm-referenced context. In that context, the ranking or comparative interpretation of candidates is the goal. Points along the scale of such a test do not immediately have implications for what a test taker knows and can do, that is, attaching an interpretation to a score or score range. The idea of criterion-referenced (Glaser 1963) measurement emerged in the 1960s and was quickly adopted as an alternative conception to norm-referenced testing, especially in the context of school-based testing. Today it is common (Linn and Gronlund 2000) to talk about standards-based assessments to mean assessments that have been developed following a framework that describes the content to be assessed such that scores on the test can be interpreted with respect to what students know and can do. Such interpretations can be assigned to a single score or, more commonly, a range of scores by means of a process called standard setting (Cizek and Bunch 2007; Hambleton and Pitoniak 2006), where panels of experts examine the items or performance on the items to determine what students in those score regions know and can do.

NAEP had from its inception a standards-based orientation. The initial implementation of NAEP in the 1960s, led by Ralph Tyler, did not report scores, but rather performance on specific items, and did not include constructed responses. When writing was first introduced in the late 1960s, the scoring methodology was holistic (Mullis 1980, p. 2). However, the methodology was not found adequate for NAEP purposes and instead the method of primary traits was developed for the second NAEP writing assessment in 1974 (Cooper 1977, p. 11; Lloyd-Jones 1977). The inapplicability of holistic scoring to NAEP measurement purposes is given by Mullis (1980):

> NAEP needed to report performance levels for particular writing skills, and the rank ordering did not readily provide this information. Also, NAEP for its own charge of measuring change over time, as well as for users interested in comparisons with national results, needed a scoring system that could be replicated, and this is difficult to do with holistic scoring. (p. 3)

The criterion-referenced rationale that Mullis advocated was very much aligned with the standards-based orientation of NAEP. According to Bourque (2009), "by the mid-1980s, states began to realize that better reporting mechanisms were needed to measure student progress" (p. 3). A policy group was established, the National Assessment Governing Board (NAGB), to direct NAEP, and shortly thereafter the "Board agreed to adopt three achievement levels (Basic, Proficient, and Advanced) for each grade and subject area assessed by NAEP" (Bourque 2009, p. 3).

With respect to writing, Mullis (1984) noted,

> For certain purposes, the most efficient and beneficial scoring system may be an adaptation or modification of an existing system. For example, the focused holistic system used by the Texas Assessment Program … can be thought of as a combination of the impressionistic holistic and primary trait scoring systems. (p. 18)

To this day, the method used by NAEP to score writing samples is a modified holistic method called focused holistic (H. Persky, personal communication, January 25,

2011; see also, Persky 2012) that seems to have first originated in Texas around 1980 (Sachse 1984).

Holistic scoring also evolved within admissions testing for different reasons, albeit in the same direction. N. Elliot (2005, p. 228) gives Paul Ramsey at ETS credit for instituting a "modified holistic" method to mean that the scoring was accompanied by detailed scoring guides. In 1992 the College Board's English Composition Test (which would become SAT Writing) began using scoring guides as well. The rationale, however, was different, namely, comparability:[13]

> We need a scoring guide for the SAT Writing test because, unlike the ECT [English Composition Test] which gives an essay once a year, the SAT will be given 5 times a year and scoring of each administration must be comparable to scoring of other administrations. Other tests, like TOEFL, which give an essay several times a year use a scoring guide like this. (Memorandum from Marylyn Sudlow to Ken Hartman, August 6, 1992)

Clearly the approach to scoring of constructed responses had implications for score meaning and score comparability. However, the psychometric support for constructed responses was limited, at least compared with the support available for multiple-choice tests. Psychometric research at ETS since the 1950s was initially oriented to dichotomously scored items; a historical account can be found in Carlson and von Davier (Chap. 5, this volume). Fred Lord's work (Lord 1952) was critical for developing a broadly applicable psychometric framework, item response theory (IRT), that would eventually include ordered polytomously scored items (Samejima 1969),[14] a needed development to accommodate constructed responses. Indeed, IRT provided the psychometric backbone for developing the second generation of NAEP (Messick et al. 1983), including the incorporation of polytomously scored constructed-response items at a time when to do so in large-scale testing was rare. (For a detailed discussion of the ETS contributions to psychometric theory and software in support of constructed-response formats, see Carlson and von Davier, Chap. 5, this volume.)

The sense of error of measurement within IRT, as represented by the idea of an information function (Birnbaum 1968), was conditional and sample independent (in a certain sense), an improvement over the conception of error in classical test theory, which was global and sample specific. IRT introduced explicitly the idea that the error or measurement was not constant at all ability levels, although it did not allow for the identification of sources of error. Concurrent developments outside the IRT sphere made it possible to begin teasing out the contribution of the scoring process to score reliability (Ebel 1951; Finlayson 1951; Lindquist 1953), culminating in generalizability theory (Cronbach et al. 1972). Such analyses were useful for

[13] Interestingly, the rationale underlying Sudlow's memorandum is the same as the rationale for instituting the methodology of equating in the SAT itself in the 1940s (College Entrance Examination Board 1942, p. 34), namely, that the SAT would be administered more than once per year and the two within-year testing populations could not be assumed to be equivalent as a year-to-year population might be.

[14] Fumiko Samejima was at ETS during the 1960s, invited by Fred Lord. A full account can be found in Wainer and Robinson (2007).

characterizing what portion of the error variability was due to different sources, among them lack of reader agreement. Bock et al. (2002), however, proposed a solution to incorporate that framework into IRT whereby the conditional standard error of measurement derived from IRT could be partitioned to identify the portion due to the rating process. (Briggs and Wilson 2007, provide for a more elaborate integration of IRT and generalizability theory.)

As had been recognized by Edgeworth (1890), readers can differ in the stringency of the scores they assign and such disagreements contribute to the error of measurement. Henry Braun[15] appears to have been the first one at ETS to introduce the idea of rater calibration, described earlier by Paul (1981), as an approach to compensate for systematic disagreements among raters. The logic of the approach was described as follows (Braun 1988): "This new approach involves appropriately adjusting scores in order to remove the noise contributed by systematic sources of variation; for example, a reader consistently assigning higher grades than the typical reader. Such adjustments are akin to an equating process" (p. 2).

The operational implementation of the idea would prove challenging, however. To implement the idea economically, specialized data collection designs were necessary and needed to be embedded in the operational scoring process over several days. The effects estimated from such an analysis are then used to adjust the raw scores. Along the same lines, Longford (1994) also studied the possibility of adjusting scores by taking into account rater severity and consistency.

An alternative to adjusting scores retrospectively is to identify those readers who appear to be unusually severe or lenient so that they can receive additional training. Bejar (1985) experimented with approaches to identify "biased" readers by means of multivariate methods in the Test of Spoken English. Myford et al. (1995) approached the problem of rater severity by applying FACETS (Linacre 2010), an extension of the IRT Rasch model that includes rater parameters, as well as the parameters for test takers and items.

18.1.2 Conclusion

When ETS was formed, the pragmatics of increasingly large scale testing together with psychometric considerations set a barrier to the use of constructed-response formats, which was viewed as unreliability due to inadequate interrater agreement. Carl Brigham, chief developer of the SAT, was also a strong proponent of more direct measures, but a solution to the scoring problem eluded him. After Brigham's death, there appeared to be no strong proponent of the format, at least not within the College Board, nor in the initial years of ETS. Without Brigham to push the point, and the strong undercurrent against constructed responses illustrated by Huddleston's (1954) perspective that writing skills do not merit their own construct, the prospects

[15] Henry Braun was vice president for research management from 1990 to 1999.

for constructed-response testing seemed dire. However, the ETS staff also included writing scholars such as Paul Diederich and Fred Godshalk, and because of them, and others, ultimately there was significant progress in solving the interrater agreement challenge with the emergence of holistic scoring. That method, which was also widely accepted outside of ETS paved the way for an increase in the use of essays. However, as we will see in the next sections, much more was needed for constructed-response formats to become viable.

18.2 Validity

Making progress on the scoring of constructed responses was critical but far from sufficient to motivate a wider reliance on constructed-response formats. Such formats necessarily require longer response times, which means fewer items can be administered in a given time, threatening *score* reliability. The conception of validity prevailing in the mid-twentieth century emphasized predictive validity, which presented a challenge for the adoption of constructed-response formats since their characteristic lower score reliability would attenuate predictive validity. The evolution of validity theory would be highly relevant to decisions regarding the use of response format, as we will see shortly. Research at ETS played a key role and was led by Samuel Messick, who not only would argue, along with others, for a unitary—as opposed to a so-called Trinitarian—conception of validity (Guion 1980) (consisting of content, criterion and construct "validities") but also, as important, for the relevance of such a unitary conception of validity to *educational* measurement. First, it is informative to review briefly the historical background.

The notion that eventually came to be known as *content validity*, and was seen as especially relevant to educational testing, probably has its roots in the idea of the sampling of items as a warrant for score interpretation. That notion was proposed early on by Robert C. Tryon as a reaction to the factor analytic conception of individual differences that prevailed at the time. Tryon (1935) argued,

> The significant fact to observe about mental measurement is that, having marked out by definition some domain for testing, the psychologist chooses as *a method of measurement* one which indicates that he knows *before giving the test to any subjects* a great deal about the nature of the factors which cause individual differences in the domain. The method is that of *sampling* behavior, and it definitely presupposes that for any defined domain there exists a *universe* of causes, or factors, or components determining individual differences. Each test-item attempts to 'tap' one or more of these components. (p. 433, emphasis in the original)

Tryon was on track with respect to assessment design by suggesting that the assessment developer should know much about what is to be tested "*before giving the test to any subject*," therefore implying the need to explicate what is to be measured in some detail as a first step in the design of an assessment (a principle fully fleshed out in ECD, Mislevy et al. 2003, much later). However, his rejection of the prevailing factor analytic perspective advocated by the prominent psychologists of the day (Spearman 1923; Thurstone 1926) was probably responsible for the lack of acceptance of his perspective.[16] Among the problems raised about the sampling perspective as a warrant to score interpretation was that, in principle, it seemed to require the preexistence of a universe of items, so that random samples could be taken from it. Such an idea presupposes some means of defining the universe of items. The resistance to the idea was most vocally expressed by Jane Loevinger (1965), who could not envision how to explicate such universes. Nevertheless, the relevance of sampling in validation was affirmed by Cronbach (1980) and Kane (1982), although not as a sufficient consideration, even though the link back to Tryon was lost along the way.

What appears to have been missed in Tryon's argument is that he intended the universe of items to be isomorphic with a "*universe* of factors, causes, or components determining individual differences" (p. 433), which would imply a crossing of content and process in the creation of a universe of items. Such an idea foreshadows notions of validity that would be proposed many decades later, specifically notions related to construct representation (Embretson 1983). Instead, in time, the sampling perspective became synonymous with content validity (Cronbach 1971): "Whether the operations that finally constitute the test correspond to the *specified universe* is the question of content validity" (p. 452, emphasis added). The idea of a universe was taken seriously by Cronbach (although using for illustration an example from social psychology, which, interestingly, implies a constructed-response test design):

> For observation of sociability, the universe specification presumably will define a category of "social acts" to be tallied and a list of situations in which observations are to be made. Each observation ought to have validity as a sample from this universe. (p. 452)

While sampling considerations evolved into content validity, and were thought to be especially applicable to educational (achievement) testing (Kane 2006), the predictive or criterion notion of "validity" dominated from 1920 to 1950 (Kane 2006) and served to warrant the use of tests for selection purposes, which in an educational context meant admissions testing. The research at ETS described earlier on writing assessment took place in that context. The predictive view presented a major hurdle to the use of constructed-response formats because, in a predictive context, it is natural to evaluate any modifications to the test, such as adding constructed-response formats, with respect to *increases* in prediction (Breland 1983):

> Because of the expense of direct assessments of writing skill, a central issue over the years has been whether or not an essay adds significantly to the measurement accuracy provided by other available measures-the high school record, objective test scores, or other information. (p. 14)

Breland provided a meta-analysis of writing assessment research showing the incremental prediction of writing samples over measures consisting only of

[16]The caution about factor analysis was expressed many decades later by Sam Messick (1972): "these concerns [about factor analysis] could lead to a marked skepticism about the construct validity of empirically derived factors as fundamental dimensions of behavioral processes" (p. 358).

multiple-choice items. Although he presented a fairly compelling body of evidence, a cost-conscious critic could have argued that the increases in prediction could just as easily have been obtained more economically by lengthening the multiple-choice component.

The third conception of validity is construct validity, dating back to the mid-twentieth-century seminal paper introducing the term (Cronbach and Meehl 1955). In that paper, validation is seen as a process that occurs after the assessment has been completed, although the process is driven by theoretical expectations. However, Cronbach and Meehl did not suggest that those expectations should be used in developing the test itself. Instead, such theoretical expectations were to be used to locate the new test within a nomological network of relationships among theoretically relevant variables and scores. At the time Cronbach and Meehl were writing, developing a test was a matter of writing items as best one could and then pretesting them. The items that did not survive were discarded. In effect, the surviving items were the de facto definition of the construct, although whether it was the intended construct could not be assumed until a conclusion could be reached through validation. In the wrong hands, such an ad hoc process could converge on the wrong test.[17] Loevinger (1957) argued that "the dangers of pure empiricism in determining the content of a test should not be underestimated" (p. 657) and concluded that

> there appears to be no convincing reason for ignoring content nor for considering content alone in determining the validity of a test or individual items. The problem is to find a coherent set of operations permitting utilization of content together with empirical considerations. (p. 658)

Clearly Loevinger considered content important, but the "coherent set of operations" she referred to was missing at the time, although it would appear soon as part of the cognitive science revolution that was beginning to emerge in the 1950s.[18]

Toward the end of that decade, another important article was published that would have important repercussions for the history of research on constructed-response formats. D. T. Campbell and Fiske (1959) made an important distinction: "For the justification of novel trait measures, for the validation of test interpretation, or for the establishment of construct validity, *discriminant* validation as well as convergent validation is required" (p. 81, emphasis in the original).

The paper is significant for contrasting the evidentiary basis for *and* against a psychometric claim.[19] In addition, the paper formalizes the notion of *method*

---

[17] Of course, in the right hands, the approach could also converge on a very effective instrument. Two of the most highly regarded assessments developed during the 20th century, the Minnesota Multiphasic Personality Inventory (MMPI) and the Strong Vocational Interest Blank, later to become the Strong–Campbell, were developed in this fashion.

[18] Miller (2003), a major leader of the revolution, provides a historical account of cognitive science.

[19] Toulmin's (1958) model of argument was published at around the same time. Toulmin stressed the role of counterarguments and rebuttals of claims or conclusions. Toulmin's model figured prominently in the subsequent evolution of validation (Kane 2006) and in assessment design (Mislevy et al. 2003). Karl Popper's (1959/1992) book, *The Logic of Scientific Discovery,* also appeared in 1959. Popper stressed the importance of falsifying theories, a concept that can be

variance, which would surface later in research about constructed-response formats, especially evaluating the measurement equivalence of multiple-choice and constructed-response formats.

As can be seen, the 1950s was a contentious and productive decade in the conceptual development of testing. Importantly, the foregoing discussion about the nature of validity did *not* take place at ETS. Nevertheless, it is highly relevant to the chapter: These developments in validity theory may have even been seen as tangential to admissions tests,[20] which represented the vast majority of ETS operations at the time. In that context, the normative interpretations of scores together with predictive validity were the accepted practice.

As mentioned earlier, in 1963 a most influential paper was published by Glaser, proposing an alternative approach to score interpretation and assessment design in an educational setting, namely, by *reference* to the level of proficiency within a very well-defined content domain. Glaser's intent was to provide an alternative to normative interpretations since norms were less relevant in the context of individualized instruction.[21] Whereas norms provide the location of a given score in a distribution of scores, a criterion-referenced interpretation was intended to be more descriptive of the test taker's skills than a normative interpretation. The criterion-referenced approach became aligned early on with the idea of mastery testing (Hambleton and Novick 1973), whereby the objective of measurement was to determine whether a student had met the knowledge requirements associated with a learning objective.

Criterion-referenced tests were thought to yield more actionable results in an educational context not by considering a score as a deviation from the mean of a distribution, the normative interpretation, but by locating the score within an interval whereby all scores in that interval would have a similar interpretation. In the simplest form, this meant determining a cut score that would define the range of pass scores and the range for fail scores, with pass implying mastery. To define those intervals, cut scores along the score scale needed to be decided on first. However, as noted by Zieky (1995), the methodology for setting such cut scores had not yet emerged. In retrospect, it is clear that if the deviation from a mean was not adequate for score interpretation, locating a score within an interval would not necessarily help either; much more was needed. In fact, reflecting on his 1963 paper, Glaser (1994) noted that "systematic techniques needed to be developed to more adequately identify and describe the components of performance, and to determine the relative weighting of these components with respect to a given task" (p. 9).

---

applied to challenge assertions about the validity of scores. Messick (1989) discussed Toulmin and Popper at length.

[20]An examination of the titles of research published in ETS's first decades clearly emphasizes predictive validity. However, consequential implications of testing or test bias appeared in the mid-1960s with the work of T. Anne Cleary (1966) and even earlier (Turnbull 1949). Also, Gulliksen's (1950) idea of intrinsic validity, cited by Cronbach and Meehl (1955), is a rare exception on early validity theorizing at ETS, to be followed some years later by the seminal theoretical work of Messick. See Kane and Bridgeman Chap. 16, (this volume) for a comprehensive description of Messick's work and a historical review of validity theory at ETS more generally.

[21]Glaser credits Ebel (1962), who was vice president at ETS at the time, with a similar idea.

The "components of performance" that Glaser thought needed to be developed echoed both Tryon's earlier "components determining individual differences" and the "coherent set of operations permitting utilization of content" that Loevinger called for. That is, there had been an implied consensus all along as to a key ingredient for test meaning, namely, identifying the underlying sources of variability in test performance, which meant a deeper understanding of the response process itself.[22]

18.2.1 Validity Theory at ETS

With the benefit of hindsight, it seems that by 1970, the conception of validity remained divided, consisting of different "validities," which had significant implications for the use of constructed-response formats in education. Three developments were needed to further that use:

- With criterion (and especially, predictive) validity as a primary conception of validity, economics would delay wider use of constructed-response formats. Replacing the Trinitarian view with a unitary view was needed to avoid associating the different "validities" with specific testing contexts.
- Even under a unitary view, the costs of constructed-response formats would remain an obstacle. An expansion of the unitary conception was necessary to *explicitly* give the evidential *and* consequential aspects of validity *equal* footing. By doing so, the calculus for the deployment of constructed-response formats would balance monetary cost with the (possibly intangible) benefits of using the format.
- As alluded to earlier, by and large, the broader discussion of validity theory was not directed at educational achievement testing. Thus, the third needed development was to make the evolution of validity theory applicable to educational testing.

These developments were related and *enormous*. Unlike the earlier evolution of validity, which had taken place outside of ETS, Sam Messick dedicated two decades to explicating the unitary view, bringing its evidential and consequential aspects more into line with one another, and making the view relevant, if not central, to educational testing. These advances, arguably, were essential to wider use of constructed-response formats in education.

Calls for a unitary view in the form of construct validity began early on. Messick (1989) quoted Loevinger that, "since predictive, concurrent, and content validities are all essentially ad hoc, construct validity is the whole of validity from a scientific point of view" (p. 17). Messick elaborated that idea, stating that, "almost any kind

[22] Of course, generalizability theory had been under development (Rajaratnam et al. 1965) during the 1960s, and it was concerned with components of *observed* score variability. It distinguishes between components of variability that attenuate the interpretation of a score, that is, error variability, and true score variability, summarizing the results into a generalizability coefficient. It does not address the understanding of the response process.

of information about a test can contribute to an understanding of construct validity, but the contribution becomes stronger if the degree of fit of the information with the theoretical rationale underlying score interpretation is explicitly evaluated" (p. 17). That is, Messick stressed the need for a theoretical rationale to integrate the different sources of validity evidence.

Importantly, Messick's (1980, 1989) unitary view explicitly extended to the consequences of test use, with implications for the use of constructed-response formats. Although the message was not well received in some quarters (Kane 2006, p. 54), it was in others. For example, Linn et al. (1991) argued,

> If performance-based assessments are going to have a chance of realizing the potential that the major proponents in the movement hope for, it will be essential that the consequential basis of validity be given much greater prominence among the criteria that are used for judging assessments. (p. 17)

By the 1990s, there had been wider acceptance that consequential evidence was relevant to validity. But that acceptance was one of the later battles that needed to be fought. The relevance of the unitary view to educational testing needed to be established first. In 1975, Messick wondered, "Why does educational measurement, by and large, highlight comparative interpretations, whether with respect to norms or to standards,[23] and at the same time play down construct interpretations?" (p. 957).

This question was raised in reaction to the predominance that criterion-referenced testing had acquired by the 1970s, Among the possible answers Messick (1975) proposed for the absence of construct interpretations was the "legacy of behaviorism and operationism that views desired behaviors as ends in themselves with little concerns for the *processes* that produce them" (p. 959, emphasis added). That speculation was later corroborated by Lorie Shepard (1991) who found that, for the most part, state testing directors had a behaviorist conception of student learning.

The positive attitude toward behaviorism among state testing directors is informative because the so-called cognitive revolution had been under way for several decades. Although its relevance was recognized early on, its impact on testing practice was meager. Susan Embretson, who was not associated with ETS, recognized those implications (Whitely and Dawis 1974).[24] In an important paper, Embretson integrated ideas from cognitive science into testing and psychometric theory by building on Loevinger's argument and layering a cognitive perspective on it. Embretson (1983) proposed the term *construct representation* to describe the extent to which performance on a test is a function of mental processes hypothesized to underlie test performance. An approach to documenting construct representation is modeling the difficulty of items as a function of variables representing the response process and knowledge hypothesized to underlie performance.[25] Modeling of item

---

[23] In the quotation, by "standards," he meant criterion referencing.

[24] Susan Embretson published as Susan Whitely earlier in her career.

[25] For example, the classic item type based on verbal analogies was thoroughly reanalyzed from a cognitive perspective (Bejar et al. 1991; Pellegrino and Glaser 1980; Sternberg 1977; Whitely and Dawis 1974) with the goal of understanding the variability in the difficulty of the items as a function of the process and knowledge assumed to be involved in analogical reasoning.

difficulty is well suited to multiple-choice items, but less so for items requiring a constructed response since there would typically be fewer of them in any given test. Nevertheless, the concept is equally applicable, as Messick (1994) noted with specific reference to performance assessment: "Evidence should be sought that the presumed sources of task complexity are indeed reflected in task performance and that the complex skill is captured in the test scores with minimal construct underrepresentation" (p. 20).

Embretson's construct representation fit well with Messick's calls for a fuller understanding of the response process as a source of validity evidence. But Messick (1990) also understood that format had the potential to introduce irrelevancies:

> Inferences must be tempered by recognizing that the test not only samples the task universe but casts the sampled tasks in a test format, thereby raising the specter of context effects or irrelevant method [i.e., format] variance possibly distorting test performance vis-a-vis domain performance. (p. 9)

Independently of the evolution of validity theory that was taking place, the calls for direct and authentic forms of assessment never stopped, as evidenced by the work on portfolio assessments at ETS (Camp 1993; Gentile 1992; Myford and Mislevy 1995) and elsewhere. Following the period of "minimum competency testing" in the 1980s there were calls for testing higher order forms of educational achievement (Koretz and Hamilton 2006), including the use of so-called authentic assessments (Wiggins 1989). The deployment of highly complex forms of assessment in the early 1990s was intended to maximize the positive educational consequences of constructed-response formats and avoid the negative consequences of the multiple-choice format, such as teaching to the narrow segment of the curriculum that a multiple-choice test would represent. However, despite the appeal of constructed-response formats, such forms of assessment still needed to be evaluated from a validity perspective encompassing both evidential and consequential considerations. As Messick (1994) noted,

> some aspects of all testing, even performance testing, may have adverse as well as beneficial educational consequences. And if both positive and negative aspects, whether intended or unintended, are not meaningfully addressed in the validation process, then the concept of validity loses its force as a social value. (p. 22)

Indeed, following the large-scale deployment of performance assessments in K–12 in the 1990s (Koretz and Hamilton 2006), it became obvious that overcoming the design challenges would take time. Although the assessments appeared to have positive effects on classroom practice, the assessments did not meet technical standards, especially with respect to score reliability. As a result, the pendulum swung back to the multiple-choice format (Koretz and Hamilton 2006, p. 535).

Not surprisingly, after the long absence of constructed-response formats from educational testing, the know-how for using such formats was not fully developed. Reintroducing such formats would require additional knowledge and a technological infrastructure that would make the format affordable.

18.2.2 Conclusion

Arguably, the predictive conception of validity prevalent through most of the twentieth century favored the multiple-choice format. The evolution of validity into a more unitary concept was not seen initially as relevant to educational measurement. Samuel Messick thought otherwise and devoted two decades to explicate the relevance of a unitary conception, incorporating along the way consequential, not just evidentiary, considerations, which was critical to reasoning about the role of response format in educational measurement.

18.3 The Interplay of Constructs and Technology

The evolution of validity theory may have been essential to providing a compelling rationale for the use of constructed-response formats. However, the cost considerations in an educational setting are still an issue, especially in an educational context: According to Koretz and Hamilton (2006), "concerns about technical quality and costs are likely to dissuade most states from relying heavily on performance assessments in their accountability systems … particularly when states are facing heavy testing demands and severe budget constraints" (p. 536).

An important contribution by ETS to the development of constructed-response formats has been to take advantage of technological developments for educational and professional testing purposes. Among the most salient advances are the following:

- using computers to deploy constructed-response formats that expand construct coverage
- taking advantage of technology to enable more efficient human scoring
- pioneering research on automated scoring in a wide range of domains to improve cost effectiveness and further leverage the computer as a delivery medium

If the scanner enabled the large-scale use of multiple-choice tests, the advent of the computer played a similar role in enabling the large-scale use of constructed-response formats.[26] Incorporating technological advances into operational testing had been common practice at ETS almost from inception (Traxler 1951, 1954). However, a far more visionary perspective was apparent at the highest levels of the organization. In 1951, ETS officer William Turnbull coined the term, tailored testing (Lord 1980, p. 151); that is, the idea of adapting the test to the test taker.[27] Some years later, as the organization's executive vice president, he elaborated on it (Turnbull 1968):

[26] For perhaps the most complete history of the scanner and how it impacted testing, see Russell (2006, pp. 36–47).

[27] "Tailoring" a test was not a totally new idea, in the sense that Binet was practicing it at the turn of the twentieth century. Also, Cowden (1946) at Princeton University used sequential sampling, a method associated with quality control, as a test design (see Weiss and Betz 1973; Wood 1973). In

> The next step should be to provide examinations in which the individual questions are contingent on the student's responses to previous questions. If you will permit the computer to raise its ugly tapes, I would like to put forward the prospect of an examination in which, for each examinee, the sequence of questions is determined by his response to items earlier in the sequence. The questions will be selected to provide the individual student with the best opportunity to display his own profile of talent and accomplishment, without wasting time on tasks either well below or well beyond his level of developed ability along any one line. Looking farther down this same path, one can foresee a time when such *tailor-made tests* will be part and parcel of the school's instructional sequence; when the results will be accumulated and displayed regularly as a basis for instruction and guidance; and when the pertinent elements of the record will be banked as a basis for such major choice points as the student's selection of a college. (p. 1428, emphasis added)

Although Turnbull was not addressing the issue of format, his interest in computer-based testing is relevant to the eventual wider use of constructed-response formats, which perhaps would not have been feasible in the absence of computer-based testing. (The potential of microcomputers for testing purposes was recognized early at ETS; Ward 1984.) That an officer and future president of ETS would envision in such detail the use of computers in testing could have set the stage for an *earlier* use of computers for test delivery than might otherwise have been the case. And, if as Fowles (2012) argued, computer delivery was in part responsible for the adoption of writing in postsecondary admissions tests like the GRE General Test, then it is possible that the early adoption of computer delivery by ETS accelerated that process.[28] The transition to computer delivery started with what was later named the *ACCUPLACER®* test, a placement test consisting entirely of multiple-choice items developed for the College Board. It was first deployed in 1985 (Ward 1988). It is an important first success because it opened the door for other tests to follow.[29]

Once computer delivery was successfully implemented, it would be natural for other ETS programs to look into the possibility. Following the deployment of ACCUPLACER, the GRE General Test was introduced in 1992 (Mills and Steffen 2000). The 1992 examination was an adaptive test consisting of multiple-choice sections for Verbal Reasoning, Quantitative Reasoning, and Analytical Reasoning. However, the Analytical Reasoning measure was replaced in 2002 by the Analytical

---

fact, an experiment in the adaptive administration of the Stanford–Binet was reported as early as 1947 (Hutt 1947) and Hick (1951) shortly thereafter presented the essence of all the components of adaptive testing as we understand the term today.

[28] The contributions of Martha Stocking (1942–2006) in this process should be acknowledged. She was hired by Fred Lord in the 1960s and was soon making contributions to adaptive testing on her own (Stocking 1969). She made many contributions to adaptive testing over her career, especially in the area of controlling the exposure of individual test items (Stocking and Swanson 1993).

[29] Although it is entirely possible that while Turnbull may have been a visionary and could have encouraged Fred Lord to think about the idea of adaptive testing, Turnbull, apparently, was not involved in the decisions leading to the implementation of adaptive testing. ACCUPLACER (Ward 1988), the first ETS-produced adaptive test, was deployed in 1985 some years after Turnbull had resigned as president of ETS in 1981, according to Bill Ward (personal communication, July 6, 2010), who was the main developer of ACCUPLACER at ETS.

Writing section, consisting of two prompts: an issue prompt (45 minutes with a choice between two prompts) and an argument prompt (30 minutes).

The transition to computer delivery in 1992 and the addition of writing in 2002 appear to have flowed seamlessly, but in fact, the process was far more circuitous. The issue and argument prompts that composed the Analytical Writing measure were a significant innovation in assessment design and an interesting example of serendipity, the interplay of formats, technology, and attending to the consequences of testing.

Specifically, the design of the eventual GRE Analytical Writing measure evolved from the GRE Analytical Reasoning (multiple-choice) measure, which was itself a major innovation in the assessment of reasoning (Powers and Dwyer 2003). The Analytical Reasoning measure evolved by including and excluding different item types. In its last incarnation, it consisted of two multiple-choice item types, analytical reasoning and logical reasoning. The logical reasoning item type called for evaluating plausible conclusions, determining missing premises, finding the weakness of a conclusion, and so on (Powers and Dwyer 2003, p. 19). The analytical reasoning item type presented a set of facts and rules or restrictions. The test taker was asked to ascertain the relationships permissible among those facts, and to judge what was necessary or possible under the given constraints (Chalifour and Powers 1989).

Although an extensive program of research supported the development of the Analytical Reasoning measure, it also presented several challenges *especially under computer delivery*. In particular, performance on the logical reasoning items correlated highly with the verbal reasoning items, whereas performance on the analytical reasoning items correlated highly with quantitative reasoning items (Powers and Enright 1987), raising doubts about the construct it assessed. Moreover, no conclusive validity evidence for the measure as a whole was found when using an external criterion (Enright and Powers 1991). The ambiguous construct underpinnings of the Analytical Reasoning measure were compounded by the presence of speededness (Bridgeman and Cline 2004), which was especially harmful under computer delivery. Given the various challenges encountered by the Analytical Reasoning measure, it is no surprise that it ultimately was replaced by the Analytical Writing measure, which offered a well-balanced design.

The issue prompt has roots in the pedagogy of composition. As D'Angelo (1984) noted, textbooks dating back to the nineteenth century distinguish four genre: narration, description, exposition, and argumentation. Argumentation was defined as "the attempt to *persuade* others of the truth of a proposition" (p. 35, emphasis added). There is less precedent, if any, for the GRE argument prompt, which presents the task of *critiquing* an argument. The germ for the idea of an argument-critique prompt was planted during efforts to better prepare minority students for the GRE Analytical Reasoning measure, specifically, the logical reasoning item type (Peter Cooper, personal communication, November 27, 2013):

> The Logical Reasoning items … took the form of a brief stimulus passage and then one or more questions with stems such as "Which of the following, if true, weakens the argument?," "The argument above rests on which of the following assumptions," and so forth,

> with five options. At a workshop in Puerto Rico, a student commented that he would prefer questions that allowed him to comment on the argument in his own terms, not just pick an answer someone else formulated to a question someone else posed. I thought to myself, "Interesting concept … but be careful what you wish for" and did nothing for a couple of years, until [Graduate Management Admission Test] GMAT … told us in the summer of 1993 that it wanted to add a constructed-response measure, to be operational by October 1994, that would get at analytical reasoning—i.e., not just be another writing measure that rewarded fluency and command of language, although these would matter as well. Mary Fowles had discussed "Issue"-like prototypes with the [Graduate Management Admission Council] GMAC's writing advisory committee, which liked the item type but seemed to want something more "analytical" if possible. I recalled the student's comment and thought that a kind of constructed-response Logical Reasoning item could pair well with the Issue-type question to give a complementary approach to analytical writing assessment: In one exercise, students would make their own argument, developing a position on an issue, and in the other exercise they would critically evaluate the line of reasoning and use of evidence in an argument made by someone else. Both kinds of skills are important in graduate-level work.

Mary Fowles (2012) picked up the story from there: "What caused this seemingly rapid introduction of direct writing assessment for admission to graduate and professional programs?" (pp. 137–138). She cited factors such as the "growing awareness [of the relationship] between thinking and writing"; the availability of the computer as a delivery medium, which "enabled most examinees to write more fluently" and "streamlined the process of collecting written responses"; and "essay testing programs [that] now had the advantage of using automated scoring" (pp. 137–138).

Although the genesis of the argument prompt type came from attempts to help prepare students of diverse backgrounds for the multiple-choice GRE Analytical Reasoning section, the analytical writing measure comprising issue and argument prompts was used first by the GMAT. In 1994, that measure was offered in paper-and-pencil form, and then moved to computer when GMAT converted to an adaptive test in 1997. The GRE first used the measure as a stand-alone test (the GRE Writing Assessment) in 1999 and incorporated it into the General Test in 2002, as noted earlier.

The transition to computer delivery in the 1990s was not limited to the GRE and GMAT. The *TOEFL*® test transitioned as well. It evolved from a test conceived in the 1960s to a measure rooted in the communicative competence construct (Canale and Swain 1980; Duran et al. 1987). The earlier efforts to bolster TOEFL by introducing stand-alone writing and speaking tests—the Test of Written English (*TWE*® test) and the Test of Spoken English (*TSE*® test)—were seen as stopgap measures that led to an "awkward" situation for the "communication of score meaning" (C. A. Taylor and Angelis 2008, p. 37). Importantly, communicative competence called for evidence of proficiency in productive skills, which meant the assessment of writing *and* speaking proficiency in academic settings. In the case of speaking, these requirements meant that ultimately complex multimodal tasks were needed where students would read or listen to a stimulus and provide a spoken response. The construct of communicative competence was unpacked in frameworks corresponding to the four skills thought to compose it: reading (Enright et al. 2000), listening (Bejar et al. 2000), writing (Cumming et al. 2000), and speaking (Butler et al. 2000). The frameworks served as the basis for experimentation, after which the blueprint for the test was set (Pearlman 2008a).

Computer delivery would prove critical to implementing such an ambitious test, especially the measurement of the productive skills. The inclusion of writing was relatively straightforward because there was already experience from GRE and GMAT. In fact, when it first transitioned to computer in 1998, TOEFL CBT used the TWE prompt as either a typed or handwritten essay. Nevertheless, there were still significant challenges, especially technological and assessment design challenges. The assessment of computer-delivered *speaking* on an international scale was unprecedented, especially considering the test security considerations.[30] The first generation of computer delivery that had served GRE and TOEFL CBT was less than ideal for effectively and securely delivering an international test administered every week. For one, testing that required speaking had the potential to interfere with other test takers. In addition, the quality of the speech captured needed to be high in all test centers to avoid potential construct-irrelevant variance. These requirements meant changes at the test centers, as well as research on the best microphones to capture spoken responses. On the back end, written and spoken responses needed to be scored quickly to comply with a turnaround of no more than 10 days. These requirements influenced the design of the next-generation test delivery system at ETS, iBT (Internet-based testing), and when the latest version of TOEFL was released in 2005, it was called the *TOEFL iBT*® test (Pearlman 2008a).

In addition to the technological challenges of delivering and scoring a secure speaking test, there were several assessment design challenges. To accommodate the international volume of test takers, it was necessary to administer the test 50 times a year. Clearly, the forms from week to week needed to be sufficiently different to prevent subsequent test takers from being able to predict the content of the test. The central concept was that of reusability, a key consideration in ECD, which was implemented by means of item templates (Pearlman 2008a).

18.3.1 Computer-Mediated Scoring

Once tests at ETS began to transition to computer delivery, computer-mediated scoring became of interest. Typically, faculty, in the case of educational tests, or practitioners, in the case of professional assessments, would congregate at a central location to conduct the scoring. As volume grew, best practices were developed, especially in writing (Baldwin 2004), and more generally (Baldwin et al. 2005; McClellan 2010). However, the increase in testing volumes called for better utilization of technology in the human scoring process.

[30] The IELTS assessment includes a speaking test but, unlike TOEFL, is administered locally by live examiners. Pearson currently offers a competitor to the TOEFL tests that includes writing and speaking measures that are scored solely by computer.

Perhaps anticipating the imminence of larger volumes, and the increasing availability of computers, there was experimentation with "remote scoring" fairly early[31] (Breland and Jones 1988). In the Breland and Jones study, the essays were distributed via courier to the raters at home. The goal was to evaluate whether solo scoring was feasible compared to centralized or conference scoring. Not surprisingly this form of remote scoring was not found to be as effective as conference scoring. All the affordances of computer technology were not taken advantage of until a few years later (Bejar and Whalen 2001; Driscoll et al. 1999; Kuntz et al. 2006).

The specialized needs of NAEP motivated a somewhat different use of the computer to mediate the human scoring process. In the early 1990s the NAEP program started to include state samples, which led to large increases in the volume of constructed responses. Such responses were contained in a single booklet for each student. To avoid potential scoring bias that would result from a single reader scoring all the constructed responses from a given student, a system was developed where the responses would be physically clipped and scanned separately. The raters would then score the scanned responses displayed on a terminal, with each response for a student routed to a different rater. The scoring of NAEP constructed responses was carried out by a subcontractor (initially NCS, and then Pearson after it acquired NCS) under direction from ETS. Scoring was centralized (all the raters were at the same location), but computer images of the work product were presented on the screen and the rater entered a score that went directly into a database.[32]

18.3.2 Automated Scoring

Though technology has had an impact on human scoring, a more ambitious idea was to automate the scoring of constructed responses. Page, a professor at the University of Connecticut, first proposed the idea for automated scoring of essays (Page 1966). It was an idea ahead of its time, because for automated scoring to be maximally useful, the responses need to be in digital form to begin with; digital test delivery was some decades away. However, as the computer began to be used for test delivery, even if it was limited to multiple-choice items, it was natural to study how the medium might be leveraged for constructed-response scoring purposes. Henry Braun, then vice president for research management, posed precisely that question (personal communication, July 9, 2014). Although a statistician by training, he was familiar with the literature on expert systems that had proliferated by the

[31] Even earlier, in the 1950s, Paul Diederich experimented enthusiastically with "lay readers," or "college-educated housewives" (Burke 1961, p. 258).

[32] Interestingly, during the same period, there was much activity at ETS related to scanning technology that had been developed for processing financial aid applications, led by Keith Reid-Green (1990). In fact, the seasonal nature of financial aid applications meant that the scanners ETS had could be used in support of other work, such as NAEP. However, in the end, the NAEP directors opted to use an external vendor.

1980s as a means of aiding and even automating expert judgment. In contrast to earlier research on actuarial judgment (Bejar et al. 2006), where the clinician and a regression equation were compared, in expert systems the role of the computer is more ambitious and consists of both analyzing an object (e.g., a doctor's course of treatment for a patient, an architectural design) and, based on that analysis, making a decision about the object, such as assigning a score level.

Randy Bennett took the lead at ETS in exploring the technology for scoring constructed responses in concert with theory about the relevant constructs, including mathematics (Bennett and Sebrechts 1996; Bennett et al. 1999, 2000a; Sandene et al. 2005; Sebrechts et al. 1991, 1996), computer science (Bennett and Wadkins 1995), graphical items (Bennett et al. 2000a; b), and formulating hypotheses (Bennett and Rock 1995). The scoring of mathematics items has reached a significant level of maturity (Fife 2013), as has the integration of task design and automated scoring (Graf and Fife 2012).

Much of the research on automated scoring was experimental, in the sense that actual applications needed to await the delivery of tests by computer. One ETS client, the National Council of Architectural Registration Boards (NCARB), was seriously considering on its own the implications of technology for the profession. The software used in engineering and architecture, computer-assisted design (CAD), was transitioning during the 1980s from minicomputers to desktop computers. A major implication of that transition was that the cost of the software came down significantly and became affordable to an increasingly larger number of architecture firms, thereby changing, to some extent, the entry requirements for the profession. Additionally, the Architectural Registration Examination introduced in 1983 was somewhat unwieldy, consisting of many parts that required several years to complete, since they could not all be taken together over the single testing window that was made available every June. A partnership between ETS and NCARB was established to transition the test to computer delivery and allow continuous testing, revise the content of the test, and take advantage of computer delivery, including automated scoring.

Management of the relationship between ETS and NCARB was housed in ETS's Center for Occupational and Professional Assessment (COPA), led by vice president Alice Irby, who was aware of the research on the utilization of computers for test delivery and scoring under Henry Braun. A project was initiated between ETS and NCARB that entailed developing new approaches to adaptive testing with multiple-choice items in a licensing context (Lewis and Sheehan 1990; Sheehan and Lewis 1992) and that had the more ambitious goal of delivering and scoring on computer the parts of the examination that required the demonstration of design skills.

The paper-and-pencil test used to elicit evidence of design skills included a very long design problem that took some candidates up to 14 hours to complete. Scoring such a work product was a challenge even for the practicing architects, called jurors. The undesirability of a test consisting of a single item from a psychometric perspective was not necessarily understood by the architects. However, they had realized that a single-item test could make it difficult for the candidate to recover from an early wrong decision. That insight led to an assessment consisting of smaller constructed-response design tasks that required demonstrations of competence in several aspects of architectural practice (Bejar 2002; Bejar and Braun 1999). The process of narrowing the test design to smaller tasks was informed by practice analyses intended to identify the knowledge, skills, and abilities (so-called KSAs) required of architects, and their importance. This information was used to construct the final test blueprint, although many other considerations entered the decision, including considerations related to interface design and scorability (Bennett and Bejar 1998).

Reconceptualizing the examination to better comply with psychometric and technological considerations was a first step. The challenge of delivering and scoring the architectural designs remained. The interface and delivery, as well as supervising the engineering of the scoring engines, was led by Peter Brittingham, while the test development effort was led by Dick Devore. The scoring approach was conceived by Henry Braun (Braun et al. 2006) and Bejar (1991). Irv Katz contributed a cognitive perspective to the project (Katz et al. 1998). The work led to operational implementation in 1997, possibly the first high-stakes operational application of automated scoring.[33]

While ETS staff supported research on automated scoring in several domains, perhaps the ultimate target was essays, especially in light of their increasing use in high-volume testing programs. Research on automated scoring of textual responses began at ETS as part of an explicit effort to leverage the potential of technology for assessment. However, the first thorough evaluation of the feasibility of automated essay scoring was somewhat fortuitous and was carried out as a collaboration with an external partner. In the early 1990s, Nancy Petersen heard Ellis B. Page discuss his system, PEG, for scoring essays[34] at an AERA reception. Petersen suggested to Page the possibility of evaluating the system in a rigorous fashion using essays from 72 prompts taken from the *PRAXIS*® program, which had recently begun to collect essays on computer. The report (Page and Petersen 1995) was optimistic about the feasibility of automated scoring but lacked detail on the functioning of the scoring system. Based on the system's relatively positive performance, there was discussion between ETS and Page regarding a possible licensing of the system for nonoperational use, but the fact that Page would not fully reveal[35] the details of the system motivated ETS to invest further in its own development and research on automated scoring of essays. That research paid off relatively quickly since the system developed, the *e-rater*® engine, was put into operation in early 1999 to score GMAT

[33]The National Board of Medical Examiners (NBME) had also wanted to use automated scoring as part of its licensing test and had a project with that goal at about the same time the ETS and NCARB project was underway. Staff from both projects met informally over the years to exchange information. The NBME examination with automated scoring of Step 3 (Primum Computer Case Simulations) became operational in 2000 (P. Harik, personal communication, July 14, 2014), backed by a considerable body of research (Clauser 2000; Clauser et al. 2002; Clyman et al. 1995).

[34]The system is currently owned by Measurement Incorporated.

[35]According to Kaplan et al. (1995), the only feature that was revealed was essay length.

essays (Burstein et al. 1998). The system has continued to evolve (Attali and Burstein 2006; Burstein et al. 2004; Burstein et al. 2013) and has become a major ETS asset. Importantly, the inner workings of e-rater are well documented (Attali and Burstein 2006; Quinlan et al. 2009), and disclosed through patents.

The e-rater engine is an example of scoring based on linguistic analysis, which is a suitable approach for essays (Deane 2006). While the automated scoring of essays is a major accomplishment, many tests rely on shorter textual responses, and for that reason approaches to the scoring of short textual responses have also been researched. The basic problem of short-answer scoring is to account for the multiple ways in which a correct answer can be expressed. The scoring is then a matter of classifying a response, however expressed, into a score level. In the simplest case, the correct answer requires reference to a single concept, although in practice a response may require more than one concept. Full credit is given if all the concepts are present in the response, although partial credit is also possible if only some of the concepts are offered.

Whereas the score humans would assign to an essay can be predicted from linguistic features that act as correlates of writing quality, in the case of short responses, there are fewer correlates on which to base a prediction of a score. In a sense, the scoring of short responses requires an actual understanding of the *content* of the response so that it can be then be classified into a score level. The earliest report on short-answer scoring at ETS (Kaplan 1992) was an attempt to infer a "grammar" from a set of correct and incorrect responses that could be used to classify future responses. The approach was subsequently applied to scoring a computer-delivered version of a task requiring the generation of hypotheses (Kaplan and Bennett 1994). A more refined approach to short-answer scoring, relying on a more robust linguistic representation of responses, was proposed by Burstein et al. (1999), although it was not applied further.

As the complexities of scoring short answers became better understood, the complexity and sophistication of the approach to scoring grew as well. The next step in this evolution was the *c-rater*™ automated scoring engine (Leacock and Chodorow 2003).[36] The system was motivated by a need to lower the scoring load of teachers. Unlike earlier efforts, c-rater requires a *model* of the correct answer such that scoring a response is a matter of deciding whether it matches the model response. Developing such a model is not a simple task given the many equivalent ways of expressing the same idea. One of the innovations introduced by c-rater was to provide an interface to model the ideal response. In effect, a model response is defined by a set of possible paraphrases of the correct answer that are then represented in canonical or standard form. To evaluate whether a given response is in the set requires linguistic processing to deal with spelling and other issues so that the student response can be recast into the same canonical form as the model. The actual scoring is a matter of matching the student response against the model, guided by a set of linguistic rules. Because student responses can contain many spelling and

[36] The system was developed under an ETS subsidiary, ETS Technologies, which was ultimately folded back into R&D at ETS.

grammatical errors, the matching process is "fairly forgiving" (Leacock and Chodorow 2003, p. 396). The c-rater engine was evaluated in studies for NAEP (Sandene et al. 2005), and in other studies has been found useful for providing feedback to students (Attali 2010; Attali and Powers 2008, 2010). The most recent evaluation of the c-rater approach (Liu et al. 2014) took advantage of some refinements introduced by Sukkarieh and Bolge (2008). O. L. Liu et al. (2014) concluded that c-rater cannot replace human scores, although it has shown promise for use in low-stakes settings.

One limitation of c-rater is scalability. A scoring model needs to be developed for each question, a rather laborious process. A further limitation is that it is oriented to scoring responses that are verbal. However, short answers potentially contain numbers, equations, and even drawings.[37]

More recent approaches to short answer scoring have been developed including one referred to as Henry ML. Whereas c-rater makes an attempt to understand the response by identifying the presence of concepts, these newer approaches evaluate low-level aspects of the response, including "sparse features" like word and character n-grams, as well as "dense features" that compare the semantic similarity of a response to responses with agreed upon-scores (Liu et al. 2016; Sakaguchi et al. 2015).

The foregoing advances were followed by progress in the scoring of spoken responses. An automated approach had been developed during the 1990s by the Ordinate Corporation based on "low-entropy" tasks, such as reading a text aloud (Bernstein et al. 2000). The approach was, however, at odds with the communicative competence perspective that was by then driving the thinking of TOEFL developers. ETS experimented with automated scoring of high-entropy spoken responses (Zechner, Bejar, & Hemat, 2007). That is, instead of reading a text aloud, the tasks called for responses that were relatively extemporaneous and therefore more in line with a communicative perspective. The initial experimentation led rather quickly to an approach that could provide more comprehensive coverage of the speaking construct (Zechner et al. 2007b, 2009a). The current system, known as the *SpeechRater*SM service, is used to score the *TOEFL Practice Online* (*TPO*™) test, which is modeled after the speaking component of the TOEFL. Efforts continue to further expand the construct coverage of the scoring engine by integrating additional aspects of speaking proficiency, such as content accuracy and discourse coherence (Evanini et al. 2013; Wang et al. 2013; Yoon et al. 2012). Additionally, the scope of applicability has been expanded beyond English as a second language (ESL) to also include the assessment of oral reading proficiency for younger students by means of low-entropy tasks (Zechner et al. 2009b, 2012). Importantly, the same underlying engine is used in this latter case, which argues well for the potential of that engine to support multiple types of assessments.

[37] The Smarter Balanced consortium, for example, field tested in 2014 such item types (Smarter Balanced Assessment Consortium 2014).

18.3.3 Construct Theory and Task Design

Technology was as important to the adoption of constructed-response formats as it was for the multiple-choice format, where the scanner made it possible to score large volumes of answer sheets. However, much more was needed in the case of constructed-response formats besides technology. Invariably, progress was preceded or accompanied by work on construct definition.

18.3.3.1 Writing

The publication that may have been responsible for the acceptance of holistic scoring (Godshalk et al. 1966) was, in fact, an attempt to empirically define the writing construct. Over the years, many other efforts followed, with various emphases (Breland 1983; Breland and Hart 1994; Breland et al. 1984, 1987). Surveys of graduate faculty identified written argumentation, both constructing and critiquing arguments, as an important skill for success in graduate school (Enright and Gitomer 1989). Summaries of research through 1999 (Breland et al. 1999) show convergence on various issues, especially the importance of defining the construct, and then designing the test accordingly to cover the intended construct, while simultaneously avoiding construct-irrelevant variance. In the case of the GMAT and GRE,[38] a design consisting of two prompts, creating and evaluating arguments, emerged after several rounds of research (Powers et al. 1999a). The design remains in GMAT and GRE.

Writing was partially incorporated into the TOEFL during the 1980s in the form of the TWE. It was a single-prompt "test." A history of the test is provided by Stansfield (1986a). With plans to include writing in the revised TOEFL, more systematic research among English language learners began to emerge, informed by appropriate theory (Hamp-Lyons and Kroll 1997). Whereas the distinction between issue and argument is thought to be appropriate for GRE and GMAT, in the case of TOEFL the broader construct of communicative competence has become the foundation for the test. With respect to writing, a distinction is made between an independent and an integrated prompt. The latter requires the test takers to refer to a document they read as part of the prompt. (See TOEFL 2011, for a brief history of the TOEFL program.)

Understandably, much of the construct work on writing has emphasized the postsecondary admissions context. However, in recent years, K-12 education reform efforts have increasingly incorporated test-based accountability approaches (Koretz and Hamilton 2006). As a result, there has been much reflection about the nature of school-based testing. The research initiative known as CBAL (Cognitively Based Assessment *of*, *for*, and *as* Learning) serves as an umbrella for experimentation on

[38] Today, the GMAT is administered and developed under the auspices of the Graduate Management Admissions Council (GMAC). It was originally developed at ETS for the GMAC and shared item types and staff with the GRE program. The interest in incorporating writing in the GMAT dates back to at least the mid-1980s (Owens 2006).

next-generation K–12 assessments. Under this umbrella, the writing construct has expanded to acknowledge the importance of other skills, specifically reading and critical thinking, and the developmental trajectories that underlie proficiency (Deane 2012; Deane and Quinlan 2010; Deane et al. 2008, 2012). In addition to expanding the breadth of the writing construct, recent work has also emphasized depth by detailing the nature of the evidence to be sought in student writing, especially argumentative writing (Song et al. 2014). Concomitant advances that would enable automated scoring for rich writing tasks have also been put forth (Deane 2013b).

18.3.3.2 Speaking

The assessment of speaking skills has traditionally taken place within an ESL context. The TSE (Clark and Swinton 1980) was the first major test of English speaking proficiency developed at ETS. Nevertheless, Powers (1984) noted that among the challenges facing the development of speaking measures were construct definition and cost. With respect to construct definition, a major conference was held at ETS in the 1980s (Stansfield 1986b) to discuss the relevance of communicative competence for the TOEFL. Envisioning TOEFL from that perspective was a likely outcome of the conference (Duran et al. 1987). Evidence of the acceptance of the communicative competence construct can be seen in its use to validate TSE scores (Powers et al. 1999b), and in the framework for incorporating a speaking component in a revised TOEFL (Butler et al. 2000). The first step in the development of an operational computer-based speaking test was the TOEFL Academic Speaking Test (TAST), a computer-based test intended to familiarize TOEFL test takers with the new format. TAST was introduced in 2002 and served to refine the eventual speaking measure included in TOEFL iBT. Automated scoring of speaking as discussed above, could help to reduce costs, but is not yet sufficiently well developed (Bridgeman et al. 2012). The *TOEIC*® Speaking and Writing test followed the TOEFL (Pearlman 2008b) in using ECD for assessment design (Hines 2010) as well as in the inclusion of speaking (Powers 2010; Powers et al. 2009).

18.3.3.3 Mathematics

Constructed-response items have been standard in the AP program since inception and were already used in NAEP by 1990 (Braswell and Kupin 1993). The SAT relied on multiple-choice items for much of its history (Lawrence et al. 2002) but also introduced in the 1990s a simple constructed-response format, the grid-in item, that allowed students to enter numeric responses. Because of the relative simplicity of numeric responses, they could be recorded on a scannable answer sheet, and therefore scored along with the multiple-choice responses. Various construct-related considerations motivated the introduction of the grid-in format, among them the influence of the standards produced by the National Council of Teachers of

Mathematics (Braswell 1992) but also considerations about the response process. For example, Bridgeman (1992) argued that in a mathematics context, the multiple-choice format could provide the student inadvertent hints and also make it possible to arrive at the right answers by reasoning backward from the options. He evaluated the SAT grid-in format with GRE items and concluded that the multiple-choice and grid-in versions of GRE items behaved very similarly. Following the adoption of the grid-in format in the SAT, a more comprehensive examination of mathematics item formats that could serve to elicit quantitative skills was undertaken, informed by advances in the understanding of mathematical cognition and a maturing computer-based infrastructure (Bennett and Sebrechts 1997; Bennett et al. 1997, 1999, 2000a, b; Sandene et al. 2005; Sebrechts et al. 1996). More recently, the mathematics strand of the CBAL initiative has attempted to unpack mathematical proficiency by means of competency models, the corresponding constructed-response tasks (Graf 2009), and scoring approaches (Fife 2013).

18.3.3.4 History[39]

A design innovation introduced by the AP history examinations was the document-based question (DBQ). Such questions require the test taker to incorporate, in a written response, information from one or more historical documents.[40] The idea for the format was based on input from a committee member who had visited libraries in England and saw that there were portfolios of primary historical documents, which apparently led to the DBQ. The DBQ was first used with the U.S. History examination, and the European History examination adopted the format the following year, as did World History when it was introduced in 2002. The scoring of document-based responses proved to be a challenge initially, but since its rationale was so linked to the construct, the task has remained.

18.3.3.5 Interpersonal Competence

Interpersonal competence has been identified as a twenty-first-century educational skill (Koenig 2011) as well as a workforce skill (Lievens and Sackett 2012). The skill was assessed early on at ETS by Larry Stricker (Stricker 1982; Stricker and Rock 1990) in a constructed-response format by means of videotaped stimuli, a relatively recent invention at the time. The recognition of the affordances of

---

[39] This section is based on an interview conducted on May 12, 2011, with Despina Danos, a senior Advanced Placement assessment developer.

[40] Although it is safe to say that assessments in the 1960s did not flow from a comprehensive framework or the explication of the target constructs, the current construct statement for AP History includes the skill of "crafting historical arguments from historical evidence" (College Board 2011, p. 8).

technology appears to have been the motivation for the work (Stricker 1982): "The advent of videotape technology raises new possibilities for assessing interpersonal competence because videotape provides a means of portraying social situations in a comprehensive, standardized, and economical manner" (p. 69).

18.3.3.6 Professional Assessments

Historically, ETS tests have been concerned with aiding the transition to the next educational level and, to a lesser extent, with tests designed to certify professional knowledge. Perhaps the earliest instance of this latter line of work is the "in-basket test" developed by Frederiksen et al. (1957). Essentially, the in-basket format is used to simulate an office environment where the test taker plays the role of school principal or business executive, for example. The format was used in an extended study concerned with measurement of the administrative skills of school principals in a simulated school (Hemphill et al. 1962). Apart from the innovative constructed-response format, the assessment was developed following what, in retrospect, was a very sophisticated assessment design approach. First, a job analysis was conducted to identify the skills required of an elementary school principal. In addition, the types of problems an elementary school principal is confronted with were identified and reduced to a series of incidents. This led to a universe of potential items by combining the problems typically confronted with the skills assumed to be required to perform as a principal based on the best research at the time (Hemphill et al. 1962, p. 47). Three skills were assumed to be (a) technical, (b) human, and (c) conceptual. The four facets of the jobs were taken to be (a) improving educational opportunity, (b) obtaining and developing personnel, (c) maintaining effective interrelationships with the community, and (d) providing and maintaining funds and facilities. The crossing of skill and facets led to a 4 × 3 matrix. Items were then written for each cell.

While the research on the assessment of school principals was highly innovative, ETS also supported the assessment of school personnel with more traditional measures. The first such assessment was bequeathed to the organization when the American Council on Education transferred the National Teacher Examination (NTE) in 1948 to the newly founded ETS. However, in the early 1980s, under President Greg Anrig[41] a major rethinking of teacher testing took place and culminated in the launching, in 1993, of the *PRAXIS SERIES*® tests. The *PRAXIS I*® and *PRAXIS II*® tests were concerned with content and pedagogical knowledge measured by multiple-choice items, as well as some types of constructed-response tasks. However, the *PRAXIS III*® tests were concerned with classroom performance and involved observing teachers in situ, a rather sharp departure from traditional mea-

[41] Greg Anrig was ETS's third president from 1981 until his death in 1993.

surement approaches. Although classroom observation has long been used in education, PRAXIS III appears to be among the first attempts to use *observations-as-measurement* in a classroom context. The knowledge base for the assessment was developed over several years (Dwyer 1994) and included scoring rubrics and examples of the behavior that would be evidence of the different skills required of teachers. The PRAXIS III work led to the Danielson *Framework for Teaching*,[42] which has served as the foundation for school-leader evaluations of teachers in many school districts, as well as for video-based products concerned with evaluation,[43] including those of the MET project (Bill and Melinda Gates Foundation 2013).

Whereas PRAXIS III was oriented toward assessing beginning teachers, ETS was also involved with the assessment of master teachers as part of a joint project with the National Board of Professional Teaching Standards (NBPTS). The goal of the assessment was to certify the expertise of highly accomplished practitioners. Pearlman (2008a, p. 88) described the rich history as "a remarkable journey of design, development, and response to empirical evidence from practice and use," including the scoring of complex artifacts. Gitomer (2007) reviewed research in support of NBPTS.

COPA was devoted to developing assessments for licensing and certification in fields outside education. In addition to that of architects mentioned earlier, COPA also considered the licensing of dental hygienists (Cameron et al. 2000; Mislevy et al. 1999, 2002b), which was one of the earliest applications of the ECD framework that will be discussed next.

18.3.3.7 Advances in Assessment Design Theory

For most of the twentieth century, there did not exist a *comprehensive* assessment design framework that could be used to help manage the complexity of developing assessments that go beyond the multiple-choice format. Perhaps this was not a problem because such assessments were relatively few and any initial design flaws could be remedied over time. However, several factors motivated the use of more ambitious designs, including the rapid technological innovations introduced during the second half of the twentieth century, concerns about the levels of achievement and competitiveness of U.S. students, the continued interest in forms of assessment beyond the multiple-choice item, and educational reform movements that have emphasized test-based accountability. A systematic approach to the design of complex assessments was needed, including ones involving the use of complex constructed responses.

---

[42] http://danielsongroup.org/framework/

[43] http://www.teachscape.com/

ECD is rooted in validity theory. Its genesis (Mislevy et al. 2006) is in the following quote from Messick (1994) concerning assessment design, which, he argued,

> would begin by asking what complex of knowledge, skills, and other attributes should be assessed, presumably because they are tied to explicit or implicit objectives of instruction or are otherwise valued by society. Next, *what behaviors or performances should reveal those constructs*, and *what task or situations should elicit those behaviors*? Thus, the nature of the construct guides the selection or construction of relevant tasks as well as the rational development of construct-based scoring criteria and rubrics. (p. 17, emphasis added)

ECD is a fleshing out of the quote into a comprehensive framework consisting of interlocking models. The *student model* focuses on describing the test taker, whereas the *evidence model* focuses on the nature and analysis of the responses. The evidence model passes its information to the student model to update the characterization of what the examinee knows and can do. Finally, the *task model* describes the items. Thus, if the goal is to characterize the students' communicative competence, an analysis of the construct is likely to identify writing and speaking skills as components, which means the student model should include characterizations of these student skills. With that information in hand, the details of the evidence model can be fleshed out: What sort of student writing and speaking performance or behavior constitutes evidence of students' writing and speaking skills? The answer to that question informs the task models, that is, what sorts of tasks are required to elicit the necessary evidence? ECD is especially useful in the design of assessments that call for constructed responses by requiring the behavior that constitutes relevant evidence of writing and speaking skills, for example, to be detailed and then prescribing the task attributes that would elicit that behavior. The evidence model, apart from informing the design of the tasks, is also the basis for scoring the responses (Mislevy et al. 2006).

ECD did not become quickly institutionalized at ETS, as Zieky (2014) noted. Nevertheless, over time, the approach has become widely used. Its applications include science (Riconscente et al. 2005), language (Mislevy and Yin 2012), professional measurement (Mislevy et al. 1999), technical skills (Rupp et al. 2012), automated scoring (Williamson et al. 2006), accessibility (Hansen and Mislevy 2008; T. Zhang et al. 2010), and task design and generation (Huff et al. 2012; Mislevy et al. 2002a). It has also been used to different degrees in the latest revisions of several ETS tests, such as TOEFL (Pearlman 2008b), in revisions of the College Board's AP tests (Huff and Plake 2010), and by the assessment community more generally (Schmeiser and Welch 2006, p. 313). Importantly, ECD is a broad design methodology that is not limited to items as the means of eliciting evidence. Games and simulations are being used with increasing frequency in an educational context, and ECD is equally applicable in both cases (Mislevy 2013; Mislevy et al. 2014, 2016).

18.3.4 Conclusion

It is clear that at ETS the transition to computer-based test delivery began early on and was sustained. The transition had an impact on constructed responses by enabling their use earlier than might have otherwise been the case. As Table 18.1 shows, online essay scoring and automated essay scoring were part of the transition for three major admissions tests: GMAT, GRE and TOEFL.

18.4 School-Based Testing

Although postsecondary admissions tests have been the main form of operational testing at ETS, school-based testing has been and continues to be an important focus. The Sequential Tests of Educational Progress (STEP) was an early ETS product in this domain. At one point it included a writing test that consisted of multiple-choice questions and an essay,[44] although it is no longer extant. By contrast, ETS involvement in two major twentieth-century school-based assessments, the *Advanced Placement Program®* examinations and the NAEP assessments, as well as in state assessments has grown. Constructed-response formats have played a major role, especially in AP and NAEP. In addition, the CBAL initiative has been prominent in recent years. They are discussed further in this section.

18.4.1 Advanced Placement

While the use of constructed responses encountered resistance at ETS in the context of admissions testing, the same was not true for the AP program, introduced in the mid-1950s. From the start, the AP program was oriented to academically advanced students who would be going to college, specifically to grant college credit or advanced placement by taking an examination. The seeds for the program were two reports (Lacy 2010), one commissioned by Harvard president James Bryant Conant (Committee on the Objectives of a General Education in a Free Society 1945), the other (General Education in School and College 1952) also produced at Harvard. These reports led to a trial of the idea in an experiment known as the Kenyon Plan.[45]

The eventual acquisition of the program by the College Board was not a given. Valentine (1987) noted that "Bowles [College Board president at the time] was not

[44] For a review, see Croon Davis et al. (1959).

[45] ETS was involved in the development and scoring of the Kenyon Plan before College Board agreed to take the program (Valentine 1987, p. 84).

Table 18.1 Writing assessment milestones for GMAT, GRE and TOEFL tests

Milestone	GMAT	GRE	TOEFL
When was writing introduced?	GMAT Analytic Writing Assessment (AWA) was introduced in the paper testing program by ETS in October 1994. The test consisted of one issue and one argument prompt.	GRE introduced the stand-alone GRE Writing Assessment in 1999, which consisted of one issue and one argument prompt (students were given a choice between two issue prompts). In 2002 the Analytical Writing measure replaced the Analytical reasoning section and became part of the GRE General Test. In 2011, the choice of issue prompts was removed and prompts variants were introduced.	Test of Written English portion of the paper-based TOEFL test was introduced in 1986 at selected administrations of TOEFL. The 1998 TOEFL CBT writing task consisted of a choice of handwritten or keyed essay, essentially the same task as the Test of Written English portion of the paper-based TOEFL test. A writing measure consisting of integrated and independent prompts was introduced with the release of TOEFL iBT in 2006.
When was computer-based/adaptive testing introduced?	GMAT switched entirely from paper-and-pencil to on-demand CAT in October 1997.	GRE switched to on-demand CAT in 1992 and abandoned CAT in favor of MST in 2011.	TOEFL CBT on-demand testing was introduced in 1998. The CBT essay score was combined with the Structure selected-response subsection to report out on a Structure Writing section score. Listening and Structure were adaptive.
When was online scoring deployed?	Under on-demand testing, scores need to be reported on an ongoing basis and that, in turn, requires continuous scoring. The Online Scoring Network (OSN) was developed for that purpose and was first used operationally in October 1997 for GMAT essays.	OSN has been used to score GRE essays since 1999 when the GRE Writing Assessment was introduced.	Online scoring was deployed when TOEFL CBT was launched in 1998.

(continued)

Table 18.1 (continued)

Milestone	GMAT	GRE	TOEFL
When was automated scoring introduced?	Automated scoring with e-rater as a contributory score started in January 1999.	e-rater scoring as a check score was introduced in 2008.	e-rater started as contributory score to independent writing for TOEFL iBT beginning July 2009; contributory score for integrated writing began November 2010.

Note. *CAT* = computer adaptive testing, *CBT* = computer based testing, *MST* = multistage testing

sure that taking the program was in the Board's interest" (p. 85). Initially, some of the AP examinations were entirely based on constructed responses, although eventually all, with the exception of Studio Art, included a mix of constructed-response and multiple-choice items. A program publication, *An Informal History of the AP Readings 1956–1976* (Advanced Placement Program of the College Board 1980), provides a description of the scoring process early in the program's history.

Interestingly, in light of the ascendancy of the multiple-choice format during the twentieth century, the use of constructed responses in AP does not appear to have been questioned. Henry Dyer, a former ETS vice president (1954–1972), seems to have been influential in determining the specifications of the test (Advanced Placement Program of the College Board 1980, p. 2). Whereas Dyer did not seem to have been opposed to the use of constructed responses in the AP program, he was far more skeptical of their value in the context of another examination being conceived at about the same time, the Test of Developed Ability.[46] In discussing the creation of that test, Dyer (1954) noted that

> there may be one or two important abilities which are measureable only through some type of free response question. If an examining committee regards such abilities as absolutely vital in its area, it should attempt to work out one or two free response questions to measure them. Later on, we shall use the data from the tryouts to determine whether the multiple-choice sections of the test do not in fact measure approximately the same abilities as the free

[46] The Test of Developed Ability is a nearly forgotten test. It is relevant to this chapter since in its original conception, it employed constructed responses. Henry Chauncey was a champion for the test at ETS (Lemann 1999, p. 95). According to N. Elliot (2005, p. 149), citing Henry Dyer, Frank Bowles, president of College Board, proposed the test as early as 1949. Bowles thought that there were "changes coming in the kind of tests that would be suitable for college admission." The Test of Developed Ability was designed to measure achievement, in contrast to the SAT, which was oriented, at the time, toward measuring ability. The design of the Test of Developed Ability called for constructed responses, which presented a major scoring hurdle and may have been one of the reasons the test never became operational. According to Lemann (1999), the projected cost of the test was six dollars, as opposed to three dollars for the SAT. This work transpired during the 1950s when some individuals thought there should be an alternative to the SAT that was more achievement oriented. In fact, such an alternative led to the founding of ACT in 1959, led by Lindquist (see, N. Elliot 2014, p.246). For further discussion of the Test of Developed Ability, see N. Elliot (2005, p. 148; 2014, p.292) and Lemann (1999).

response sections. If they do, the free response section will be dropped, if not, they will be retained. (p. 7)

Thus, there was realization that the AP program was unique with respect to other tests, in part because of its use of constructed responses. In *An Informal History of the AP Readings 1956–76* (Advanced Placement Program of the College Board 1980), it was noted that

> neither the setting nor the writing of essay examination was an innovation. The ancient Chinese reputedly required stringent written examinations for high government offices 2,500 years ago. European students have long faced pass-or-perish examinations at the end of their courses in the Lycée, Gymnasium, or British Secondary system. In this country, from 1901 to 1925, the College Board Comprehensives helped to determine who would go to the best colleges. But the Advanced Placement Program was new, and in many ways unique. (p. 2)

As the College Board's developer and administrator for the AP program, ETS has conducted much research to support it. The contributions focused on fairness (e.g., Breland et al. 1994; Bridgeman et al. 1997; Dorans et al. 2003; Stricker and Ward 2004), scoring (e.g., Braun 1988; Burstein et al. 1997; Coffman and Kurfman 1968; Myford and Mislevy 1995; Zhang et al. 2003), psychometrics (e. g., Bridgeman et al. 1996a, b; Coffman and Kurfman 1966; Lukhele et al. 1994; Moses et al. 2007), and validity and construct considerations (e. g., Bennett et al. 1991; Bridgeman 1989; Bridgeman and Lewis 1994).

18.4.2 Educational Surveys[47]

As noted earlier, NAEP has been a locus of constructed-response innovation at ETS. NAEP was managed by the Education Commission of the States until 1983 when ETS was awarded the contract to operate it. With the arrival of NAEP, ETS instituted matrix sampling, along with IRT (Messick et al. 1983); both had been under development at ETS under Fred Lord,[48] and both served to undergird a new approach to providing the "Nation's Report Card" in several subjects, with extensive use of constructed-response formats. To NAEP's credit, explicating the domain of knowledge to be assessed by means of "frameworks" had been part of the assessment development process from inception. Applebee (2007) traced the writing framework back to 1969. Even before that date, however, formal frameworks

[47] For a fuller discussion of educational surveys see Beaton and Barone (Chap. 8, this volume) and Kirsch et al. (Chap. 9, this volume).

[48] Lord (1965) credits William Turnbull with the essence of the idea of item sampling and Robert L. Ebel with its application to norming. An implication of item and matrix sampling for constructed-response formats is that they make it possible to administer a large number of items, without any one student responding to a long test, by assigning subsets of items to different students. The idea can be leveraged for school-based testing (Bejar and Graf 2010; Bock and Mislevy 1988).

providing the rationale for content development were well documented (Finley and Berdie 1970). The science framework refers early on to "inquiry skills necessary to solve problems in science, specifically the ability to recognize scientific hypotheses" (p. 14). The assessment of inquiry skills has since become standard in science assessment but was only recently implemented operationally with the redesigned AP exams.

NAEP has been the source of multiple content and psychometric innovations (Mazzeo et al. 2006), including the introduction of mixed-format assessment consisting of both multiple choice and the large-scale use of constructed-response items. Practical polytomous IRT was developed in a NAEP context as documented by Carlson and von Davier (Chap. 5, this volume), and, as described earlier, NAEP introduced innovations concerned with the scoring of written responses. Finally, ETS continues to collaborate with NAEP in the exploration of technological advances to testing (Bennett et al. 2010).[49] The transition to digital delivery is underway as of this writing. In fact, the 2017 writing assessment was administered on tablets supplied by NAEP and research into the use of mixed-format adaptive in mathematics has also been carried out (Oranje et al. 2014).

18.4.3 Accountability Testing

The start of K–12 testing in the United States dates back to the nineteenth century, when Horace Mann, an educational visionary, introduced several innovations into school testing, among them the use of standardized (written constructed-response) tests (U.S. Congress and Office of Technology Assessment 1992, chapter 4). The innovations Mann introduced were, in part, motivated by a perception that schools were not performing as well as could be expected. Such perceptions have endured and have continued to fuel the debate about the appropriate use of tests in K–12. More recently, the Nation at Risk report (National Commission on Excellence in Education 1983) warned that "the educational foundations of our society are presently being eroded by a rising tide of mediocrity that threatens our very future as a Nation and a people" (para. 1). Similarly, the linking of the state of education to the nation's economic survival[50] was behind one effort in the early 1990s (U.S. Department of Labor and Secretary's Commission on Achieving Necessary Skills 1991; known as SCANS), and it had significant implications for the future of testing. As Linn (1996) noted, the system of assessment expected to emerge from the SCANS effort and to be linked to instruction, "would require direct appraisals of student performance" (p. 252) and would serve to promote the measured skills.

[49] NAEP has set up a website on technology-based assessments: http://nces.ed.gov/nationsreportcard/tba/

[50] The concerns continue and have been described as a "perfect storm" (Kirsch et al. 2007).

The calls for direct assessment that promotes learning joined the earlier Frederiksen (1984) assault on the multiple-choice item type, which had been heard loudly and clearly, judging by the number of citations to that article.[51] The idea of authentic assessment (Wiggins 1989) as an alternative to the standardized multiple-choice test, took hold among many educators, and several states launched major performance-based assessments. ETS participated in exploring these alternatives, especially portfolio assessment (Camp 1985, 1993), including in their evaluation in at least one state, California (Thomas et al. 1998).

Stetcher (2010) provided a detailed review of the different state experiments in the early 1990s in Vermont, Kentucky, Maryland, Washington, California, and Connecticut. A summary of a conference (National Research Council 2010, p. 36) noted several factors that led to the demise of these innovative programs:

- Hurried implementation made it difficult to address scoring, reliability, and other issues.
- The scientific foundation required by these innovative assessments was lacking.
- The cost and burden to the school was great, and questions were raised as to whether they were worth it.
- There were significant political considerations, including cost, time, feasibility of implementation, and conflicts in purpose among constituencies.

Not surprisingly, following this period of innovation, there was a return to the multiple-choice format. Under the No Child Left Behind (NCLB) legislation,[52] the extent of federally mandated testing increased dramatically and once again the negative consequences of the predominant use of multiple-choice formats were raised. In response, a research initiative was launched at ETS known as CBAL (Bennett and Gitomer 2009).[53] Referring to the circumstances surrounding accountability testing under NCLB, Bennett and Gitomer noted,

> In the United States, the problem is … an accountability assessment system with at least two salient characteristics. The first characteristic is that there are now significant consequences for students, teachers, school administrators, and policy makers. The second characteristic is, paradoxically, very limited educational value. This limited value stems from the fact that our accountability assessments typically reflect a shallow view of proficiency defined in terms of the skills needed to succeed on relatively short and, too often, quite artificial test items (i.e., with little direct connection to real-world contexts). (p. 45)

The challenges that needed to be overcome to develop tests based on a deeper view of student achievement were significant and included the fact that more meaningful tests would require constructed-response formats to a larger degree, which required a means of handling the trade-off between reliability and time. As Linn and Burton (1994), and many others, have reminded us regarding constructed-response tests, "a substantial number of tasks will still be needed to have any reasonable level of confidence in making a decision that an individual student has or has not met the

[51] As of August 2014, it had been cited 639 times.

[52] http://www.ed.gov/policy/elsec/leg/esea02/index.html

[53] A website describing the CBAL initiative can be found at https://www.ets.org/cbal

standard" (p. 10). Such a test could not reasonably be administered in a single seating. A system was needed in which tests would be administered at more than one occasion. A multi-occasion testing system raises methodological problems of its own, as was illustrated by the California CLAS assessment (Cronbach et al. 1995). Apart from methodological constraints, increasing testing time could be resented, unless the tests departed from the traditional mold and actually promoted, not just probed, learning. This meant that the new assessment needed to be an integral part of the educational process. To help achieve that goal, a theory of action was formulated (Bennett 2010) to link the attributes of the envisioned assessment system to a set of hypothesized action mechanisms leading to improved student learning. (Of course, a theory of action *is* a theory, and whether the theory is valid is an empirical question.)

Even with a vision of an assessment system, and a rationale for how such a vision would lead to improved student learning, considerable effort is required to explicate the system and to leverage technology to make such assessments scalable and affordable. The process entailed the formulation of competency models for specific domains, including reading (Sheehan and O'Reilly 2011), writing (Deane et al. 2012), mathematics (Graf 2009), and science (Liu et al. 2013); the elaboration of constructs, especially writing (Song et al. 2014); and innovations in automated scoring (Deane 2013a, b; Fife 2013) and task design (Bennett 2011; Sheehan and O'Reilly 2011).

The timing of the CBAL system coincided roughly with the start of a new administration in Washington that had educational plans of its own, ultimately cast as the Race to the Top initiative.[54] The assessments developed under one portion of the Race to the Top initiative illustrate a trend toward the use of significant numbers of items requiring constructed responses. In addition, technology is being used more extensively, including adaptive testing by the Smarter Balanced Assessment Consortium, and automated scoring by some of its member states.

18.4.4 Conclusion

Admissions testing has been the primary business at ETS for most of its existence. Constructed-response formats were resisted for a long time in that context, although in the end they were incorporated. By contrast, the same resistance was not encountered in some school assessments, where they were used from the start in the AP program as well as in the NAEP program. The CBAL initiative has continued and significantly expanded that tradition by conceiving of instructionally rich computer-based tasks grounded in scientific knowledge about student learning.

[54] The tone for the initiative was oriented to increased reliance on constructed-response formats, as noted by President Obama: "And I'm calling on our nation's governors and state education chiefs to develop standards and assessments that don't simply measure whether students can fill in a bubble on a test" (White House 2009).

18.5 Validity and Psychometric Research Related to Constructed-Response Formats

The foregoing efforts occurred in the context of a vigorous validity and psychometric research program over several decades in support of constructed-response formats. It is beyond the scope of this chapter to review the literature resulting from that effort. However, the scope of the research is noteworthy and is briefly and selectively summarized below.

18.5.1 Construct Equivalence

The choice between multiple-choice or constructed-response format, or a mix of the two, is an important design question that is informed by whether the two formats function in similar ways. The topic has been approached conceptually and empirically (Bennett et al. 1990, 1991; Bridgeman 1992; Enright et al. 1998; Katz et al. 2000; Messick 1993; Wainer and Thissen 1993; Ward 1982; Ward et al. 1980).

18.5.2 Predictive Validity of Human and Computer Scoring

The predictive validity of tests based on constructed responses scored by humans and computers has not been studied extensively. A study (Powers et al. 2002) appears to be one of the few on the subject. More recently, Bridgeman (2016) showed the impressive psychometric predictive power of the GRE and TOEFL writing assessments.

18.5.3 Equivalence Across Populations and Differential Item Functioning

The potential incomparability of the evidence elicited by different test formats has fairness implications and not surprisingly has received much attention (e.g., Breland et al. 1994; Bridgeman and Rock 1993; Dorans 2004; Dorans and Schmitt 1993; Schmitt et al. 1993; Zwick et al. 1993, 1997). The challenges of differential item functioning across language groups have also been addressed (Xi 2010). Similarly, the role of different response formats when predicting external criterion measures has been investigated (Bridgeman and Lewis 1994), as have the broader implications of format for the admissions process (Bridgeman and McHale 1996).

18.5.4 Equating and Comparability

The use of constructed-response formats presents many operational challenges. For example, ensuring the comparability of scores from different forms is equally applicable to tests comprising constructed-response items as it is for multiple-choice tests. The primary approach to ensuring score comparability is through equating (Dorans et al. 2007), a methodology that had been developed for multiple-choice tests. As the use of constructed-response formats has grown, there has been an increase in research concerning equating of tests composed entirely, or partly, of constructed responses (Kim and Lee 2006; Kim and Walker 2012; Kim et al. 2010). Approaches to achieving comparability without equating, which rely instead on designing *tasks* to be comparable, have also been studied (Bejar 2002; Bridgeman et al. 2011; Golub-Smith et al. 1993).

18.5.5 Medium Effects

Under computer delivery, task presentation and the recording of responses is very different for multiple-choice and constructed-response items. These differences could introduce construct-irrelevant variance due to the testing medium. The investigation of that question has received significant attention (Gallagher et al. 2002; Horkay et al. 2006; Mazzeo and Harvey 1988; Powers et al. 1994; Puhan et al. 2007; Wolfe et al. 1993).

18.5.6 Choice

Students' backgrounds can influence their interest and familiarity with the topics presented in some types of constructed-response items, which can lead to an unfair assessment. The problem can be compounded by the fact that relatively few constructed-response questions can be typically included in a test since responding to them is more time consuming. A potential solution is to let students choose from a set of possible questions rather that assigning the same questions to everyone. The effects of choice have been investigated primarily in writing (Allen et al. 2005; Bridgeman et al. 1997; Lukhele et al. 1994) but also in other domains (Powers and Bennett 1999).

18.5.7 *Difficulty Modeling*

The difficulty of constructed-response items and the basis for, and control of, variability in difficulty have been studied in multiple domains, including mathematics (Katz et al. 2000), architecture (Bejar 2002), and writing (Bridgeman et al. 2011; Joe et al. 2012).

18.5.8 *Diagnostic and Formative Assessment*

Diagnostic assessment is a broad topic that has much in common with formative assessment because in both cases it is expected that the provided information will lead to actions that will enhance student learning. ETS contributions in this area have included the development of psychometric models to support diagnostic measurement based on constructed responses. Two such developments attempt to provide a psychometric foundation for diagnostic assessments. Although these efforts are not explicitly concerned with constructed responses, they support such assessments by accommodating polytomous responses. One approach is based on Bayesian networks (Almond et al. 2007), whereas the second approach follows a latent variable tradition (von Davier 2013).

18.6 Summary and Reflections

The multiple-choice item format is an early-twentieth-century American invention. Once the format became popular following its use in the Army Alpha and SAT, it became difficult for constructed-response formats to regain a foothold. The psychometric theory that also emerged in the early twentieth century emphasized score reliability and predictive validity. Those emphases presented further hurdles. The interest in constructed-response formats, especially to assess writing skills, did not entirely die, however. In fact, there was early research at ETS that would be instrumental in eventually institutionalizing constructed-response formats, although it was a journey of nearly 50 years. The role of ETS in that process has been significant. The chapter on performance assessment by Suzanne Lane and Clement Stone (Lane and Stone 2006) in *Educational Measurement* is an objective measure. Approximately 20% of the chapter's citations were to publications authored by ETS staff. This fact is noteworthy, because the creation of an ETS-like organization had been objected to by Carl Brigham on the grounds that an organization that produced tests would work to preserve the status quo, with little incentive to pursue innovation. As he noted in a letter to Conant (cited in Bennett, Chap. 1, this volume):

> one of my complaints against the proposed organization is that although the word research will be mentioned many times in its charter, the very creation of powerful machinery to do

> more widely those things that are now being done badly will stifle research, discourage new developments, and establish existing methods, and even existing tests, as the correct ones. (p. 6)

His fears were not unreasonable in light of what we know today about the potential for lack of innovation in established organizations (Dougherty and Hardy 1996). However, according to Bennett (2005), from its inception, the ETS Board of Trustees heeded Brigham's concerns, as did the first ETS president (from 1947 until 1970), Henry Chauncey. That climate was favorable to conducting research that would address how to improve and modernize existing tests.[55] Among the many areas of research were investigations related to the scoring of writing. That early research led to a solution to what has been the long-standing problem of operationally scoring essays with acceptable scoring reliability.

Even if the scoring agreement problem was on its way to being solved, it was still the case that tasks requiring a longer constructed response would also take more time and that therefore fewer items could be administered in a given period. With predictive validity as the key metric for evaluating the "validity" of scores, the inclusion of constructed-response tasks continued to encounter resistance. An exception to this trend was the AP program, which relied on constructed-response tasks from its inception. There was also pioneering work on constructed-response assessments early in ETS's history (Frederiksen et al. 1957; Hemphill et al. 1962). However, in both of these cases, the context was very different from the admissions testing case that represented the bulk of ETS business.

Thus a major development toward wider use of constructed-response formats was the evolution of validity theory away from an exclusive focus on predictive considerations. Messick's (1989) work was largely dedicated to expanding the conception of validity to include not only the psychometric attributes of the test, the evidentiary aspect of validation, but also the repercussions that the use of the test could have, the consequential aspect. This broader view did not necessarily endorse the use of one format over another but provided a framework in which constructed-response formats had a greater chance for acceptance.

With the expansion of validity, the doors were opened a bit more, although costs and scalability considerations remained. These considerations were aided by the transition of assessment from paper to computer. The transition to computer-delivered tests at ETS that started in 1985 with the deployment of ACCUPLACER set the stage for the transition of other tests—like the GMAT, GRE, and TOEFL—to digital delivery and the expansion of construct coverage and constructed-response formats, especially for writing and eventually speaking.

---

[55]A good example can be seen in the evolution of the GRE, a test owned by ETS. In 1992, ETS introduced adaptive testing in the GRE by building on research by Fred Lord and others. In 2011, the GRE was revised again to include, among other changes, a different form of adaptive testing that has proven more robust than the earlier approach. The current adaptive testing approach, a multistage design (Robin et al. 2014), was experimented with much earlier at ETS (Angoff and Huddleston 1958), was extensively researched (Linn et al. 1969), and has since proven to be preferable in an on-demand admissions testing context.

Along the way, there was abundant research and implementation of results in response to the demands resulting from the incorporation of expanded constructs and the use of the computer to support those demands. For example, the psychometric infrastructure for mixed-format designs, including psychometric modeling of polytomous responses, was first used in 1992 (Campbell et al. 1996, p. 113) and developed at ETS. The use of constructed-response formats also required an efficient means of scoring responses captured from booklets. ETS collaborated with subcontractors in developing the necessary technology, as well as developing control procedures to monitor the quality of scoring. Online scoring systems were also developed to accommodate the transition to continuous administration that accompanied computer-based testing. Similarly, automated scoring was first deployed operationally in 1997, when the licensing test for architects developed by ETS became operational (Bejar and Braun 1999; Kenney 1997). The automated scoring of essays was first deployed operationally in 1999 when it was used to score GMAT essays.

Clearly, by the last decade of the twentieth century, the fruits of research at ETS around constructed-response formats were visible. The increasingly ambitious assessments that were being conceived in the 1990s stimulated a rethinking of the assessment design process and led to the conception of ECD (Mislevy et al. 2003). In addition, ETS expanded its research agenda to include the role of assessment in instruction and forms of assessment that, in a sense, are beyond format. Thus the question is no longer one of choice between formats but rather whether an assessment that is grounded in relevant science can be designed, produced, and deployed. That such assessments call for a range of formats and response types is to be expected. The CBAL initiative represents ETS's attempt to conceptualize assessments that can satisfy the different information needs of K–12 audiences with state-of-the-art tasks grounded in the science of student learning, while taking advantage of the latest technological and methodological advances. Such an approach seems necessary to avoid the difficulties that accountability testing has encountered in the recent past.

18.6.1 What Is Next?

If, as Alphonse De Lamartine (1849) said, "history teaches us everything, including the future" (p. 21), what predictions about the future can be made based on the history just presented? Although for expository reasons I have laid the history of constructed-response research at ETS as a series of sequential hurdles that appear to have been solved in an orderly fashion, in reality it is hard to imagine how the story would have unfolded at the time that ETS was founded. While there were always advocates of the use of constructed-response formats, especially in writing, Huddleston's views that writing was essentially verbal ability, and therefore could be measured with multiple-choice verbal items, permeated decision making at ETS.

Given the high stakes associated with admissions testing and the technological limitations of the time, in retrospect, relying on the multiple-choice format arguably was the right course of action from both the admissions committee's point of view and from the student's point of view. It is well known that James Bryant Conant instituted the use of the SAT at Harvard for scholarship applicants (Lemann 2004), shortly after his appointment as president in 1933, based on a recommendation by his then assistant Henry Chauncey, who subsequently became the first president of ETS.[56] Conant was motivated by a desire to give students from more diverse backgrounds an opportunity to attend Harvard, which in practice meant giving students from other than elite schools a chance to enroll. The SAT, with its curriculum agnostic approach to assessment, was fairer to students attending public high schools than the preparatory-school-oriented essay tests that preceded it. That is, the consequential aspect of validation may have been at play much earlier than Messick's proposal to incorporate consequences of test use as an aspect of validity, and it could be in this sense partially responsible for the long period of limited use into which the constructed-response format fell.

However well-intentioned the use of the multiple-choice format may have been, Frederiksen (1984) claimed that it represented the "real test bias." In doing so, he contributed to fueling the demand for constructed-response forms of assessment. It is possible to imagine that the comforts of what was familiar, the multiple-choice format, could have closed the door to innovation, a fear that had been expressed by Carl Brigham more generally.[57] For companies emerging in the middle of the twentieth century, a far more ominous danger was the potential disruptions that could accrue from a transition to the digital medium that would take place during the second half of the century. The Eastman Kodak Company, known for its film and cameras, is perhaps the best known example of the disruption that the digital medium could bring: It succumbed to the digital competition and filed for bankruptcy in 2012. However, this is not a classic case of being disrupted out of existence,[58] because Kodak invented the first digital camera! The reasons for Kodak's demise are far more nuanced and include the inability of the management team to figure out in time how to operate in a hybrid digital and analog world (Chopra 2013). Presumably a different management team could have successfully transitioned the company to a digital world.[59]

In the testing industry, by contrast, ETS not only successfully navigated the digital transition, *it actually led the transition* to a digital testing environment with the

---

[56] In 1938, Chauncey and Conant "persuaded all the College Board schools to use the SAT as the main admissions tests for scholarship applicants" (Lemann 2004, p. 8), and in 1942, the written College Board exams were suspended and the SAT became the admissions tests for all students, not just scholarship applicants.

[57] The tension between innovation and "operate and maintain" continues to this day (T. J. Elliot 2014).

[58] Both Foster (1986) and Christensen (1997) discussed the potential of established companies to fall prey to nimble innovating companies.

[59] In the leadership literature (Howell and Avolio 1993), a distinction is made between two styles of leadership: transactional and transformational. Transactional leaders aim to achieve their goals through accountability, whereas transformational leaders aim to achieve their goals by being charismatic and inspirational. It is easy to imagine that different prevailing types of leadership are better or worse at different points in the life of a company.

launching of ACCUPLACER in 1985.[60] The transition to adaptive testing must have been accompanied by a desire to innovate and explore how technology could be used in testing, because in reality, there were probably few compelling business or even psychometric reasons to launch a computer-based placement test in the 1980s. Arguably, the early transition made it possible for ETS to eventually incorporate constructed-response formats into its tests *sooner*, even if the transition was not even remotely motivated by the use of constructed-response formats. By building on a repository of research related to constructed-response formats motivated by validity and fairness considerations, the larger transition to a digital ecosystem did not ultimately prove to be disruptive at ETS and instead made it possible to take advantage of the medium, finally, to deploy assessments containing constructed-response formats and to envision tests as integral to the educational process rather than purely technological add-ons.

In a sense, the response format challenge has been solved: Admissions tests now routinely include constructed-response items, and the assessments developed to measure the Common Core State Standards also include a significant number of constructed-response items. Similarly, NAEP, which has included constructed-response items for some time, is making the transition to digital delivery via tablets. ETS has had a significant role in the long journey. From inception, there was a perspective at ETS that research is critical to an assessment organization (Chauncey, as cited by Bennett, Chap. 1, this volume). Granted that the formative years of the organization were in the hands of enlightened and visionary individuals, it appears that the research that supported the return of constructed-response formats was not prescribed from above but rather the result of *intra*preneurship,[61] or individual researchers largely pursuing their own interests.[62] If this is the formula that worked in the past, it could well continue to work in the future, if we believe De Lamartine. Of course, Santayana argued that history is always written wrong and needs to be rewritten. Complacency about the future, therefore, is not an option—it will still need to be constructed.

Acknowledgments The chapter has greatly benefited from the generous contributions of many colleagues at ETS. Peter Cooper, Robby Kantor, and Mary Fowles shared generously their knowledge of writing assessment. Despina Danos, Alice Sims-Gunzenhauser, and Walt MacDonald did likewise with respect to the history of the Advanced Placement program. Mel Kubota and Marylyn Sudlow shared their knowledge of the history of writing within the SAT admissions program. John Barone, Ina Mullis, and Hillary Persky were my primary sources for NAEP's history. Pan Mollaun and Mary Schedl shared their knowledge on the history of TOEFL. Sydell Carlton generously provided a firsthand account of the milestone study "Factors in Judgments of Writing Ability." And Mike Zieky's long tenure at ETS served to provide me with useful historical background throughout the project. Karen McQuillen and Jason Wagner provided access to the nooks and crannies of the Brigham Library and greatly facilitated tracking down sources. The chronology reported in Table 18.1 would have been nearly impossible to put together without the assistance of Jackie

[60] ETS developed ACCUPLACER for the College Board.

[61] The term that Pinchot (1987) introduced relatively recently to describe within-company entrepreneurship.

[62] N. Elliot (2005, p. 183) wrote instead about "lone wolves."

Briel, Gary Driscoll, Roby Kantor, Fred McHale, and Dawn Piacentino; I'm grateful for their patience. Within the ETS Research division, I benefited from input by Keelan Evanini, Irv Katz, Sooyeon Kim, and Klaus Zechner.

Other individuals were also generous contributors, especially Bert Green, Mary Pommerich, Dan Segal, and Bill Ward, who shared their knowledge of the development of adaptive testing; Nancy Petersen contributed on the origins of automated scoring at ETS. I'm especially grateful to Norbert Elliot for his encouragement. His *On a Scale: A Social History of Writing Assessment in America* was an invaluable source in writing this chapter. He also offered valuable suggestions and sources, many of which I was able to incorporate.

Randy Bennett, Brent Bridgeman, and Don Powers, by virtue of being major contributors to the knowledge base on constructed-response formats at ETS, were ideal reviewers and provided many valuable suggestions in their respective reviews. Doug Baldwin and Larry Stricker offered valuable criticism on an earlier draft. Finally, the manuscript has benefited greatly from the thorough editorial review by Randy Bennett, Jim Carlson, and Kim Fryer. However, it should be clear that I remain responsible for any flaws that may remain.

References

Advanced Placement Program of the College Board. (1980). *An informal history of the AP readings 1956–1976*. New York: The College Board.

Allen, N. L., Holland, P. W., & Thayer, D. T. (2005). Measuring the benefits of examinee-selected questions. *Journal of Educational Measurement, 42*, 27–51. https://doi.org/10.1111/j.0022-0655.2005.00003.x

Almond, R. G., DiBello, L. V., Moulder, B., & Zapata-Rivera, J.-D. (2007). Modeling diagnostic assessments with Bayesian networks. *Journal of Educational Measurement, 44*, 341–359. https://doi.org/10.1111/j.1745-3984.2007.00043.x

Anderson, H. A., & Traxler, A. E. (1940). The reliability of the reading of an English essay test: A second study. *The School Review, 48*(7), 521–530.

Angoff, W. H., & Huddleston, E. M. (1958). *The multi-level experiment: A study of a two-level test system for the College Board Scholastic Aptitude Test* (Statistical Report No. SR-58-21). Princeton: Educational Testing Service.

Applebee, A. N. (2007). Issues in large-scale writing assessment: Perspectives from the National Assessment of Educational Progress. *Journal of Writing Assessment, 3*(2), 81–98.

Attali, Y. (2010). Immediate feedback and opportunity to revise answers: Application of a graded response IRT model. *Applied Psychological Measurement, 38*, 632–644.

Attali, Y., & Burstein, J. (2006). Automated essay scoring with *e-rater* V.2. *Journal of Technology, Learning, and Assessment, 4*(3). Retrieved from http://www.jtla.org/

Attali, Y., & Powers, D. E. (2008). *Effect of immediate feedback and revision on psychometric properties of open-ended GRE subject test items* (Research Report No. RR-08-21). Princeton: Educational Testing Service. http://dx.doi.org/10.1002/j.2333-8504.2008.tb02107.x

Attali, Y., & Powers, D. E. (2010). Immediate feedback and opportunity to revise answers to open-ended questions. *Educational and Psychological Measurement, 70*(1), 22–35. https://doi.org/10.1177/0013164409332231

Baker, F. B. (1971). Automation of test scoring, reporting and analysis. In R. L. Thorndike (Ed.), *Educational measurement* (2nd ed., pp. 202–236). Washington, DC: American Council on Education.

Baldwin, D. (2004). A guide to standardized writing assessment. *Educational Leadership, 62*(2), 72–75.

Baldwin, D., Fowles, M., & Livingston, S. (2005). *Guidelines for constructed-responses and other performance assessments*. Princeton: Educational Testing Service.

Bejar, I. I. (1985). *A preliminary study of raters for the Test of Spoken English* (TOEFL Research Report No. 18). Princeton: Educational Testing Service. http://dx.doi.org/10.1002/j.2330-8516.1985.tb00090.x

Bejar, I. I. (1991). A methodology for scoring open-ended architectural design problems. *Journal of Applied Psychology, 76*, 522–532. https://doi.org/10.1037/0021-9010.76.4.522

Bejar, I. I. (2002). Generative testing: From conception to implementation. In S. H. Irvine & P. C. Kyllonen (Eds.), *Item generation for test development* (pp. 199–218). Mahwah: Erlbaum.

Bejar, I. I. (2012). Rater cognition: Implications for validity. *Educational Measurement: Issues and Practice, 31*(3), 2–9. https://doi.org/10.1111/j.1745-3992.2012.00238.x

Bejar, I. I., & Braun, H. I. (1999). *Architectural simulations: From research to implementation: Final report to the National Council of Architectural Registration Boards* (Research Memorandum No. RM-99-02). Princeton: Educational Testing Service.

Bejar, I. I., & Graf, E. A. (2010). Updating the duplex design for test-based accountability in the twenty-first century. *Measurement: Interdisciplinary Research & Perspective, 8*(2), 110–129. https://doi.org/10.1080/15366367.2010.511976

Bejar, I. I., & Whalen, S. J. (2001). *U.S. Patent No. 6,295,439*. Washington, DC: Patent and Trademark Office.

Bejar, I. I., Chaffin, R., & Embretson, S. E. (1991). *Cognitive and psychometric analysis of analogical problem solving*. New York: Springer. https://doi.org/10.1007/978-1-4613-9690-1

Bejar, I. I., Douglas, D., Jamieson, J., Nissan, S., & Turner, J. (2000). *TOEFL 2000 listening framework: A working paper* (TOEFL Monograph Series No. 19). Princeton: Educational Testing Service.

Bejar, I. I., Williamson, D. M., & Mislevy, R. J. (2006). Human scoring. In D. M. Williamson, R. J. Mislevy, & I. I. Bejar (Eds.), *Automated scoring of complex tasks in computer-based testing* (pp. 49–82). Mahwah: Erlbaum.

Bennett, R. E. (2005). *What does it mean to be a nonprofit educational measurement organization in the 21st century?* Princeton: Educational Testing Service.

Bennett, R. E. (2010). Cognitively Based Assessment of, for, and as Learning (CBAL): A preliminary theory of action for summative and formative assessment. *Measurement: Interdisciplinary Research & Perspective, 8*(2), 70–91. https://doi.org/10.1080/15366367.2010.508686

Bennett, R. E. (2011). Formative assessment: A critical review. *Assessment in Education: Principles, Policy & Practice, 18*(1), 5–25. https://doi.org/10.1080/0969594X.2010.513678

Bennett, R. E., & Bejar, I. I. (1998). Validity and automated scoring: It's not only the scoring. *Educational Measurement: Issues and Practice, 17*(4), 9–16. https://doi.org/10.1111/j.1745-3992.1998.tb00631.x

Bennett, R. E., & Gitomer, D. H. (2009). Transforming K–12 assessment: Integrating accountability testing, formative assessment and professional support. In C. Wyatt-Smith & J. Cumming (Eds.), *Educational assessment in the 21st century* (pp. 43–61). New York: Springer. https://doi.org/10.1007/978-1-4020-9964-9_3

Bennett, R. E., & Rock, D. A. (1995). Generalizability, validity, and examinee perceptions of a computer-delivered formulating-hypothesis test. *Journal of Educational Measurement, 32*, 19–36. https://doi.org/10.1111/j.1745-3984.1995.tb00454.x

Bennett, R. E., & Sebrechts, M. M. (1996). The accuracy of expert-system diagnoses of mathematical problem solutions. *Applied Measurement in Education, 9*, 133–150. https://doi.org/10.1207/s15324818ame0902_3

Bennett, R. E., & Sebrechts, M. M. (1997). A computer-based task for measuring the representational component of quantitative proficiency. *Journal of Educational Measurement, 34*, 64–77. https://doi.org/10.1111/j.1745-3984.1997.tb00507.x

Bennett, R. E., & Ward, W. C. (Eds.). (1993). *Construction versus choice in cognitive measurement*. Hillsdale: Erlbaum.

Bennett, R. E., & Wadkins, J. R. J. (1995). Interactive performance assessment in computer science: The Advanced Placement Computer Science (APCS) Practice System. *Journal of Educational Computing Research, 12*(4), 363–378. https://doi.org/10.2190/5WQA-JB0J-1CM5-4H50

Bennett, R. E., Rock, D. A., Braun, H. I., Frye, D., Spohrer, J. C., & Soloway, E. (1990). The relationship of expert-system scored constrained free-response items to multiple-choice and open-ended items. *Applied Psychological Measurement, 14*, 151–162. https://doi.org/10.1177/014662169001400204

Bennett, R. E., Rock, D. A., & Wang, M. (1991). Equivalence of free-response and multiple-choice items. *Journal of Educational Measurement, 28*, 77–92. https://doi.org/10.1111/j.1745-3984.1991.tb00345.x

Bennett, R. E., Steffen, M., Singley, M. K., Morley, M., & Jacquemin, D. (1997). Evaluating an automatically scorable, open-ended response type for measuring mathematical reasoning in computer-adaptive tests. *Journal of Educational Measurement, 34*, 162–176. https://doi.org/10.1111/j.1745-3984.1997.tb00512.x

Bennett, R. E., Morley, M., Quardt, D., Rock, D. A., Singley, M. K., Katz, I. R., & Nhouyvanishvong, A. (1999). Psychometric and cognitive functioning of an under-determined computer-based response type for quantitative reasoning. *Journal of Educational Measurement, 36*, 233–252. https://doi.org/10.1111/j.1745-3984.1999.tb00556.x

Bennett, R. E., Morley, M., & Quardt, D. (2000a). Three response types for broadening the conception of mathematical problem solving in computerized tests. *Applied Psychological Measurement, 24*, 294–309. https://doi.org/10.1177/01466210022031769

Bennett, R. E., Morley, M., Quardt, D., & Rock, D. A. (2000b). Graphical modeling: A new response type for measuring the qualitative component of mathematical reasoning. *Applied Measurement in Education, 13*(3), 303–322. https://doi.org/10.1207/S15324818AME1303_5

Bennett, R. E., Persky, H., Weiss, A. R., & Jenkins, F. (2010). Measuring problem solving with technology: A demonstration study for NAEP. *Journal of Technology, Learning and Assessment, 8*(8). Retrieved from http://escholarship.bc.edu/jtla/vol8/8

Bernstein, J., de Jong, J., Pisoni, D., & Townshend, D. (2000). Two experiments in automatic scoring of spoken language proficiency. In P. Delcloque (Ed.), *Proceedings of the InSTIL2000: Integrating speech technology in learning* (pp. 57–61). Dundee: University of Abertay.

Bill and Melinda Gates Foundation. (2013). *Ensuring fair and reliable measures of effective teaching: Culminating findings from the MET project's three-year study*. Seattle: Author.

Birnbaum, A. (1968). Some latent trait models and their use in inferring an examinee's ability. In F. M. Lord & M. R. Novick (Eds.), *Statistical theories of mental test scores* (pp. 395–479). Reading: Addison-Wesley.

Bock, R. D., & Mislevy, R. J. (1988). Comprehensive educational assessment for the states: The Duplex Design. *Educational Evaluation and Policy Analysis, 10*(2), 89–105. https://doi.org/10.1177/014662102237794

Bock, R. D., Brennan, R. L., & Muraki, E. (2002). The information in multiple ratings. *Applied Psychological Measurement, 26*, 364–375.

Bourque, M. L. (2009). *A history of NAEP achievement levels: Issues, implementation, and impact 1989–2009*. Washington, DC: National Assessment Governing Board.

Braswell, J. S. (1992). Changes in the SAT in 1994. *Mathematics Teacher, 85*(1), 16–21.

Braswell, J., & Kupin, J. (1993). Item formats for assessment in mathematics. In R. E. Bennett & W. C. Ward (Eds.), *Construction versus choice in cognitive measurement* (pp. 167–182). Hillsdale: Erlbaum.

Braun, H. I. (1988). Understanding scoring reliability: Experiments in calibrating essay readers. *Journal of Educational Statistics, 13*(1), 1–18. https://doi.org/10.2307/1164948

Braun, H. I., Bejar, I. I., & Williamson, D. M. (2006). Rule-based methods for automated scoring: Applications in a licensing context. In D. M. Williamson, R. J. Mislevy, & I. I. Bejar (Eds.), *Automated scoring of complex tasks in computer-based testing* (pp. 83–122). Mahwah: Erlbaum.

Breland, H. M. (1983). *The direct assessment of writing skill: A measurement review* (College Board Report No. 83-6). Princeton: Educational Testing Service.

Breland, H. M., & Hart, F. M. (1994). *Defining legal writing: An empirical analysis of the legal memorandum* (Research Report No. 93-06). Newtown: Law School Admission Council.

Breland, H. M., & Jones, R. J. (1988). *Remote scoring of essays* (College Board Report No. 88-03). New York: College Entrance Examination Board.

Breland, H. M., Grandy, J., Rock, D., & Young, J. W. (1984). *Linear models of writing assessments* (Research Memorandum No. RM-84-02). Princeton: Educational Testing Service.

Breland, H. M., Camp, R., Jones, R. J., Morris, M. M., & Rock, D. A. (1987). *Assessing writing skill*. New York: College Entrance Examination Board.

Breland, H. M., Danos, D. O., Kahn, H. D., Kubota, M. Y., & Bonner, M. W. (1994). Performance versus objective testing and gender: An exploratory study of an Advanced Placement history examination. *Journal of Educational Measurement, 31*, 275–293. https://doi.org/10.1111/j.1745-3984.1994.tb00447.x

Breland, H. M., Bridgeman, B., & Fowles, M. (1999). *Writing assessment in admission to higher education: Review and framework* (Research Report No. RR-99-03). Princeton: Educational Testing Service. http://dx.doi.org/10.1002/j.2333-8504.1999.tb01801.x

Brennan, R. L. (2001). *Generalizability theory*. New York: Springer. https://doi.org/10.1007/978-1-4757-3456-0

Bridgeman, B. (1989). *Comparative validity of multiple-choice and free-response items on the Advanced Placement examination in biology* (Research Report No. RR-89-01). Princeton: Educational Testing Service. http://dx.doi.org/10.1002/j.2330-8516.1989.tb00327.x

Bridgeman, B. (1992). A comparison of quantitative questions in open-ended and multiple-choice formats. *Journal of Educational Measurement, 29*, 253–271. https://doi.org/10.1111/j.1745-3984.1992.tb00377.x

Bridgeman, B. (2016). Can a two-question test be reliable and valid for predicting academic outcomes? *Educational Measurement: Issues and Practice, 35*(4), 21–24. https://doi.org/10.1111/emip.12130

Bridgeman, B., & Cline, F. (2004). Effects of differentially time-consuming tests on computer-adaptive test scores. *Journal of Educational Measurement, 41*, 137–148. https://doi.org/10.1111/j.1745-3984.2004.tb01111.x

Bridgeman, B., & Lewis, C. (1994). The relationship of essay and multiple-choice scores with grades in college courses. *Journal of Educational Measurement, 31*, 37–50. https://doi.org/10.1111/j.1745-3984.1994.tb00433.x

Bridgeman, B., & McHale, F. J. (1996). *Gender and ethnic group differences on the GMAT Analytical Writing Assessment* (Research Report No. RR-96-02). Princeton: Educational Testing Service. http://dx.doi.org/10.1002/j.2333-8504.1996.tb01680.x

Bridgeman, B., & Rock, D. A. (1993). Relationships among multiple-choice and open-ended analytical questions. *Journal of Educational Measurement, 30*, 313–329.

Bridgeman, B., Morgan, R., & Wang, M.-M. (1996a). *Reliability of Advanced Placement examinations* (Research Report No. RR-96-03). Princeton: Educational Testing Service. http://dx.doi.org/10.1002/j.2333-8504.1996.tb01681.x

Bridgeman, B., Morgan, R., & Wang, M.-M. (1996b). *The reliability of document-based essay questions on Advanced Placement history examinations* (Research Report No. RR-96-05). Princeton: Educational Testing Service. http://dx.doi.org/10.1002/j.2333-8504.1996.tb01683.x

Bridgeman, B., Morgan, R., & Wang, M. (1997). Choice among essay topics: Impact on performance and validity. *Journal of Educational Measurement, 34*, 273–286. https://doi.org/10.1111/j.1745-3984.1997.tb00519.x

Bridgeman, B., Trapani, C., & Bivens-Tatum, J. (2011). Comparability of essay question variants. *Assessing Writing, 16*(4), 237–255. https://doi.org/10.1016/j.asw.2011.06.002

Bridgeman, B., Powers, D. E., Stone, E., & Mollaun, P. (2012). TOEFL iBT speaking test scores as indicators of oral communicative language proficiency. *Language Testing, 29*(1), 91–108. https://doi.org/10.1177/0265532211411078

Briggs, D. C., & Wilson, M. (2007). Generalizability in item response modeling. *Journal of Educational Measurement, 44*, 131–155. https://doi.org/10.1111/j.1745-3984.2007.00031.x

Brunswik, E. (1952). The conceptual framework of psychology. In *International encyclopedia of unified science* (Vol. 1, No. 10). Chicago: University of Chicago Press.

Burke, V. M. (1961). A candid opinion on lay readers. *The English Journal, 50*(4), 258–264. https://doi.org/10.2307/810913
Burstein, J., Kaplan, R., Wolff, S., & Lu, C. (1997). *Automatic scoring of Advanced Placement Biology essays* (Research Report No. RR-97-22). Princeton: Educational Testing Service. http://dx.doi.org/10.1002/j.2333-8504.1997.tb01743.x
Burstein, J., Braden-Harder, L., Chodorow, M., Hua, S., Kaplan, B., Kukich, K., et al. (1998). *Computer analysis of essay content for automated score prediction: A prototype automated scoring system for GMAT Analytical Writing Assessment essays* (Research Report No. RR-98-15). Princeton: Educational Testing Service. http://dx.doi.org/10.1002/j.2333-8504.1998.tb01764.x
Burstein, J., Wolff, S., & Lu, C. (1999). Using lexical semantic techniques to classify free responses. In N. Ide & J. Veronis (Eds.), *The depth and breadth of semantic lexicons* (pp. 227–244). Dordrecht: Kluwer Academic. https://doi.org/10.1007/978-94-017-0952-1_11
Burstein, J., Chodorow, M., & Leacock, C. (2004). Automated essay evaluation. *AI Magazine, 25*(3), 27–36.
Burstein, J., Tetreault, J., & Madnani, N. (2013). The e-rater automated essay scoring system. In M. D. Shermis & J. Burstein (Eds.), *Handbook of automated essay evaluation: Current application and new directions* (pp. 55–67). New York: Routledge.
Butler, F. A., Eignor, D., Jones, S., McNamara, T., & Suomi, B. K. (2000). *TOEFL 2000 speaking framework: A working paper* (Research Memorandum No. RM-00-06). Princeton: Educational Testing Service.
Cameron, C. A., Beemsterboer, P. L., Johnson, L. A., Mislevy, R. J., Steinberg, L. S., & Breyer, F. J. (2000). A cognitive task analysis for dental hygiene. *Journal of Dental Education, 64*(5), 333–351.
Camp, R. (1985). The writing folder in post-secondary assessment. In P. J. A. Evans (Ed.), *Directions and misdirections in English evaluation* (pp. 91–99). Ottawa: Canadian Council of Teachers of English.
Camp, R. (1993). The place of portfolios in our changing views of writing assessment. In R. E. Bennett & W. C. Ward (Eds.), *Construction versus choice in cognitive measurement* (pp. 183–212). Hillsdale: Erlbaum.
Campbell, D. T., & Fiske, D. W. (1959). Convergent and discriminant validation by the multitrait-multimethod matrix. *Psychological Bulletin, 56*, 81–105. https://doi.org/10.1037/h0046016
Campbell, J. R., Donahue, P. L., Reese, C. M., & Phillips, G. W. (1996). *NAEP 1994 Reading report card for the nation and the states*. Washington, DC: National Center for Education Statistics.
Canale, M., & Swain, M. (1980). Theoretical bases of communicative approaches to second language teaching and testing. *Applied Linguistics, 1*, 1–47. https://doi.org/10.1093/applin/1.1.1
Chalifour, C., & Powers, D. E. (1989). The relationship of content characteristics of GRE analytical reasoning items to their difficulties and discriminations. *Journal of Educational Measurement, 26*, 120–132. https://doi.org/10.1111/j.1745-3984.1989.tb00323.x
Chopra, A. (2013). *The dark side of innovation. St.* Johnsbury: Brigantine.
Christensen, C. M. (1997). *The innovator's dilemma: When new technologies cause great firms to fail.* Boston: Harvard Business School Press.
Cizek, G. J., & Bunch, M. B. (2007). *Standard setting: A guide to establishing and evaluating performance standards on tests*. Thousand Oaks: Sage. https://doi.org/10.4135/9781412985918
Clark, J. L. D., & Swinton, S. S. (1980). *The Test of Spoken English as a measure of communicative ability in English-medium instructional settings* (TOEFL Research Report No. 07). Princeton: Educational Testing Service. http://dx.doi.org/10.1002/j.2333-8504.1980.tb01230.x
Clauser, B. E. (2000). Recurrent issues and recent advances in scoring performance assessments. *Applied Psychological Measurement, 24*, 310–324. https://doi.org/10.1177/01466210022031778
Clauser, B. E., Kane, M. T., & Swanson, D. B. (2002). Validity issues for performance-based tests scored with computer-automated scoring systems. *Applied Measurement in Education, 15*(4), 413–432.

Cleary, A. T. (1966). *Test bias: Validity of the Scholastic Aptitude Test for Negro and White students in integrated colleges* (Research Bulletin No. RB-66-31). Princeton: Educational Testing Service. http://dx.doi.org/10.1002/j.2333-8504.1966.tb00529.x

Clyman, S. G., Melnick, D. E., & Clauser, B. E. (1995). Computer-based simulations. In E. L. Mancall & E. G. Bashook (Eds.), *Assessing clinical reasoning: The oral examination and alternative methods* (pp. 139–149). Evanston: American Board of Medical Specialties.

Coffman, W. E. (1971). Essay examinations. In R. L. Thorndike (Ed.), *Educational measurement* (2nd ed., pp. 271–302). Washington, DC: American Council on Education. https://doi.org/10.3102/00028312005001099

Coffman, W. E., & Kurfman, D. G. (1966). *Single score versus multiple score reading of the American History Advanced Placement examination (Research Bulletin No. RB-66-22).* Princeton: Educational Testing Service. http://dx.doi.org/10.1002/j.2333-8504.1966.tb00359.x

Coffman, W. E., & Kurfman, D. G. (1968). A comparison of two methods of reading essay examinations. *American Educational Research Journal, 5*(1), 99–107.

College Entrance Examination Board. (2011). *AP World History: Course and exam description.* New York: Author.

College Entrance Examination Board. (1942). *Forty-second annual report of the Executive Secretary*. New York: Author.

Committee on the Objectives of a General Education in a Free Society. (1945). *General education in a free society: Report of the Harvard committee*. Cambridge, MA: Harvard University Press.

Cooper, C. R. (1977). Holistic evaluation of writing. In C. R. Cooper & L. Odell (Eds.), *Evaluating writing* (pp. 3–33). Urbana: National Council of Teachers of English.

Coward, A. F. (1950). *The method of reading the Foreign Service Examination in English composition* (Research Bulletin No. RB-50-57). Princeton: Educational Testing Service. http://dx.doi.org/10.1002/j.2333-8504.1950.tb00680.x

Coward, A. F. (1952). The comparison of two methods of grading English compositions. *Journal of Educational Research, 46*(2), 81–93. https://doi.org/10.1080/00220671.1952.10882003

Cowden, D. J. (1946). An application of sequential sampling to testing students. *Journal of the American Statistical Association, 41*(236), 547–556. https://doi.org/10.1080/01621459.1946.10501897

Crisp, V. (2010). Towards a model of the judgment processes involved in examination marking. *Oxford Review of Education, 36*, 1–21. https://doi.org/10.1080/03054980903454181

Cronbach, L. J. (1971). Test validation. In R. L. Thorndike (Ed.), *Educational measurement* (2nd ed., pp. 443–507). Washington, DC: American Council on Education.

Cronbach, L. J. (1980). Validity on parole: How can we go straight? In W. B. Schrader (Ed.), *New directions for testing and measurement: Measuring achievement over a decade (Proceedings of the 1979 ETS Invitational Conference)* (pp. 99–108). San Francisco: Jossey-Bass.

Cronbach, L. J., & Meehl, P. E. (1955). Construct validity in psychological tests. *Psychological Bulletin, 52*, 281–302. https://doi.org/10.1037/h0040957

Cronbach, L. J., Gleser, G. C., Nanda, H., & Rajaratnam, N. (1972). *The dependability of behavioral measurements: Theory of generalizability for scores and profiles*. New York: John Wiley.

Cronbach, L. J., Linn, R. L., Brennan, R. L., & Haertel, E. (1995). Generalizability analysis for educational assessments. *Educational and Psychological Measurement, 57*(3), 373–399. https://doi.org/10.1177/0013164497057003001

Croon Davis, C., Stalnaker, J. M., & Zahner, L. C. (1959). Review of the Sequential Tests of Educational Progress: Writing. In O. K. Buros (Ed.), *The fifth mental measurement yearbook.* Lincoln: Buros Institute of Mental Measurement.

Cumming, A., Kantor, R., Powers, D. E., Santos, T., & Taylor, C. (2000). *TOEFL 2000 writing framework: A working paper* (TOEFL Monograph Series No. 18). Princeton: Educational Testing Service.

Cumming, A., Kantor, R., & Powers, D. E. (2002). Decision making while rating ESL/EFL writing tasks: A descriptive framework. *The Modern Language Journal, 86*(1), 67–96. https://doi.org/10.1111/1540-4781.00137
D'Angelo, F. (1984). Nineteenth-century forms/modes of discourse: A critical inquiry. *College Composition and Communication, 35*(1), 31–42. https://doi.org/10.2307/357678
Dawes, R. M., & Corrigan, B. (1974). Linear models in decision making. *Psychological Bulletin, 81*, 95–106. https://doi.org/10.1037/h0037613
Deane, P. (2006). Strategies for evidence identification through linguistic assessment of textual responses. In D. M. Williamson, R. J. Mislevy, & I. I. Bejar (Eds.), *Automated scoring of complex tasks in computer-based testing* (pp. 313–371). Mahwah: Erlbaum.
Deane, P. (2012). Rethinking K–12 writing assessment. In N. Elliot & L. Perelman (Eds.), *Writing assessment in the 21st century: Essays in honor of Edward M. White* (pp. 87–100). New York: Hampton Press.
Deane, P. (2013a). Covering the construct: An approach to automated essay scoring motivated by a socio-cognitive framework for defining literacy skills. In M. D. Shermis & J. Burstein (Eds.), *Handbook of automated essay evaluation: Current application and new directions* (pp. 298–312). New York: Routledge.
Deane, P. (2013b). On the relation between automated essay scoring and modern views of the writing construct. *Assessing Writing, 18*(1), 7–24. https://doi.org/10.1016/j.asw.2012.10.002
Deane, P., & Quinlan, T. (2010). What automated analyses of corpora can tell us about students' writing skills. *Journal of Writing Research, 2*(2), 151–177. https://doi.org/10.17239/jowr-2010.02.02.4
Deane, P., Odendahl, N., Quinlan, T., Fowles, M., Welsh, C., & Bivens-Tatum, J. (2008). *Cognitive models of writing: Writing proficiency as a complex integrated skill* (Research Report No. RR-08-55). Princeton: Educational Testing Service. http://dx.doi.org/10.1002/j.2333-8504.2008.tb02141.x
Deane, P., Sabatini, J., & Fowles, M. (2012). Rethinking k-12 writing assessment to support best instructional practices. In C. Bazerman, C. Dean, J. Early, K. Lunsford, S. Null, P. Rogers, & A. Stansell (Eds.), *International advances in writing research: Cultures, places, measures* (pp. 81–102). Anderson: The WAC Clearinghouse and Parlor Press.
De Lamartine, A. (1849). *History of the French Revolution of 1848*. London: Bohn.
Diederich, P. B. (1957). *The improvement of essay examinations* (Research Memorandum No. RM-57-03). Princeton: Educational Testing Service.
Diederich, P. B., French, J. W., & Carlton, S. T. (1961). *Factors in judgments of writing ability* (Research Bulletin No. RB-61-15). Princeton: Educational Testing Service. http://dx.doi.org/10.1002/j.2333-8504.1961.tb00285.x
Dorans, N. J. (2004). Using subpopulation invariance to assess test score equity. *Journal of Educational Measurement, 41*, 43–68. https://doi.org/10.1111/j.1745-3984.2004.tb01158.x
Dorans, N. J., & Schmitt, A. P. (1993). Constructed response and differential item functioning. In R. E. Bennett & W. C. Ward (Eds.), *Construction versus choice in cognitive measurement: Issues in constructed response, performance testing, and portfolio assessment* (pp. 135–166). Hillsdale: Erlbaum.
Dorans, N. J., Holland, P. W., Thayer, D. T., & Tateneni, K. (2003). Invariance of score linking across gender groups for three Advanced Placement Program examinations. In N. J. Dorans (Ed.), *Population invariance of score linking: Theory and applications to Advanced Placement program examinations* (Research Report No. RR-03-27; pp. 79–118). Princeton: Educational Testing Service. http://dx.doi.org/10.1002/j.2333-8504.2003.tb01919.x
Dorans, N. J., Holland, P., & Pommerich, M. (Eds.). (2007). *Linking and aligning scores and scales*. New York: Springer. https://doi.org/10.1007/978-0-387-49771-6
Dougherty, D., & Hardy, C. (1996). Sustained product innovation in large, mature organizations: Overcoming innovation-to-organization problems. *Academy of Management Journal, 39*(5), 1120–1153. https://doi.org/10.2307/256994

Downey, M. T. (1965). *Ben D. Wood, educational reformer*. Princeton: Educational Testing Service.

Driscoll, G., Hatfield, L. A., Johnson, A. A., Kahn, H. D., Kessler, T. E., Kuntz, D., et al. (1999). *U.S. Patent No. 5,987,302*. Washington, DC: U.S. Patent and Trademark Office.

Du Bois, P. H. (1970). *A history of psychological testing*. Boston: Allyn and Bacon.

Duran, R. P., Canale, M., Penfield, J., Stansfield, C. W., & Liskin-Gasparro, J. E. (1987). TOEFL from a communicative viewpoint on language proficiency: A working paper. In R. Freedle & R. P. Duran (Eds.), *Cognitive and linguistic analyses of test performance* (pp. 1–150). Norwood: Ablex.

Dwyer, C. A. (1994). *Development of the knowledge base for the PRAXIS III: Classroom performance assessments assessment criteria*. Princeton: Educational Testing Service.

Dyer, H. S. (1954, January 5). *A common philosophy for the Test of Developed Ability*. Princeton: Educational Testing Service.

Ebel, R. L. (1951). Estimation of the reliability of ratings. *Psychometrika, 16*, 407–424. https://doi.org/10.1007/BF02288803

Ebel, R. L. (1962). Content standard test scores. *Educational and Psychological Measurement, 22*(1), 15–25.

Edgeworth, F. Y. (1890). The element of chance in competitive examinations. *Journal of the Royal Statistical Society, 53*(4), 460–475.

Elbow, P., & Yancey, K. B. (1994). On the nature of holistic scoring: An inquiry composed on e-mail. *Assessing Writing, 1*(1), 91–107. https://doi.org/10.1016/1075-2935(94)90006-X

Elliot, N. (2005). *On a scale: A social history of writing assessment in America*. New York: Lang.

Elliot, N. (2014). *Henry Chauncey: An American life*. New York: Lang.

Elliot, T. J. (2014). Escaping gravity: Three kinds of knowledge as fuel for innovation in an operate and maintain company. In K. Pugh (Ed.), *Smarter innovation: Using interactive processes to drive better business results* (pp. 3–12). Peoria: Ark Group.

Embretson, S. E. (1983). Construct validity: Construct representation versus nomothetic span. *Psychological Bulletin, 93*, 179–197. https://doi.org/10.1037/0033-2909.93.1.179

Enright, M. K., & Gitomer, D. (1989). *Toward a description of successful graduate students* (Research Report No. RR-89-09). Princeton: Educational Testing Service. http://dx.doi.org/10.1002/j.2330-8516.1989.tb00335.x

Enright, M. K., & Powers, D. E. (1991). *Validating the GRE Analytical Ability Measure against faculty ratings of analytical reasoning skills* (Research Report No. RR-90-22). Princeton: Educational Testing Service. http://dx.doi.org/10.1002/j.2333-8504.1990.tb01358.x

Enright, M. K., Rock, D. A., & Bennett, R. E. (1998). Improving measurement for graduate admissions. *Journal of Educational Measurement, 35*, 250–267. https://doi.org/10.1111/j.1745-3984.1998.tb00538.x

Enright, M. K., Grabe, W., Mosenthal, P., Mulcahy-Ernt, P., & Schedl, M. (2000). *TOEFL 2000 reading framework: A working paper* (TOEFL Monograph Series No. 17). Princeton: Educational Testing Service.

Evanini, K., Xie, S., & Zechner, K. (2013). *Prompt-based content scoring for automated spoken language assessment*. Paper presented at the Eighth Workshop on the Innovative Use of NLP for Building Educational Applications, Atlanta.

Fife, J. H. (2013). *Automated scoring of mathematics tasks in the Common Core Era: Enhancements to m-rater™ in support of CBAL mathematics and the Common Core Assessments* (Research Report No. RR-13-26). Princeton: Educational Testing Service. http://dx.doi.org/10.1002/j.2333-8504.2013.tb02333.x

Finlayson, D. S. (1951). The reliability of the marking of essays. *British Journal of Educational Psychology, 21*, 126–134. https://doi.org/10.1111/j.2044-8279.1951.tb02776.x

Finley, C., & Berdie, F. S. (1970). *The National Assessment approach to exercise development*. Ann Arbor, MI: National Assessment of Educational Progress. Retrieved from ERIC database. (ED 067402)

Foster, R. N. (1986). *Innovation: The attacker's advantage*. New York: Summit Books. https://doi.org/10.1007/978-3-322-83742-4

Fowles, M. (2012). Writing assessments for admission to graduate and professional programs: Lessons learned and a note for the future. In N. Elliot & L. Perelman (Eds.), *Writing assessment in the 21st century: Essays in honor of Edward M. White* (pp. 135–148). New York: Hampton Press.

Frederiksen, N. (1984). The real test bias: Influences of testing on teaching and learning. *American Psychologist, 39*, 193–202. https://doi.org/10.1037/0003-066X.39.3.193

Frederiksen, N., Saunders, D. R., & Wand, B. (1957). The in-basket test. *Psychological Monographs, 71*(438). https://doi.org/10.1037/h0093706

Freedman, S. W., & Calfee, R. C. (1983). Holistic assessment of writing: Experimental design and cognitive theory. In P. Mosenthal, L. Tamor, & S. A. Walmsley (Eds.), *Research on writing: Principles and methods* (pp. 75–98). New York: Longman.

Gallagher, A., Bennett, R. E., Cahalan, C., & Rock, D. A. (2002). Validity and fairness in technology-based assessment: Detecting construct-irrelevant variance in an open-ended, computerized mathematics task. *Educational Assessment, 8*(1), 27–41. https://doi.org/10.1207/S15326977EA0801_02

General Education in School and College. (1952). *A committee report by members of the faculties of Andover, Exeter, Lawrenceville, Harvard, Princeton, and Yale*. Cambridge: Harvard University Press.

Gentile, C. A. (1992). *NAEP's 1990 pilot portfolio study: Exploring methods for collecting students' school-based writing*. Washington, DC: National Center for Education Statistics.

Gigerenzer, G., & Goldstein, D. G. (1996). Reasoning the fast and frugal way: Models of bounded rationality. *Psychological Review, 103*(4), 650–669. https://doi.org/10.1037/0033-295X.103.4.650

Gitomer, D. H. (2007). *The impact of the National Board for Professional Teaching Standards: A review of the research* (Research Report No. RR-07-33). Princeton: Educational Testing Service. http://dx.doi.org/10.1002/j.2333-8504.2007.tb02075.x

Glaser, R. (1963). Instructional technology and the measurement of learning outcomes: Some questions. *American Psychologist, 18*, 519–521. https://doi.org/10.1037/h0049294

Glaser, R. (1994). Criterion-referenced tests: Part I. Origins. *Educational Measurement: Issues and Practice, 13*(4), 9–11. https://doi.org/10.1111/j.1745-3992.1994.tb00562.x

Godshalk, F. I., Swineford, F., & Coffman, W. E. (1966). *The measurement of writing ability*. New York: College Entrance Examination Board.

Golub-Smith, M., Reese, C., & Steinhaus, K. (1993). *Topic and topic type comparability on the Test of Written English* (TOEFL Research Report No. 42). Princeton: Educational Testing Service. http://dx.doi.org/10.1002/j.2333-8504.1993.tb01521.x

Graf, E. A. (2009). *Defining mathematics competency in the service of cognitively based assessment for Grades 6 through 8* (Research Report No. RR-09-42). Princeton: Educational Testing Service. http://dx.doi.org/10.1002/j.2333-8504.2009.tb02199.x

Graf, E. A., & Fife, J. (2012). Difficulty modeling and automatic item generation of quantitative items: Recent advances and possible next steps. In M. Gierl & S. Haladyna (Eds.), *Automated item generation: Theory and practice* (pp. 157–180). New York: Routledge.

Gulliksen, H. (1950). *Theory of mental tests*. New York: John Wiley. https://doi.org/10.1037/13240-000

Guion, R. M. (1980). On Trinitarian doctrines of validity. *Professional Psychology-Research and Practice, 11*(3), 385–398. https://doi.org/10.1037/0735-7028.11.3.385

Haertel, E. H. (2006). Reliability. In R. L. Brennan (Ed.), *Educational measurement* (4th ed., pp. 65–110). Westport: American Council on Education and Praeger.

Hambleton, R. K., & Novick, M. R. (1973). Toward an integration of theory and method for criterion referenced tests. *Journal of Educational Measurement, 10*, 159–170. https://doi.org/10.1111/j.1745-3984.1973.tb00793.x

Hambleton, R. K., & Pitoniak, M. (2006). Setting performance standards. In R. L. Brennan (Ed.), *Educational measurement* (4th ed., pp. 433–470). Westport: American Council on Education and Praeger.

Hammond, K. R., & Stewart, T. R. (2001). *The essential Brunswik: Beginnings, explications, applications*. New York: Oxford University Press.

Hammond, K. R., Hursch, C. J., & Todd, F. J. (1964). Analyzing the components of clinical inference. *Psychological Review, 71*(6), 438–456. https://doi.org/10.1037/h0040736

Hamp-Lyons, L., & Kroll, B. (1997). *TOEFL 2000—Writing: Composition, community, and assessment* (TOEFL Monograph Series No. MS-05). Princeton: Educational Testing Service.

Hansen, E. G., & Mislevy, R. J. (2008). *Design patterns for improving accessibility for test takers with disabilities* (Research Report No. RR-08-49). Princeton: Educational Testing Service. http://dx.doi.org/10.1002/j.2333-8504.2008.tb02135.x

Heller, J. I., Sheingold, K., & Myford, C. M. (1998). Reasoning about evidence in portfolios: Cognitive foundations for valid and reliable assessment. *Educational Assessment, 5*(1), 5–40. https://doi.org/10.1207/s15326977ea0501_1

Hemphill, J. K., Griffiths, D. E., & Frederiksen, N. (1962). *Administrative performance and personality: A study of the principal in a simulated elementary school*. New York: Teachers College.

Hick, W. E. (1951). Information theory and intelligence tests. *British Journal of Psychology 4(3)*, 157–164.

Hines, S. (2010). *Evidence-centered design: The TOEIC speaking and writing tests* (TOEIC Compendium Study No. 7). Princeton: Educational Testing Service.

Horkay, N., Bennett, R. E., Allen, N., Kaplan, B., & Yan, F. (2006). Does it matter if I take my writing test on computer? An empirical study of mode effects in NAEP. *Journal of Technology, Learning, and Assessment, 5*(2), 1–49.

Howell, J., & Avolio, B. J. (1993). Transformational leadership, transactional leadership, locus of control, and support for innovation: Key predictors of consolidated-business-unit performance. *Journal of Applied Psychology, 78*(6), 891–902. https://doi.org/10.1037/0021-9010.78.6.891

Hubin, D. R. (1988). *The Scholastic Aptitude Test: Its development and introduction, 1900–1947*. Retrieved from http://darkwing.uoregon.edu/~hubin/

Huddleston, E. M. (1954). The measurement of writing ability at the college-entrance level: Objective vs. subjective testing techniques. *Journal of Experimental Education, 22*(3), 165–207. https://doi.org/10.1080/00220973.1954.11010477

Huff, K., & Plake, B. S. (2010). Evidence-centered assessment design in practice. *Applied Measurement in Education, 23*(4), 307–309. https://doi.org/10.1080/08957347.2010.510955

Huff, K., Alves, C., Pellegrino, J., & Kaliski, P. (2012). Using evidence-centered design task models in automatic item generation. In M. Gierl & S. Haladyna (Eds.), *Automated item generation: Theory and practice* (pp. 102–118). New York: Routledge.

Huot, B. (1990). Reliability, validity, and holistic scoring: What we know and what we need to know. *College Composition and Communication, 41*(2), 201–213. https://doi.org/10.2307/358160

Huot, B., & Neal, M. (2006). Writing assessment: A techno-history. In C. A. MacArthur, S. Graham, & J. Fitzgerald (Eds.), *Handbook of writing research* (pp. 417–432). New York: Guilford Press.

Hutt, M. L. (1947). A clinical study of "consecutive" and "adaptive" testing with the revised Stanford-Binet. *Journal of Consulting Psychology, 11*, 93–103. https://doi.org/10.1037/h0056579

Joe, J. N., Park, Y. S., Brantley, W., Lapp, M., & Leusner, D. (2012, April). *Examining the effect of prompt complexity on rater behavior: A mixed-methods study of GRE Analytical Writing Measure Argument prompts*. Paper presented at the annual meeting of the National Council of Measurement in Education, Vancouver, British Columbia, Canada.

Jones, V., & Brown, R. H. (1935). Educational tests. *Psychological Bulletin, 32*(7), 473–499. https://doi.org/10.1037/h0057574

Kane, M. T. (1982). A sampling model for validity. *Applied Psychological Measurement, 6*, 125–160. https://doi.org/10.1177/014662168200600201

Kane, M. T. (2006). Validation. In R. L. Brennan (Ed.), *Educational measurement* (4th ed., pp. 17–64). Westport: American Council on Education and Praeger.

Kaplan, R. (1992). *Using a trainable pattern-directed computer program to score natural language item responses* (Research Report No. RR-91-31). Princeton: Educational Testing Service. http://dx.doi.org/10.1002/j.2333-8504.1991.tb01398.x.

Kaplan, R. M., & Bennett, R. E. (1994). *Using the free-response scoring tool to automatically score the formulating-hypotheses item* (Research Report No. RR-94-08). Princeton: Educational Testing Service. http://dx.doi.org/10.1002/j.2333-8504.1994.tb01581.x

Kaplan, R. M., Burstein, J., Trenholm, H., Lu, C., Rock, D., Kaplan, B., & Wolff, C. (1995). *Evaluating a prototype essay scoring procedure using off-the shelf software* (Research Report No. RR-95-21). Princeton: Educational Testing Service. https://dx.doi.org/10.1002/j.2333-8504.1995.tb01656.x

Katz, I. R., Martinez, M. E., Sheehan, K. M., & Tatsuoka, K. K. (1998). Extending the rule space methodology to a semantically-rich domain: Diagnostic assessment in architecture. *Journal of Educational and Behavioral Statistics, 23*(3), 254–278. https://doi.org/10.3102/10769986023003254

Katz, I. R., Bennett, R. E., & Berger, A. E. (2000). Effects of response format on difficulty of SAT-Mathematics items: It's not the strategy. *Journal of Educational Measurement, 37*, 39–57. https://doi.org/10.1111/j.1745-3984.2000.tb01075.x

Kenney, J. F. (1997). New testing methodologies for the Architect Registration Examination. *CLEAR Exam Review, 8*(2), 23–28.

Kim, S., & Lee, W.-C. (2006). An extension of four IRT linking methods for mixed-format tests. *Journal of Educational Measurement, 43*, 53–76. https://doi.org/10.1111/j.1745-3984.2006.00004.x

Kim, S., & Walker, M. (2012). Determining the anchor composition for a mixed-format test: Evaluation of subpopulation invariance of linking functions. *Applied Measurement in Education, 25*(2), 178–195. https://doi.org/10.1080/08957347.2010.524720

Kim, S., Walker, M. E., & McHale, F. J. (2010). Comparisons among designs for equating mixed-format tests in large-scale assessments. *Journal of Educational Measurement, 47*, 36–53. https://doi.org/10.1111/j.1745-3984.2009.00098.x

Kirsch, I., Braun, H. I., Yamamoto, K., & Sum, A. (2007). *America's perfect storm: Three forces changing our nation's future*. Princeton: Educational Testing Service.

Koenig, J. A. (2011). *Assessing 21st century skills: Summary of a workshop*. Washington, DC: National Academies Press.

Koretz, D., & Hamilton, L. S. (2006). Testing for accountability in K–12. In R. L. Brennan (Ed.), *Educational measurement* (4th ed., pp. 531–578). Westport: American Council on Education and Praeger.

Kuntz, D., Cody, P., Ivanov, G., & Perlow, J. E. (2006). *U.S. Patent No. 8,374,540 B2*. Washington, DC: Patent and Trademark Office.

Lacy, T. (2010). Examining AP: Access, rigor, and revenue in the history of the Advanced Placement program. In D. R. Sadler, G. Sonnert, R. H. Tai, & K. Klopfenstein (Eds.), *AP: A critical examination of the Advanced Placement program* (pp. 17–48). Cambridge, MA: Harvard Education Press.

Lane, E. S., & Stone, C. A. (2006). Performance assessment. In R. L. Brennan (Ed.), *Educational measurement* (4th ed., pp. 387–431). Westport: American Council on Education and Praeger.

Lawrence, I., Rigol, G., Van Essen, T., & Jackson, C. A. (2002). *A historical perspective on the SAT 1926–2001* (Research Report No. 2002-7). New York: College Entrance Examination Board.

Leacock, C., & Chodorow, M. (2003). c-rater: Scoring of short-answer questions. *Computers and the Humanities, 37*(4), 389–405. https://doi.org/10.1023/A:1025779619903

Lemann, N. (1999). *The big test: The secret history of the American meritocracy*. New York: Farrar, Straus, and Giroux.

Lemann, N. (2004). A history of admissions testing. In R. Zwick (Ed.), *Rethinking the SAT: The future of standarized testing in univesity admissions* (pp. 5–14). New York: RoutledgeFalmer.

Lewis, C., & Sheehan, K. (1990). Using Bayesian decision theory to design a computerized mastery test. *Applied Psychological Measurement, 14*, 367–386. https://doi.org/10.1177/014662169001400404

Lievens, F., & Sackett, P. R. (2012). The validity of interpersonal skills assessment via situational judgment tests for predicting academic success and job performance. *Journal of Applied Psychology, 97*(2), 460–468.

Linacre, J. M. (2010). Facets Rasch measurement, version 3.67.1 [Computer software]. Chicago: Winsteps.com.

Lindquist, E. F. (1953). *Design and analysis of experiments in psychology and education*. Boston: Houghton Mifflin.

Linn, R. L. (1996). Work readiness assessment: Questions of validity. In B. L. Resnick & J. G. Wirt (Eds.), *Linking school and work: Roles for standards and assessment* (pp. 245–266). San Francisco: Jossey-Bass.

Linn, R. L., & Burton, E. (1994). Performance-based assessment: Implications of task specificity. *Educational Measurement: Issues and Practice, 13*(1), 5–8. https://doi.org/10.1111/j.1745-3992.1994.tb00778.x

Linn, R. L., & Gronlund, N. E. (2000). *Measurement and assessment in teaching* (8th ed.). Upper Saddle River: Prentice Hall.

Linn, R. L., Rock, D. A., & Cleary, T. A. (1969). The development and evaluation of several programmed testing methods. *Educational and Psychological Measurement, 29*(1), 29–146. https://doi.org/10.1177/001316446902900109

Linn, R. L., Baker, E. L., & Dunbar, S. B. (1991). Complex performance based assessment: Expectations and validation criteria. *Educational Researcher, 20*(8), 15–21. https://doi.org/10.3102/0013189X020008015

Liu, L., Rogat, A., & Bertling, M. (2013). *A CBAL science model of cognition: Developing a competency model and learning progressions to support assessment development* (Research Report No. RR-13-29). Princeton: Educational Testing Service. http://dx.doi.org/10.1002/j.2333-8504.2013.tb02336.x

Liu, O. L., Brew, C., Blackmore, J., Gerard, L., Madhok, J., & Linn, M. C. (2014). Automated scoring of constructed-response science items: Prospects and obstacles. *Educational Measurement: Issues and Practice, 33*(2), 19–28. https://doi.org/10.1111/emip.12028

Liu, O. L., Rios, J. A., Heilman, M., Gerard, L., & Linn, M. C. (2016). Validation of automated scoring of science assessments. *Journal of Research in Science Teaching, 53*, 215–233. https://doi.org/10.1002/tea.21299

Lloyd-Jones, R. (1977). Primary trait scoring. In C. R. Cooper & L. Odell (Eds.), *Evaluating writing: Describing, measuring, judging* (pp. 33–66). Urban: National Council of Teachers of English.

Loevinger, J. (1957). Objective tests as instruments of psychological theory. *Psychological Reports, 3*, 653–694. https://doi.org/10.2466/pr0.1957.3.3.635

Loevinger, J. (1965). Person and population as psychometric concepts. *Psychological Review, 72*, 143–155. https://doi.org/10.1037/h0021704

Longford, N. T. (1994). Reliability of essay rating and score adjustment. *Journal of Educational and Behavioral Statistics, 19*(3), 171–200. https://doi.org/10.2307/1165293

Lord, F. M. (1952). A theory of test scores. *Psychometrika Monograph, 17*(7).

Lord, F. M. (1965). *Item sampling in test theory and in research design* (Research Bulletin No. RB-65-22). Princeton: Educational Testing Service. http://dx.doi.org/10.1002/j.2333-8504.1965.tb00968.x

Lord, F. M. (1980). *Applications of item-response theory to practical testing problems*. Hillsdale: Erlbaum.

Lukhele, R., Thissen, D., & Wainer, H. (1994). On the relative value of multiple-choice, constructed response, and examinee-selected items on two achievement tests. *Journal of Educational Measurement, 31*, 234–250. https://doi.org/10.1111/j.1745-3984.1994.tb00445.x

Lumley, T. (2002). Assessment criteria in a large-scale writing test: What do they really mean to the raters? *Language Testing, 19*(3), 246–276. https://doi.org/10.1191/0265532202lt230oa

Mariano, L. T. (2002). *Information accumulation, model selection and rater behavior in constructed response assessments* (Unpublished doctoral dissertation). Pittsburgh: Carnegie Mellon University.

Mazzeo, J., & Harvey, A. L. (1988). *The equivalence of scores from automated and conventional educational and psychological tests: A review of the literature.* (Report No. 88-8). New York: College Entrance Examination Board.

Mazzeo, J., Lazer, S., & Zieky, M. J. (2006). Monitoring educational progress with group-score assessments. In R. L. Brennan (Ed.), *Educational measurement* (4th ed., pp. 661–699). Westport: American Council on Education and Praeger.

McClellan, C. A. (2010). *Constructed-response scoring—Doing it right* (R&D Connections No. 13). Princeton: Educational Testing Service.

Meehl, P. E. (1954). *Clinical versus statistical prediction: A theoretical analysis and a review of evidence.* Minneapolis: University of Minnesota Press. https://doi.org/10.1037/11281-000

Messick, S. (1972). Beyond structure: In search of functional models of psychological process. *Psychometrika, 37*, 357–375. https://doi.org/10.1007/BF02291215

Messick, S. (1975). The standard problem: Meaning and values in measurement and evaluation. *American Psychologist, 30*(10), 955–966. https://doi.org/10.1037/0003-066X.30.10.955

Messick, S. (1980). Test validity and the ethics of assessment. *American Psychologist, 35*(11), 1012–1027. https://doi.org/10.1037/0003-066X.35.11.1012

Messick, S. (1989). Validity. In R. L. Linn (Ed.), *Educational measurement* (3rd ed., pp. 13–103). New York: Macmillian.

Messick, S. (1990). *Validity of test interpretation and use* (Research Report No. RR-90-11). Princeton: Educational Testing Service. http://dx.doi.org/10.1002/j.2333-8504.1990.tb01343.x

Messick, S. (1993). Trait equivalence as construct validity of score interpretation across multiple methods of measurement. In R. E. Bennett & W. C. Ward (Eds.), *Construction versus choice in cognitive measurement: Issues in constructed response, performance testing, and portfolio assessment* (pp. 61–74). Hillsdale: Erlbaum.

Messick, S. (1994). The interplay of evidence and consequences in the validation of performance assessments. *Educational Researcher, 23*(2), 13–23. https://doi.org/10.3102/0013189X023002013

Messick, S., & Kogan, N. (1966). Personality consistencies in judgment: Dimensions of role constructs. *Multivariate Behavioral Research, 1*, 165–175. https://doi.org/10.1207/s15327906mbr0102_3

Messick, S., Beaton, A. E., Lord, F. M., Baratz, J. C., Bennett, R. E., Duran, R. P., et al. (1983). *National assessment of educational progress reconsidered: A new design for a new era* (NAEP Report No. 83-1). Princeton: Educational Testing Service.

Miller, G. (2003). The cognitive revolution: A historical perspective. *Trends in Cognitive Sciences, 7*(3), 141–144. https://doi.org/10.1016/S1364-6613(03)00029-9

Mills, C. N., & Steffen, M. (2000). The GRE computer adaptive test: Operational issues. In W. J. van der Linden & C. A. W. Glas (Eds.), *Computerized adaptive testing: Theory and practice* (pp. 75–99). Amsterdam, the Netherlands: Kluwer. https://doi.org/10.1007/0-306-47531-6_4

Mislevy, R. J. (2013). Evidence-centered design for simulation-based assessment. *Military Medicine, 178*(Suppl. 1), 107–114. https://doi.org/10.7205/MILMED-D-13-00213

Mislevy, R. J., & Yin, C. (2012). Evidence-centered design in language testing. In G. Fulcher & F. Davidson (Eds.), *The Routledge handbook of language testing* (pp. 208–222). London: Routledge.

Mislevy, R. J., Steinberg, L. S., Breyer, F. J., Almond, R. G., & Johnson, L. (1999). A cognitive task analysis with implications for designing simulation-based performance assessment. *Computers in Human Behavior, 15*(3–4), 335–374. https://doi.org/10.1016/S0747-5632(99)00027-8

Mislevy, R. J., Steinberg, L. S., & Almond, R. G. (2002a). On the roles of task model variables in assessment design. In S. H. Irvine & P. C. Kyllonen (Eds.), *Item generation for test development* (pp. 97–128). Mahwah: Erlbaum.

Mislevy, R. J., Steinberg, L. S., Breyer, F. J., Almond, R. G., & Johnson, L. (2002b). Making sense of data from complex assessments. *Applied Measurement in Education, 15*(4), 363–390. https://doi.org/10.1016/S0747-5632(99)00027-8

Mislevy, R. J., Steinberg, L. S., & Almond, R. G. (2003). On the structure of educational assessments. *Measurement: Interdisciplinary Research and Perspectives, 1*(1), 3–62. https://doi.org/10.1207/S15366359MEA0101_02

Mislevy, R. J., Steinberg, L., Almond, R. G., & Lucas, J. F. (2006). Concepts, terminology, and basic models of evidence-centered design. In D. M. Williamson, R. J. Mislevy, & I. I. Bejar (Eds.), *Automated scoring of complex tasks in computer-based testing* (pp. 49–82). Mahwah: Erlbaum.

Mislevy, R. J., Oranje, A., Bauer, M. I., von Davier, A., Hao, J., Corrigan, S., et al. (2014). *Psychometrics considerations in game-based assessment*. Retrieved from http://www.instituteofplay.org/wp-content/uploads/2014/02/GlassLab_GBA1_WhitePaperFull.pdf

Mislevy, R. J., Corrigan, S., Oranje, A., Dicerbo, K., Bauer, M. I., von Davier, A. A., & Michael, J. (2016). Psychometrics for game-based assessment. In F. Drasgow (Ed.), *Technology and testing: Improving educational and psychological measurement* (pp. 23–48). Washington, DC: National Council on Measurement in Education.

Moses, T. P., Yang, W.-L., & Wilson, C. (2007). Using kernel equating to assess item order effects on test scores. *Journal of Educational Measurement, 44*, 157–178. https://doi.org/10.1111/j.1745-3984.2007.00032.x

Mullis, I. V. S. (1980). *Using the primary trait system for evaluating writing*. Washington, DC: National Assessment of Educational Progress.

Mullis, I. V. S. (1984). Scoring direct writing assessments: What are the alternatives? *Educational Measurement: Issues and Practice, 3*(1), 16–18. https://doi.org/10.1111/j.1745-3992.1984.tb00728.x

Myers, A. E., McConville, C. B., & Coffman, W. E. (1966). Simplex structure in the grading of essay tests. *Educational and Psychological Measurement, 26*(1), 41–54.

Myford, C. M., & Mislevy, R. J. (1995). *Monitoring and improving a portfolio assessment system* (CSE Technical Report No. 402). Los Angeles: National Center for Research on Evaluation, Standards, and Student Testing.

Myford, C. M., Marr, D. B., & Linacre, J. M. (1995). *Reader calibration and its potential role in equating for the Test of Written English* (Research Report No. RR-95-40). Princeton: Educational Testing Service. http://dx.doi.org/10.1002/j.2333-8504.1995.tb01674.x

Nagy, W. E., & Scott, J. (2000). Vocabulary processes. In M. Kamil, P. Mosenthal, & P. D. Pearson (Eds.), *Handbook of reading research* (Vol. 3, pp. 269–284). Mahwah: Erlbaum.

National Commission on Excellence in Education. (1983). *A nation at risk: The imperative for educational reform, a report to the Nation and the Secretary of Education, United States Department of Education*. Retrieved from the Department of Education website: http://www2.ed.gov/pubs/NatAtRisk/index.html

National Research Council. (2010). *Best practices for state assessment systems. Part 1, Summary of a workshop*. Washington, DC: National Academies Press. Retrieved from http://www.nap.edu/catalog.php?record_id=12906

Norton, L. S. (1990). Essay-writing: What really counts? *Higher Education, 20*, 411–442. https://doi.org/10.1007/BF00136221

Oranje, A., Mazzeo, J., Xu, X., & Kulick, E. (2014). An adaptive approach to group-score assessments. In D. Yan, A. A. von Davier, & C. Lewis (Eds.), *Computerized multistage testing: Theory and applications* (pp. 371–390). New York: CRC Press.

Owens, K. M. (2006). *Use of the GMAT Analytical Writing Assessment: Past and present* (Research Report No. 07-01). McLean: GMAC.

Page, E. B. (1966). The imminence of grading essays by computer. *Phi Delta Kappan, 47*, 238–243.

Page, E. B., & Petersen, N. S. (1995). The computer moves into essay grading: Updating the ancient test. *Phi Delta Kappan, 76*, 561–565.

Patz, R. J., Junker, B. W., Johnson, M. S., & Mariano, L. T. (2002). The hierarchical rater model for rated test items and its application to large-scale educational assessment data. *Journal of Educational and Behavioral Statistics in Medicine, 27*, 341–384. https://doi.org/10.3102/10769986027004341

Paul, S. R. (1981). Bayesian methods for calibration of examiners. *British Journal of Mathematical and Statistical Psychology, 34*, 213–223. https://doi.org/10.1111/j.2044-8317.1981.tb00630.x

Pearlman, M. (2008a). Finalizing the test blueprint. In C. A. Chapelle, M. Enright, & J. Jamieson (Eds.), *Building a validity argument for the Test of English as a Foreign Language* (pp. 227–258). New York: Routledge.

Pearlman, M. (2008b). The design architecture of NBPTS certification assessments. In L. Ingvarson & J. Hattie (Eds.), *Advances in program evaluation: Vol 11. Assessing teachers for professional certification: The first decade of the National Board of Professional Teaching Standards* (pp. 55–91). Bingley: Emerald Group. https://doi.org/10.1016/S1474-7863(07)11003-6

Pellegrino, J. W., & Glaser, R. (1980). Components of inductive reasoning. In P. A. Federico, R. E. Snow, & W. E. Montague (Eds.), *Aptitude, learning, and instruction: Cognitive process analyses* (pp. 177–217). Hillsdale: Erlbaum.

Persky, H. (2012). Writing assessment in the context of the National Assessment of Educational Progress. In N. Elliot & L. Perelman (Eds.), *Writing assessment in the 21st century: Essays in honor of Edward M. White* (pp. 69–86). New York: Hampton Press.

Pinchot, G. (1987). Innovation through intrapreneuring. *Research Management, 30*(2), 14–19.

Popper, K. R. (1992). *The logic of scientific discovery*. London: Routledge. (Original work published 1959).

Powers, D. E. (1984). *Considerations for developing measures of speaking and listening* (Research Report No. RR-84-18). Princeton: Educational Testing Service. http://dx.doi.org/10.1002/j.2330-8516.1984.tb00058.x

Powers, D. E. (2010). *The case for a comprehensive, four-skills assessment of English-language proficiency* (R&D Connections No. 14). Princeton: Educational Testing Service.

Powers, D. E., & Bennett, R. E. (1999). Effects of allowing examinees to select questions on a test of divergent thinking. *Applied Measurement in Education, 12*(3), 257–279. https://doi.org/10.1207/S15324818AME1203_3

Powers, D. E., & Dwyer, C. A. (2003). *Toward specifying a construct of reasoning* (Research Memorandum No. RM-03-01). Princeton: Educational Testing Service.

Powers, D. E., & Enright, M. K. (1987). Analytical reasoning skills in graduate study: Perceptions of faculty in six fields. *Journal of Higher Education, 58*(6), 658–682. https://doi.org/10.2307/1981103

Powers, D. E., Fowles, M. E., Farnum, M., & Ramsey, P. (1994). Will they think less of my handwritten essay if others word process theirs? Effects of essay scores of intermingling handwritten and word-processed essays. *Journal of Educational Measurement, 31*, 220–233. https://doi.org/10.1111/j.1745-3984.1994.tb00444.x

Powers, D. E., Kubota, M. Y., Bentley, J., Farnum, M., Swartz, R., & Willard, A. E. (1998). *Qualifying readers for the Online Scoring Network: Scoring argument essays* (Research Report No. RR-98-28). Princeton: Educational Testing Service. http://dx.doi.org/10.1002/j.2333-8504.1998.tb01777.x

Powers, D. E., Fowles, M. E., & Welsh, C. (1999a). *Further validation of a writing assessment for graduate admissions* (Research Report No. RR-99-18). Princeton: Educational Testing Service. http://dx.doi.org/10.1002/j.2333-8504.1999.tb01816.x

Powers, D. E., Schedl, M. A., Leung, S. W., & Butler, F. (1999b). Validating the revised Test of Spoken English against a criterion of communicative success. *Language Testing, 16*(4), 399–425.

Powers, D. E., Burstein, J. C., Chodorow, M. S., Fowles, M. E., & Kukich, K. (2002). Comparing the validity of automated and human scoring of essays. *Journal of Educational Computing Research, 26*(4), 407–425. https://doi.org/10.2190/CX92-7WKV-N7WC-JL0A

Powers, D. E., Kim, H.-J., Yu, F., Weng, V. Z., & VanWinkle, W. H. (2009). *The TOEIC speaking and writing tests: Relations to test-taker perceptions of proficiency in English* (TOEIC Research Report No. 4). Princeton: Educational Testing Service.
Puhan, P., Boughton, K., & Kim, S. (2007). Examining differences in examinee performance in paper and pencil and computerized testing. *Journal of Technology, Learning, and Assessment, 6*(3). Retrieved from http://www.jtla.org/
Pula, J. J., & Huot, B. A. (1993). A model of background influences on holistic raters. In M. M. Williamson & B. A. Huot (Eds.), *Validating holistic scoring for writing assessment: Theoretical and empirical foundations* (pp. 237–265). Cresskill: Hampton Press.
Quinlan, T., Higgins, D., & Wolff, S. (2009). *Evaluating the construct-coverage of e-rater* (Research Report No. RR-09–01). Princeton: Educational Testing Service. http://dx.doi.org/10.1002/j.2333-8504.2009.tb02158.x
Rajaratnam, N., Cronbach, L. J., & Gleser, G. C. (1965). Generalizability of stratified-parallel tests. *Psychometrika, 30*(1), 39–56. https://doi.org/10.1007/BF02289746
Reid-Green, K. S. (1990). A high speed image processing system. *IMC Journal, 26*(2), 12–14.
Riconscente, M. M., Mislevy, R. J., & Hamel, L. (2005). *An introduction to PADI task templates* (Technical Report No. 3). Menlo Park: SRI International.
Robin, F., Steffen, M., & Liang, L. (2014). The implementation of the GRE Revised General Test as a multistage test. In D. Yan, A. A. von Davier, & C. Lewis (Eds.), *Computerized multistage testing: Theory and applications* (pp. 325–341). Boca Raton: CRC Press.
Rupp, A. A., Levy, R., Dicerbo, K. E., Crawford, A. V., Calico, T., Benson, M., et al. (2012). Putting ECD into practice: The interplay of theory and data in evidence models within a digital learning environment. *Journal of Educational Data Mining, 4*, 1–102.
Russell, M. (2006). *Technology and assessment: The tale of two interpretations*. Greenwich: Information Age.
Sachse, P. P. (1984). Writing assessment in Texas: Practices and problems. *Educational Measurement: Issues and Practice, 3*(1), 21–23. https://doi.org/10.1111/j.1745-3992.1984.tb00731.x
Sakaguchi, K., Heilman, M., & Madnani, N. (2015). Effective feature integration for automated short answer scoring. In *Proceedings of the 2015 conference of the North American Chapter of the association for computational linguistics: Human language technologies* (pp. 1049–1054). Retrieved from https://aclweb.org/anthology/N/N15/N15-1000.pdf
Samejima, F. (1969). Estimation of latent ability using a response pattern of graded scores. *Psychometrika, 34* (4, Whole Pt. 2). https://doi.org/10.1007/BF03372160
Sandene, B., Horkay, N., Bennett, R., Allen, N., Braswell, J., Kaplan, B., & Oranje, A. (2005). *Online assessment in mathematics and writing: Reports from the NAEP Technology-Based Assessment project* (NCES 2005-457). Washington, DC: U.S. Government Printing Office.
Schmeiser, C. B., & Welch, C. J. (2006). Test development. In R. L. Brennan (Ed.), *Educational measurement* (4th ed., pp. 307–353). Westport: American Council on Education and Praeger.
Schmitt, A., Holland, P., & Dorans, N. J. (1993). Evaluating hypotheses about differential item functioning. In P. Holland & H. Wainer (Eds.), *Differential item functioning* (pp. 281–315). Hillsdale: Erlbaum.
Sebrechts, M. M., Bennett, R. E., & Rock, D. A. (1991). Agreement between expert-system and human raters on complex constructed-response quantitative items. *Journal of Applied Psychology, 76*, 856–862. https://doi.org/10.1037/0021-9010.76.6.856
Sebrechts, M. M., Enright, M. K., Bennett, R. E., & Martin, K. (1996). Using algebra word problems to assess quantitative ability: Attributes, strategies, and errors. *Cognition and Instruction, 14*(3), 285–343. https://doi.org/10.1207/s1532690xci1403_2
Sheehan, K. M., & Lewis, C. (1992). Computerized mastery testing with nonequivalent testlets. *Applied Psychological Measurement, 16*, 65–76. https://doi.org/10.1177/014662169201600108
Sheehan, K. M., & O'Reilly, T. (2011). *The CBAL reading assessment: An approach for balancing measurement and learning goals* (Research Report No. RR-11-21). Princeton: Educational Testing Service. http://dx.doi.org/10.1002/j.2333-8504.2011.tb02257.x

Shepard, L. A. (1991). Psychometricians' beliefs about learning. *Educational Researcher, 20*(7), 2–16. https://doi.org/10.3102/0013189X020007002

Shohamy, E., Gordon, C. M., & Kraemer, R. (1992). The effect of raters background and training on the reliability of direct writing tests. *Modern Language Journal, 76*(1), 27–33. https://doi.org/10.1111/j.1540-4781.1992.tb02574.x

Smarter Balanced Assessment Consortium. (2014). *Smarter Balanced scoring guide for selected short-text mathematics items* (Field Test 2014). Retrieved from http://www.smarterbalanced.org/wordpress/wp-content/uploads/2014/10/Smarter-Balanced-Scoring-Guide-for-Selected-Short-Text-Mathematics-Items.pdf

Song, Y., Heilman, M., Klebanov Beigman, B., & Deane, P. (2014, June). *Applying argumentation schemes for essay scoring*. Paper presented at the First Workshop on Argumentation Mining, Baltimore, Maryland. https://doi.org/10.3115/v1/W14-2110

Sparks, J. R., Song, Y., Brantley, W., & Liu, O. L. (2014). *Assessing written communication in higher education: Review and recommendations for next-generation assessment* (Research Report No. RR-14-37). Princeton: Educational Testing Service. http://dx.doi.org/10.1002/ets2.12035

Spearman, C. (1923). *The nature of "intelligence" and the principles of cognition.* New York: Macmillan.

Stanley, J. C. (1971). Reliability. In R. L. Thorndike (Ed.), *Educational measurement* (2nd ed., pp. 356–442). Washington, DC: American Council on Education.

Stansfield, C. (1986a). A history of the Test of Written English: The developmental year. *Language Testing, 3*, 224–334. https://doi.org/10.1177/026553228600300209

Stansfield, C. (1986b). *Toward communicative competence testing: Proceedings of the second TOEFL Invitational Conference* (TOEFL Research Report No. 21). Princeton: Educational Testing Service.

Sternberg, R. J. (1977). Component processes in analogical reasoning. *Psychological Review, 84*(4), 353–378. https://doi.org/10.1037/0033-295X.84.4.353

Stetcher, B. (2010). *Performance assessment in an era of standards-based educational accountability*. Stanford: Stanford University, Stanford Center for Opportunity Policy in Education.

Stocking, M. (1969). *Short tailored tests* (Research Bulletin No. RB-69-63). Princeton: Educational Testing Service. http://dx.doi.org/10.1002/j.2333-8504.1969.tb00741.x

Stocking, M. L., & Swanson, L. (1993). A method for severely constrained item selection in adaptive testing. *Applied Psychological Measurement, 17*, 277–292. https://doi.org/10.1177/014662169301700308

Stricker, L. J. (1982). Interpersonal competence instrument: Development and preliminary findings. *Applied Psychological Measurement, 6*, 69–81. https://doi.org/10.1177/014662168200600108

Stricker, L. J., & Rock, D. A. (1990). Interpersonal competence, social intelligence, and general ability. *Personality and Individual Differences, 11*(8), 833–839. https://doi.org/10.1016/0191-8869(90)90193-U

Stricker, L. J., & Ward, W. C. (2004). Stereotype threat, inquiring about test takers' ethnicity and gender, and standardized test performance. *Journal of Applied Social Psychology, 34*(4), 665–693. https://doi.org/10.1111/j.1559-1816.2004.tb02564.x

Sukkarieh, J. Z., & Bolge, E. (2008). Leveraging c-rater's automated scoring capability for providing instructional feedback for short constructed responses. In P. Woolf, E. Aimeur, R. Nkambou, & S. Lajoie (Eds.), *Intelligent tutoring systems: Vol. 5091. Proceedings of the 9th international conference on intelligent tutoring systems, 2008, Montreal, Canada* (pp. 779–783). New York: Springer. https://doi.org/10.1007/978-3-540-69132-7_106

Suto, I., & Nadas, R. (2009). Why are some GCSE examination questions harder to mark accurately than others? Using Kelly's Repertory Grid technique to identify relevant question features. *Research Papers in Education, 24*(3), 335–377. https://doi.org/10.1080/02671520801945925

Suto, I., Nadas, R., & Bell, J. (2009). Who should mark what? A study of factors affecting marking accuracy in a biology examination. *Research Papers in Education, 26*, 21–51. https://doi.org/10.1080/02671520902721837

Taylor, C. A., & Angelis, P. (2008). The evolution of the TOEFL. In C. A. Chapelle, M. Enright, & J. Jamieson (Eds.), *Bulding a validity argument for the Test of English as a Foreign Language* (pp. 27–54). New York: Routledge. https://doi.org/10.1007/BF02288939

Taylor, C. W. (1947). A factorial study of fluency in writing. *Psychometrika, 12*, 239–262. https://doi.org/10.1007/BF02288939

Thomas, W. H., Storms, B. A., Sheingold, K., Heller, J. I., Paulukonis, S. T., & Nunez, A. M. (1998). *California Learning Assessment System: Portfolio assessment research and development project: Final report*. Princeton: Educational Testing Service.

Thurstone, L. L. (1926). *The nature of intelligence*. New York: Harcourt, Brace.

TOEFL. (2011). TOEFL program history. *TOEFL iBT Research Insight, 1*(6).

Torgerson, W. S., & Green, B. F. (1952). A factor analysis of English essay readers. *Journal of Educational Psychology, 43*(6), 354–363. https://doi.org/10.1037/h0052471

Toulmin, S. E. (1958). *The uses of argument*. New York: Cambridge University Press.

Traxler, A. E. (1951). Administering and scoring the objective test. In E. F. Lindquist (Ed.), *Educational measurement* (pp. 329–416). Washington, DC: American Council on Education.

Traxler, A. E. (1954). Impact of machine and devices on developments in testing and related fields. In *Proceedings of the 1953 invitational conference on testing problems* (pp. 139–146). Princeton: Educational Testing Service.

Tryon, R. C. (1935). A theory of psychological components—An alternative to "mathematical factors." *Psychological Review, 42*, 425–454. https://doi.org/10.1037/h0058874

Tucker, L. R. (1964). A suggested alternative formulation in the developments by Hursch, Hammond, and Hursch, and by Hammond, Hursch, and Todd. *Psychological Review, 71*(6), 528–530. https://doi.org/10.1037/h0047061

Turnbull, W. W. (1949). Influence of cultural background on predictive test scores. In *Proceedings of the ETS invitational conference on testing problems* (pp. 29–34). Princeton: Educational Testing Service.

Turnbull, W. W. (1968). Relevance in testing. *Science, 160*, 1424–1429. https://doi.org/10.1126/science.160.3835.1424

U.S. Congress & Office of Technology Assessment. (1992). *Testing in American schools: Asking the right questions* (No. OTA-SET-519). Washington, DC: U.S. Government Printing Office.

U.S. Department of Labor, Secretary's Commission on Achieving Necessary Skills. (1991). *What work requires of schools: A SCANS report for America 2000*. Washington, DC: Author.

Valentine, J. A. (1987). *The College Board and the school curriculum: A history of the College Board's influence on the substance and standards of American education, 1900–1980*. New York: College Entrance Examination Board.

Vaughan, C. (1991). Holistic assessment: What goes on in the rater's mind? In L. Hamp-Lyons (Ed.), *Assessing second language writing in academic contexts* (pp. 111–125). Norwood: Albex.

von Davier, M. (2013). The DINA model as a constrained general diagnostic model: Two variants of a model equivalency. *British Journal of Mathematical and Statistical Psychology, 67*(1), 49–71. https://doi.org/10.1111/bmsp.12003

Wainer, H., & Robinson, D. H. (2007). Fumiko Samejima. *Journal of Educational and Behavioral Statistics, 32*(2), 206–222. https://doi.org/10.3102/1076998607301991

Wainer, H., & Thissen, D. (1993). Combining multiple-choice and constructed response test scores: Toward a Marxist theory of test construction. *Applied Measurement in Education, 6*(2), 103–118. https://doi.org/10.1207/s15324818ame0602_1

Wang, X., Evanini, K., & Zechner, K. (2013). Coherence modeling for the automated assessment of spontaneous spoken responses. In *Proceedings of NAACL-HLT 2013* (pp. 814–819). Atlanta: Association of Computational Lingustics.

Ward, W. C. (1982). A comparison of free-response and multiple-choice forms of verbal aptitude tests. *Applied Psychological Measurement, 6*, 1–11. https://doi.org/10.1177/014662168200600101

Ward, W. C. (1984). Using microcomputers to administer tests. *Educational Measurement: Issues and Practice, 3*(2), 16–20. https://doi.org/10.1111/j.1745-3992.1984.tb00744.x

Ward, W. C. (1988). The College Board computerized placement tests: An application of computerized adaptive testing. *Machine-Mediated Learning, 2*, 271–282.

Ward, W. C., Frederiksen, N., & Carlson, S. B. (1980). Construct validity of free-response and machine scorable forms of a test of scientific thinking. *Journal of Educational Measurement, 17*, 11–29. https://doi.org/10.1111/j.1745-3984.1980.tb00811.x

Weiss, D. J., & Betz, N. E. (1973). *Ability measurement: Conventional or adaptive?* (Research Report No. 73-1). Minneapolis: University of Minnesota.

White, E. M. (1984). Holisticism. *College Composition and Communication, 35*(4), 400–409. https://doi.org/10.2307/357792

White House. (2009, March 10). *Remarks by the president to the Hispanic Chamber of Commerce on a complete and competitive American education*. Retrieved from https://www.whitehouse.gov/the_press_office/Remarks-of-the-President-to-the-United-States-Hispanic-Chamber-of-Commerce/

Whitely, S. E., & Dawis, R. V. (1974). Effects of cognitive intervention on latent ability measured from analogy items. *Journal of Educational Psychology, 66*, 710–717.

Wiggins, G. (1989). A true test: Toward more authentic and equitable assessment. *Phi Delta Kappan, 70*, 703–713.

Williamson, D. M., Mislevy, R. J., & Bejar, I. I. (Eds.). (2006). *Automated scoring of complex tasks in computer-based testing*. Mahwah: Erlbaum.

Wolfe, E. W. (1997). The relationship between essay reading style and scoring proficiency in a psychometric scoring system. *Assessing Writing, 4*(1), 83–106. https://doi.org/10.1016/S1075-2935(97)80006-2

Wolfe, E. W., Bolton, S., Feltovich, B., & Welch, C. (1993). *A comparison of word-processed and handwritten essays from a standardized writing assessment* (Research Report No. 93-8). Iowa City: American College Testing.

Wolfe, E. W., Kao, C. W., & Ranney, M. (1998). Cognitive differences in proficient and nonproficient essay scorers. *Written Communication, 15*(4), 465–492. https://doi.org/10.1177/0741088398015004002

Wolfe, E. W., Matthews, S., & Vickers, D. (2010). The effectiveness and efficiency of distributed online, regional online, and regional face-to-face training for writing assessment raters. *Journal of Technology, Learning, and Assessment, 10*(1.) Retrieved from http://www.jtla.org/

Wong, K. F. E., & Kwong, J. Y. Y. (2007). Effects of rater goals on rating patterns: Evidence from an experimental field study. *Journal of Applied Psychology, 92*(2), 577–585. https://doi.org/10.1037/0021-9010.92.2.577

Wood, R. (1973). Response-contingent testing. *Review of Educational Research, 43*(4), 529–544. https://doi.org/10.3102/00346543043004529

Xi, X. (2010). How do we go about investigating test fairness? *Language Testing, 27*(2), 147–170. https://doi.org/10.1177/0265532209349465

Yoon, S.-Y., Bhat, S., & Zechner, K. (2012). Vocabulary profile as a measure of vocabulary sophistication. In *Proceedings of the 7th Workshop on Innovative Use of NLP for Building Educational Applications, NAACL-HLT* (pp. 180–189). Stroudsburg: Association for Computational Linguistics.

Zechner, K., Bejar, I. I., & Hemat, R. (2007a). *Toward an understanding of the role of speech recognition in nonnative speech assessment* (TOEFL iBT Research Series No. 2). Princeton: Educational Testing Service.

Zechner, K., Higgins, D., & Xi, X. (2007b, October). *SpeechRater: A construct-driven approach to scoring spontaneous non-native speech*. Paper presented at the 2007 Workshop of the International Speech Communication Association (ISCA) Special Interest Group on Speech and Language Technology in Education (SLaTE), Farmington.

Zechner, K., Higgins, D., Xi, X., & Williamson, D. (2009a). Automatic scoring of non-native spontaneous speech in tests of spoken English. *Speech Communication, 51*, 883–895. https://doi.org/10.1016/j.specom.2009.04.009

Zechner, K., Sabatini, J., & Chen, L. (2009b). Automatic scoring of children's read-aloud text passages and word lists. In *Proceedings of the NAACL-HLT Workshop on Innovative Use of NLP for Building Educational Applications* (pp. 10–18). Stroudsburg: Association for Computational Linguistics.
Zechner, K., Evanini, K., & Laitusis, C. (2012, September). *Using automatic speech recognition to assess the reading proficiency of a diverse sample of middle school students.* Paper presented at the Interspeech Workshop on Child, Computer Interaction, Portland.
Zhang, L. Y., Powers, D. E., Wright, W., & Morgan, R. (2003). *Applying the Online Scoring Network (OSN) to Advanced Placement Program (AP) tests* (Research Report No. RR-03-12). Princeton: Educational Testing Service. http://dx.doi.org/10.1002/j.2333-8504.2003.tb01904.x
Zhang, T., Mislevy, R. J., Haertel, G., Javitz, H., & Hansen, E. G. (2010). *A Design pattern for a spelling assessment for students with disabilities* (Technical Report No. 2). Menlo Park: SRI.
Zieky, M. J. (1995). A historical perspective on setting standards. In L. Crocker & M. J. Zieky (Eds.), *Joint conference on standard setting for large-scale assessments* (Vol. 2, pp. 1–38). Washington, DC: National Assessment Governing Board and National Center for Education Statistics.
Zieky, M. J. (2014). An introduction to the use of evidence-centered design in test development. *Educational Psychology, 20*, 79–88. https://doi.org/10.1016/j.pse.2014.11.003
Zwick, R., Donoghue, J. R., & Grima, A. (1993). Assessment of differential item functioning for performance tasks. *Journal of Educational Measurement, 30*, 233–251. ps://doi.org/10.1111/j.1745-3984.1993.tb00425.x
Zwick, R., Thayer, D. T., & Mazzeo, J. (1997). Descriptive and inferential procedures for assessing differential item functioning in polytomous items. *Applied Measurement in Education, 10*, 321–344. https://doi.org/10.1207/s15324818ame1004_2

Chapter 19
Advancing Human Assessment: A Synthesis Over Seven Decades

Randy E. Bennett and Matthias von Davier

This book has documented the history of ETS's contributions to educational research and policy analysis, psychology, and psychometrics. We close the volume with a brief synthesis in which we try to make more general meaning from the diverse directions that characterized almost 70 years of work.

Synthesizing the breadth and depth of the topics covered over that time period is not simple. One way to view the work is across time. Many of the book's chapters presented chronologies, allowing the reader to follow the path of a research stream over the years. Less evident from these separate chronologies was the extent to which multiple streams of work not only coexisted but sometimes interacted.

From its inception, ETS was rooted in Henry Chauncey's vision of describing individuals through broad assessment of their capabilities, helping them to grow and society to benefit (Elliot 2014). Chauncey's conception of broad assessment of capability required a diverse research agenda.

Following that vision, his research managers assembled an enormous range of staff expertise. Only through the assemblage of such expertise could one bring diverse perspectives and frameworks from many fields to a problem, leading to novel solutions.

In the following sections, we summarize some of the key research streams evident in different time periods, where each period corresponds to roughly a decade. Whereas the segmentation of these time periods is arbitrary, it does give a general

---

This work was conducted while M. von Davier was employed with Educational Testing Service.

R.E. Bennett (✉) • M. von Davier
Educational Testing Service, Princeton, NJ, USA
e-mail: rbennett@ets.org

R.E. Bennett, M. von Davier (eds.), *Advancing Human Assessment*,
Methodology of Educational Measurement and Assessment,
DOI 10.1007/978-3-319-58689-2_19

sense of the progression of topics across time.[1] Also somewhat arbitrary is the use of publication date as the primary determinant of placement into a particular decade. Although the work activity leading up to publication may well have occurred in the previous period, the result of that activity and the impact that it had was typically through its dissemination.

19.1 The Years 1948–1959

19.1.1 Psychometric and Statistical Methodology

As will be the case for every period, a very considerable amount of work centered on theory and on methodological development in psychometrics and statistics. With respect to the former, the release of Gulliksen's (1950) *Theory of Mental Tests* deserves special mention for its codification of classical test theory. But more forward looking was work to create a statistically grounded foundation for the analysis of test scores, a latent-trait theory (Lord 1952, 1953). This direction would later lead to the groundbreaking development of item response theory (IRT; Lord and Novick 1968), which became a well-established part of applied statistical research in domains well beyond education and is now an important building block of generalized modeling frameworks, which connect the item response functions of IRT with structural models (Carlson and von Davier, Chap. 5, this volume). Green's (1950a, b) work can be seen as an early example that has had continued impact not commonly recognized. His work pointed out how latent structure and latent-trait models are related to factor analysis, while at the same time placing latent-trait theory into the context of latent class models. Green's insights had profound impact, reemerging outside of ETS in the late 1980s (de Leeuw and Verhelst 1986; Follman 1988; Formann 1992; Heinen 1996) and, in more recent times, at ETS in work on generalized latent variable models (Haberman et al. 2008; Rijmen et al. 2014).

In addition to theoretical development, substantial effort was focused on methodological development for, among other purposes, the generation of engineering solutions to practical scale-linking problems. Examples include Karon and Cliff's (1957) proposal to smooth test-taker sample data before equating, a procedure used today by most testing programs that employ equipercentile equating (Dorans and Puhan, Chap. 4, this volume); Angoff's (1953) method for equating test forms by using a miniature version of the full test as an external anchor; and Levine's (1955) procedures for linear equating under the common-item, nonequivalent-population design.

[1] In most cases, citations included as examples of a work stream were selected based on their discussion in one of the book's chapters.

19.1.2 Validity and Validation

In the 2 years of ETS's beginning decade, the 1940s, and in the 1950s that followed, great emphasis was placed on predictive studies, particularly for success in higher education. Studies were conducted against first-semester performance (Frederiksen 1948) as well as 4-year academic criteria (French 1958). As Kane and Bridgeman (Chap. 16, this volume) noted, this emphasis was very much in keeping with conceptions of validity at the time, and it was, of course, important to evaluating the meaning and utility of scores produced by the new organization's operational testing programs. However, also getting attention were studies to facilitate trait interpretations of scores (French et al. 1952). These interpretations posited that response consistencies were the result of test-taker dispositions to behave in certain ways in response to certain tasks, dispositions that could be investigated through a variety of methods, including factor analysis. Finally, the compromising effects of construct-irrelevant influences, in particular those due to coaching, were already a clear concern (Dear 1958; French and Dear 1959).

19.1.3 Constructed-Response Formats and Performance Assessment

Notably, staff interests at this time were not restricted to multiple-choice tests because, as Bejar (Chap. 18, this volume) pointed out, the need to evaluate the value of additional methods was evident. Work on constructed-response formats and performance assessment was undertaken (Ryans and Frederiksen 1951), including development of the in-basket test (Fredericksen et al. 1957), subsequently used throughout the world for job selection, and a measure of the ability to formulate hypotheses as an indicator of scientific thinking (Frederiksen 1959). Research on direct writing assessment (e.g., through essay testing) was also well under way (Diederich 1957; Huddleston 1952; Torgerson and Green 1950).

19.1.4 Personal Qualities

Staff interests were not restricted to the verbal and quantitative abilities underlying ETS's major testing programs, the Scholastic Aptitude Test (the *SAT*® test) and the *GRE*® General Test. Rather, a broad investigative program on what might be termed *personal qualities* was initiated. Cognition, more generally defined, was one key interest, as evidenced by publication of the Kit of Selected Tests for Reference Aptitude and Achievement Factors (French 1954). The Kit was a compendium of marker assessments investigated with sufficient thoroughness to make it possible to use in factor analytic studies of cognition such that results could be more directly compared across studies. Multiple reference measures were provided for each factor, including measures of abilities in the reasoning, memory, spatial, verbal, numeric, motor, mechanical, and ideational fluency domains.

In addition, substantial research targeted a wide variety of other human qualities. This research included personality traits, interests, social intelligence, motivation, leadership, level of aspiration and need for achievement, and response styles (acquiescence and social desirability), among other things (French 1948, 1956; Hills 1958; Jackson and Messick 1958; Melville and Frederiksen 1952; Nogee 1950; Ricciuti 1951).

19.2 The Years 1960–1969

19.2.1 Psychometric and Statistical Methodology

If nothing else, this period was notable for the further development of IRT (Lord and Novick 1968). That development is one of the major milestones of psychometric research. Although the organization made many important contributions to classical test theory, today psychometrics around the world mainly uses IRT-based methods, more recently in the form of generalized latent variable models. One of the important differences from classical approaches is that IRT properly grounds the treatment of categorical data in probability theory and statistics. The theory's modeling of how responses statistically relate to an underlying variable allows for the application of powerful methods for generalizing test results and evaluating the assumptions made. IRT-based item functions are the building blocks that link item responses to underlying explanatory models (Carlson and von Davier, Chap. 5, this volume). Leading up to and concurrent with the seminal volume *Statistical Theories of Mental Test Scores* (Lord and Novick 1968), Lord continued to make key contributions to the field (Lord 1965a, b, 1968a, b).

In addition to the preceding landmark developments, a second major achievement was the invention of confirmatory factor analysis by Karl Jöreskog (1965, 1967, 1969), a method for rigorously evaluating hypotheses about the latent structure underlying a measure or collection of measures. This invention would be generalized in the next decade and applied to the solution of a great variety of measurement and research problems.

19.2.2 Large-Scale Survey Assessments of Student and Adult Populations

In this period, ETS contributed to the design and conducted the analysis of the Equality of Educational Opportunity Study (Beaton and Barone, Chap. 8, this volume). Also of note was that, toward the end of the decade, ETS's long-standing program of longitudinal studies began with initiation of the Head Start Longitudinal Study (Anderson et al. 1968). This study followed a sample of children from before preschool enrollment through their experience in Head Start, in another preschool, or in no preschool program.

19.2.3 Validity and Validation

The 1960s saw continued interest in prediction studies (Schrader and Pitcher 1964), though noticeably less than in the prior period. The study of construct-irrelevant factors that had concentrated largely on coaching was less evident, with interest emerging in the phenomenon of test anxiety (French 1962). Of special note is that, due to the general awakening in the country over civil rights, ETS research staff began to focus on developing conceptions of equitable treatment of individuals and groups (Cleary 1968).

19.2.4 Constructed-Response Formats and Performance Assessment

The 1960s saw much investigation of new forms of assessment, including in-basket performance (Frederiksen 1962; L. B. Ward 1960), formulating-hypotheses tasks (Klein et al. 1969), and direct writing assessment. As described by Bejar (Chap. 18, this volume), writing assessment deserves special mention for the landmark study by Diederich et al. (1961) documenting that raters brought "schools of thought" to the evaluation of essays, thereby initiating interest in the investigation of rater cognition, or the mental processes underlying essay grading. A second landmark was the study by Godshalk et al. (1966) that resulted in the invention of holistic scoring.

19.2.5 Personal Qualities

The 1960s brought a very substantial increase to work in this area. The work on cognition produced the 1963 "Kit of Reference Tests for Cognitive Factors" (French et al. 1963), the successor to the 1954 "Kit." Much activity concerned the measurement of personality specifically, although a range of related topics was also investigated, including continued work on response styles (Damarin and Messick 1965; Jackson and Messick 1961; Messick 1967), the introduction into the social–psychological literature of the concept of prosocial (or altruistic) behavior (Bryan and Test 1967; Rosenhan 1969; Rosenhan and White 1967), and risk taking (Kogan and Doise 1969; Kogan and Wallach 1964; Wallach et al. 1962). Also of note is that this era saw the beginnings of ETS's work on cognitive styles (Gardner et al. 1960; Messick and Fritzky 1963; Messick and Kogan 1966). Finally, a research program on creativity began to emerge (Skager et al. 1965, 1966), including Kogan's studies of young children (Kogan and Morgan 1969; Wallach and Kogan 1965), a precursor to the extensive line of developmental research that would appear in the following decade.

19.2.6 Teacher and Teaching Quality

Although ETS had been administering the National Teachers Examination since the organization's inception, relatively little research had been conducted around the evaluation of teaching and teachers. The 1960s saw the beginnings of such research, with investigations of personality (Walberg 1966), values (Sprinthall and Beaton 1966), and approaches to the behavioral observation of teaching (Medley and Hill 1967).

19.3 The Years 1970–1979

19.3.1 Psychometric and Statistical Methodology

Causal inference was a major area of research in the field of statistics generally in this decade, and that activity included ETS. Rubin (1974b, 1976a, b, c, 1978) made fundamental contributions to the approach that allows for evaluating the extent to which differences observed in experiments can be attributed to effects of underlying variables.

More generally, causal inference as treated by Rubin can be understood as a missing-data and imputation problem. The estimation of quantities under incomplete-data conditions was a chief focus, as seen in work by Rubin (1974a, 1976a, b) and his collaborators (Dempster et al. 1977), who created the expectation-maximization (EM) algorithm, which has become a standard analytical method used not only in estimating modern psychometric models but throughout the sciences. As of this writing, the Dempster et al. (1977) article had more than 45,000 citations in Google Scholar.

Also falling under causal inference was Rubin's work on matching. Matching was developed to reduce bias in causal inferences using data from nonrandomized studies. Rubin's (1974b, 1976a, b, c, 1979) work was central to evaluating and improving this methodology.

Besides landmark contributions to causal inference, continued development of IRT was taking place. Apart from another host of papers by Lord (1970, 1973, 1974a, b, 1975a, b, 1977), several applications of IRT were studied, including for linking test forms (Marco 1977; see also Carlson and von Davier, Chap. 5, this volume). In addition, visiting scholars made seminal contributions as well. Among these contributions were ones on testing the Rasch model as well as on bias in estimates (Andersen 1972, 1973), ideas later generalized by scholars elsewhere (Haberman 1977).

Finally, this period saw Karl Jöreskog and colleagues implement confirmatory factor analysis (CFA) in the LISREL computer program (Jöreskog and van Thillo 1972) and generalize CFA for the analysis of covariance structures (Jöreskog 1970), path analysis (Werts et al. 1973), simultaneous factor analysis in several populations (Jöreskog 1971), and the measurement of growth (Werts et al. 1972). Their inventions, particularly LISREL, continue to be used throughout the social sciences within the general framework of structural equation modeling to pose and evaluate psychometric, psychological, sociological, and econometric theories and the hypotheses they generate.

19.3.2 Large-Scale Survey Assessments of Student and Adult Populations

Worthy of note were two investigations, one of which was a continuation from the previous decade. That latter investigation, the Head Start Longitudinal Study, was documented in a series of program reports (Emmerich 1973; Shipman 1972; Ward 1973). Also conducted was the National Longitudinal Study of the High School Class of 1972 (Rock, Chap. 10, this volume).

19.3.3 Validity and Validation

In this period, conceptions of validity, and concerns for validation, were expanding. With respect to conceptions of validity, Messick's (1975) seminal paper "The Standard Problem: Meaning and Values in Measurement and Evaluation" called attention to the importance of construct interpretations in educational measurement, a perspective largely missing from the field at that time. As to validation, concerns over the effects of coaching reemerged with research finding that two quantitative item types being considered for the SAT were susceptible to short-term preparation (Evans and Pike 1973), thus challenging the College Board's position on the existence of such effects. Concerns for validation also grew with respect to test fairness and bias, with continued development of conceptions and methods for investigating these issues (Linn 1973, 1976; Linn and Werts 1971).

19.3.4 Constructed-Response Formats and Performance Assessment

Relatively little attention was given to this area. An exception was continued investigation of the formulating-hypotheses item type (Evans and Frederiksen 1974; Ward et al. 1980).

19.3.5 Personal Qualities

The 1970s saw the continuation of a significant research program on personal qualities. With respect to cognition, the third version of the "Factor Kit" was released in 1976: the "Kit of Factor-Referenced Cognitive Tests" (Ekstrom et al. 1976). Work on other qualities continued, including on prosocial behavior (Rosenhan 1970, 1972) and risk taking (Kogan et al. 1972; Lamm and Kogan 1970; Zaleska and Kogan 1971). Of special note was the addition to the ETS staff of Herman Witkin and colleagues, who significantly extended the prior decade's work on cognitive styles (Witkin et al. 1974, 1977; Zoccolotti and Oltman 1978). Work on kinesthetic aftereffect (Baker et al. 1976, 1978, 1979) and creativity (Frederiksen and Ward 1978; Kogan and Pankove 1972; Ward et al. 1972) was also under way.

19.3.6 Human Development

The 1970s saw the advent of a large work stream that would extend over several decades. This work stream might be seen as a natural extension of Henry Chauncey's interest in human abilities, broadly conceived; that is, to understand human abilities, it made sense to study from where those abilities emanated. That stream, described in detail by Kogan et al. (Chap. 15, this volume), included research in many areas. In this period, it focused on infants and young children, encompassing their social development (Brooks and Lewis 1976; Lewis and Brooks-Gunn 1979), emotional development (Lewis 1977; Lewis et al. 1978; Lewis and Rosenblum 1978), cognitive development (Freedle and Lewis 1977; Lewis 1977, 1978), and parental influences (Laosa 1978; McGillicuddy-DeLisi et al. 1979).

19.3.7 Educational Evaluation and Policy Analysis

One of the more notable characteristics of ETS research in this period was the emergence of educational evaluation, in good part due to an increase in policy makers' interest in appraising the effects of investments in educational interventions. This work, described by Ball (Chap. 11, this volume), entailed large-scale evaluations of television programs like *Sesame Street* and *The Electric Company* (Ball and Bogatz 1970, 1973) and early computer-based instructional systems like PLATO and TICCIT (Alderman 1978; Murphy 1977), as well as a wide range of smaller studies (Marco 1972; Murphy 1973). Some of the accumulated wisdom gained in this period was synthesized in two books, the *Encyclopedia of Educational Evaluation* (Anderson et al. 1975) and *The Profession and Practice of Program Evaluation* (Anderson and Ball 1978).

Alongside the intense evaluation activity was the beginning of a work stream on policy analysis (see Coley et al., Chap. 12, this volume). That beginning concentrated on education finance (Goertz 1978; Goertz and Moskowitz 1978).

19.3.8 Teacher and Teaching Quality

Rounding out the very noticeable expansion of research activity that characterized the 1970s were several lines of work on teachers and teaching. One line concentrated on evaluating the functioning of the National Teachers Examination (NTE; Quirk et al. 1973). A second line revolved around observing and analyzing teaching behavior (Quirk et al. 1971, 1975). This line included the Beginning Teacher Evaluation Study, the purpose of which was to identify teaching behaviors effective in promoting learning in reading and mathematics in elementary schools, a portion of which was conducted by ETS under contract to the California Commission for Teacher Preparation and Licensing. The study included extensive classroom observation and analysis of the relations among the observed behaviors, teacher characteristics, and student achievement (McDonald and Elias 1976; Sandoval 1976). The final line of research concerned college teaching (Baird 1973; Centra 1974).

19.4 The Years 1980–1989

19.4.1 Psychometric and Statistical Methodology

As was true for the 1970s, in this decade, ETS methodological innovation was notable for its far-ranging impact. Lord (1980) furthered the development and application of IRT, with particular attention to its use in addressing a wide variety of testing problems, among them parameter estimation, linking, evaluation of differential item functioning (DIF), and adaptive testing. Holland (1986, 1987), as well as Holland and Rubin (1983), continued the work on causal inference, further developing its philosophical and epistemological foundations, including exploration of a longstanding statistical paradox described by Lord (1967).[2] An edited volume, *Drawing Inferences From Self-Selected Samples* (Wainer 1986), collected work on these issues.

Rubin's work on matching, particularly propensity score matching, was a key activity through this dccadc. Rubin (1980a), as well as Rosenbaum and Rubin (1984, 1985), made important contributions to this methodology. These widely cited publications outlined approaches that are frequently used in scientific research when experimental manipulation is not possible.

Building on his research of the previous decade, Rubin (1980b, c) developed "multiple imputation," a statistical technique for dealing with nonresponse by generating random draws from the posterior distribution of a variable, given other variables. The multiple imputations methodology forms the underlying basis for several major group-score assessments (i.e., tests for which the focus of inference is on population, rather than individual, performance), including the National Assessment of Educational Progress (NAEP), the Programme for International Student Assessment (PISA), and the Programme of International Assessment of Adult Competencies (PIAAC; Beaton and Barone, Chap. 8, this volume; Kirsch et al., Chap. 9, this volume).

Also of note was the emergence of DIF as an important methodological research focus. The standardization method (Dorans and Kulick 1986), and the more statistically grounded Mantel and Haenszel (1959) technique proposed by Holland and Thayer (1988), became stock approaches used by operational testing programs around the world for assessing item-level fairness. Finally, the research community working on DIF was brought together for an invited conference in 1989 at ETS.

Although there were a large number of observed-score equating studies in the 1980s, one development stands out in that it foreshadowed a line of research undertaken more than a decade later. The method of kernel equating was introduced by Holland and Thayer (1989) as a general procedure that combines smoothing,

---

[2] Lord's (1967) paradox refers to the situation, in observational studies, in which the statistical treatment of posttest scores by means of different corrections using pretest scores (i.e., regression vs. posttest minus pretest differences) can lead to apparent contradictions in results. This phenomenon is related to regression artifacts (D. T. Campbell and Kenny, 1999; Eriksson and Haggstrom, 2014).

modeling, and transforming score distributions. This combination of statistical procedures was intended to provide a flexible tool for observed-score equating in a nonequivalent-groups anchor-test design.

19.4.2 Large-Scale Survey Assessments of Student and Adult Populations

ETS was first awarded the contract for NAEP in 1983 after evaluating previous NAEP analytic procedures and releasing *A New Design for a New Era* (Messick et al. 1983). The award set the stage for advances in assessment design and psychometric methodology, including extensions of latent-trait models that employed covariates. These latent regression models used maximum likelihood methods to estimate population parameters from observed item responses without estimating individual ability parameters for test takers (Mislevy 1984, 1985). Many of the approaches developed for NAEP were later adopted by other national and international surveys, including the Progress in International Reading Literacy Study (PIRLS), the Trends in International Mathematics and Science Study (TIMSS), PISA, and PIAAC. These surveys are either directly modeled on NAEP or are based on other surveys that were themselves NAEP's direct derivates.

The major design and analytic features shared by these surveys include (a) a balanced incomplete block design that allows broad coverage of content frameworks, (b) use of modern psychometric methods to link across the multiple test forms covering this content, (c) integration of cognitive tests and respondent background data using those psychometric methods, and (d) a focus on student (and adult) populations rather than on individuals as the targets of inference and reporting.

Two related developments should be mentioned. The chapters by Kirsch et al. (Chap. 9, this volume) and Rock (Chap. 10, this volume) presented in more detail work on the 1984 Young Adult Literacy Study (YALS) and the 1988 National Educational Longnitudinal Study, respectively. These studies also use multiple test forms and advanced psychometric methods based on IRT. Moreover, YALS was the first to apply a multidimensional item response model (Kirsch and Jungeblut 1986).

19.4.3 Validity and Validation

The 1980s saw the culmination of Messick's landmark unified model (Messick 1989), which framed validity as a unitary concept. The highlight of the period, Messick's chapter in *Educational Measurement*, brought together the major strands of validity theory, significantly influencing conceptualization and practice throughout the field.

Also in this period, research on coaching burgeoned in response to widespread public and institutional user concerns (see Powers, Chap. 17, this volume). Notable was publication of *The Effectiveness of Coaching for the SAT: Review and Reanalysis of Research From the Fifties to the FTC* (Messick 1980), though many other studies were also released (Alderman and Powers 1980; Messick 1982; Powers 1985; Powers and Swinton 1984; Swinton and Powers 1983). Other sources of construct-irrelevant variance were investigated, particularly test anxiety (Powers 1988). Finally, conceptions of fairness became broader still, motivated by concerns over the flagging of scores from admissions tests that were administered under nonstandard conditions to students with disabilities; these concerns had been raised most prominently by a National Academy of Sciences panel (Sherman and Robinson 1982). Most pertinent was the 4-year program of research on the meaning and use of such test scores for the SAT and GRE General Test that was initiated in response to the panel's report. Results were summarized in the volume *Testing Handicapped People* by Willingham et al. (1988).

19.4.4 Constructed-Response Formats and Performance Assessment

Several key publications highlighted this period. Frederiksen's (1984) *American Psychologist* article "The Real Test Bias: Influences of Testing on Teaching and Learning" made the argument for the use of response formats in assessment that more closely approximated the processes and outcomes important for success in academic and work environments. This classic article anticipated the K–12 performance assessment movement of the 1990s and its 2010 resurgence in the Common Core Assessments. Also noteworthy were Breland's (1983) review showing the incremental predictive value of essay tasks over multiple-choice measures at the postsecondary level and his comprehensive study of the psychometric characteristics of such tasks (Breland et al. 1987). The Breland et al. volume included analyses of rater agreement, generalizability, and dimensionality. Finally, while research continued on the formulating-hypotheses item type (Ward et al. 1980), the investigation of portfolios also emerged (Camp 1985).

19.4.5 Personal Qualities

Although investigation of cognitive style continued in this period (Goodenough et al. 1987; Messick 1987; Witkin and Goodenough 1981), the death of Herman Witkin in 1979 removed its intellectual leader and champion, contributing to its decline. This decline coincided with a drop in attention to personal qualities research more generally, following a shift in ETS management priorities from the very clear think tank orientation of the 1960s and 1970s to a greater focus on research to assist existing testing programs and the creation of new ones. That focus remained centered largely on traditional academic abilities, though limited research proceeded on creativity (Baird and Knapp 1981; Ward et al. 1980).

19.4.6 Human Development

Whereas the research on personal qualities noticeably declined, the work on human development remained vibrant, at least through the early part of this period, in large part due to the availability of external funding and staff members highly skilled at attracting it. With a change in management focus, the reassignment of some developmental staff to other work, and the subsequent departure of the highly prolific Michael Lewis, interest began to subside. Still, this period saw a considerable amount and diversity of research covering social development (Brooks-Gunn and Lewis 1981; Lewis and Feiring 1982), emotional development (Feinman and Lewis 1983; Lewis and Michalson 1982), cognitive development (Lewis and Brooks-Gunn 1981a, b; Sigel 1982), sexual development (Brooks-Gunn 1984; Brooks-Gunn and Warren 1988), development of Chicano children (Laosa 1980a, 1984), teenage motherhood (Furstenberg et al. 1987), perinatal influences (Brooks-Gunn and Hearn 1982), parental influences (Brody et al. 1986; Laosa 1980b), atypical development (Brinker and Lewis 1982; Brooks-Gunn and Lewis 1982), and interventions for vulnerable children (Brooks-Gunn et al. 1988; Lee et al. 1988).

19.4.7 Educational Evaluation and Policy Analysis

As with personal qualities, the evaluation of educational programs began to decline during this period. In contrast to the work on personal qualities, evaluation activities had been almost entirely funded through outside grants and contracts, which diminished considerably in the 1980s. In addition, the organization's most prominent evaluator, Samuel Ball, departed to take an academic appointment in his native Australia. The work that remained investigated the effects of instructional software like the IBM Writing to Read program (Murphy and Appel 1984), educational television (Murphy 1988), alternative higher education programs (Centra and Barrows 1982), professional training (Campbell et al. 1982), and the educational integration of students with severe disabilities (Brinker and Thorpe 1984).

Whereas funding for evaluation was in decline, support for policy analysis grew. Among other things, this work covered finance (Berke et al. 1984), teacher policy (Goertz et al. 1984), education reform (Goertz 1989), gender equity (Lockheed 1985), and access to and participation in graduate education (Clewell 1987).

19.4.8 Teacher and Teaching Quality

As with program evaluation, the departure of key staff during this period resulted in diminished activity, with only limited attention given to the three dominant lines of research of the previous decade: functioning of the NTE (Rosner and Howey 1982), classroom observation (Medley and Coker 1987; Medley et al. 1981), and college teaching (Centra 1983). Of particular note was Centra and Potter's (1980) article "School and Teacher Effects: An Interrelational Model," which proposed an early structural model for evaluating input and context variables in relation to achievement.

19.5 The Years 1990–1999

19.5.1 Psychometric and Statistical Methodology

DIF continued to be an important methodological research focus. In the early part of the period, an edited volume, *Differential Item Functioning*, was released based on the 1989 DIF conference (Holland and Wainer 1993). Among other things, the volume included research on the Mantel–Haenszel (1959) procedure. Other publications, including on the standardization method, have had continued impact on practice (Dorans and Holland 1993; Dorans et al. 1992). Finally, of note were studies that placed DIF into model-based frameworks. The use of mixture models (Gitomer and Yamamoto 1991; Mislevy and Verhelst 1990; Yamamoto and Everson 1997), for example, illustrated how to relax invariance assumptions and test DIF in generalized versions of item response models.

Among the notable methodological book publications of this period was *Computer Adaptive Testing: A Primer*, edited by Wainer et al. (1990). This volume contained several chapters by ETS staff members and their colleagues.

Also worthy of mention was research on extended IRT models, which resulted in several major developments. Among these developments were the generalized partial credit model (Muraki 1992), extensions of mixture IRT models (Bennett et al. 1991; Gitomer and Yamamoto 1991; Yamamoto and Everson 1997), and models that were foundational for subsequent generalized modeling frameworks. Several chapters in the edited volume *Test Theory for a New Generation of Tests* (Frederiksen et al. 1993) described developments around these extended IRT models.

19.5.2 Large-Scale Survey Assessments of Student and Adult Populations

NAEP entered its second decade with the new design and analysis methodology introduced by ETS. Articles describing these methodological innovations were published in a special issue of the *Journal of Educational Statistics* (Mislevy et al. 1992b; Yamamoto and Mazzeo 1992). Many of these articles remain standard references, used as a basis for extending the methods and procedures of group-score assessments. In addition, Mislevy (1991, 1993a, b) continued work on related issues.

A significant extension to the large-scale assessment work was a partnership with Statistics Canada that resulted in development of the International Adult Literacy Survey (IALS). IALS collected data in 23 countries or regions of the world, 7 in 1994 and an additional 16 in 1996 and 1998 (Kirsch et al., Chap. 9, this volume). Also in this period, ETS research staff helped the International Association for the Evaluation of Educational Achievement (IEA) move the TIMSS 1995 and 1999 assessments to a more general IRT model, later described by Yamamoto and Kulick (2002). Finally, this period saw the beginning of the Early Childhood Longitudinal Study, Kindergarten Class of 1998–1999 (ECLS-K), which followed students through the eighth grade (Rock, Chap. 10, this volume).

19.5.3 Validity and Validation

Following the focus on constructs advocated by Messick's (1989) chapter, the 1990s saw a shift in thinking that resulted in concerted attempts to ground assessment design in domain theory, particularly in domains in which design had been previously driven by content frameworks. Such theories often offered a deeper and clearer description of the cognitive components that made for domain proficiency and the relationships among the components. A grounding in cognitive-domain theory offered special advantages for highly interactive assessments like simulations because of the expense involved in their development, which increased dramatically without the guidance provided by theory for task creation and scoring. From Messick (1994a), and from work on an intelligent tutoring system that combined domain theory with rigorous probability models (Gitomer et al. 1994), the foundations of evidence-centered design (ECD) emerged (Mislevy 1994, 1996). ECD, a methodology for rigorously reasoning from assessment claims to task development, and from item responses back to claims, is now used throughout the educational assessment community as a means of creating a stronger validity argument a priori.

During this same period, other investigators explored how to estimate predictive validity coefficients by taking into account differences in grading standards across college courses (Ramist et al. 1994). Finally, fairness for population groups remained in focus, with continued attention to admissions testing for students with disabilities (Bennett 1999) and release of the book *Gender and Fair Assessment* by Willingham and Cole (1997), which comprehensively examined the test performance of males and females to identify potential sources of unfairness and possible solutions.

19.5.4 Constructed-Response Formats and Performance Assessment

At both the K–12 and postsecondary levels, interest in moving beyond multiple-choice measures was widespread. ETS work reflected that interest and, in turn, contributed to it. Highlights included Messick's (1994a) paper on evidence and consequences in the validation of performance assessments, which provided part of the conceptual basis for the invention of ECD, and publication of the book *Construction Versus Choice in Cognitive Measurement* (Bennett and Ward 1993), framing the breadth of issues implicated in the use of non-multiple-choice formats.

In this period, many aspects of the functioning of constructed-response formats were investigated, including construct equivalence (Bennett et al. 1991; Bridgeman 1992), population invariance (Breland et al. 1994; Bridgeman and Lewis 1994), and effects of allowing test takers choice in task selection (Powers and Bennett 1999). Work covered a variety of presentation and response formats, including formulating hypotheses (Bennett and Rock 1995), portfolios (Camp 1993; LeMahieu et al. 1995), and simulations for occupational and professional assessment (Steinberg and Gitomer 1996).

Appearing in this decade were ETS's first attempts at automated scoring, including of computer science subroutines (Braun et al. 1990), architectural designs (Bejar 1991), mathematical step-by-step solutions and expressions (Bennett et al. 1997; Sebrechts et al. 1991), short-text responses (Kaplan 1992), and essays (Kaplan et al. 1995). By the middle of the decade, the work on scoring architectural designs had been implemented operationally as part of the National Council of Architectural Registration Board's Architect Registration Examination (Bejar and Braun 1999). Also introduced at the end of the decade into the Graduate Management Admission Test was the *e-rater*® automated scoring engine, an approach to automated essay scoring (Burstein et al. 1998). The e-rater scoring engine continues to be used operationally for the GRE General Test Analytical Writing Assessment, the *TOEFL*® test, and other examinations.

19.5.5 Personal Qualities

Interest in this area had been in decline since the 1980s. The 1990s brought an end to the cognitive styles research, with only a few publications released (Messick 1994b, 1996). Some research on creativity continued (Bennett and Rock 1995; Enright et al. 1998).

19.5.6 Human Development

As noted, work in this area also began to decline in the 1980s. The 1990s saw interest diminish further with the departure of Jeanne Brooks-Gunn, whose extensive publications covered an enormous substantive range. Still, a significant amount of research was completed, including on parental influences and beliefs (Sigel 1992), representational competence (Sigel 1999), the distancing model (Sigel 1993), the development of Chicano children (Laosa 1990), and adolescent sexual, emotional, and social development (Brooks-Gunn 1990).

19.5.7 Education Policy Analysis

This period saw the continuation of a vibrant program of policy studies. Multiple areas were targeted, including finance (Barton et al. 1991), teacher policy (Bruschi and Coley 1999), education reform (Barton and Coley 1990), education technology (Coley et al. 1997), gender equity (Clewell et al. 1992), education and the economy (Carnevale 1996; Carnevale and DesRochers 1997), and access to and participation in graduate education (Ekstrom et al. 1991; Nettles 1990).

19.5.8 Teacher and Teaching Quality

In this period, a resurgence of interest occurred due to the need to build the foundation for the *PRAXIS®* program, which replaced the NTE. An extensive series of surveys, job analyses, and related studies was conducted to understand the knowledge, skills, and abilities required for newly licensed teachers (Reynolds et al. 1992; Tannenbaum 1992; Tannenbaum and Rosenfeld 1994). As in past decades, work was done on classroom performance (Danielson and Dwyer 1995; Powers 1992), some of which supplied the initial foundation for the widely used *Framework for Teaching Evaluation Instrument* (Danielson 2013).

19.6 The Years 2000–2009

19.6.1 Psychometric and Statistical Methodology

The first decade of the current century saw increased application of Bayesian methods in psychometric research, in which staff members continued ETS's tradition of integrating advances in statistics with educational and psychological measurement. Among the applications were posterior predictive checks (Sinharay 2003), a method not unlike the frequentist resampling and resimulation studied in the late 1990s (M. von Davier 1997), as well as the use of Bayesian networks to specify complex measurement models (Mislevy et al. 2000). Markov chain Monte Carlo methods were employed to explore the comprehensive estimation of measurement and structural models in modern IRT (Johnson and Jenkins 2005) but, because of their computational requirements, currently remain limited to small- to medium-sized applications.

Alternatives to these computationally demanding methods were considered to enable the estimation of high-dimensional models, including empirical Bayes methods and approaches that utilized Monte Carlo integration, such as the stochastic EM algorithm (M. von Davier and Sinharay 2007).

These studies were aimed at supporting the use of explanatory IRT applications taking the form of a latent regression that includes predictive background variables in the structural model. Models of this type are used in the NAEP, PISA, PIAAC, TIMSS, and PIRLS assessments, which ETS directly or indirectly supported. Sinharay and von Davier (2005) also presented extensions of the basic numerical integration approach to data having more dimensions. Similar to Johnson and Jenkins (2005), who proposed a Bayesian hierarchical model for the latent regression, Li et al. (2009) examined the use of hierarchical linear (or multilevel) extensions of the latent regression approach.

The kernel equating procedures proposed earlier by Holland and Thayer (1989; also Holland et al. 1989) were extended and designs for potential applications were described in *The Kernel Method of Test Equating* by A. A. von Davier, Holland, and Thayer (2004). The book's framework for observed-score equating encapsulates several well-known classical methods as special cases, from linear to equipercentile approaches.

A major reference work was released, titled *Handbook of Statistics: Vol. 26. Psychometrics* and edited by Rao and Sinharay (2006). This volume contained close to 1200 pages and 34 chapters reviewing state-of-the-art psychometric modeling. Sixteen of the volume's chapters were contributed by current or former ETS staff members.

The need to describe test-taker strengths and weaknesses has long motivated the reporting of subscores on tests that were primarily designed to provide a single score. Haberman (2008) presented the concept of proportional reduction of mean squared errors, which allows an evaluation of whether subscores are technically defensible. This straightforward extension of classical test theory derives from a formula introduced by Kelley (1927) and provides a tool to check whether a subscore is reliable enough to stand on its own or whether the true score of the subscore under consideration would be better represented by the observed total score. (Multidimensional IRT was subsequently applied to this issue by Haberman and Sinharay 2010, using the same underlying argument.)

Also for purposes of better describing test-taker strengths and weaknesses, generalized latent variable models were explored, but with the intention of application to tests designed to measure multiple dimensions. Apart from the work on Bayesian networks (Mislevy and Levy 2007; Mislevy et al. 2003), there were significant extensions of approaches tracing back to the latent class models of earlier decades (Haberman 1988) and to the rule space model (Tatsuoka 1983). Among these extensions were developments around the reparameterized unified model (DiBello et al. 2006), which was shown to partially alleviate the identification issues of the earlier unified model, as well as around the general diagnostic model (GDM; M. von Davier 2008a). The GDM was shown to include many standard and extended IRT models, as well as several diagnostic models, as special cases (M. von Davier 2008a, b). The GDM has been successfully applied to the *TOEFL iBT*® test, PISA, NAEP, and PIRLS data in this as well as in the subsequent decade (M. von Davier 2008a; Oliveri and von Davier 2011, 2014; Xu and von Davier 2008). Other approaches later developed outside of ETS, such as the log-linear cognitive diagnostic model (LCDM; Henson et al. 2009), can be directly traced to the GDM (e.g., Rupp et al. 2010) and have been shown to be a special case of the GDM (von Davier 2014).

19.6.2 Large-Scale Survey Assessments of Student and Adult Populations

As described by Rock (Chap. 10, this volume), the Early Childhood Longitudinal Study continued through much of this decade, with the last data collection in the eighth grade, taking place in 2007. Also, recent developments in the statistical procedures used in NAEP were summarized and future directions described (M. von Davier et al. 2006).

A notable milestone was the Adult Literacy and Lifeskills (ALL) assessment, conducted in 2003 and 2006–2008 (Kirsch et al., Chap. 9, this volume). ALL was a household-based, international comparative study designed to provide participating countries with information about the literacy and numeracy skills of their adult populations. To accomplish this goal, ALL used nationally representative samples of 16- to 65-year-olds.

In this decade, ETS staff members completed a multicountry feasibility study for PISA of computer-based testing in multiple languages (Lennon, Kirsch, von Davier, Wagner, and Yamamoto 2003) and a report on linking and linking stability (Mazzeo and von Davier 2008).

Finally, in 2006, ETS and IEA established the IEA/ETS research institute (IERI), which promotes research on large-scale international skill surveys, publishes a journal, and provides training around the world through workshops on statistical and psychometric topics (Wagemaker and Kirsch 2008).

19.6.3 Validity and Validation

In the 2000s, Mislevy and colleagues elaborated the theory and generated additional prototypic applications of ECD (Mislevy et al. 2003, 2006), including proposing extensions of the methodology to enhance accessibility for individuals from special populations (Hansen and Mislevy 2006). Part of the motivation behind ECD was the need to more deeply understand the constructs to be measured and to use that understanding for assessment design. In keeping with that motivation, the beginning of this period saw the release of key publications detailing construct theory for achievement domains, which feed into the domain analysis and modeling aspects of ECD. Those publications concentrated on elaborating the construct of communicative competence for the TOEFL computer-based test (CBT), comprising listening, speaking, writing, and reading (Bejar et al. 2000; Butler et al. 2000; Cumming et al. 2000; Enright et al. 2000). Toward the end of the period, the Cognitively Based Assessment of, for, and as Learning (*CBAL*®) initiative (Bennett and Gitomer 2009) was launched. This initiative took a similar approach to construct definition as TOEFL CBT but, in CBAL's case, to the definition of English language arts and mathematics constructs for elementary and secondary education.

At the same time, the communication of predictive validity results for postsecondary admissions tests was improved. Building upon earlier work, Bridgeman and colleagues showed how the percentage of students who achieved a given grade point average increased as a function of score level, a more easily understood depiction than the traditional validity coefficient (Bridgeman et al. 2008). Also advanced was the research stream on test anxiety, one of several potential sources of irrelevant variance (Powers 2001).

Notable too was the increased attention given students from special populations. For students with disabilities, two research lines dominated, one related to testing and validation concerns that included but went beyond the postsecondary admissions focus of the 1980s and 1990s (Ekstrom and Smith 2002; Laitusis et al. 2002), and the second on accessibility (Hansen et al. 2004; Hansen and Mislevy 2006; Hansen et al. 2005). For English learners, topics covered accessibility (Hansen and Mislevy 2006; Wolf and Leon 2009), accommodations (Young and King 2008), validity frameworks and assessment guidelines (Pitoniak et al. 2009; Young 2009), and instrument and item functioning (Martiniello 2009; Young et al. 2008).

19.6.4 Constructed-Response Formats and Performance Assessment

Using ECD, several significant computer-based assessment prototypes were developed, including for NAEP (Bennett et al. 2007) and for occupational and professional assessment (Mislevy et al. 2002). The NAEP Technology-Rich Environments project was significant because assessment tasks involving computer simulations were administered to nationally representative samples of students and because it included an analysis of students' solution processes. This study was followed by NAEP's first operational technology-based component, the Interactive Computer Tasks, as part of the 2009 science assessment (U.S. Department of Education, n.d.-a). Also of note was the emergence of research on games and assessment (Shute et al. 2008, 2009).

With the presentation of constructed-response formats on computer came added impetus to investigate the effect of computer familiarity on performance. That issue was explored for essay tasks in NAEP (Horkay et al. 2006) as well as for the entry of complex expressions in mathematical reasoning items (Gallagher et al. 2002).

Finally, attention to automated scoring increased considerably. Streams of research on essay scoring and short-text scoring expanded (Attali and Burstein 2006; Leacock and Chodorow 2003; Powers et al. 2002; Quinlan et al. 2009), a new line on speech scoring was added (Zechner et al. 2007, 2009), and publications were released on the grading of graphs and mathematical expressions (Bennett et al. 2000).

19.6.5 Personal Qualities

Although it almost disappeared in the 1990s, ETS's interest in this topic reemerged following from the popularization of so-called noncognitive constructs in education, the workplace, and society at large (Goleman 1995). Two highly visible topics accounted for a significant portion of the research effort, one being emotional intelligence (MacCann and Roberts 2008; MacCann et al. 2008; Roberts et al. 2006) and the other stereotype threat (Stricker and Bejar 2004; Stricker and Ward 2004), the notion that concern about a negative belief as to the ability of one's demographic group might adversely affect test performance.

19.6.6 Human Development

With the death of Irving Sigel in 2006, the multidecade history of contributions to this area ended. Before his death, however, Sigel continued to write actively on the distancing model, representation, parental beliefs, and the relationship between research and practice generally (Sigel 2000, 2006). Notable in this closing period was publication of his coedited *Child Psychology in Practice*, volume 4 of the *Handbook of Child Psychology* (Renninger and Sigel 2006).

19.6.7 Education Policy Analysis

Work in this area increased considerably. Several topics stood out for the attention given them. In elementary and secondary education, the achievement gap (Barton 2003), gender equity (Coley 2001), the role of the family (Barton and Coley 2007), and access to advanced course work in high school (Handwerk et al. 2008) were each examined. In teacher policy and practice, staff examined approaches to teacher preparation (Wang et al. 2003) and the quality of the teaching force (Gitomer 2007b).

With respect to postsecondary populations, new analyses were conducted of data from the adult literacy surveys (Rudd et al. 2004; Sum et al. 2002), and access to graduate education was studied (Nettles and Millett 2006). A series of publications by Carnevale and colleagues investigated the economic value of education and its equitable distribution (Carnevale and Fry 2001, 2002; Carnevale and Rose 2000). Among the many policy reports released, perhaps the highlight was *America's Perfect Storm* (Kirsch et al. 2007), which wove labor market trends, demographics, and student achievement into a social and economic forecast that received international media attention.

19.6.8 Teacher and Teaching Quality

Notable in this period were several lines of research. One centered on the functioning and impact of the certification assessments created by ETS for the National Board of Professional Teaching Standards (Gitomer 2007a; Myford and Engelhard 2001), which included the rating of video-recorded classroom performances. A second line more generally explored approaches for the evaluation of teacher effectiveness and teaching quality (Gitomer 2009; Goe et al. 2008; Goe and Croft 2009) as well as the link between teaching quality and student outcomes (Goe 2007). Deserving special mention was Braun's (2005) report "Using Student Progress to Evaluate Teachers: A Primer on Value-Added Models," which called attention to the problems with this approach. Finally, a third work stream targeted professional development, including enhancing teachers' formative assessment practices (Thompson and Goe 2009; Wylie et al. 2009).

19.7 The Years 2010–2016

19.7.1 Psychometric and Statistical Methodology

Advances in computation have historically been an important driver of psychometric developments. In this period, staff members continued to create software packages, particularly for complex multidimensional analyses. One example was software for the operational use of multidimensional item response theory (MIRT) for simultaneous linking of multiple assessments (Haberman 2010). Another example was software for the operational use of the multidimensional discrete latent-trait model for IRT (and MIRT) calibration and linking (M. von Davier and Rost 2016). This software is used extensively for PIAAC and PISA.

Whereas software creation has constituted a continued line of activity, research on how to reduce computational burden has also been actively pursued. Of note in this decade was the use of graphical modeling frameworks to reduce the calculations required for complex multidimensional estimation. Rijmen (2010) as well as Rijmen et al. (2014) showed how these advances can be applied in large-scale testing applications, producing research software for that purpose. On a parallel track, von Davier (2016) described the use of all computational cores of a workstation or server to solve measurement problems in many dimensions more efficiently and to analyze the very large data sets coming from online testing and large-scale assessments of national or international populations.

In the same way as advances in computing have spurred methodological innovation, those computing advances have made the use of new item response types more feasible (Bejar, Chap. 18, this volume). Such response types have, in turn, made new analytic approaches necessary. Research has examined psychometric models and latent-trait estimation for items with multiple correct choices, self-reports using anchoring vignettes, data represented as multinomial choice trees, and responses collected from interactive and simulation tasks (Anguiano-Carrasco et al. 2015; Khorramdel and von Davier 2014), in the last case including analysis of response time and solution process.

Notable methodological publications collected in edited volumes in this period covered linking (von Davier 2011), computerized multistage testing (Yan et al. 2014), and international large-scale assessment methodology (Rutkowski et al. 2013). In addition, several contributions appeared by ETS authors in a three-volume handbook on IRT (Haberman 2016; von Davier and Rost 2016). Chapters by other researchers detail methods and statistical tools explored while those individuals were at ETS (e.g., Casabianca and Junker 2016; Moses 2016; Sinharay 2016).

19.7.2 Large-Scale Survey Assessments of Student and Adult Populations

In this second decade of the twenty-first century, the work of many research staff members was shaped by the move to computer-based, large-scale assessment. ETS became the main contractor for the design, assessment development, analysis, and project management of both PIAAC and PISA. PIAAC was fielded in 2012 as a multistage adaptive test (Chen et al. 2014b). In contrast, PISA 2015 was administered as a linear test with three core domains (science, mathematics, and reading), one innovative assessment domain (collaborative problem solving), and one optional domain (financial literacy).

NAEP also fielded computer-based assessments in traditional content domains and in domains that would not be suitable for paper-and-pencil administration. Remarkable were the delivery of the 2011 NAEP writing assessment on computer (U.S. Department of Education, n.d.-b) and the 2014 Technology and Engineering Literacy assessment (U.S. Department of Education, n.d.-c). The latter assessment contained highly interactive simulation tasks involving the design of bicycle lanes and the diagnosis of faults in a water pump. A large pilot study exploring multistage adaptive testing was also carried out (Oranje and Ye 2013) as part of the transition of all NAEP assessments to administration on computers.

Finally, ETS received the contract for PISA 2018, which will also entail the use of computer-based assessments in both traditional and nontraditional domains.

19.7.3 Validity and Validation

The work on construct theory in achievement domains for elementary and secondary education that was begun in the prior decade continued with publications in the English language arts (Bennett et al. 2016; Deane et al. 2015; Deane and Song 2015; Sparks and Deane 2015), mathematics (Arieli-Attali and Cayton-Hodges 2014; Graf 2009), and science (Liu et al. 2013). These publications detailed the CBAL competency, or domain, models and their associated learning progressions, that is, the pathways most students might be expected to take toward domain competency. Also significant was the Reading for Understanding project, which reformulated and exemplified the construct of reading comprehension for the digital age (Sabatini and O'Reilly 2013). Finally, a competency model was released for teaching (Sykes and Wilson 2015), intended to lay the foundation for a next generation of teacher licensure assessment.

In addition to domain modeling, ETS's work in validity theory was extended in several directions. The first direction was through further development of ECD, in particular its application to educational games (Mislevy et al. 2014). A second direction resulted from the arrival of Michael Kane, whose work on the argument-based approach added to the research program very substantially (Kane 2011, 2012,

2016). Finally, fairness and validity were combined in a common framework by Xi (2010).

Concerns for validity and fairness continued to motivate a wide-ranging research program directed at students from special populations. For those with disabilities, topics included accessibility (Hansen et al. 2012; Stone et al. 2016), accommodations (Cook et al. 2010), instrument and item functioning (Buzick and Stone 2011; Steinberg et al. 2011), computer-adaptive testing (Stone et al. 2013; Stone and Davey 2011), automated versus human essay scoring (Buzick et al. 2016), and the measurement of growth (Buzick and Laitusis 2010a, b). For English learners, topics covered accessibility (Guzman-Orth et al. 2016; Young et al. 2014), accommodations (Wolf et al. 2012a, b), instrument functioning (Gu et al. 2015; Young et al. 2010), test use (Lopez et al. 2016; Wolf and Farnsworth 2014; Wolf and Faulkner-Bond 2016), and the conceptualization of English learner proficiency assessment systems (Hauck et al. 2016; Wolf et al. 2016).

19.7.4 Constructed-Response Formats and Performance Assessment

As a consequence of growing interest in education, the work on games and assessment that first appeared at the end of the previous decade dramatically increased (Mislevy et al. 2012, 2014, 2016; Zapata-Rivera and Bauer 2012).

Work on automated scoring also grew substantially. The focus remained on response types from previous periods, such as essay scoring (Deane 2013a, b), short answer scoring (Heilman and Madnani 2012), speech scoring (Bhat and Yoon 2015; Wang et al. 2013), and mathematical responses (Fife 2013). However, important new lines of work were added. One such line, made possible by computer-based assessment, was the analysis of keystroke logs generated by students as they responded to essays, simulations, and other performance tasks (Deane and Zhang 2015; He and von Davier 2015, 2016; Zhang and Deane 2015). This analysis began to open a window into the processes used by students in problem solving. A second line, also made possible by advances in technology, was conversation-based assessment, in which test takers interact with avatars (Zapata-Rivera et al. 2014). Finally, a work stream was initiated on "multimodal assessment," incorporating analysis of test-taker speech, facial expression, or other behaviors (Chen et al. 2014a, c).

19.7.5 Personal Qualities

While work on emotional intelligence (MacCann et al. 2011; MacCann et al. 2010; Roberts et al. 2010), and stereotype threat (Stricker and Rock 2015) continued, this period saw a significant broadening to a variety of noncognitive constructs and their applications. Research and product development were undertaken in education (Burrus et al. 2011; Lipnevich and Roberts 2012; Oliveri and Ezzo 2014) as well as for the workforce (Burrus et al. 2013; Naemi et al. 2014).

19.7.6 Education Policy Analysis

Although the investigation of economics and education had diminished due to the departure of Carnevale and his colleagues, attention to a wide range of policy problems continued. Those problems related to graduate education (Wendler et al. 2010), minority representation in teaching (Nettles et al. 2011), developing and implementing teacher evaluation systems (Goe et al. 2011), testing at the pre-K level (Ackerman and Coley 2012), achievement gaps in elementary and secondary education (Barton and Coley 2010), and parents opting their children out of state assessment (Bennett 2016).

A highlight of this period was the release of two publications from the ETS Opportunity Project. The publications, "Choosing Our Future: A Story of Opportunity in America" (Kirsch et al. 2016) and "The Dynamics of Opportunity in America" (Kirsch and Braun 2016), comprehensively analyzed and directed attention toward issues of equality, economics, and education in the United States.

19.7.7 Teacher and Teaching Quality

An active and diverse program of investigation continued. Support was provided for testing programs, including an extensive series of job analyses for revising PRAXIS program assessments (Robustelli 2010) as well as work toward the development of new assessments (Phelps and Howell 2016; Sykes and Wilson 2015). The general topic of teacher evaluation remained a constant focus (Gitomer and Bell 2013; Goe 2013; Turkan and Buzick 2016), including continued investigation into implementing it through classroom observation (Casabianca et al. 2013; Lockwood et al. 2015; Mihaly and McCaffrey 2014) and value-added modeling (Buzick and Jones 2015; McCaffrey 2013; McCaffrey et al. 2014). Researchers also explored the impact of teacher characteristics and teaching practices on student achievement (Liu et al. 2010), the effects of professional development on teacher knowledge (Bell et al. 2010), and the connection between teacher evaluation and professional learning (Goe et al. 2012). One highlight of the period was release of the fifth edition of AERA's *Handbook of Research on Teaching* (Gitomer and Bell 2016), a comprehensive reference for the field. A second highlight was *How Teachers Teach: Mapping the Terrain of Practice* (Sykes and Wilson 2015), which, as noted earlier, laid out a conceptualization of teaching in the form of a competency model.

19.8 Discussion

As the previous sections might suggest, the history of ETS research is marked by both constancy and changes in focus. The constancy can be seen in persistent attention to problems at the core of educational and psychological measurement. Those problems have centered on developing and improving the psychometric and statistical methodology that helps connect observations to inferences about individuals, groups, and institutions. In addition, the problems have centered on evaluating those inferences—that is, the theory, methodology, and practice of validation.

The changes in focus across time have occurred both within these two persistently pursued areas and among those areas outside of the measurement core. For example, Kane and Bridgeman (Chap. 16, this volume) documented in detail the progression that has characterized ETS's validity research, and multiple chapters did the same for the work on psychometrics and statistics. In any event, the emphasis given these core areas remained strong throughout ETS's history.

As noted, other areas experienced more obvious peaks and valleys. Several of these areas did not emerge as significant research programs in their own right until considerably after ETS was established. That characterization would be largely true, for example, of human development (beginning in the 1970s), educational evaluation (1970s), large-scale assessment/adult literacy/longitudinal studies (1970s), and policy analysis (1980s), although there were often isolated activities that preceded these dates. Once an area emerged, it did not necessarily persist, the best examples being educational evaluation, which spanned the 1970s to 1980s, and human development, which began at a similar time point, declined through the late 1980s and 1990s, and reached its denouement in the 2000s.

Still other areas rose, fell, and rose again. Starting with the founding of ETS, work on personal qualities thrived for three decades, all but disappeared in the 1980s and 1990s, and returned by the 2000s close to its past levels, but this time with the added focus of product development. The work on constructed-response formats and performance assessment also began early on and appeared to go dormant in the 1970s, only to return in the 1980s. In the 1990s, the emphasis shifted from a focus on paper-and-pencil measurement to presentation and scoring by computer.

What drove the constancy and change over the decades? The dynamics were most likely due to a complex interaction among several factors. One factor was certainly the influence of the external environment, including funding, federal education policy, public opinion, and the research occurring in the field. That environment, in turn, affected (and was affected by) the areas of interest and expertise of those on staff who, themselves, had impact on research directions. Finally the interests of the organization's management were affected by the external environment and, in turn, motivated actions that helped determine the staff composition and research priorities.

Aside from the changing course of research over the decades, a second striking characteristic is the vast diversity of the work. At its height, this diversity arguably rivaled that found in the psychology and education departments of major research universities anywhere in the world. Moreover, in some areas—particularly in psychometrics and statistics—it was often considerably deeper.

This breadth and depth led to substantial innovation, as this chapter has highlighted and the prior ones have detailed. That innovation was often highly theoretical—as in Witkin and Goodenough's (1981) work on cognitive styles, Sigel's (1990) distancing theory, Lord and Novick's (1968) seminal volume on IRT, Messick's (1989) unified conception of validity, Mislevy's (1994, 1996) early work on ECD, Deane et al.'s (2015) English language arts competency model, and Sykes and Wilson's (2015) conceptions of teaching practice. But that innovation was also very often practical—witness the in-basket test (Frederiksen et al. 1957), LISREL (Jöreskog and van Thillo 1972), the EM algorithm (Dempster et al. 1977), Lord's (1980) "Applications of Item Response Theory to Practical Testing Problems," the application of Mantel–Haenszel to DIF (Holland and Thayer 1988), the plausible-values solution to the estimation of population performance in sample surveys (Mislevy et al. 1992a), and e-rater (Burstein et al. 1998). These innovations were not only useful but *used*, in all the preceding cases widely employed in the measurement community, and in some cases used throughout the sciences.

Of no small consequence is that ETS innovations—theory and practical development—were employed throughout the organization's history to support, challenge, and improve the technical quality of its testing programs. Among other things, the challenges took the form of a continuing program of validity research to identify and address construct-irrelevant influences, for example, test anxiety, coaching, stereotype threat, lack of computer familiarity, English language complexity in content assessments, and accessibility—which might unfairly affect the performance of individuals and groups.

A final observation is that research was used not only for the generation of theory and of practical solutions in educational and psychological studies but also for helping government officials and the public address important policy problems. The organization's long history of contributions to informing policy are evident in its roles with respect to the Equality of Educational Opportunity Study (Beaton 1968); the evaluation of *Sesame Street* (Ball and Bogatz 1970); the Head Start, early childhood, and high school longitudinal studies; the adult literacy studies; NAEP, PISA, and PIAAC; and the many policy analyses of equity and opportunity in the United States (Kirsch et al. 2007; Kirsch and Braun 2016).

We close this chapter, and the book, by returning to the concept of a nonprofit measurement organization as outlined by Bennett (Chap. 1, this volume). In that conception, the organization's raison d'être is public service. Research plays a fundamental role in realizing that public service obligation to the extent that it helps advance educational and psychological measurement as a field, acts as a mechanism for enhancing (and routinely challenging) the organization's testing programs, and helps contribute to the solution of big educational and social challenges. We would assert that the evidence presented indicates that, taken over its almost 70-year history, the organization's research activities have succeeded in filling that fundamental role.

References

Ackerman, D. J., & Coley, R. J. (2012). *State pre-K assessment policies: Issues and status* (Policy Information Report). Princeton: Educational Testing Service.

Alderman, D. L. (1978). *Evaluation of the TICCIT computer-assisted instructional system in the community college*. Princeton: Educational Testing Service.

Alderman, D. L., & Powers, D. E. (1980). The effects of special preparation on SAT verbal scores. *American Educational Research Journal, 17*, 239–251. https://doi.org/10.3102/00028312017002239

Andersen, E. B. (1972). *A computer program for solving a set of conditional maximum likelihood equations arising in the Rasch model for questionnaires* (Research Memorandum No. RM-72-06). Princeton: Educational Testing Service.

Andersen, E. B. (1973). A goodness of fit test for the Rasch model. *Psychometrika, 38*, 123–140. https://doi.org/10.1007/BF02291180

Anderson, S. B., & Ball, S. (1978). *The profession and practice of program evaluation*. San Francisco: Jossey-Bass.

Anderson, S. B., Beaton, A. E., Emmerich, W., & Messick, S. J. (1968). *Disadvantaged children and their first school experiences: ETS-OEO Longitudinal Study—Theoretical considerations and measurement strategies* (Program Report No. PR-68-04). Princeton: Educational Testing Service.

Anderson, S. B., Ball, S., & Murphy, R. T. (Eds.). (1975). *Encyclopedia of educational evaluation: Concepts and techniques for evaluating education and training programs*. San Francisco: Jossey-Bass.

Angoff, W. H. (1953). *Equating of the ACE psychological examinations for high school students* (Research Bulletin No. RB-53-03). Princeton: Educational Testing Service. http://dx.doi.org/10.1002/j.2333-8504.1953.tb00887.x

Anguiano-Carrasco, C., MacCann, C., Geiger, M., Seybert, J. M., & Roberts, R. D. (2015). Development of a forced-choice measure of typical-performance emotional intelligence. *Journal of Psychoeducational Assessment, 33*, 83–97. http://dx.doi.org/10.1002/10.1177/0734282914550387

Arieli-Attali, M., & Cayton-Hodges, G. A. (2014). *Expanding the CBAL mathematics assessments to elementary grades: The development of a competency model and a rational number learning progression* (Research Report No. RR-14-08). Princeton: Educational Testing Service. http://dx.doi.org/10.1002/ets2.12008

Attali, Y., & Burstein, J. (2006). Automated essay scoring with e-rater V.2. *Journal of Technology, Learning, and Assessment, 4*(3). Retrieved from http://www.jtla.org/

Baird, L. L. (1973). Teaching styles: An exploratory study of dimensions and effects. *Journal of Educational Psychology, 64*, 15–21. https://doi.org/10.1037/h0034058

Baird, L. L., & Knapp, J. E. (1981). *The inventory of documented accomplishments for graduate admissions: Results of a field trial study of its reliability, short-term correlates, and evaluation* (Research Report No. RR-81-18). Princeton: Educational Testing Service. http://dx.doi.org/10.1002/j.2333-8504.1981.tb01253.x

Baker, A. H., Mishara, B. L., Kostin, I. W., & Parker, L. (1976). Kinesthetic aftereffect and personality: A case study of issues involved in construct validation. *Journal of Personality and Social Psychology, 34*, 1–13. https://doi.org/10.1037/0022-3514.34.1.1

Baker, A. H., Mishara, B. L., Parker, L., & Kostin, I. W. (1978). When "reliability" fails, must a measure be discarded?—The case of kinesthetic aftereffect. *Journal of Research in Personality, 12*, 262–273. https://doi.org/10.1016/0092-6566(78)90053-3

Baker, A. H., Mishara, B. L., Kostin, I. W., & Parker, L. (1979). Menstrual cycle affects kinesthetic aftereffect, an index of personality and perceptual style. *Journal of Personality and Social Psychology, 37*, 234–246. https://doi.org/10.1037/0022-3514.37.2.234

Ball, S., & Bogatz, G. A. (1970). *The first year of* Sesame Street*: An evaluation* (Program Report No. PR-70-15). Princeton: Educational Testing Service.

Ball, S., & Bogatz, G. A. (1973). *Reading with television: An evaluation of* The Electric Company (Program Report No. PR-73-02). Princeton: Educational Testing Service.

Barton, P. E. (2003). *Parsing the achievement gap: Baselines for tracking progress* (Policy Information Report). Princeton: Educational Testing Service.

Barton, P. E., & Coley, R. J. (1990). *The education reform decade* (Policy Information Report). Princeton: Educational Testing Service.

Barton, P. E., & Coley, R. J. (2007). *The family: America's smallest school* (Policy Information Report). Princeton: Educational Testing Service.

Barton, P. E., & Coley, R. J. (2010). *The Black–White achievement gap: When progress stopped* (Policy Information Report). Princeton: Educational Testing Service.

Barton, P. E., Coley, R. J., & Goertz, M. E. (1991). *The state of inequality* (Policy Information Report). Princeton: Educational Testing Service.

Beaton, A. E. (1968). *The computer techniques used in the equality of Educational Opportunity Survey* (Research Memorandum No. RM-68-16). Princeton: Educational Testing Service.

Bejar, I. I. (1991). A methodology for scoring open-ended architectural design problems. *Journal of Applied Psychology, 76*, 522–532. https://doi.org/10.1037/0021-9010.76.4.522

Bejar, I. I., & Braun, H. I. (1999). *Architectural simulations: From research to implementation—Final report to the National Council of Architectural Registration Boards* (Research Memorandum No. RM-99-02). Princeton: Educational Testing Service.

Bejar, I. I., Douglas, D., Jamieson, J., Nissan, S., & Turner, J. (2000). *TOEFL 2000 listening framework: A working paper* (TOEFL Monograph Series Report No. 19). Princeton: Educational Testing Service.

Bell, C. A., Wilson, S. M., Higgins, T., & McCoach, D. B. (2010). Measuring the effects of professional development on teacher knowledge: The case of developing mathematical ideas. *Journal for Research in Mathematics Education, 41*, 479–512.

Bennett, R. E. (1999). Computer-based testing for examinees with disabilities: On the Road to generalized accommodation. In S. Messick (Ed.), *Assessment in higher education: Issues of access, quality, student development, and public policy* (pp. 181–191). Mahwah: Erlbaum.

Bennett, R. E. (2016). *Opt out: An examination of issues* (RR-16-13). Princeton, NJ: Educational Testing Service. https://doi.org/10.1002/ets2.12101

Bennett, R. E., & Gitomer, D. H. (2009). Transforming K-12 assessment: Integrating accountability testing, formative assessment, and professional support. In C. Wyatt-Smith & J. Cumming (Eds.), *Educational assessment in the 21st century* (pp. 43–61). New York: Springer. https://doi.org/10.1007/978-1-4020-9964-9_3

Bennett, R. E., & Rock, D. A. (1995). Generalizability, validity, and examinee perceptions of a computer-delivered Formulating-Hypotheses test. *Journal of Educational Measurement, 32*, 19–36. https://doi.org/10.1111/j.1745-3984.1995.tb00454.x

Bennett, R. E., & Ward, W. C. (Eds). (1993). *Construction vs. choice in cognitive measurement: Issues in constructed response, performance testing, and portfolio assessment*. Hillsdale, NJ: Erlbaum.

Bennett, R. E., Rock, D. A., & Wang, M. (1991). Equivalence of free-response and multiple-choice items. *Journal of Educational Measurement, 28*, 77–92. https://doi.org/10.1111/j.1745-3984.1991.tb00345.x

Bennett, R. E., Deane, P., & van Rijn, P. W. (2016). From cognitive-domain theory to assessment practice. *Educational Psychologist, 51*, 82–107. https://doi.org/10.1080/00461520.2016.1141683

Bennett, R. E., Morley, M., & Quardt, D. (2000). Three response types for broadening the conception of mathematical problem solving in computerized tests. *Applied Psychological Measurement, 24*, 294–309. https://doi.org/10.1177/01466210022031769

Bennett, R. E., Persky, H., Weiss, A. R., & Jenkins, F. (2007). *Problem solving in technology-rich environments: A report from the NAEP Technology-Based Assessment Project* (NCES 2007–466). Washington, DC: National Center for Education Statistics.

Bennett, R. E., Steffen, M., Singley, M. K., Morley, M., & Jacquemin, D. (1997). Evaluating an automatically scorable, open-ended response type for measuring mathematical reasoning in computer-adaptive tests. *Journal of Educational Measurement, 34*, 163–177. https://doi.org/10.1111/j.1745-3984.1997.tb00512.x

Braun, H. I. (2005). *Using student progress to evaluate teachers: A primer on value-added models (PIC-VAM)*. Princeton, NJ: Educational Testing Service.

Braun, H. I., Bennett, R. E., Frye, D, & Soloway, E. (1990). Scoring constructed responses using expert systems. *Journal of Educational Measurement, 27*, 93–108. https://doi.org/10.1111/j.1745-3984.1990.tb00736.x

Berke, J. S., Goertz, M. E., & Coley, R. J. (1984). *Politicians, judges, and city schools: Reforming school finance in New York*. New York: Russell Sage Foundation.

Bhat, S., & Yoon, S-Y. (2015). Automatic assessment of syntactic complexity for spontaneous speech scoring. *Speech Communication, 67*, 42–57. https://doi.org/10.1016/j.specom.2014.09.005

Breland, H. M. (1983). *The direct assessment of writing skill: A measurement review* (Report No. 83–6). New York: College Entrance Examination Board.

Breland, H. M., Camp, R., Jones, R. J., Morris, M. M., & Rock, D. A. (1987). Assessing writing skill. New York: College Entrance Examination Board.

Breland, H. M., Danos, D. O., Kahn, H. D., Kubota, M. Y., & Bonner, M. W. (1994). Performance versus objective testing and gender: An exploratory study of an Advanced Placement history examination. *Journal of Educational Measurement, 31*(4), 275–293. https://doi.org/10.1111/j.1745-3984.1994.tb00447.x

Bridgeman, B. (1992). A comparison of quantitative questions in open-ended and multiple-choice formats. *Journal of Educational Measurement, 29*, 253–271. https://doi.org/10.1111/j.1745-3984.1992.tb00377.x

Bridgeman, B., & Lewis, C. (1994). The relationship of essay and multiple-choice scores with grades in college courses. *Journal of Educational Measurement, 31*, 37–50. https://doi.org/10.1111/j.1745-3984.1994.tb00433.x

Bridgeman, B., Burton, N., & Pollack, J. (2008). Predicting grades in college courses: A comparison of multiple regression and percent succeeding approaches. *Journal of College Admission, 199*, 19–25.

Brinker, R. P., & Lewis, M. (1982). Discovering the competent handicapped infant: A process approach to assessment and intervention. *Topics in Early Childhood Special Education, 2*(2), 1–16. https://doi.org/10.1177/027112148200200205

Brinker, R. P., & Thorpe, M. E. (1984). *Evaluation of the integration of severely handicapped students in education and community settings* (Research Report No. RR-84-11). Princeton: Educational Testing Service. http://dx.doi.org/10.1002/j.2330-8516.1984.tb00051.x

Brody, G. H., Pellegrini, A. D., & Sigel, I. E. (1986). Marital quality and mother–child and father–child interactions with school-aged children. *Developmental Psychology, 22*, 291–296. https://doi.org/10.1037/0012-1649.22.3.291

Brooks, J., & Lewis, M. (1976). Infants' responses to strangers: Midget, adult, and child. *Child Development, 47*, 323–332. https://doi.org/10.2307/1128785

Brooks-Gunn, J. (1984). The psychological significance of different pubertal events in young girls. *Journal of Early Adolescence, 4*, 315–327. https://doi.org/10.1177/0272431684044003

Brooks-Gunn, J. (1990). Adolescents as daughters and as mothers: A developmental perspective. In I. E. Sigel & G. H. Brody (Eds.), *Methods of family research: Biographies of research projects* (Vol. 1, pp. 213–248). Hillsdale: Erlbaum.

Brooks-Gunn, J., & Hearn, R. P. (1982). Early intervention and developmental dysfunction: Implications for pediatrics. *Advances in Pediatrics, 29*, 497–527.

Brooks-Gunn, J., & Lewis, M. (1981). Infant social perceptions: Responses to pictures of parents and strangers. *Developmental Psychology, 17*, 647–649. https://doi.org/10.1037/0012-1649.17.5.647

Brooks-Gunn, J., & Lewis, M. (1982). Temperament and affective interaction in handicapped infants. *Journal of Early Intervention, 5*, 31–41. https://doi.org/10.1177/105381518200500105

Brooks-Gunn, J., & Warren, M. P. (1988). The psychological significance of secondary sexual characteristics in nine- to eleven-year-old girls. *Child Development, 59*, 1061–1069. https://doi.org/10.2307/1130272

Brooks-Gunn, J., McCormick, M. C., & Heagarty, M. C. (1988). Preventing infant mortality and morbidity: Developmental perspectives. *American Journal of Orthopsychiatry, 58*, 288–296. https://doi.org/10.1111/j.1939-0025.1988.tb01590.x

Bruschi, B. A., & Coley, R. J. (1999). *How teachers compare: The prose, document, and quantitative literacy of America's teachers* (Policy Information Report). Princeton: Educational Testing Service.

Bryan, J. H., & Test, M. A. (1967). Models and helping: Naturalistic studies in aiding behavior. *Journal of Personality and Social Psychology, 6*, 400–407. https://doi.org/10.1037/h0024826

Burrus, J., MacCann, C., Kyllonen, P. C., & Roberts, R. D. (2011). Noncognitive constructs in K-16: Assessments, interventions, educational and policy implications. In P. J. Bowman & E. P. S. John (Eds.), *Readings on equal education: Diversity, merit, and higher education—Toward a comprehensive agenda for the twenty-first century* (Vol. 25, pp. 233–274). New York: AMS Press.

Burrus, J., Jackson, T., Xi, N., & Steinberg, J. (2013). *Identifying the most important 21st century workforce competencies: An analysis of the Occupational Information Network (O*NET)* (Research Report No. RR-13-21). Princeton: Educational Testing Service. http://dx.doi.org/10.1002/j.2333-8504.2013.tb02328.x

Burstein, J., Braden-Harder, L., Chodorow, M., Hua, S., Kaplan, B. A., Kukich, K.,. .. Wolff, S. (1998). *Computer analysis of essay content for automated score prediction: A prototype automated scoring system for GMAT Analytical Writing Assessment essays* (Research Report No. RR-98-15). Princeton: Educational Testing Service. http://dx.doi.org/10.1002/j.2333–8504.1998.tb01764.x

Butler, F. A., Eignor, D., Jones, S., McNamara, T., & Suomi, B. K. (2000). *TOEFL 2000 speaking framework: A working paper* (Research Memorandum No. RM-00-06). Princeton: Educational Testing Service.

Buzick, H. M., & Jones, N. D. (2015). Using test scores from students with disabilities in teacher evaluation. *Educational Measurement: Issues and Practice, 34*(3), 28–38. https://doi.org/10.1111/emip.12076

Buzick, H. M., & Laitusis, C. (2010a). *A summary of models and standards-based applications for grade-to-grade growth on statewide assessments and implications for students with disabilities* (Research Report No. RR-10-14). Princeton: Educational Testing Service. https://doi.org/10.1002/j.2333-8504.2010.tb02221.x

Buzick, H. M., & Laitusis, C. C. (2010b). Using growth for accountability: Measurement challenges for students with disabilities and recommendations for research. *Educational Researcher, 39*, 537–544. https://doi.org/10.3102/0013189X10383560

Buzick, H. M., & Stone, E. (2011). *Recommendations for conducting differential item functioning (DIF) analyses for students with disabilities based on previous DIF studies* (Research Report No. RR-11-34). Princeton: Educational Testing Service. http://dx.doi.org/10.1002/j.2333-8504.2011.tb02270.x

Buzick, H. M., Oliveri, M. E., Attali, Y., & Flor, M. (2016). Comparing human and automated essay scoring for prospective graduate students with learning disabilities and/or ADHD. *Applied Measurement in Education, 29*, 161–172. http://dx.doi.org/10.1080/08957347.2016.1171765

Camp, R. (1985). The writing folder in post-secondary assessment. In P. J. A. Evans (Ed.), *Directions and misdirections in English evaluation* (pp. 91–99). Ottawa: Canadian Council of Teachers of English.

Camp, R. (1993). The place of portfolios in our changing views of writing assessment. In R. E. Bennett & W. C. Ward (Eds.), *Construction versus choice in cognitive measurement* (pp. 183–212). Hillsdale: Erlbaum.

Campbell, D. T., & Kenny, D. A. (1999). *A primer on regression artifacts*. New York: Guilford Press.

Campbell, J. T., Esser, B. F., & Flaugher, R. L. (1982). *Evaluation of a program for training dentists in the care of handicapped patients* (Research Report No. RR-82-52). Princeton: Educational Testing Service.

Carnevale, A. P. (1996). Liberal education and the new economy. *Liberal Education, 82*(2), 1–8.

Carnevale, A. P., & Desrochers, D. M. (1997). The role of community colleges in the new economy. *Community College Journal, 67*(5), 25–33.

Carnevale, A. P., & Fry, R. A. (2001). Economics, demography and the future of higher education policy. In *Higher expectations: Essays on the future of postsecondary education* (pp. 13–26). Washington, DC: National Governors Association.

Carnevale, A. P., & Fry, R. A. (2002). The demographic window of opportunity: College access and diversity in the new century. In D. E. Heller (Ed.), *Condition of access: Higher education for lower income students* (pp. 137–151). Westport: Praeger.

Carnevale, A. P., & Rose, S. J. (2000). Inequality and the new high-skilled service economy. In J. Madrick (Ed.), *Unconventional wisdom: Alternative perspectives on the new economy* (pp. 133–156). New York: Century Foundation Press.

Casabianca, J. M., & Junker, B. (2016). Multivariate normal distribution. In W. J. van der Linden (Ed.), *Handbook of item response theory* (Vol. 2, pp. 35–46). Boca Raton: CRC Press. https://doi.org/10.1177/0013164413486987

Casabianca, J. M., McCaffrey, D. F., Gitomer, D. H., Bell, C. A., Hamre, B. K., & Pianta, R. C. (2013). Effect of observation mode on measures of secondary mathematics teaching. *Educational and Psychological Measurement, 73*, 757–783.

Centra, J. A. (1974). The relationship between student and alumni ratings of teachers. *Educational and Psychological Measurement, 34*, 321–325. https://doi.org/10.1177/001316447403400212

Centra, J. A. (1983). Research productivity and teaching effectiveness. *Research in Higher Education, 18*, 379–389. https://doi.org/10.1007/BF00974804

Centra, J. A., & Barrows, T. S. (1982). *An evaluation of the University of Oklahoma advanced programs: Final report* (Research Memorandum No. RM-82-03). Princeton: Educational Testing Service.

Centra, J. A., & Potter, D. A. (1980). School and teacher effects: An interrelational model. *Review of Educational Research, 50*, 273–291. https://doi.org/10.3102/00346543050002273

Chen, L., Feng, G., Joe, J. N., Leong, C. W., Kitchen, C., & Lee, C. M. (2014a). Towards automated assessment of public speaking skills using multimodal cues. In *Proceedings of the 16th International Conference on Multimodal Interaction* (pp. 200–203). New York: ACM. http://dx.doi.org/10.1145/2663204.2663265

Chen, H., Yamamoto, K., & von Davier, M. (2014b). Controlling multistage testing exposure rates in international large-scale assessments. In D. Yan, A. A. von Davier, & C. Lewis (Eds.), *Computerized multistage testing: Theory and applications* (pp. 391–409). Boca Raton: CRC Press.

Chen, L., Yoon, S.-Y., Leong, C. W., Martin, M., & Ma, M. (2014c). An initial analysis of structured video interviews by using multimodal emotion detection. In *Proceedings of the 2014 Workshop on Emotion Representation and Modelling in Human-Computer-Interaction-Systems* (pp. 1–6). New York: Association of Computational Linguistics. https://doi.org/10.1145/2668056.2668057

Cleary, T. A. (1968). Test bias: Prediction of grades of Negro and White students in integrated colleges. *Journal of Educational Measurement, 5*, 115–124. https://doi.org/10.1111/j.1745-3984.1968.tb00613.x

Clewell, B. C. (1987). *Retention of Black and Hispanic doctoral students* (GRE Board Research Report No. 83-4R). Princeton: Educational Testing Service. http://dx.doi.org/10.1002/j.2330-8516.1987.tb00214.x

Clewell, B. C., Anderson, B. T., & Thorpe, M. E. (1992). *Breaking the barriers: Helping female and minority students succeed in mathematics and science*. San Francisco: Jossey-Bass.

Coley, R. J. (2001). *Differences in the gender gap: Comparisons across racial/ethnic groups in education and work* (Policy Information Report). Princeton: Educational Testing Service.

Coley, R. J., Cradler, J., & Engel, P. (1997). *Computers and classrooms: The status of technology in U.S. schools* (Policy Information Report). Princeton: Educational Testing Service.

Cook, L. L., Eignor, D. R., Sawaki, Y., Steinberg, J., & Cline, F. (2010). Using factor analysis to investigate accommodations used by students with disabilities on an English-language arts assessment. *Applied Measurement in Education, 23*, 187–208. https://doi.org/10.1080/08957341003673831

Cumming, A., Kantor, R., Powers, D. E., Santos, T., & Taylor, C. (2000). *TOEFL 2000 writing framework: A working paper* (TOEFL Monograph Series No. TOEFL-MS-18). Princeton: Educational Testing Service.

Damarin, F., & Messick, S. (1965). *Response styles and personality variables: A theoretical integration of multivariate research* (Research Bulletin No. RB-65-10). Princeton: Educational Testing Service. http://dx.doi.org/10.1002/j.2333-8504.1965.tb00967.x

Danielson, C. (2013). *Framework for teaching evaluation instrument.* Princeton: The Danielson Group.

Danielson, C., & Dwyer, C. A. (1995). How PRAXIS III® supports beginning teachers. *Educational Leadership, 52*(6), 66–67.

Deane, P. (2013a). Covering the construct: An approach to automated essay scoring motivated by socio-cognitive framework for defining literacy skills. In M. D. Shermis & J. Burstein (Eds.), *Handbook of automated essay evaluation: Current application and new directions* (pp. 298–312). New York: Routledge.

Deane, P. (2013b). On the relation between automated essay scoring and modern views of the writing construct. *Assessing Writing, 18*(1), 7–24. https://doi.org/10.1016/j.asw.2012.10.002

Deane, P., & Song, Y. (2015). *The key practice, discuss and debate ideas: Conceptual framework, literature review, and provisional learning progressions for argumentation* (Research Report No. RR-15-33). Princeton: Educational Testing Service. http://dx.doi.org/10.1002/ets2.12079

Deane, P., & Zhang, M. (2015). *Exploring the feasibility of using writing process features to assess text production skills* (Research Report No. RR-15-26). Princeton: Educational Testing Service. http://dx.doi.org/10.1002/ets2.12071

Deane, P., Sabatini, J. P., Feng, G., Sparks, J. R., Song, Y., Fowles, M. E.,… Foley, C. (2015). *Key practices in the English language arts (ELA): Linking learning theory, assessment, and instruction* (Research Report No. RR-15-17). Princeton: Educational Testing Service. http://dx.doi.org/10.1002/ets2.12063

Dear, R. E. (1958). *The effects of a program of intensive coaching on SAT scores* (Research Bulletin No. RR-58-05). Princeton: Educational Testing Service. http://dx.doi.org/10.1002/j.2333-8504.1958.tb00080.x

de Leeuw, J., & Verhelst, N. (1986). Maximum likelihood estimation in generalized Rasch models. *Journal of Educational and Behavioral Statistics, 11*, 183–196. https://doi.org/10.3102/10769986011003183

Dempster, A. P., Laird, N. M., & Rubin, D. B. (1977). Maximum likelihood from incomplete data via the EM algorithm (with discussion). *Journal of the Royal Statistical Society, Series B, 39*, 1–38.

DiBello, L. V., Roussos, L. A., & Stout, W. (2006). Review of cognitively diagnostic assessment and a summary of psychometric models. In C. R. Rao & S. Sinharay (Eds.), *Handbook of statistics: Vol 26. Psychometrics* (pp. 979–1030). Amsterdam: Elsevier. https://doi.org/10.1016/s0169-7161(06)26031-0

Diederich, P. B. (1957). *The improvement of essay examinations* (Research Memorandum No. RM-57-03). Princeton: Educational Testing Service.

Diederich, P. B., French, J. W., & Carlton, S. T. (1961). *Factors in judgments of writing ability* (Research Bulletin No. RB-61-15). Princeton: Educational Testing Service. http://dx.doi.org/10.1002/j.2333-8504.1961.tb00286.x

Dorans, N. J., & Holland, P. W. (1993). DIF detection and description: Mantel–Haenszel and standardization. In P. W. Holland & H. Wainer (Eds.), *Differential item functioning* (pp. 35–66). Hillsdale: Erlbaum.

Dorans, N. J., & Kulick, E. (1986). Demonstrating the utility of the standardization approach to assessing unexpected differential item performance on the Scholastic Aptitude Test. *Journal of Educational Measurement, 23*, 355–368. https://doi.org/10.1111/j.1745-3984.1986.tb00255.x

Dorans, N. J., Schmitt, A. P., & Bleistein, C. A. (1992). The standardization approach to assessing comprehensive differential item functioning. *Journal of Educational Measurement, 29*, 309–319. https://doi.org/10.1111/j.1745-3984.1992.tb00379.x

Ekstrom, R. B., & Smith, D. K. (Eds.). (2002). *Assessing individuals with disabilities in educational, employment, and counseling settings*. Washington, DC: American Psychological Association. https://doi.org/10.1037/10471-000

Ekstrom, R. B., French, J. W., & Harman, H. H. (with Dermen, D.). (1976). *Manual for kit of factor-referenced cognitive tests*. Princeton: Educational Testing Service.

Ekstrom, R. B., Goertz, M. E., Pollack, J. C., & Rock, D. A. (1991). *Undergraduate debt and participation in graduate education: The relationship between educational debt and graduate school aspirations, applications, and attendance among students with a pattern of full-time continuous postsecondary education* (GRE Research Report No. 86-5). Princeton: Educational Testing Service. http://dx.doi.org/10.1002/j.2333-8504.1982.tb01330.x

Elliot, N. (2014). *Henry Chauncey: An American life*. New York: Lang.

Emmerich, W. (1973). *Disadvantaged children and their first school experiences: ETS–Head Start longitudinal study—Preschool personal–social behaviors: Relationships with socioeconomic status, cognitive skills, and tempo* (Program Report No. PR-73-33). Princeton: Educational Testing Service.

Enright, M. K., Grabe, W., Mosenthal, P., Mulcahy-Ernt, P., & Schedl, M. (2000). *TOEFL 2000 reading framework: A working paper* (TOEFL Monograph Series No. MS-17). Princeton: Educational Testing Service.

Enright, M. K., Rock, D. A., & Bennett, R. E. (1998). Improving measurement for graduate admissions. *Journal of Educational Measurement, 35*, 250–267. https://doi.org/10.1111/j.1745-3984.1998.tb00538.x

Eriksson, K., & Haggstrom, O. (2014). Lord's paradox in a continuous setting and a regression artifact in numerical cognition research. *PLoS ONE, 9*(4). https://doi.org/10.1371/journal.pone.0095949

Evans, F. R., & Frederiksen, N. (1974). Effects of models of creative performance on ability to formulate hypotheses. *Journal of Educational Psychology, 66*, 67–82. https://doi.org/10.1037/h0035808

Evans, F. R., & Pike, L. W. (1973). The effects of instruction for three mathematics item formats. *Journal of Educational Measurement, 10*, 257–272. https://doi.org/10.1111/j.1745-3984.1973.tb00803.x

Feinman, S., & Lewis, M. (1983). Social referencing at ten months: A second-order effect on infants' responses to strangers. *Child Development, 50*, 848–853. https://doi.org/10.2307/1129892

Fife, J. H. (2013). *Automated scoring of mathematics tasks in the Common Core Era: Enhancements to m-rater™ in support of CBAL mathematics and the Common Core Assessments* (Research Report No. RR-13-26). Princeton: Educational Testing Service. http://dx.doi.org/10.1002/j.2333-8504.2013.tb02333.x

Follman, D. (1988). Consistent estimation in the Rasch model based on nonparametric margins. *Psychometrika, 53*, 553–562. https://doi.org/10.1007/BF02294407

Formann, A. K. (1992). Linear logistic latent class analysis for polytomous data. *Journal of the American Statistical Association, 87*, 476–486. https://doi.org/10.1080/01621459.1992.10475229

Frederiksen, N. O. (1948). *The prediction of first term grades at Hamilton College* (Research Bulletin No. RB-48-02). Princeton: Educational Testing Service. http://dx.doi.org/10.1002/j.2333-8504.1948.tb00867.x

Frederiksen, N. (1959). *Development of the test "Formulating Hypotheses": A progress report* (Office of Naval Research Technical Report No. NR-2338[00]). Princeton: Educational Testing Service.

Frederiksen, N. O. (1962). Factors in in-basket performance. *Psychological Monographs: General and Applied, 76*(22), 1–25. https://doi.org/10.1037/h0093838

Frederiksen, N. (1984). The real test bias: Influences of testing on teaching and learning. *American Psychologist, 39*, 193–202. https://doi.org/10.1037/0003-066X.39.3.193

Frederiksen, N., & Ward, W. C. (1978). Measures for the study of creativity and scientific problem-solving. *Applied Psychological Measurement, 2*, 1–24. https://doi.org/10.1177/014662167800200101

Frederiksen, N., Mislevy, R. J., & Bejar, I. I. (Eds.). (1993). *Test theory for a new generation of tests*. Hillsdale: Erlbaum.

Frederiksen, N., Saunders, D. R., & Wand, B. (1957). The in-basket test. *Psychological Monographs: General and Applied, 71*(9), 1–28. https://doi.org/10.1037/h0093706

Freedle, R., & Lewis, M. (1977). Prelinguistic conversations. In M. Lewis & L. A. Rosenblum (Eds.), *The origins of behavior: Vol. 5. Interaction, conversation, and the development of language* (pp. 157–185). New York: Wiley.

French, J. W. (1948). The validity of a persistence test. *Psychometrika, 13*, 271–277. https://doi.org/10.1007/BF02289223

French, J. W. (1954). *Manual for Kit of Selected Tests for Reference Aptitude and Achievement Factors*. Princeton: Educational Testing Service.

French, J. W. (1956). *The effect of essay tests on student motivation* (Research Bulletin No. RB-56-04). Princeton: Educational Testing Service. http://dx.doi.org/10.1002/j.2333-8504.1956.tb00060.x

French, J. W. (1958). Validation of new item types against four-year academic criteria. *Journal of Educational Psychology, 49*, 67–76. https://doi.org/10.1037/h0046064

French, J. W. (1962). Effect of anxiety on verbal and mathematical examination scores. *Educational and Psychological Measurement, 22*, 555–567. https://doi.org/10.1177/001316446202200313

French, J. W., & Dear, R. E. (1959). Effect of coaching on an aptitude test. *Educational and Psychological Measurement, 19*, 319–330. https://doi.org/10.1177/001316445901900304

French, J. W., Tucker, L. R., Newman, S. H., & Bobbitt, J. M. (1952). A factor analysis of aptitude and achievement entrance tests and course grades at the United States Coast Guard Academy. *Journal of Educational Psychology, 43*, 65–80. https://doi.org/10.1037/h0054549

French, J. W., Ekstrom, R. B., & Price, L. A. (1963). *Manual for Kit of Reference Tests for Cognitive Factors*. Princeton: Educational Testing Service.

Furstenberg, F. F., Jr., Brooks-Gunn, J., & Morgan, S. P. (1987). *Adolescent mothers in later life*. New York: Cambridge University Press. https://doi.org/10.1017/CBO9780511752810

Gallagher, A., Bennett, R. E., Cahalan, C., & Rock, D. A. (2002). Validity and fairness in technology-based assessment: Detecting construct-irrelevant variance in an open-ended computerized mathematics task. *Educational Assessment, 8*, 27–41. https://doi.org/10.1207/S15326977EA0801_02

Gardner, R. W., Jackson, D. N., & Messick, S. (1960). Personality organization in cognitive controls and intellectual abilities. *Psychological Issues, 2*(4, Whole No. 8). https://doi.org/10.1037/11215-000

Gitomer, D. H. (2007a). *The impact of the National Board for Professional Teaching Standards: A review of the research* (Research Report No. RR-07-33). Princeton: Educational Testing Service. https://doi.org/10.1002/j.2333-8504.2007.tb02075.x

Gitomer, D. H. (2007b). *Teacher quality in a changing policy landscape: Improvements in the teacher pool* (Policy Information Report). Princeton: Educational Testing Service.

Gitomer, D. H. (2009). *Measurement issues and assessment for teaching quality*. Los Angeles: Sage. https://doi.org/10.4135/9781483329857

Gitomer, D. H., & Bell, C. A. (2013). Evaluating teachers and teaching. In K. F. Geisinger, B. A. Bracken, J. F. Carlson, J.-I. C. Hansen, N. R. Kuncel, S. P. Reise, & M. C. Rodriguez (Eds.), *APA handbook of testing and assessment in psychology: Vol. 3. Testing and assessment in school psychology and education* (pp. 415–444). Washington, DC: American Psychological Association. https://doi.org/10.1037/14049-020

Gitomer, D. H., & Bell, C. A. (2016). *Handbook of research on teaching* (5th ed.). Washington, DC: American Educational Research Association. https://doi.org/10.3102/978-0-935302-48-6

Gitomer, D. H., & Yamamoto, K. (1991). *Performance modeling that integrates latent trait and class theory* (Research Report No. RR-91-01). Princeton: Educational Testing Service. http://dx.doi.org/10.1002/j.2333-8504.1991.tb01367.x

Gitomer, D. H., Steinberg, L. S., & Mislevy, R. J. (1994). *Diagnostic assessment of troubleshooting skill in an intelligent tutoring system* (Research Report No. RR-94-21-ONR). Princeton: Educational Testing Service. http://dx.doi.org/10.1002/j.2333-8504.1994.tb01594.x

Godshalk, F. I., Swineford, F., & Coffman, W. E. (1966). *The measurement of writing ability*. New York: College Entrance Examination Board.

Goe, L. (2007). *The link between teacher quality and student outcomes: A research synthesis* (NCCTQ Report). Washington, DC: National Comprehensive Center for Teacher Quality.

Goe, L. (2013). Can teacher evaluation improve teaching? *Principal Leadership, 13*(7), 24–29.

Goe, L., & Croft, A. (2009). *Methods of evaluating teacher effectiveness* (NCCTQ Research-to-Practice Brief). Washington, DC: National Comprehensive Center for Teacher Quality.

Goe, L., Bell, C., & Little, O. (2008). *Approaches to evaluating teacher effectiveness: A research synthesis* (NCCTQ Report). Washington, DC: National Comprehensive Center for Teacher Quality.

Goe, L., Biggers, K., & Croft, A. (2012). *Linking teacher evaluation to professional development: Focusing on improving teaching and learning* (NCCTQ Research & Policy Brief). Washington, DC: National Comprehensive Center for Teacher Quality.

Goe, L., Holdheide, L., & Miller, T. (2011). *A practical guide to designing comprehensive teacher evaluation systems: A tool to assist in the development of teacher evaluation systems* (NCCTQ Report). Washington, DC: National Comprehensive Center for Teacher Quality.

Goertz, M. E. (1978). *Money and education: Where did the 400 million dollars go? The impact of the New Jersey Public School Education Act of 1975*. Princeton: Educational Testing Service.

Goertz, M. E. (1989). *What Americans study* (Policy Information Report). Princeton: Educational Testing Service.

Goertz, M. E., & Moskowitz, J. (1978). *Plain talk about school finance*. Washington, DC: National Institute of Education.

Goertz, M. E., Ekstrom, R., & Coley, R. (1984). *The impact of state policy on entrance into the teaching profession*. Princeton: Educational Testing Service.

Goleman, D. (1995). *Emotional intelligence: Why it can matter more than IQ*. New York: Bantam.

Goodenough, D. R., Oltman, P. K., & Cox, P. W. (1987). The nature of individual differences in field dependence. *Journal of Research in Personality, 21*, 81–99. https://doi.org/10.1016/0092-6566(87)90028-6

Graf, E. A. (2009). *Defining mathematics competency in the service of cognitively based assessment for Grades 6 through 8* (Research Report No. RR-09-02). Princeton: Educational Testing Service. http://dx.doi.org/10.1002/j.2333-8504.2009.tb02199.x

Green, B. F., Jr. (1950a). *A general solution for the latent class model of latent structure analysis* (Research Bulletin No. RB-50-38). Princeton: Educational Testing Service. http://dx.doi.org/10.1002/j.2333-8504.1950.tb00917.x

Green, B. F., Jr. (1950b). *Latent structure analysis and its relation to factor analysis* (Research Bulletin No. RB-50-65). Princeton: Educational Testing Service. http://dx.doi.org/10.1002/j.2333-8504.1950.tb00920.x

Gu, L., Lockwood, J., & Powers, D. E. (2015). *Evaluating the TOEFL Junior® Standard Test as a measure of progress for young English language learners* (Research Report No. RR-15-22). Princeton: Educational Testing Service. http://dx.doi.org/10.1002/ets2.12064

Gulliksen, H. (1950). *Theory of mental tests*. New York: Wiley. https://doi.org/10.1037/13240-000

Guzman-Orth, D., Laitusis, C., Thurlow, M., & Christensen, L. (2016). *Conceptualizing accessibility for English language proficiency assessments* (Research Report No. RR-16-07). Princeton: Educational Testing Service. https://doi.org/10.1002/ets2.12093

Haberman, S. J. (1977). Maximum likelihood estimates in exponential response models. *Annals of Statistics, 5*, 815–841. https://doi.org/10.1214/aos/1176343941

Haberman, S. J. (1988). A stabilized Newton–Raphson algorithm for log-linear models for frequency tables derived by indirect observation. *Sociological Methodology, 18*, 193–211. https://doi.org/10.2307/271049

Haberman, S. J. (2008). When can subscores have value? *Journal of Educational and Behavioral Statistics, 33*, 204–229. https://doi.org/10.3102/1076998607302636

Haberman, S. (2010). *Limits on the accuracy of linking* (Research Report No. RR-10-22). Princeton: Educational Testing Service. http://dx.doi.org/10.1002/j.2333-8504.2010.tb02229.x

Haberman, S. J. (2016). Exponential family distributions relevant to IRT. In W. J. van der Linden (Ed.), *Handbook of item response theory* (Vol. 2, pp. 47–69). Boca Raton: CRC Press.

Haberman, S. J., & Sinharay, S. (2010). Reporting of subscores using multidimensional item response theory. *Psychometrika, 75*, 209–227. https://doi.org/10.1007/s11336-010-9158-4

Haberman, S. J., von Davier, M., & Lee, Y.-H. (2008). *Comparison of multidimensional item response models: Multivariate normal ability distributions versus multivariate polytomous ability distributions* (Research Report No. RR-08-45). Princeton: Educational Testing Service. http://dx.doi.org/10.1002/j.2333-8504.2008.tb02131.x

Handwerk, P., Tognatta, N., Coley, R. J., & Gitomer, D. H. (2008). *Access to success: Patterns of Advanced Placement participation in U.S. high schools* (Policy Information Report). Princeton: Educational Testing Service.

Hansen, E. G., & Mislevy, R. J. (2006). Accessibility of computer-based testing for individuals with disabilities and English language learners within a validity framework. In M. Hricko & S. L. Howell (Eds.), *Online and distance learning: Concepts, methodologies, tools, and applications* (pp. 214–261). Hershey: Information Science.

Hansen, E. G., Forer, D. C., & Lee, M. J. (2004). *Toward accessible computer-based tests: Prototypes for visual and other disabilities* (Research Report No. RR-04-25). Princeton: Educational Testing Service. http://dx.doi.org/10.1002/j.2333-8504.2004.tb01952.x

Hansen, E. G., Mislevy, R. J., Steinberg, L. S., Lee, M. J., & Forer, D. C. (2005). Accessibility of tests for individuals with disabilities within a validity framework. *System, 33*(1), 107–133. https://doi.org/10.1016/j.system.2004.11.002

Hansen, E. G., Laitusis, C. C., Frankel, L., & King, T. C. (2012). Designing accessible technology-enabled reading assessments: Recommendations from teachers of students with visual impairments. *Journal of Blindness Innovation and Research, 2*(2). http://dx.doi.org/10.5241/2F2-22

Hauck, M. C., Wolf, M. K., & Mislevy, R. (2016). *Creating a next-generation system of K–12 English learner language proficiency assessments* (Research Report No. RR-16-06). Princeton: Educational Testing Service. http://dx.doi.org/10.1002/ets2.12092

He, Q., & von Davier, M. (2015). Identifying feature sequences from process data in problem-solving items with n-grams. In L. A. van der Ark, D. M. Bolt, W.-C. Wang, J. A. Douglas, & S.-M. Chow (Eds.), *Quantitative psychology research* (pp. 173–190). Madison: Springer International. https://doi.org/10.1007/978-3-319-19977-1_13

He, Q., & von Davier, M. (2016). Analyzing process data from problem-solving items with n-grams: Insights from a computer-based large-scale assessment. In Y. Rosen, S. Ferrara, & M. Mosharraf (Eds.), *Handbook of research on technology tools for real-world skill development* (Vol. 2, pp. 749–776). Hershey: Information Science Reference. http://dx.doi.org/10.4018/978-1-4666-9441-5.ch029

Heilman, M., & Madnani, N. (2012). Discriminative edit models for paraphrase scoring. In *Proceedings of the first Joint Conference on Lexical and Computational Semantics* (pp. 529–535). Stroudsburg: Association of Computational Linguistics.

Heinen, T. (1996). *Latent class and discrete latent trait models: Similarities and differences*. Thousand Oaks: Sage.

Henson, R., Templin, J., & Willse, J. (2009). Defining a family of cognitive diagnosis models using log-linear models with latent variables. *Psychometrika, 74*, 191–210. https://doi.org/10.1007/s11336-008-9089-5

Hills, J. R. (1958). Needs for achievement, aspirations, and college criteria. *Journal of Educational Psychology, 49*, 156–161. https://doi.org/10.1037/h0047283

Holland, P. W. (1986). Statistics and causal inference. *Journal of the American Statistical Association, 81*, 945–970. https://doi.org/10.1080/01621459.1986.10478354

Holland, P. W. (1987). *Which comes first, cause or effect?* (Research Report No. RR-87-08). Princeton: Educational Testing Service. http://dx.doi.org/10.1002/j.2330-8516.1987.tb00212.x

Holland, P. W., & Rubin, D. B. (1983). On Lord's paradox. In H. Wainer & S. Messick (Eds.), *Principals of modern psychological measurement: A festschrift for Frederic M. Lord* (pp. 3–25). Hillsdale: Erlbaum.

Holland, P. W., & Thayer, D. T. (1988). Differential item performance and the Mantel–Haenszel procedure. In H. Wainer & H. I. Braun (Eds.), *Test validity* (pp. 129–145). Hillsdale: Erlbaum.
Holland, P. W., & Thayer, D. T. (1989). *The kernel method of equating score distributions* (Program Statistics Research Technical Report No. 89-84). Princeton: Educational Testing Service. http://dx.doi.org/10.1002/j.2330-8516.1989.tb00333.x
Holland, P. W., & Wainer, H. (Eds.). (1993). *Differential item functioning*. Hillsdale: Erlbaum.
Holland, P. W., King, B. F., & Thayer, D. T. (1989). *The standard error of equating for the kernel method of equating score distributions* (Research Report No. RR-89-06). Princeton: Educational Testing Service. http://dx.doi.org/10.1002/j.2330-8516.1989.tb00332.x
Horkay, N., Bennett, R. E., Allen, N., Kaplan, B., & Yan, F. (2006). Does it matter if I take my writing test on computer? An empirical study of mode effects in NAEP. *Journal of Technology, Learning and Assessment, 5*(2) Retrieved from http://ejournals.bc.edu/ojs/index.php/jtla/
Huddleston, E. M. (1952). *Measurement of writing ability at the college-entrance level: Objective vs. subjective testing techniques* (Research Bulletin No. RB-52-07). Princeton: Educational Testing Service. http://dx.doi.org/10.1002/j.2333-8504.1952.tb00925.x
Jackson, D. N., & Messick, S. (1958). Content and style in personality assessment. *Psychological Bulletin, 55*, 243–252. https://doi.org/10.1037/h0045996
Jackson, D. N., & Messick, S. (1961). Acquiescence and desirability as response determinants on the MMPI. *Educational and Psychological Measurement, 21*, 771–790. https://doi.org/10.1177/001316446102100402
Johnson, M. S., & Jenkins, F. (2005). *A Bayesian hierarchical model for large-scale educational surveys: An application to the National Assessment of Educational Progress* (Research Report No. RR-04-38). Princeton: Educational Testing Service. http://dx.doi.org/10.1002/j.2333-8504.2004.tb01965.x
Jöreskog, K. G. (1965). Testing a simple structure hypothesis in factor analysis. *Psychometrika, 31*, 165–178. https://doi.org/10.1007/BF02289505
Jöreskog, K. G. (1967). Some contributions to maximum likelihood factor analysis. *Psychometrika, 32*, 443–482. https://doi.org/10.1007/BF02289658
Jöreskog, K. G. (1969). A general approach to confirmatory maximum likelihood factor analysis. *Psychometrika, 34*, 183–202. https://doi.org/10.1007/BF02289343
Jöreskog, K. G. (1970). A general method for analysis of covariance structures. *Biometrika, 57*, 239–251. https://doi.org/10.1093/biomet/57.2.239
Jöreskog, K. G. (1971). Simultaneous factor analysis in several populations. *Psychometrika, 36*, 409–426. https://doi.org/10.1007/BF02291366
Jöreskog, K. G., & van Thillo, M. (1972). *LISREL: A general computer program for estimating a linear structural equation system involving multiple indicators of unmeasured variables* (Research Bulletin No. RB-72-56). Princeton: Educational Testing Service. http://dx.doi.org/10.1002/j.2333-8504.1972.tb00827.x
Kane, M. (2011). The errors of our ways. *Journal of Educational Measurement, 48*, 12–30. https://doi.org/10.1111/j.1745-3984.2010.00128.x
Kane, M. (2012). Validating score interpretations and uses. *Language Testing, 29*, 3–17. https://doi.org/10.1177/0265532211417210
Kane, M. (2016). Validation strategies: Delineating and validating proposed interpretations and uses of test scores. In S. Lane, M. R. Raymond, & T. M. Haladyna (Eds.), *Handbook of test development* (2nd ed., pp. 64–80). New York: Routledge.
Kaplan, R. M. (1992). *Using a trainable pattern-directed computer program to score natural language item responses* (Research Report No. RR-91-31). Princeton: Educational Testing Service. http://dx.doi.org/10.1002/j.2333-8504.1991.tb01398.x
Kaplan, R. M., Burstein, J., Trenholm, H., Lu, C., Rock, D., Kaplan, B., & Wolff, C. (1995). *Evaluating a prototype essay scoring procedure using off-the shelf software* (Research Report No. RR-95-21). Princeton: Educational Testing Service. http://dx.doi.org/10.1002/j.2333-8504.1995.tb01656.x
Karon, B. P., & Cliff, R. H., & (1957). *The Cureton–Tukey method of equating test scores* (Research Bulletin No. RB-57-06). Princeton: Educational Testing Service. http://dx.doi.org/10.1002/j.2333-8504.1957.tb00072.x

Kelley, T. (1927). *Interpretation of educational measurements*. Yonkers: World Book.
Khorramdel, L., & von Davier, M. (2014). Measuring response styles across the big five: A multiscale extension of an approach using multinomial processing trees. *Multivariate Behavioral Research, 49*, 2, p. 161–177. http://dx.doi.org/10.1080/00273171.2013.866536
Kirsch, I., & Braun, H. (Eds.). (2016). *The dynamics of opportunity in America*. New York: Springer. https://doi.org/10.1007/978-3-319-25991-8
Kirsch, I. S., & Jungeblut, A. (1986). *Literacy: Profiles of America's young adults* (NAEP Report No. 16-PL-01). Princeton: National Assessment of Educational Progress.
Kirsch, I. S., Braun, H., Yamamoto, K., & Sum, A. (2007). *America's perfect storm: Three forces changing our nation's future* (Policy Information Report). Princeton: Educational Testing Service.
Kirsch, I., Braun, H., Lennon, M. L., & Sands, A. (2016). *Choosing our future: A story of opportunity in America*. Princeton: Educational Testing Service.
Klein, S. P., Frederiksen, N., & Evans, F. R. (1969). Anxiety and learning to formulate hypotheses. *Journal of Educational Psychology, 60*, 465–475. https://doi.org/10.1037/h0028351
Kogan, N., & Doise, W. (1969). Effects of anticipated delegate status on level of risk taking in small decision-making groups. *Acta Psychologica, 29*, 228–243. https://doi.org/10.1016/0001-6918(69)90017-1
Kogan, N., & Morgan, F. T. (1969). Task and motivational influences on the assessment of creative and intellective ability in children. *Genetic Psychology Monographs, 80*, 91–127.
Kogan, N., & Pankove, E. (1972). Creative ability over a five-year span. *Child Development, 43*, 427–442. https://doi.org/10.2307/1127546
Kogan, N., & Wallach, M. A. (1964). *Risk taking: A study in cognition and personality*. New York: Holt, Rinehart, and Winston.
Kogan, N., Lamm, H., & Trommsdorff, G. (1972). Negotiation constraints in the risk-taking domain: Effects of being observed by partners of higher or lower status. *Journal of Personality and Social Psychology, 23*, 143–156. https://doi.org/10.1037/h0033035
Laitusis, C. C., Mandinach, E. B., & Camara, W. J. (2002). *Predictive validity of SAT I Reasoning Test for test-takers with learning disabilities and extended time accommodations* (Research Report No. RR-02-11). Princeton: Educational Testing Service. https://doi.org/10.1002/j.2333-8504.2002.tb01878.x
Lamm, H., & Kogan, N. (1970). Risk taking in the context of intergroup negotiation. *Journal of Experimental Social Psychology, 6*, 351–363. https://doi.org/10.1016/0022-1031(70)90069-7
Laosa, L. M. (1978). Maternal teaching strategies in Chicano families of varied educational and socioeconomic levels. *Child Development, 49*, 1129–1135. https://doi.org/10.2307/1128752
Laosa, L. M. (1980a). Maternal teaching strategies and cognitive styles in Chicano families. *Journal of Educational Psychology, 72*, 45–54. https://doi.org/10.1037/0022-0663.72.1.45
Laosa, L. M. (1980b). Maternal teaching strategies in Chicano and Anglo-American families: The influence of culture and education on maternal behavior. *Child Development, 51*, 759–765. https://doi.org/10.2307/1129462
Laosa, L. M. (1984). Ethnic, socioeconomic, and home language influences upon early performance on measures of abilities. *Journal of Educational Psychology, 76*, 1178–1198. https://doi.org/10.1037/0022-0663.76.6.1178
Laosa, L. M. (1990). Psychosocial stress, coping, and development of Hispanic immigrant children. In F. C. Serafica, A. I. Schwebel, R. K. Russell, P. D. Isaac, & L. B. Myers (Eds.), *Mental health of ethnic minorities* (pp. 38–65). New York: Praeger.
Leacock, C., & Chodorow, M. (2003). *c-rater*: Scoring of short-answer questions. *Computers and the Humanities, 37*, 389–405. https://doi.org/10.1023/A:1025779619903
Lee, V. E., Brooks-Gunn, J., & Schnur, E. (1988). Does Head Start work? A 1-year follow-up comparison of disadvantaged children attending Head Start, no preschool, and other preschool programs. *Developmental Psychology, 24*, 210–222. https://doi.org/10.1037/0012-1649.24.2.210
LeMahieu, P. G., Gitomer, D. H., & Eresh, J. A. T. (1995). Portfolios in large-scale assessment: Difficult but not impossible. *Educational Measurement: Issues and Practice, 14*(3), 11–28. https://doi.org/10.1111/j.1745-3992.1995.tb00863.x

Lennon, M. L., Kirsch, I. S., von Davier, M., Wagner, M., & Yamamoto, K. (2003). *Feasibility study for the PISA ICT Literacy Assessment: Report to Network A* (ICT Literacy Assessment Report). Princeton: Educational Testing Service.

Levine, R. (1955). *Equating the score scales of alternate forms administered to samples of different ability* (Research Bulletin No. RB-55-23). Princeton: Educational Testing Service. http://dx.doi.org/10.1002/j.2333-8504.1955.tb00266.x

Lewis, M. (1977). Early socioemotional development and its relevance to curriculum. *Merrill-Palmer Quarterly, 23*, 279–286.

Lewis, M. (1978). Attention and verbal labeling behavior in preschool children: A study in the measurement of internal representations. *Journal of Genetic Psychology, 133*, 191–202. https://doi.org/10.1080/00221325.1978.10533377

Lewis, M., & Brooks-Gunn, J. (1979). *Social cognition and the acquisition of self.* New York: Plenum Press. https://doi.org/10.1007/978-1-4684-3566-5

Lewis, M., & Brooks-Gunn, J. (1981a). Attention and intelligence. *Intelligence, 5*, 231–238. https://doi.org/10.1016/S0160-2896(81)80010-4

Lewis, M., & Brooks-Gunn, J. (1981b). Visual attention at three months as a predictor of cognitive functioning at two years of age. *Intelligence, 5*, 131–140. https://doi.org/10.1016/0160-2896(81)90003-9

Lewis, M., & Feiring, C. (1982). Some American families at dinner. In L. Laosa & I. Sigel (Eds.), *Families as learning environments for children* (pp. 115–145). New York: Plenum Press. https://doi.org/10.1007/978-1-4684-4172-7_4

Lewis, M., & Michalson, L. (1982). The measurement of emotional state. In C. E. Izard & P. B. Read (Eds.), *Measuring emotions in infants and children* (Vol. 1, pp. 178–207). New York: Cambridge University Press.

Lewis, M., & Rosenblum, L. A. (Eds.). (1978). *Genesis of behavior: Vol. 1. The development of affect.* New York: Plenum Press.

Lewis, M., Brooks, J., & Haviland, J. (1978). Hearts and faces: A study in the measurement of emotion. In M. Lewis & L. A. Rosenblum (Eds.), *Genesis of behavior: Vol. 1. The development of affect* (pp. 77–123). New York: Plenum Press. https://doi.org/10.1007/978-1-4684-2616-8_4

Li, D., Oranje, A., & Jiang, Y. (2009). On the estimation of hierarchical latent regression models for large-scale assessments. *Journal of Educational and Behavioral Statistics, 34*, 433–463. https://doi.org/10.3102/1076998609332757

Linn, R. L. (1973). Fair test use in selection. *Review of Educational Research, 43*, 139–161. https://doi.org/10.3102/00346543043002139

Linn, R. L. (1976). In search of fair selection procedures. *Journal of Educational Measurement, 13*, 53–58. https://doi.org/10.1111/j.1745-3984.1971.tb00898.x

Linn, R. L., & Werts, C. E. (1971). Considerations for studies of test bias. *Journal of Educational Measurement, 8*, 1–4. https://doi.org/10.1111/j.1745-3984.1971.tb00898.x

Lipnevich, A. A., & Roberts, R. D. (2012). Noncognitive skills in education: Emerging research and applications in a variety of international contexts. *Learning and Individual Differences, 22*, 173–177. https://doi.org/10.1016/j.lindif.2011.11.016

Liu, O. L., Lee, H. S., & Linn, M. C. (2010). An investigation of teacher impact on student inquiry science performance using a hierarchical linear model. *Journal of Research in Science Teaching, 47*, 807–819. https://doi.org/10.1002/tea.20372

Liu, L., Rogat, A., & Bertling, M. (2013). *A CBAL science model of cognition: Developing a competency model and learning progressions to support assessment development* (Research Report No. RR-13-29). Princeton: Educational Testing Service. http://dx.doi.org/10.1002/j.2333-8504.2013.tb02336.x

Lockheed, M. E. (1985). Women, girls and computers: A first look at the evidence. *Sex Roles, 13*, 115–122. https://doi.org/10.1007/BF00287904

Lockwood, J. R., Savitsky, T. D., & McCaffrey, D. F. (2015). Inferring constructs of effective teaching from classroom observations: An application of Bayesian exploratory factor analysis without restrictions. *Annals of Applied Statistics, 9*, 1484–1509. https://doi.org/10.1214/15-AOAS833

Lopez, A. A., Pooler, E., & Linquanti, R. (2016). *Key issues and opportunities in the initial identification and classification of English learners* (Research Report No. RR-16-09). Princeton: Educational Testing Service. http://dx.doi.org/10.1002/ets2.12090

Lord, F. M. (1952). A theory of test scores. *Psychometrika Monograph, 17*(7).

Lord, F. M. (1953). An application of confidence intervals and of maximum likelihood to the estimation of an examinee's ability. *Psychometrika, 18*, 57–76. https://doi.org/10.1007/BF02289028

Lord, F. M. (1965a). An empirical study of item-test regression. *Psychometrika, 30*, 373–376. https://doi.org/10.1007/BF02289501

Lord, F. M. (1965b). A note on the normal ogive or logistic curve in item analysis. *Psychometrika, 30*, 371–372. https://doi.org/10.1007/BF02289500

Lord, F. M. (1967). A paradox in the interpretation of group comparisons. *Psychological Bulletin, 68*, 304–335. https://doi.org/10.1037/h0025105

Lord, F. M. (1968a). An analysis of the Verbal Scholastic Aptitude Test using Birnbaum's three-parameter logistic model. *Educational and Psychological Measurement, 28*, 989–1020. https://doi.org/10.1177/001316446802800401

Lord, F. M. (1968b). *Some test theory for tailored testing* (Research Bulletin No. RB-68-38). Princeton: Educational Testing Service. http://dx.doi.org/10.1002/j.2333-8504.1968.tb00562.x

Lord, F. M. (1970). Estimating item characteristic curves without knowledge of their mathematical form. *Psychometrika, 35*, 43–50. https://doi.org/10.1007/BF02290592

Lord, F. M. (1973). Power scores estimated by item characteristic curves. *Educational and Psychological Measurement, 33*, 219–224. https://doi.org/10.1177/001316447303300201

Lord, F. M. (1974a). Estimation of latent ability and item parameters when there are omitted responses. *Psychometrika, 39*, 247–264. https://doi.org/10.1007/BF02291471

Lord, F. M. (1974b). Individualized testing and item characteristic curve theory. In D. H. Krantz, R. C. Atkinson, R. D. Luce, & P. Suppes (Eds.), *Contemporary developments in mathematical psychology* (Vol. 2, pp. 106–126). San Francisco: Freeman.

Lord, F. M. (1975a). The "ability" scale in item characteristic curve theory. *Psychometrika, 40*, 205–217. https://doi.org/10.1007/BF02291567

Lord, F. M. (1975b). *Evaluation with artificial data of a procedure for estimating ability and item characteristic curve parameters* (Research Bulletin No. RB-75-33). Princeton: Educational Testing Service. http://dx.doi.org/10.1002/j.2333-8504.1975.tb01073.x

Lord, F. M. (1977). Practical applications of item characteristic curve theory. *Journal of Educational Measurement, 14*, 117–138. https://doi.org/10.1111/j.1745-3984.1977.tb00032.x

Lord, F. M. (1980). *Applications of item response theory to practical testing problems*. Mahwah: Erlbaum.

Lord, F. M., & Novick, M. R. (1968). *Statistical theories of mental test scores*. Reading: Addison-Wesley.

MacCann, C., & Roberts, R. D. (2008). New paradigms for assessing emotional intelligence: Theory and data. *Emotions, 8*, 540–551. https://doi.org/10.1037/a0012746

MacCann, C., Schulze, R., Matthews, G., Zeidner, M., & Roberts, R. D. (2008). Emotional intelligence as pop science, misled science, and sound science: A review and critical synthesis of perspectives from the field of psychology. In N. C. Karafyllis & G. Ulshofer (Eds.), *Sexualized brains: Scientific modeling of emotional intelligence from a cultural perspective* (pp. 131–148). Cambridge, MA: MIT Press.

MacCann, C., Wang, L., Matthews, G., & Roberts, R. D. (2010). Emotional intelligence and the eye of the beholder: Comparing self- and parent-rated situational judgments in adolescents. *Journal of Research in Personality, 44*, 673–676. https://doi.org/10.1016/j.jrp.2010.08.009

MacCann, C., Fogarty, G. J., Zeidner, M., & Roberts, R. D. (2011). Coping mediates the relationship between emotional intelligence (EI) and academic achievement. *Contemporary Educational Psychology, 36*, 60–70. https://doi.org/10.1016/j.cedpsych.2010.11.002

Mantel, N., & Haenszel, W. (1959). Statistical aspects of the analysis of data from retrospective studies of disease. *Journal of the National Cancer Institute, 22*, 719–748.

Marco, G. L. (1972). *Impact of Michigan 1970–71 Grade 3 Title I reading programs* (Program Report No. PR-72-05). Princeton: Educational Testing Service.

Marco, G. L. (1977). Item characteristic curve solutions to three intractable testing problems. *Journal of Educational Measurement, 14*, 139–160. https://doi.org/10.1111/j.1745-3984.1977.tb00033.x

Martiniello, M. (2009). Linguistic complexity, schematic representations, and differential item functioning for English language learners in math tests. *Educational Assessment, 14*, 160–179. https://doi.org/10.1080/10627190903422906

Mazzeo, J., & von Davier, M. (2008). (2008). Review of the Programme for International Student Assessment (PISA) test design: Recommendations for fostering stability in assessment results. *Education Working Papers EDU/PISA/GB, 28*, 23–24.

McCaffrey, D. F. (2013). *Will teacher value-added scores change when accountability tests change?* (Carnegie Knowledge Network Brief No. 8). Retrieved from http://www.carnegieknowledgenetwork.org/briefs/valueadded/accountability-tests/

McCaffrey, D. F., Han, B., & Lockwood, J. R. (2014). Using auxiliary teacher data to improve value-added: An application of small area estimation to middle school mathematics teachers. In R. W. Lissitz & H. Jiao (Eds.), *Value added modeling and growth modeling with particular application to teacher and school effectiveness* (pp. 191–217). Charlotte: Information Age.

McDonald, F. J., & Elias, P. (1976). *Beginning teacher evaluation study, Phase 2: The effects of teaching performance on pupil learning* (Vol. 1, Program Report No. PR-76-06A). Princeton: Educational Testing Service.

McGillicuddy-DeLisi, A. V., Sigel, I. E., & Johnson, J. E. (1979). The family as a system of mutual influences: Parental beliefs, distancing behaviors, and children's representational thinking. In M. Lewis & L. A. Rosenblum (Eds.), *Genesis of behavior: Vol. 2. The child and its family* (pp. 91–106). New York: Plenum Press.

Medley, D. M., & Coker, H. (1987). The accuracy of principals' judgments of teacher performance. *Journal of Educational Research, 80*, 242–247. https://doi.org/10.1080/00220671.1987.10885759

Medley, D. M., & Hill, R. A. (1967). Dimensions of classroom behavior measured by two systems of interaction analysis. *Educational Leadership, 26*, 821–824.

Medley, D. M., Coker, H., Lorentz, J. L., Soar, R. S., & Spaulding, R. L. (1981). Assessing teacher performance from observed competency indicators defined by classroom teachers. *Journal of Educational Research, 74*, 197–216. https://doi.org/10.1080/00220671.1981.10885311

Melville, S. D., & Frederiksen, N. (1952). Achievement of freshmen engineering students and the Strong Vocational Interest Blank. *Journal of Applied Psychology, 36*, 169–173. https://doi.org/10.1037/h0059101

Messick, S. (1967). The psychology of acquiescence: An interpretation of research evidence. In I. A. Berg (Ed.), *Response set in personality assessment* (pp. 115–145). Chicago: Aldine.

Messick, S. (1975). The standard problem: Meaning and values in measurement and evaluation. *American Psychologist, 30*, 955–966. https://doi.org/10.1037/0003-066X.30.10.955

Messick, S. (1980). *The effectiveness of coaching for the SAT: Review and reanalysis of research from the fifties to the FTC*. Princeton: Educational Testing Service.

Messick, S. (1982). Issues of effectiveness and equity in the coaching controversy: Implications for educational testing and practice. *Educational Psychologist, 17*, 67–91. https://doi.org/10.1080/00461528209529246

Messick, S. (1987). Structural relationships across cognition, personality, and style. In R. E. Snow & M. J. Farr (Eds.), *Aptitude, learning, and instruction: Vol. 3. Conative and affective process analysis* (pp. 35–75). Hillsdale: Erlbaum.

Messick, S. (1989). Validity. In R. Linn (Ed.), *Educational measurement* (3rd ed., pp. 13–103). New York: Macmillan.

Messick, S. (1994a). The interplay of evidence and consequences in the validation of performance assessments. *Educational Researcher, 23*(2), 13–23. https://doi.org/10.3102/0013189X023002013

Messick, S. (1994b). The matter of style: Manifestations of personality in cognition, learning, and teaching. *Educational Psychologist, 29*, 121–136. https://doi.org/10.1207/s15326985ep2903_2

Messick, S. (1996). Bridging cognition and personality in education: The role of style in performance and development. *European Journal of Personality, 10*, 353–376. https://doi.org/10.1002/(SICI)1099-0984(199612)10:5<353::AID-PER268>3.0.CO;2-G

Messick, S., Beaton, A. E., & Lord, F. (1983). *National Assessment of Educational Progress: A new design for a new era*. Princeton: Educational Testing Service.

Messick, S., & Fritzky, F. J. (1963). Dimensions of analytic attitude in cognition and personality. *Journal of Personality, 31*, 346–370. https://doi.org/10.1111/j.1467-6494.1963.tb01304.x

Messick, S., & Kogan, N. (1966). Personality consistencies in judgment: Dimensions of role constructs. *Multivariate Behavioral Research, 1*, 165–175. https://doi.org/10.1207/s15327906mbr0102_3

Mihaly, K., & McCaffrey, D. F. (2014). Grade-level variation in observational measures of teacher effectiveness. In T. J. Kane, K. A. Kerr, & R. C. Pianta (Eds.), *Designing teacher evaluation systems: New guidance from the Measures of Effective Teaching Project* (pp. 9–49). San Francisco: Jossey-Bass.

Mislevy, R. J. (1984). Estimating latent distributions. *Psychometrika, 49*, 359–381. https://doi.org/10.1007/BF02306026

Mislevy, R. J. (1985). Estimation of latent group effects. *Journal of the American Statistical Association, 80*, 993–997. https://doi.org/10.1080/01621459.1985.10478215

Mislevy, R. J. (1991). Randomization-based inference about latent variables from complex samples. *Psychometrika, 56*, 177–196. https://doi.org/10.1007/BF02294457

Mislevy, R. J. (1993a). Should "multiple imputations" be treated as "multiple indicators"? *Psychometrika, 58*, 79–85. https://doi.org/10.1007/BF02294472

Mislevy, R. J. (1993b). Some formulas for use with Bayesian ability estimates. *Educational and Psychological Measurement, 53*, 315–328.

Mislevy, R. (1994). Evidence and inference in educational assessment. *Psychometrika, 59*, 439–483. https://doi.org/10.1177/0013164493053002002

Mislevy, R. J. (1996). Test theory reconceived. *Journal of Educational Measurement, 33*, 379–416. https://doi.org/10.1111/j.1745-3984.1996.tb00498.x

Mislevy, R. J., & Levy, R. (2007). Bayesian psychometric modeling from an evidence centered design perspective. In C. R. Rao & S. Sinharay (Eds.), *Handbook of statistics: Vol. 26. Psychometrics* (pp. 839–865). Amsterdam: Elsevier. http://dx.doi.org/10.1016/S0169-7161(06)26026-7

Mislevy, R. J., & Verhelst, N. (1990). Modeling item responses when different subjects employ different solution strategies. *Psychometrika, 55*, 195–215. https://doi.org/10.1007/BF02295283

Mislevy, R. J., Beaton, A. E., Kaplan, B., & Sheehan, K. M. (1992a). Estimating population characteristics from sparse matrix samples of item responses. *Journal of Educational Measurement, 29*, 133–161. https://doi.org/10.1111/j.1745-3984.1992.tb00371.x

Mislevy, R., Johnson, E., & Muraki, E. (1992b). Scaling procedures in NAEP. *Journal of Educational and Behavioral Statistics, 17*, 131–154. https://doi.org/10.2307/1165166

Mislevy, R. J., Almond, R. G., Yan, D., & Steinberg, L. S. (2000). *Bayes nets in educational assessment: Where do the numbers come from?* (CSE Technical Report No. 518). Los Angeles: UCLA CRESST.

Mislevy, R. J., Steinberg, L. S., Breyer, F. J., Almond, R. G., & Johnson, L. (2002). Making sense of data from complex assessments. *Applied Measurement in Education, 15*, 363–389. https://doi.org/10.1207/S15324818AME1504_03

Mislevy, R. J., Steinberg, L. S., & Almond, R. G. (2003). On the structure of educational assessments. *Measurement: Interdisciplinary Research and Perspectives, 1*, 3–62. https://doi.org/10.1207/S15366359MEA0101_02

Mislevy, R. J., Steinberg, L., Almond, R. G., & Lucas, J. F. (2006). Concepts, terminology, and basic models of evidence-centered design. In D. M. Williamson, R. J. Mislevy, & I. I. Bejar (Eds.), *Automated scoring of complex tasks in computer-based testing* (pp. 49–82). Mahwah: Erlbaum.

Mislevy, R. J., Behrens, J. T., Dicerbo, K. E., & Levy, R. (2012). Design and discovery in educational assessment: Evidence-centered design, psychometrics, and educational data mining. *Journal of Educational Data Mining, 4*(1), 11–48.

Mislevy, R. J., Oranje, A., Bauer, M. I., von Davier, A., Hao, J., Corrigan, S., et al. (2014). *Psychometric considerations in game-based assessment*. Redwood City: GlassLab.
Mislevy, R. J., Corrigan, S., Oranje, A., DiCerbo, K., Bauer, M. I., von Davier, A., & John, M. (2016). Psychometrics and game-based assessment. In F. Drasgow (Ed.), *Technology and testing: Improving educational and psychological measurement* (pp. 23–48). New York: Routledge.
Moses, T. (2016). Loglinear models for observed-score distributions. In W. J. van der Linden (Ed.), *Handbook of item response theory* (Vol. 2, pp. 71–85). Boca Raton: CRC Press.
Muraki, E. (1992). A generalized partial credit model: Application of an EM algorithm. *Applied Psychological Measurement, 16*, 159–176. https://doi.org/10.1177/014662169201600206
Murphy, R. T. (1973). *Adult functional reading study* (Program Report No. PR-73-48). Princeton: Educational Testing Service.
Murphy, R. T. (1977). *Evaluation of the PLATO 4 computer-based education system: Community college component*. Princeton: Educational Testing Service.
Murphy, R. T. (1988). *Evaluation of* Al Manaahil*: An original Arabic children's television series in reading* (Research Report No. RR-88-45). Princeton: Educational Testing Service. http://dx.doi.org/10.1002/j.2330-8516.1988.tb00301.x
Murphy, R. T., & Appel, L. R. (1984). *Evaluation of the Writing to Read instructional system, 1982–1984*. Princeton: Educational Testing Service.
Myford, C. M., & Engelhard, G., Jr. (2001). Examining the psychometric quality of the National Board for Professional Teaching Standards Early Childhood/Generalist Assessment System. *Journal of Personnel Evaluation in Education, 15*, 253–285. https://doi.org/10.1023/A:1015453631544
Naemi, B. D., Seybert, J., Robbins, S. B., & Kyllonen, P. C. (2014). *Examining the WorkFORCE® Assessment for Job Fit and Core Capabilities of FACETS* (Research Report No. RR-14-32). Princeton: Educational Testing Service. http://dx.doi.org/10.1002/ets2.12040
Nettles, M. T. (1990). *Black, Hispanic, and White doctoral students: Before, during, and after enrolling in graduate school* (Minority Graduate Education Project Report No. MGE-90-01). Princeton: Educational Testing Service.
Nettles, M., & Millett, C. (2006). *Three magic letters: Getting to Ph.D.* Baltimore: Johns Hopkins University Press.
Nettles, M. T., Scatton, L. H., Steinberg, J. H., & Tyler, L. L. (2011). *Performance and passing rate differences of African American and White prospective teachers on PRAXIS examinations* (Research Report No. RR-11-08). Princeton: Educational Testing Service. http://dx.doi.org/10.1002/j.2333-8504.2011.tb02244.x
Nogee, P. (1950). *A preliminary study of the "Social Situations Test"* (Research Memorandum No. RM-50-22). Princeton: Educational Testing Service.
Oliveri, M. E., & Ezzo, C. (2014). The role of noncognitive measures in higher education admissions. *Journal of the World Universities Forum, 6*(4), 55–65.
Oliveri, M. E., & von Davier, M. (2011). Investigation of model fit and score scale comparability in international assessments. *Psychological Test and Assessment Modeling, 53*, 315–333.
Oliveri, M. E., & von Davier, M. (2014). Toward increasing fairness in score scale calibrations employed in international large-scale assessments. *International Journal of Testing, 14*(1), 1–21. http://dx.doi.org/10.1080/15305058.2013.825265
Oranje, A., & Ye, J. (2013). Population model size, bias, and variance in educational survey assessments. In L. Rutkowski, M. von Davier, & D. Rutkowski (Eds.), *Handbook of international large scale assessments* (pp. 203–228). Boca Raton: CRC Press.
Phelps, G., & Howell, H. (2016). Assessing mathematical knowledge for teaching: The role of teaching context. *The Mathematics Enthusiast, 13*(1), 52–70.
Pitoniak, M. J., Young, J. W., Martiniello, M., King, T. C., Buteux, A., & Ginsburgh, M. (2009). *Guidelines for the assessment of English language learners*. Princeton: Educational Testing Service.
Powers, D. E. (1985). Effects of test preparation on the validity of a graduate admissions test. *Applied Psychological Measurement, 9*, 179–190. https://doi.org/10.1177/014662168500900206

Powers, D. E. (1988). Incidence, correlates, and possible causes of test anxiety in graduate admissions testing. *Advances in Personality Assessment, 7*, 49–75.

Powers, D. E. (1992). *Assessing the classroom performance of beginning teachers: Educators' appraisal of proposed evaluation criteria* (Research Report No. RR-92-55). Princeton: Educational Testing Service. https://doi.org/10.1002/j.2333-8504.1992.tb01487.x

Powers, D. E. (2001). Test anxiety and test performance: Comparing paper-based and computer-adaptive versions of the Graduate Record Examinations (GRE) General Test. *Journal of Educational Computing Research, 24*, 249–273. https://doi.org/10.2190/680W-66CR-QRP7-CL1F

Powers, D. E., & Bennett, R. E. (1999). Effects of allowing examinees to select questions on a test of divergent thinking. *Applied Measurement in Education, 12*, 257–279. https://doi.org/10.1207/S15324818AME1203_3

Powers, D. E., & Swinton, S. S. (1984). Effects of self-study for coachable test item types. *Journal of Educational Psychology, 76*, 266–278. https://doi.org/10.1037/0022-0663.76.2.266

Powers, D. E., Burstein, J. C., Chodorow, M. S., Fowles, M. E., & Kukich, K. (2002). Comparing the validity of automated and human scoring of essays. *Journal of Educational Computing Research, 26*, 407–425. https://doi.org/10.2190/CX92-7WKV-N7WC-JL0A

Quinlan, T., Higgins, D., & Wolff, S. (2009). *Evaluating the construct-coverage of the e-rater scoring engine* (Research Report No. RR-09-01). Princeton: Educational Testing Service. http://dx.doi.org/10.1002/j.2333-8504.2009.tb02158.x

Quirk, T. J., Steen, M. T., & Lipe, D. (1971). Development of the Program for Learning in Accordance with Needs Teacher Observation Scale: A teacher observation scale for individualized instruction. *Journal of Educational Psychology, 62*, 188–200. https://doi.org/10.1037/h0031144

Quirk, T. J., Trismen, D. A., Nalin, K. B., & Weinberg, S. F. (1975). The classroom behavior of teachers during compensatory reading instruction. *Journal of Educational Research, 68*, 185–192. https://doi.org/10.1080/00220671.1975.10884742

Quirk, T. J., Witten, B. J., & Weinberg, S. F. (1973). Review of studies of the concurrent and predictive validity of the National Teacher Examinations. *Review of Educational Research, 43*, 89–113. https://doi.org/10.3102/00346543043001089

Ramist, L., Lewis, C., & McCamley-Jenkins, L. (1994). *Student group differences in predicting college grades: Sex, language, and ethnic groups* (Research Report No. RR-94-27). Princeton: Educational Testing Service. http://dx.doi.org/10.1002/j.2333-8504.1994.tb01600.x

Rao, C. R., & Sinharay, S. (Eds.). (2006). *Handbook of statistics: Vol. 26. Psychometrics*. Amsterdam: Elsevier.

Renninger, K. A., & Sigel, I. E. (Eds.). (2006). *Handbook of child psychology: Vol. 4. Child psychology in practice* (6th ed.). New York: Wiley.

Reynolds, A., Rosenfeld, M., & Tannenbaum, R. J. (1992). *Beginning teacher knowledge of general principles of teaching and learning: A national survey* (Research Report No. RR-92-60). Princeton: Educational Testing Service. http://dx.doi.org/10.1002/j.2333-8504.1992.tb01491.x

Ricciuti, H. N. (1951). *A comparison of leadership ratings made and received by student raters (Research Memorandum No. RM-51-04)*. Princeton: Educational Testing Service.

Rijmen, F. (2010). Formal relations and an empirical comparison among the bi-factor, the testlet, and a second-order multidimensional IRT model. *Journal of Educational Measurement, 47*, 361–372. https://doi.org/10.1111/j.1745-3984.2010.00118.x

Rijmen, F., Jeon, M., von Davier, M., & Rabe-Hesketh, S. (2014). A third order item response theory model for modeling the effects of domains and subdomains in large-scale educational assessment surveys. *Journal of Educational and Behavioral Statistics, 38*, 32–60. https://doi.org/10.3102/1076998614531045

Roberts, R. D., MacCann, C., Matthews, G., & Zeidner, M. (2010). Emotional intelligence: Toward a consensus of models and measures. *Social and Personality Psychology Compass, 4*, 821–840. https://doi.org/10.1111/j.1751-9004.2010.00277.x

Roberts, R. D., Schulze, R., O'Brien, K., McCann, C., Reid, J., & Maul, A. (2006). Exploring the validity of the Mayer–Salovey–Caruso Emotional Intelligence Test (MSCEIT) with established emotions measures. *Emotions, 6*, 663–669. https://doi.org/10.1037/1528-3542.6.4.663

Robustelli, S. L. (2010). *Validity evidence to support the development of a licensure assessment for entry-level teachers: A job-analytic approach* (Research Memorandum No. RM-10-10). Princeton: Educational Testing Service.

Rosenbaum, P. R., & Rubin, D. B. (1984). Reducing bias in observational studies using subclassification on the propensity score. *Journal of the American Statistical Association, 79*, 516–524. https://doi.org/10.1080/01621459.1984.10478078

Rosenbaum, P. R., & Rubin, D. B. (1985). The bias due to incomplete matching. *Biometrics, 41*, 103–116. https://doi.org/10.2307/2530647

Rosenhan, D. (1969). Some origins of concern for others. In P. H. Mussen, J. Langer, & M. V. Covington (Eds.), *Trends and issues in developmental psychology* (pp. 134–153). New York: Holt, Rinehart, and Winston.

Rosenhan, D. (1970). The natural socialization of altruistic autonomy. In J. Macaulay & L. Berkowitz (Eds.), *Altruism and helping behavior: Social psychological studies of some antecedents and consequences* (pp. 251–268). New York: Academic Press.

Rosenhan, D. L. (1972). Learning theory and prosocial behavior. *Journal of Social Issues, 28*, 151–163. https://doi.org/10.1111/j.1540-4560.1972.tb00037.x

Rosenhan, D., & White, G. M. (1967). Observation and rehearsal as determinants of prosocial behavior. *Journal of Personality and Social Psychology, 5*, 424–431. https://doi.org/10.1037/h0024395

Rosner, F. C., & Howey, K. R. (1982). Construct validity in assessing teacher knowledge: New NTE interpretations. *Journal of Teacher Education, 33*(6), 7–12. https://doi.org/10.1177/002248718203300603

Rubin, D. B. (1974a). Characterizing the estimation of parameters in incomplete-data problems. *Journal of the American Statistical Association, 69*, 467–474.

Rubin, D. B. (1974b). Estimating causal effects of treatments in randomized and nonrandomized studies. *Journal of Educational Psychology, 66*, 688–701. https://doi.org/10.1037/h0037350

Rubin, D. B. (1976a). Noniterative least squares estimates, standard errors, and F-tests for analyses of variance with missing data. *Journal of the Royal Statistical Society, Series B, 38*, 270–274.

Rubin, D. B. (1976b). Comparing regressions when some predictor values are missing. *Technometrics, 18*, 201–205. https://doi.org/10.1080/00401706.1976.10489425

Rubin, D. B. (1976c). Inference and missing data. *Biometrika, 63*, 581–592. https://doi.org/10.1093/biomet/63.3.581

Rubin, D. B. (1978). Bayesian inference for causal effects: The role of randomization. *Annals of Statistics, 6*, 34–58. https://doi.org/10.1214/aos/1176344064

Rubin, D. B. (1979). Using multivariate matched sampling and regression adjustment to control bias in observational studies. *Journal of the American Statistical Association, 74*, 318–328. https://doi.org/10.1080/01621459.1979.10482513

Rubin, D. B. (1980a). Bias reduction using Mahalanobis-metric matching. *Biometrics, 36*, 293–298. https://doi.org/10.2307/2529981

Rubin, D. B. (1980b). *Handling nonresponse in sample surveys by multiple imputations*. Washington, DC: U.S. Department of Commerce, Bureau of the Census.

Rubin, D. B. (1980c). Illustrating the use of multiple imputations to handle nonresponse in sample surveys. *42nd Session of the International Statistical Institute, 1979*(Book 2), 517–532.

Rudd, R., Kirsch, I., & Yamamoto, K. (2004). *Literacy and health in America* (Policy Information Report). Princeton: Educational Testing Service.

Rupp, A., Templin, J., & Henson, R. (2010). *Diagnostic measurement: Theory, methods, and applications*. New York: Guilford Press.

Rutkowski, L., von Davier, M., & Rutkowski, D. (Eds.). (2013). *Handbook of international large-scale assessment: Background, technical issues, and methods of data analysis*. London: Chapman and Hall.

Ryans, D. G., & Frederiksen, N. (1951). Performance tests of educational achievement. In E. F. Lindquist (Ed.), *Educational measurement* (pp. 455–494). Washington, DC: American Council on Education.

Sabatini, J., & O'Reilly, T. (2013). Rationale for a new generation of reading comprehension assessments. In B. Miller, L. E. Cutting, & P. McCardle (Eds.), *Unraveling reading comprehension: Behavioral, neurobiological and genetic components* (pp. 100–111). Baltimore: Paul H. Brooks.

Sandoval, J. (1976). *Beginning Teacher Evaluation Study: Phase II. 1973–74. Final report: Vol. 3. The evaluation of teacher behavior through observation of videotape recordings* (Program Report No. PR-76-10). Princeton: Educational Testing Service.

Schrader, W. B., & Pitcher, B. (1964). *Adjusted undergraduate average grades as predictors of law school performance* (Law School Admissions Council Report No. LSAC-64-02). Princeton: LSAC.

Sebrechts, M. M., Bennett, R. E., & Rock, D. A. (1991). Agreement between expert-system and human raters on complex constructed-response quantitative items. *Journal of Applied Psychology, 76*, 856–862. https://doi.org/10.1037/0021-9010.76.6.856

Sherman, S. W., & Robinson, N. M. (Eds.). (1982). *Ability testing of handicapped people: Dilemma for government, science, and the public*. Washington, DC: National Academy Press.

Shipman, V. C. (Ed.). (1972). *Disadvantaged children and their first school experiences: ETS–Head Start longitudinal study* (Program Report No. 72-27). Princeton: Educational Testing Service.

Shute, V. J., Ventura, M., Bauer, M. I., & Zapata-Rivera, D. (2008). *Monitoring and fostering learning through games and embedded assessments* (Research Report No. RR-08-69). Princeton: Educational Testing Service. https://doi.org/10.1002/j.2333-8504.2008.tb02155.x

Shute, V. J., Ventura, M., Bauer, M., & Zapata-Rivera, D. (2009). Melding the power of serious games and embedded assessment to monitor and foster learning. In U. Ritterfeld, M. J. Cody, & P. Vorderer (Eds.), *Serious games: Mechanisms and effects* (pp. 295–321). New York: Routledge.

Sigel, I. E. (1982). The relationship between parental distancing strategies and the child's cognitive behavior. In L. M. Laosa & I. E. Sigel (Eds.), *Families as learning environments for children* (pp. 47–86). New York: Plenum Press. https://doi.org/10.1007/978-1-4684-4172-7_2

Sigel, I. E. (1990). Journeys in serendipity: The development of the distancing model. In I. E. Sigel & G. H. Brody (Eds.), *Methods of family research: Biographies of research projects: Vol. 1. Normal families* (pp. 87–120). Hillsdale: Erlbaum.

Sigel, I. E. (1992). The belief–behavior connection: A resolvable dilemma? In I. E. Sigel, A. V. McGillicuddy-DeLisi, & J. J. Goodnow (Eds.), *Personal belief systems: The psychological consequences for children* (2nd ed., pp. 433–456). Hillsdale: Erlbaum.

Sigel, I. E. (1993). The centrality of a distancing model for the development of representational competence. In R. R. Cocking & K. A. Renninger (Eds.), *The development and meaning of psychological distance* (pp. 141–158). Hillsdale: Erlbaum.

Sigel, I. E. (1999). Approaches to representation as a psychological construct: A treatise in diversity. In I. E. Sigel (Ed.), *Development of mental representation: Theories and applications* (pp. 3–12). Mahwah: Erlbaum.

Sigel, I. (2000). Educating the Young Thinker model, from research to practice: A case study of program development, or the place of theory and research in the development of educational programs. In J. L. Roopnarine & J. E. Johnson (Eds.), *Approaches to early childhood education* (3rd ed., pp. 315–340). Upper Saddle River: Merrill.

Sigel, I. E. (2006). Research to practice redefined. In K. A. Renninger & I. E. Sigel (Eds.), *Handbook of child psychology: Vol. 4. Child psychology in practice* (6th ed., pp. 1017–1023). New York: Wiley.

Sinharay, S. (2003). *Practical applications of posterior predictive model checking for assessing fit of common item response theory models* (Research Report No. RR-03-33). Princeton: Educational Testing Service. http://dx.doi.org/10.1002/j.2333-8504.2003.tb01925.x

Sinharay, S. (2016). Bayesian model fit and model comparison. In W. J. van der Linden (Ed.), *Handbook of item response theory* (Vol. 2, pp. 379–394). Boca Raton: CRC Press.

Sinharay, S., & von Davier, M. (2005). *Extension of the NAEP BGROUP program to higher dimensions* (Research Report No. RR-05-27). Princeton: Educational Testing Service. http://dx.doi.org/10.1002/j.2333-8504.2005.tb02004.x

Skager, R. W., Schultz, C. B., & Klein, S. P. (1965). Quality and quantity of accomplishments as measures of creativity. *Journal of Educational Psychology, 56*, 31–39. https://doi.org/10.1037/h0021901

Skager, R. W., Schultz, C. B., & Klein, S. P. (1966). Points of view about preference as tools in the analysis of creative products. *Perceptual and Motor Skills, 22*, 83–94. https://doi.org/10.2466/pms.1966.22.1.83

Sparks, J. R., & Deane, P. (2015). *Cognitively based assessment of research and inquiry skills: Defining a key practice in the English language arts* (Research Report No. RR-15-35). Princeton: Educational Testing Service. https://doi.org/10.1002/ets2.12082

Sprinthall, N. A., & Beaton, A. E. (1966). Value differences among public high school teachers using a regression model analysis of variance technique. *Journal of Experimental Education, 35*, 36–42. https://doi.org/10.1080/00220973.1966.11010982

Steinberg, L. S., & Gitomer, D. H. (1996). Intelligent tutoring and assessment built on an understanding of a technical problem-solving task. *Instructional Science, 24*, 223–258. https://doi.org/10.1007/BF00119978

Steinberg, J., Cline, F., & Sawaki, Y. (2011). *Examining the factor structure of a state standards-based science assessment for students with learning disabilities* (Research Report No. RR-11-38). Princeton: Educational Testing Service. https://doi.org/10.1002/j.2333-8504.2011.tb02274.x

Stone, E., & Davey, T. (2011). *Computer-adaptive testing for students with disabilities: A review of the literature* (Research Report No. RR-11-32). Princeton: Educational Testing Service. https://doi.org/10.1002/j.2333-8504.2011.tb02268.x

Stone, E., Cook, L. L., & Laitusis, C. (2013). *Evaluation of a condition-adaptive test of reading comprehension for students with reading-based learning disabilities* (Research Report No. RR-13-20). Princeton: Educational Testing Service. https://doi.org/10.1002/j.2333-8504.2013.tb02327.x

Stone, E., Laitusis, C. C., & Cook, L. L. (2016). Increasing the accessibility of assessments through technology. In F. Drasgow (Ed.), *Technology and testing: Improving educational and psychological measurement* (pp. 217–234). New York: Routledge.

Stricker, L. J., & Bejar, I. (2004). Test difficulty and stereotype threat on the GRE General Test. *Journal of Applied Social Psychology, 34*, 563–597. https://doi.org/10.1111/j.1559-1816.2004.tb02561.x

Stricker, L. J., & Rock, D. A. (2015). An "Obama effect" on the GRE General Test? *Social Influence, 10*, 11–18. https://doi.org/10.1080/15534510.2013.878665

Stricker, L. J., & Ward, W. C. (2004). Stereotype threat, inquiring about test takers' ethnicity and gender, and standardized test performance. *Journal of Applied Social Psychology, 34*, 665–693. https://doi.org/10.1111/j.1559-1816.2004.tb02561.x

Sum, A., Kirsch, I., & Taggart, R. (2002). *The twin challenges of mediocrity and inequality: Literacy in the U.S. from an international perspective* (Policy Information Report). Princeton: Educational Testing Service.

Swinton, S. S., & Powers, D. E. (1983). A study of the effects of special preparation on GRE analytical scores and item types. *Journal of Educational Psychology, 75*, 104–115. https://doi.org/10.1037/0022-0663.75.1.104

Sykes, G., & Wilson, S. M. (2015). *How teachers teach: Mapping the terrain of practice*. Princeton: Educational Testing Service.

Tannenbaum, R. J. (1992). *A job analysis of the knowledge important for newly licensed (certified) general science teachers* (Research Report No. RR-92-77). Princeton: Educational Testing Service. https://doi.org/10.1002/j.2333-8504.1992.tb01509.x

Tannenbaum, R. J., & Rosenfeld, M. (1994). Job analysis for teacher competency testing: Identification of basic skills important for all entry-level teachers. *Educational and Psychological Measurement, 54*, 199–211. https://doi.org/10.1177/0013164494054001026

Tatsuoka, K. K. (1983). Rule space: An approach for dealing with misconceptions based on item response theory. *Journal of Educational Measurement, 20*, 345–354. https://doi.org/10.1111/j.1745-3984.1983.tb00212.x

Thompson, M., & Goe, L. (2009). *Models for effective and scalable teacher professional development* (Research Report No. RR-09-07). Princeton: Educational Testing Service. https://doi.org/10.1002/j.2333-8504.2009.tb02164.x

Torgerson, W. S., & Green, B. F. (1950). *A factor analysis of English essay readers* (Research Bulletin No. RB-50-30). Princeton: Educational Testing Service. http://dx.doi.org/10.1002/j.2333-8504.1950.tb00470.x

Turkan, S., & Buzick, H. M. (2016). Complexities and issues to consider in the evaluation of content teachers of English language learners. *Urban Education, 51*, 221–248. https://doi.org/10.1177/0042085914543111

U.S. Department of Education. (n.d.-a). *Technical notes on the interactive computer and hands-on tasks in science*. Retrieved from http://www.nationsreportcard.gov/science_2009/ict_tech_notes.aspx

U.S. Department of Education. (n.d.-b). *2011 writing assessment*. Retrieved from http://www.nationsreportcard.gov/writing_2011/

U.S. Department of Education. (n.d.-c). *2014 technology and engineering literacy (TEL) assessment*. Retrieved from http://www.nationsreportcard.gov/tel_2014/

von Davier, A. A. (Ed.). (2011). *Statistical models for test equating, scaling and linking*. New York: Springer. https://doi.org/10.1007/978-0-387-98138-3

von Davier, A. A., Holland, P. W., & Thayer, D. T. (2004). *The Kernel method of test equating*. New York: Springer. https://doi.org/10.1007/b97446

von Davier, M. (1997). Bootstrapping goodness-of-fit statistics for sparse categorical data: Results of a Monte Carlo study. *Methods of Psychological Research Online, 2*(2), 29–48.

von Davier, M. (2008a). A general diagnostic model applied to language testing data. *British Journal of Mathematical and Statistical Psychology, 61*, 287–307. https://doi.org/10.1348/000711007X193957

von Davier, M. (2008b). The mixture general diagnostic model. In G. R. Hancock & K. M. Samuelsen (Eds.), *Advances in latent variable mixture models* (pp. 255–274). Charlotte: Information Age.

von Davier, M. (2014). *The log-linear cognitive diagnostic model (LCDM) as a special case of the general diagnostic model (GDM), Research Report No. RR-14-40*. Princeton: Educational Testing Service. https://doi.org/10.1002/ets2.12043

von Davier, M. (2016). *High-performance psychometrics: The parallel-E parallel-M algorithm for generalized latent variable models* (Research Report No. 16-34). Princeton: Educational Testing Service.

von Davier, M., & Rost, J. (2016). Logistic mixture-distribution response models. In W. J. van der Linden (Ed.), *Handbook of item response theory* (Vol. 1, pp. 391–406). Boca Raton: CRC Press.

von Davier, M., & Sinharay, S. (2007). An importance sampling EM algorithm for latent regression models. *Journal of Educational and Behavioral Statistics, 32*, 233–251. https://doi.org/10.3102/1076998607300422

von Davier, M., Sinharay, S., Oranje, A., & Beaton, A. (2006). Statistical procedures used in the National Assessment of Educational Progress (NAEP): Recent developments and future directions. In C. R. Rao & S. Sinharay (Eds.), *Handbook of statistics: Vol. 26. Psychometrics* (pp. 1039–1055). Amsterdam: Elsevier. http://dx.doi.org/10.1016/S0169-7161(06)26032-2

Wagemaker, H., & Kirsch, I. (2008). Editorial. In D. Hastedt & M. von Davier (Eds.), *Issues and methodologies in large scale assessments* IERI monograph series, Vol. 1, pp. 5–7). Hamburg: IERInstitute.

Wainer, H. (Ed.). (1986). *Drawing inferences from self-selected samples*. New York: Springer. https://doi.org/10.1007/978-1-4612-4976-4

Wainer, H., Dorans, N. J., Green, B. F., Mislevy, R. J., Steinberg, L., & Thissen, D. (1990). Future challenges. In H. Wainer, N. J. Dorans, R. Flaugher, B. F. Green, R. J. Mislevy, L. Steinberg, & D. Thissen (Eds.), *Computer adaptive testing: A primer* (pp. 233–270). Hillsdale: Erlbaum.

Walberg, H. J. (1966). *Personality, role conflict, and self-conception in student teachers* (Research Bulletin No. RB-66-10). Princeton: Educational Testing Service.

Wallach, M. A., & Kogan, N. (1965). *Modes of thinking in young children: A study of the creativity–intelligence distinction*. New York: Holt, Rinehart, and Winston.

Wallach, M. A., Kogan, N., & Bem, D. J. (1962). Group influence on individual risk taking. *Journal of Abnormal and Social Psychology, 65*, 75–86. https://doi.org/10.1037/h0044376

Wang, A. H., Coleman, A. B., Coley, R. J., & Phelps, R. P. (2003). *Preparing teachers around the world* (Policy Information Report). Princeton: Educational Testing Service.

Wang, X., Evanini, K., & Zechner, K. (2013). Coherence modeling for the automated assessment of spontaneous spoken responses. In *Proceedings of NAACL-HLT 2013* (pp. 814–819). Atlanta: Association of Computational Linguistics.

Ward, L. B. (1960). The business in-basket test: A method of assessing certain administrative skills. *Harvard Business Review, 38*, 164–180.

Ward, W. C. (1973). *Disadvantaged children and their first school experiences: ETS–Head Start Longitudinal Study—Development of self-regulatory behaviors* (Program Report No. PR-73-18). Princeton: Educational Testing Service.

Ward, W. C., Frederiksen, N., & Carlson, S. B. (1980). Construct validity of free-response and machine-scorable forms of a test. *Journal of Educational Measurement, 17*, 11–29. https://doi.org/10.1111/j.1745-3984.1980.tb00811.x

Ward, W. C., Kogan, N., & Pankove, E. (1972). Incentive effects in children's creativity. *Child Development, 43*, 669–676. https://doi.org/10.2307/1127565

Wendler, C., Bridgeman, B., Cline, F., Millett, C., Rock, J., Bell, N., & McAllister, P. (2010). *The path forward: The future of graduate education in the United States*. Princeton: Educational Testing Service.

Werts, C. E., Jöreskog, K. G., & Linn, R. L. (1972). A multitrait–multimethod model for studying growth. *Educational and Psychological Measurement, 32*, 655–678. https://doi.org/10.1177/001316447203200308

Werts, C. E., Jöreskog, K. G., & Linn, R. L. (1973). Identification and estimation in path analysis with unmeasured variables. *American Journal of Sociology, 78*, 1469–1484. https://doi.org/10.1086/225474

Willingham, W. W., & Cole, N. S. (1997). *Gender and fair assessment*. Mahwah: Erlbaum.

Willingham, W. W., Ragosta, M., Bennett, R. E., Braun, H. I., Rock, D. A., & Powers, D. E. (Eds.). (1988). *Testing handicapped people*. Boston: Allyn and Bacon.

Witkin, H. A., & Goodenough, D. R. (1981). *Cognitive styles: Essence and origins*. New York: International Universities Press.

Witkin, H. A., Moore, C. A., Oltman, P. K., Goodenough, D. R., Friedman, F., Owen, D. R., & Raskin, E. (1977). Role of the field-dependent and field-independent cognitive styles in academic evolution: A longitudinal study. *Journal of Educational Psychology, 69*, 197–211. https://doi.org/10.1037/0022-0663.69.3.197

Witkin, H. A., Price-Williams, D., Bertini, M., Christiansen, B., Oltman, P. K., Ramirez, M., & van Meel, J. M. (1974). Social conformity and psychological differentiation. *International Journal of Psychology, 9*, 11–29. https://doi.org/10.1080/00207597408247089

Wolf, M. K., & Farnsworth, T. (2014). English language proficiency assessments as an exit criterion for English learners. In A. J. Kunnan (Ed.), *The companion to language assessment: Vol. 1. Abilities, contexts, and learners. Part 3, assessment contexts* (pp. 303–317). Wiley-Blackwell: Chichester.

Wolf, M. K., & Faulkner-Bond, M. (2016). Validating English language proficiency assessment uses for English learners: Academic language proficiency and content assessment performance. *Educational Measurement: Issues and Practice, 35*(2), 6–18. https://doi.org/10.1111/emip.12105

Wolf, M. K., & Leon, S. (2009). An investigation of the language demands in content assessments for English language learners. *Educational Assessment, 14*, 139–159. https://doi.org/10.1080/10627190903425883

Wolf, M. K., Kao, J. C., Rivera, N. M., & Chang, S. M. (2012a). Accommodation practices for English language learners in states' mathematics assessments. *Teachers College Record, 114*(3), 1–26.

Wolf, M. K., Kim, J., & Kao, J. (2012b). The effects of glossary and read-aloud accommodations on English language learners' performance on a mathematics assessment. *Applied Measurement in Education, 25*, 347–374. https://doi.org/10.1080/08957347.2012.714693

Wolf, M. K., Guzman-Orth, D., & Hauck, M. C. (2016). *Next-generation summative English language proficiency assessments for English learners: Priorities for policy and research* (Research Report No. RR-16-08). Princeton: Educational Testing Service. https://doi.org/10.1002/ets2.12091

Wylie, E. C., Lyon, C. J., & Goe, L. (2009). *Teacher professional development focused on formative assessment: Changing teachers, changing schools* (Research Report No. RR-09-10). Princeton: Educational Testing Service. http://dx.doi.org/10.1002/j.2333-8504.2009.tb02167.x

Xi, X. (2010). How do we go about investigating test fairness? *Language Testing, 27*, 147–170. https://doi.org/10.1177/0265532209349465

Xu, X., & von Davier, M. (2008). Linking for the general diagnostic model. In D. Hastedt & M. von Davier (Eds.), *Issues and methodologies in large scale assessments* IERI monograph series, Vol. 1, pp. 97–111). Hamburg: IERInstitute.

Yamamoto, K., & Everson, H. T. (1997). Modeling the effects of test length and test time on parameter estimation using the HYBRID model. In J. Rost & R. Langeheine (Eds.), *Applications of latent trait class models in the social sciences* (pp. 89–99). New York: Waxmann.

Yamamoto, K., & Kulick, E. (2002). Scaling methodology and procedures for the TIMSS mathematics and science scales. In M. O. Martin, K. D. Gregory, & S. E. Stemler (Eds.), *TIMSS 1999 technical report* (pp. 259–277). Chestnut Hill: Boston College.

Yamamoto, K., & Mazzeo, J. (1992). Item response theory scale linking in NAEP. *Journal of Educational Statistics, 17*, 155–173. https://doi.org/10.2307/1165167

Yan, D., von Davier, A. A., & Lewis, C. (Eds.). (2014). *Computerized multistage testing: Theory and applications*. Boca Raton: CRC Press.

Young, J. W. (2009). A framework for test validity research on content assessments taken by English language learners. *Educational Assessment, 14*, 122–138. https://doi.org/10.1080/10627190903422856

Young, J. W., & King, T. C. (2008). *Testing accommodations for English language learners: A review of state and district policies* (Research Report No. RR-08-48). Princeton: Educational Testing Service. https://doi.org/10.1002/j.2333-8504.2008.tb02134.x

Young, J. W., Cho, Y., Ling, G., Cline, F., Steinberg, J., & Stone, E. (2008). Validity and fairness of state standards-based assessments for English language learners. *Educational Assessment, 13*, 170–192. https://doi.org/10.1080/10627190802394388

Young, J. W., Steinberg, J., Cline, F., Stone, E., Martiniello, M., Ling, G., & Cho, Y. (2010). Examining the validity of standards-based assessments for initially fluent students and former English language learners. *Educational Assessment, 15*, 87–106. https://doi.org/10.1080/10627197.2010.491070

Young, J. W., King, T. C., Hauck, M. C., Ginsburgh, M., Kotloff, L. J., Cabrera, J., & Cavalie, C. (2014). *Improving content assessment for English language learners: Studies of the linguistic modification of test items* (Research Report No. RR-14-23). Princeton: Educational Testing Service. https://doi.org/10.1002/ets2.12023

Zaleska, M., & Kogan, N. (1971). Level of risk selected by individuals and groups when deciding for self and for others. *Sociometry, 34*, 198–213. https://doi.org/10.2307/2786410

Zapata-Rivera, D., & Bauer, M. (2012). Exploring the role of games in educational assessment. In M. C. Mayrath, J. Clarke-Midura, D. H. Robinson, & G. Schraw (Eds.), *Technology-based assessments for 21st century skills: Theoretical and practical implications from modern research* (pp. 149–171). Charlotte: Information Age.

Zapata-Rivera, D., Jackson, T., & Katz, I. R. (2014). Authoring conversation-based assessment scenarios. In R. A. Sottilare, A. C. Graesser, X. Hu, & K. Brawner (Eds.), *Design recommendations for intelligent tutoring systems* (pp. 169–178). Orlando: U.S. Army Research Laboratory.

Zechner, K., Bejar, I. I., & Hemat, R. (2007). *Toward an understanding of the role of speech recognition in nonnative speech assessment* (Research Report No. RR-07-02). Princeton: Educational Testing Service. http://dx.doi.org/10.1002/j.2333-8504.2007.tb02044.x

Zechner, K., Higgins, D., Xi, X., & Williamson, D. (2009). Automatic scoring of non-native spontaneous speech in tests of spoken English. *Speech Communication, 51*, 883–895. https://doi.org/10.1016/j.specom.2009.04.009

Zhang, M., & Deane, P. (2015). *Process features in writing: Internal structure and incremental value over product features* (Research Report No. RR-15-27). Princeton: Educational Testing Service. http://dx.doi.org/10.1002/ets2.12075

Zoccolotti, P., & Oltman, P. K. (1978). Field dependence and lateralization of verbal and configurational processing. *Cortex, 14*, 155–163. https://doi.org/10.1016/S0010-9452(78)80040-9

Author Index

R.E. Bennett, M. von Davier (eds.), *Advancing Human Assessment*,
Methodology of Educational Measurement and Assessment,
DOI 10.1007/978-3-319-58689-2

C

D

Q

R

X

Y

Z

Subject Index

R.E. Bennett, M. von Davier (eds.), *Advancing Human Assessment*, Methodology of Educational Measurement and Assessment,
DOI 10.1007/978-3-319-58689-2